Culture History and Convergent Evolution

Vertebrate Paleobiology and Paleoanthropology Series

Edited by

Eric Delson
Vertebrate Paleontology, American Museum of Natural History
New York, NY, USA

Eric J. Sargis
Yale University
Department of Anthropology, New Haven, CT, USA

Focal topics for volumes in the series will include systematic paleontology of all vertebrates (from agnathans to humans), phylogeny reconstruction, functional morphology, Paleolithic archaeology, taphonomy, geochronology, historical biogeography, and biostratigraphy. Other fields (e.g., paleoclimatology, paleoecology, ancient DNA, total organismal community structure) may be considered if the volume theme emphasizes paleobiology (or archaeology). Fields such as modeling of physical processes, genetic methodology, nonvertebrates or neontology are out of our scope.

Volumes in the series may either be monographic treatments (including unpublished but fully revised dissertations) or edited collections, especially those focusing on problem-oriented issues, with multidisciplinary coverage where possible.

More information about this series at http://www.springer.com/series/6978

Culture History and Convergent Evolution

Can We Detect Populations in Prehistory?

Edited by

Huw S. Groucutt

Extreme Events Research Group, Max Planck Institute for Chemical Ecology, Jena, Germany;

Department of Archaeology, Max Planck Institute for the Science of Human History, Jena, Germany;

Max Planck Institute for Biogeochemistry, Jena, Germany

Springer

Editor
Huw S. Groucutt
Extreme Events Research Group
Max Planck Institute for Chemical Ecology
Jena, Germany

Department of Archaeology
Max Planck Institute for the Science of Human History
Jena, Germany

Max Planck Institute for Biogeochemistry
Jena, Germany

ISSN 1877-9077 ISSN 1877-9085 (electronic)
Vertebrate Paleobiology and Paleoanthropology
ISBN 978-3-030-46125-6 ISBN 978-3-030-46126-3 (eBook)
https://doi.org/10.1007/978-3-030-46126-3

Cover image: Handaxe from Nefud Desert. Copyright: Palaeodeserts Project

This Springer imprint is published by the registered company Springer Nature Switzerland AG
The registered company address is: Gewerbestrasse 11, 6330 Cham, Switzerland

Preface

Archaeology uses patterns of similarities and differences in material culture to construct narratives about the human past. Its methodologies to do this borrow from both the natural and social sciences, and over time diverse schools of thought have developed. Crudely put, some see culture (learned behaviours) as absolutely paramount in determining the form of things like stone tools, while others emphasize 'pragmatic' factors such as the use of different kinds of raw material. Other perspectives sit between these extremes, such as those emphasizing the centrality of variation in mobility strategies in determining the nature of the archaeological record.

At the root of these different approaches and perspectives are diverse ways of understanding the character and meaning of similarities and differences in the archaeological record. The aim of this book is not to 'heal' and reconcile the diverse approaches used in archaeology, but rather to explore one of the fundamental building blocks for all perspectives. 'Convergent evolution' in this setting refers to the independent evolution of particular forms of material culture, as opposed to their spread with the movement of people or ideas. Convergent evolution is therefore the opposite side of the coin to cultural transmission. While cultural transmission theory has been widely discussed in the literature in recent years, rather less has been made of convergent evolution.

Most, or hopefully all, archaeologists would accept that convergent evolution characterizes at least some elements of the archaeological record: the question is, how much? If we cannot satisfactorily answer this question then accounts of the past—be they of highly 'cultural-historical' character or biologically-derived cladistics-based perspectives—will build a potential ticking time bomb into their DNA.

When I first pitched the idea of this book I soon discovered a similar book was about to come out: the excellent *Convergent Evolution in Stone-Tool Technology* (2018, MIT Press), edited by Michael J. O'Brien and colleagues. The convergent evolution of a book on convergent evolution highlights the importance of this topic in the contemporary research climate. As O'Brien and colleagues' book focuses very much on lithic technology and is somewhat weighted towards chapters looking at the Americas, I decided to take a slightly different tack. While still primarily relating to lithic technologies, which constitute the overwhelming body of data for the human past, I sought to connect this field of study with the notion of 'populations'.

The reasons for emphasizing this notion of 'population' are three-fold. Firstly, the idea of 'populations' offers a bridge between subject areas. It is, for example, an absolutely central notion in genetics, so it is important to think about what we can (and cannot) say about human populations in prehistory. Secondly, archaeology is gradually coming to the realization that population dynamics (demography) are central to understanding long-term processes in the human past. For example, genetic studies often assume panmixia (random mating), but archaeology and related disciplines can provide information on population structure which can allow models to be refined. Thirdly, in archaeological accounts, populations are often seen as being central and are often linked with particular forms of material culture. Thus, for example, one can read about the 'Aterians', the 'Gravettians', the 'Nubians', etc. Yet at root, these vaguely defined 'populations' are effectively just guesses based on patterns observed in archaeological data. Yet, what that patterning *means* is not self-evident. What does it mean if we find a particular kind of lithic technology in one area for 100,000 years? No straightforward framework exists to link the kind of long-term patterning visible in the archaeological record with the existence of 'populations' as commonly understood in social or biological ways. I think that linking thinking about populations with considerations of convergent evolution can offer a useful way to orient our thoughts about the past. If we are to understand populations in prehistory, then we require both solid theory and practice which allow us to distinguish convergent evolution of material culture from cultural transmission.

I have deliberately not sought to act as a heavy-handed editor. The numerous themes explored—such as the causes of variability in the archaeological record and the character and recognition of populations in the past—are both complex topics and ones which can be approached from very different perspectives. My aim was to highlight diverse theoretical and methodological approaches to these themes. If this book encourages researchers to consider the role of convergent evolution more carefully, then I will consider it to have been a success. Failure to address this topic will arguably damn the relevance of archaeology and particularly areas such as lithic analysis. Conversely, if we are able to develop sensible and balanced perspectives and methodologies, then our field can grow into a mature science. Many recent accounts of human evolution and prehistory are heavily biological in character. If we are to come to balanced perspectives it is up to us archaeologists to emphasize the importance of culture, and doing so means getting to grips with convergent evolution and the recognition of populations in prehistory.

I thank all of the contributors to this volume, the dozens of peer-reviewers, and Eric Delson (Springer VERT Series editor). All played important roles in shaping this volume. I hope that readers find it a useful and enjoyable book.

Jena, Germany
January 2020

Huw S. Groucutt

Contents

Contributors

* Indicates primary address

R. Alexander Bentley Department of Anthropology, University of Tennessee, Knoxville, TN, USA

Dries Cnuts FNRS, Bruxelles, Belgium

Katja Douze APA, Department of Genetics and Evolution, University of Geneva, Geneva, Switzerland

Huw S. Groucutt *Extreme Events Research Group, Max Planck Institute for Chemical Ecology, Jena, Germany;
Department of Archaeology, Max Planck Institute for the Science of Human History, Jena, Germany;
Max Planck Institute for Biogeochemistry, Jena, Germany

Marina Igreja LARC-DGPC, Ministry of Culture of Portugal/ENVARCH, Cibio-Inbio, University of Porto, Porto, Portugal

Danielle A. Macdonald Department of Anthropology, The University of Tulsa, Tulsa, OK, USA

Alex Mackay *Centre for Archaeological Science, School of Earth and Environmental Sciences, University of Wollongong, Wollongong, Australia;
Department of Archaeology, University of Cape Town, South Africa

Lisa A. Maher Department of Anthropology, University of California, Berkeley, CA, USA

John McNabb Department of Archaeology, University of Southampton, Southampton, UK

Michael J. O'Brien *Department Communication, History, and Philosophy, Texas A&M University-San Antonio, San Antonio, USA;
Department of Anthropology, University of Missouri, Columbia, USA

Guillaume Porraz *CNRS, UMR 7269-LAMPEA, Aix-Marseille Université, Aix-en-Provence, France;
Evolutionary Studies Institute, University of the Witwatersrand, Johannesburg, South Africa

Kathryn L. Ranhorn Institute of Human Origins, School of Human Evolution and Social Change, Arizona State University, Tempe, AZ, USA

Natasha Reynolds UMR 5199 PACEA, Université de Bordeaux, Bâtiment B8, Pessac Cedex, France

Veerle Rots *TraceoLab/Prehistory, University of Liège, Liège, Belgium;
FNRS, Bruxelles, Belgium

Patrick Schmidt *Department of Early Prehistory and Quaternary Ecology, Eberhard Karls University of Tübingen, Tübingen, Germany;
Department of Geosciences, Applied Mineralogy, Eberhard Karls University of Tübingen, Tübingen, Germany

Stephen Shennan Institute of Archaeology, University College London, London, UK

Ceri Shipton *Centre of Excellence for Australian Biodiversity and Heritage, Australian National University, Canberra, Australia;
Institute of Archaeology, University College London, 31–34 Gordon Square, Bloomsbury, London WC1H 0PH, UK

Michael J. Shott Department of Anthropology, University of Akron, Akron, OH, USA

Enza Elena Spinapolice Dipartimento di Scienze dell'Antichità, Sapienza Università di Roma, Rome, Italy

Aaron Jonas Stutz Department of Anthropology, Emory University, Atlanta, GA, USA

Christian A. Tryon Department of Anthropology, University of Connecticut, Storrs, CT, USA

Manuel Will *Department of Early Prehistory and Quaternary Ecology, University of Tübingen, Tübingen, Germany;
Gonville and Caius College, University of Cambridge, Cambridge, UK;
PAVE Research Group, Department of Archaeology and Anthropology, University of Cambridge, Cambridge, UK

Chapter 1
Into the Tangled Web of Culture-History and Convergent Evolution

Huw S. Groucutt

Keywords Convergence • Archaeology • Demography • Material culture • Populations

> *"The question "how common is convergence?" remains unanswered and may be unanswerable. Our examples indicate that even the minimum detectable levels of convergence are often high and we conclude that at all levels convergence has been greatly underestimated."* (Moore and Willmer 1997, p. 1)

Background and Context

The themes explored in this book revolve around the related areas of convergent (independent) evolution of particular forms of material culture, the notion and recognition of populations in prehistory, and issues of taxonomy (such as 'technocomplexes' and 'industries') that archaeologists debate as the subject moves (generally) beyond culture-historical interpretations. Another recent volume explored convergent evolution in lithic technologies (O'Brien et al. 2018a). My aim here is to complement such research and push it into debates on 'populations' and archaeological taxonomy across space and time. In the first part of this introduction I describe the background and context of this volume. I subsequently describe the individual chapters.

H. S. Groucutt (✉)
Extreme Events Research Group, Max Planck Institute for Chemical Ecology, 8 Hans-Knöll-Straße 8, 07745 Jena, Germany
e-mail: hgroucutt@ice.mpg.de

Department of Archaeology, Max Planck Institute for the Science of Human History, Kahlaische Straße 10, 07745 Jena, Germany

Max Planck Institute for Biogeochemistry, 10 Hans-Knöll-Straße, 07745 Jena, Germany

Firstly, some comments on definitions. By convergent evolution we mean the appearance of the same (or very similar) features of material culture in different places due to their independent invention. This is opposed to similarities reflecting either population movement or the spread of ideas by cultural diffusion. These poles are often described using the biologically-rooted terms of 'analogy' (convergent evolution) and 'homology' (similarities due to relatedness). While 'homology' in the archaeological record reflects relatedness and connectivity, and therefore offers insights into how human societies moved through space and changed through time, convergent evolution has the potential to severely disrupt and complicate these narratives, by falsely implying connections that never occurred.

Convergent evolution is common in biology, and has been much discussed (e.g. McGhee 2011). As McGhee pointed out, while Darwin concluded *On the Origin of Species* by stating "from so simple a beginning endless forms most beautiful…have been, and are being evolved" (1859, p. 490), it is actually doubtful whether "endless forms" is a particularly accurate way of looking at things. Similar features have repeatedly evolved, in diverse lineages, over millions of years. Ultimately, convergent evolution is so frequent in biology because of 'evolutionary constraint', "that is, the number of evolutionary pathways available to life is in fact not endless, but is quite restricted" (McGhee 2011, p. xi). Some scholars distinguish 'parallel evolution', which can be seen as a special example of convergent evolution where a trait emerges in two groups from the same ancestral state, rather than truly independently (McGhee 2011, p. 3). In 'normal' convergent evolution, two different traits evolve into two traits which are similar/the same, whereas in parallel evolution one trait evolves into two similar traits, which both appear independently, but from a common base. It is also possible for convergence to stem from 'reverse evolution', when a trait evolves to a similar condition to earlier in its lineage, and where the original form may be preserved in other branches. For the

H. S. Groucutt (ed.), *Culture History and Convergent Evolution: Can We Detect Populations in Prehistory?*,
Vertebrate Paleobiology and Paleoanthropology, https://doi.org/10.1007/978-3-030-46126-3_1

present purpose, 'parallel evolution' and 'reverse evolution' can be considered variants of general convergent evolution. In future examples, however, it may be useful to distinguish these different processes in cultural settings.

The importance of convergent evolution in biological evolution is clear. Eyes—to give one particularly iconic example—have appeared independently at least 49 times (McGhee 2011). But what about convergent evolution in cultural evolution? When it comes to early prehistoric archaeological evidence we are mostly dealing with stone tool data. There are powerful reasons to expect convergent evolution here. Unlike forms of material culture where recycling is possible (e.g. metal), lithic reduction is a one-way process. And the nature of this one-way process is constrained by specific fracture mechanics, rooted in physical processes that are fairly well understood (e.g. Cotterell and Kamminga 1987, 1990; Tostevin 2012; Lin et al. 2018). As Eren et al. (2018, p. 70) put it, you "cannot strike a spherical flake". The 'design space' of possible stone tool technologies is actually reasonably small, and the chances of convergent evolution therefore high. Similarly, early humans would have used tools for a limited number of tasks—such as cutting and scraping—and this relative homogeneity of function would also likely have been a powerful driver for the re-invention of particular aspects of material culture. If something as complex as agriculture was invented independently in several parts of the world at broadly the same time, then surely different ways of producing stone tools are also likely to demonstrate myriad examples of convergent evolution. And we can imagine different forms of convergent evolution: such as those which are similar by random chance and those which are similar due to strong 'pragmatic' factors (such as microlithic technology, as explored by Clarkson and colleagues (2018).

'Culture-history' refers to a distinct archaeological tradition that has interpreted and understood archaeological 'cultures' as corresponding to distinct human populations that have long associations with particular regions. This understanding of the archaeological record has emphasized distinct ethnic identities with simple, linear histories. There are often modern political aspects driving the promotion of culture-historical accounts, such as nationalism. Culture-historical approaches tend to be built on inductivism, and downplay the diversity of factors structuring the archaeological record and its complicated relationship with identity and genetic structure. Convergent evolution, for example, is a serious impediment to notions that finding a particular form of pot, or stone tool, automatically equates to a particular group of people. While many archaeologists have developed processual and post-processual approaches to studying the past, culture-history remains a significant school of thought in archaeology, albeit often somewhat disguised and packaged in different ways. The contemporary idea of the 'Nubian Complex', for example, is classically culture-historical in its construction (see Groucutt 2020). Various publications have explored the character and history of culture-historical interpretations in archaeology (e.g. Lyman et al. 1997; Trigger 2006).

Finally, the notion of 'population' deserves some thought. The idea of population is central in genetics, and is also widely used in various human sciences (e.g. Krieger 2012; Kreager et al. 2015a). The term population comes to us from the Latin *populus,* which refers "generally to people living in a state, but more particularly to citizens" (Kreager et al. 2015b, p. 25). In that sense the origin of the word population is a group of something in a particular place.

Since the time of Aristotle and other early scholars, the meaning of the term has changed and developed. In biology the term population tends to be used to describe a geographically and temporally delineated group of individuals belonging to some higher-order taxonomic category (such as a species). In its application in genetics, there is much more complexity than the old idea of the *populus*. There is, for example, the notion that over relevant timescales, populations display panmixia, that is all individuals in a population mate at random with other individuals in the population (e.g. Wright 1984). The reality of course is that to varying degrees, mating in populations is not random. This leads to some interesting ideas, and it has been claimed that "were sufficiently detailed data available, every individual would be a 'population'" (Lawson 2015, p. 109). In genetics, programs such as STRUCTURE can be used to statistically define populations. This is a rather different perspective on the notion of populations than those used in subjects such as history and geography. More discussion needs to occur on how different disciplines are using the concept of populations, and explore the extent to which groups defined by different disciplines/datasets actually correlate with each other. Given growing arguments for admixture in hominin evolution, do traditional definitions of population remain useful? Such questions require much thought and discussion.

In human evolutionary studies, the role of demography (that is, broadly, the study of the sizes, 'shapes' and connectivity of populations) has been prominent in some narratives (e.g. Powell et al. 2009). A recent study on the evolution of *Homo sapiens* by Scerri and colleagues (2018) emphasizes the importance of population structure in the evolution of our species. Diverse datasets are consistent with this model, while many previous analyses have assumed (or at least implied) panmixia (i.e. random mating between individuals in a population) at vast scales (such as the whole of Africa). Modeling high levels of population structure is complex and computationally demanding, but nevertheless is emerging as being essential to understanding the human past. Things become even less clear when poorly-defined populations are glued to poorly-defined aspects of variability in the archaeological record described in terms 'industries' or 'technocomplexes'. Overall, Krieger (2012, p. 635) is surely correct that the "population sciences need to expand and deepen their theorizing about who and what makes populations". This should be a significant aspect in prehistoric archaeology and related fields over the coming years.

Archaeology on the Rocks

Most of the contributions in this book focus on variation in lithic (stone tool) assemblages. This should hardly come as a surprise given that in their temporal span, abundance and preservability, lithics dominate the early prehistoric archaeological record.

Lithic analysis is strongly divided between different schools and research traditions. While there is methodological disagreement, we would also do well to recognize that sometimes there is more commonality than we realize. Lithic analysis as a whole has generally moved from 'static' (often typological) to 'dynamic' (often technological) notions. This could also be seen as a move from a focus on objects to a focus on processes. Dynamic perspectives vary in form considerably. Despite claims to the contrary, many similar ideas are found in the Anglo-world 'reduction sequence' approach and in the French-originated chaîne opératoire school (e.g. Shott 2003). While combined analyses with different approaches and backgrounds are rare, those which do exist are highly significant and deserve careful study (e.g. Scerri et al. 2016) Views of lithic reduction as 'dynamic' also come in many forms, such as Dibble's (e.g. 1987, 1995a) notion of scraper form changing with retouch intensity. My personal perspective is that all approaches to lithic analyses have good and bad points. We should focus on taking what works well to answer particular questions, and not on defending particular research traditions that we happened to have been raised in.

We can think about lithics in terms of the 'static' forms that we find in archaeological sites, and the 'dynamic' processes which produced them. The way to distinguish similar forms may be to differentiate the processes that produced those forms. Yet, conversely, detailed analysis of static forms may be the way to distinguish between superficially similar processes producing them. Just as the distinction between 'stylistic' and 'functional' features is blurry, so the distinction between static and dynamic conceptions breaks down under analysis.

Traditionally, lithic analysis and archaeology in general focused on describing forms, such as in typological frameworks. In the later twentieth century this tended to fade somewhat as the field shifted towards a focus on process. Yet with the current rise of geometric morphometrics (GMM) approaches, we are in essence seeing a return to a focus on object form (in the guise of detailed thought on shape), albeit at a higher level. I believe that care is needed with 'morphocentrism', as Charboneau (2018) put it. Charboneau (2018) pointed out that these approaches have been useful in some ways, but also have limitations, such as their inability to deal with novel forms (rather than subtle variation). Likewise, even to the extent to which 'shape' is important, it is complicated by factors such as differential reduction intensity and does not correlate with transmission mechanisms in as pure a way as something like a gene does. Aside from issues such as whether lithics have any meaningful 'landmarks' for GMM analysis, we must question whether highly detailed analyses of subtle variations in shape are going to provide satisfactory answers to the questions we wish to ask. Yet, we should celebrate the focus on data, objectivity and replicability in such approaches.

One cannot deny the dynamism bought to archaeological accounts by process-centered accounts (such as those of the chaîne opératoire school), yet we should also recognize their limitations. How does one quantify and compare inferred processes (and any higher-level interpretation requires inference)? In my opinion a focus on process means that the archaeological record increasingly consists of a series of highly detailed assemblage studies—with abundant refits, use-wear analyses and so on—which float in space and time and are poorly articulated with other assemblages. Answering the key questions in human evolutionary studies require comparisons across time and space, and comparing processes has not proven easy. Building comparative frameworks should be at the heart of methodological development.

In my opinion then, we have to find a balance between studying forms and processes. There are ways this can be done, such as the framework developed by Tostevin (2012) and extended into multivariate analyses by Scerri (e.g. Scerri et al. 2014). We need strong theoretical frameworks. And we have to separate historical contingencies from essential characteristics. The tendency of chaîne opératoire advocates to reject quantification, for instance, can be seen as a reaction against the pseudo-quantification (e.g. cumulative graphs) of Bordes (Soressi and Geneste 2011). Likewise, it is possible to believe in 'evolutionary archaeology' and find particular methods such as cladistics problematic. More broadly, 'tree-like' models of the past may seem intuitive, but it is not clear how accurate or helpful they are (Groucutt et al. 2015a; Scerri et al. 2018). Human prehistory is not like a family tree.

It is also interesting to observe ongoing debates about methods in other fields. For example, there is currently a debate raging in the study of dinosaurs. Baron and colleagues (2017) proposed a major revision of the dinosaur cladistic structure (phylogeny). Others have criticized this argument and present alternative cladograms (e.g. Langer et al. 2017). The point is that, despite decades of study, the field is currently up in the air. As Benton (2019, p. 83) puts it, "this might sound shocking, or an indictment of the cladistic method". He defends cladistics, seeing the only alternative as "assertion and guesswork". This example should surely give us food for thought on the utility of cladistics. It is possible that the dinosaurs rapidly split into different groups early in their evolution, and this, along with convergent evolution, makes distinguishing the deep relationships challenging. These debates should warn us against the view that because cladistics 'works' in biology then it will work in archaeology. Likewise, criticisms of the chaîne opératoire approach (e.g. Dibble 1995b; Tostevin 2011) should be carefully evaluated by proponents of that school, not simply ignored.

While critics of 'evolutionary archaeology' have long argued that there is so much horizontal 'blending' that it is illusory to imagine cultural change in a tree-like manner, it is arguably true that as Buchanan and colleagues (2018, p. 275) put it "numerous studies have now shown that blending is not more prevalent in culture than biology". However, such views can be taken in different ways. For example, while I agree that the dominant mode of cultural transmission is vertical inheritance, the importance of occasional horizontal transfer and convergent evolution (invention) should also not be underestimated. I have no problem with 'evolutionary' processes, and while I think the dominant themes of the archaeological record do reflect inheritance, it is never the less true that if one day a lion decides it wants to be a donkey, it is probably going to be disappointed when the sun sets, whereas if a blade-maker wants a handaxe, they may well be able to achieve their wish. Just as with evolution, gradualism (or even stasis), may dominate in terms of time spans, yet sudden changes and convergent evolution cannot be downplayed. Key aspects of material culture may appear precisely at moments of major transition and turbulence. One can measure thousands of flakes in multiple dimensions, but if a fundamentally new element emerges—groundstone technology, for example—that data is not really going to help to clarify things. My point is not that 'evolutionary'/ phylogenetic approaches to lithic analysis are not useful, but simply that, in the grand sweep of time and space, they are perhaps not sufficient. With particular questions, in particular study areas, they have their uses.

Numerous examples in this volume outline the limitations of a 'morphocentric' approach. If we think, for example, of the 'Clactonian' assemblages discussed by McNabb (2020), they are both preceded by and followed by 'Acheulean' assemblages featuring iconic handaxes. Understanding the place of the Clactonian is not going to be helped by detailed analyses of handaxes, as these are not a significant part of Clactonian assemblages (if they are present at all). So it is only by rounded technological analyses and consideration of chronostratigraphic issues that such themes can be addressed.

Context and Chronology

Let us step back from the details of different analytical frameworks for studying lithic assemblages. One thing I would like to emphasize is that we should always consider the extent to which technologies/assemblages actually are similar, before we move into complex discussions on why they are similar. For example, different assemblages from the Levantine Middle Paleolithic are often described as being 'the same'. This is discussed as an interesting example because these assemblages are made by both Neanderthals and *Homo sapiens*, and thus the apparent similarity in lithics through time feeds into negative views on whether lithics can actually tell us anything at all. My view is that there is actually no time-transgressive 'Levantine Mousterian'. Assemblages are broadly similar, as we would expect from closely related groups occupying the same kind of environment, and within the 'mode-3' world of subtle variations in core reduction methods and tool types. In my opinion, when one looks in detail the Middle Paleolithic assemblages of the Levant are not 'the same' at all (e.g. Groucutt et al. 2019). McGhee (2018, p. 27) goes to considerable lengths describing the apparent mystery of convergent evolution in the Levantine Middle Paleolithic, which he sees as a very exciting example of "iterative evolution that is both parallel and convergent". In my opinion, this unnecessarily complicates the situation.

Before we move into sophisticated analyses we should make sure that our starting premise is actually meaningful. The same could be said of purported similarities in 'microlithic technologies' in southern and eastern Africa and South Asia (Mellars et al. 2013). Yet, again here it should be pointed out that on multiple levels of technology, as well as other aspects such as chronology, these examples are simply not very similar (e.g. Groucutt et al. 2015a, b; Lewis 2017; Clarkson et al. 2018).

Numerous other debates and issues swirl around the areas we are discussing. For example, correlations between form and function remain very problematic. This is, for example, explored by Douze and colleagues (2020) in their consideration of 'points'; a category which subsumes a lot of technological and morphological variability. The importance of dating is paramount, particularly chronometric dating. And in fact, it could be argued that many of the problems we face today reflect chronometric issues. Unfortunately, archaeology has a propensity to appeals to authority, and if one is firmly wedded to an idea it is possible to argue why 'x' site is too far away to be important or why 'y' dating sample should be ignored. We should of course be critical of all chronometric age estimates, but where they seem to be reliable, they should be central to our accounts. Without time, the archaeological record is a mess.

As numerous papers in this book argue, good chronological control is absolutely vital. The development and refinement of absolute dating techniques is central, and should be a continued target for funding and research. Single amino acid radiocarbon dating, for example, promises a major advance (e.g. Devièse et al. 2018). For earlier periods the use of techniques such as optically stimulated luminescence often results in prodigious error ranges, yet they are often the only option available. Understanding the strengths and weakness of different techniques and age estimates is central. For example, it is important to emphasize that a minimum age is a minimum age, not an approximation of a specific age. It is also important for archaeologists to improve their understandings of the strengths and weaknesses of different models, not simply to act as passive consumers. One need not spend months measuring grains of quartz in the dark to understand that the kind of model chosen has a fundamental impact of optically stimulated luminescence dating techniques. If a paper therefore gives minimal information—such as not saying whether the central age model, minimum age model or finite mixture model was used—its results should not be relied on. And it should be made clear that such practices are not acceptable. Likewise, as Reynolds (2020) discusses, it is important to understand the outlines of different radiocarbon pre-treatment methods. Attempting to remove contamination from collagen, for example, is not straightforward. Many published radiocarbon estimates may be wrong. And it is often very hard to distinguish contamination from sources such as humic acids in soils. If at all possible, multiple, independent dating techniques should be used to give more certainty. Techniques such as Bayesian modelling are useful, but if the underlying dates are flawed they will only create artificial certainty. Finally, as Reynolds (2020) argues it is precisely by weaving together technological and chrono-stratigraphic information that we can build reliable frameworks. If a site represents a remarkable exception from an otherwise clear pattern, it may well be published in a high-impact journal, but it may also be the result of problems such as incorrect dating.

Converging

Being able to distinguish convergent evolution from other mechanisms is crucial in understanding the meaning of the archaeological record. The example of the long-lived Acheulean is interesting here. While likely to be provocative, Shipton (2020) makes a strong case that the Acheulean does not represent the repeated invention of similar technologies, but rather is a genuine cultural tradition. Shipton (2020) uses comparisons of lithic data, experimental knapping, and consideration of chronometric age estimates to make his argument. Whatever one thinks of his conclusion, Shipton (2020) is surely right to highlight these kinds of approaches as the way to address the issue of convergent evolution.

Other chapters in this volume highlight numerous examples of actual or probable convergent evolution. In doing so they join many examples already discussed in the literature. The presence of stone tool making in several distantly related species of primates likely reflects convergent evolution (e.g. Carvalho and Beardmore-Herd 2019). In more recent periods numerous examples exist. For example, most researchers think the view that 'Solutreans' crossed the Atlantic to settle the Americas (e.g. Stanford and Bradley 2012) is highly unlikely. It is much more likely that superficial similarities—particularly 'overshot flaking'—reflect convergent evolution due to similar technological repertoires (bifacial flaking) (e.g. Eren et al. 2013, 2014). Other examples that many archaeologists would accept as representing convergent evolution include the origin of Levallois technology (e.g. Adler et al. 2014) through to particular retouched tool forms such as tanged/pedunculated tools (e.g. Scerri 2012) and fluted points (e.g. Charpentier et al. 2002). As a result of both the spread of these examples in space and time and their occurring across different aspects of the reduction process we should always test a null hypothesis of convergent evolution. Whether certain technologies represent convergent evolution or not is currently the subject of considerable debate. For example, 'Nubian Levallois' technology has been argued to represent a very strong culture-historical signal (discussed in Groucutt 2020), yet numerous authors have suggested that convergent evolution probably best explains the distribution of Nubian Levallois technology (e.g. Groucutt et al. 2015a; Will et al. 2015; Clarkson et al. 2018; Eren et al. 2018).

While the reality of convergent evolution in the archaeological record is recognized by effectively all researchers, it should be pointed out that this introduces the notion of sliding scale in thinking about the importance of convergent evolution. I, for one, think there is a lot of patterning in the archaeological record and that this primarily reflects underlying population dynamics. Many dubious claims for particular models are based on single lithic types, and are easily dispensed with when one looks at the situation objectively. Convergent evolution seems to generally apply to individual elements of lithic assemblages, and so a rounded evaluation of different transmission processes needs to take a whole assemblage view. Just looking at a particular core reduction method, or a particular retouched tool form is unlikely to give very clear solutions and are very vulnerable to convergent evolution (e.g. Groucutt 2020; Will and Mackay 2020).

Perhaps harder to deal with, perhaps, are putative instances of 'false negatives'. Tryon and Ranhorn (for example) point out that population connected by ancestry/cultural transmission, could actually go on to produce quite different lithics, because of factors such as different raw material variability. This difference could be mistaken for a lack of cultural transmission. Likewise, Stutz (2020) explores the possibility that contemporaneous technological variability—such as between different kinds of Upper Palaeolithic entity in the Levant—may not reflect the existence of different populations, but rather the same populations behaving different in different places due to ecological, demographic and mobility gradients. Stutz (2020) suggests that the same could apply with 'Bohunician' and 'Szeletian' assemblages in east-central Europe. While I do not doubt that regional patterns in the archaeological record mean *something*, it is really not clear what that something is. And it almost certainly not a one to one match with distinct and homogenous populations. Similarities between assemblages do not necessarily mean cultural transmission/shared histories, but then neither do differences necessarily mean a lack of cultural transmission/shared histories. I will discuss below ways in which we might get around this, but in short the key is surely multi-scale evaluations of what we mean by 'similar' and 'different', as well as detailed contextual understanding (e.g. chronology).

Population Thinking

We still do not know very clearly what variation in lithic technology or other aspects of the archaeological record mean in relation to populations. What, for instance, does the earlier heterogeneity of Levantine Epipaleolithic entities compared to a later homogenization with the 'Natufian' actually mean in terms of human population dynamics (Maher and Macdonald 2020)? And while on the Epipaleolithic example, what do different proportions of microlith forms tell us? This is how the record of the terminal Pleistocene of the area has been structured by modern archaeologists, but what is the actual social or behavioral meaning of making one microlith shape over another?

Several chapters offer insights into the nature and theorization of populations in early prehistory. Groucutt explores the simplistic equation of 'Nubian Levallois technology = Nubian Complex = the Nubians'. In such cases it seems clear from objective evaluations that even the lithic arguments do not stand up to scrutiny, let alone demographic/social interpretations of those claims. In other cases, however, there are clear patterns in the archaeological record, but what do these mean? What does it mean that MSA people in North Africa west of the Nile often made tanged/pedunculated tools, for tens and tens of thousands of years in the Late Pleistocene (Scerri 2017)? We currently have little grasp on the kinds of learning and social dynamics that could explain such phenomena.

To me, it seems clear that we need to move beyond outdated ad hoc 'techcomplexes' which lack coherent definition (e.g. some are defined by a core reduction method, some by a retouched tool form, some by the absence of certain features, etc.), let alone to use these as proxies for populations. The criticisms of 'named stone tool industries' have been presented in both mostly theoretical (Shea 2014; Scerri 2017) and in quantified 'practical' senses (Scerri et al. 2014). But what is to replace them? My inclination is that we have to more towards more continuous perspectives, such as the clouds of attributes states explored by Scerri and colleagues (2014; see also Mackay et al. 2014). How these relate to populations remains complicated, of course, but doing this at least moves towards an objective and data-centered approach, instead of collapsing huge complexity into simple words like 'Aterian'. And objectively characterizing the nature of lithic variability is something that we can do. In the context of the Middle Stone Age, for example, the 'Comparative Analysis of Middle Stone Age Artefacts' (CosMSAfrica) project has recently been launched in an effort to standardize methodologies across Africa (Will et al. 2019).

The final area when it comes to 'population thinking' reflects the need to think through the dynamics and implications of 'admixed populations'. Growing evidence indicates frequent admixture between distinct populations in the Pleistocene (although we must of course always be aware of how much this reflects the model parameters input by researchers). Stutz (2020) suggests that it was precisely an admixed *Homo sapiens*/Neanderthal population which made the Upper Paleolithic transition in Southwest Asia. This is another reminder that population dynamics, in their various forms and scales, are central to accounts of the Pleistocene.

Diversification

Most chapters in this volume concern stone tools, and ways of thinking about them. But that is not the exclusive focus of the book (e.g. Shennan 2020; Schmidt 2020). Reynolds (2020) highlights that understanding the European Upper Paleolithic record takes us beyond the realm of lithics alone, and into a world of personal ornaments, osseous technologies and so on. I think this is an important point, and we have to avoid the lure of 'lithics for lithics' sake arguments. The strongest narratives, in my opinion, come from cross-cutting lithic information with other kinds of datasets.

Several chapters (e.g. Reynolds 2020; Shennan 2020) discuss genetic evidence. In the case of Europe, there is now enough ancient DNA to begin to outline some very interesting processes. It is fascinating to see that sometimes archaeology and genetics suggest congruent narratives, whereas other times they do not. For example, genetic evidence suggests major population turnover during the Late Upper Paleolithic, yet there is not a clear archaeological signal of this (Reynolds 2020). As ever greater numbers of ancient genomes are sequenced, there is a growing possibility to weave together biological and cultural narratives.

It is also important to think about the nature of processes which lie 'behind' variability in lithics. For example, it is very interesting to think about the notion of learning. If populations can be represented by particular forms of material culture, then leaning is the process by which these patterns are perpetuated. Interesting discussions have been published on learning, distinguishing emulation and imitation, and so on (e.g. Bentley 2018; Wilkins 2018). What kind of transmission mechanisms meant that similar forms, such as the bifacial technologies discussed by Shipton (2020), were transmitted for hundreds of thousands of years? The simple knapping experiment Shipton (2020) conducts highlights the potential of this kind of experiment to guide our thoughts here. The interesting hypothesis from Shipton's knapping study can be tested by larger experiments in the future. When it comes to learning we would do well to build on Tostevin's (2012) ideas on visibility—which focusses on the parts of lithic technological systems that require intimate social connection. Stutz (2020), for example, builds from this notion. There are diverse ways in which we can think about issues of learning and teaching. For example, Maher and Macdonald (2020) explore the notion of 'communities of practice' as a way to link assemblages with wider social structures and lifeways of Epipaleolithic human groups.

The Chapters

The chapters in this volume are organized in broadly chronological order, and cover a variety of topics, locations, and time periods. Here I will briefly summarize each of the chapters.

Shipton (2020) explores a topic of perennial interest, the Acheulean, characterized by iconic large cutting tools such as handaxes. The chapter explores whether the Acheulean represents repeated convergent evolution of similar forms, or rather if it spread from a single origin. Shipton does this in three ways. Firstly, he reports an anecdotal yet very interesting knapping experiment, to highlight that biface knapping is hard to invent, but easy to transmit. Secondly, he compares handaxe and cleaver elongation to highlight regional technological differences. Finally, he discusses the oldest ages for Acheulean sites in different regions, highlighting that current evidence supports an East African origin followed by a spread. Shipton's chapter highlights the important issue of understanding how East Asian bifaces are similar and different to the Acheulean assemblages west of the Movius line. He argues that they are in fact quite different, and that this represents an example of parallelism, both being invented from a common Oldowan base. Clearly, much remains to be done in understanding the world of the Acheulean, and Shipton indicates some of the ways in which we can go about improving our current knowledge.

McNabb (2020) explores the Clactonian, a technocomplex known from several sites in southeast England, dating to MIS 11 (i.e. ca. 400 ka). In contrast to both older and younger assemblages in the region, the Clactonian lacks large cutting tools such as handaxes, and is instead characterized by a simple core and flake technology. McNabb discuses different interpretations of the Clactonian, which makes an interesting case study of the meaning of archaeologically defined entities (technocomplexes, industries, phases, or whatever else one wishes to call them). As usual, McNabb does a good job of situating the research which led to the definition of the Clactonian in terms of its historical context. Changing views on the Clactonian have ranged from culture-historical perspectives to the very pragmatic (e.g. raw material factors). McNabb provides both a summary of the Clactonian in itself, but also a useful case study for the interconnected issues which this book is focused on. Future research will turn up surprises in poorly explored parts of the world, so it is important that lessons are learned from the areas where relatively large amounts of research have been conducted, such as northwest Europe.

Groucutt (2020) explores the example of the 'Nubian Complex'. Following recent findings of assemblages in southern Arabia characterized by 'Nubian Levallois' technology there has been some enthusiasm for the idea that this kind of technology is diagnostic of the 'Nubian Complex' which is seen as a culture-historical signal for the spread of a Northeast African population. Groucutt explores the history of the Nubian Complex, highlighting the problems and contradictions of the various definitions. In the end the evidence suggests that Nubian Levallois technology is found over such a huge temporal and spatial scale that the notion of the Nubian Complex is not helpful. Nubian Levallois technology is argued to provide a very interesting example of convergent evolution, probably repeatedly re-invented from a common background of 'normal' preferential Levallois technology.

Spinapolice (2020) explores the character and meaning of lithic variability in the Middle Stone Age of East Africa. As a region containing some of the most famous sites in Pleistocene archaeology, various perspectives have been taken on this record: some have seen it as signaling the origin of our species, some have argued that East Africa contained key 'refugia', and so on. To consider such narratives it is important to have a clear understanding of the archaeological record of the area. Spinapolice (2020) provides a useful synthesis of the East African record, and highlights the complex and diverse ways in which lithic evidence can be related to social and demographic dynamics. This chapter reminds us of the importance of rooting archaeological accounts in terms of anthropology. Spinapolice introduces the idea of Significant Technological Units as a way to look at the Middle Stone Age record, and explores the example of bladelet production.

Will and Mackay (2020) present an evaluation of convergent evolution in Africa over the last 300,000 years. Their multi-scale approach offers refreshing insights, and highlights that convergence is both common and scale-dependent. Because of this, great care has to be taken in building narratives of population movement based on material culture (and here they build on their influential paper reporting Nubian Levallois technology in South Africa, thereby giving a strong example of convergent evolution [Will et al. 2015]). Nearly all lithic analysis would agree that convergent evolution *sometimes* occurred. Will and Mackay (2020) explore the important question: how frequent is convergence in lithic technology? To answer that requires clear definitions and strong theory, as well as an objective evaluation of the archaeological record. Will and Mackay (2020) argue that convergence is more common with certain technologies than others. Backed microliths, for example, seem to clearly show convergent origins (see also Clarkson et al. 2018). On the other hand, they argue that balanced perspectives are needed, and that clearly processes such as diffusion and migration did play a role in Stone Age Africa, and that we can see genuine examples of spatially and temporally specific technological features.

Douze and colleagues (2020) present a combined approach to understanding 'points' from Bushman Rockshelter in South Africa. Points have often been seen as a diagnostic feature of the Middle Stone Age, yet as Douze and colleagues discuss, definitions of what a point is have been highly variable. Given that points have been a key part of arguments for regionalization in the Middle Stone Age (e.g. Clark 1988), it is important to think about the character and diversity of point production. Douze and colleagues findings show the potential of combining detailed technological analyses with usewear and residue analyses to understand the archaeological record. Their results suggest that points at Bushman Rock Shelter were typically used for cutting and scraping tasks. Given the lack of high-resolution data it is difficult to distinguish regional trends in technologies such as point production. In South Africa, *relatively* intensively studied compared to the rest of the continent, arguments can be made about patterning in the archaeological record, but for vast areas of Africa so little is known that we currently have little idea on the extent to which 'point production ' repeatedly emerged by convergent evolution, as opposed to cultural transmission. Douze and colleagues (2020) show how modern 'techno-functional' analyses should be conducted. And even if one then wants to conduct highly detailed quantitative analyses of artefacts, this can be seen as an addition, not an alternative to this kind of approach.

Tryon and Ranhorn (2020) present an interesting case study of the role of raw material types in considerations of lithic technology, typology and metrics, and the way in which this can impact considerations of similarities and differences in the archaeological record which then feed into models of populations and behavior. They look at various East African assemblage from the Acheulean to the Holocene, and in particular emphasize the impact of the use of quartz by knappers. This is both in terms of the behavior of the knappers, and in terms of interpretation by contemporary lithic analysts. For knappers, certain raw materials impose constraints on the forms than can be produced. From the perspective of the lithic analyst, they argue that recent approaches such as those of Tostevin (2012), and the multivariate development of this approach (Scerri et al. 2014), cannot be easily applied to quartz dominated assemblages. For example, Tryon and Ranhorn (2020; see also Ranhorn 2017) describe difficulties such as reading scar patterns of quartz lithics from sites like Nasera. Likewise, the bipolar technology which is frequent in East Africa is not a significant element of the technological repertoire in areas studied Tostevin (2012) and Scerri et al. (2014) and therefore the methodologies that these authors developed. These are clearly issues that we need to take seriously and discuss if we are to develop reliable and replicable comparative frameworks. Finally, I think that Tryon and Ranhorn's (2020) chapter shows the benefits of thinking about themes through time instead of limiting oneself to a certain realm (the 'MSA' for example). That is something I would like to see more of in the future.

Stutz (2020) offers a novel perspective on a perennial topic of interest, the Middle to Upper Paleolithic transition. This transition has often been explained in terms of a 'human revolution' or, in some unclear way, the migration of *Homo sapiens* out of Africa. These narratives have never been particularly convincing, and Stutz (2020) outlines an alternative model where the transition reflected behavioral change in admixed *Homo sapiens*/Neanderthal populations. Stutz (2020) builds this perspective from considerations of niche construction and biocultural evolutionary dynamics. Instead of the earliest appearance of the Upper Paleolithic in a region being a simple marker of the appearance of *Homo sapiens*, Stutz argues that the timeframe is better characterized as a long-term process unfolding as the result of admixture between Neanderthals and *Homo sapiens*. The Upper Palaeolithic emerges as a result of economic intensification within this new admixed population.

Reynolds (2020) considers the European Upper Paleolithic record, in both general terms and with a case study of Mid Upper Paleolithic Russia. This is a particularly pertinent area given the theme of this book, as it features numerous taxonomic units that are commonly associated with distinct populations (the Gravettians, the Solutreans, etc). However, as Reynolds (2020) argues, these taxonomic units are historically contingent and the theoretical bridge from these poorly defined entities to purported populations is very problematic. To move beyond the current situation Reynolds proposes an approach combining detailed understanding of the chronology of assemblages (which she refers to here as 'the warp') and detailed comparisons of material culture ('the weft'). By weaving these two parts together we can construct a reliable chronocultural framework. Of course, some of what Reynolds (2020) discusses has a specific relevance to the European Upper Paleolithic. However, the basic themes of the need for secure chrono-stratigraphy and for a bottom up rather than top down (i.e. techcomplex) approach apply to the entire archaeological record. Likewise, Reynolds (2020) discussion of radiocarbon dating is important, and deserves to be carefully considered.

Maher and Macdonald (2020) take us to the world of the Epipaleolithic Levant. This is typically studied in terms of being a precursor to agricultural communities, yet as Maher and Macdonald explore it also a fascinating period in its own right, without having to be seen as the precursor to something. Maher and Macdonald (2020) emphasize the importance of technology in understanding the Epipaleolitic of the Levant. They highlights the notion of 'communities of practice'. This situates lithic technology in terms of wider social practices and lifeways. Whether it is an approach one favors or not, Maher and Macdonald (2020) give a very clear description of a chaîne opératoire approach to lithic analysis. They remind us that the study of lithics should be aimed at trying to elucidate human lifeways and societies, not as an end in itself. They report the recovery of an incredible ca. three million lithics from the site of Kharaneh IV in eastern Jordan, of which they have analyzed about 10%. Given such vast numbers, we are reminded of the importance of stone tools in early human societies and the powerful information they can therefore surely provide.

Shott (2020) explores North American points. Points have been central to the construction of the North American archaeological record, yet as Shott explores, they can be seen from multiple perspectives. Shott (2020) explores questions such as how one point type changed to another and why did certain forms last for long periods. Such approaches emphasize the need for deep levels of theory, to transcend simplistic traditional views. What defines a 'point' (see also Douze et al. 2020)? To Shott, points in the Americas are mostly projectile tips, but beyond that embody high levels of variation. This variation offers a way to cut through the simplistic division of the record into 'cultures'. For Shott (2020) building a meaningful understanding of points takes various forms, from geometric morphometric analyses through to considerations of the appropriate time scales at which to consider points (i.e. beyond the scale of ethnographic observation).

O'Brien and Bentley (2020) explore the colonization of North America, with a particular focus on learning. They seek to explore novel ways of considering the notion of 'populations'. Several pre-Clovis assemblages are now increasingly accepted, but it is with Clovis that we see the first widespread human presence in the area. O'Brien and Bentley give a very clear and useful discussion of various issues relating to homology and analogy in material culture, and argue the best way to distinguish them is using cladistics. O'Brien and Bentley (2020) describe learning (particularly social learning) as the basis of cultural transmission. They explore the idea of fitness landscapes as applied to culture. Through these notions they explore variability in Clovis technologies—which they emphasize can be seen both in terms of overall shape, and in specific aspects such as scar patterns. They explore how as Clovis technology spread across America some aspects changed while others did not, which they interpret as reflecting different levels of learning/transmission.

Schmidt (2020) represents a very different approach to the others in this book, exploring heat treatment of stone used for stone tool production, rather than details of lithic technology as most chapters address. Heat treatment has been celebrated as an early example of complex behavior in humans, yet it has been debated how exactly humans were heating the rocks and the implications these behaviors have for planning, cognition, etc. Schmidt (2020) compares different examples of heat treatment across time and space, from the Middle Stone Age of South Africa to the Paleo-Indian period of North America. He explores how the specifics of heat treatment varied in these different settings—with the stone sometimes being buried, sometimes not, and the temperature of the fire varying. Schmidt (2020) argues that this behavioral diversity indicates repeated convergent evolution of heat treatment in different settings. While the study of early prehistory remains focused on stone tool reduction technology, Schmidt (2020) reminds us that finding additional lines of research on early human behavior offers very exciting avenues for future research. As great as lithics are, it is incumbent on us to continually seek to develop innovative new techniques to study the past.

Finally, Shennan (2020) explores the themes of style, function and cultural transmission. Shennan (2020) summarizes some of the seminal debates in archaeology, such as the Bordes-Binford debate, and brings us to the major contemporary topics of research in cultural evolutionary studies. Shennan (2020) outlines useful ways forward. These include the formation of clear and testable models, for example testing a null hypothesis of 'isolation by distance' can be a very useful approach (see also Scerri et al. 2014, 2018). Shennan and colleagues (2015) have previously demonstrated the utility of this approach in relation to Neolithic pottery and ornaments. Shennan (2020) outlines ways in which hypotheses on prehistoric cultural evolution and relationships can be tested using ancient DNA evidence.

Conclusion

My main aim with this book is to provoke questions, while offering few answers. The issues involved are too deep and fundamental for simple and immediate solutions. It seems evident that the definitions of many of the terms we commonly use and the ways we employ them—such as populations and cultures—are at best poorly defined and theorized, and at worst positively confusing and unhelpful. Without developing the crucial 'scaffolding' that such concepts should facilitate, the ever-growing mass of data is going to lack secure anchoring.

Things used to be so simple in archaeology, when it could be declared that "typological similarity is an indicator of cultural relatedness" (Willey 1953, p. 363). Over the following decades the naivety of such views became clear (e.g. Binford 1968; Clarke 1968). Yet, in my opinion, archaeology is still struggling to deal with convergent evolution. And, in their different ways, both 'evolutionary archaeology' and 'chaîne opératoire' approaches (to rather sloppily characterize two relative research poles) have struggled with the recognition and consideration of convergent evolution. Archaeologists have tended to argue, in effect, that the more similar things are then the less likely convergent evolution is (e.g. Kroeber 1931; Clarke 1968). I do not think this perspective has worked, and we now need to develop better ways of thinking about convergent evolution. Convergent evolution does not just matter in terms of understanding how people made different forms of material culture, but because of the wider implications in terms of recognizing social structures and populations. Casting light on the demography of early humans is now crucial in advancing our understanding of human evolution (e.g. Scerri et al. 2018), and a failure to develop ways to recognize convergent evolution means building a time-bomb into the use of cultural data.

In my opinion, strong arguments on convergent evolution will come from threefold analyses of archaeological data (such as lithics) using sophisticated and objective techniques to explore patterns of similarity and difference in multiple independent areas of material culture (e.g. core reduction methods, forms of retouched tools, etc.) and ideally include both lithic and non-lithic data, experimental studies (such as knapping experiments), and chronometric dating of archaeological sites. Various combinations of these three elements can be found in the chapters of this book. Fire can be described in terms of a triangle (heat, fuel, oxygen). If any one side is removed or dampened, the fire goes out. Likewise, with studies of human prehistory, archaeological data, experimental studies and chronometric dating stand in symbiotic relationship and neglecting one means an entire argument can collapse.

In my opinion, methodological plurality is a good thing. Methods should be used if they help to answer questions, not just because they were used by earlier researchers. Surely progress will come from combining the 'static' but objective analysis of objects with the 'dynamic' but sometimes rather abstract study of processes. In terms of higher-level theoretical frameworks, we need approaches which can clarify both gradual processes of change (or even stasis) and sudden transitions and inventions. We need a body of methods and theory which are unique to archaeology (within the wider field of paleoanthropology), and not simply attempts at wholesale import of methods from other areas (e.g. biology, ethnography). We should also try and recognize our biases, and move beyond a focus on individual know how and arguments from authority, towards more objective and quantified perspectives (see e.g. O'Brien et al. 2018b, p. 16). Likewise, we should emphasize the study of questions, not just using a new method because it is new. The methodological and theoretical pitfalls are numerous, but if we can work through these issues we can make material culture relevant to understanding some of the biggest questions in the study of human evolution and prehistory. Building a solid understanding of convergent evolution will firmly embed our subject as a science, while failure to do so will mean that we drift into storytelling while the scientific research is done by others.

References

Adler, D. S., Wilkinson, K. N., Blockley, S., Mark, D. F., Pinhasi, R., Schmidt-Magee, B. A., et al. (2014). Eearly Levallois technology and the Lower to Middle Paleolithic transition in the southern Caucasus. *Science, 345,* 1609–1613.

Baron, M. G., Norman, D. B., & Barrett, P. M. (2017). A new hypothesis of dinosaur relationships and early dinosaur evolution. *Nature, 543,* 501–506.

Bentley, R. A. (2018). The transparency of imitation versus emulation in the Middle Paleolithic. In M. J. O'Brien, B. Buchanan, & M. I. Eren (Eds.), *Convergent evolution in stone-tool technology* (pp. 47–59). Cambridge, MA: MIT Press.

Benton, M. J. (2019). *The dinosaurs rediscovered: How a scientific revolution is rewriting history*. London: Thames and Hudson.

Binford, L. R. (1968). Archaeological perspectives. In S. R. Binford & L. R. Binford (Eds.), *New perspectives in archaeology* (pp. 5–32). New York: Aldine.

Buchanan, B., Eren, M. I., & O'Brien, M. J. (2018). Assessing the likelihood of convergence among North American projectile-point types. In M. J. O'Brien, B. Buchanan, & M. I. Eren (Eds.), *Convergent evolution in stone-tool technology* (pp. 275–287). Cambridge, MA: MIT Press.

Carvalho, C., & Beardmore-Herd, M. (2019). Technological origins: Primate perspectives and early hominin tool use in Africa. In *Oxford research encyclopedia of African history*. Oxford University Press: Oxford. http://dx.doi.org/10.1093/acrefore/9780190277734.013.75.

Charboneau, M. (2018). Technical constraints on the convergent evolution of technologies. In M. J. O'Brien, B. Buchanan, & M. I. Eren (Eds.), *Convergent evolution in stone-tool technology* (pp. 73–89). Cambridge, MA: MIT Press.

Charpentier, V., Inizan, M. L., & Féblot-Augustins, J. (2002). Fluting in the Old World: The Neolithic projectile points of Arabia. *Lithic Technology, 27,* 39–46.

Clark, J. D. (1988). The Middle Stone Age of East Africa and the beginnings of regional identity. *Journal of World Prehistory, 2,* 235–305.

Clarke, D. L. (1968). *Analytical archaeology*. London: Methuen.

Clarkson, C., Hiscock, P., Mackay, A., & Shipton, C. (2018). Small, sharp and standardized: Global convergence in backed-microlith technology. In M. J. O'Brien, B. Buchanan, & M. I. Eren (Eds.), *Convergent evolution in stone-tool technology* (pp. 175–200). Cambridge, MA: MIT Press.

Cotterell, B., & Kamminga, J. (1987). The formation of flakes. *American Antiquity, 52,* 675–708.

Cotterell, B., & Kamminga, J. (1990). *Mechanics of pre-industrial technology: An introduction to the mechanics of ancient and traditional material culture*. Cambridge: Cambridge University Press.

Darwin, C. (1859). *On the origin of species by means of natural selection; or the preservation of favoured races in the struggle for life*. London: Murray.

Devièse, T., Comeskey, D., McCullagh, J., Bronk-Ramsey, C., & Higham, T. (2018). New protocol for compound-specific radiocarbon analysis of archaeological bones. *Rapid Communications in Mass Spectrometry, 32,* 373–379.

Dibble, H. L. (1987). The interpretation of Middle Paleolithic scraper morphology. *American Antiquity, 52,* 109–117.

Dibble, H. L. (1995a). Middle Paleolithic scraper reduction: Background, clarification, and review of the evidence to date. *Journal of Archaeological Method and Theory, 2,* 299–368.

Dibble, H. L. (1995b). Biache Saint-Vaast, Level IIa: A comparison of analytical approaches. In H. L. Dibble & O. Bar-Yosef (Eds.), *The definition and interpretation of Levallois technology* (pp. 93–116). Madison: Prehistory Press.

Douze, K., Igreja, M., Rots, V., Cnuts, D., & Porraz, G. (2020). Technology and function of Middle Stone Age points. Insights from a combined approach at Bushman Rock Shelter, South Africa. In H. Groucutt (Ed.), *Culture history and convergent evolution: Can we detect populations in prehistory?* (pp. 127–141). Cham, Switzerland: Springer.

Eren, M. I., Patten, R. J., O'Brien, M. J., & Meltzer, D. J. (2013). Refuting the technological cornerstone of the Ice-Age Atlantic crossing hypothesis. *Journal of Archaeological Science, 40,* 2934–2941.

Eren, M. I., Patten, R. J., O'Brien, M. J., & Meltzer, D. (2014). More on the rumor of "Intentional overshot flaking" and the purported ice-age Atlantic crossing. *Lithic Technology, 39,* 55–63.

Eren, M. I., Buchanan, B., & O'Brien, M. (2018). Why convergence should be a potential hypothesis for the emergence and occurrence of stone-tool form and production processes: An illustration using replication. In M. J. O'Brien, B. Buchanan, & M. I. Eren (Eds.), *Convergent evolution in stone-tool technology* (pp. 61–71). Cambridge, MA: MIT Press.

Groucutt, H. S. (2020). Culture and convergence: The curious case of the Nubian Complex. In H. Groucutt (Ed.), *Culture history and convergent evolution: Can we detect populations in prehistory?* (pp. 55–86). Cham, Switzerland: Springer.

Groucutt, H. S., Petraglia, M. D., Bailey, G., Scerri, E. M. L., Parton, A., et al. (2015a). Rethinking the dispersal of *Homo sapiens* out of Africa. *Evolutionary Anthropology, 24,* 149–164.

Groucutt, H. S., Scerri, E. M. L., Lewis, L., Clark-Balzan, L., & Blinkhorn, J. (2015b). Stone tool assemblages and models for the dispersal of *Homo sapiens* out of Africa. *Quaternary International, 382,* 8–30.

Groucutt, H. S., Scerri, E. M. L., Stringer, C., & Petraglia, M. D. (2019). Skhul lithic technology and the dispersal of *Homo sapiens* into Southwest Asia. *Quaternary International, 510,* 30–52.

Kreager, P., Winney, B., Ulijaszek, S., & Capelli, C. (Eds.). (2015a). *Population in the human sciences: Concepts, models, evidence*. Oxford: Oxford University Press.

Kreager, P., Winney, B., Ulijaszek, S., & Capelli, C. (2015b). An introduction. In P. Kreager, B. Winner, S. Ulijaszek, & C. Capelli (Eds.), *Population in the human sciences: Concepts, models, evidence* (pp. 3–51). Oxford: Oxford University Press.

Krieger, N. (2012). Who and what is a "population"? Historical debates, current controversies, and implications for understanding "population health" and rectifying health inequities. *The Milbank Quarterly, 90,* 634–681.

Kroeber, A. L. (1931). Historical reconstruction of culture growths and organic evolution. *American Anthropologist, 33,* 149–156.

Langer, M. C., Ezcurra, M. D., Rauhut, O. W. M., Benton, M. J., Knol, F., McPhee, B. W., et al. (2017). Untangling the dinosaur family tree. *Nature, 551,* E1–E3.

Lawson, D. J. (2015). Populations in statistical genetic modelling and inference. In P. Kreager, B. Winner, S. Ulijaszek, & C. Capelli (Eds.), *Population in the human sciences: Concepts, models, evidence* (pp. 108–130). Oxford: Oxford University Press.

Lewis, L. (2017). *Early microlithic technologies and behavioural variability in Southern Africa and South Asia*. British Archaeological Reports International Series S2861, BAR, Oxford.

Lyman, R. L., O'Brien, M. J., & Dunnell, R. C. (1997). *The rise and fall of culture history*. New York: Plenum.

Lin, W., Rezek, Z., & Dibble, H. L. (2018). Experimental design and experimental inference in stone artefact archaeology. *Journal of Archaeological Method and Theory, 25,* 663–688.

Mackay, A., Stewart, B., & Chase, B. M. (2014). Coalescence and fragmentation in the Late Pleistocene archaeology of Southernmost Africa. *Journal of Human Evolution, 72,* 71–98.

Maher, L. A., & Macdonald, D. A. (2020). Communities of interaction: Tradition and learning in stone tool production through the lens of the Epipaleolithic of Khareh IV, Jordan. In H. Groucutt (Ed.), *Culture history and convergent evolution: Can we detect populations in prehistory?* (pp. 213–243). Cham, Switzerland: Springer.

McGhee, G. R. (2011). *Convergent evolution: Limited forms most beautiful*. Cambridge, MA: MIT Press.

McGhee, G. R. (2018). Limits on the possible forms of stone tools: A perspective from convergent biological evolution. In M. J. O'Brien, B. Buchanan, & M. I. Eren (Eds.), *Convergent evolution in stone-tool technology* (pp. 23–46). Cambridge, MA: MIT Press.

McNabb, J. (2020). Problems and pitfalls in understanding the Clactonian. In H. Groucutt (Ed.), *Culture history and convergent evolution: Can we detect populations in prehistory?* (pp. 29–53). Cham, Switzerland: Springer.
Mellars, P., Gori, K. C., Carr, M., Soares, P. A., & Richards, M. B. (2013). Genetic and archaeological perspectives on the initial modern human colonization of southern Asia. *Proceedings of the National Academy of Sciences USA, 110,* 10699–10704.
Moore, J., & Willmer, P. (1997). Convergent evolution in invertebrates. *Biological Reviews, 72,* 1–60.
O'Brien, M. J., & Bentley, R. A. (2020). Learning strategies and population dynamics during the Pleistocene colonization of North America. In H. Groucutt (Ed.), *Culture history and convergent evolution: Can we detect populations in prehistory?* (pp. 261–281). Cham, Switzerland: Springer.
O'Brien, M. J., Buchanan, B., & Eren, M. I. (Eds.). (2018a). *Convergent evolution in stone-tool technology.* Cambridge, MA: MIT Press.
O'Brien, M. J., Buchanan, B., & Eren, M. I. (Eds.). (2018b). Issues in archaeological studies of convergence. In M. J. O'Brien, B. Buchanan, & M. I. Eren (Eds.), *Convergent evolution in stone-tool technology* (pp. 3–20). Cambridge, MA: MIT Press.
Ranhorn, K. L. (2017). *Cultural transmission and lithic technology in Middle Stone Age Eastern Africa.* Unpublished Ph.D. thesis, The George Washington University.
Reynolds, N. (2020). Threading the weft, testing the warp: Population concepts and the European Upper Palaeolithic chronocultural framework. In H. Groucutt (Ed.), *Culture history and convergent evolution: Can we detect populations in prehistory?* (pp. 187–212). Cham, Switzerland: Springer.
Powell, A., Shennan, S., & Thomas, M. G. (2009). Late Pleistocene demography and the appearance of modern human behavior. *Science, 324,* 1298–1301.
Scerri, E. M. L. (2012). A new stone tool assemblage revisited: Reconsidering the 'Aterian' in Arabia. *Proceedings of the Seminar for Arabian Studies, 42,* 357–370.
Scerri, E. M. L. (2017). The North African Middle Stone Age and its place in recent human evolution. *Evolutionary Anthropology, 26,* 119–135.
Scerri, E. M. L., Drake, N. A., Jennings, R., & Groucutt, H. S. (2014). Earliest evidence for the structure of *Homo sapiens* populations in Africa. *Quaternary Science Reviews, 101,* 207–216.
Scerri, E. M. L., Blinkhorn, J., Gravina, B., & Delagnes, A. (2016). Can lithic attribute analyses identify discrete reduction trajectories? A quantitative study using refitting lithic constellations. *Journal of Archaeological Method and Theory, 23,* 669–691.
Scerri, E. M. L., Thomas, M. G., Manica, A., Gunz, P., Stock, J., Stringer, C. B., et al. (2018). Did our species evolve in subdivided populations spread across Africa, and why does it matter? *Trends in Ecology and Evolution, 33,* 582–594.
Schmidt, P. (2020). Culture, environmental adaptation or specific problem solving? On convergence and innovation dynamics related to techniques used for stone heat treatment. In H. Groucutt (Ed.), *Culture history and convergent evolution: Can we detect populations in prehistory?* (pp. 283–290). Cham, Switzerland: Springer.
Shea, J. J. (2014). Sink the Mousterian? Named stone tool industries (NASTIES) as obstacles to investigating hominin evolutionary relationships in the Later Middle Palaeolithic Levant. *Quaternary International, 350,* 169–179.
Shennan, S. (2020). Style, function and cultural transmission. In H. Groucutt (Ed.), *Culture history and convergent evolution: Can we detect populations in prehistory?* (pp. 291–298). Cham, Switzerland: Springer.
Shennan, S. J., Crema, E., & Kerig, T. (2015). Isolation-by-distance, homophily, and "core" vs. "package" cultural evolution models in prehistoric Europe. *Evolution and Human Behavior, 36,* 103–109.
Shipton, C. (2020). The unity of Acheulean culture. In H. Groucutt (Ed.), *Culture history and convergent evolution: Can we detect populations in prehistory?* (pp. 13–27). Cham, Switzerland: Springer.
Shott, M. J. (2003). Chaîne Opératoire and reduction sequence. *Lithic Technology, 28,* 95–105.
Shott, M. J. (2020). Toward a theory of the point. In H. Groucutt (Ed.), *Culture history and convergent evolution: Can we detect populations in prehistory?* (pp. 245–259). Cham, Switzerland: Springer.
Soressi, M., & Geneste, J. M. (2011). The history and efficacy of the chaîne opératoire approach to lithic analysis: Studying techniques to reveal past societies in an evolutionary perspective. *Paleoanthropology, 2011,* 250–334.
Spinapolice, E. E. (2020). Lithic variability and cultures in the East African Middle Stone Age. In H. Groucutt (Ed.), *Culture history and convergent evolution: Can we detect populations in prehistory?* (pp. 87–102). Cham, Switzerland: Springer.
Stanford, D., & Bradley, B. (2012). *Across Atlantic ice: The origin of America's Clovis culture.* Berkley: University of California Press.
Stutz, A. J. (2020). The Middle-Upper Paleolithic transition: A long-term biocultural effect of anatomically modern human dispersal. In H. Groucutt (Ed.), *Culture history and convergent evolution: Can we detect populations in prehistory?* (pp. 157–186). Cham, Switzerland: Springer.
Tostevin, G. B. (2011). Levels of theory and social practise in the reduction sequence and chaîne opératoire methods of lithic analysis. *Paleoanthropology, 2011,* 351–375.
Tostevin, G. B. (2012). *Seeing lithics: A middle-range theory for testing cultural transmission in the Pleistocene.* Oxford: Oxbow Books.
Tryon, C. A., & Ranhorn, K. L. (2020). Raw material and regionalization in Stone Age East Africa. In H. Groucutt (Ed.), *Culture history and convergent evolution: Can we detect populations in prehistory?* (pp. 143–156). Cham, Switzerland: Springer.
Trigger, B. T. (2006). *A history of archaeological theory* (2nd ed.). Cambridge: Cambridge University Press.
Will, M., & Mackay, A. (2020). A matter of space and time: How frequent is convergence in lithic technology in the African archaeological record over the last 300 kyr? In H. Groucutt (Ed.), *Culture history and convergent evolution: Can we detect populations in prehistory?* (pp. 103–126). Cham, Switzerland: Springer.
Will, M., Mackay, A., & Philips, N. (2015). Implications of Nubian-like core reduction systems in southern Africa for the identification of early modern human dispersals. *PLoS ONE, 10,* e0131824.
Will, M., Tryon, C., Shaw, M., Scerri, E. M. L., Ranhorn, K., Pargeter, J., et al. (2019). Comparitive analysis of Middle Stone Age artifacts. *Evolutionary Anthropology, 28,* 57–59.
Willey, G. R. (1953). Archaeological theories and interpretations: New World. In A. L. Kroeber (Ed.), *Anthropology today* (pp. 361–385). Chicago: University of Chicago Press.
Wilkins, J. (2018). The point is the point: Emulative social learning and weapon manufacture in the Middle Stone Age of South Africa. In M. J. O'Brien, B. Buchanan, & M. I. Eren (Eds.), *Convergent evolution in stone-tool technology* (pp. 153–173). Cambridge, MA: MIT Press.
Wright, S. (1984). *Evolution and the genetics of populations.* Chicago: University of Chicago Press.

Chapter 2
The Unity of Acheulean Culture

Ceri Shipton

Abstract This chapter examines the issue of whether the Acheulean is a genuine homologous cultural entity, descended via a chain of social reproduction from a common 'ancestor', or whether it was a technological phase that was repeatedly independently invented. An anecdotal experiment is used to determine the relative ease of inventing biface knapping from scratch, versus transmitting it with one bout of social observation. Handaxe and cleaver elongation is compared between East African and Indian Acheulean assemblages to determine if there are systematic differences that might reflect different lineages of social transmission. The age of the first appearance of the Acheulean in various parts of the world is modelled to determine if spread from a single source or independent inventions best fits the timing of its distribution. The issue of whether Pleistocene bifaces from East Asia are homologous with the Acheulean or were independently invented is examined by comparing the extent of bifacial shaping between East Asian and western Acheulean assemblages. The chapter concludes with the following contentions. Acheulean bifaces are hard to invent, or even emulate, but easy to imitate. Pleistocene East Asian bifaces are an example of parallelism; that is, not *de novo* independent invention, but invention from the same Oldowan substrate as the Acheulean. The western Acheulean is however a coherent cultural entity that seems to have spread from a single source region, and with regionally consistent variations suggesting it was maintained through social transmission.

Keywords Cleaver • Handaxe • Invention • Movius Line • Parallelism • Transmission

C. Shipton (✉)
Centre of Excellence for Australian Biodiversity and Heritage, Australian National University, Canberra, Australia
e-mail: ceri.shipton@anu.edu.au

Institute of Archaeology, University College London, 31–34 Gordon Square, Bloomsbury, London WC1H 0PH, UK

Introduction

Acheulean sites have been found at locations separated by over 80° of latitude (Aldhouse-Green et al. 2012; Lotter and Kuman 2017), and in time by over 1.5 million years (Beyene et al. 2013; Benito-Calvo et al. 2014). Such an enormous geographical and temporal range transcends different hominin species and makes the idea that the Acheulean is a single cultural entity, tracing its roots to a common ancestor, seem improbable. For our own species it is not until the modern era of global transport and communication that the same artifact types become so widespread across the globe, and now there is typically rapid turnover of forms rather than the almost interminable persistence of the Acheulean.

One explanation for Acheulean ubiquity, is that it was in part genetically determined (Corbey et al. 2016); a hypothesis which myself and others have critiqued elsewhere (Wynn and Gowlett 2018; Shipton and Nielsen 2018; Hosfield et al. 2018). Here, I wish to address the more plausible alternative explanations that the characteristic bifacial artifact forms of the Acheulean were repeatedly independently invented (Tennie et al. 2016, 2017), or that at least the Asian, European, and African Acheulean traditions arose independently (Barsky et al. 2018; Gallotti 2016).

Acheulean assemblages the world over have a diagnostic feature in common: the presence of bifacially shaped artifact forms. In particular, a tear-drop shaped form with a long cutting edge around much of its perimeter—the handaxe; and, also for most Acheulean assemblages, a form with a broad unretouched bit as its cutting edge—the cleaver. Bifacial flaking is one of the simplest ways to remove multiple flakes from a flattish stone, with bifacial forms present from the Oldowan (de la Torre 2004). It is the easiest way to shape a stone: bifacial flaking being the most common method used to shape stone tools throughout prehistory (Inizian et al. 1983), and with the most elaborate stone artifact shapes in prehistory being bifacial (e.g. Carballo 2007).

H. S. Groucutt (ed.), *Culture History and Convergent Evolution: Can We Detect Populations in Prehistory?*,
Vertebrate Paleobiology and Paleoanthropology, https://doi.org/10.1007/978-3-030-46126-3_2

Handaxe-like bifaces crop up repeatedly in later prehistory (e.g. Moore 2003; Brumm and Moore 2012; Brumm and Rainey 2015), so the possibility that Acheulean ones were also independently invented needs serious consideration. Cleavers on the other hand present a very different case, being a very specific tool (Inizian et al. 1983), with few or no parallels at other points in prehistory. Bifacial flaking on cleavers is often limited to a few marginal scars to regularize the outline shape, with the principal cutting edge typically the unretouched straight edge of a flake blank. The only potential example of non-Acheulean cleavers is the localized Middle Paleolithic of the Vasco-Cantabrian region in western Europe (Thiébaut et al. 2012; Utrilla et al. 2015; Deschamps 2017). However, the Vasco-Cantabrian Middle Paleolithic, previously thought to be a late regionalization, is now known to have its origins in Marine Isotope Stage 5 (Deschamps 2017; Álvarez-Alonso 2014), while the Iberian Acheulean persists until Marine Isotope Stage 6 (Rios-Garaizar et al. 2011; Álvarez-Alonso 2014). There may then have been long term continuity of bifacial forms from the Acheulean to the Middle Paleolithic, as unusually for Europe, cleavers were common in the Iberian Acheulean.

This chapter will argue that the Acheulean is a unitary cultural phenomenon, explained by strong social transmission rather than repeated independent invention (convergent evolution). The results of an anecdotal experiment on the social transmission of Acheulean-like biface knapping will be presented; Acheulean biface elongation in Africa and India will be compared to test for cultural divergence versus stochastic variation; the timing of the first appearance of the Acheulean in various parts of the world will be assessed to see if a diffusion or independent invention model is a better fit; finally the phenomenon of shaping of para-Acheulean bifaces in East Asia will be explored as a possible example of convergence.

An Anecdotal Experiment of Biface Transmission

An experiment by Geribàs and colleagues (2010) compared handaxe knapping between experts and complete novices with no prior experience of knapping. Drawing on the Geribàs experiment, this section looks at the difference one bout of social transmission can make to a novice's ability to make handaxes. A naïve subject with no prior experience of knapping was asked to make two handaxes; firstly without any prior knowledge apart from what the final form should look like, and secondly after having seen the process demonstrated once.

The subject, JP, despite being the girlfriend of the author, knew nothing of the process of making stone tools, but was shown a pointy handaxe made by expert knapper Chris Clarkson and asked to replicate it with no other verbal instructions. JP was given a glove, a leather pad, and a copper bopper hammer, and told to select a piece of Norfolk flint from a pile containing a variety of shapes, all suitable for knapping, but not necessarily suitable for making handaxes.

Similar to the novices in the Geribàs experiment, JP's principal approach was to strike the clast in the secant plane, i.e. to bash it on the ends rather than work in from the sides (Fig. 2.1). Although JP's clast was ultimately too thick to have ever been made into a handaxe, the Geribàs study indicates that, even with an appropriate clast thickness, one of the reasons naïve people fail to make handaxes is because of their focus on the secant plane and their failure to identify acute angles.

The other striking thing about JP's attempts to knap was the variety of methods attempted. She began by picking up another clast and attempting indirect percussion; she then tried direct percussion; then she rested the core on the hard floor; then she moved a large quartzite cobble to use as an anvil (Fig. 2.2); before finally giving up. The Geribàs et al. experiment also found that novices frequently used the ground and anvils as supports. JP's exploring of different percussive techniques in a single knapping bout was impressive, but not having the knowledge to identify appropriate angles and platforms for flaking, she was not able to strike more than a handful of flakes from the clast.

After this failed attempt, the author then knapped a handaxe with JP watching, but no verbal instructions were provided. During this demonstration there was a key moment of realization: "ah, you hit it from the other side". JP then attempted her second handaxe, choosing a much flatter clast, identifying acute angles, and working in from the edges. Remarkably, with this one bout of imitative social transmission, she was able to produce a handaxe that would be immediately recognizable archaeologically (Fig. 2.3). She flaked it around the entire perimeter leaving no trace of the original clast and created a globular butt and sharp cutting edge.

Of course, with a single subject this is only a pilot study and the conclusions must be regarded as highly tentative, but it is encouraging that some findings of the Geribàs et al. experiment were repeated. The failure of the initial attempt here, despite the variety of percussion methods employed, suggests that even Oldowan style freehand percussion, may not be as easy to invent as Tennie et al. hypothesize. The striking improvement after just one bout of watching another knapper suggests that imitation is also a far more efficient way of learning to knap handaxes than emulation and trial and error. If there was motivation to make them, handaxes could have spread rapidly between hominins in social contact.

Fig. 2.1 The first attempt by novice knapper JP to replicate a handaxe without having seen it done before. Nowhere on the piece has a bifacial edge been established. Note the copper bopper hammer has left marks across all surfaces due to heavy but ineffective strikes, and the base of the clast (bottom right) exhibits extensive battering damage. The scale is in centimeters

Acheulean Biface Elongation

Elongation (length to width ratio) is one of the principal ways in which Acheulean bifaces vary between assemblages (Shipton 2013; Callow 1986; Wynn and Tierson 1990). In this section biface elongation is compared between East African and Indian assemblages to test between the zone of latent solutions and social transmission models of Acheulean ubiquity. If handaxes and cleavers were repeatedly independently invented (Tennie et al. 2017), site-wise variation in elongation should be random at the continental scale, with no systematic difference between East Africa and India. If

Fig. 2.2 JP's varied method of percussion on her first attempt to make a handaxe. Top left—indirect percussion; top right—direct percussion; bottom left—using the floor as a support; bottom right—on-anvil percussion. Note that she is striking the clast in the secant plane

handaxes and cleavers were socially transmitted artifact types, then we should expect local grouping between assemblages as the result of regional traditions.

Measurements on East African and Indian bifaces were obtained from samples collected for previous studies (Shipton 2013, 2016, 2018), with the addition of a small sample of 21 bifaces from Kalambo Falls housed in the British Museum. The East African assemblages in the sample were Olduvai Gorge Bed II and Bed IV; Kariandusi; Isenya; Kalambo Falls; and Olorgesailie CL1-1, Member 6/7, and Upper Member 1. The Indian assemblages in the sample were Isampur Quarry; Teggihalli II; Singi Talav;

Fig. 2.3 JP's second attempt at making a handaxe, after having seen the process demonstrated once. Note the piece is flaked around the entire perimeter and has the characteristic globular butt and elongate cutting edge of a handaxe and would be archaeologically recognizable as such. The scale is in centimeters

Chirki-Nevasa; Morgaon; Bhimbetka; and Patpara. Assemblages with sample sizes of handaxes or cleavers of less than ten were excluded.

Tables 2.1 and 2.2 summarize the data and Figs. 2.4 and 2.5 show the pattern of variation in elongation both within and between sites for East Africa and India. For both biface types, East African assemblages tend to be more elongate than Indian assemblages. One-way ANOVAs confirmed the heterogeneity in assemblage mean elongation for both handaxes (df = 322, F = 10.494, P < 0.001) and cleavers (df = 252, F = 6.405, P < 0.001). The Indian assemblage Chirki-Nevasa has more elongate handaxes than the East African assemblage Olorgesailie Upper Member 1, and more elongate cleavers than the East African assemblage Kalambo Falls. But, the Chirki-Nevasa bifaces are still less elongate on average than all other East African assemblages. Notably, the differences in elongation are apparent between both classic Acheulean sites from either region, such as Kariandusi and Morgaon, as well as sites from the end of the Acheulean, such as Kalambo Falls and Patpara.

Some possible explanations for this geographic pattern are differences in rock type and blank form driving the differences in elongation. However, both the Indian and East African assemblages include examples that were invariably made on lava flakes such as the cleavers from Morgaon and Olorgesailie CL1-1, with an equal variances t-test confirming the significance of the difference between these two (df = 52, t = 3.222, P = 0.02). Likewise, both the Indian and East African assemblages include examples invariably made on quartzite flakes, such as the cleavers from Kalambo Falls and Bhimbetka, with an equal variances t-test confirming the significance of the difference between these two (df = 44, t = 2.901, P = 0.06). Differences in reduction intensity between East Africa and India might be invoked to explain these differences in elongation (cf. McPherron 1999). However, reduction intensity has been shown to have only a subtle influence on handaxe shape (Shipton and Clarkson 2015b); and in the case of cleavers, as their cutting edge is typically formed from the unretouched edge of the flake blank, they were by definition not resharpened (Shipton and Clarkson 2015a). Discounting systematic differences between East Africa and India in reduction intensity, or the influence of blank and rock type on biface elongation, we are left with the explanation that the differences in elongation arose due to divergent cultural traditions between these two Acheulean regions.

Table 2.1 Elongation values (length/width) for various East African and Indian Acheulean handaxe assemblages

Site	N	Minimum	Lower quartile	Mean	Upper quartile	Maximum
Isenya	18	1.77	1.87	2.05	2.2	2.37
Kalambo Falls	10	1.58	1.77	1.86	1.95	2.13
Olduvai Gorge Bed IV	41	1.52	1.74	1.85	1.95	3.07
Olduvai Gorge Bed II	30	1.41	1.75	1.85	2.01	2.59
Olorgesailie Member 6/7	16	1.69	1.74	1.84	1.93	2.12
Kariandusi	68	1.35	1.65	1.73	1.83	2.04
Chirki-Nevasa	21	1.38	1.59	1.72	1.82	2.32
Olorgesailie Upper Member 1	17	1.29	1.57	1.66	1.82	1.93
Isampur Quarry	47	1.01	1.52	1.66	1.84	2.29
Singi Talav	28	1.25	1.48	1.65	1.81	2.04
Teggihalli II	12	1.14	1.37	1.5	1.67	1.82
Patpara	20	1.22	1.3	1.46	1.59	1.71

Table 2.2 Elongation values (length/width) for various East African and Indian Acheulean cleaver assemblages

Site	N	Minimum	Lower quartile	Mean	Upper quartile	Maximum
Olorgesailie Member 6/7	19	1.41	1.65	1.76	1.91	2.05
Isenya	16	1.43	1.53	1.66	1.8	1.9
Olorgesailie CL1-1	25	1.29	1.56	1.65	1.75	1.99
Kariandusi	12	1.04	1.44	1.64	1.79	2.09
Olduvai Gorge Bed IV	11	1.34	1.45	1.62	1.7	1.88
Chirki-Nevasa	11	1.32	1.35	1.62	1.83	2.19
Kalambo Falls	11	1.34	1.44	1.59	1.73	1.91
Morgaon	31	1.16	1.32	1.47	1.63	1.92
Bhimbetka	36	1.19	1.33	1.44	1.52	1.8
Teggihalli II	19	1.28	1.3	1.43	1.52	1.75
Isampur Quarry	38	0.95	1.27	1.42	1.54	2.02

The First Appearance of the Acheulean

There remains the possibility that, while they represent social traditions, Acheulean bifaces were independently invented in various regions including East Africa and India (Barsky et al. 2018). Immediately prior to the emergence of the Acheulean, the hominin occupied world stretched the length of the African continent (Balter et al. 2008; Sahnouni et al. 2002), into Asia as far as north as the Lesser Caucasus (Ferring et al. 2011), and east into India (Gaillard et al. 2016; Dennell et al. 1988; Malassé et al. 2016) and China (Han et al. 2017; Hou and Zhao 2010; Li et al. 2017; Zhu et al. 2018). If stone tool using populations occupied this vast territory and the Acheulean was easy to invent, we should expect its emergence soon after the first appearance of hominins in a region. Alternatively, if the Acheulean was only invented once and spread from that source as a single tradition, we should expect a pattern of younger ages of first appearance the farther afield one moves from the source.

There have been claims for a very early appearance of the Acheulean in Armenia 1.85 Ma at the site of Karakhach (Trifonov et al. 2016). However, as illustrated, the three possible artifacts are not convincing as Acheulean bifaces (see Fig. 12 in Trifonov et al. 2016); they are extensively rolled and do not appear to have been shaped. Notwithstanding Karakhach, the three oldest Acheulean sites, with ages of 1.7–1.75 Ma, are Konso-Gardula (Beyene et al. 2013), Kokiselei (Lepre et al. 2011), and Olduvai Gorge FLK West (Diez-Martín et al. 2014): all located in East Africa, less than 1000 km apart. The earliest Acheulean sites in southern Africa are dated to 1.6–1.4 Ma (Gibbon et al. 2009; Chazan et al. 2008; Herries and Shaw 2011). The earliest sites in the Levant date to a similar 1.6–1.4 Ma timeframe (Ginat et al. 2003; Martínez-Navarro et al. 2012; Tchernov 1988), and there is one Acheulean site in India with a 1.5 Ma age (Pappu et al. 2011). The earliest date for the Acheulean on the Atlantic Coast of north-western Africa is around 1 Ma (Raynal et al. 2001). Moving further afield into Europe, the Acheulean does not appear until after 1 Ma (Moncel et al. 2013; Vallverdu et al. 2014). Notably, in all these regions there are older non-Acheulean assemblages, so its spread does not reflect the first arrival of stone knapping hominins in an area.

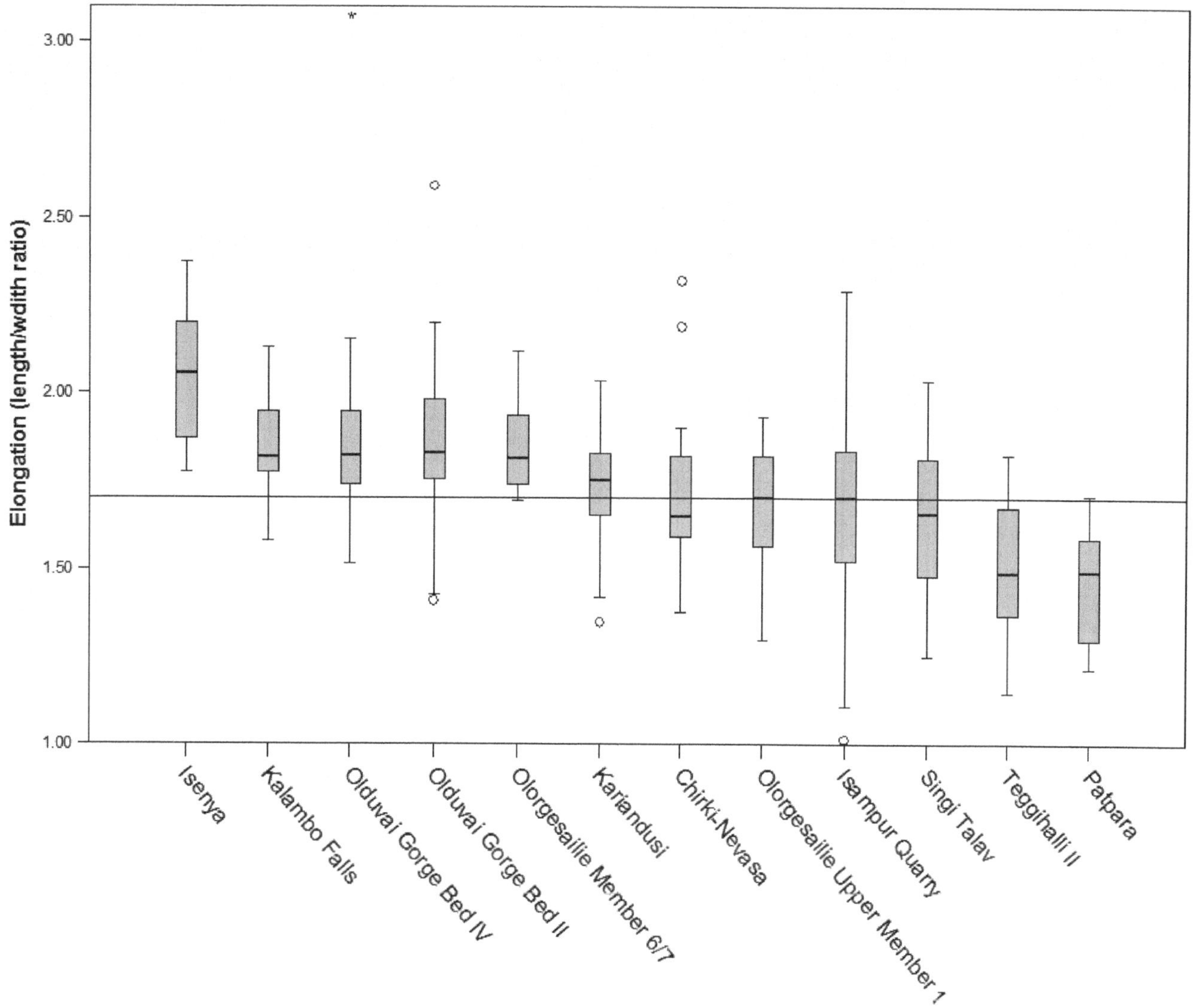

Fig. 2.4 Elongation in Acheulean handaxes for selected East African and Indian assemblages, ordered by mean values. Note that East African assemblages tend to sit above the reference line at 1.7 while Indian assemblages tend to sit below it

Date range estimates are large for sites of this age, particularly outside of East Africa where radiometric dating of volcanic eruptions is usually not possible. However, on current evidence it seems that there is time on the order of 100,000–200,000 years for the Acheulean to have spread from its East African homeland to southern Africa and the Levant, and perhaps as far as India; with several hundred thousand more years for the Acheulean to reach Europe. Table 2.3 shows the above sites alongside an approximate as-the-crow-flies distance from Kokiselei. If the Acheulean were repeatedly independently invented, we would expect there to be no relationship between the age of the first appearance of the Acheulean in a region and its distance from a putative East African source. A linear regression analysis was conducted of the data in Table 2.3 to test this. Attirampakkam was not included in the following analysis as the as-the-crow flies distance goes unrealistically across the Indian Ocean; it is also the only site dated by the relatively experimental technique of cosmogenic nuclides and the only site where the age estimate has not yet been corroborated by another within 200,000 years and 4000 km. The regression analysis (df = 9, F = 40.614, $P < 0.001$) indicates that there is in fact a strong relationship between the distance from Kokiselei and the age of the first appearance of the Acheulean in a region, with an R squared value of 0.835. The most parsimonious interpretation for the appearance of the Acheulean first in East Africa, later in southern Africa and the Levant, and much later still in Europe, is that it was a single tradition which spread through social transmission.

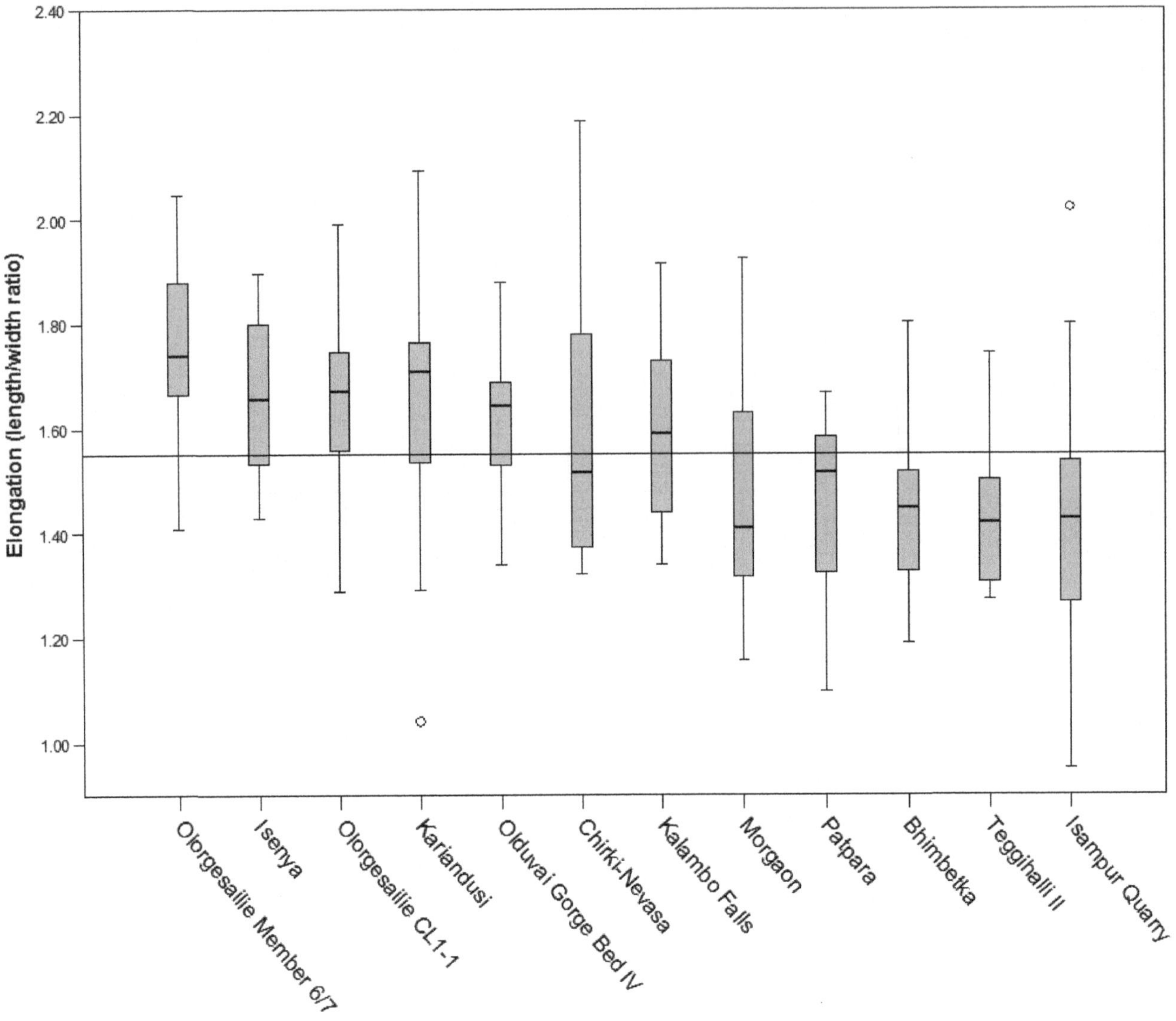

Fig. 2.5 Elongation in Acheulean cleavers for selected East African and Indian assemblages, ordered by mean values. Note that East African assemblages tend to sit above the reference line at 1.55, while Indian assemblages tend to sit below it

Table 2.3 The age estimates of the earliest Acheulean sites in East Africa, southern Africa, north-western Africa, the Middle East, India, and Europe and their as-the-crow-flies distance from Kokiselei. Age estimates are in million years and distance is approximated to the nearest 5 km

Site	Distance from Kokiselei	Age	References
Kokiselei	0	1.75	Lepre et al. (2011)
Konso-Gardula	235	1.75	Beyene et al. 2013)
Olduvai	775	1.7	Diez-Martín et al. (2014)
Sterkfontein	3460	1.4	Herries and Shaw (2011)
Rietputs Formation	3795	1.57	Gibbon et al. 2009)
ʻUbeidiya	3180	1.4	Tchernov (1988)
Nahal Zihor	2860	1.5	Ginat et al. 2003)
Thomas Quarry	5550	1	Raynal et al. (2001)
Attirampakkam	4940	1.5	Pappu et al. (2011)
Quipar	5360	0.9	Scott and Gibert (2009)
La Noira	5760	0.7	Moncel et al. (2013)

The Movius Line

Perhaps the most pertinent issue when it comes to the Acheulean and convergence is the Movius Line. In an early review of the Lower Paleolithic cultures of Asia, Hallam Movius (1948) noted that whereas bifaces are prevalent in India and areas to the west, they are rare or absent in East Asia. In the intervening years, several discoveries of Pleistocene biface assemblages in East Asia have purported to dissolve the Movius Line. The principal areas where such bifaces have been reported are the Bose (Baise) Basin in southern China (Hou et al. 2000), the Luonan Basin and Danjiankou region in central China (Wang 2005; Li et al. 2014), Dingcun on the Loess Plateau in northern China (Yang et al. 2014), and the Imjin-Hantan River Basin on the Korean Peninsula (Norton et al. 2006). Much has been written on whether these bifaces belong to the Acheulean tradition or are an example of convergence in Pleistocene hominin behaviour (e.g. Petraglia and Shipton 2008; Wang et al. 2012; Li et al. 2014; Lycett and Bae 2010).

Several distinctions between East Asian and western Acheulean bifaces are apparent. First, even in the above-mentioned areas of East Asia, bifaces occur at extremely low densities. In the Danjiankou region for example the maximum number of bifaces excavated per square meter at a site is 0.027 (Li et al. 2014), whereas in the western Acheulean densities of over 1 are common and over 10 is not unheard of (Méndez-Quintas et al. 2018). Second, both univariate and geometric morphometric studies have shown that East Asian bifaces tend to be both absolutely and relatively (to width) thicker than those of the Acheulean, and as a consequence heavier (Shipton and Petraglia 2010; Wang et al. 2012; Kuman et al. 2016). There are exceptions to this pattern, with the Danjiankou and Luonan bifaces falling in the range of Acheulean variation. A third distinguishing feature of East Asian bifaces is the dearth of cleavers (Corvinus 2004), which are a common biface type in India, Africa, Iberia, and some Middle Eastern sites. While there have been claims for cleavers in East Asia, for the most part these do not conform to the classic Acheulean cleaver where the bit is formed by the intersection of a dorsal flake scar and the termination of the large flake blank. An exception to this is again the Luonan Basin bifaces (Petraglia and Shipton 2008). The fourth distinction between East Asian and western Acheulean bifaces is the degree to which they have been bifacially shaped. Many of the purported handaxes from East Asia are in fact unifacial (Li et al. 2014; Hou et al. 2000). Absolute numbers of flake scars are low for East Asian bifaces in comparison to those of the Acheulean (Li et al. 2014), with marginal trimming to regularize the edge not apparent (Kuman et al. 2014).

Here the fourth of these distinctions between East Asian and western Acheulean bifaces is explored in more detail. Shaping is assessed through two measures, the bifaciality index (the ratio of the number of scars on the more flaked surface to the less flaked surface), and the scar density index (the number of scars per unit of surface area).

In the original publication on Bose, comparison with the bifaciality index from Olorgesailie was used to show that they fall within the range of Acheulean variation for this variable (Hou et al. 2000). However, the sample from Olorgesailie contained a large proportion of cleavers where much of the shaping, including the crucial large scar that will form the bit, is done prior to the striking of the flake blank and therefore would not be measured by the bifaciality index. To reassess shaping in the Bose large cutting tools, their bifaciality index (Hou et al. 2000) was compared with six handaxe assemblages, two from Africa (including Olorgesailie) (Shipton 2018), two from Europe (Shipton and Clarkson 2015b), and two from India (Shipton 2016). Aside from Olorgesailie, the other five assemblages were chosen for their comparability to Bose, where the large cutting tools are primarily made on cobbles of coarse-grained rocks such as sandstone, quartzite, and quartz, although flake-made large tools and finer grained chert also feature in the Bose assemblage. The six assemblages included a sample of the quartz handaxes from Olduvai Gorge (multiple Beds), a sample of phonolite handaxes from Olorgesailie (multiple Members), quartzite handaxes from Singi Talav, basalt cobble and flake handaxes from Chirki, flint cobble handaxes from Swanscombe, and chert cobble and flake handaxes from Broom.

Table 2.4 shows that most of the large cutting tools from Bose are in fact unifacial, with a negligible bifaciality index. Discounting these, even the bifacial large cutting tools from Bose are at the lowermost end of the range of Acheulean handaxe variation for the bifaciality index, although the differences between the Bose bifaces and some of these Acheulean assemblages are not statistically significant. Further details on the Bose bifaces are necessary to evaluate the degree to which they were shaped, but, given that a majority of the large tools are unifacial, it seems they are not comparable to the Acheulean where bifacially shaped artifacts typically form the dominant class of large tool (Gowlett 2015).

In the Danjiankou region the majority of large tools have at least some bifacial working (Li et al. 2014), so they are a potential candidate for Acheulean-like shaping. Published data on the Scar Density Index (Clarkson 2013), the number of flake scars per unit area of a piece of knapped stone, is available for the Danjiankou bifaces (Li et al. 2015). The Scar Density Index is a measure of reduction intensity, and

Table 2.4 The Bifaciality Index of six handaxe assemblages and the large cutting tools from Bose

Site	N	Mean	SD
Olduvai quartz	32	0.73335	0.16138
Olorgesailie	43	0.74789	0.17214
Singi Talav	28	0.80218	0.12777
Chirki	21	0.74117	0.15291
Swanscombe	34	0.76019	0.13712
Broom	29	0.81187	0.12893
Bose bifacial	35	0.68	0.22
Bose unifacial	64	0.03	0.11

for shaped tools such as bifaces indicates the amount of knapping that went in to creating the form that entered the archaeological record (Shipton and Clarkson 2015b). If the Danjiankou bifaces were shaped to the same extent as Acheulean ones, we should expect comparable levels of reduction intensity. Scar Density values were compared between the Danjiankou bifacial large cutting tools (unifacial ones were excluded from the analysis) and handaxes from Acheulean assemblages that have elements of blank and rock type in common with Danjiankou, such as the use of trachyte and other igneous rocks, quartz rich metamorphic rocks, and cobble and flake blanks. These assemblages were Singi Talav and Chirki from India, and Isenya, Kariandusi, Olduvai Gorge Beds II and IV, and Olorgesailie Members 1 and 6/7 from East Africa.

Figure 2.6 shows the variation in Scar Density Index (SDI) values for the Acheulean assemblages and both terraces from which the Danjiankou artifacts were recovered. Both Danjiankou assemblages have markedly lower SDI values than the Acheulean assemblages, with a one-way ANOVA test showing there was significant heterogeneity in this sample (df = 294, F = 22.352, $P < 0.001$). The Danjiankou bifaces even have lower SDI values than Olduvai Gorge Bed II, one of the oldest Acheulean assemblages where there was relatively little shaping, with an equal variances t-test showing that this pattern was significant at the $P = 0.005$ level (df = 124, t = 2.875). Figure 2.7 shows a selection of Danjiankou bifaces with relatively high SDI values alongside a range of Acheulean ones, to illustrate the limited amount of flaking that went into creating the former.

While a number of researchers have sought to abandon the Movius Line (e.g. Dennell 2016), it remains an important distinction between the western Acheulean tradition with high densities of intensively flaked and relatively thin bifaces, often including cleavers; and the sporadic East Asian examples of thick and cortical Pleistocene bifaces. The Luonan bifaces are more similar to the western Acheulean than any other assemblage currently known from East Asia and require further investigation. In relation to convergence, two explanations are possible for the general pattern of the Movius Line. The East Asian bifaces may be an example of parallelism; an independent invention of large bifacial tools, but from the same Oldowan substrate as the Acheulean was invented from in East Africa. In the latter, flake production from bifacial (discoidal) cores was an established feature of the hominin knapping repertoire prior to the invention of the Acheulean (e.g. de la Torre et al. 2008; Stout et al. 2010). Alternatively, East Asian bifaces may represent the dispersal of the Acheulean into East Asia but with the loss of some aspects of biface knapping due to the founder effect (Stout 2011). This might explain the loss of cleavers and the lack of biface thinning, but it does not explain the sporadic distribution of East Asian bifaces, or the low levels of reduction intensity, lower even than the very early Acheulean from Olduvai Gorge Bed II. Notwithstanding Luonan, the most parsimonious explanation for the East Asian Pleistocene bifaces is that they were independently invented, and given their, patchy distribution, possibly more than once.

Conclusion

The vast temporal and geographical extent of the Acheulean raises the possibility that it was not a single cultural entity, but a technological phase that was repeatedly independently invented. The experiment conducted by Geribàs and colleagues (2010) and the anecdotal experiment reported here, suggest that it is not easy to invent *de novo* or even emulate biface knapping. Furthermore, with the capacity for imitation and overimitation in our own species, the anecdotal experiment reported here indicates that biface knapping is easily transmitted via social transmission.

Mark Nielsen and I have argued on the basis of generalized means-end correspondence in multiple-step manufacturing sequences, and, the repeated localized occurrence of arbitrary variations in the most complex manufacturing sequences; that a propensity for imitation and even overimitation were features of Acheulean hominin behavior (Shipton 2010; Nielsen 2012; Shipton and Nielsen 2015; Shipton in press). If these mechanisms for robust social transmission were operating during the Acheulean, its

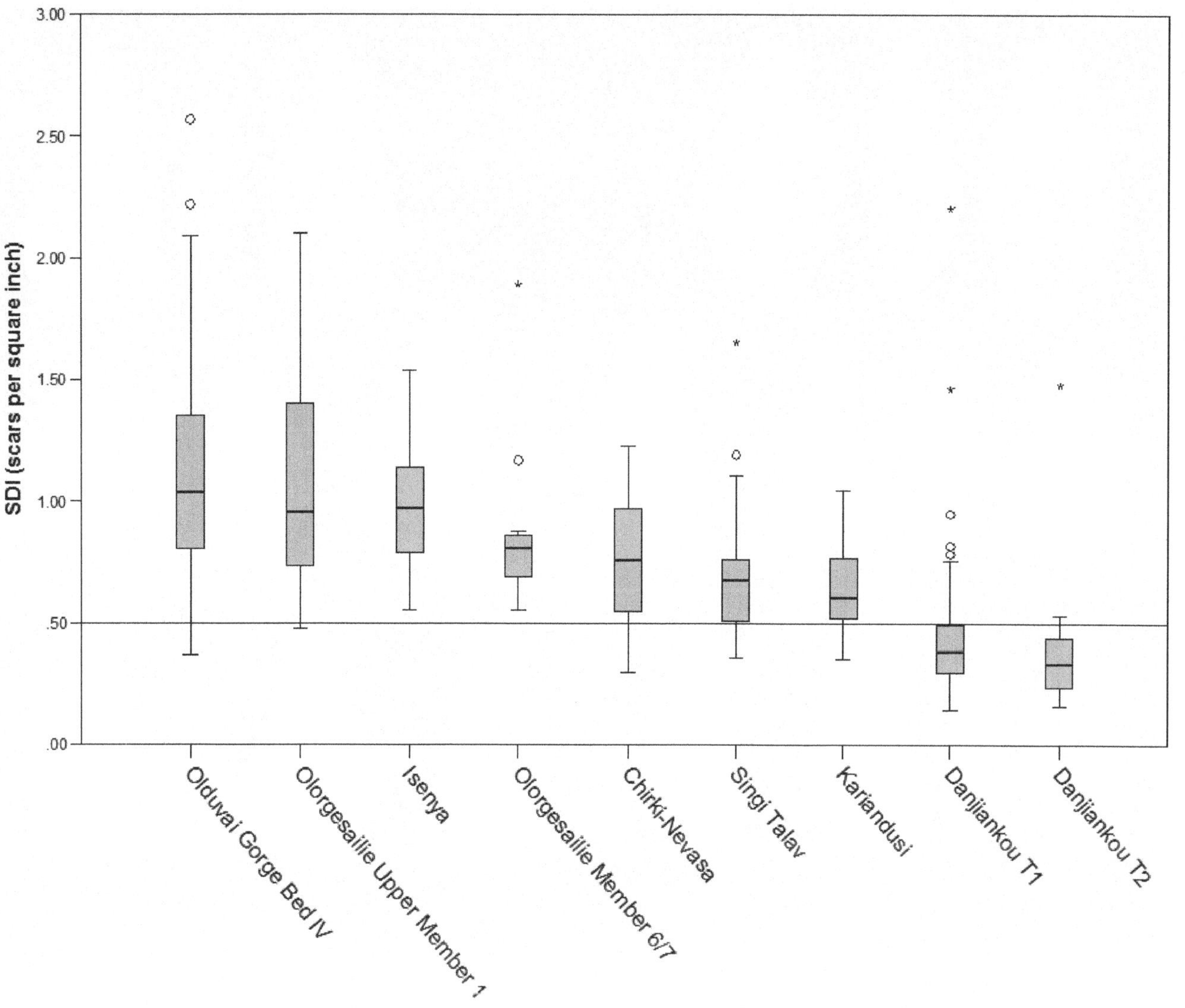

Fig. 2.6 Variation in Scar Density Index for seven Acheulean assemblages and two from the Danjiankou region. Assemblages are ordered by mean Scar Density Index (SDI) value. The reference line is at 0.485

artifact forms could have been maintained indefinitely and spread from a single source over a large area. The possibility that handaxes and cleavers were repeatedly invented from scratch without a prior knapping tradition seems remote for an animal even with the baseline levels of social transmission seen throughout the great apes, let alone if they had propensities to imitate and overimitate like our own species. A strong convergence argument to explain the ubiquity of the Acheulean can therefore be rejected.

This does not preclude the possibility of independent invention of Acheulean bifaces from a baseline Oldowan knapping tradition. Such parallelism indeed appears to have been operating with the emergence of bifacial large cutting tools in the Lower Paleolithic of East Asia. These bifaces, although similar in some respects to those of the Acheulean, are distinguished from them by their low density and patchy occurrence, their relative thickness, the dearth of cleavers, and, as demonstrated above, the low levels of shaping. East Asian bifaces thus provide us with models of what non-Acheulean large cutting tool assemblages look like.

When it comes to the western Acheulean several factors point to it being a single cultural phenomenon, rather than being invented in multiple places penecontemporaneously. Firstly, there is the specificity of cleavers as a tool type, which, unlike handaxe-like forms, do not recur at other points in prehistory. Secondly, regionally consistent differences are maintained over the course of the Acheulean (Wynn and Tierson 1990; Vaughan 2001; Lycett and Gowlett 2008) that are not easily explained by reduction intensity, rock type variation, or blank type variation. These differences suggest these biface forms were socially transmitted over extremely long periods. Thirdly, improvements

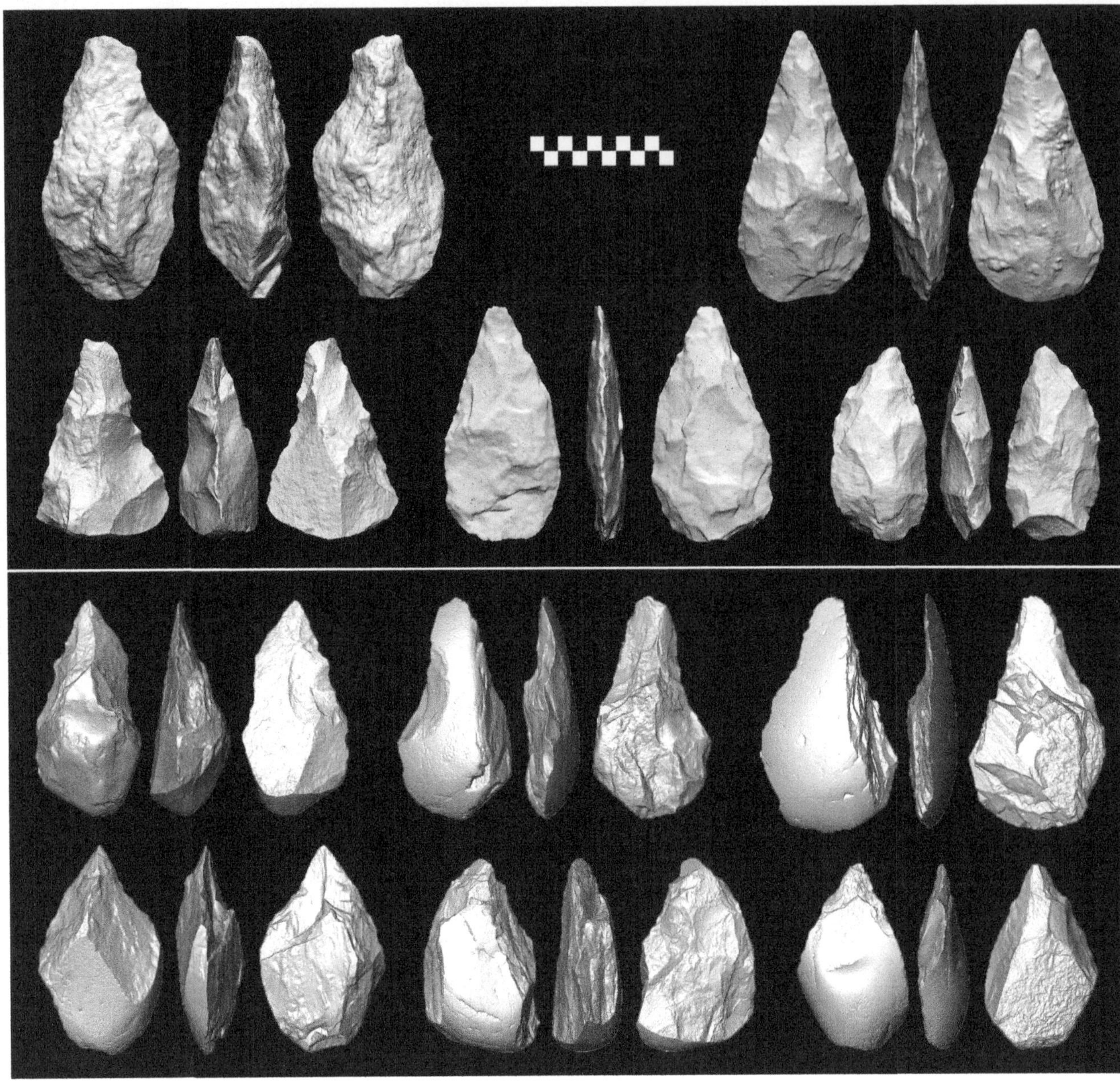

Fig. 2.7 Acheulean handaxes with a range of Scar Density Index (SDI) values (top) shown alongside a selection of Danjiankou bifaces with relatively high SDI values (bottom). Acheulean bifaces are from Olduvai Gorge Bed II (top left) (SDI = 0.55), Isenya (top right) (SDI = 1.13), Chirki (left) (SDI = 0.68), Olduvai Gorge Bed IV (middle) (SDI = 1.36), and Olorgesailie Member 6/7 (right) (SDI = 1.16). Note that a large proportion of the surface area of all the Danjiankou bifaces is still cortical. Lower part of the figure adapted from Li et al. (2015). The scale is in centimeters

over time in Acheulean knapping skill (Shipton 2013, 2018; Schick and Toth 2017; Chazan 2015), suggests it was a tradition that was maintained and improved upon. Fourthly, the three oldest Acheulean sites are all to be found in East Africa, and the date of the first appearance of the Acheulean in other parts of the world is consistent with a model of dispersal or diffusion from this source. Therefore, the contention of this chapter is that the Acheulean was indeed a unitary cultural phenomenon.

Acknowledgments I thank Huw Groucutt for the invitation to write this chapter, JP for agreeing to participate in the study, Chris Clarkson for providing the model handaxe, and three anonymous reviewers for constructive feedback on a previous draft of the chapter.

References

Aldhouse-Green, S., Peterson, R., & Walker, E. (2012). *Neanderthals in Wales: Pontnewydd and the Elwy Valley Caves*. Oxford: Oxbow Books.

Álvarez-Alonso, D. (2014). First Neanderthal settlements in northern Iberia: The Acheulean and the emergence of Mousterian technology in the Cantabrian region. *Quaternary International, 326,* 288–306.

Balter, V., Blichert-Toft, J., Braga, J., Telouk, P., Thackeray, F., & Albarède, F. (2008). U-Pb dating of fossil enamel from the Swartkrans Pleistocene hominid site, South Africa. *Earth and Planetary Science Letters, 267,* 236–246.

Barsky, D., Carbonell, E., & Ramos, R. S. (2018). Diversity and multiplicity in the Asian Acheulian. *L'Anthropologie, 122,* 59–73.

Benito-Calvo, A., Barfod, D. N., McHenry, L. J., & de la Torre, I. (2014). The geology and chronology of the Acheulean deposits in the Mieso area (East-Central Ethiopia). *Journal of Human Evolution, 76,* 26–38.

Beyene, Y., Katoh, S., WoldeGabriel, G., Hart, W. K., Uto, K., Sudo, M., et al. (2013). The characteristics and chronology of the earliest Acheulean at Konso, Ethiopia. *Proceedings of the National Academy of Sciences, 110,* 1584–1591.

Brumm, A., & Moore, M. W. (2012). Biface distributions and the Movius Line: A Southeast Asian perspective. *Australian Archaeology, 74,* 34–46.

Brumm, A., & Rainey, A. (2015). The Acheulean downunder: Modern human 'handaxes' from the Barkly Tableland of Northern Australia. *Lithics-The Journal of the Lithic Studies Society, 32,* 30–49.

Callow, P. (1986). A comparison of British and French Acheulian bifaces. In S. Collcutt (Ed.), *The Palaeolithic of Britain and its nearest neighbours: Recent trends* (pp. 3–7). Sheffield: J.R. Collis.

Carballo, D. M. (2007). Implements of state power: Weaponry and martially themed obsidian production near the Moon Pyramid, Teotihuacan. *Ancient Mesoamerica, 18,* 173–190.

Chazan, M. (2015). Technological trends in the Acheulean of Wonderwerk Cave, South Africa. *African Archaeological Review, 32,* 701–728.

Chazan, M., Ron, H., Matmon, A., Porat, N., Goldberg, P., Yates, R., et al. (2008). Radiometric dating of the Earlier Stone Age sequence in excavation I at Wonderwerk Cave, South Africa: Preliminary results. *Journal of Human Evolution, 55,* 1–11.

Clarkson, C. (2013). Measuring core reduction using 3D flake scar density: A test case of changing core reduction at Klasies River Mouth, South Africa. *Journal of Archaeological Science, 40,* 4348–4357.

Corbey, R., Jagich, A., Vaesen, K., & Collard, M. (2016). The acheulean handaxe: More like a bird's song than a beatles' tune? *Evolutionary Anthropology: Issues, News, and Reviews, 25,* 6–19.

Corvinus, G. (2004). Homo erectus in East and Southeast Asia, and the questions of the age of the species and its association with stone artifacts, with special attention to handaxe-like tools. *Quaternary International, 117,* 141–151.

de la Torre, I. (2004). The Omo revisited: Evaluating the technological skills of Pliocene hominids. *Current Anthropology, 45,* 439–465.

de la Torre, I., Mora, R., & Martínez-Moreno, J. (2008). The early Acheulean in Peninj (Lake Natron, Tanzania). *Journal of Anthropological Archaeology, 27,* 244–264.

Dennell, R. W. (2016). Life without the Movius line: The structure of the east and Southeast Asian Early Palaeolithic. *Quaternary International, 400,* 14–22.

Dennell, R. W., Rendell, H., & Hailwood, E. (1988). Early tool-making in Asia: Two-million-year-old artefacts in Pakistan. *Antiquity, 62,* 98–106.

Deschamps, M. (2017). Late Middle Palaeolithic assemblages with flake cleavers in the western Pyrenees: The Vasconian reconsidered. *Quaternary International, 433,* 33–49.

Diez-Martín, F., Uribelarrea, D., Baquedano, E., Mark, D., Mabulla, A., Fraile, C., et al. (2014). The origin of the Acheulean: The 1.7 million-year-old site of FLK West, Olduvai Gorge (Tanzania). *Scientific Reports, 5,* 17839–17839.

Ferring, R., Oms, O., Agustí, J., Berna, F., Nioradze, M., Shelia, T., et al. (2011). Earliest human occupations at Dmanisi (Georgian Caucasus) dated to 1.85–1.78 Ma. *Proceedings of the National Academy of Sciences, 108,* 10432–10436.

Gaillard, C., Singh, M., Malassé, A. D., Bhardwaj, V., Karir, B., Kaur, A., et al. (2016). The lithic industries on the fossiliferous outcrops of the Late Pliocene masol formation, Siwalik frontal range, northwestern India (Punjab). *Comptes Rendus Palevol, 15,* 341–357.

Gallotti, R. (2016). The East African origin of the Western European Acheulean technology: Fact or paradigm? *Quaternary International, 411,* 9–24.

Geribàs, N., Mosquera, M., & Vergès, J. M. (2010). What novice knappers have to learn to become expert stone toolmakers. *Journal of Archaeological Science, 37,* 2857–2870.

Gibbon, R. J., Granger, D. E., Kuman, K., & Partridge, T. C. (2009). Early Acheulean technology in the Rietputs Formation, South Africa, dated with cosmogenic nuclides. *Journal of Human Evolution, 56,* 152–160.

Ginat, H., Zilberman, E., & Saragusti, I. (2003). Early pleistocene lake deposits and Lower Paleolithic finds in Nahal (wadi) Zihor, Southern Negev desert, Israel. *Quaternary Research, 59,* 445–458.

Gowlett, J. (2015). Variability in an early hominin percussive tradition: The Acheulean versus cultural variation in modern chimpanzee artefacts. *Philosophical Transactions of the Royal Society B, 370,* 20140358.

Han, F., Bahain, J.-J., Deng, C., Boëda, É., Hou, Y., Wei, G., et al. (2017). The earliest evidence of hominid settlement in China: Combined electron spin resonance and uranium series (ESR/U-series) dating of mammalian fossil teeth from Longgupo cave. *Quaternary International, 434,* 75–83.

Herries, A. I., & Shaw, J. (2011). Palaeomagnetic analysis of the Sterkfontein palaeocave deposits: Implications for the age of the hominin fossils and stone tool industries. *Journal of Human Evolution, 60,* 523–539.

Hosfield, R., Cole, J., & McNabb, J. (2018). Less of a bird's song than a hard rock ensemble. *Evolutionary Anthropology: Issues, News, and Reviews, 27,* 9–20.

Hou, Y., Potts, R., Baoyin, Y., Zhengtang, G., Deino, A., Wei, W., et al. (2000). Mid-Pleistocene Acheulean-like stone technology of the Bose basin, South China. *Science, 287,* 1622–1626.

Hou, Y., & Zhao, L. (2010). An archeological view for the presence of early humans in China. *Quaternary International, 223,* 10–19.

Inizian, M.-L., Reduron-Ballinger, M., Roche, H., & Tixier, J. (1983). *Technology and terminology of knapped stone (Préhistoire de la Pierre Taillée).* Nanterre: Cercle de Recherches et d'Etudes Préhistorique.

Kuman, K., Li, C., & Li, H. (2014). Large cutting tools in the Danjiangkou Reservoir Region, central China. *Journal of Human Evolution, 76,* 129–153.

Kuman, K., Li, H., & Li, C. (2016). Large cutting tools from the Danjiangkou Reservoir Region, central China: Comparisons and contrasts with western and South Asian Acheulean. *Quaternary International, 400,* 58–64.

Lepre, C. J., Roche, H., Kent, D. V., Harmand, S., Quinn, R. L., Brugal, J.-P., et al. (2011). An earlier origin for the Acheulian. *Nature, 477,* 82–85.

Li, H., Kuman, K., & Li, C. (2015). Quantifying the reduction intensity of handaxes with 3D technology: A pilot study on handaxes in the Danjiangkou Reservoir region, central China. *PLoS ONE, 10,* e0135613.

Li, H., Li, C., & Kuman, K. (2014). Rethinking the "Acheulean" in East Asia: Evidence from recent investigations in the Danjiangkou Reservoir Region, central China. *Quaternary International, 347,* 163–175.

Li, H., Li, C., & Kuman, K. (2017). Longgudong, an Early Pleistocene site in Jianshi, South China, with stratigraphic association of human teeth and lithics. *Science China Earth Sciences, 60,* 452–462.

Lotter, M. G., & Kuman, K. (2017). The Acheulean in South Africa, with announcement of a new site (Penhill Farm) in the lower Sundays River Valley, Eastern Cape Province, South Africa. *Quaternary International*, 43–65.

Lycett, S. J., & Bae, C. J. (2010). The Movius Line controversy: The state of the debate. *World Archaeology, 42,* 521–544.

Lycett, S. J., & Gowlett, J. A. (2008). On questions surrounding the Acheulean 'tradition'. *World Archaeology, 40,* 295–315.

Malassé, A. D., Singh, M., Karir, B., Gaillard, C., Bhardwaj, V., Moigne, A.-M., et al. (2016). Anthropic activities in the fossiliferous Quranwala Zone, 2.6 Ma, Siwaliks of Northwest India, historical context of the discovery and scientific investigations. *Comptes Rendus Palevol, 15,* 295–316.

Martínez-Navarro, B., Belmaker, M., & Bar-Yosef, O. (2012). The Bovid assemblage (Bovidae, Mammalia) from the Early Pleistocene site of'Ubeidiya, Israel: Biochronological and paleoecological implications for the fossil and lithic bearing strata. *Quaternary International, 267,* 78–97.

McPherron, S. P. (1999). Ovate and pointed handaxe assemblages: Two points make a line. *Préhistoire Européenne, 14,* 9–32.

Méndez-Quintas, E., Santonja, M., Pérez-González, A., Duval, M., Demuro, M., & Arnold, L. (2018). First evidence of an extensive Acheulean large cutting tool accumulation in Europe from Porto Maior (Galicia, Spain). *Scientific Reports, 8,* 3082.

Moncel, M.-H., Despriée, J., Voinchet, P., Tissoux, H., Moreno, D., Bahain, J.-J., et al. (2013). Early evidence of Acheulean settlement in northwestern Europe-La Noira Site, a 700 000 year-old occupation in the center of France. *PLoS ONE, 8,* e75529.

Moore, M. W. (2003). Australian Aboriginal biface reduction techniques on the Georgina River, Camooweal, Queensland. *Australian Archaeology, 56,* 22–34.

Movius, H. L. (1948). The Lower Palaeolithic cultures of southern and eastern Asia. *Transactions of the American Philosophical Society, 38,* 329–420.

Nielsen, M. (2012). Imitation, pretend play, and childhood: Essential elements in the evolution of human culture? *Journal of Comparative Psychology, 126,* 170–181.

Norton, C. J., Bae, K., Harris, J. W., & Lee, H. (2006). Middle Pleistocene handaxes from the Korean peninsula. *Journal of Human Evolution, 51,* 527–536.

Pappu, S., Gunnell, Y., Akhilesh, K., Braucher, R., Taieb, M., Demory, F., et al. (2011). Early Pleistocene presence of Acheulian hominins in south India. *Science, 331*(6024), 1596–1599.

Petraglia, M. D., & Shipton, C. (2008). Large cutting tool variation west and east of the Movius Line. *Journal of Human Evolution, 55,* 962–966.

Raynal, J.-P., Alaoui, F. S., Geraads, D., Magoga, L., & Mohi, A. (2001). The earliest occupation of North-Africa: The Moroccan perspective. *Quaternary International, 75,* 65–75.

Rios-Garaizar, J., Garate, D., Gómez-Olivencia, A., Iriarte-Avilés, E., Aranburu-Artano, A., Arceredillo-Alonso, D., et al. (2011). The Lower to Middle Palaeolithic transition in northern Iberia: New data from Arlanpe Cave. *Antiquity, 85*(329).

Sahnouni, M., Hadjouis, D., Van Der Made, J., Canals, A., Medig, M., Belahrech, H., et al. (2002). Further research at the Oldowan site of Ain Hanech, North-eastern Algeria. *Journal of Human Evolution, 43,* 925–937.

Schick, K., & Toth, N. (2017). Acheulean Industries of the Early and Middle Pleistocene, Middle Awash, Ethiopia. *L'Anthropologie, 121,* 451–491.

Scott, G. R., & Gibert, L. (2009). The oldest hand-axes in Europe. *Nature, 461,* 82–85.

Shipton, C. (2010). Imitation and shared intentionality in the Acheulean. *Cambridge Archaeological Journal, 20*(02), 197–210.

Shipton, C. (2013). *A million years of hominin sociality and cognition: Acheulean bifaces in the Hunsgi-Baichbal Valley, India.* BAR International Series. Oxford: Archaeopress.

Shipton, C. (2016). Hierarchical organization in the Acheulean to Middle Palaeolithic transition at Bhimbetka, India. *Cambridge Archaeological Journal, 26,* 601–618.

Shipton, C. (2018). Biface knapping skill in the East African Acheulean: Progressive trands and random walks. *African Archaeological Review, 35,* 107–131.

Shipton, C. (in press). The evolution of social transmission in the Acheulean. In K. Overmann & F. L. Coolidge (Eds.), *Squeezing minds from stones.* Oxford: Oxford University Press.

Shipton, C., & Clarkson, C. (2015a). Flake scar density and handaxe reduction intensity. *Journal of Archaeological Science: Reports, 2,* 169–175.

Shipton, C., & Clarkson, C. (2015b). Handaxe reduction and its influence on shape: An experimental test and archaeological case study. *Journal of Archaeological Science: Reports, 3,* 408–419.

Shipton, C., & Nielsen, M. (2015). Before cumulative culture. *Human Nature, 26,* 331–345.

Shipton, C., & Nielsen, M. (2018). The acquisition of knapping skill in the Acheulean. In L. Di Paolo & F. Di Vincenzo (Eds.), *Evolution of primate social cognition* (pp. 283–297). Dordrecht: Springer.

Shipton, C., & Petraglia, M. D. (2010). Inter-continental variation in Acheulean bifaces. In C. Norton & D. Braun (Eds.), *Asian paleoanthropology* (pp. 49–55). Dordrecht: Springer.

Stout, D. (2011). Stone toolmaking and the evolution of human culture and cognition. *Philosophical Transactions of the Royal Society B: Biological Sciences, 366,* 1050–1059.

Stout, D., Semaw, S., Rogers, M. J., & Cauche, D. (2010). Technological variation in the earliest Oldowan from Gona, Afar, Ethiopia. *Journal of Human Evolution, 58,* 474–491.

Tchernov, E. (1988). The age of'Ubeidiya Formation (Jordan Valley, Israel) and the earliest hominids in the Levant. *Paléorient*, 63–65.

Tennie, C., Braun, D. R., Premo, L., & McPherron, S. P. (2016). The Island test for cumulative culture in the Paleolithic. In M. N. Haidle, N. Conard, & M. Bolus (Eds.), *The nature of culture* (pp. 121–133). Dordrecht: Springer.

Tennie, C., Premo, L., Braun, D. R., & McPherron, S. P. (2017). Early stone tools and cultural transmission. *Current Anthropology, 58,* 652–672.

Thiébaut, C., Mourre, V., Chalard, P., Colonge, D., Coudenneau, A., Deschamps, M., et al. (2012). Lithic technology of the final Mousterian on both sides of the Pyrenees. *Quaternary International, 247,* 182–198.

Trifonov, V., Lyubin, V., Belyaeva, E., Lebedev, V., Trikhunkov, Y. I., Tesakov, A., et al. (2016). Stratigraphic and tectonic settings of early Paleolithic of north-west Armenia. *Quaternary International, 420,* 178–198.

Utrilla, P., Mazo, C., & Domingo, R. (2015). Fifty thousand years of prehistory at the cave of Abauntz (Arraitz, Navarre): A nexus point between the Ebro Valley, Aquitaine and the Cantabrian Corridor. *Quaternary International, 364,* 294–305.

Vallverdu, J., Saladié, P., Rosas, A., Huguet, R., Cáceres, I., Mosquera, M., Garcia-Tabernero, A., Estalrrich, A., Lozano-Fernández, I., Pineda-Alcalá, A., & Carrancho, Á. (2014). Age and date for early arrival of the Acheulian in Europe (Barranc de la Boella, la Canonja, Spain). *PLoS One, 9*(7), e103634.

Vaughan, C. D. (2001). A million years of style and function: Regional and temporal variation in Acheulean handaxes. In T. D. Hurt & G. F. Rakita (Eds.), *Style and function: Conceptual issues in evolutionary archaeology* (pp. 141–163). Westport, CT: Bergin and Garvey.

Wang, S. (2005). *Perspectives on hominid behaviour and settlement patterns: A story of the Lower Palaeolithic sites in the Luonan Basin, China*. BAR International Series. Oxford: Archaeopress.

Wang, W., Lycett, S. J., von Cramon-Taubadel, N., Jin, J. J., & Bae, C. J. (2012). Comparison of handaxes from Bose Basin (China) and the western Acheulean indicates convergence of form, not cognitive differences. *PLoS ONE, 7,* e35804.

Wynn, T., & Gowlett, J. (2018). The handaxe reconsidered. *Evolutionary Anthropology: Issues, News, and Reviews, 27,* 21–29.

Wynn, T., & Tierson, F. (1990). Regional comparison of the shapes of later Acheulean handaxes. *American Anthropologist, 92,* 73–84.

Yang, S.-X., Huang, W.-W., Hou, Y.-M., & Yuan, B.-Y. (2014). Is the Dingcun lithic assembly a "chopper-chopping tool industry", or "Late Acheulian"? *Quaternary International, 321,* 3–11.

Zhu, Z., Dennell, R., Huang, W., Wu, Y., Qiu, S., Yang, S., et al. (2018). Hominin occupation of the Chinese Loess Plateau since about 2.1 million years ago. *Nature, 559,* 608–612.

Chapter 3
Problems and Pitfalls in Understanding the Clactonian

John McNabb

Abstract The Clactonian is a stone tool industry dating to MIS 11 and found in southern England. Its maker is currently thought to be *Homo heidelbergensis*, a hominin species known to make handaxes in Britain before and after the Clactonian. Currently, neither direct nor proxy dating techniques are able to establish a clear contemporaneity between the Clactonian and the Acheulean. Clactonian technology is a basic one, in that it is dedicated to the production of sharp edges. Clactonian sites are usually located near to water bodies or rivers where raw material (flint) is present. Its contemporary interpretation is influenced by the culture-historical approach, prevalent in the 1920s and 1930s when the Clactonian was first identified. This paper briefly reviews the historical context of the industry. It then places modern interpretations in the broader contemporary Middle Pleistocene chrono-stratigraphic and environmental context. Although no overarching new interpretation of the Clactonian is offered, convergent evolution can be seen to explain some of the patterns seen in Clactonian knapping technology. However, it cannot explain the Clactonian phenomenon itself. Some important points for future consideration in the Clactonian debate are presented.

Keywords Acheulean • Lithic • Middle Pleistocene • Non-handaxe • Paleolithic • Stone tool • Technology

Introduction

The Clactonian is a non-handaxe stone tool industry found in Britain and dated to the early part of Marine Isotope Stage (MIS) 11, c. 364–427 ka (McNabb 2007).

J. McNabb (✉)
Department of Archaeology, University of Southampton, Southampton, UK

It is a perfect subject for a volume on culture history and convergent evolution as it was born of, and understood through, the restrictive legacy of imperial culture history. By this I mean the global theory building of white male European archaeologists who, during the Edwardian and post-Edwardian periods, perpetuated beliefs in the universal validity of hierarchies of racial cultural and gender superiority. Although the age of empires was ending after the First World War, the Victorian mind-set that generated it soldiered on and still sought to impose a Eurocentric outlook on the world's Prehistory. This was empowered by a misreading of natural selection. Selection, it was believed, worked on organisms to improve them; a successful adaptation was a perfect adaptation. The perfectibility of adaptation applied to cultures as well people. This expressed itself in the early twentieth century through a variation on social Darwinism as Prehistory and human origins were sublimated to explore the divisions that separated living peoples. The co-evolution of humans and their material culture could expose the deep roots of racial differentiation and so explain why some races appeared more developed than others.

That legacy still influences our understanding of the Clactonian today. Ultimately, culture history puts boundaries around things to define and distinguish them. This made them easy to slot into developmental sequences. The Clactonian was about such boundaries, expressed through stone tools and stratigraphy—a 'stone and strat' debate. The hominin makers of the Clactonian were a relatively late insertion into the discussion.

The Clactonian exerts a strange fascination over those who study it. Its devotees harass each other with minutiae, exaggerate ambiguities and fire off contradictions from entrenched positions like soldiers from by-gone wars. It is a mirror for most of the problems that beset archaeology as a whole, at the scale of interpreter as well as interpretation. I will not describe in much detail the background to the Clactonian as a culture-historical phenomenon, I have done that elsewhere (McNabb 2007), and good reviews of the literature are

H. S. Groucutt (ed.), *Culture History and Convergent Evolution: Can We Detect Populations in Prehistory?*,
Vertebrate Paleobiology and Paleoanthropology, https://doi.org/10.1007/978-3-030-46126-3_3

provided by others (Bridgland 1994; White 2000; Pettitt and White 2012; Wenban-Smith 2013). Instead I will briefly draw out a few aspects of the Clactonian's life-history that, in my opinion, reflect its historicized character.

My intention is not to offer a new interpretation of the Clactonian as I do not have one. Instead I will review the current contexts of the Clactonian debate. However, I will offer some personal insights, data from my own research, and observations on differing aspects of the debate along the way. I have been studying the Clactonian for thirty-five years and I still don't get it; but I still love it.

Birth pangs and early childhood, pre-World War One—up to the late 1920s. Underpinning the theory building of this period was the notion that human and cultural evolution developed through a series of stages, and those stages were globally ubiquitous. The baleful influence of both Piltdown (Spencer 1990) and the longer running eolith controversy (O'Connor 2003; Sommer 2004; McNabb 2012), focused attention on Western Europe. Here was the oldest evidence for human evolution, apart from *Pithecanthropus*. The European races were the most advanced because they'd had longer to evolve. Archaeology and fossil evidence combined to demonstrate that. The Javanese *Pithecanthropus* was too ape-like and foreign! Even if our history began in Asia, as some believed, it was only after our ancestors arrived in Europe that the really significant advances began. Prehistory was co-opted to confirm what Europeans already knew to be true (McNabb 2012), namely, that they were better than everybody else and their ancestry proved it.

Hazzledine Warren, the father of the Clactonian (O'Connor 2006), was collecting artefacts from the West Cliff foreshore at Clacton-on-Sea before the First World War (Warren 1911, 1912). He identified a non-handaxe assemblage that he originally correlated with a similar one from Mesvin in Belgium (Warren 1922, 1923, 1924a). However in 1926 he renamed it the Clactonian (Warren 1926; McNabb 1996a).

During the 1920s Warren promoted his belief that the Clactonian was a nodule-tool non-handaxe industry. This was based on his views of what did and did not constitute a flake core. In modern terms, knapping by alternate flaking creates more acute edges. However, knapping by parallel flaking from flat platforms usually leaves a much higher angle between striking platform and flaking face. Warren saw the former as making a tool edge, while the latter was flake production. This was a dichotomy he maintained throughout his life and it directly influenced his interpretation of the Clactonian as a whole.

The idea of such a simple unsophisticated industry at the root of cultural evolution fitted the evolutionary theories of the time. Warren discounted eoliths. Arguments throughout the 1920s veered between mono-glacialists vs. multi-glacialists, and whether there was one evolving hominin species or a number of them, and how cultural evolution mapped onto these, see (Warren 1924b) for a good example of the complexity surrounding this topic (O'Connor 2007). An unshakeable belief shared by all researchers was that the passage of time had to be marked by progressive evolutionary development, what I have elsewhere characterized as the idea of progressive time (McNabb 1996a). As a simple and unsophisticated assemblage the Clactonian became a necessary starting point in any developmental sequence.

Growing up and maturity, the late 1920s until after the Second World War. By the end of the 1920s a consensus had emerged. The multi-glacialists held the ascendancy based on the four-fold glacial sequence worked out for the Alps (Penck and Bruckner 1909). This framework neatly divided up Pleistocene/progressive time into successive glacial and interglacial slots, demanding cultural progress in each sequential inter-glacial. Several evolving parallel lineages of hominins (phyla) each with their own diagnostic culture/assemblage type were fitted into this geological framework. The hominins of the Acheulean handaxe phylum were distinguished from the hominins of the Clactonian/Mousterian phylum. Each successive inter-glacial period saw advancements in material culture and in the characteristic signature tool of each phylum; so each phylum's *type fossil*, went through developmental stages as did its maker.

The 1930s were the *La Belle Époque* of culture history with the *fossile directeur* of the parallel phyla acting as dating techniques (Trigger 1989). Warren's nodule tool Clactonian was now reinterpreted by the Abbé Breuil as a flake-tool industry and once again sat at the root of an evolving phylum, in this case the Mousterian of the Neanderthals (Breuil 1932). Breuil epitomized the 1930s zeitgeist, and as the world's first globally famous prehistorian his influence was far reaching. British versions of his culture historical framework by Kenneth Oakley (King and Oakley 1936; Oakley and Leakey 1937) and T. T. Paterson (Paterson 1940, 1940–1941) were reactions to its details not to its broader principles.

After World War Two—towards retirement. The years after the Second World War saw a growing uncertainty in the understanding of the Clactonian's cultural and chrono-stratigraphic context. In this post-imperial period, as former colonies and protectorates sought their independence, there was an increasing focus on more local/regional archaeologies. There was something unsavory about the imposition of Eurocentric frameworks on the rest of the world, and the whole culture-historical project was now seen as a contributor to the racial and cultural underpinnings of Nazi genocide. The pathway to culture history had ended at the gates of Auschwitz. But what was there to replace it? Back to a single evolving phylum?

O'Connor (2007) is definitive on the post-war period. The 1950s saw two key developments. The first was the application of pollen analysis to British Middle Pleistocene sites (Pike and Godwin 1953) and its use as a relative dating framework (West 1956; Turner 1970). In Britain it took the place of the *fossile directeur* of material culture. Secondly, the publication of Louis Leakey's magisterial volume on Olduvai Gorge (Leakey 1951). While written in the 1930s in the culture historical context it charted a linear developmental cultural evolution from the non-handaxe chopper tool Oldowan through a series of more progressive Acheulean handaxe assemblages. Even if the detail of each successive Acheulean stage was now doubted (West and McBurney 1954), the notion of progressive time evidently still applied to handaxes. The real power of Olduvai was the demonstration of progressive time all at one site. It made the piecemeal European sequences seem parochial.

Warren published his last views on the Clactonian in the early 1950s, and his final chrono-stratigraphic interpretation a few years later (Warren 1951, 1955). The Clactonian was still a nodule tool assemblage but now it was linked to a major post-war development, the influential concept of cultural provinces by Hallam Movius (Movius 1948). This was as much to do with the East-West power blocks of the cold-war as with hominin territoriality (Trigger 1989; McNabb 1996a). The Clactonian sat squarely and uncomfortably in the African, Indian and European 'Acheulean family of handaxe cultures' along with a handful of other non-handaxe assemblages.

Warren died in 1959 but the Clactonian already had a new champion, John Wymer, who followed the fashion of a more nationally orientated interpretative focus. He too was promoting an Oldowan-like core-tool Clactonian (Wymer 1956). This may have owed something to Leakey's conception of the Oldowan. In 1959 and 1960 Leakey was arguing the Oldowan was 600 ka (Morell 1995). The next year, the new Potassium-Argon technique revealed it to be 1.75 Ma (Leakey et al. 1961). Suddenly any similarities between the Oldowan and the Clactonian seemed coincidental—more convergent evolution than culture history.

From the 1960s through to the 1980s there was little change in the understanding of the Clactonian. Its interpretation as a distinctive cultural entity was not questioned, but its context was a puzzle. In this post-colonial era it was still historically constituted, its boundaries established and maintained by a culture historical mind-set now in disrepute. I think of this as the late-classical phase of the Clactonian's life-history (Wymer 1968, 1974, 1985b; Roe 1981).

Palynology and handaxe typology (Pike and Godwin 1953; West and McBurney 1954; Wymer 1956) underpinned a generally progressive sequence from Clactonian to Early Acheulean (if present), and then to Middle Acheulean (the bulk of UK handaxe sites), and finally to Late Acheulean with Levallois. Whether the Clactonian was ancestral to, or just predated the Acheulean, was left unanswered (Wymer 1956, 1968, 1974; Roe 1968, 1981; Waechter 1973). There was little attempt to contextualize the Clactonian in its broader European or global context. A notable exception was that by Desmond Collins (Collins 1969).

A local industry for local archaeologists—the Clactonian in old age and demise. The first sustained attack on the reality of the Clactonian was by Milla Ohel (Ohel 1979) who asserted that measurements of typological and technological features on flakes and debitage from the Clactonian were identical to those from the Acheulean. In effect he quantified the physical evidence of knapping. One explanation for this similarity was that the non-handaxe sites were preparatory areas for the roughing out of handaxes, finished off by soft hammers elsewhere. The idea drew stinging responses from Wymer and the British archaeologists (see comments to Ohel's paper in Current Anthropology). Wymer's views on the Clactonian continued to reflect the late-classical outlook throughout the 1980s (Wymer 1985a, 1988)

The theoretical debates that engaged archaeology in the 1990s left the British Lower Paleolithic largely untouched. However. one important influence was the gradual 'humanizing' of the subject. Almost up until the late 1980s British Lower Paleolithic archaeology remained a purely stone and strat debate. The broader subject area of human origins had reacted to the theory of the Later Prehistorians by turning more to primate and chimp-based models of cognition and behavior. Behavior became the new culture. An important first step in this for the Clactonian was a paper by Steve Mithen (Mithen 1994) arguing that physical environment conditioned hominin learning behavior.

Subsequently, my own work (very much stone and strat) comparing the knapping technology of hard hammer core working in the Clactonian and Acheulean established that both assemblage types flaked their cores in the same way (McNabb 1992a). While this observation was accepted by most archaeologists my broader denial of the Clactonian was not, it also drew the fire of the late-classical establishment (see Wenban-Smith 2013; Pettitt and White 2012 on this). The late 1990s saw the Clactonian debate stall for lack of new data and ideas.

The Clactonian franchise—a modern reboot. Around 2000 those new ideas arrived, initiating a new cycle of Clactonian interpretations rooted in demography and population movements—a continuation of the humanizing of the subject. These arose from the work of Mark White, David Bridgland and Danielle Schreve. Beginning in 2000 Schreve and White (White 2000; White and Schreve 2000) linked hominin occupation in the UK to those times when it was actually possible to cross from the Continent to Britain because of lowered sea-level (Meijer and Preece 1995). This idea has been subsequently developed (Bridgland and White 2015; White 2015; White and Bridgland 2018) to show that patterns in handaxe typology are linked to potential changes

Table 3.1 The forest history of the Hoxnian (Ho) pollen stage, now equated with MIS11c. Based on the environment around the kettle lakes at Marks Tey (Turner 1970) and at Hoxne (West 1956)

HoIV	Post Temperate—the mixed oak forest all but disappears and birch and pine are the dominant trees species; there is a stronger presence of grasses and non-tree pollen	HoIVb. At Marks Tey the forests are still more common than grasslands, but the grasses now dominate over the shrubby crowberries. Birch is somewhat more frequent than pine.
		HoIVa. The crowberry appears at Mark Tey, a shrub well adapted to open heaths and birch and pine, particularly the latter, make marked rises in frequency.
HoIII	The Late Temperate when the mixed oak forest declines and hornbeam and fir tree forests replace earlier ones	Ho IIIb. At Marks Tey the fir trees show a marked rise in this stage and the alders continue steadily. Low frequencies of oak and yew.
		HoIIIa. At Marks Tey the firs appear but are low in numbers, the alders and hazels are still present, the hazels especially.
HoII	Early Temperate—mixed oak forests, dense with little overall pollen from open area tree species	HoIIc. At Marks Tey alder declines in frequency from the previous phase but remain present, while at Hoxne they increase locally. At Marks Tey yew and particularly hazel trees are now dominant, with Elm slightly more frequent. Lime trees may have been common in the Hoxne forests. At Marks Tey the elms and yews decline as the phase continues. Middle-end of this phase at Marks Tey and Hoxne sees a dramatic fall in tree pollen and a rise in open country species. This is the NAP—the non-arboreal pollen phase.
		HoIIb. The oak trees begin a slow decline in frequency and alder now dominates at Marks Tey. Lime trees and hazel become more common too. At Hoxne the replacement of birch by oak was relatively quick. The end of this stage at Hoxne is marked by a fluctuation of forest with a slight increase in open country.
		HoIIa. Oak rises markedly at Marks Tey to distinguish the beginning of this zone, the birch and pine trees decline in frequency. At Hoxne a birch forest is present and open grassland is declining.
HoI	Pre-Temperate	Birch is the dominant tree at Marks Tey, but pine trees increase as this phase progresses, and there is a rise in elm and oak towards the end of the phase.
1An	The Late Glacial	The recovery after the retreat of the Anglian (MIS 12) ice sheets. Open country and grasslands. At Hoxne the sea buckthorn, a deciduous shrub, dominates. The climate is cool but warming and the trees are beginning to return.

in hominin populations. Wenban-Smith also noted a similar structure to the Middle Pleistocene record (Wenban-Smith 1998). Ficron and cleaver making hominins arrive in Britain from MIS 9, replacing earlier hominins who made different kinds of axes in MIS 11. Both are preceded by Clactonian groups, early in each interglacial, who subsequently disappear before the Acheuleans arrive.

Ashton's (2018) recent views on the Clactonian also come under the umbrella of demography-driven culture. Populations of handaxe making hominins drift westwards into central Europe where a lack of suitable raw materials results in the knowledge of handaxe making being lost. This is the source area for further population movements at the beginning of MIS 11. These hominins cross the land bridge into Britain where their core and flake technology represents the Clactonian. The stability of MIS 11 landscapes and the utility of the Clactonian tool kit do not require any further development. It 'ain't broke and don't need fixing', so they carry on using it as it is. Local landscapes of habit develop as traditional knowledge is almost fossilized through continued utility; cultural stability predicated upon predictable resources, in turn rooted in landscape stability. It is only when climate induced disruption produces landscape unpredictability that things change. This is a trigger for new pulses of hominin migration with handaxe makers coming to replace the small and isolated Clactonian groups. Intriguingly, the non-arboreal pollen phase (NAP) a widespread period of deforestation within pollen zone HoIIc (Tye et al. 2016), see Table 3.1, is suggested as a catalyst for the end of the Clactonian in Britain.

Rounding off the 'neo-culturalists' is Wenban-Smith's views on the Clactonian-Acheulean relationship (Wenban-Smith 1998, 2013). In this configuration of the data, and rooted in the interpretation of the Southfleet elephant site (see below), Wenban-Smith argues for the in-situ evolution of the Acheulean from the Clactonian in Britain while high sea-levels separate Britain from the Continent. He queries the evidence for reconnections and migration opportunities prior to the MIS 11b stadial (but see Tye et al. 2016). Landscape is also a driver here, but it is a shift in behavior within changing landscapes that operationalizes the change from a static and locally configured cultural repertoire (Clactonian) to a landscape of expectation where handaxes are moved around in anticipation of encounters with animal resources. If I read him right, the monotonous Clactonian hunters relied on there being flint somewhere locally, whereas the Acheulean hunters were taking the tools to the job and leaving nothing to chance. At its root the Clactonian/Acheulean dichotomy is one of forward planning and the kit to match it; a landscape of places rather than more open spaces. Isolated from population incursions prior to MIS 11b the Clactonian develops into the Acheulean as hominins begin to understand landscapes in new ways. This is a local case of in-situ evolution (convergent evolution?) based on geographical isolation. Wenban-Smith's argument is set within a nuanced understanding of how archaeologists' appreciation of landscape is itself an artefact of a partial and discontinuous record—like poorly stacked slabs of Emmental cheese where the voids no longer quite match up.

The Clactonian World and Its Sites

Traditionally the British Clactonian was present at four sites; Swanscombe in Kent, Clacton-on-Sea in Essex, Barnham St Gregory in Suffolk, and Little Thurrock in Essex, see Fig. 3.1. The first two sites comprised a number of different

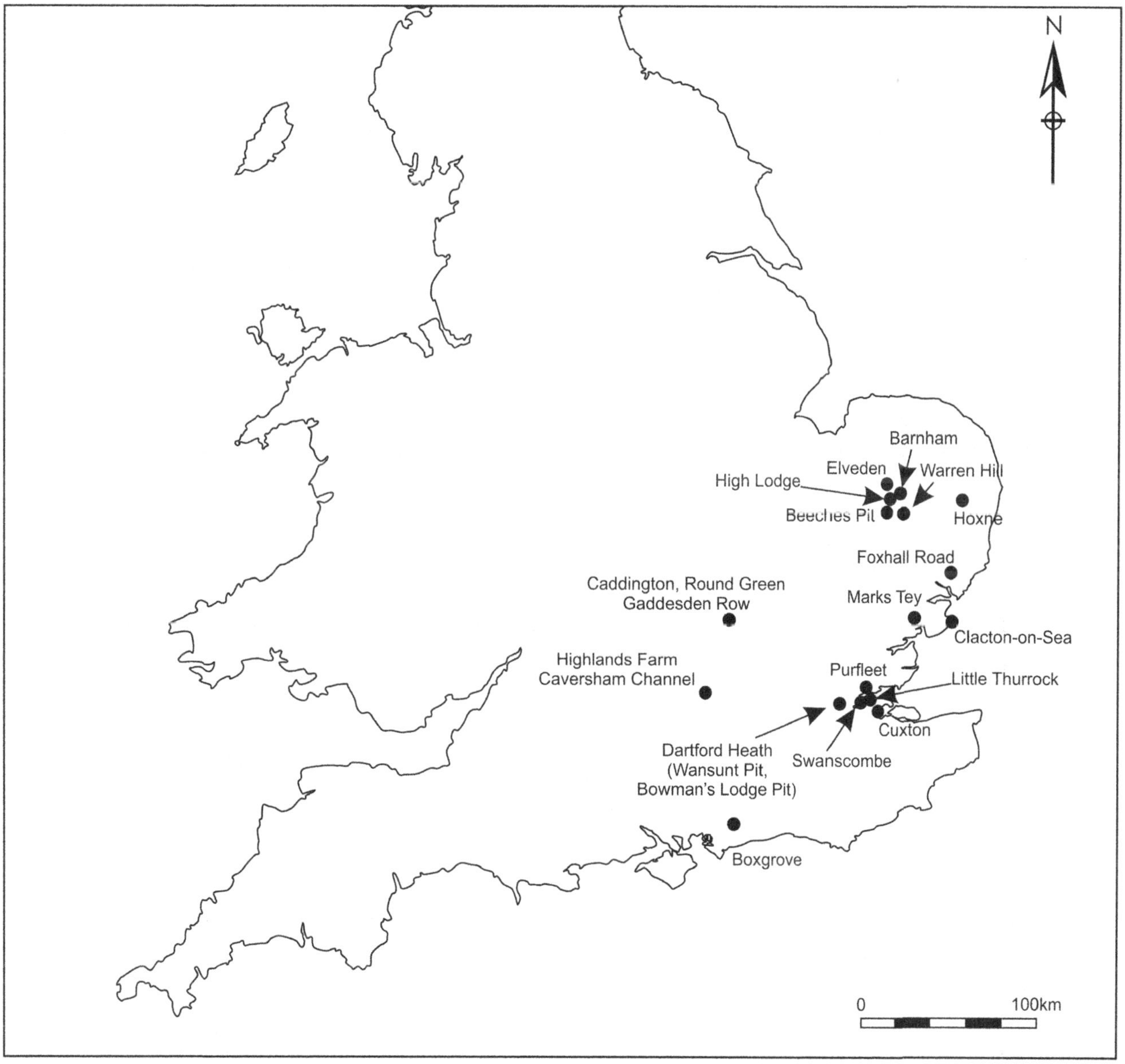

Fig. 3.1 Map showing Clactonian sites and other locations mentioned in the text

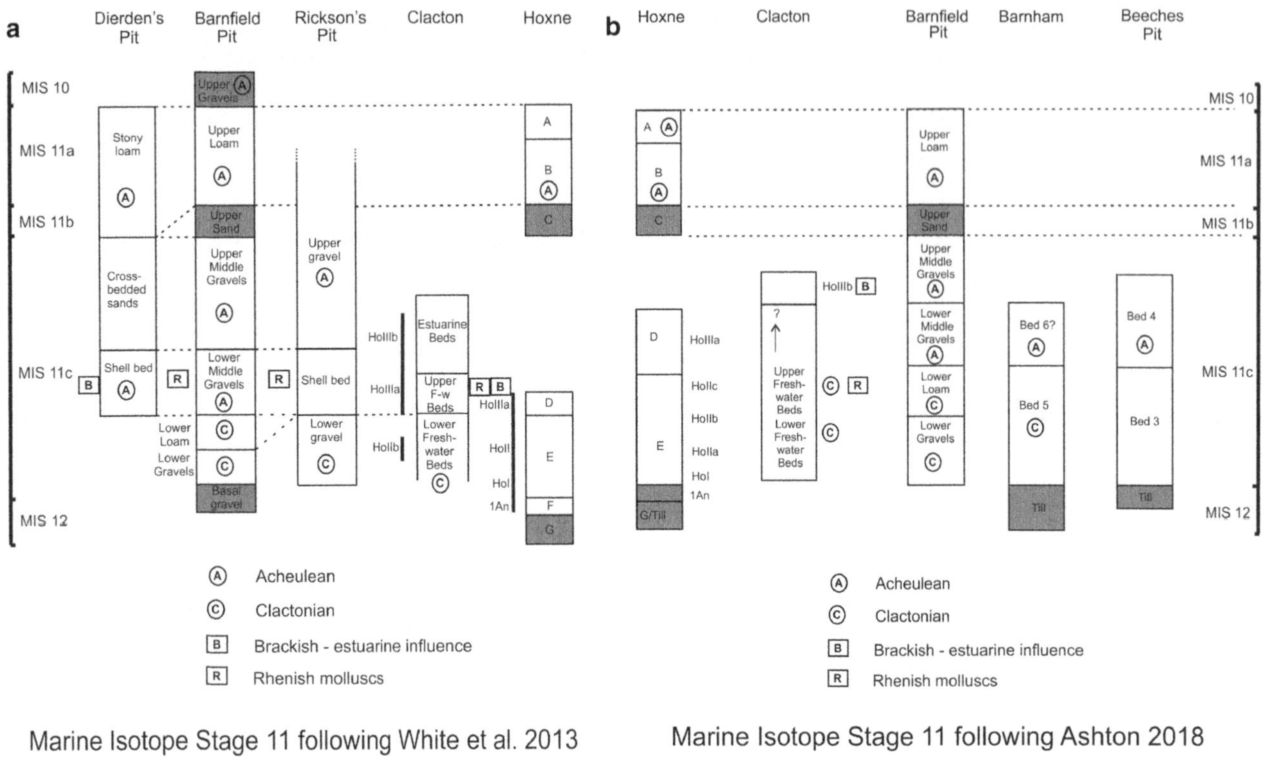

Fig. 3.2 Two slightly different interpretations of the MIS 11 interglacial and where, in relation to certain climate driven phases, particular stratigraphic layers and sites are situated. Redrawn and slightly amended from the originals

localities. They were dated by pollen, fauna, artefact typology and stratigraphy to the first temperate period after the Anglian glaciation, the Hoxnian interglacial (Ho), later equated with MIS 11 364–427 ka.

Since the publication of my overview of the Clactonian (McNabb 2007) there have been considerable changes in our understanding of the environmental history of MIS 11. The publication of the revised stratigraphy for Hoxne (Figs. 3.1 and 3.2) by Ashton et al. (2008), and the recognition that the classic Hoxnian pollen diagram in Table 3.1 does not span the whole of the MIS 11 interglacial (Ashton et al. 2008), critically altered our impression of hominin occupation in the Swanscombe interglacial as it is sometimes called. Fieldwork by Conway on the continuous stratigraphic succession at Barnfield Pit, Swanscombe (Conway 1996a, b), represented an early recognition that a simple cold–warm–cold Hoxnian climatic cycle as suggested by the pollen data (Turner 1970; Wymer 1974) was an oversimplification. Of particular note is the work by Preece and colleagues on the 'Rhenish suite' (Kerney 1971; Preece et al. 2007; White et al. 2013). This is a group of terrestrial molluscs which track the re-connection of the Thames to the Rhine (or more likely the Scheldt) in periods of lowered sea level. Since different components of this mollusc fauna appear sequentially at different times, the Rhenish fauna serves to link geographically distant sites in a relative dating framework (Preece et al. 2006; Ashton 2018).

The classic four phase Hoxnian interglacial pollen diagram, Table 3.1, is now taken to represent the expansion, and contraction of interglacial forests in MIS 11c, the first warm phase of the Swanscombe interglacial, see Table 3.2. Following this a stadial (MIS 11b) is recognized, succeeded by a return to more temperate but cooler conditions (MIS 11a) toward the end of the interglacial. It is against this climatic pulse-beat that Nick Ashton, Mark White and their respective colleagues (Ashton 2018; White and Bridgland 2018) map the incoming and outgoing movements of Middle Pleistocene hominins from the Continent and back. Although there is some disagreement over the correlation of specific layers, there is a consensus over the broad development of the interglacial and its climatic history. Two similar, but slightly contradictory schemes are shown in Fig. 3.2.

The interpretive difference between the two schemes is not that significant for our purposes. The Clactonian of the Swanscombe interglacial is mostly confined to pollen phase HoIIb of the Early Temperate pollen zone of MIS 11c; in other words the middle and warmest part of this warm climatic episode when oak and alder forests were at their densest. With the exception of Barnham (see below), the Clactonian is confined to the banks of one river, the Thames,

Table 3.2 A synthesis of the ideas of Mark White, Nick Ashton, Richard Preece and their colleagues drawn from references cited in the text. The Acheulean of MIS 11a is characterized by assemblages which show a higher frequency of S twists on their handaxes' edges

MIS 11 stage	Pollen (Ho) stage	Archaeological sites	
		Clactonian sites	Acheulean sites
11a			Swanscombe Upper Loam, Dierden's Pit, Rickson's Pit, Hoxne Upper Industry, Hoxne Lower Industry Bowman's Lodge and Wansunt Loam, Pearson's Pit stony Loam, Greenhythe stony loam, Elveden ovates, Foxhall Road
11b			
11c	HoIV		
	HoIIIb		Upper Middle Gravels
	HoIIIa		Lower Middle Gravels Beeches Pit bed 4, ?Barnham Area IV
	HoIIc		?Barnham Area IV prior to NAP in HoIIc
	HoIIb	Clacton gravels and marls Swanscombe LG and LL, Southfleet elephant (phase 6) Barnham Area I	
	HoIIa	?Rolled flakes and cores in lowest gravels at Swanscombe and Clacton ↕ ?	
	HoI		
Late Anglian			

in one section of its reach – the Lower Thames Valley. It is an island in time and space with handaxes predating it (e.g. Highlands Farm and other sites in the Caversham Channel in the late-Anglian Black Park Terrace, Fig. 3.1), and post-dating it, Table 3.2. However, establishing a contemporaneity between any Clactonian and Acheulean sites in HoII is currently challenging.

It is tempting to think of the Clactonian's world view as one of paths and animal trackways through otherwise densely wooded terrain; a linear perception along rivers and trails joining one place to another. Although difficult to navigate, the forests and under canopy may not have been impenetrable (Ashton et al. 2006). Animal trails through the bush will have led to rivers and water holes, and pollen analysis at a number of sites suggest areas of open country, perhaps maintained by herds of feeding herbivores (Wenban-Smith 2013). Such a restricted perception of the world might explain the Clactonian's apparent simplicity and static character. Following such a heuristic conception, social worlds would have been limited to immediate kin and others in your group. A chance encounter with strangers along a game trail would have been a tense and dangerous time; this is the world of Gamble's local hominin networks (Gamble 1999). Desmond Collins suggested a forest connection for the Clactonian, as did Steve Mithen (Collins 1969; Mithen 1994). Although Clactonian sites in MIS 11 have not been found in open environments or away from water/rivers, Acheulean ones certainly have (McNabb and Ashton 1995).

Clacton-on-Sea. Currently five localities are known from here all of which have provided archaeological data, but crucially a number of them have provided environmental data which has proved important in the chrono-stratigraphic placement of these sites. A detailed summary of their environmental and archaeological contributions is provided by Bridgland (Bridgland 1994; Bridgland et al. 1999) and McNabb (McNabb 2007), and a suggested correlation of the various Clacton localities is presented in Fig. 3.3.

The West Cliff site was Hazzledine Warren's original collecting locality (Warren 1911, 1922, 1923, 1955; McNabb 1992b) and represents the fullest expression of Middle Pleistocene deposits at Clacton. Boreholes through the cliff and the foreshore deposits (Pike and Godwin 1953) enabled the organic rich beach exposures of the surface of the Clacton channel to be located within HoIIb. The organic muds stained the artefacts a characteristic black colour. It was on this material that Warren developed his interpretation of the Clactonian as a nodule-tool assemblage. Subsequently, Warren collected extensively from the foreshore at Lion Point (Warren 1932, 1951) where the stratigraphic succession appears greatly compressed (McNabb 1996b). As the westward most extension of the channel deposits at Clacton, erosion has removed most of the deposits from this area. This assemblage was published in 1951 and as a larger

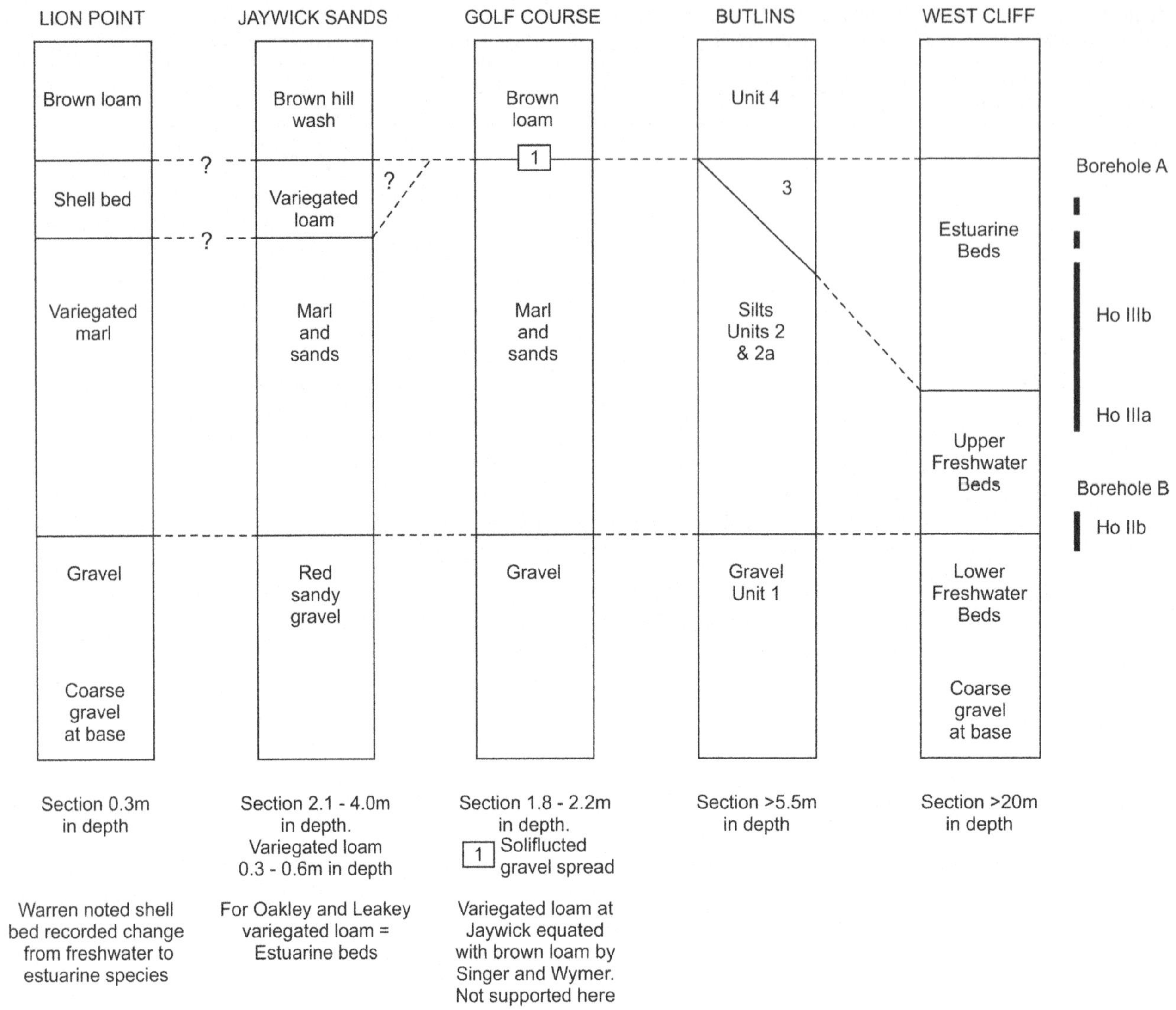

Fig. 3.3 A schematic representation of the five main localities at Clacton-on-Sea in Essex. Bed depths and relationships are not shown to scale

assemblage than that from West Cliff replaced it as the definitive Clactonian type-assemblage. As the West Cliff was the measure against which all Clactonian was judged before World War Two, so Lion Point became the measure after it.

In addition to earlier work, three archaeological investigations have been conducted at Clacton; Kenneth Oakley and Mary Leakey in 1934 at Jaywick Sands (Oakley and Leakey 1937) who also cut a few test pits into the Lion Point foreshore; Ronald Singer and John Wymer at the Golf Course (Singer et al. 1973); and there was a careful watching brief and sampling programme afforded by the demolition of the old Butlin's Holiday Camp, directed by David Bridgland (Bridgland et al. 1999).

The jewel in the Clacton crown remains the Golf Course excavation sampling a part of the southern bank of the river. In-situ knapping is attested via refits and micro-wear. Hominins came to a gravel bar and made cores and flakes on the pebbles and cobbles they found there. Whereas Wymer interpreted the site as being in primary context and only slightly disturbed (Singer et al. 1973; Wymer 1985b), I interpreted the in-situ material as one episode among many in a longer continuum of aggradation and occupation along the river's bank (McNabb 1992a, 2007).

Swanscombe. At this famous locality there are four traditional locations which contribute to the Middle Pleistocene history of the area; the Barnfield Pit, Rickson's Pit, New Craylands Lane Pit (aka Craylands Lane) and Dierden's Pit at Ingress Vale. All are within a few kilometres of each other and sample a succession of deposits now considered to represent the full extent of the Swanscombe interglacial—MIS 11c-11a, see Fig. 3.4. The reinterpretations of White et al. (2013) have played a significant part in improving our understanding of the complicated area-wide stratigraphy at

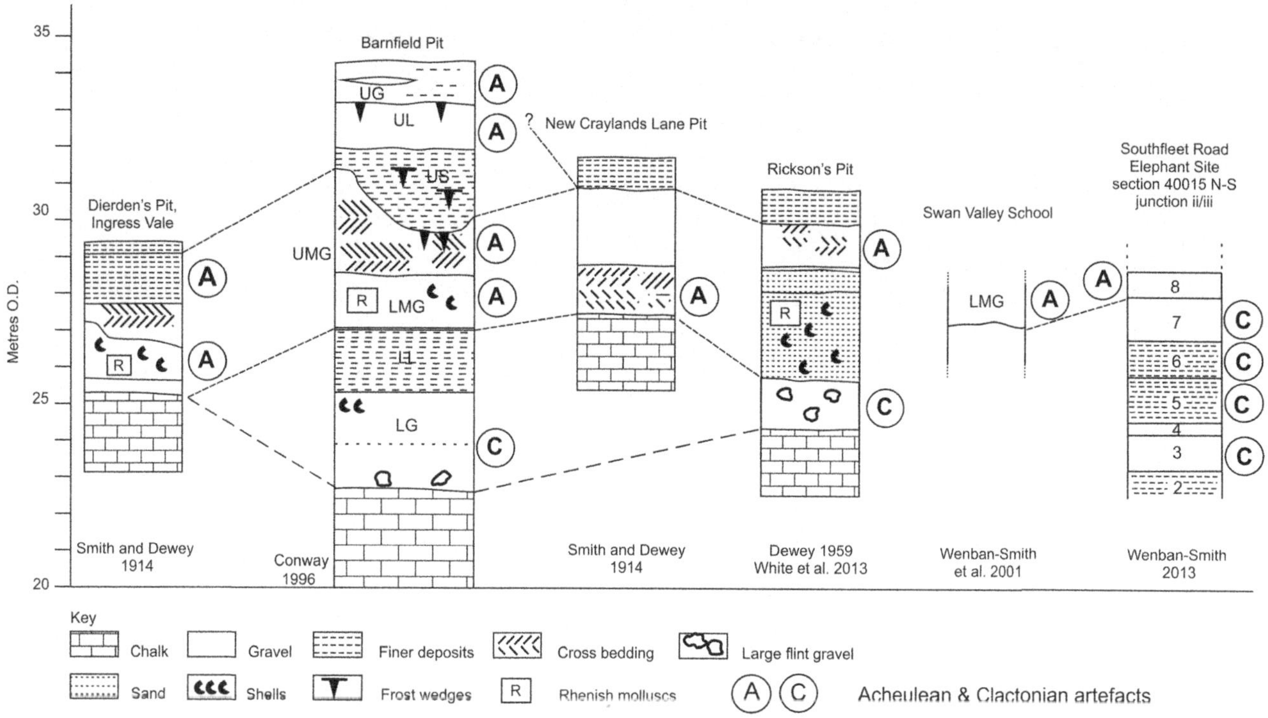

Fig. 3.4 Schematic depiction of suggested relationships between different sedimentary units from the main archaeological localities in the vicinity of Swanscombe and Ebbsfleet, Kent. Drawn to scale

Swanscombe. Crucially, new exposures and excavations by Francis Wenban-Smith at the Swan Valley School site (Wenban-Smith and Bridgland 2001), and at the spectacular Southfleet Road Elephant site (Wenban-Smith 2013), have extended our understanding of the regional wide significance of the depositional succession preserved in the Barnfield Pit. Complete histories and detailed stratigraphy's are given in McNabb (McNabb 2007), Bridgland (Bridgland 1994), and White (White 2000).

The Barnfield Pit contains a complete history of the interglacial at a single locality—Fig. 3.4. The Clactonian is confined to the Lower Gravel channel, a wide single thread river channel that sinuously flowed past wooded slopes, and to the succeeding Lower Loam which was a marsh. Both were the subject of careful excavations before the First World War (Smith and Dewey 1913, 1914) when the core and flake character of their stone tool assemblage was established. Further work by artefact collector's reinforced the Clactonian character of the deposits in the inter-war years (Chandler 1928–9, 1931; Marston 1937, 1942). But the discovery of the Swanscombe skull by A. T. Marston in 1935 and 1936 in the handaxe rich Upper Middle Gravels focused attention away from the basal units, and it was not until the late 1960s and early 1970s that excavations resumed in the Clactonian deposits under John d'Arcy Waechter (Waechter et al. 1970; Conway et al. 1996).

An important point which is not often given enough air time is the recognition by Conway (Conway 1996a, b) of a 'resting' phase or non-aggradational period half way up the Lower Gravels (between Lower Gravel units 3 and 2). For a while the river either dried up, moved elsewhere, or began eroding its bed. It divides the lower half of the Lower Gravels (Waechter's Units 4 and 3) from the upper half (units 2, 1 and the midden). Other interruptions to the depositional sequence also suggest periods of time when the river was not flowing—such as the midden developed on the top of the Lower Gravels (on the surface of unit 1). This was originally thought to be a refuse collection of hominin food bones, but turned out to be a natural scour surface. In the overlying marshy Lower Loam the section drawings clearly indicate temporary (seasonal?) land surfaces associated with small streams. On one such surface Waechter's team excavated a Clactonian butchery event (McNabb 2007)—the Lower Loam knapping floor. All that remained of the carcass was the head of a large extinct fallow deer (*Dama dama clactoniana*) with its impressive antler rack still attached. Presumably the head had been removed to make the butchery of the body itself easier. Adjacent to the skull and antlers were a series of refitting flakes, large and small, a large flake used as a wedge, possibly to sever the spinal cord, and a flaked flake (see note 1) which had been repeatedly re-sharpened. Sadly the remainder of the carcass was never found.

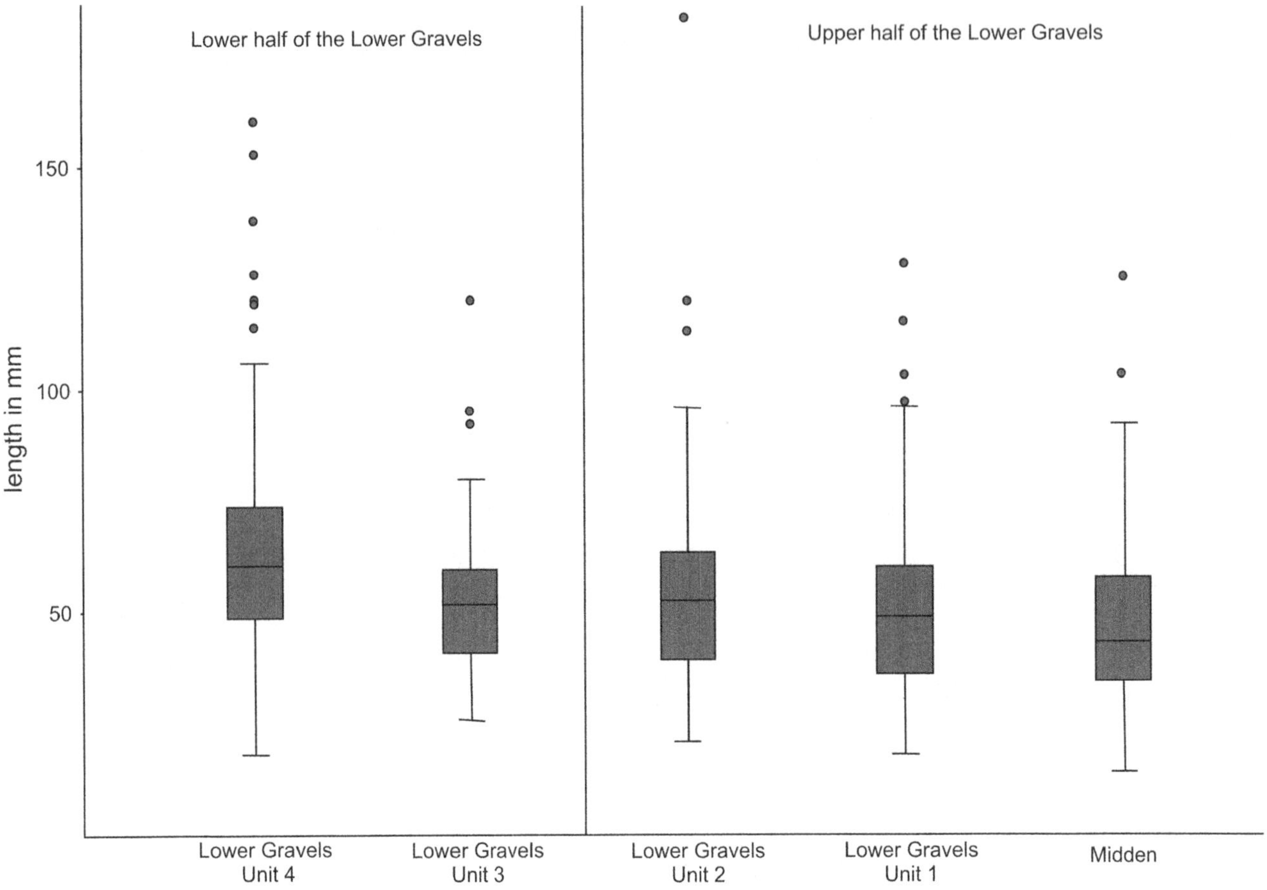

Fig. 3.5 Box and whisker plots of artefact length for the full depth of the Lower Gravels at the Barnfield Pit Swanscombe. Data from the Waechter excavation, 1968–1971. The upward fining is demonstrated in the median value for each unit's distribution becoming smaller (4 = 61.0 mm, 3 = 52.0 mm, 2 = 53.0 mm, 1 = 49.0 mm, midden = 43.0 mm), and the slight decrease in range up the sequence. Kruskal Wallis test comparing the combined length in mm data for units 4 and 3 vs. units 2 to the midden found a statistically significant difference between the two distributions ($N = 710$, $H = 31.962$, $df = 1$, $p \leq 0.01$)

Emphasizing the further passage of time, a soil developed on the surface of the Lower Loam. It too contained Clactonian artefacts. On its surface were animal footprint trails. This was an open land surface when the river did not flow over this spot.

All of the above implies that the Clactonian in the Lower Gravels and Lower Loam at the Barnfield accumulated over a considerable period of time. The Barnfield Pit is probably the best evidence for the duration of the Clactonian in HoIIb. Figure 3.5 uses artefact length to demonstrate this. The lower half of the Lower Gravels is an upward fining sequence; the river gradually losing its power to transport bigger gravel clasts. Although the river regains its energy somewhat (Fig. 3.5 Unit 2) in the upper half of the Lower Gravel, this too is a gradual upward fining sequence as clasts become steadily smaller.

The Southfleet Road elephant site (Wenban-Smith 2013) is a critical site and something of a game changer in Clactonian studies. The front half of an extinct species of elephant (straight tusked elephant—*Palaeoloxodon antiquus*) was recovered during construction (the back half sadly destroyed). The beast died on a dry land surface (Fig. 3.4—phase 6) with a still or sluggish water body nearby, and forest as well as open country in the vicinity. As interpreted by Wenban-Smith a group of between 4 and 13 Clactonian hominins butchered the carcass possibly over a few days, and possibly over more than one visit. In-situ knapping scatters were present next to the carcass (assemblage 6.3), and close by on slightly higher ground (assemblage 6.1). The site was buried and preserved by clays which indicate a rise in the level of the local stream/waterbody. The area was then overlain by slope wash deposits (phase 7) which also contained a Clactonian assemblage interpreted as sweepings from the adjacent environment and broadly contemporary with the Clactonian of phase 6.

There are marked parallels between the deer butchery episode of the Lower Loam knapping floor and that of the elephant at Southfleet Road. A number of different flint nodules are brought to the carcass. They were knapped by hammer stones. Sharp edged flakes and flaked flakes were made and used to dismember the carcass. At Barnfield, they took all the cores away with them. In assemblage 6.3 four cores were left, but upwards of thirteen pieces of raw material may have originally been schlepped into the site (Wenban-Smith 2013); anything up to nine were brought to the Lower Loam knapping floor.

As Wenban-Smith notes, an important observation here is the recognition of an invariant Clactonian signature at three different landscape scales; at the level of the activity specific butchery knapping scatter (assemblage 6.3), at the level of the immediate surrounding area (assemblage 6.1), and finally at the scale of the broader landscape (artefacts in phase 7). It is difficult not to see this as a product of habitual behavior. In addition, the Southfleet Road site extends the regional significance of the Clactonian at Swanscombe. The Lower Gravel channel is present at Barnfield Pit and then again at Rickson's Pit, Fig. 3.4. The environmental data convincingly supports the elephant site as being contemporary with the Lower Gravel/Lower Loam (but cannot be pinned down further). The Clactonian hunters were therefore ranging to the south of the main Thames channel as well as along its banks.

Barnham St Gregory. This was an important location in the Clactonian site catalogue during the inter-war years. It was one of only two sites (Barnfield being the other) where the Acheulean/Clactonian cultural stratigraphic succession was clearly demonstrated; the Acheulean overlay and therefore succeeded the Clactonian. The site also extended the regional significance of the Clactonian outside of the Thames Valley, Fig. 3.1. This Suffolk site, according to its excavator T. T. Paterson (Paterson 1937, 1942, 1945) showed a developmental sequence as the Clactonian evolved and its knappers interchanged technological know-how with handaxe makers. Although this late imperial scale interpretative framework was not widely taken up, the Clactonian interpretation remained unchallenged and was further bolstered by excavations by John Wymer in the late 1970s (Wymer 1985b). However in the 1990s this was disputed by excavations led by Nick Ashton from the British Museum (Ashton et al. 1994, 1998). Hominins came to a river bank-side locality and exploited a lag gravel of flint cobbles along its edge making a core and flake assemblage. Further along the same cobble band the excavators recovered a handaxe assemblage also within the gravel, indicating a contemporaneity between the two assemblages. Initially, this was interpreted as evidence for the two being part of a single continuum of hominin behavior, with each assemblage interpreted as a reaction to changing local environments rather than two distinct cultural phenomena. However a return to the site by the British Museum team has added new data (Ashton et al. 2016). Although the handaxe and the core and flake assemblages are still interpreted as within the same cobble band, careful stratigraphic work has demonstrated that the core and flake assemblage had been buried by silt before the handaxe assemblage was made. In other words the former predates the latter, a return to the traditional interpretations of the site that would have pleased John Wymer enormously (see also Wenban-Smith 2013 on this).

The Clactonian deposits at Barnham are currently associated with a *D. ruderatus* fauna, the earlier of the two potentially chrono-stratigraphic mollusc species from the Rhenish suite (above). No mollusc suite is as yet equated with the chronologically later Acheulean at the site (R. Preece pers. comm.). These data suggest a broad contemporaneity (HoIIb) between the Clactonian outside of the Thames Valley, and that within it.

Little Thurrock. This is one of only two Clactonian sites described since World War Two (the Southfleet Road elephant site being the other). Artefacts from a 'working floor' at Little Thurrock were first described by the great George Worthington Smith (Smith 1894; McNabb 1992a). The current site was originally found by John Wymer's father in 1910 and published by the younger Wymer in the late 1950s (Wymer 1957), see Fig. 3.1. It is a channel margin site, its non-handaxe assemblage being the sweepings of bank side occupation upstream. The gravel is the very feather edge of the northern bank of the Thames river. It dates to the end of a glacial phase or the beginning of an interglacial and is overlain by a brick earth deposit—fully temperate sands and silts, also marking the river's feather edge.

Much controversy has centered on whether a single gravel spread is present, or two distinct ones at slightly different altitudes. The gravel spread at 15 m OD represents the Clactonian site and has been excavated on three occasions (Wymer 1957; Snelling 1964; McNabb 1992a; Bridgland and Harding 1993), and sections were carefully observed by Bernard Conway and Richard West among others (West 1969; Wymer 1985b; Bridgland 1994). The current consensus is that the two gravels are part of a single eroded river margin.

What is key here is that the site is no longer assumed to be MIS 11 and so contemporary with the other Clactonian sites. A major revision by David Bridgland of the Thames river terrace stratigraphy has found widespread acceptance (Bridgland 1994), and Little Thurrock now dates to late MIS 10 or earliest MIS 9 and the overlying brickearth to temperate MIS 9. This underpins the new interpretations of the Clactonian as described above (White 2015). Wenban-Smith (2013) following West (1969) raises the spectre of reworking from higher and older deposits to dispute the MIS 9 age.

Purfleet and Cuxton, see Fig. 3.1. The models of hominin occupation presented by Mark White and colleagues, and which are based upon the new chrono-stratigraphic position of Little Thurrock have roped in two other sites, also dated to late MIS 10/early MIS 9, which have been interpreted as Clactonian on the basis of a lack of handaxes. Mark White and I continue to dispute the relevance of these sites, though their age is not contentious (McNabb 2007; Pettitt and White 2012).

At 15 Rochester Road, Cuxton, c. 5 cubic meters of sediment produced 118 flakes and cores (Cruse et al. 1987) from a lower gravel unit. Handaxes and thinning flakes were absent. The overlying gravel bed, separated by an erosion phase contained handaxes. It is likely that this higher unit is related to the gravel from the Cuxton Rectory site, some thirty meters away, a well-known handaxe locality (Tester 1965). Equally small but prolific excavations by Wenban Smith a little further along Rochester Road found a continuation of this same handaxe gravel sitting on bedrock (Wenban-Smith 2004). This suggests that the lower core and flake material found by Cruse at 15 Rochester Road may be some localized channel predating the handaxe gravel; the recognition of an erosion phase separating the two implies some time separated them. My concern is that the lower half of this excavated area is too small to be certain of a genuine core and flake assemblage. If handaxes were infrequent at some sites (which they often are) then a keyhole excavation in someone's driveway would likely miss them. Paul Callow who analysed the lithics (in Cruse et al. 1987) was adamant the small collection should not be called Clactonian.

I have similar concerns with the lowest gravels at the well excavated site of Purfleet. This is another key site whose sedimentary depth may span either the earliest temperate phase of the MIS 9 interglacial, or its full span (Schreve et al. 2002; Bridgland et al. 2013). Here the basal unit is considered a downstream equivalent of the Little Thurrock gravel and so dates to late MIS 10 or early MIS 9. From this deposit just over a hundred flakes and cores have been recovered from a ~500 m long excavated face. White's contention (White and Bridgland 2018) is that this is a sufficiently large enough area to be sure of the character of the assemblage, the opposite situation to Cuxton. I agree, but continuing with the heurism of secondary context Acheulean sites with very low handaxe frequencies, a density of only five artefacts per square meter may not be enough to pick up the Acheulean character of the site despite its section length. As I pointed out long ago, an Acheulean hunter who brings a handaxe to butcher a carcass, makes a few sharp flakes from a core to help, and then takes the handaxe away with her has left a Clactonian site behind.

Just to be clear, I do not dispute the core and flake character of Cuxton or Purfleet, they are currently non-biface assemblages, I just wonder whether they should be called Clactonian.

What Exactly is the Clactonian?

To answer this it is time to turn to the data on individual Clactonian sites and their lithic assemblages. I do not intend to describe in detail the various elements that make up these assemblages, or the method of analysis applied as these have already been fully published elsewhere (McNabb 2007). Table 3.3 gives a selective summary of the basic data.

A generalized Clactonian *chaîne opératoire* is presented in Fig. 3.6. It must be admitted at the outset it is not a particularly exciting one. A series of flint clasts of different sizes found on fluvial gravel banks and bars are knapped by a number of very basic hard hammer direct percussion techniques (alternate flaking, parallel flaking and single removals, usually in various combinations) to produce flakes of varying sizes. A small number of these are then picked out and either flaked again (flaked flakes and their spalls) or modified into a small series of retouched tool forms. Additionally, Keeley's micro-wear analysis at Clacton-on-Sea, Swanscombe and elsewhere (Keeley 1980), showed that a substantial proportion of unretouched flake edges were often employed in a variety of cutting, scraping, wedging and shaping activities. In terms of physical appearance, the character of the retouch can be grouped under three headings —(a) those where a sharp edge predominates (unretouched edges, flaked flakes and their spalls), (b) evenly retouched edges (scrapers of various sorts and lengths of scraper retouch), and (c) irregular or unevenly retouched edges (retouched notches, various denticulates and irregular lengths of retouch either locally present or unevenly distributed across a flake's edge). Additionally, a small number of retouched tools show a combination of two different kinds of edge modification on the same flake. These are labelled multiple tools. Usually this is a flaked flake accompanied by another form of edge modification.

In terms of frequencies it is the sharp-edged group that predominate at every Clactonian site. This is almost wholly based on the occurrence of flaked flakes and their spalls. In the absence of micro-wear it must be assumed that sharp edged flakes were also used. The Southfleet Road elephant site (Wenban-Smith 2013) has provided powerful support for this observation. Wenban-Smith independently interprets the Clactonian as an assemblage type characterized by sharp edges and flaked flakes, with many of the flakes and flaked flakes from the knapping scatter next to the elephant demonstrating macroscopic use wear damage.

On the banks of the Thames, Clacton-on-Sea. Here I will focus on only three localities; the Warren collection from Lion Point (including the handful of artefacts from Oakley and Leakey's 1934 test trenches on the Lion Point foreshore), the 1934 Oakley and Leakey trenches and test pits at Jaywick Sands, and the Singer and Wymer excavation at the Golf Course, comprising the gravels and the overlying finer grained marls, see Table 3.3.

The Lion Point material is an assemblage of largish flakes and chunky looking cores, whose character may reflect Hazzledine Warren's ability to see artefacts in the mud of low tide as much as anything else. Figure 3.7 compares the

Table 3.3 Frequencies of core knapping techniques on a selection of non-PCT (non-prepared core technology/Levallois) hard hammer flake cores from UK Clactonian sites, a selection of non-handaxe assemblages from Europe, and from UK Acheulean sites. Data from McNabb (2007), Cole (2011), Fluck (2011). Data used with kind permission. Codes (A1, A2 etc. refer to artefact categories in McNabb 2007)

Cores with no fixed perimeter and no fixed flaking face (type A)								Cores with a fixed perimeter but no fixed flaking face (type B)	
Assemblage type	Site	Alternate flaking (A1)	Alternate and parallel flaking (A2)	Parallel flaking (A3 and A4)	Single flakes (A5)	Mixed techniques (A6)	Other non-PCT cores (A7)	Centripetal bi-convex (B1 and B2)	Other (B3)
UK Clactonian	Clacton—Lion Point Warren and Oakley-Leakey 1934 (MIS 11)	128	18	17	4	44	4	15	0
	Clacton—Jaywick and Oakley-Leakey 1934 (MIS 11)	11	0	0	3	6	1	1	0
	Clacton—Golf Course, Gravel (MIS 11)	27	4	7	10	14	3	0	0
	Clacton—Golf Course, Marl (MIS 11)	7	1	0	2	3	0	0	0
	Swanscombe—Chandler, Lower Gravels (MIS 11)	34	8	0	3	16	0	2	2
	Swanscombe—Waechter Lower Gravels, units 4 and 3 MIS 11)	18	4	3	3	9	6	2	0
	Swanscombe—Waechter Lower Gravels units, 2 and 1 and midden (MIS 11)	14	3	1	2	8	1	2	0
	Swanscombe—Rickson's Pit, L. S.B. Leakey 1934	7	0	0	2	14	0	1	0
	Little Thurrock—Bridgland and Wymer (MIS 10/9)	2	1	0	2	1	0	0	0
European non-handaxe assemblages	Vértesszőlős Site I, all raw materials, Hungary, (MIS 13)	84	100	57	83	64	1	1	0
	La Micoque, France, levels 1 and 2, (MIS 11)	22	4	1	13	16	0	0	0
	Bolomor, Spain, Levels 15–17, (MIS 9-8)	3	3	0	4	6	0	0	0

(continued)

Table 3.3 (continued)

Cores with no fixed perimeter and no fixed flaking face (type A)								Cores with a fixed perimeter but no fixed flaking face (type B)	
Assemblage type	Site	Alternate flaking (A1)	Alternate and parallel flaking (A2)	Parallel flaking (A3 and A4)	Single flakes (A5)	Mixed techniques (A6)	Other non-PCT cores (A7)	Centripetal bi-convex (B1 and B2)	Other (B3)
UK Acheulean	High Lodge—Sieveking C2, D, E (MIS 13)	6	47	8	10	20	7	4	0
	Hoxne upper Industry, West Cutting, Singer and Wymer, (MIS 11)	2	3	1	0	2	4	0	0
	Hoxne lower Industry, West Cutting, (Singer and Wymer), unrolled (MIS 11)	5	2	3	1	4	2	1	0
	Cuxton Tester (MIS 9 or 7)	4	7	2	3	8	1	0	2
	Swanscombe, Waechter, lower Middle Gravels (MIS 11)	4	0	1	1	3	1	2	0
	Elveden, BM, excavations, all units (MIS 11)	14	2	5	5	9	1	0	0

length of cores, unretouched whole flakes and retouched flakes for each of the Middle Pleistocene Clacton locations considered here. The larger general size of Lion Point is readily apparent. A Kruskal-Wallis H test comparing the artefact length data reveals a statistically significant difference between the four localities (N = 1537, H = 343.699, df = 4, p $\leq$ 0.01), and pairwise comparisons (data not presented) clearly show that it is Lion Point that differs from the others. The larger size of the Lion Point data contributed significantly to the culture historical conception of the Clactonian being large crude and technologically simplistic (McNabb 1992a, 2007).

Despite the size difference the Lion Point *chaîne opératoire* closely follows that of the other Clacton localities. Comparing the Warren collection with Singer and Wymer's assemblage from the Golf Course gravel (the closest in sample size) the *chaîne opératoire* shows a basic similarity in approach to flaking and assemblage composition. The same knapping techniques are present in both. The alternate and mixed techniques predominate at both sites with alternate being the most frequent. But those cores worked by parallel flaking, or parallel in combination with another technique, are far more common at Lion Point. Perhaps this reflects Warren's views on the differences between cores and nodule tools. He may well have been collecting or keeping those which he thought highlighted the differences between the two.

Pebble-sized and smaller/medium cobble-sized cores dominate all the Clacton assemblages, and overwhelmingly retain various amounts of cortex indicating the cores could have been flaked further. The vast majority fall into the range of having 4–12 flake scars, and those cores which show more ($\geq$ 13 scars) and have no cortex at all are very infrequent. Extensive reduction of cores was not a hall-mark of the Clacton knappers. Warren's assemblage contains cores with a fixed margin (type B cores—fixed margin but no preferential flaking face), which, curiously, are not present in the gravel and marl of the Golf Course.

The whole flakes from these cores were grouped under the Toth system (Toth 1985) of classifying flakes in relation to the presence or absence (and amount) of cortex on the butt and dorsal surface (primary data in McNabb 1992a). Statistical tests showed significant differences in the frequency of occurrence of the different Toth types between the various Clacton localities (Chi-square = 42.150, df = 15, p $\leq$ 0.01), but this is not surprising since the river's bed load contains a mixture of bank side sweepings (Jaywick), larger material

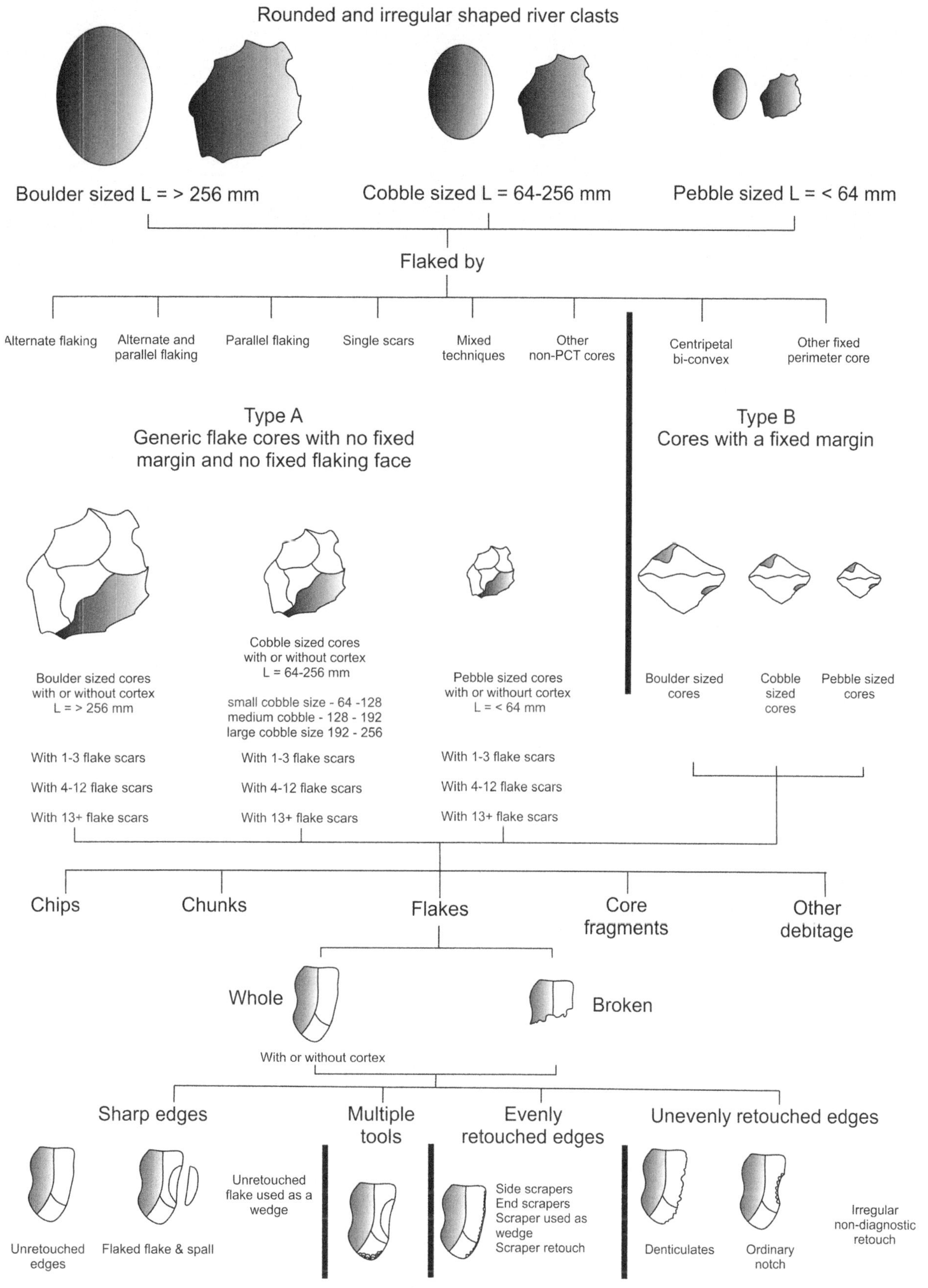

Fig. 3.6 The Clactonian *chaîne opératoire*

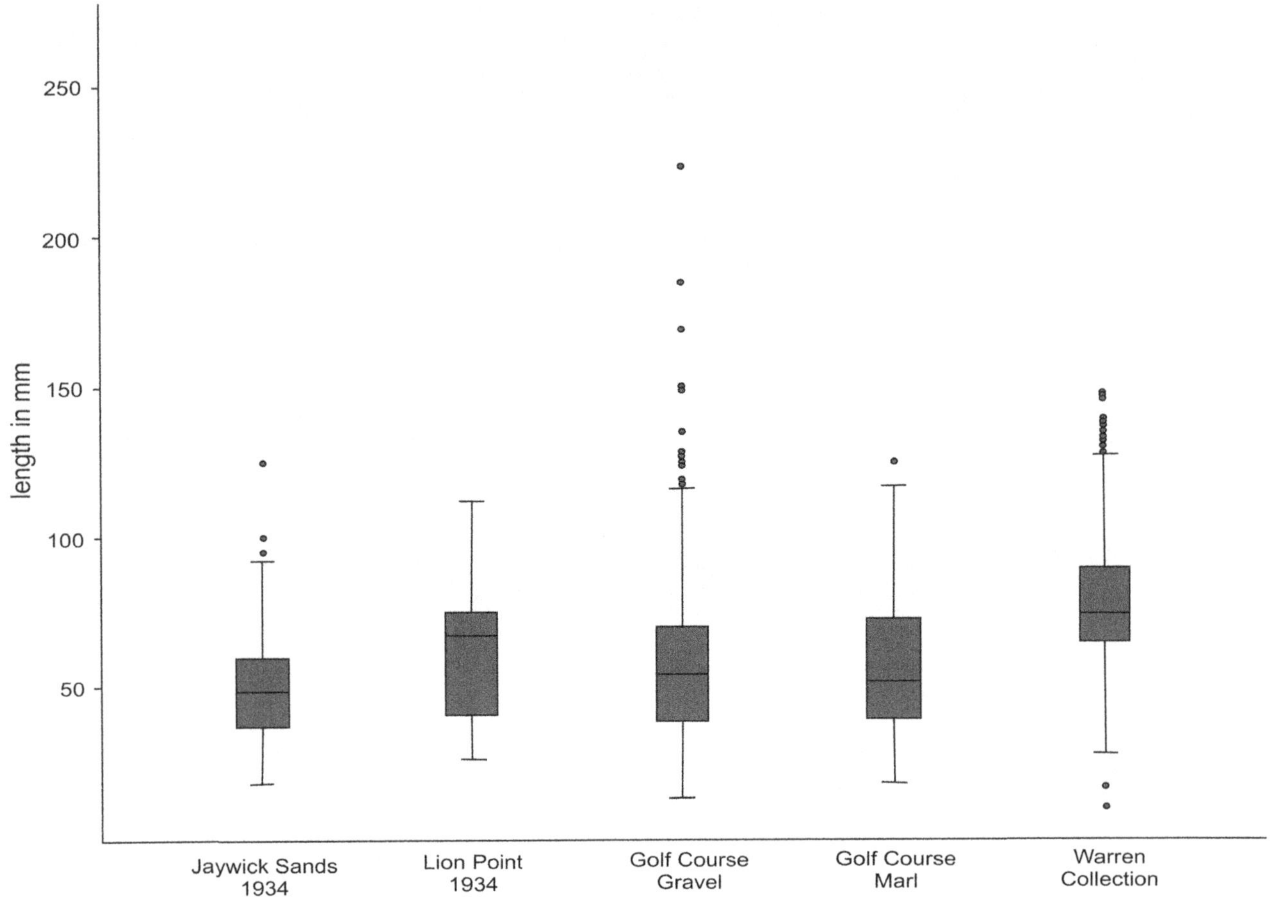

Fig. 3.7 Box and whisker plots of artefact length for the main localities at Clacton-on-Sea. Warren collection refers to Warren's collecting activities at Lion Point and Lion Point 1934 refers to the material recovered by Oakley and Leakey's excavations on the Lion Point foreshore in that year

from deeper in the channel (Lion Point, see below), and disturbed primary context occupation (Golf Course gravel). Toth types 5 (partial cortex on the dorsal) and to a lesser extent Toth type 2 (cortex on dorsal and butt) predominate in each locality, and their frequency maps on to the core data, reinforcing the observation that the majority of the cores could have been flaked further, as does the much lower frequencies of Type 6 (no cortex at all).

In terms of the uses to which the flakes were put, at all the Clacton localities the flaked flakes and their spalls are by far the most frequent form of flake modification, with multiple tools, denticulates and notches, and irregular or non-diagnostic patches of retouch on flake edges, providing a low frequency background of flake tools.

It is worth remembering that Warren was collecting from beach deposits at low tide. Oakley and Leakey (1937) believed the Lion Point exposures to be from a deeper part of the channel (thalweg) than their Jaywick artefacts. These came from the upper part of the gravels and sampled from the north bank to mid-stream. Their trenches were not bottomed to the channel's bed. Wymer's 1969–1970 test pits and main excavation sampled the full depth of the southern branch of the channel where the main river split into two around an island of London Clay, to the east of Jaywick. The parity in the Clacton knapping, and its products, sampling different parts of the Middle Pleistocene Thames, at different locations along the river is therefore indicative of a strongly persistent behavioural signal. Artefact frequencies at these sites vary, as would be expected, but the *chaîne opératoire* does not.

On the banks of the Thames, Swanscombe, the Barnfield Pit. As at Clacton, an older collected assemblage has influenced the perception of the archaeology from here. The Chandler collection (Chandler 1928–9, 1931) is a particularly biased assemblage in my opinion. Its highly selected character is clearly emphasized in the comparative length data in Fig. 3.8, with Warren's own highly selected Lion Point assemblage added to highlight this. Again, the larger size of the Chandler artefacts actively contributed to a perception of the Clactonian being older and cruder because small artefacts were indicative of finesse and sophistication in the culture historical mindset.

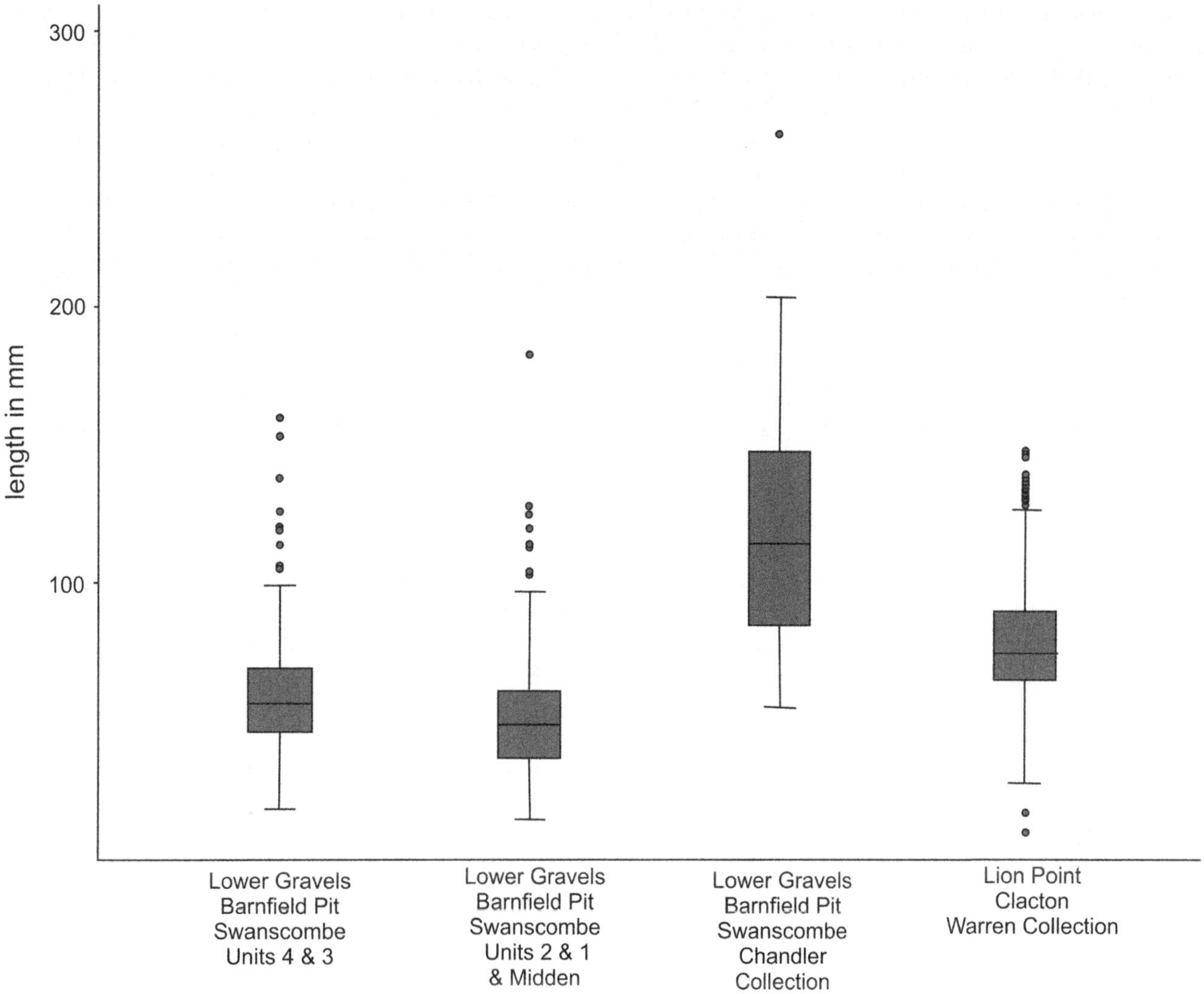

Fig. 3.8 Box and whisker plots of artefact length for the Lower Gravels at Barnfield Pit, Swanscombe as compared with classic artefact collections which heavily influence the traditional perception of the Clactonian

Following the *chaîne opératoire* however, the frequency of occurrence of knapping techniques between the Chandler collection, and the bottom half of the Lower Gravels vs. their upper half found no statistically meaningful division between the three (Chi-square = 15.640, df = 10, p = 0.110). So although the size of the artefacts retained by Chandler are not representative of Clactonian flaking, the knapping techniques are. That a consistent knapping practice is present in the lower and upper parts of the Lower Gravels is demonstrated by a further Chi-squared test only on the knapping technique data for the lower part vs the upper part (Fig. 3.8). A non-statistically significant difference was present (Chi square = 2.859, df = 5, p = 0.722).

As at the Clacton localities, partially cortical flake cores in the pebble to small cobble-sized ranges predominate. Again, the more exhaustively flaked cores which lack cortex are noticeable only by their infrequency, and scar counts also fall well within the 4–12 frequency group. Flake cores with a fixed margin but no permanent flaking face occur occasionally. The Toth categories for whole flakes show a similar predominance of the type 5 followed by the type 2 flakes, both consistent with the partial reduction of cortical river clasts. The type 6 flakes with no cortex occur less frequently, though a higher than expected frequency is present in the upper half of the Lower Gravels.

As far as the retouch is concerned the pattern at the Barnfield Pit parallels that of the Clacton localities. The sharp-edged morphology is the most frequently occurring type of modification to a flake's edge (flaked flakes and their spalls), with multiple tools, scraper edges, denticulates and notches occasionally present.

On the banks of the Thames, Little Thurrock. Combining the Bridgland excavation (Bridgland and Harding 1993) and that of Wymer (Wymer 1957), which is curated in the British

Museum under the name 'Institute of Archaeology Collection', the assemblage has a curiously low core frequency—none from Bridgland and only six from Wymer. A single type B core is illustrated by Wymer (1957, Fig. 5.1) but is no longer part of the collection. Artefacts in the British Museum marked Grays (especially in red enamel), Gtk and Gtx, and variations on this, represent material collected by Wymer on visits before his main excavation occurred (Wymer letter to McNabb, dated 16/8/1988).

Concerns about the site's age notwithstanding the *chaîne opératoire* is remarkably similar to the other Clactonian sites—in other words unremarkable. The core frequency is too small to be meaningful, but its knapping patterns are characterized by alternate and by parallel flaking as elsewhere. The cores are in the pebble and small cobble size range with an even mixture of 1–3 and 4–12 flake scars. Not surprisingly the Toth flake categories 5 and 2 predominate, in that order of frequency, and flaked flakes are the most common form of retouch, with retouched notches a poor second.

Sadly, I do not have comparable data for Barnham.

The contested stones. There are a small number of artefacts found in Clactonian assemblages that really should not be there according to the classic definition. Yet there they are. People have gone to extraordinary lengths to discount them, exploiting the smallest ambiguities, usually because their own interpretative position requires them to disappear. Some are illustrated in Fig. 3.9. The handaxe in Fig. 3.9a was carefully dug out of the base of the Lower Gravel by A. T. Marston. Its provenance is unambiguous (McNabb 1996b). A second handaxe from the Lower Gravel was also found by him. Both can be classified as classic handaxes.

Non-classic handaxes are bifacially thinned and shaped tools which retain the idea of a handaxe and its cutting edge without conforming to its physical appearance (Ashton and McNabb 1994). That illustrated in Fig. 3.9b came from the top 20 cm of the Lower Gravel at Barnfield Pit, recovered during the Waechter excavation. Its provenance is clearly marked as such and it is in exactly the same condition as the rest of the artefacts from the spit. Mark Newcomer (Ohel 1979) believed the site note book recorded its provenance as being from the base of the gravels. Newcomer was simply wrong about this—a trick of memory a decade later. There were no note books for the lower part of the sequence. A patch of Lower Middle Gravel in the baulk above this square contained no artefacts so the little non-classic could not have fallen in either. Its provenance is clear, excavated from square A2 in the top of the Lower Gravels in 1970. The non-classic handaxe in Fig. 3.9c comes from the Lion Point channel. Wenban-Smith discounts this and others from here because they may have been derived from older deposits on- or off-shore (Wenban-Smith 2013). This is certainly possible elsewhere. But at Lion Point it is not. On-shore, any older deposits have long since been eroded away (Bridgland et al. 1999), and off-shore only the Middle Pleistocene Clacton channel is preserved.

Figure 3.9d (bottom left) is a chopping tool from the base of the Lower Gravels. There are three variations on this artefact type in the British Lower Paleolithic. Firstly, there are those with a zig-zag cutting edge made by hard hammer alternate flaking. They may well be cores. They are found in handaxe and non-handaxe assemblages. Secondly, there are examples with carefully shaped edges, whose working looks like handaxe thinning, or the extensive semi-abrupt retouch applied to the making of well-made scrapers (examples are found in Bowman's Lodge and Caddington). These are associated exclusively with handaxe assemblages (McNabb 1992a). Then there is the sole example in Fig. 3.9d bottom left. It has a careful fan-shaped cutting edge, but shows none of the thinning present on Acheulean examples. As with the non-classic handaxes, it's a rare Clactonian example of a properly shaped tool.

As I have noted elsewhere (2007) the presence of a few classic handaxes or shaped tools found in isolation in Clactonian deposits does not change anything, they are certainly not grounds for denying the Clactonian as I once claimed (McNabb 1992a; Ashton and McNabb 1994). Likewise, the presence of a few non-classics does not affect the overall interpretation of the Clactonian as a core and flake dominated assemblage. The key point here is one made by Wenban-Smith (2013) and by older archaeologists like John Wymer in his reply to Milla Ohel (Ohel 1979). No Clactonian locality retains any evidence of *handaxe manufacture*—of the *extensive* shaping and thinning of a bifacial edge. Even the non-classics do not violate this. The presence of non-classics could support those interpretations that see the Acheulean emerge, locally, from the Clactonian. It is the addition of the concept of thinning and shaping that actually characterizes the Acheulean's materiality.

Scrapers are rare in Clactonian assemblages. When a length of flake edge is modified by retouch it is usually abrupt and non-invasive. Denticulate scrapers are sometimes cited (when a continuous length of edge is denticulated). A dramatic conformation of this was discovered in section cleaning in the Lower Gravels Swanscombe in 1995, see Fig. 3.9e. The well-made scrapers seen in Acheulean assemblages like Bowman's Lodge (Tester 1951, 1976), Hoxne Upper Industry (Singer et al. 1993) or High Lodge (Ashton et al. 1992), here taken to be Acheulean, do not occur in the Clactonian (McNabb 1992b). The presence of such scrapers is more appropriate in Acheulean assemblages. They are conceptual templates (Ashton and McNabb 1994), like handaxes and equally carefully shaped—ideas of tools held in the memory, a product of social learning. It is significant that at the landscape level of Southfleet Road, proper scrapers only appear with the Acheulean in phase 7 (Wenban-Smith 2013).

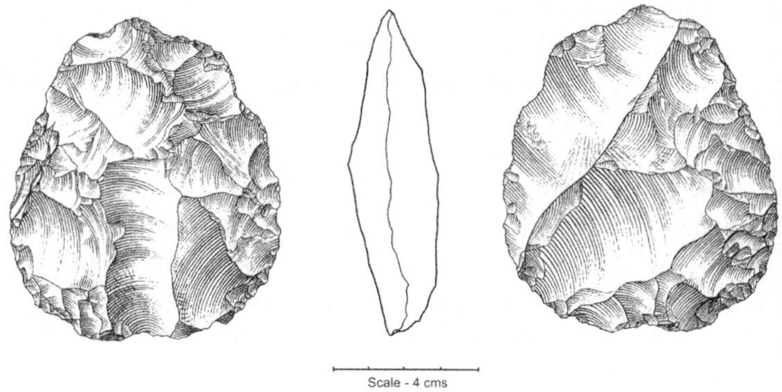

3.9a. The Black Ovate
A classic handaxe from the base
of the Lower Gravels

3.9c. Non-classic handaxe
from Lion Point, Clacton. Warren
Collection

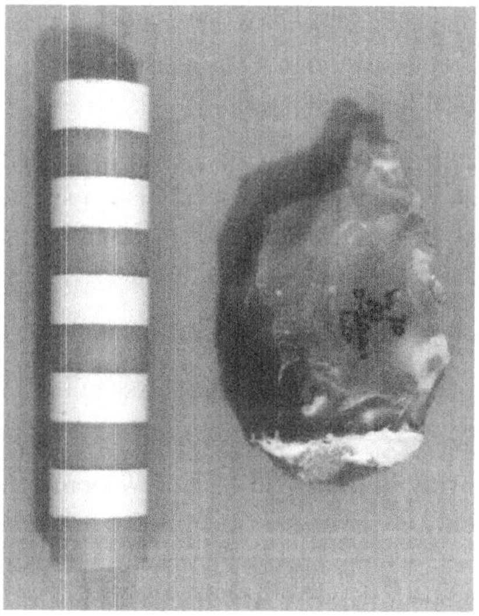

3.9b. Non-classic handaxe
clearly marked as coming from
the top 20 cms of the Lower
Gravel.
Waechter excavation

3.9d. Top - chopping tool, *sensu* Warren, a
zig-zag cutting edge made by hard hammer flake
removals, Bowman's Lodge. Bottom right -
chopping tool made by careful bifacial thinning,
Bowman's Lodge. Bottom left - fan shaped
chopping tool made by hard hammer bifacial flaking
but not thinned, Lower Gravels, Barnfield Pit.

3.9e. Denticulated scraper
on a big flake. Recovered during
section cleaning in the Lower
Gravels in 1995

Fig. 3.9 A selection of anomalies from Clactonian sites. Shaped tools from an assemblage type that does not normally include them

Summary. In answering the question that headed this section, the MIS 11 Clactonian is a stone tool assemblage type knapped to make and maximize the frequency of sharp edges, with a much lower emphasis on other activities requiring a different edge character. Butchery and carcass processing are obvious possibilities, as are the making of wooden spears or other organic artefacts. Clactonian knappers exploited local lithic resources along river banks and water bodies. They brought the raw material to the job and knapped it on the spot. The Clactonian is therefore an expedient response for the need for cutting and for shaping things with sharp edges. The shaping and thinning of classic handaxes is not part of the repertoire, though occasional non-classic handaxes are. Perhaps they were attempts to copy the artefacts of the handaxe makers who occasionally passed through their territories, sometimes losing a handaxe on route, or older artefacts found exposed in the landscape? Or perhaps non-classics are best understood as infrequent experiments with new kinds of cutting edge within a mind-set otherwise focused on making sharp edges which could be left behind.

The Clactonian is clearly present in the warmest and most densely forested part of MIS 11c (HoIIb). There is a strong link with water and river banks, particularly where suitable flint resources are close by. On the molluscan evidence, Barnham should continue to be called Clactonian. The non-biface assemblage from MIS 9 Little Thurrock has strong technological links (especially in core working) to the MIS 11c Clactonian. It is a salient point that only two Clactonian sites have been reported since the Second World War, Little Thurrock and Southfleet, and one of those was actually discovered before the First World War. The Clactonian as a non-handaxe tradition in the UK is hardly prolific.

How should we interpret the Clactonian?

In 2007 I predicted that the pendulum of interpretation was beginning to swing back towards cultural explanations (McNabb 2007). That swing is now complete. The neo-culturalists have yet to define what *they* mean by culture in the British Lower Paleolithic, but it is clear that, as with the old culture historians, it involves drawing sharp boundaries around different taxonomic units, and then explaining their persistence over time by reference to differing traditions of inherited knowledge. But this is not a re-hash of the old-fashioned culture history. It is not underpinned by concepts of progressive development in parallel lineages of culturally differentiated hominins.

Mirroring the Clactonian debate, Mary Leakey explained the relationship between the Oldowan and the Acheulean by reference to a model in which the former underwent in-situ local evolution (Developed Oldowan). Hybridization eventually explained the disappearance of the Oldowan/Developed Oldowan as its makers themselves became handaxe makers. De la Torre and Mora note that this is almost a blueprint for Breuil's ideas from the golden age of culture history (de la Torre and Mora 2014). Breuil (1932) and Paterson (1945) among others explained the demise of the Clactonian by reference to this. De la Torre and Mora point out that Mary Leakey dug at Clacton in the early 1930s and with Kenneth Oakley produced a British variation on parallel phyla in opposition to that of Breuil (see above). The elements of that interpretation underpinned her later Olduvai Gorge work, unlike that of Louis Leakey who had argued for a single evolving tradition from non-handaxe to handaxe at Olduvai, in a sense harking back to the pre-parallel phyla era with its emphasis on unilinear progress.

In effect the neo-culturalists are recycling the explanations of the culture-historians, but without signing up to the same theoretical underpinnings:-

- the Clactonian arrives from elsewhere and locally evolves into the Acheulean (local development model).
- the Acheulean arrives from elsewhere and locally replaces the Clactonian (replacement model).

The third historically constituted explanation would be

- the Clactonian is absorbed into the Acheulean by cultural fusion (hybridization model).

This latter differs from the local development model in that it explicitly requires two identifiably different cultural strands to be contemporary, and then to fuse to form a distinct transitional industry which still retains recognizable elements from each parent. Neither of these pre-requisites are present in the British MIS 11 archaeological record, hence the hybridization model is not widely discussed.

In the local development model a single hominin lineage is usually implied, but in the replacement model one or more hominin lineages may be invoked. Contemporary hominin species were implicit in the parallel phyla of Breuil and others in the 1930s, and were used by Mary Leakey to explain her Oldowan/Developed Oldowan vs. Acheulean dichotomy (Leakey 1971; de la Torre and Mora 2014), Contemporaneity remains a distinct possibility in explaining European core and flake vs Acheulean industries (Carbonell et al. 1999). However, Ashton and colleagues sound clear warnings about premature associations between species and particular tool traditions while the status of *Homo heidelbergensis* remains under review (Ashton et al. 2016).

On the other hand some of the explanations of the Clactonian that emerged during its more senior years (as above) owe little to the historical debates.

- the Clactonian is the Acheulean since its makers lost the knowledge of handaxes while migrating over long distances.
- the Clactonian is the Acheulean, it always was, however the cultural paradigm has prevented recognition of this. An approach rooted in comparative lithic technology reveals this bias.
- the Clactonian and the Acheulean are contemporary, although the resolution in the record is not fine enough to show this yet. This prevents us from seeing the core-and-flake vs. handaxe dichotomy as differential use of the landscape. Prey is attracted to water where tool stone is close by. In such circumstances handaxes are unnecessary. Handaxes are made only when away from such locales and hunters must range further afield (activity differentiation explanations *sensu* Ohel). A forest/river bank technology vs. an open-country one (*sensu* Collins or Mithen). Variations on such explanations are being invoked to account for the contemporaneity of handaxe and non-handaxe assemblages at Olduvai Gorge (Uribelarrea et al. 2017), and landscape as a factor was suggested for Olduvai by the geological work of Hay (1976).

Finally, there is convergent evolution which has not, as far as I am aware, ever been explicitly applied to the Clactonian. Long ago (McNabb 1992a, 2007) I suggested that the Clactonian represented a knapping repertoire of flaking practices that could be common to any hominins utilising basic hard hammer knapping. Two of my students, Hannah Fluck and James Cole (Cole 2011; Fluck 2011) independently came to the same conclusion. Data from their respective theses is presented in Table 3.3 and shows that so-called Clactonian flaking patterns are those shared by any hominins knapping irregular nodules with hard hammers, including those who make handaxes. To be clear on this point, alternate and parallel flaking, and single removals are an approach to the problem of making flakes outside of the more structured prepared core working techniques. All stone tool industries, from Prehistoric to Historic times and in every part of the world utilize them in some way, and even prepared core working will utilize them in shaping the core to the point that the predetermined debitage product will be struck. In the Clactonian (and British Acheulean) they are applied to the reduction of irregularly shaped and sized, flint nodules, but they also appear in the Oldowan of South Africa and East Africa (personal observation) although applied in different ways.

One potential implication of such a view is that Clactonian materiality is not actually cultural—it's the inevitable result of pragmatic needs. Real cultural knowledge would be confined to the understanding of how to survive in a landscape—all the tricks and dodges that hunters learned and passed on as they managed to last from one season to the next. Real traditions of inherited knowledge only come when there is something to be passed on that can only be acquired by learning from others, something that cannot be re-invented by convergent evolution. In such a view materiality is only cultural from the Acheulean onwards (see also Shipton 2020), when thinning and shaping become part of the inherited repertoire of hominins.

In this sense the Clactonian as a reflection of basic knapping practices could be seen as a product of endless convergent evolution when the need for flakes and sharp edges presented itself; a common solution to a common problem. However, it does not explain the Clactonian as a phenomenon in time and space. It does not explain why Clactonian knappers did not make classic handaxes. To explore this we must look beyond convergent evolution—to what cultural knowledge really was, and how Lower Paleolithic hominins learnt.

Currently I suspect the answer to the Clactonian will be found either in a better understanding of how knowledge was passed on from one generation to the next, or in the identification of genuine Clactonian knapping traditions on the continent.

The latter would allow us to support or falsify the in-situ vs replacement models, as well as comprehend how the Clactonian can return at the MIS 10/9 boundary at Little Thurrock and Purfleet (White and Schreve 2000). In reviewing Fluck's data, Ashton et al. (2016) note 108 non-handaxe assemblages were reported by her from Europe, but only 14 have more than 50 artefacts, and only 4 are dated to MIS 11. Fluck is clear that the European non-biface assemblages are different when compared to the Clactonian; raw materials account for some of these differences while in other cases the retouched tool components differ. It is worth reiterating that Fluck's extensive study of the non-biface assemblages of Western Europe found no substantial evidence for a Clactonian-like tradition in MIS 11, 10 or 9, but at the same time confirmed the presence of genuine non-handaxe assemblages.

If the neo-culturalists wish to support the distinctiveness of their cultural taxonomic units I believe they must also engage with the mechanisms hominin used to perpetuate distinctiveness—the transfer of genuine cultural knowledge across the generations. Palmer and colleagues defined culture as simply 'knowledge from the ancestors', in other words *any* information passed down from one generation to the next (Palmer et al. 2005). I believe that the way knowledge is transferred will influence what can be learnt. We should recall that *Homo heidelbergensis* had a brain a third smaller than our own, and their life histories may have been somewhat different (Nowell and White 2010; Hopkinson et al. 2013) although reconstructing this is notoriously difficult. Young Heidelbergs may have matured more quickly and experienced much shorter periods of childhood and adolescence (terms used *sensu lato*). These are key periods for modern humans when language, social skills and adult social behaviour are learnt. We should be very wary of assuming the capacity to acquire, enhance and then pass on knowledge was similar to our own. We should be wary of assuming a capacity for culture as we understand it, despite evidence for complex social behaviours such as communal hunting at Atapuerca (Rodríguez-Hidalgo et al. 2017).

An interesting departure from the normative view of cultural transmission has been suggested by Plunkett and White in discussing the handaxe site of Foxhall Road (White and Plunkett 2004). They link inter-generational social learning with role models rather than invariant bodies of knowledge about how things should be done. Such a concept draws its inspiration from primate studies (as above). As mooted earlier, knowledge from the ancestors in the Acheulean would entail learning how to survive and succeed in both the social and physical worlds, as well as how to make shaped tools (socially inherited knowledge about thinning and shaping, the capacity to mentalize concepts such as handaxe shape), which in the presumed absence of fully modern linguistic capability would involve learning at least by imitation—process copying with a knowledge of the end state (Stade 2017). Incorporating this with White and Plunkett's notion of different mentors in a social group, means the knappers in Acheulean groups need not all conform to a single standard of practice and end-product. Different mentors could have different ways of doing things and make different looking handaxes. This explains why at Foxhall Road, Caddington, Round Green, and Beeches Pit, all primary- or near-primary context UK Acheulean sites (Fig. 3.1), you get such a variety of handaxe shapes at the same site. Social groups with different role models will produce diverse material culture, and small groups may generate and sustain such diverse practices (Hopkinson et al. 2013), but not pass them on to the larger population.

In such a scenario Boxgrove and other Acheulean sites with high levels of conformity in handaxe shape could have been the product of a small isolated group, late in the MIS 13 interglacial. Its role model/s monotonously churn out the same thing, because there is little stimulus for change. Knowledge from the ancestors for them is in the use of soft hammers and tranchets, and cutting edges all the way around the axe, but they rarely vary their repertoire.

And the Clactonian? It cannot be explained by convergent evolution alone, its spatio-temporal extent requires some behavioural/cultural underpinning. Was its culturally inherited repertoire of social knowledge restricted to learning about how to survive in the outside world? Imitation learning by a pack hunting predator? In which case its knapping technology did not need not be too sophisticated if its adaptive niche remained stable (*sensu* Ashton), or it targeted parts of the landscape where it knew affordances were already present (*sensu* Wenban-Smith). There would be no need to endlessly reinvent a simple technology for making sharp edges that once learnt by imitation learning was never forgotten. Alternatively, the pressing need for sharp edges may well have triggered convergent evolution when the need arose.

Then again perhaps the Clactonian was the product of a particular hominin lineage, one that did not possess a sufficiently developed enough theory of mind (Dunbar et al. 2014) to enable hominins to recognise others as individuals like themselves, and so could not have mentors who could pass on social knowledge about material culture that could only be learned from other individuals? Perhaps.

Answers to these questions will go a long way to understanding the Clactonian.

Note

1. The term flaked flakes (McNabb 1992a, 2007; Conway et al. 1996) is a descriptive one used in order to side-step the difficult issue of whether these artefacts are cores on flakes, or retouched tools. They may well be both at different times and different places. The term is therefore a more neutral one which describes a technological practice with no explanatory inferences.

References

Ashton, N. (2018). Landscapes of habit and persistent places during MIS 11 in Europe: A return journey from Britain. In M. Pope, J. McNabb, & C. Gamble (Eds.), *Crossing the human threshold: Dynamic transformation and persistent places during the Middle Pleistocene* (pp. 142–164). Abingdon Oxford: Routledge.

Ashton, N., Cook, J., & Rose, J. (Eds.). (1992). *High Lodge; Excavations by G. de G. Sieveking 1962–1968 and J. Cook 1988*. London: British Museum Press.

Ashton, N., Lewis, S. G., & Parfitt, S. (Eds.). (1998). *Excavations at the lower palaeolithic site at East Farm, Barnham, 1989–1994. Occasional Papers British Museum 125*. London: Trustees of the British Museum.

Ashton, N., Lewis, S. G., Parfitt, S., & White, M. (2006). Riparian landscapes and human habitat preferences during the Hoxnian (MIS 11) interglacial. *Journal of Quaternary Science, 21,* 497–505.

Ashton, N., Lewis, S. G., Parfitt, S. A., Davis, R. J., & Stringer, C. B. (2016). Handaxe and non-handaxe assemblages during marine isotope stage 11 in North Western Europe. Recent investigations at Barnham Sufolk, UK. *Journal of Quaternary Science, 31,* 837–843.

Ashton, N., Lewis, S. G., Parfitt, S. A., Penkman, K. E. H., & Coope, R. (2008). New evidence for complex climate change in MIS 11 from Hoxne, Suffolk, UK. *Quaternary Science Reviews, 27,* 625–668.

Ashton, N., & McNabb, J. (1994). Bifaces in perspective. In N. Ashton, & A. David (Eds.), *Stories in stone. Lithic studies society occasional paper 4.* (pp. 182–191). Oxford: The Lithics Studies Society.

Ashton, N., McNabb, J., Irving, B., Lewis, S. G., & Parfitt, S. (1994). Contemoraneity of Clactonian and Acheulean flint industries at Barnham, Suffolk. *Antiquity, 68,* 585–589.

Breuil, H. (1932). Les Industries a Éclats du Paléolithique Ancien, 1.—Le Clactonien. *Préhistoire, 1,* 125–190.

Bridgland, D. R. (1994). *Quaternary of the Thames*. London: Chapman and Hall.

Bridgland, D. R., Field, M. H., Holmes, J. A., McNabb, J., Preece, R. C., Selby, I., et al. (1999). Middle pleistocene interglacial thames-medway deposits at Clacton-on-Sea, England: Reconsideration of the biostratigraphical and environmental context of the type Clactonian palaeolithic industry. *Quaternary Science Reviews, 18,* 109–146.

Bridgland, D. R., & Harding, P. (1993). Middle pleistocene thames terrace deposits at Globe Pit, Little Thurrock, and their contained clactonian industry. *Proceedings of the Geologists' Association, 104,* 263–283.

Bridgland, D. R., Harding, P., Allen, P., Candy, I., Cherry, C., Horne, D., et al. (2013). An enhanced record of MIS 9 environments, geochronology and geoarchaeology; Data from construction of the high speed 1 (London-Channel Tunnel) rail-link and other recent investigations at Purfleet, Essex, UK. *Proceedings of the Geologists' Association, 124,* 417–476.

Bridgland, D. R., & White, M. J. (2015). Chronological variations in Handaxes: Patterns detected from fluvial archives in North-Western Europe. *Journal of Quaternary Science, 30,* 623–638.

Carbonell, E., Mosquera, M., Rodríguez, X. P., Sala, R., & Van der Made, J. (1999). Out of Africa: The dispersal of the earliest technical systems reconsidered. *Journal of Anthropological Archaeology, 18,* 119–136.

Chandler, R. H. (1928–9). On the Clactonian industry of Swanscombe. *Proceedings of the Prehistoric Society of East Anglia, 6,* 79–116.

Chandler, R. H. (1931). Preliminary notice. Types of Clactonian implements at Swanscombe. *Proceedings of the Prehistoric Society of East Anglia, 6,* 377–378.

Cole, J. (2011). *Hominin cognitive and behavioural complexity in the pleistocene: Assessment through identity, intentionality and visual display*. Unpublished Ph.D, University of Southampton.

Collins, D. (1969). Culture traditions and environment of early man. *Current Anthropology, 10,* 267–316.

Conway, B. (1996a). Chapter 6. The geology outside the national nature reserve, 1968–1972. In B. Conway, J. McNabb, & N. Ashton (Eds.), *Excavations at Barnfield Pit, Swanscombe, 1968–1972, occasional papers British museum 94* (pp. 67–89). London: Trustees British Museum.

Conway, B. (1996b). Chapter 8. The stratigraphy and chronology of the pleistocene deposits of Barnfield Pit, Swanscombe. In B. Conway, J. McNabb, & N. Ashton. (Eds.), *Excavations at Barnfield Pit, Swanscombe, 1968–1972. Occasional papers British museum 94* (pp. 117–136). London: Trustees of the British Museum.

Conway, B., McNabb, J., & Ashton, N. (Eds.). (1996). *Excavations at Barnfield Pit, Swanscombe, 1968–1972. Occasional papers British museum 94*. London: Trustees of the British Museum.

Cruse, R. J., Bridgland, D. R., Callow, P., Currant, A., Hubbard, R. N. L. B., Debenham, N. C., et al. (1987). Further investigation of the Acheulean site at Cuxton. *Archaeologia Cantiana, 104,* 39–81.

de la Torre, I., & Mora, R. (2014). The transition to the Acheulean in East Africa: An assessment of paradigms and evidence from Olduvai Gorge (Tanzania). *Journal of Archaeological Method and Theory, 214,* 781–823.

Dunbar, R., Gamble, C., & Gowlett, J. (2014). *Thinking Big. How the evolution of social life shaped the human mind*. London: Thames and Hudson.

Fluck, H. L. (2011). *Non-biface assemblages in middle pleistocene Western Europe. A comparative study*. Unpublished Ph.D, University of Southampton.

Gamble, C. (1999). *The palaeolithic societies of Europe*. Cambridge: Cambridge University Press.

Hay, R. L. (1976). *Geology of the Olduvai Gorge*. London: University of California Press Berkeley.

Hopkinson, T., Nowell, A., & White, M. J. (2013). Life histories, metapopulation ecology, and innovation in the Acheulean. *PalaeoAnthropology*. https://doi.org/10.4207/PA.2013.ART80.

Keeley, L. H. (1980). *Experimental determination of stone tool uses*. Chicago: University of Chicago Press.

Kerney, M. P. (1971). Interglacial deposits in Barnfield Pit, Swanscombe, and their Molluscan Fauna. *Journal of the Geological Society of London, 127,* 69–86.

King, W. B. R., & Oakley, K. P. (1936). The pleistocene succession in the lower parts of the Thames Valley. *Proceedings of the Prehistoric Society, 2,* 52–76.

Leakey, L. S. B. (1951). *Olduvai Gorge*. Cambridge: Cambridge University Press.

Leakey, L. S. B., Evernden, J. F., & Curtis, G. H. (1961). Age of Bed I, Olduvai. *Nature, 191,* 478–479.

Leakey, M. D. (1971). *Olduvai Gorge: Excavations in Beds I and II 1960–1963*. Cambridge: Cambridge University Press.

Marston, A. T. (1937). The Swanscombe skull. *Journal of the Royal Anthropological Institute, 67,* 339–406.

Marston, A. T. (1942). Flint industries of the high terrace at Swanscombe. *Proceedings of the Geologists' Association, 53,* 106.

McNabb, J. (1992a). *The Clactonian: British lower palaeolithic flint technology in biface and non-biface assemblages*. Unpublished Ph. D. thesis. University of London.

McNabb, J. (1992b). The cutting edge, Bifaces in the Clactonian. *Lithics, 13,* 4–10.

McNabb, J. (1996a). 4. Through the looking glass: An historical perspective on archaeological research at Barnfield Pit, Swanscombe, ca. 1900–1964. In B. Conway, J. McNabb, & N. Ashton (Eds.), *Excavations at Barnfield Pit, Swanscombe, 1968–1972. Occasional papers British museum 94* (pp. 31–51). London: Trustees of the British Museum.

McNabb, J. (1996b). More from the cutting edge: Further discoveries of Clactonian bifaces. *Antiquity, 70,* 428–436.

McNabb, J. (2007). *The British lower palaeolithic: stones in contention*. London: Routledge.

McNabb, J. (2012). *Dissent with Modification. Human origins, palaeolithic archaeology and evolutionary anthropology in Britain 1859–1901*. Oxford: Archaeopress.

McNabb, J., & Ashton, N. (1995). Thoughtful flakers. *Cambridge Archaeological Journal, 5,* 289–298.

Meijer, T., & Preece, R. C. (1995). Malacological evidence relating to the insularity of the British Isles during the quaternary. In R. C. Preece (Ed.), *Island Britain: A quaternary perspective* (pp. 89–110). London: Geological Society. Special Publications 96.

Mithen, S. (1994). Technology and society during the middle Pleistocene: Hominid group size, social learning and industrial variability. *Cambridge Archaeological Journal, 4,* 3–32.

Morell, V. (1995). *Ancestral passions*. New York: Touchstone (for Simon and Schuster).

Movius, H. (1948). The lower palaeolithic cultures of South-Eastern and Eastern Asia. *Transactions of the American Philosophical Society, 38,* 329–420.

Nowell, A., & White, M. J. (2010). Growing up in the Middle Pleistocene: Life history strategies and their relationship to Acheulean industries. In A. Nowell & I. Davidson (Eds.), *Stone tools and the evolution of human cognition* (pp. 67–82). Boulder: University Press of Colorado.

O'Connor, A. (2003). Geology, Archaeology, and 'the Raging Vortex of the "Eolith" Controversy'. *Proceedings of the Geologists' Association, 114,* 255–262.

O'Connor, A. (2006). Samuel Hazzledine Warren and the construction of a chronological framework for the British quaternary in the early twentieth century. *Proceedings of the Geologists' Association, 117,* 41–52.

O'Connor, A. (2007). *Finding time for the old stone age. A history of palaeolithic archaeology and quaternary geology in Britain, 1860–1960.* Oxford: Oxford University Press.

Oakley, K. P., & Leakey, M. (1937). Report on excavations at Jaywick Sands, Essex (1934). *Proceedings of the Prehistoric Society, 3,* 217–260.

Ohel, M. Y. (1979). The Clactonian: An independent complex or an integral part of the Acheulean? *Current Anthropology, 20,* 685–713.

Palmer, C. T., Coe, K., & Wadley, R. L. (2005). On tools and traditions. *Current Anthropology, 46,* 459–460.

Paterson, T. T. (1937). Studies on the Palaeolithic succession in England. *Proceedings of the Prehistoric Society, 3,* 87–135.

Paterson, T. T. (1940a). Geology and early man. *Nature, 146,* 12–15.

Paterson, T. T. (1940–1941). On a World correlation of the Pleistocene. *Transactions of the Royal Society of Edinburgh, 60,* 373–425.

Paterson, T. T. (1942). Lower Palaeolithic man in the Cambridge District. Unpublished Ph.D thesis, University of Cambridge.

Paterson, T. T. (1945). Core, culture and complex in the old stone age. *Proceedings of the Prehistoric Society, 11,* 1–19.

Penck, A., & Bruckner, E. (1909). *Die Alpen im Eiszeitalter.* Leipzig: Tachnitz.

Pettitt, P., & White, M. J. (2012). *British palaeolithic: Hominin societies at the edge of the pleistocene world.* Oxford: Routledge.

Pike, K., & Godwin, H. (1953). The interglacial at Clacton-on-Sea, Essex. *Quarterly Journal of the Geological Society of London, 108,* 261–272.

Preece, R. C., Gowlett, J. A. J., Parfitt, S. A., Bridgland, D. R., & Lewis, S. G. (2006). Humans in the Hoxnian: Habitat, context and fire use at beeches pit, West Stow, Suffolk, UK. *Journal of Quaternary Science, 21,* 485–496.

Preece, R. C., Parfitt, S. A., Bridgland, D. R., Lewis, S. G., Rowe, P. J., Atkinson, T. C., et al. (2007). Terrestrial environments during MIS 11: Evidence from the Palaeolithic Site at West Stow, Suffolk, UK. *Quaternary Science Reviews, 26,* 1236–1300.

Rodríguez-Hidalgo, A., Saladi, P., Oll, A., Arsuaga, J. L., Bermúdez de Castro, J. M., & Carbonell, E. (2017). Human predatory behavior and the social implications of communal hunting based on evidence from the TD10.2 Bison bone bed at Gran Dolina (Atapuerca, Spain). *Journal of Human Evolution, 105,* 89–122.

Roe, D. A. (1968). British lower and middle palaeolithic handaxe groups. *Proceedings of the Prehistoric Society, 34,* 1–82.

Roe, D. A. (1981). *The lower and middle palaeolithic periods in Britain.* London: Routledge and Kegan Paul.

Schreve, D. C., Bridgland, D. R., Allen, P., Blackford, J. J., Gleed-Owen, C. P., Griffiths, H. I., et al. (2002). Sedimentology, palaeontology, and archaeology of late middle Pleistocene River Thames Terrace Deposits at Purfleet, Essex, UK. *Quaternary Science Reviews, 21,* 1423–1464.

Shipton, C. (2020). The unity of Acheulean culture. In H. Groucutt (Ed.), *Culture history and convergent evolution: Can we detect populations in prehistory?* (pp. 13–27). Switzerland: Springer, Cham.

Singer, R., Gladfelter, B. G., & Wymer, J. (Eds.). (1993). *The lower palaeolithic site at Hoxne, England.* Chicago: Chicago University Press.

Singer, R., Wymer, J., Gladfelter, B. G., & Wolff, R. G. (1973). Excavation of the Clactonian industry at the Golf Course, Clacton-on-Sea, Essex. *Proceedings of the Prehistoric Society, 39,* 6–74.

Smith, R. A., & Dewey, H. (1913). Stratification at Swanscombe: Report on Excavations Made on Behalf of the British Museum and H.M. *Geological Survey. Archaeologia, 64,* 177–204.

Smith, R. A., & Dewey, H. (1914). The high terrace of the thames: Report on excavations made on behalf of the British Museum and H.M. Geological survey in 1913. *Archaeologia, 65,* 187–212.

Smith, W. G. (1894). *Man the primeval savage. His haunts and relics from the Hill-Tops of Bedfordshire to Blackwall.* London: Stanford.

Snelling, A. J. R. (1964). Excavations at the Globe Pit, Little Thurrock, Grays, Essex. *Essex Naturalist, 31,* 199–208.

Sommer, M. (2004). Eoliths as evidence for human origins? The British Context. *Historical and Philosphical Life Sciences, 26,* 209–241.

Spencer, F. (1990). *Piltdown a scientific forgery.* London: Natural History Museum Publications.

Stade, C. M. (2017). *Lithic morphological variability as a proxy for palaeolithic linguistic ability: A knapping training study exploring cultural transmission, theory of mind and language.* Unpublished Ph.D, Southampton.

Tester, P. J. (1951). Palaeolithic flint implements from the Bowman's Lodge Gravel Pit, Dartford Heath. *Archaeologia Cantiana, 63,* 122–134.

Tester, P. J. (1965). An Acheulean site at cuxton. *Archaeologia Cantiana, 80,* 30–60.

Tester, P. J. (1976). Further consideration of the Bowman's lodge industry. *Archaeologia Cantiana, 91,* 29–39.

Toth, N. (1985). The Oldowan reassessed: A close look at early stone artefacts. *Journal of Archaeological Science, 12,* 101–120.

Trigger, B. (1989). *A history of archaeological thought.* Cambridge: Cambridge University Press.

Turner, C. (1970). The middle pleistocene deposits at Marks Tey, Essex. *Philosophical Transactions of the Royal Society London. Series B, 257,* 373–440.

Tye, G. J., Sherriff, J., Candy, I., Coxon, P., Palmer, A., McClymont, E. L., et al. (2016). The $d^{18}O$ stratigraphy of the Hoxnian lacustrine sequence at Marks Tey, Essex, UK: Implications for the climatic structure of MIS 11 in Britain. *Journal of Quaternary Science, 31,* 75–92.

Uribelarrea, D., Martín-Perea, D., Díez-Martín, F., Sánchez-Yustos, P., Domínguez-Rodrigo, M., Baquedano, E., et al. (2017). A reconstruction of the paleolandscape during the earliest Acheulian of FLK West: The co-existence of Oldowan and Acheulian industries during lowermost Bed II (Olduvai Gorge, Tanzania). *Palaeogeography, Palaeoclimatology, Palaeoecology, 488,* 50–58.

Waechter, J. d. A. (1973). The late middle Acheulean industries in the Swanscombe Area. In D. E. Strong (Ed.), *Archaeological theory and practice* (pp. 67–86). London: Seminar Press.

Waechter, J. d. A., Newcomer, M. H., & Conway, B. (1970). Swanscombe 1970. *Proceedings of the royal anthropological institute of Great Britain and Ireland,* 43–64.

Warren, S. H. (1911). On a palaeolithic (?) wooden spear. *Quarterly Journal of the Geological Society of London, 67,* xcix.

Warren, S. H. (1912). Palaeolithic remains from Clacton-on-Sea, Essex. *Essex Naturalist, 17,* 15.

Warren, S. H. (1922). The Mesvinian industry of Clacton-on-Sea, Essex. *Proceedings of the Prehistoric Society of East Anglia, 3,* 1–6.

Warren, S. H. (1923). The *Elephas-Antiquus* Bed of Clacton-on-Sea. *Quarterly Journal of the Geological Society of London, 79,* 634.

Warren, S. H. (1924a). The elephant Bed of Clacton-on-Sea. *Essex Naturalist, 21,* 32–40.

Warren, S. H. (1924b). Pleistocene classifications. *Proceedings of the Geologists' Association, 35,* 265–282.
Warren, S. H. (1926). The classification of the lower palaeolithic with especial reference to Essex. In A. F. Ravenshear (Ed.), *The south eastern naturalist. Transactions of the south-eastern union of scientific societies, volume 31. Including the proceedings of the thirty first annual congress held at Colchester in 1926* (pp. 38–50).
Warren, S. H. (1932). The palaeolithic industries of the Clacton and Dovercourt Districts. *Essex Naturalist, 24,* 1–29.
Warren, S. H. (1951). The Clacton flint industry: A new interpretation. *Proceedings of the Geologists' Association, 62,* 107–135.
Warren, S. H. (1955). The Clacton (Essex) channel deposits. *Quarterly Journal of the Geological Society of London, 111,* 287–307.
Wenban-Smith, F. (1998). Clactonian and Acheulean industries in Britain; Their chronology and significance reconsidered. In N. Ashton, F. Healy, & P. Pettitt (Eds.), *Stone age archaeology. Essays in honour of John Wymer. The lithic studies society occasional papers 6* (pp. 90–97). Oxford: Lithics Studies Society.
Wenban-Smith, F. (2004). Handaxe typology and lower palaeolithic cultural development: Ficrons, cleavers and two giant handaxes from cuxton. *Lithics, 25,* 11–21.
Wenban-Smith, F. (Ed.). (2013). *The Ebbsfleet Elephant: Excavations at Southfleet Road, Swanscombe in Advance of High Speed 1, 2003–4. Oxford Archaeology Monograph, 20.* Oxford: Oxford Archaeology.
Wenban-Smith, F., & Bridgland, D. R. (2001). Palaeolithic archaeology at the Swan Valley Community School, Swanscombe, Kent. *Proceedings of the Prehistoric Society, 67,* 219–259.
West, R. G. (1956). The quaternary deposits at Hoxne, Suffolk. *Philosophical Transactions of the Royal Society London. Series B, 239,* 265–356.
West, R. G. (1969). Pollen analysis from interglacial deposits at Aveley and Grays, Essex. *Proceedings of the Geologists' Association, 80,* 271–282.
West, R. G., & McBurney, C. M. B. (1954). The quaternary deposits at Hoxne, Suffolk, and their archaeology. *Proceedings of the Prehistoric Society, 20,* 131–154.
White, M. J. (2000). The Clactonian question: On the interpretation of core-and-flake assemblages in the British lower palaeolithic. *Journal of World Prehistory, 14,* 1–63.
White, M. J. (2015). 'Dancing to the rhythms of the Biotidal zone': Settlement history and culture history in middle pleistocene Britain. In F. Coward, R. Hosfield, M. Pope, & F. Wenban-Smith (Eds.), *Settlement, society and cognition in human evolution* (pp. 154–173). Cambridge: Cambridge University Press.
White, M. J., & Bridgland, D. R. (2018). Thresholds in lithic technology and human behaviour during MIS 9 in Britain. In M. Pope, J. McNabb, & C. Gamble (Eds.), *Crossing the human threshold. Dynamic transformations and persistent places during the middle pleistocene* (pp. 165–191). Abingdon Oxford: Routledge.
White, M. J., & Plunkett, S. (Eds.). (2004). *Miss Layard excavates: A palaeolithic site at Foxhall Road, Ipswich, 1903–1905.* Liverpool: Western Academic and Specialist Press.
White, M. J., & Schreve, D. C. (2000). Insular Britain–Peninsula Britain: Palaeogeography, colonisation and settlement history of lower palaeolithic Britain. *Proceedings of the Prehistoric Society, 66,* 1–28.
White, T. S., Preece, R. C., & Whittaker, J. E. (2013). Molluscan and ostracod successions from Dierden's Pit, Swanscombe: Insights into the fluvial history, sea-level record and human occupation of the Hoxnian Thames. *Quaternary Science Reviews, 70,* 73–90.
Wymer, J. (1956). Palaeoliths from gravel of the ancient channel between Caversham and Henley at highlands near Henley. *Proceedings of the Prehistoric Society, 22,* 29–36.
Wymer, J. (1957). A Clactonian flint industry at Little Thurrock, Grays, Essex. *Proceedings of the Geologists' Association, 68,* 159–177.
Wymer, J. (1968). *Lower palaeolithic archaeology in Britain.* London: John Baker Publishers Ltd.
Wymer, J. (1974). Clactonian and Acheulean industries in Britain: Their chronology and significance. *Proceedings of the Geologists' Association, 85,* 391–421.
Wymer, J. (1985a). Early man in Britain: Time and change. *Modern Geology, 9,* 261–272.
Wymer, J. (1985b). *Palaeolithic sites of East Anglia.* Norwich: Geo Books.
Wymer, J. (1988). Palaeolithic archaeology and the British quaternary sequence. *Quaternary Science Reviews, 7,* 79–98.

Chapter 4
Culture and Convergence: The Curious Case of the Nubian Complex

Huw S. Groucutt

Abstract 'Nubian Levallois' lithic technology has been found from South Africa to India, it occurs sporadically over a period of more than two hundred thousand years, and it appears to be associated with at least two hominin species. Despite this, proponents of the 'Nubian Complex' argue that this technocomplex—often, but not exclusively, defined by the presence of Nubian Levallois technology—offers a strong culture historical signal. This argument claims that the Nubian Complex is an originally Northeast African entity, dating to Marine Isotope Stage 5, and that by tracing the distribution of Nubian Levallois technology it is possible to trace the spread of *Homo sapiens* from Northeast Africa. In light of these bold claims, it is important to test the reality and usefulness of the Nubian Complex idea. In this paper I review the history of the Nubian Complex, evaluate sites assigned to it, and consider the characteristics and significance of Nubian Levallois technology. This review suggests that the original reasons for defining the Nubian Complex were flawed, definitions of it are overly-variable and inconsistent, and that the concept is driving misleading models that are actively harming interpretations of the record. It should therefore be abandoned. Perhaps the most telling criticism of the Nubian Complex is that even its proponents do not agree on which sites should be included (e.g. Bir Tarfawi). I explore the possibility that Nubian Levallois technology—which should be disentangled from the culture-historical concept of the 'Nubian Complex'—represents a case of convergent evolution and identify avenues for future research. This reorientation facilitates insights into the behavioral significance of Nubian Levallois technology, in terms of factors such as standardization and mobility strategies.

Keywords Convergent evolution • Culture-history • Levallois • Lithics • Middle Paleolithic • Middle Stone Age • Nubian Levallois • Regionalization

Introduction

The so-called 'Nubian Complex' offers a fascinating case study for researchers interested in culture history, the meaning of variability in lithic technology, convergent evolution and the philosophy of science. As shall be explored below, the original reasons for defining the Nubian Complex have arguably been negated and the entity has been so variably defined as to be seemingly meaningless. Despite this, the idea of the Nubian Complex has persisted, arguably demonstrating a common tendency in archaeology towards confirmation bias (Nickerson 1998). The aim of this paper is to test the notion of the Nubian Complex; whether it really existed, and whether it is helpful in understanding human prehistory.

To its proponents, the Nubian Complex reflects the long-term existence of a Northeast African population defined by a particular form of material culture (e.g. Van Peer 1998), as well as subsequently demonstrating the dispersal of our species into Southwest Asia (e.g. Rose et al. 2011). However, two elements are being conflated in the 'Nubian' debate. Firstly, there is the notion of the Nubian Complex, as a technocomplex with culture-historical overtones (Tables 4.1 and 4.2). Secondly, there is the existence of a kind of Levallois reduction method, generally called the 'Nubian Levallois' method (see Fig. 4.1) (Table 4.3). Previous attempts by this author to instead refer to this

H. S. Groucutt (✉)
Extreme Events Research Group, Max Planck Institute for Chemical Ecology, Hans-Knöll-Straße 8, 07745 Jena, Germany
e-mail: hgroucutt@ice.mpg.de

Department of Archaeology, Max Planck Institute for the Science of Human History, Kahlaische Straße 10, 07745 Jena, Germany

Max Planck Institute for Biogeochemistry, 10 Hans-Knöll-Straße, 07745 Jena, Germany

H. S. Groucutt (ed.), *Culture History and Convergent Evolution: Can We Detect Populations in Prehistory?*,
Vertebrate Paleobiology and Paleoanthropology, https://doi.org/10.1007/978-3-030-46126-3_4

Table 4.1 Key definitions of the Nubian Complex (NC) presented in chronological order of their publication

Author(s)	Key features	Chronology of NC	Geography
Van Peer (1998)	Long term co-existence of the Lower Nile Valley Complex (LNVC) and the Nubian Complex (NC). NC "characterised by the use of so-called Nubian Levallois methods" and tools including "bifacial foliates, Nubian end scrapers, Nazlet Khater points, and truncated-faceted pieces" (p. 120). LNVC (incorrectly) guessed to be contemporary with NC.	Late Middle Pleistocene (ca. 240 ka) to the transition to the Upper Paleolithic, ca. 40 ka.	Lower Nile Valley, Eastern Sahara, Red Sea Mountains.
Vermeersch (2001)	Follows above idea of the NC, that is seen here as a mostly MIS 5 entity, although Nubian technology is said to occur in the Middle Pleistocene. He adds that the NC "seems to contain a true blade component" (105). Vermeersch includes the Taramsan in NC.	130–50 ka, but Nubian technology occurring earlier.	As above.
Van Peer and Vermeersch (2000)	NC sites have both Levallois flake and point production, the latter including the Nubian method (p. 48). Retouched tools are "bifacial foliates, various retouched point types including Mousterian and Nazlet Khater points, and truncated-facetted pieces" plus "side scrapers, denticulates, and…. Upper Palaeolithic types" (p. 49).	Late Middle Pleistocene to around 60 ka in Nile Valley, younger elsewhere.	As above, plus hints at East Africa connections.
Vermeersch and Van Peer (2002)	They present (p. 355) a slightly modified but basically similar narrative to that of Van Peer (1998), reflecting work on material from sites such as Nazlet Khater. NC people "frequently used the Nubian Levallois method" (p. 356). In this view, Taramsan not part of the NC.	Early NC: 200–120 ka Late NC: 120–70 ka.	NE Africa.
Van Peer (2004)	In early NC, Nubian technology and bifacial foliates. Late NC lacks bifacial component.	As above.	NE Africa.
Van Peer and Vermeersch (2007)	Essentially as earlier publications, but note geographic expansion.	Particularly MIS 5, continues to ca. 60 ka.	NE Africa, East Africa, Yemen.
Van Peer et al. (2010)	Echoes earlier definitions, but is more complex. NC *sensu stricto* is said to date to MIS 5a…but Nubian reduction is said to occur in activity phase 1 (Middle Pleistocene) at Taramsa 1. And likewise sites lacking Nubian Levallois cores are said to belong to NC (p. 240).	*Sensu stricto* MIS 5a, *sensu lato* MIS 5, *sensu latissimo* Middle Pleistocene to MIS 4/3.	Nile Valley, East Africa, Yemen.
Rose et al. (2011)	NC "industries are distinguished by a characteristic and highly standardized method of preferential Levallois reduction…Nubian core technology… recognized by its triangular/sub-triangular shaped cores and a specific opposed platform preparation of the primarily working surface." (p. 1).	MIS 5.	NE Africa, East Africa, Arabia.
Rots et al. (2011)	With 'Lupemban' and 'Sangoan' roots, NC is an amalgamation of MIS 5 industries. Within this looser definition, the NC of Van Peer (1998) is restricted to later MIS 5. In addition, they argue that both percussive and projectile technologies are "characteristic features" of the NC (p. 658).	Middle Pleistocene roots, but Early NC: early MIS 5, late NC: late MIS 5.	NE Africa.
Usik et al. (2013)	"Afro-Arabian Nubian Technocomplex" contains "the African and Arabian Nubian Traditions, which, in turn, consist of technologically related lithic industries that are distinguished by the presence of the Nubian Levallois core reduction strategy" (p. 244). Little focus on other factors such as retouched tools. They also use term "Nilotic Nubian Tradition" (p. 245), that seems to encompass more of a Van Peer (1998) long chronology view.	Early NC: MIS 5e, Late NC: MIS 5a, but these occur within a Middle Pleistocene to MIS 3 Nilotic Nubian Tradition.	Northeast Africa, East Africa, southwest Arabia.

(continued)

Table 4.1 (continued)

Author(s)	Key features	Chronology of NC	Geography
Crassard and Hilbert (2013)	It appears that to them a NC site is one with Nubian Levallois technology. Nubian Levallois Technology is "typically northeast African" (p. 1) and the "origin of Nubian Levallois technology lies in Africa" (p. 9). They argue that distinctions between Type 1 and Type 2 cores (and Type 1/2) should be abandoned and that there is high variability within Levallois reduction methods.	Mostly MIS 5, but also to MIS 3… Although they mistakenly show an MIS 8 to 6 core.	Northeast Africa, East Africa, Arabia.
Rose and Marks (2014)	NC "characterized by Nubian Levallois cores and, to a lesser extent, classic centripetal Levallois cores". Some sites they assign to NC do not include Nubian cores. Industries such as the Khormusan and Taramsan are said to be descended from, but not part of, the NC (p. 61), yet later they are described part of the Late NC (p. 74).	Mostly MIS 5. They are unclear if post-MIS 5 assemblages belong to NC or not.	NE Africa and Arabia.
Schmidt et al. (2015)	NC is northeast African and is "characterised technologically by the Nubian Levallois methods for points and the classical Levallois method for flake production"	?	NE Africa.
Van Peer (2016)	The hallmark of NC is the "Nubian reduction strategy…Its origins can be found in Lupemban foliate façonnage" (p. 151). Early NC has "thin leaf-shaped foliates and so called "Nubian endscrapers" (p. 153), these are absent in late NC, which also has high levels of Type 1 reduction. Despite his own definition, Van Peer assigns Haua Fteah (Pre-Aurignacian) and Jebel Faya (Assemblage C) to the NC, despite neither having Nubian Levallois technology!	'By MIS 5e' to MIS 5a…but 'Nubian Aterian' occurring in MIS 4	Specific to NE Africa, yet also found in East Africa, Arabia (e.g. Jebel Faya), etc.
Hilbert et al. (2016)	NC is a "group of technologically related industries that might share a common origin and occupy a discrete geographic and temporal range" (1). "Nubian Levallois technology is the defining characteristic" of NC" (1). Despite being distinctive and diagnostic, they suggest a "strict delineation between the Nubian and the Levallois "constructed" reduction system (*sensu* Boëda et al. 1998), is problematic" (7).	MIS 5	NE Africa, East Africa Levant, Arabia.
Hilbert et al. (2017).	NC "is a Middle Stone Age (in Africa) and Middle Palaeolithic (in Arabia) technocomplex primarily distinguished by the use of the Nubian Levallois method, which is a highly specific approach to point manufacture. Nubian core technology is a regional variant of the preferential Levallois method of point production" (p. 78).	120–55 ka	North Africa, East Africa, Arabia.

technology—which involves the production of Levallois products with generally convergent margins from cores with the key feature of having a 'median distal ridge'—in a more morphological and less geographically-suggestive way as 'beaked Levallois' reduction (e.g. Groucutt et al. 2015a), following Seligman's (1921) original definition of this technology, have met with silence and so the term 'Nubian Levallois' will be used here. I will return to the technological theme later in the paper, after exploring the history of the Nubian Complex as a culture-historical entity and the stratigraphic, chronological and technological characteristics of sites assigned to the Nubian Complex and sites where Nubian Levallois technology has been reported.

For those who believe in the reality and utility of the Nubian Complex—by which they often mean the presence of Nubian Levallois technology—it offers a culture historical signal that is unusually clear for this early period of human prehistory. Other issues aside, this presupposes that convergent evolution is a negligible factor in human prehistory. As numerous examples in the present volume, as well as another recent volume on the topic (O'Brien et al. 2018) amply demonstrate, this is a highly dubious assumption. The

Table 4.2 Chronometric age estimates for sites with actual or probable Nubian Levallois technology, organized approximately youngest to oldest

Assemblage	Ages and notes
Affad-23, Sudan	OSL estimates of ca. 16–15 ka. Some of the assemblage shows Nubian Levallois characteristics. Ref: Osypińska and Osypiński (2016).
BP177, Sudan	OSL ages of ca. 27–17 ka for upper horizon, and ca. 65–25 ka for lower horizons. The assemblages contain abundant Nubian Levallois cores (both Type 1 and Type 2, as well as bifacial foliates). Ref: Masojć (2010), Masojć et al. (2017).
Taramsa 1 burial, Egypt.	OSL estimates range from ca. 110–21 ka. Little detailed information on OSL. Samples from layers close to skeleton produced MIS 4 and 5 estimates, samples from within the skull itself give ages from 22.4 ± 1.3 ka (TS-2) to 30.2 ± 3.2 ka (TS-5), with an average of 24.3 ± 3.2 ka. Nubian Levallois artefacts in stratigraphically complex deposits around skeleton. Ref: Van Peer et al. (2010, p. 224).
Taramsa 1, Cc12-13, Egypt	Radiocarbon date of 38,100 ± 1,400 bp (OxA-2602) associated with assemblage displaying a mix of Levallois and Taramsa reduction, at least some of the former of Nubian Levallois character: Ref: Van Peer et al., (2010, p. 169 and p. 183).
Mochena Borago, Ethiopia	Upper T group, C14 dates of ca. 50–45 ka, contains Type 1 and Type 2 Nubian Levallois cores. However, due to small size of charcoal pieces modern pretreatment was not used, there is therefore a risk of contamination and incorrect age estimation (e.g. perhaps underestimates). Ref: Brandt et al. (2017).
Level 1, Boker Tachtit, Israel	Existing C14 dates of 50–47 ka are problematic as they lack modern pre-treatment, and dates are at limits of radiocarbon dating. Cores display a Nubian character (e.g. Volkman 1989, p. 92), as several authors have argued (e.g. Clark 1988; Belfer-Cohen and Goring-Morris 2009), while to others IUP cores are 'Nubian-like' (Nishiaki 2018). Hilbert et al. (2016) see Boker Tachtit cores as having similar morphology to Nubian Levallois cores, but belonging to different reduction sequences. Some argue that early Boker Tachtit technology is already Upper Palaeolithic (non-Levallois) (e.g. Meignen 2012). Ref for dating: Marks (1983).
'Taramsan' assemblages from Taramsa 1, Egypt	Phase IV deposits at Taramsa 1 have minimum OSL estimates as young as 39,500 ± 3,800 bp (897/2), but the authors argue that estimates of 56,900 ± 6,900 bp (897/8) and 55,800 ± 5,200 bp (897/1) date the assemblage. Little information is given on the OSL estimates so it is hard to evaluate their reliability. Proponents of the Nubian Complex oscillate on whether the Taramsan belongs to the Nubian Complex or not. Ref: Van Peer et al. (2010).
Nazlet Safaha 2, Egypt	Assemblages contain use of the 'Safahan' method, which can be seen as very closely related to Nubian Levallois technology, as it involves prepared of a Nubian central ridge before this is removed by a removal from distal. An OSL estimate of 59.8 ± 6.6 ka (OxL-898/1) is argued to correlate with occupation phase. Ref: Van Peer et al. (2002, p. 173).
Katoati, India	OSL ages associated with Nubian Levallois cores of 48 ± 11 ka (unit S4, but this age is inverted, and is beneath an estimate of 61 ± 9 ka, an age of ca 60 ka therefore seems more likely) and of 95.6 ± 13.1 ka in unit S8. Ref: Blinkhorn et al. (2015).
Uan Tabu, Libya	OSL estimate of 61 ± 10 ka on this Saharan tanged tool assemblage, which includes Nubian Levallois technology. Ref: Martini et al. (1998).
2004, Sudan	Assemblage with two Type 1 Nubian Levallois cores (1.6% of cores) stratigraphically above the site of ANW-3 which has preliminary U-series ages of ca. 65 ka. Ref: Rose and Marks (2014).
Taramsa 1, Cc 04, Egypt	Lithics above layer with OSL estimate of 73 ± 8.3 ka (OSL 897/9). Small assemblage of four cores, two of which are Nubian, plus a foliate. Ref: Van Peer et al. (2010).
ANW-3, Sudan	Unpublished u-series estimates on wood of ca. 65 to 62.5 ka. Assemblage has seven Type 1 Nubian Levallois cores (1.54% of cores). Ref: Rose and Marks (2014).
Taramsa 1, Cc 10, Egypt	Small assemblage containing Nubian technology, just below OSL samples from same unit of 58.9 ± 9.4 ka (793/3) and 79.2 ± 5.2 ka (TAM-4). Ref: Van Peer et al. (2010).
Wasariya Rusinga, Kenya	Type 1 Nubian Levallois core. Dated to between 100 and 35 ka by radiocarbon and tephrochonology. Ref: Tryon et al. (2012).
Taramsa 1, Cc07, Egypt	Ca. 80 to 40 ka age on a small assemblage containing two Type 1 Nubian cores, with an estimate from the same deposit as the assemblage of 73.0 ± 8.3 ka (OSL 897/9), but one of 40.5 ± 4.6 ka (897/10) from an underlying deposit. Little information is available on these OSL samples and the stratigraphy is complex. Ref: Van Peer et al. (2010, pp. 77–80).
34A, Sudan	Age of just over 84 ka suggested by preliminary (unpublished) U-series estimate on teeth from the site of 1017 which overlies 34A. Marks (1968b) reported only one Type 1 Nubian core was found at 34A. Ref: Rose and Marks (2014).
Aduma, Ethiopia	In deposits dated to 100–80 ka by multiple u-series and luminescence estimates, a small number of Nubian Levallois cores were reported. However, re-examination of the material by the author leads me to question this characterization. Ref: Yellen et al. (2005).

(continued)

Table 4.2 (continued)

Assemblage	Ages and notes
Keraswanin, Kenya	At least one Type 1 Nubian Levallois core recovered from the surface, but directly associated with the Wakondo Tuff which is dated to ca. 100 ka, based on U-series estimates of this deposit reported by Beverly et al. (2015) and Blegen et al. (2015) at Rusinga Island. Ref: Blegen et al. (2018).
Sodmein Cave, Egypt	Age of ca. 135–75 ka for Nubian Complex layers. Little information is available on the site or its lithic assemblages. From Van Peer et al. (1996a) it seems that only 23 lithics were found in MP-5, and these were scattered. A single Type 2 Nubian Levallois core is reported (Mercier et al. 1999). The correlation between lithics and dated hearths is unclear. Mercier et al. (1993) produced ages from 109 ± 8 ka (95/96) to 127 ± 10 ka (93/489), with average of 118 ± 8 ka. New TL and OSL age estimates on heated chert by Schmidt et al. (2015) have a greater scatter, from 86.9 ± 9.4 ka (SodTL2) to 121.2 ± 14.8 ka (SodTL0). Unpublished OSL estimates (pIRIR290 dating of feldspars) of between ~70 ka (layers G1 and G2) and ~110 ka (layer I) (Klasen et al. 2018).
Taramsa 1 Cc38, Egypt	OSL ages of 117 ± 10.5 ka (OSL 897/4) and 88.8 ± 9.5 ka (OSL 897/3) bracket the lithic assemblage, but little information is given on these OSL estimates. Other techniques suggested ages of >180 ka (AAR, AAL-7003) and >49,0000 (OxA-4038) with C14. Cc38 is a small (n = 368) assemblage with unclear number of Nubian Levallois cores, but not many as there are only six Levallois cores in total. Refits show diverse reduction systems. The apparently Nubian technology here is very atypical (p. 76) and would probably not be classified as Nubian Levallois by some authors. Ref: Van Peer et al. (2010, pp. 55–57).
Ain Difla, Jordan	Age of ca. 180–90 ka is indicated by TL, U series and ESR ages. TL and early uptake ESR estimates converge on an earlier MIS 5 age. Some artefacts certainly seem to have Nubian Levallois characteristics. Ref: Clark et al., (1997, e.g. p. 85).
ETH-72-6, Gademotta-Kulkuletti, Ethiopia	Lithic assemblage with Nubian Levallois technology (Wendorf and Schild 1974; Douze and Delagnes 2016) bracketed between Aliyo Tuff/Kibish Formation dated to 104 ± 1.0 ka/99.8 ± 1.0 ka and the Unit D Gademotta Formation dated to 183 ± 10 ka. Refs: Morgan and Renne (2008); Brown et al. (2012).
Skhul, Israel	U-series/ESR and TL dates suggest a chronology of ca. 120 ka. Little has been published on the lithic assemblages from the site, but analysis of a small part of the excavated material by Groucutt et al. (2019) found a Nubian Levallois-like element was present. Ref: Grün et al. (2005).
Mata'na G, Kharga, Egypt	U-series estimates on tufa deposits give a minimum age of 103 ± 14 ka for lithics including some Nubian Levallois forms reported by Caton Thompson (1952), with an underlying maximum age of 127.9 ± 13 ka. Refs: Smith et al. (2004, 2007).
Wadi Midauwara, Kharga, Egypt	Maximum age of 124.8 ± 4 ka (U-series on tufa) associated with a small collection of artefacts collected over a large area, including both Type 1 and Type 2 Nubian Levallois cores. Ref: Smith et al. (2007).
8-B-11, Sai Island, Sudan	A maximum age of ca. 150 ka suggested by underlying OSL estimates of 223 ± 19 ka (OSL-3), 182 ± 20 (OSL-2) and 152 ± 10 ka (OSL-1). Very little information has been published on the purportedly Nubian Complex assemblage from this site (no illustrations, no basic typology, etc.) so the reader has to rely on faith that the excavators' attribution to the Nubian Complex is meaningful. Lack of publication on this material may suggest that the situation is more complex than suggested. Ref: Van Peer et al. (2003).
Aybut Al Auwal (TH-59), Dhofar, Oman	Minimum age of ca. 107 ka suggested by OSL estimates of 106.6 ± 9 ka (AYB1-OSL1) and 107 ± 9 ka (AYB-OSL2). Dates are on fluvial sediments containing a few redeposited lithics—two of which are purportedly 'diagnostic'—and should only be seen as minimum age estimates. The discoverers have since described the artefacts being in a "secondary" position (Rose et al. 2018, p. 168). Ref: Rose et al. (2011).
Bulaq Wadi 3, locus 1, Kharga, Egypt	Minimum ages of 114.4 ± 4.2 ka (U-series on tufa) for artefacts collected by Caton Thompson (1952) which include Nubian Levallois forms. According to Smith and colleagues, the artefacts are redeposited. Ref: Smith et al. (2007).
Taramsa 1 Cc 05, Egypt	An age of more than 117 ± 10.5 ka (OSL 897/4) and perhaps more than 165.5 ± 17.8 ka suggested by OSL estimates, on which little information is provided, for an assemblage including a Nubian Type 1 component, in an assemblage generally dominated by discoidal reduction. The stratigraphy is complex, and chronology assumes that the authors have correctly understood the stratigraphy. Ref: Van Peer et al. (2010, pp. 225, 282).
Taramsa 1, Cc 17, Egypt.	Dating and stratigraphy as for Cc05 discussed above. Refits from Cc17 show Type 2 Nubian Levallois reduction occurred in this probably Middle Pleistocene assemblage. Ref: Van Peer et al. (2010).
ETH-72-1, Gademotta-Kulkuletti, Ethiopia	Assemblage between tuffs dated to 183 ± 10 ka (Unit D: Gademotta Formation) and 276 ± 4 ka/280 ± 8 ka (Unit 10: Gademotta Formation). This provides a good example of the ambiguity of assigning artefacts/assemblages to the Nubian Complex. Van Peer and Vermeersch (2000, p. 57) states that cores from the site "do seem to be of Nubian type". A core from here was shown by Crassard and Hilbert (2013, see their Fig. 11), although it is incorrectly labelled as being from Aduma. Ref: Morgan and Renne (2008).

(continued)

Table 4.2 (continued)

Assemblage	Ages and notes
Umm al-Sha'al, Al Kharj, Saudi Arabia	One of the five artefacts found scatted in Unit 3 at a depth of 170–180 cm below the surface is described as being a broken preferential Levallois core of the "Nubian Tradition". It was found close to sampling points for two inverted OSL estimates of 194 ± 21 ka (AKE31-1) and 253 ± 27 ka (AKE31-2). These OSL estimates are affected by dose rate uncertainties and some of the grains seem to be partially bleached, leading the authors to describe these estimates as "problematic". Ref: Crassard et al. (2018).
ETH-72-B, Gademotta-Kulkuletti, Ethiopia	Assemblage beneath tuff dated to 276 ± 4 ka/280 ±8 ka (Unit 10: Gademotta Formation). Van Peer and Vermeersch (2000, p. 56) state that artefact illustrations from the site "strongly suggest" the presence of Nubian technology. Ref: Morgan and Renne (2008).

prevalence of convergent evolution should hardly be surprising given high levels of convergent evolution even in biological systems (e.g. McGhee 2011; Bergey et al. 2018), where horizontal information transfer is less prominent than in cultural contexts.

Setting aside for the moment the morpho-technological distinctiveness of Nubian Levallois technology, it is important to note that previous research has arguably already falsified the Nubian Complex, or at least raised significant grounds for caution. In the comments on Van Peer's (1998) article which marked the real birth of the Nubian Complex, Schild (1998) raised pertinent criticisms. Later, Kleindienst (2006, p. 22) succinctly argued that "the term "Nubian Complex" masks variability rather than aiding communication about the clustering of typological and technological traits". In a large-scale comparative study of the actual technological variability in North Africa (rather than comparing perceived types of assemblages), involving hundreds of thousands of attribute states on stone tools from across North Africa broadly dating to MIS 5, Scerri and colleagues (2014) explored patterns of similarity and difference in lithic technology. They did so without *a priori* expectations. The basic conclusion of this study was that assemblages were most different from assemblages furthest away from them, except where they were connected by paleohydrological corridors. This environmentally rooted perspective offers a very different view from one in which all assemblages can be assigned to an *ad hoc* defined technocomplex, or named artefact industry (NASTIES), as Shea (2014) put it. In the case of North Africa, assemblages from northeast Africa traditionally assigned to the 'Aterian' are more similar to nearby sites assigned to the 'Nubian Complex' than they are to other 'Aterian' sites in northwest Africa (Scerri et al. 2014). In other words, variability correlates with geography rather than the traditional named industries, some of which are defined based on core reduction methods, some on the presence of particular retouched tool forms, and so on. Scerri et al. (2014) found the same pattern in independent statistical tests across cores, flakes and retouched tools. These patterns of similarities and differences, reflecting constellations of attribute states, offer a much richer and more nuanced view than perspectives based on the presence or absence of a single typological category. These attribute-based, quantified data suggest that the term 'Nubian Complex' is simply not a meaningful or helpful way of characterizing the archaeological record of Northeast Africa.

Understanding the reality or otherwise of the Nubian Complex has important practical bearings for archaeology. If, for example, all evidence to the contrary is ignored and one decides that 'Nubian' Levallois technology is diagnostic of the Nubian Complex which equates with a group of people, then significant shortcuts can be made in research. So when selective (i.e. biased) collections are made (e.g. Crassard and Hilbert 2013; Goder Goldberger et al. 2017) which focus on collecting 'diagnostic' forms—diagnostic because these studies start with the conclusion that Nubian Levallois technology is 'diagnostic'—this is, I would argue, a problem. Of course, if you conclude in advance that a particular technology is diagnostic of a time period, this is an understandable focus. The problem is that, as discussed below, Nubian Levallois technology actually does not offer a strong temporal and cultural signal. Convergent evolution seems to clearly explain at least some of the patterning of Nubian Levallois technology seen in the archaeological record (e.g. Will et al. 2015), and therefore making selective collections of artefacts is highly problematic, as it removes the context of these artefacts. The aim here is to objectively explore the evidence for the Nubian Complex and for the spatial and temporal distribution and characteristics of Nubian Levallois technology. Whatever one's perspective, it is important to note that incredibly few apparently Nubian Complex sites have reliable age estimates and that little information has been published on the lithic assemblages of key sites.

Over the following pages I will outline the history of the idea of the Nubian Complex, then describe relevant sites in Northeast Africa, the rest of Africa, Arabia, the Levant and then India, before finally considering Nubian Levallois technology from a technological rather than culture-historical perspective. The reader may wish to refer back to the tables as they read through this chapter, outlining the highly variable definitions of the Nubian Complex through

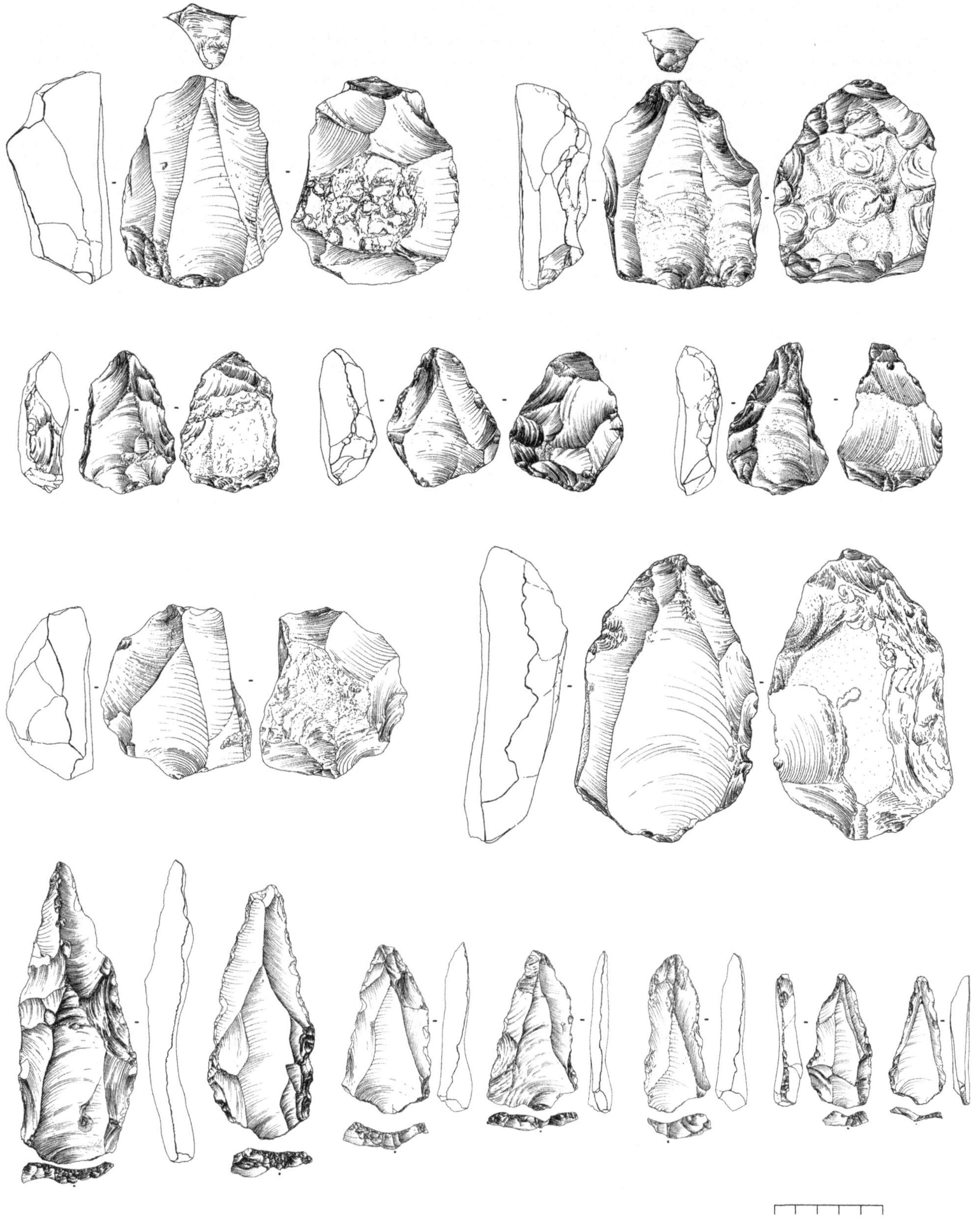

Fig. 4.1 Illustrations of Nubian Levallois cores (top three rows) and associated Levallois products (bottom row) from various sites in Dhofar, Oman (modified from Rose et al. 2011, Creative Commons Attribution). Scale in cm

Table 4.3 Definitions of Nubian Levallois technology and related observations

Seligman (1921) gave the first description of Nubian Levallois technology. Several cores from Egypt and a diacritical illustration are shown. He suggests that such cores have a 'type length' of ~8–10 cm (p. 123), which can be seen a precursor to the later notion of standardization. Seligman sees Nubian Levallois cores as being developed from tortoise (i.e. Levallois) cores and used as a tool, a "heavy drawing or dragging point" which he proposes to call a "tortoise point" (p. 123) because they are Levallois cores on which the distal end is shaped like the beak of a tortoise. While Seligman was wrong in seeing these cores primarily as tools, he recognized their distinctive morphology and manufacture, and describes them as forms with distinctive distal ends shaped by preparatory removals which create a median distal ridge (p. 126). Seligman's mistake was to see the distal striking platform as being formed after, rather than before, the distal preparatory removals. Seligman noted that beaked Levallois cores occur in low frequencies in the European Paleolithic record.
Vignard (1930) described Nubian Levallois cores from Egypt as point cores, emphasizing the removal of two large flakes or blades from the distal end of the core, before the preferential removal was struck (p. 312). He notes that similar forms are found in France.
Montet (1957) Illustrated Nubian Levallois cores from Egypt, described as Levallois point cores. Also Nubian Levallois points are depicted and said to be similar to those described by Vignard (1930).
Guichard and Guichard (1965). Type 1 Nubian defined as "A Levallois point core characterized by a special technique…the main phases of the technique are as follows: (a) flaking off the periphery of a plaquette to obtain an oval thick core; (b) at the distal (and pointed end), the removal of two elongated flakes or Levallois blades with negative bulbs close to each other; the traces of removal of both blades are delineated by a short ridge approximately in the core symmetry axis; (c) towards the base, retouches starting from the denudation described (b) apparently meant to smooth the edges so as not to injure the worker's hand; (d) the preparation of a striking platform…(e) the removal of a Levallois point" (pp. 68/69). They claim that this core form "was designed according to a precise, unvarying pattern, elaborated in every detail" but also noted that "this particular type is relatively rare" (p. 69). Type 2 Nubian Levallois is described as a "Levallois core, often of triangular shape, or sub-triangular, give pointed flakes or points of Levallois technique which are not, in a strict-sense, Levallois points because they are not debited along a ridge" (p. 69). It was noted that Nubian Levallois cores are "often prepared on a plaquette", and that "the distal end presents a protruding point, a real rostrum, the striking platform at the opposite end is well prepared" (p. 69). They noted that Nubian Levallois cores are often associated with overshot preferential removals. They state that previously Nubian Levallois cores had been found in various parts of Africa and that "both type 1 and type 2 cores are never entirely missing from European Paleolithic sites (as for example, the Saint-Cirq Magdalenian…), not to mention the Grand-Pressigny Neolithic." (p. 99).
Bordes (1980) Discusses the Guichards' definition, and states that he thinks Nubian core should only be used for Type 2 cores (p. 47). Type 1 are just a slightly modified version of a non-Nubian Levallois point core. This is an interesting statement, as more recent accounts (e.g. Van Peer 1992) tend to emphasize the distinctiveness of Type 1 cores, with Type 2 blending into 'normal' centripetal Levallois cores. In this view, Nubian Levallois cores can both be said to merely be a slight variant on standard Levallois cores, and to be present in settings including the French Neolithic.
Van Peer (1992) states that Type 1 cores have "been noticed since long (sic) in Egyptian Paleolithic assemblages. It was described systematically by Guichard and Guichard (1965, pp. 68–69). He repeats the definition of the Guichards but adds the following points, for Type 1, "here is a combination of a system of transversal ridges in the proximal part of the upper surface and a central ridge in the distal part. This pattern of ridges does not produce points in a morphological sense (triangular flakes), but rather pointed flakes" (Van Peer 1992, p. 41). For Type 2, he adds that according to the Guichards' definition, "the method should rather be integrated as a subvariety of the classical method for flakes. Indeed, the scar pattern implements a system of transversal ridges. One might, however, argue that the transversal disposition of scars creates a *de facto* central ridge which will give way to a pointed flake, and, therefore, the method has to be kept among the point methods (*ibid.*).
Vermeersch et al. (1990) describe Nubian Type 1 cores as a "more or less triangular nodule" with a central ridge shaped by "two intersecting elongated flakes from the distal end" (p. 81). They add that the medial-ridge preparation only occurred at the distal end of cores, hence the products are more like "pointed flakes, rather than real points" (p. 81). Type 2 cores are said to be triangular with a distal end prepared from the margins to create pointed removals, however "in many instances, however, it is hard to distinguish these cores from 'classical' Levallois cores, except for their triangular shape" (p. 81).
Chiotti et al. (2009) from their work in Egypt, suggested that "While Nubian Type 1, Nubian Type 2, and radial Levallois cores are morphologically distinct, our data provide evidence that these three core types are related and thus should not be interpreted as being discrete entities" (p. 310). Type 1 "exhibit distal preparation of the core surface, usually through the removal of two flakes…These removals frequently originate from a platform surface that is relatively high on the core surface, and the two removals are not dissimilar from large burin blows along the side of the cores", while Type 2, shaped by "removals from the lateral edges, which can lead to an almost centripetal preparation". At most sites both Type 1 and Type 2 cores were found, and many with both types of preparation, with cores being of similar sizes, which they suggest means different core forms "cannot be interpreted as being behaviorally distinctive, or as different types". They describe how Nubian cores can become non-Nubian cores, and vice versa. They also argue that "it was not clear that "pointedness" was the intended results for flakes removed from Nubian cores" (p. 311). While Nubian cores tend to produce more pointed flakes than 'standard' centripetal preferential cores, the distinction is not absolute.
Van Peer et al. (2010) follow Guichard and Guichard's (1965) definition and Van Peer's (1992) comments. Adds here that a 'characteristic' feature of Type 1 cores is the distal preparation of the median distal ridge is that "these removals are struck under a relatively wide angle to the core's plane of intersection". These emphasizes the "triangular plan shape of the core" (p. 49). They comment that for Type 2 cores, "there is very little reason to distinguish this as a separate Levallois method…the basis to identify such cores is mainly morphological. The pattern of preparation itself grades between that of the classical Levallois method and the Nubian type-1" (p. 50).
Rose et al. (2011) define Nubian Levallois technology as "A regional variant of the preferential Levallois method for producing points… recognized by its triangular/sub-triangular shaped cores and a specific opposed platform preparation of the primary working surface". Type 1 debitage surface shape "formed by two distal-divergent removals creating a steeply angled median distal ridge", while type 2 shaped by "bilateral shaping", although the "two methods are not mutually exclusive". The key is that Nubian cores have "highly characteristic preparation of the

(continued)

Table 4.3 (continued)

distal end of the core to create a steeply peaked triangular cross-section" (p. 1). While they argue that Nubian Levallois is very distinct, they argue that "there is overlap between Nubian Type 2 core preparation and some preferential point-producing Levallois reduction systems in the Levantine Mousterian" (p. 3).
Usik et al. (2013) argue that the Nubian Levallois method is a particular way of producing "elongated Levallois points or pointed flakes" (p. 248). Nubian cores are "highly standardized and distinguished by an array of morphological features, each of which is essential for the piece to be classified as Nubian Levallois. First and foremost is the steeply angled median distal ridge that serves to control the distal lateral convexity of the core's primary working surface". "Core morphology is another essential feature…" and has to be either "triangular, cordiform of pitched" (p. 249). Pitched shape is defined as cores with "more or less parallel elongated lateral sides with a convergent distal end". Finally, "the core must have a prepared main striking platform". Below the level of Levallois strategy and Nubian Levallois method, they distinguish the organizational systems used to form the median distal ridge: type 1 with "two distal-divergent removals, creating the steep distal ridge"… while type 2 shaping is "achieved through bilateral shaping of the primary working surface", although they are keen to emphasize that this difference reflects final core state and that knappers were "able to switch from one type to another during phases of re-preparation" (p. 250). Type 1/2 combines elements of these two forms. So they argue that the "discrete" character of the Type 1 and 2 dichotomy is exaggerated. Note that they talk about Nubian Levallois cores "grading into bidirectional cores or recurrent cores" (249). It is not clear how this can be squared with Nubian Levallois cores apparently being so distinctive. Despite aiming for a concise definition, their definition essentially comes down to having a median distal ridge (which will result in the core having a 'triangular, cordiform or pitched' shape as a result). Levallois cores tend to have prepared platforms so this part of their definition is essentially unnecessary. Note that they describe core reduction at the site of TH.383c as follows, the "first step was the formation of the distal platform" (p. 250). This differs from the description given by Guichard and Guichard (1965). Likewise, they briefly note a prominence of 'dihedral chapeau' platforms in their study region, but do not press this seemingly regionally distinct feature.
Van Peer (2016) sees the Nubian reduction strategy is a "specific manifestation of the Levallois system to produce pointed forms. Its origins can be found in Lupemban foliate façonnage. The Nubian production system is a very distinct association of discrete technical and morphological traits" (p. 151)
This paper. Nubian Levallois technology is a form of centripetally prepared preferential Levallois reduction, characterized by the shaping of a pronounced median distal ridge forming a distal 'beak', shaping the subsequent preferential removal. The shaping of this ridge can be by removals from distal ('Type 1') or lateral ('Type 2'), although these are recognized as overlapping considerably so are a matter of emphasis. This preparation specifically refers to shaping the distal ridge, not the overall debitage surface. The other features typically found with such cores (e.g. a broadly triangular shape), flow from the property of a medial distal ridge. The Nubian Levallois reduction method overlaps with and blends into other technologies.

time (Table 4.1), of chronometric age estimates for assemblages with Nubian Levallois technology (Table 4.2), and of different definitions and understandings of Nubian Levallois technology (Table 4.3).

History of the Nubian Complex

Recent discoveries of Nubian Levallois technology in Southwest Asia have led to a renewed interest in the notion of the Nubian Complex. According to its proponents (e.g. Usik et al. 2013), the Nubian Complex was 'discovered' by Jean and Geneviere Guichard in the 1960s, and then firmly codified by later studies such as the work of the Belgian Middle Egypt Prehistoric Project. These studies apparently gave the Nubian Complex firm definitions and a solid (mostly MIS 5) chronology. When Nubian Levallois technology was then discovered in Southwest Asia this was taken by proponents as indicating an early phase of the dispersal of our species out of Africa, consistent with an archaeological argument that dispersals had occurred much earlier than suggested by most genetics studies (e.g. Boivin et al. 2013; Groucutt et al. 2015a, b, c).

What is the Nubian Complex? As shown in Table 4.1, answers to this question are much more complicated than might be expected. There are two basic ideas, with variable levels of interplay between them. One is the idea advanced by Philip Van Peer in particular (e.g. Van Peer 1998, 2016), which emphasizes a broad and polythetic definition which sees the Nubian Complex as a long-lived entity defined by the variable presence of Nubian Levallois cores and perhaps other aspects such as 'Nazlet Khater' (thinned tip) points. In this view, assemblages from sites such as Haua Fteah and Jebel Faya can be assigned to the Nubian Complex (Van Peer 2016), despite the fact that Nubian Levallois technology has not been reported at either site. In the second basic view of the Nubian Complex, we are dealing with an essentially Marine Isotope Stage 5 (ca. 126–74 ka) phenomenon, defined by the presence of Nubian Levallois technology. This latter view has flowed in particular from findings in Arabia (e.g. Rose et al. 2011; Crassard and Hilbert 2013), although none of the Arabian sites have reliable chronometric age estimates, and to a lesser extent the Levant (Goder Goldberger et al. 2016). In this view if a site has Nubian Levallois technology it belongs to the Nubian Complex—although as we shall see, proponents have reacted in variable ways to the recent discovery of Nubian

Levallois technology in places as far away as South Africa. This view argues that Nubian Levallois technology was only made by *Homo sapiens*, basing the claim on a denial of convergent evolution and on the apparent association of the Taramsa burial with Nubian Levallois technology. These points will be discussed below.

Why has the notion of the Nubian Complex seemingly met with such enthusiasm? In part it reflects the fact that it is has fitted the zeitgeist of African MSA archaeology, in which regionalization and 'point' production have both been strongly emphasized (e.g. McBrearty and Brooks 2000). It also offers an alluringly simple framework. Rather than having to grapple with the complexities, and even contradictions, of technological variability (e.g. Hovers 2009; Tostevin 2012; Scerri et al. 2014), the Nubian Complex apparently offers an alluring shortcut between technological characteristics and the history of a regional population.

In reality, the Nubian Complex was born with the paper of Van Peer (1998). The original basis for defining the Nubian Complex was Van Peer's (1998) claim that the Northeast African record featured two technocomplexes that were contemporaneous. This contemporaneity was argued to have a strongly cultural character: there were Nubian Complex people who made particular kinds of artefacts, sometimes including Nubian Levallois reduction, and, occasionally, particular retouched forms such as Nazlet Khater points, and there were Lower Nile Valley Complex people who did not make these things. These two groups lived close to each other for some two hundred thousand years, yet their material culture remained distinct. One is of course reminded of previous famous work, such as Bordes (e.g. 1961) notion of distinct 'Mousterian tribes' of France. On the whole Pleistocene archaeology has moved beyond such notions.

Various earlier studies had reported and described Nubian Levallois technology—although not using that name—in Northeast Africa. Seligman (1921) offered the first clear definition of Nubian Levallois technology. Seligman called Nubian Levallois cores 'tortoise points', because he thought they were Levallois (tortoise) cores that had been distally modified to creak a 'beak', because he thought they were used as tools. Indeed, the distal morphology of Nubian Levallois cores is their defining feature, and to Seligman (1921, p. 123) "this stout, rather blunt point recalls the beak of a tortoise". Seligman's description was in many ways accurate, but his error was that he thought the distal striking platform was prepared after rather than before flake removals shaping the median distal ridge. This was an early lesson in the importance of understanding aspects such as the chronology of flake removals, not just the static shape of objects. Subsequent papers reporting Nubian Levallois technology include those of Vignard (1930), Montet (1957), Guichard and Guichard (1965, 1968), and Marks (1968a, b). These papers describe the apparent distinctiveness of Nubian Levallois technology which made it very interesting. Flowing from these works, however, was a mass of confusing and undated 'industries', rather than robust models of technological change across time and space.

The publications of Guichard and Guichard (1965, 1968) are often treated as the foundational works of the Nubian Complex (e.g. Usik et al. 2013). It should be noted that these pioneering studies—conducted in the context of the rescue work being done before flooding caused by the construction of the Aswan Dam—were later described by the Guichards themselves as "rough", "brief", and "preliminary" (Guichard and Guichard 1968). This is in no way to denigrate the important work of the Guichards, but merely to point out some of the limitations of this work given its later exaltation. They describe several kinds of artefacts as 'Nubian…' such as Nubian handaxes, Nubian sidescrapers and Nubian cores. Of these, only the latter has remained in common use. My interpretation of the Nubian handaxes and Nubian sidescrapers that the Guichards described and illustrated is that neither represents a real or meaningful group. In both cases they include morphologically and technologically diverse forms. As an aside, it is interesting to note how the 'Nubian sidescrapers' of Guichard and Guichard (1965) have evolved over time into 'Nubian endscapers' (Van Peer 2016).

To the Guichards, finding Nubian handaxes, Nubian sidescrapers and Nubian cores together represented a significant combination which could define one of their industrial groups (the 'Nubian Middle Palaeolithic'), although this combination was only recognized at three sites. They strongly emphasize that it is this combination of forms which is significant, and that Nubian Levallois cores, Nubian sidesrapers and bifacially flaked pieces "taken individually are not very significant" (Guichard and Guichard 1965, pp. 99–100). Over time these cautious words have been exaggerated. Usik et al. (2013, p. 259), for example, claim that "Nubian core technology…was the main criterion by which the Nubian Complex was initially defined (Guichard and Guichard 1965)".

Several interesting points emerging from the Guichards work are worth emphasizing. Firstly, it should be noted that Guichard and Guichard (1965) also reported Nubian Levallois cores in sites they defined as belonging to the Early Paleolithic, containing frequent large cutting tools. Secondly, Nubian Levallois cores were very rare at sites they assigned to the Nubian Middle Paleolithic. It was only at one site (420) that Nubian Levallois cores made up more than 20% of the cores. In total, only 38 Type 1 Nubian Levallois cores were identified from a total of 1,192 cores in the Middle Paleolithic assemblages. Nubian Levallois technology simply does not seem to be that prominent a part of the assemblages the Guichards studied. In fact, one of the Early Paleolithic sites (400-4) has a higher proportion of Nubian Levallois cores out of all core forms (11.9%) than the average for the Nubian Middle Paleolithic sites (8.7%), for which Nubian Levallois cores were supposedly one of the diagnostic forms! Most of the Nubian Levallois cores recovered from Middle Paleolithic assemblages were Type 2, and as will be discussed later this is potentially problematic as Type 2 Nubian Levallois reduction has widely been argued to grade into 'normal' Levallois technology.

Over the following years further work was done on assemblages containing (small numbers of) Nubian Levallois cores, and new industrial groupings posited (e.g. Marks 1968a, b). In the 'Nubian Mousterian' industry, Nubian Levallois cores occurred at variable frequencies, from 0% at site 1036, for example, 3% at site 6 up to 23.2% at 1035 and 23.4% at site 1038 (Marks 1968a). In the same book, Chmielewksi (1968) reported the site of Arkin 5, where out of 9,769 artefacts, 93 were Levallois cores, described as Type 1 Nubian Levallois cores. However, from the illustrations provided it is not clear that this description is accurate in all cases, although some are fairly clear. The assemblage also contains large numbers of large cutting tools.

It was really with the Van Peer (1998) claim for the contemporaneity of Nubian Complex and Lower Nile Valley Complex that the Nubian Complex came to be understood in a modern way. In this model two populations lived nearby but remained distinct for about two hundred thousand years. The Nubian Complex was defined so polythetically that it would prove very hard to pin down. Sites from Bir Tarfawi (Wendorf et al. 1993), for example, were assigned to the Nubian Complex, despite not having any reported Levallois technology (Van Peer 1998). This was because in the Van Peer (e.g. 1998; 2016) view, the presence of Nubian Levallois technology is not necessary for a site to belong to the Nubian Complex. Forms as diverse as bifacial foliates, Nazlet Khater points and 'truncated faceted' pieces were also apparently characteristic, but it is unclear if their individual presence is significant or if it their co-occurrence that he thinks significant (Van Peer 1998, p. 120). The Lower Nile Valley Complex was defined by things it did not contain, which is always likely to be a problematic approach. This loose and long-chronology definition of the Nubian Complex has continued to dissolve with time (Van Peer 2016). Despite the discovery that the Lower Nile Valley Complex sites are younger than the Nubian Complex sites (Van Peer 2004; Van Peer et al. 2010), the notion of the Nubian Complex has simply morphed into a slightly new model. Rots and colleagues, for instance, outline this revised view in which the Nubian Complex was "heralded" (Rots et al. 2011, p. 639) in the Middle Pleistocene, where it is a "taxonomic umbrella" for various MIS 5 industries, and where the Nubian Complex of Van Peer (1998) actually refers to a specific element of this, the Late Nubian Complex at the end of MIS 5.

Recent discoveries in Arabia and the Levant led to a renewed interest in the Nubian Complex. These recent studies have sidestepped the fact that the arguments for the original definition of the Nubian Complex have arguably been falsified, and instead emphasize Marine Isotope Stage 5 as the key period for the Nubian Complex. In these recent definitions the polythetic view of the Nubian Complex has dropped away, and instead Nubian Complex simply equates with presence of Nubian Levallois technology.

The Nubian Complex in Northeast Africa

As the apparent home of the Nubian Complex, it is important to evaluate the nature of sites assigned to the Nubian Complex in northeast Africa (for reviews of the North African MSA see Scerri 2017; Scerri and Spinapolice 2019). Research in the 1960s had identified Nubian Levallois cores in both Early and Middle Stone Age contexts along the Nile in Sudan (Guichard and Guichard 1965, 1968; Marks 1968a, b), although no chronometric dating was available. Work by the Belgian Middle Egypt Prehistory Project along the Egyptian Nile, the Bir Tarfawi and Bir Sahara project (Wendorf et al. 1993), and work at oases such Kharga and Dakhleh (e.g. Kleindienst 2006; Hawkins 2012) led to significant advances in our understanding of the prehistory of Northeast Africa.

At Bir Tarfawi and Bir Sahara in western Egypt, numerous Middle Paleolithic sites were collected and excavated, and dated to past interglacials such as MIS 5e (Wendorf et al. 1993). These localities are perhaps the key reference point for the Middle Stone Age of Northeast Africa, owing to the density of sites in a small area, and it is instructive to note that *no* Nubian Levallois technology has been reported from the numerous sites found at these localities. This lack of Nubian Levallois technology was reported by Schild (1998) in response to Van Peer (1998). Van Peer (1998) saw the Bir Tarfawi and Bir Sahara assemblages as being central to his Nubian Complex. And this has since been often repeated (e.g. Rose et al. 2011) and even described as *the* key example of the "Early Nubian" (Dibble et al. 2013, p. 195). However, the lack of Nubian Levallois cores seems to have troubled Rose and Marks (2014), who shifted their interpretation and included the Bir Tarfawi/Sahara localities in the Lower Nile Valley Complex. This is despite the fact that these assemblages are neither in the same place (i.e. near the Nile) nor same time (i.e. MIS 4/3) as other proposed Lower Nile Valley Complex sites. If this change reflects the lack of Nubian Levallois technology at Bir Tarfawi/Sahara, then it is not clear why Rose and MARKS (2014) still include some sites which also lack Nubian Levallois technology in the Nubian Complex. Rose and colleagues (2018, p. 58) seem to revert to their earlier model, and include Bir Tarfawi and Bir Sahara in the Nubian Complex. That there is a lack of agreement on whether such key localities as Bir Tarfawi/Bir Sahara can be described as Nubian Complex sites highlights the problems of the reality of this technocomplex.

Other apparently key localities such as Sai Island and Sodmein Cave have not been published in detail (see below), so the argument that these should be assigned to the Nubian Complex rest on incomplete data which cannot be meaningfully evaluated by others.

Sodmein Cave in the Red Sea Mountains of Egypt was another site emphasized by Van Peer (1998) in his definition of the Nubian Complex. Yet in the decades since then, very little has been published on the site. At Sodmein, lithics apparently occur at very low densities, and the purportedly different archaeological levels appear to blend into each other (Van Peer et al. 1996a). No clear typological counts or other detailed information has been published. We are simply told that several layers can be assigned to the Nubian Complex (Van Peer et al. 1996b). MP5, for instance, contained a Type 2 Nubian Levallois core. As seemingly only infinite radiocarbon ages are available for MP3 and MP4, the chronology of the apparent Nubian Complex at Sodmein Cave rests on material claimed to be associated with MP5. Dates from this layer cluster around MIS 5 (Table 4.2), but the relationship between the scattered lithic artefacts and the dated hearths is currently unclear. Vermeersch (2001) and Van Peer and Vermeersch (2000) both state there are two Nubian Complex levels at the site. Yet Schmidt et al. (2015) state that there are three Nubian Complex levels. Sodmein Cave is probably a key Northeast African site, but so far insufficient data has been published to clearly understand the sequence.

Two important sites in the early years of work by the Belgian team were Nazlet Khater 1 and Nazlet Safaha, both near the Nile in southern Egypt (Vemeersch 2000; 2002). Nazlet Khater 1 is an important site as, unlike most Middle Stone Age sites in Northeast Africa, it genuinely does contain a large number of Nubian Levallois cores. In the lower layer, for example, there are 19 Nubian Levallois cores (Type 1) and two classical Levallois cores (Vermeersch et al. 2002). In the Middle layer, 61 cores are described as Nubian Levallois and 20 classical Levallois cores (Vermeersch et al. 2002, p. 44). Very few retouched forms were found in these layers. In the upper layer there are 160 Type 1 Nubian Levallois cores, two Type 2 and 35 classical Levallois. In all of these assemblages Nubian Levallois methods are less frequent in terms of end products than they are in final core forms, but the site is never the less important in showing that Nubian Levallois technology can be a significant part of the repertoire. However, Nazlet Khater 1 is undated, so it is not immediately clear how it relates to other sites.

At other undated Middle Paleolithic sites in the area, Nubian Levallois cores occurred in varying frequencies, such as at El Gawanim 1 and 2, Nazlet Khater 2 and 3, and Makhdama 6, while at other sites Nubian Levallois technology was found in association with large cutting tools (Vermeersch 2000, 2002). These are all significant assemblages, but they are all undated.

The site of Nazlet Safaha is significant terms of understanding Nubian Levallois technology. An OSL estimate from Nazlet Safaha 2 of ca. 60 ka was reported by Van Peer et al. (2002) (Table 4.2). The Safahan method of core reduction was defined from the Nazlet Safaha material. In this, two elongated flakes are removed from the distal end of a core, creating a median distal ridge "not unlike the Nubian type 1 method" (Van Peer et al. 2002, p. 183). Elsewhere, Van Peer is more direct, stating that the removals from distal created a "Nubian central ridge" (Van Peer 2004, p. 220), that is "identical to that of the Nubian 1 method" (Van Peer 1991, p. 135). A large flake is then struck removing this central ridge, as a final stage in preparation before the removal of a preferential flake from the other end of the core. The key here then is the overall character of the reduction method. If, however, a core was found which was being reduced by the Safahan method but had been abandoned before the final flake was struck from the distal, it would be morphologically indistinguishable from a core being reduced by Nubian Levallois reduction. The key point here seems to be that Nubian Levallois technology is a rather slippery entity, which can appear in many guises. Also, it emphasizes that the typology of static core forms can be misleading.

The sequence at Taramsa in southern Egypt is as close to a 'type site' as there is for the Nubian Complex. Careful examination of the monograph published on Taramsa 1 (Van Peer et al. 2010), however, reveals a number of problems with the site and its interpretations. Firstly, the stratigraphy of the site is extremely complex. Prehistoric humans repeatedly dug into a gravel deposit to extract raw materials. As a result, there are numerous stratigraphic inversions. Excavation of several separate trenches through these complex deposits inevitably led to problems of correlation across the site. The excavators acknowledge this complexity, stating that "stratigraphy was the major challenge…and it took us a long time to understand the 3-dimensionality of the features, their stratigraphic succession and the nature of their formation processes…different units of anthropogenic waste deposits are often very similar in terms of their lithologic composition (Van Peer et al. 2010, p. 25). Methods which might have helped in such a situation, such as piece plotting, were "rarely done" (ibid.). In the end, the excavators "believe that we have been able to disentangle the major phases of activity" (ibid., p. 26), but in such a setting only a strong chronological framework is going to robustly clarify the sequence.

However, secondly, while OSL age estimates are presented throughout Van Peer et al. (2010), almost no information on the methods used is given, nor basic standards of data reporting met. The only information about how these results were generated is a brief personal communication note from Richard Bailey (ibid., p. 282), which basically states that 'standard methods' were used. But the actual data are not presented. The reader is not provided with simple information, such as the age model used to calculate the results. Given the lack of data provided on these OSL estimates, they should be regarded as preliminary. Even if the estimates are in themselves accurate, as charitably might be assumed, how these dates correlate with the site's complex deposits remains rather opaque.

Thirdly, the descriptions and 'industrial' designations given by Van Peer et al. (2010) are in fact much less secure when the actual lithic data is looked at. These reservations do not undermine the importance of Taramsa as a key archaeological site in the region. The problem, however, is the mistaken sense of clarity on the sequence and its implications which then feed into the wider literature. The description of the site involves a series of assemblages, which are then amalgamated into 'activity phases'. While impressive in technical descriptions—based largely on refitting—the chronological and stratigraphic complexity of the site means the sequence is arguably less clearly resolved than has been suggested.

A complete discussion of the Taramsa site is beyond the scope of this paper, and the reader is referred to the monograph on the site (Van Peer et al. 2010). Here I will simply highlight a few salient points. Firstly, Nubian Levallois technology occurs in seemingly Middle Pleistocene deposits at Taramsa (Table 4.2). Secondly, the few small purportedly MIS 5 assemblages with Nubian Levallois technology do not give an impression of a coherent 'Nubian Complex' entity. Indeed, Van Peer and colleagues (2010, p. 229) comment, apparently without irony; "it seems rather surprising that this Nubian Complex is so little represented at Taramsa". The real irony is that Taramsa is one of the flagship sites of the Nubian Complex. One of the 'Nubian Complex' assemblages, Cc38, is described as "not obviously similar to anything else in the northeast African MSA" (ibid., p. 228). Indeed, I think that many researchers would not classify the Cc38 artefacts as demonstrating Nubian Levallois technology at all.

Taramsa is also well known for the discovery of a skeleton assigned to *Homo sapiens* (Van Peer et al. 2010). Indeed, this has been argued to provide a key link between our species and the Nubian Complex (e.g. Usik et al. 2013). Lithics found near the skeleton included some described as Nubian cores (Van Peer et al. 2010, p. 218). However, given the fact that this is a probable burial in the context of a repeatedly exploited—and therefore stratigraphically complex—quarry, great care needs to be taken in firmly correlating the skeleton with artefacts in the surrounding sediments. Chronometric dating has been applied in the form of OSL dating. While some sediments near to the skeleton produced MIS 5 and 4 age estimates, sand from inside the skull of the skeleton, and therefore stratigraphically the most secure samples, produced an age of MIS 2 (Table 4.2). While the discoverers prefer an older age, the strongest luminescence data suggests that the burial is intrusive and dates to ca. 25 ka.

The site of Taramsa is arguably most important for its 'Taramsan' assemblages, rather than the purported Nubian Complex. It is unclear if those who believe in the Nubian Complex think that the Taramsan should be part of the Nubian Complex or not. While the Nubian Levallois method is supposed to be a unique and distinctive technology, Van Peer and colleagues (2010) confuse matters when they describe Taramsan blade reduction sequences/cores as having a "Nubian pattern of preparation" (e.g. p. 185). Van Peer (2004, p. 220) argued that the Taramsan includes "orthodox Nubian Levallois" reduction.

For some sites which apparently demonstrate the existence of the Nubian Complex, so little has been published on the sites and their lithic assemblage that they cannot be scrutinized. The purported Nubian Complex at Saï Island, Sudan, falls into such a category (Van Peer et al. 2004).

Other sites in Northeast Africa contain occasional examples of Nubian Levallois technology (Table 4.2). At Kharga Oasis, for instance, occasional Nubian Levallois cores and products have been identified and correlated with deposits with MIS 5 age estimates (Table 4.2). When considering the low frequency of Nubian Levallois material identified at the sites they were dating, Smith and colleagues (2007, p. 699) were led to wonder whether they were an "indigenous innovation", or what is called here convergent evolution.

From where does the supposed Nubian Complex of Northeast Africa develop? Scholars such as Van Peer and colleagues (e.g. 2010) have used terms such as Sangoan and Lupemban to describe selected northeast African assemblages, and see these as precursors to the Nubian Complex. These terms are problematic in general, but particularly in Northeast Africa (e.g. Taylor 2016; Scerri 2017). With the Sangoan, for instance, understanding a huge diversity of bifacial forms is not helped by simply categorizing them into an industry. These categorizations have been done with tiny numbers of relevant artefacts, in a weak chronological framework (e.g. Van Peer et al. 2003). Key artefact forms defining Sangoan and Lupemban assemblages further south are missing in Northeast Africa (Scerri 2017). In fact, at places such as Kharga Oasis (Hawkins et al. 2001) and Bir Tarfawi (Wendorf et al. 1993) there are early (MIS 7) Middle Stone Age assemblages characterized by Levallois technology. The notion that Nubian Levallois technology evolved from Lupemban façonnage technology (e.g. Van Peer 2016) is problematic. It is perhaps more likely that Nubian Levallois technology repeatedly appeared from a common base in Levallois reduction. Taylor (2016, p. 290) summarizes the problems with Van Peer's characterization of the Saï Island site, for which only preliminary OSL determinations and cursory technological descriptions are available. Van Peer and colleagues (2003, p. 190) originally described "Upper Sangoan" material from Saï Island, but this was later classified as "clearly Lupemban" (Van Peer et al. 2004; Van Peer and Vermeersch 2007, p. 189). Currently there is insufficient evidence to argue that a developmental sequence links the Nubian Complex as a specifically northeast African entity to the proceeding 'Sangoan' and 'Lupemban'.

The end of the Nubian Complex in Northeast Africa is as troubling as its beginning. Aside from uncertainties of proponents whether entities such as the Taramsan belong to the Nubian Complex, recent findings suggesting a very young (MIS 3 and 2) presence of Nubian Levallois technology in Sudan are also very interesting (as discussed below) (Table 4.2). This is ironic given that the Nubian Complex has repeatedly been described as an MIS 5 technocomplex (Table 4.1), as discussed above, based on often weak evidence, where dating methods are insufficiently described (e.g. Taramsa) or where limited information is available on the lithic assemblages (e.g. Sodmein). These recent findings in Sudan show young Nubian Levallois technologies, that are more securely dated than most of the purportedly MIS 5 examples.

Firstly, at Affad 23 (southern Dongola Reach, Sudan) detailed technological analysis and refitting allowed the reconstruction of several Levallois reduction methods (Osypiński and Osypiński 2016). One of these consists of a method of Levallois production matching the definition of the Nubian type 2 Method. Multiple OSL dates position these lithics at ~16–15 ka, making them some of the youngest MSA reported anywhere in Africa.

More recently, the site of BP177 from the Bayuda Desert region of Sudan and not far from Affad 23 was reported by Masojć and colleagues (2017). BP177 consists of a shallow depression at the top of a flat hilltop. Excavation revealed over 60,000 lithic artefacts in an area of 25 m^2. Two horizons of lithic artefacts were identified within a shallow sedimentary sequence. Diverse raw materials were used, including petrified wood, quartz, and chert. The lithic technology is mainly Levallois in character and includes a significant element of Nubian Levallois, while discoidal and 'classic' centripetal Levallois cores are also present. Three OSL dates give chronological information on the site (Table 4.2). The lower MSA horizon is a narrow band just above an OSL date of ca. 63 ka, while the upper MSA horizon is bracketed by dates of ca. 26 and 18 ka. The upper horizon is particularly rich (57,553 artefacts). Of the 933 cores, 483 are described as 'other/fragment' but of the remaining cores grouped into types, 25.6% are described as Nubian Levallois cores (n = 115). In the lower layer, of the identifiable cores, 26.8% are Nubian Levallois cores (n = 11). Levallois flakes and points are abundant, as are bifacial foliates and scrapers. The authors claim there are a small number of handaxes present, but the one illustrated example is perhaps just as likely to be a prepared preferential Levallois core. Likewise, in a brief original report on the site Masojć (2010) presents a few illustrations of small bifacial forms (foliates, retouched points, etc.). The combination of Nubian Type 2 reduction and the production of bifacial foliates should, according to the Van Peer scheme, put the site in the Early Nubian Complex, which apparently dates to MIS 5e. Yet, at BP177 this technology apparently dates to just ca. 62–17 ka.

Interestingly, after finding material which is 'too young' for the Nubian Complex, the same team recently reported a site which may be 'too old'. They claim that site EDAR 15 in Sudan contains a small number of lithics demonstrating Nubian Levallois technology (Masojć et al. 2019). However, the one example illustrated does not feature the criteria which define Nubian Levallois cores, so for now the characteristics of this assemblage are unclear. If it is indeed shown to contain Nubian Levallois technology then this might confirm the existence of this technology in the Middle Pleistocene, as it is located in a layer between OSL age estimates of ca. 156 ka and ca. 181 ka (Masojć et al. 2019); although the D_e distribution data which is provided for only one of the samples looks highly dispersed and so the dates may not be reliable. All other MSA material from Sudan reported by Masojć and colleagues (2019) lacks Nubian Levallois technology.

Nubian Levallois Technology Elsewhere in Africa

Just as Nubian Levallois technology is not limited to MIS 5 as has often been claimed (e.g. Rose et al. 2011), neither is it confined to Northeast Africa and, recently, Arabia. Nubian Levallois technology has been found at multiple sites in north-central and north-west Africa, in East Africa and in southern Africa. These findings have been dealt with in varying ways by proponents of the Nubian Complex. East African sites are sometimes included in the Nubian Complex, and are sometimes not (Table 4.1). In the case of North Africa away from Egypt and Sudan, Van Peer (1998, 2016) has claimed that 'Aterian' assemblages (or more neutrally, let us say tanged tool assemblages or TTAs, see Scerri 2017) were descendants of the Nubian Complex. Recent discoveries of Nubian Levallois cores in South Africa have either been ignored, denied (e.g. Goder Goldberger et al. 2016), or side-lined as being too far away to be significant (Hilbert et al. 2016). More on this below.

Findings of Nubian Levallois technology across northern Africa have long been known. Chavaillon (1973) for example illustrates Nubian Levallois cores from the site of Hassi Ouchtat in Algeria. Van Peer (1986), before he invented the Nubian Complex, reported the presence of Nubian Levallois technology at various sites in the Maghreb, such as Oued Djouf El Djemel, Ain Mansourah, Gazet Oum Ali, El Oubira and Talbelbala. Nubian Levallois technology has been reported at Adrar Bous in the central Sahara (Clark et al. 2008). At Uan Tabu in southern Libya, OSL dating suggests an age of about 61 ka for an assemblage featuring both Nubian Levallois reduction methods and the production of tanged tools (Table 4.2). Van Peer (1998, 2016) has long claimed that the Aterian was an outgrowth of the Nubian Complex. However, tanged tool assemblages occur in early MIS 5—about the time that many proponents argue the Nubian Complex appeared—and perhaps even earlier by ca. 145 ka if one accepts the weighted mean of TL estimates from Ifri N'Amar (Richter et al. 2010). These data do not support an ancestor/descendant relationship between Nubian Complex and tanged tool assemblages. The simplest reading of this data is that technological features such as tanging and Nubian Levallois reduction occur in a complex technological patchwork in the area (Scerri et al. 2014), in a way that cannot helpfully be condensed to industrial terms such as Aterian or Nubian Complex.

Nubian Levallois technology has been identified as far west as Mauritania, and its current apparent absence from West Africa south of Mauritania probably reflects a lack of research in this area as much as anything else. Pasty (1999, p. 126) describes the presence of Nubian Levallois technology at Mauritanian sites such as Arouakim (approximately 22.50 N, 11.42 W). Core forms (e.g. p. 127) are said to resemble those defined by Guichard and Guichard (1965) as Nubian Levallois. At Arouakim, tanged points are also clearly often made on Nubian Levallois points (ibid., p. 149).

Numerous East African sites contain Nubian Levallois technology. Whether these assemblages are part of the Nubian Complex seems to have troubled proponents of the Nubian Complex for many years (Table 4.1). The two key sequences here are K'One and Gademotta in Ethiopia, while a series of other sites feature variable levels of Nubian Levallois technology.

Several dense MSA assemblages were excavated at K'One (Kurashina 1978). Unfortunately, these assemblages are not chronometrically dated, but the 'locality 5 extension' assemblage is critical for its clear demonstration of Nubian Levallois technology in East Africa. Of 161 cores from the site, 49 (30.4%) are Type 1 Nubian Levallois cores, making them the most common core form at the site. And that figure only includes complete cores, so the real number is even greater. Extensive refits from the site are highly informative on the character of lithic reduction (e.g. Kurashina 1978, pp. 428–429). While to some the site is clearly part of the Nubian Complex (Rose et al. 2011), to Van Peer (2016, p. 153) the site is not a "classic" Nubian Complex assemblage, and apparently has "many idiosyncratic technical features".

Nubian Levallois cores from K'one locality 5 extension are rather diminutive in size compared to those from areas such as Dhofar (discussed below). The K'one cores have average length, width and thickness figures of 39.1, 30.2 and 12.0 mm respectively (Kurashina 1978). In regards to technology the Nubian Levallois cores feature typical characteristics for this reduction method. In other characteristics, the K'one material appears typical of other sub-Saharan MSA assemblages, with bifacially retouched points being a common tool type, for example. An analysis of part of the thousands of flakes collected by Kurashina (1978) cast light on reduction methods in the locality 5 extension assemblage, and suggested an emphasis on Levallois flake rather than point production. Nubian Levallois reduction may have been employed in a final phase of reduction (Kurashina 1978, p. 359). On the other hand, different core forms display similar size values, arguing against the hypothesis of reduction methods tracking reduction intensity at this site. Re-examination of this material using a modern methodology, and applying chronometric dating, would be extremely interesting.

The other key East African sequence in the current context is the series of MSA sites found at the Gademotta-Kulkuletti complex, also in Ethiopia. These were reported by Wendorf and Schild (1974) and have long been a classic early MSA sequence. Recent work by Douze and Delagnes (2016) has significantly added to technological understanding of the sequence. At the site of ETH-72-6 four Nubian Levallois cores were recovered (8.5% of the cores), as well as other artefacts consistent with this reduction method. Douze and Delagnes (2016) root this appearance of Nubian Levallois technology at ETH-72-6, dated to between ca. 190 and 100 ka (Table 4.2) and probably closer to the latter, in the context of a long term local developmental sequence. The context here is Levallois technology and point production, and there is no evidence of Nubian Levallois technology emerging from 'Lupemban' façonnage technology. It should also be noted that the presence of Nubian Type 1 Levallois technology in early MIS 5 or earlier does not fit with the original schema in which this technology should be a terminal MIS 5 phenomenon.

Currently there seems to be no consensus on whether the early MSA assemblages from Gademotta-Kulkuletti do or do not contain Nubian technology (Table 4.2). For example, Van Peer and Vermeersch (2000, p. 56) argue that illustrations published by Wendorf and Schild (1974) "strongly suggest" the presence of Nubian Levallois technology at ETH-72-B, a site dating to over 280 ka. Likewise, proponents of the Nubian Complex are unclear if the technology is found at ETH-72-1, dating to between ca. 180 and 280 ka. Crassard and Hilbert (2013) accidently included a drawing of a core from the site in their Fig. 11, although it is incorrectly labelled as coming from Aduma. This core was originally illustrated by Wendorf and Schild (1974; Plate XXVII). It was then redrawn for consistency with other selected illustrations, and correctly captioned, by Clark (1988, p. 259) from where Crassard and Hilbert (2013) came by the drawing, although they incorrectly cite Wendorf and Schild (1974) instead of Clark (1988). This raises important questions on how the Nubian Complex is understood. Do Crassard and Hilbert no longer think that the core is a Nubian Levallois core, or do they now think that the Nubian Complex may date as far back as MIS 8? Neither option seems particularly supportive of the argument for Nubian Levallois technology being highly distinctive and spatially and temporally restricted.

Numerous other East African sites have produced evidence for Nubian Levallois technology (Table 4.2). These include Mochena Borago (Brandt et al. 2012, 2017), where the technology occurs in a generally non-Levallois assemblage, indicating once again the variable contextual settings of Nubian Levallois technology which are simply not captured by including such material in the 'Nubian Complex'. A Type 1 Nubian Levallois core was recovered from Garba III (Mussi et al. 2014), as were two of these cores at Hargeisa in Somalia (Clark 1954). Further south, several assemblages in Kenya have produced Nubian Levallois cores. These include one associated with sediments dating to ca. 100 ka at Keraswanin (Blegen et al. 2018). Likewise, at the unpublished 'locality 92' in the Kapthurin formation, several Nubian Levallois cores were recovered (Blegen pers. comm.). A classic Nubian Type 1 Levallois core was recovered in an area of Late Pleistocene sediments at Rusinga Island (Tryon et al. 2012).

More recently discoveries of Nubian Levallois technology have been made in South Africa, in particular at two localities in western South Africa (Hallinan and Shaw 2015; Will et al. 2015), and its presence can retroactively be seen in an early publication (Samson 1968). Given the thousands of kilometers separating these sites from the purported heartland of the Nubian Complex, they are very significant in terms of distinguishing cultural transmission versus convergent evolution more widely. Proponents of the Nubian Complex have tended to simply ignore the South African findings. The exceptions are Hilbert et al. (2016, p. 2) who state that Nubian cores have been found in South Africa but that "doe (sic) to their isolated character incorporating them into the Nubian Complex is problematic, further research need (sic) to be done". So they choose to ignore them in the remainder of their paper. Goder Goldberger and colleagues (2016, p. 134) state that the South African material is merely 'Nubian-like' because of the apparent lack of a "protruding distal ridge on most of the cores". Having gone down the route of presence/absence rather than proportions, proponents of the Nubian Complex have put themselves in a difficult position. Even if "most" of the South African cores do not meet strict requirements, some do. Will and colleagues (2015) explicitly used the criteria defined by Usik et al. (2013) to define Nubian Levallois technology.

Secondly, the idea that all Northeast African and Southwest Asian examples of 'Nubian technology' are highly similar and identical to idealized depictions is itself highly dubious. Even if median distal ridges are not sharply pronounced on some of the South African cores, in other cases they clearly are (e.g. Will et al. 2015, Fig. 7g, f). The major Nubian Levallois findings have been made at currently undated, open-air sites. But correlation with the stratified sequence at Mertenhof strongly suggests a post Howiesons-Poort, i.e. probably MIS 3, chronology for the open-air sites. Both cores and flakes at Mertenhof display similar elements of technology to open air sites such as Uitspankraal 7 (UPK7).

At UKP7, systematic recording recovered 31 Nubian Levallois cores as well as multiple Levallois products and overshot flakes consistent with the cores (Will et al. 2015). Nubian Levallois cores occur as Type 1, Type 1/2 and Type 2. Will and colleagues argue that their findings strongly indicate convergent evolution of Nubian Levallois technology. Hallinan and Shaw (2015) report the site of Tweefontein. They identified at least 50 cores consistent with being Nubian Levallois cores—with a much higher incidence of Nubian Type 1 cores compared to UPK7—and 150 points. Some of the 'bifacial pieces' actually also appear to be Nubian Levallois core. Hallinan and Shaw (2015) also implicate convergent evolution in the presence of these forms in South Africa. At recent conferences several talks have shown Nubian Levallois cores at other sites in South Africa, so it is hoped that these will soon be published and chronometric and environmental contexts clarified.

The Nubian Complex in Arabia

In the last decade there has been a dramatic acceleration in research in Arabia (e.g. Petraglia and Rose 2009; Groucutt and Petraglia 2012), an area which was previously very poorly understood. The publication of dozens of Middle Paleolithic assemblages with Nubian Levallois technology in Dhofar, southern Oman, by Rose and colleagues (2011) led to a major reinvigoration of the Nubian Complex idea. To Rose and colleagues, the findings are unambiguous proof of demographic connections between southern Arabia and Northeast Africa: "we discovered a stone tool industry made by one of the earliest modern human populations on earth.' (Rose et al. 2018, p. xxii). The Dhofar 'Nubian Complex' and MIS 5 layer at Jebel Faya apparently represent "two distinct, African derived Middle Stone Age populations" (Al-Abri et al. 2012, p. 291). Below I describe the evolution of the notion of the Nubian Complex in Arabia in a primarily chronological manner.

Before looking at individual sites in more detail, it is worthwhile considering the overall nature of the findings from Dhofar. These sites have huge numbers of Nubian Levallois cores (sometimes >90% of cores); they generally occur in greater frequencies than in most Northeast African sites. Secondly, there are of course similarities between these cores and some cores (or more precisely, core reduction methods) found in places like Northeast Africa, but there are also differences. For example, Rose and colleagues (2011) themselves noted the prevalence of distinctive 'dihedral chapeau' striking platforms in Dhofar, which have not been found elsewhere. This platform type involves creating a chapeau du gendarme shape, but by two removals in a dihedral-like manner, and then subsequently applying finer faceting (see the bottom core in Fig. 4.2). Likewise, Nubian Levallois cores in Dhofar appear to be more elongated than those in Northeast Africa (Rose and Marks 2014; Hilbert et al. 2017). And there are other factors that will be discussed below. This pattern of similarities and differences suggests the need for detailed comparative studies, and robust chronostratigraphic contextualization of assemblages. So far, neither of these has been conducted.

For some years, it has been known that Nubian Levallois cores occurred in southern Arabia (Inizan and Ortlieb 1987), although they were not initially described as such. Crassard (2009) conducted detailed analysis on variability in Levallois point production method, noting similarities in some cases with Nubian Levallois technology. The diversity of methods for Levallois point production in Yemen is an extremely significant observation (Crassard 2009; Crassard and Thiébaut 2011), although one unfortunately lacking chronological control. Such diversity might be what we would expect to find as a precursor to Dhofari assemblages more dominated by Nubian Levallois reduction, if this had emerged by convergent evolution in an Arabian setting. There certainly seems to be more diversity in Levallois methods in Yemen than in neighboring Dhofar.

The discovery of hundreds of Nubian Levallois cores and associated flakes in Dhofar (Rose et al. 2011), represents a significant advance in our understanding of the prehistory of Arabia. These sites should not simply be seen as dots on a map of Nubian Complex sites. Instead, these findings should be studied and contextualized with as few assumptions as possible. Firstly, the genuine dominance of Nubian Levallois technology in Dhofar is different from the occasional presence of Nubian Levallois cores in areas such as northern Arabia and the Levant (discussed below). Secondly, the Dhofari assemblages are mostly focused in a small area around the village of Mudayy in western Dhofar. This distribution is very significant as these assemblages are located near an important spring at Mudayy (Usik et al. 2013). Therefore, rather than the Dhofari assemblages with Nubian Complex representing simply a spread of Nubian Complex bearers, the story may instead reflect convergent evolution around one of the few reliable water sources in a generally arid region. These sites are located in an area of abundant and high-quality chert, between Jebel Qara and the Empty Quarter. Even today, southern Dhofar is extremely unusual in Arabia for the rainfall it receives (and cloud cover which reduces evaporation). It is therefore quite possible to imagine a scenario where an early population spread across Arabia became restricted to areas such as Dhofar, particularly around reliable springs, where they developed a particular kind of technology either through drift or through 'adaptation' to high levels of mobility away from these occasional water sources. More on this possibility later.

Rose and colleagues (2011) reported 110 Middle Paleolithic sites with Nubian Levallois technology, varying from isolated artefacts to large assemblages. In nearly all of the assemblages studied, Type 1 Nubian Levallois reduction was dominant. Assemblages were systematically collected from grids at three sites for detailed analysis, and a selective collection made at another site. The three systematic collections allowed an objective and quantified description of the assemblages. Of the 157 cores collected at Aybut Ath Thani, 90% were Nubian Levallois cores. At Mudayy As Sodh, 78% were. While the figures for the possibly mixed palimpsest assemblage of Jebel Sanoora is 66.3%. Retouched tools associated with these assemblages were generically Middle Paleolithic, and the kinds of retouched forms emphasized by Van Peer (e.g. 1998) were not found in the Dhofari assemblages. Levallois products ('points') almost all have centripetal scar patterns, and are typically thick and broadly convergent in shape (Fig. 4.1).

As the only Arabian 'Nubian Complex' site with chronometric age estimates (OSL), it is important to evaluate the site of Aybut Al Auwal in detail. Many studies uncritically cite this as demonstrating a human presence at ca. 107 ka (e.g. Scally and Durbin 2012; Mellars et al. 2013; Grove et al. 2015; Bae et al. 2017). The reality of the site is much less clear. As with many sites in the area, Aybut Al Auwal consists of a large open-air site where eroding chert

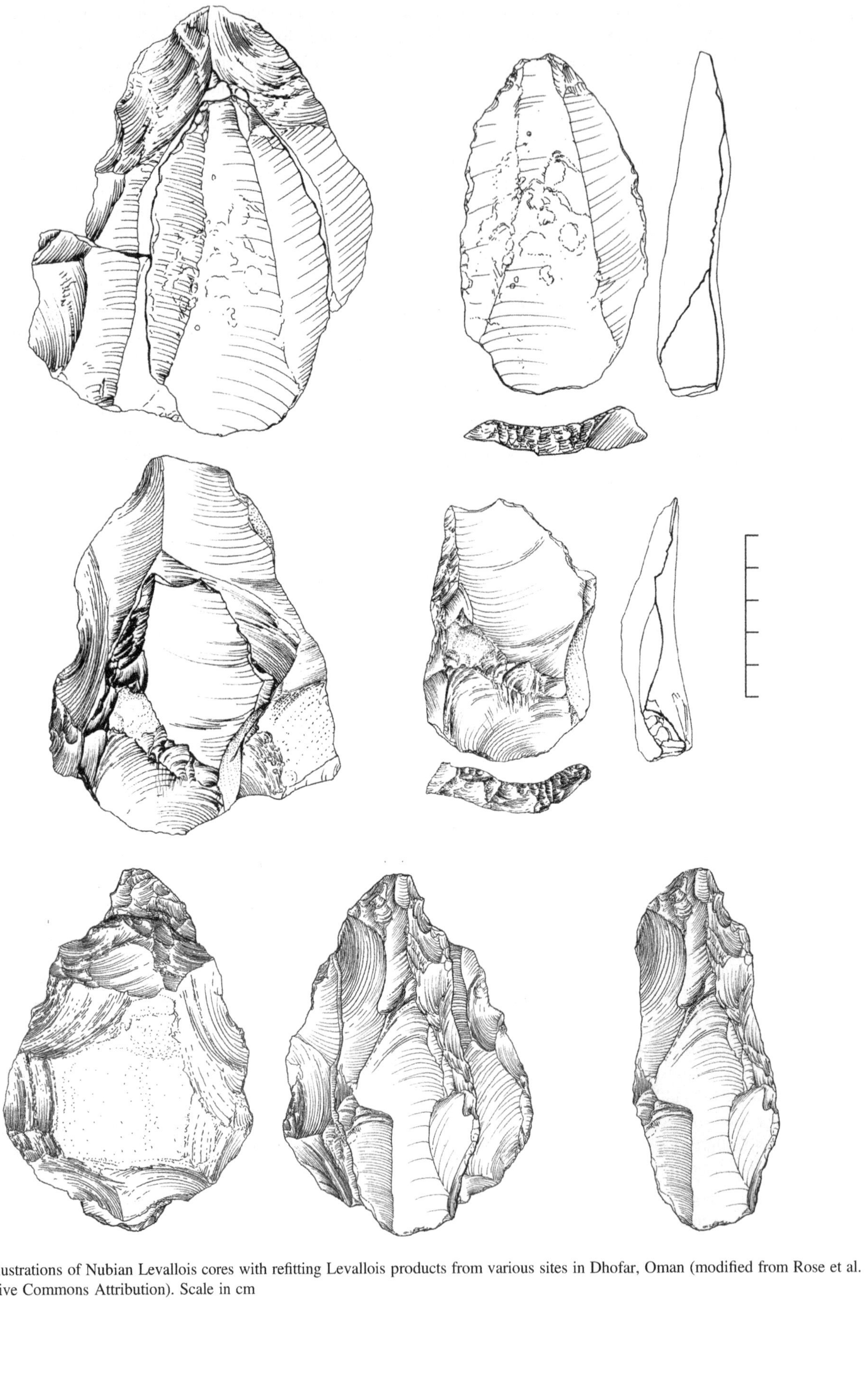

Fig. 4.2 Illustrations of Nubian Levallois cores with refitting Levallois products from various sites in Dhofar, Oman (modified from Rose et al. 2011, Creative Commons Attribution). Scale in cm

beds were repeatedly used to acquire raw material during prehistory, that was then knapped on site. The surface is covered in both Middle and Late Paleolithic artefacts. Unlike the systematic collections described above, only a "random" collection of Middle Paleolithic artefacts was made at Aybut al Auwal. The overall character of the assemblage is therefore not clear, which has not stopped later papers presenting percentages of artefact types and other proportional data from the site (Rose and Marks 2014, p. 68). Ten artefacts were extracted from a shallow deposit of fluvial sediments, resulting from an ephemeral headwater stream crossing the site. These artefacts are said to be from Unit 3 (p. 8) but the stratigraphic log also has them in Unit 2. Key to these artefacts are a single Nubian Type 1 Levallois core, and the proximal fragment of a Levallois product. Two OSL estimates from Unit 3 sediments give an average age of ca. 107 ka (Table 4.2). It is, however, critical to note that these artefacts are redeposited in the fluvial sediments. The age estimate is therefore a minimum age, by an unclear length of time. Rose et al. (2011) conflate this issue by presenting other sites (e.g. Kharga, Saï Island) in terms of minimum and maximum ages, but presenting the Aybut al Auwal dates as if they are precisely dating the human occupation. Subsequent publications continued to state that Aybut al Auwal shows that Nubian Levallois technology appeared in Dhofar in MIS 5c (e.g. Al-Abri et al. 2012). Van Peer (2016, p. 155) argues that the dates on fluvial sediments at Aybut Al Auwal date the human occupation to "mid-MIS 5 precisely". However, Rose and colleagues (2018, p. 168) admit that the artefacts are in a secondary position, so the ages are by definition minimal.

Rose et al. (2011) argue that the Aybut al Auwal assemblage, and other Dhofari findings, belong to the Late Nubian Complex in the Van Peer scheme. It will be remembered (Table 4.1), however, that in this scheme the Late Nubian Complex/Nubian Complex *sensu stricto* only emerges at the very end of MIS 5, yet at the one dated site in Dhofar, it dates to at least 107 thousand years ago.

Usik and colleagues (2013) reported on continued research in Dhofar. The number of Middle Paleolithic sites they had discovered in Dhofar had now increased to 260. Their paper is useful in many regards, and offers a clarity of definition compared to earlier studies. Even in this account, though, ambiguities slip in, such as arguing that Nubian Levallois technology 'grades into' other core types (Usik et al. 2013, p. 250).

Usik et al. (2013) also followed the procedure of reporting in detail on a small number of assemblages, although it is unclear how these assemblages were selected and how representative they are. Nevertheless, the systematic collections employed provide an excellent example of how this kind of collection should be done in these open-air sites. These localities have extremely high levels of Nubian Levallois technology. For example, at the site of TH.69, 92% of the large and systematic collection of cores are Nubian Levallois cores. Usik et al. (2013) also report sites they describe as 'Mudayyan', which they argue are younger than the 'classic Nubian' sites. These Mudayyan sites contain Nubian Levallois cores, but these occur alongside other reduction methods. Both weathering and landscape positioning suggest that these 'Mudayyan' assemblages are indeed younger than 'classic Nubian Complex' sites (Usik et al. 2013; Rose and Marks 2014; Rose et al. 2018), but it is unclear by how much. These assemblages contain 'micro-Nubian' Levallois cores, but once against things appear more blurred than is sometimes perceived, as these micro-Nubian cores "grade into" bidirectional cores with essentially flat debitage surfaces (Rose and Marks 2014).

Rose and colleagues (2018) have recently published a welcome monograph on their research in Dhofar. In terms of the Middle Paleolithic, they only discuss a few selected assemblages. As with their earlier publications, it is unclear why these and not others were selected. The authors depict the Middle Paleolithic of Arabia as being essentially entirely 'Nubian', which ignores the reality for most of the peninsula outside a small area within Dhofar. The only overall regional information given consists of maps showing the location of all sites in Dhofar. They indicate that sites vary from dense workshops to isolated individual lithics, but this kind of information is not shown on a map. The basic picture which emerges is of a dense distribution of sites near the village of Mudayy and its important springs, but with a simple 'dots on maps' approach, it is hard to understand the extent to which Nubian Levallois technology is abundant across the wider area. Another aspect which emerges from Rose et al. (2018) is the extent to which assemblages in the area are often mixed palimpsests. While they make an argument that patination and distance from raw material sources gives relative age, this is a complex issue, and variability in patination is not a simple process. It is clear from many of the illustrations given that the Dhofar MP sites have common features, including high levels of elongation, and a prominence of dihedral chapeau platforms. Unfortunately, prominence is not given to these distinctive and interesting characteristics, and instead the story of how the Nubian Complex evolved out of the 'Lupemban' is repeated.

In recent years Nubian Levallois technology has also been identified in central and northern Arabia. Crassard and Hilbert (2013) report that dozens of Middle Paleolithic sites were identified around Al Kharj in the Nejd region of Saudi Arabia. They focus on one of these sites AK-22, which they describe as a Nubian Complex site. This is a surface scatter covering an area of around 1,200 m^2. Locally outcropping quartzite is the dominant raw material at the site. Crassard and Hilbert (2013, p. 7) chose to employ a "selective sampling" of "diagnostic elements". This is of course potentially problematic, as what is and is not considered 'diagnostic' in a region where the archaeological record is very poorly understood is not something which is self-evident. The notion of something being 'diagnostic' is rather subjective and lacks clarity. In total, Crassard and Hilbert (2013) collected 177 artefacts, from a site with presumably many thousands of artefacts. Of these 123 are cores, representing diverse Levallois and non-Levallois reduction methods. Of this small and selective collection, 16 cores are described as Nubian Levallois cores, representing 13% of the collection. Given that these cores are believed by Crassard and Hilbert to be diagnostic of something and were therefore preferentially collected, it therefore seems that Nubian Levallois technology represents a small component of the AK-22 assemblage. No refits are reported, so their interpretation focusses almost entirely on the morphology of 16 cores. The attribution of some of these cores to Nubian Levallois is not always obviously justifiable—perhaps suggesting instead that Nubian Levallois cores represent an end of the range of variability of the Levallois cores which dominate the assemblage. Pieces such as AK-22-64 (p. 11) and AK 22-60 (p. 11) are not really classically Nubian in their dorsal surface morphology and scar patterning. The cores are centripetally prepared Levallois cores, with a variable presence of a median-distal ridge. Given the fact that one of the defining features of Levallois technology is the formation of lateral and distal convexity, it is not surprising that this convexity blends into a 'ridge' in some cases. It is therefore possible that Crassard and Hilbert (2013) have in essence selectively gathered Levallois cores from a huge assemblage which are broadly triangular and have a somewhat ridged median-distal surface.

The Al Kharj Nubian Levallois technology seems to show some differences to the Dhofar material—such as apparently having Safaha flakes in the assemblage, and lacking the 'dihedral chapeau' preparation found in the south. Despite this, Crassard and Hilbert (2013) see the assemblage as being part of the Nubian Complex, and give little emphasis to the differences between assemblages in different areas.

Hilbert and colleagues (2016) conducted a comparison of aspects of lithic technology from the Saudi Nejd sites to those from Dhofar, which they claim demonstrates high levels of technological homogeneity. The basic problem with this is that the Nejd material is based on biased collections. So it perhaps not surprising that artefacts collected because they looked similar to others from Dhofar are, in fact, similar. They claim to be testing a hypothesis that raw material would have had a strong influence on lithic technology—which it clearly does in some regards, and just as clearly does not in others. The hypothesis (p. 3) that they attempt to test "is that Nubian cores from Al-Kharj will show comparable patterns in respect to the specific attributes presented by Usik et al. (2013) and different from the Nubian cores found in Dhofar" and they explore the influence of raw materials. There are no null hypotheses. Despite Nubian Levallois technology being the dominant technology in Dhofar and rare at AK-22, despite dihedral chapeau du gendarme striking platforms being common in Dhofar but seemingly absent in the Al Kharj sites, despite the Al Kharj cores typically much less elongate etc., we are still told that the lithics are 'homogenous'.

According to the information shown in Hilbert et al. (2016) there are about ten sites with Nubian Levallois technology in the Al Kharj area. In their paper they report on three of these. Hilbert and colleagues (2016) state what while AK-22 was not systematically collected, the sites of AK-40 and AK-43 were. However, it seems odd that 84 of the 112 lithics from AK-22 are cores, as are 146 of the 260 lithics at AK-43. These are very unusual figures for systematic collections, unless the assemblages have undergone considerable taphonomic disruption. Seventeen out of 122 cores from AK-22 are described as being Nubian Levallois cores (it was 16 when first reported in Crassard and Hilbert 2013; likewise, Schiettecatte and colleagues [2013, p. 290] report that 122 Levallois cores were collected at AK-22, while Hilbert et al. [2016, p. 10] say there are 90). At AK-40, 28 of the 84 (33.3%) cores are said to be Nubian Levallois cores, as are 58 of 146 (39.7%) cores at AK-43.

Crassard and colleagues (2018) report the site of Umm al-Sha'al, near the Al Kharj sites described above. A single distal fragment of a core is described as belonging to the 'Nubian tradition'. OSL ages from close to this fragment indicate a chronology of MIS 7. Most of the excavated material from the site comes from the upper layers of the site, in units 4 and 5 which are dated by OSL to 71 ± 6 ka (AKE31-4) and 81 ± 9 ka (AKE31-5) respectively, as well as being above an estimate of 87 ± 10 ka (AKE31-3) for the top of the underlying unit 3. These deposits then date to the end of MIS 5 or early MIS 4, and should presumably contain a rich 'Nubian Complex' assemblage according to the model advanced by proponents of this entity. While Crassard and colleagues (2018) seem to try to make such a model fit, talking of the "distinctive medio-distal guiding ridge" on some of the late MIS 5 cores from the site, but restrain from calling anything Nubian Levallois. Likewise, one core has an opposed platform allowing recurrent bidirectional flaking which the authors argue has "in other contexts…been observed in association to (sic) the Nubian techno-complex" (p. 13). Finally, they claim that "the use of convergent mesial/distal preparation of the Levallois surface on one specimen is giving (sic) the upper assemblage from the site some Nubian affinities" (p. 15). The probably late MIS 5 lithics from Umm al-Sha'al demonstrate a focus on centripetal and unidirectional reduction, and the lack of discovery of any clearly Nubian Levallois technology is an important observation (as with other excavated sites in Arabia, one redeposited core in Dhofar aside). Cores demonstrate workshop characteristics (i.e. low reduction intensity), but the Levallois flakes recovered—perhaps made on cores subsequently removed from the locality—indicate a tendency to centripetal Levallois technology in fitting with other MIS 5 assemblages in Arabia. None of these Levallois products demonstrate a Nubian Levallois reduction method.

Hilbert and colleagues (2017) reported the presence of the Nubian Complex—that is, a few Nubian Levallois cores—in northern Arabia, near the town of Al Jawf. They report the discovery of 48 Middle Paleolithic sites in the area, using diverse Levallois and non-Levallois reduction methods. Collection methods varied, but in some cases were selective. Most of the 'sites' reported had incredibly low artefact densities, sometimes less than one artefact per square meter over large areas. They selected ten assemblages to present in their paper, with the reason for the selection of these sites over others not being clear.

To summaries the data presented from these sites, Hilbert and colleagues (2017) found a total of 14 Nubian Levallois cores, from a total of six sites (the four other sites reported did not have any). And as the collections are selective and the authors thought in advance that Nubian Levallois technology was 'diagnostic', even this represents an exaggeration of the frequency of this technology. Beyond showing that Nubian Levallois technology is present, these selective collections make it basically impossible to objectively understand the technological characteristics of the assemblages. The impression given by these assemblages is that Nubian Levallois cores represent one end of a range of variability, with various forms of Levallois cores found (see also Chiotti et al. 2009). The 'Nubian' forms are simply centripetally prepared preferential Levallois cores which have varying manifestations of median distal ridges, hardly surprising given that the imposition of lateral and distal convexity is part of the definition of Levallois technology. Hilbert et al. (2017) seem to emphasize Nubian Levallois cores because they say that other Levallois core forms are found very widely distributed. Aside from the same also being true for Nubian cores, this surely highlights the problem with simple typological categories.

The final site in our discussion of the purported Nubian Complex in Arabia comes from the recent discovery of a grand total of two apparently Nubian Levallois cores in the Adam region of northern Oman (Beshkani et al. 2017). Here we are presented with a classic conflation of the Nubian Complex and Nubian Levallois technology. The authors state "the Nubian Complex is generally divided into types one and two" (p. 1), when they mean Nubian Levallois reduction. This is a crucial distinction—to the extreme proponents of the Nubian Complex, a site does not need Nubian Levallois cores to be classed as Nubian Complex (e.g. Van Peer 2016). The two cores shown by Beshkani and colleagues (2017) are similar to those described by previous authors as Type Nubian Levallois 1 cores, but they do not have the triangular to sub-triangular shape generally found elsewhere. They argue that it is "uncertain whether the Sufrat material represents a local industry" (p. 3).

What do these findings from Arabia mean? Firstly, it is crucial to describe the reality of the record not just treat highly variable sites as identical Nubian Complex 'dots on the map'. The pattern which emerges is clearly that the record in western Dhofar, where Nubian Levallois cores are abundant, is different from the rest of the Peninsula. Away from western Dhofar, Nubian Levallois technology occurs infrequently, in variable technological settings. It is also important to note that numerous dated Middle Paleolithic assemblages do not demonstrate *any* Nubian Levallois technology. Key examples include Mundafan al Buhayrah (ca. 85 ka), the only dated MIS 5 archaeological site in southwest Arabia and the Empty Quarter (Groucutt et al. 2015b). The same goes for Jebel Faya in the far east of Arabia (Armitage et al. 2011). Likewise, MIS 5 Middle Paleolithic sites in the Nefud Desert such as Al Wusta (Groucutt et al. 2018), KAM-1 (Scerri et al. 2015), JQ-1 (Petraglia et al. 2012) and several sites which are currently in preparation for publication all lack Nubian Levallois technology. This is also the case for the numerous undated surface sites in the area (e.g. Scerri et al. 2015; Groucutt et al. 2016). These sites occur in the area which separates Dhofar from Northeast Africa, by both northern and southern routes, and the absence of Nubian Levallois technology at these sites is striking.

Nubian Levallois Technology in the Levant

Findings in Arabia have more recently been followed by the reporting of sites with Nubian Levallois technology in the southern Levant. This has been most strongly expressed by Goder Goldberger and colleagues (2016, 2017). In the H2 assemblage from the Negev highlands, Goder Golderger et al. (2016) reported the presence of four Nubian Levallois cores (2% of cores) in an assemblage of 686 lithics, including 196 cores, 46 retouched flakes and 12 Levallois flakes; and one of the three Nubian Levallois cores illustrated is rather atypical. Most of the site's Levallois cores are centripetal and bidirectionally prepared non-Nubian Levallois cores. They examined other assemblages in the Negev, Har Oded and NMR. At these site Levallois cores are rare, at around 20% of the collected cores. At Har Oded 11 cores are described as Nubian Type 1 and three as Type 1/2, while at NMR six are described as Type 1 and four as Type 2. Very diverse methods of Levallois point production are described in the original report on those sites (Bouté and Rosen 1989), one of which can be described as Nubian Levallois.

Goder Goldberger and colleagues (2016) present a subtle mutation on the Nubian Complex idea by emphasizing the idea of cultural diffusion rather than demic dispersal as a mechanism to spread Nubian Levallois technology. They claim that because Nubian Levallois cores are not found in the northern Levant and Europe (i.e. a small region at the northwestern margins of Eurasia) and because Nubian Levallois cores occur as part of a 'technological package' with other forms such as 'normal' centripetal Levallois technology, they do not represent convergent evolution. This is because they contrast early Middle Paleolithic sites in the area, which emphasize unidirectional reduction, with sites which they posit date to MIS 5 which are dominated by centripetal Levallois reduction and Nubian Levallois technology. However, given that Nubian technology is itself a form of centripetal Levallois technology, slightly modified by the addition of a median distal ridge, it is hard to see a unique and distinctive 'technological package' as the authors claim.

Subsequently, Goder Goldberger and colleagues (2017) reported the recovery of a few Nubian Levallois cores in the Arava area of Israel. They conducted a cursory survey and noted the presence of a few Nubian Levallois cores as well as other core forms, "mostly" (p. 5) of recurrent centripetal form. At the site of Nahal Paran 9 they collected four Nubian Levallois cores. While they do show the presence of Nubian Levallois cores in the study area, that is about all that can be said.

Hussain and colleagues (2015) report on recent survey in the Wadi Sabra area of southwest Jordan. They found very low density lithics in the area, and the number of artefacts they report is small (i.e. 13 cores across the whole area). Core morphology/technology is generally Levallois, but representing diverse patterns of preparation and exploitation. Products included varied forms of flakes, blades and points. At least three of the cores have affinities with Nubian Levallois technology. These authors, however, give a very useful discussion of the problems of the Nubian Complex idea, such as emphasizing the possibility of convergent evolution. They describe Nubian Levallois technology as representing "the mere lateral spectrum of preferential Levallois point production" (Hussain et al. 2015, p. 73).

These recent findings join a long record of occasional Nubian Levallois or closely related forms in the Levant (e.g. Rust 1950; Munday 1976; Ronen 1974; Clark et al. 1997). As summarized by Vermeersch (2001), in some cases artefacts from Levantine sites show clear Nubian Levallois technology. This technology tends to occur at low frequencies, and in variable technological settings. The extent of similarity in core reduction methods in the early phase of the Boker Tachtit sequence with Nubian Levallois technology has been debated (Table 4.2). At a minimum we can state that there are certainly similarities in some senses. It is likely that further studies will identify more Nubian Levallois technology in Levantine sites. Groucutt et al. (2019), for example, found that even in a single small collection of lithics from the MIS 5e site of Skhul some displayed affinities with Nubian Levallois technologies.

An interesting observation, not previously made as far as I am aware, is that the Middle Paleolithic site of Shukbah, excavated by Garrod (1942) appears to contain Nubian Levallois technology (see Callander 2004, e.g. Fig. 8.1). These occur in a setting generally similar to late Middle Paleolithic assemblages in the area, emphasizing the varied settings in which Nubian Levallois technologies occur. Examination by this author of a collection of artefacts from Shukbah at the Pitt Rivers Museum, University of Oxford, quickly found several which are consistent with Nubian Levallois technology. Crucially, hominin fossils assigned to Neanderthals were recovered from the deposits with these artefacts (Keith 1931). This suggests that Nubian Levallois technology might not be associated with a single hominin species. At other sites with Neanderthal fossils, there are cores which are also very similar to Nubian Levallois cores even if one would not fully accept them as being Nubian (e.g. Fig. 5.1 in Hovers et al. 2008). While the latter example is perhaps less 'classic', it meets the essential criteria of Usik et al. (2013) to be classified as a Nubian Levallois core. Given the choice of many proponents of the Nubian Complex to classify sites to this technocomplex based on simple presence/absence of Nubian Levallois technology, such findings should be considered carefully.

Nubian Levallois Technology in India

The final area where Nubian Levallois technology has recently been discovered is the Thar Desert of India (Blinkorn et al. 2013, 2015). Cores and flakes from both MIS 3 and MIS 5 layers from Katoati (Blinkhorn et al. 2013) meet the definitions of Nubian Levallois technology advanced by proponents of the Nubian Complex (e.g. Usik et al. 2013). Further Nubian Levallois cores have been found in landscape surveys in the area (Blinkorn et al. 2015). These are currently isolated single finds, but nevertheless demonstrate the existence of both Types 1 and 2 Nubian Levallois reduction in the area.

Proponents of the Nubian Complex have tended to sideline the Thar Desert findings, stating that because of "their isolated character…further research need (sic) to be done" (Hilbert et al. 2016, p. 2). But this is at least to admit that they agree that the Indian finds are indeed Nubian

Levallois technologies. Goder Goldberger and colleagues (2017) simply avoid the issue by describing the findings as simply "Nubian-like". They do, however, admit that these 'Nubian-like' findings may indicate convergent evolution. Of course the problem with this idea is the inference that in the Nubian 'homeland', Nubian Levallois cores are always entirely clear and homogenous. This is simply not the case, and the findings from India fit into the overall range of variability in material described as Nubian Levallois.

Nubian Levallois Technology

Over the previous pages the history of the Nubian Complex has been outlined, and sites with Nubian Levallois technology discussed. Arguably, the lack of tight spatial and temporal boundedness to Nubian Levallois technology suggests that convergent evolution is a stronger possibility than cultural transmission in driving the overall pattern of its distribution. While convergent evolution seems to be generally accepted in the case of purportedly 'outlying' areas such as South Africa, the case for convergent evolution is perhaps no less strong in areas such as East Africa, Northeast Africa and Southwest Asia. It is about the same distance (~2,500 km) from Dhofar to the Indian sites with Nubian Levallois technology as it is from Dhofar to the key Nile Valley sites. This should be borne in mind when scholars argue that the Indian sites are too "isolated" to have any bearing on the Nubian Complex (e.g. Hilbert et al. 2017). While on a map of the world, areas such as Northeast Africa and Arabia may look close together, these are vast areas and the sites are thousands of kilometers apart. In a straight line the distance between the Dhofar sites and Taramsa is the same as the distance between London and Moscow. The Red Sea means that at least another 500 km needs to be added to a terrestrial route between Taramsa—as close as there is to type site for the Nubian Complex (Van Peer 2016; Van Peer et al. 2010)—and the Dhofar sites such as Aybut Al Auwal. That 3,000 km distance also involves crossing the longest river in the world, passing through chains of mountains, and some of the driest places on the planet.

Furthermore, the reality is that in the area emphasized by proponents of the Nubian Complex the only assemblages dominated by Nubian Levallois technology are those in the vicinity of the village of Mudayy in western Dhofar and a few sites in the Nile Valley. The paucity of reliable chronometric age estimates should also be made clear. In these conditions, widespread convergent evolution is a very real possibility, and indeed should be the null hypothesis. The hypothesis of convergent evolution to explain the presence of Nubian Levallois technology across space and time can be tested by comparative analyses. To move forward it is important to separate the notion of the Nubian Complex as a culture-historical entity, defined so as to be in effect unfalsifiable (Van Peer 2016), from considerations of Nubian Levallois technology as a specific type of core reduction method.

Researchers often talk about 'Nubian cores', which gives a sense of them as unique and distinct. Of course, these are in reality a type of Levallois core (or rather, the Nubian Levallois reduction method is a form of Levallois reduction method). Recognizing the place of Nubian Levallois technology in the wider Levallois family is an important aspect in considerations of the likelihood of convergent evolution. Several points can be made on Nubian Levallois technology. Firstly, Nubian Levallois reduction can be seen as a sub-type of preferential Levallois reduction with centripetal preparation. The only real difference between 'classical centripetally prepared preferential Levallois' and 'Nubian' Levallois is the shaping of a median distal ridge. Any knapper will understand well the way the shape of the debitage surface and the pattern of arrises (ridges between flake scars) control the shape of subsequent flakes removed. Nubian Levallois technology is merely a heightened recognition of this fact; the median distal ridge is in effect a supersized arris. Pseudo-ridges commonly occur in assemblages characterized by centripetal Levallois reduction. It is easy to see how these could have been recognized, replicated, and exaggerated. Indeed, this may explain the presence of very small numbers of 'Nubian' Levallois cores in the context of generally non-Nubian Levallois assemblages. In reality, the distinctiveness and orientation of median distal ridges on 'Nubian' lithics actually varies considerably (see e.g. Van Peer et al. 2010, p. 76). The different components that Usik et al. (2013) use to define Nubian Levallois reduction are either generic for Levallois technology (e.g. preparation of a main striking platform) or are interlinked and related to the presence of the median distal ridge (e.g. core shape being broadly triangular, an often well-prepared distal striking platform). Those who have discovered Nubian Levallois artefacts in South Africa and India have cautiously described their finds as 'Nubian-like' (Blinkhorn et al. 2013; Will et al. 2015). The reality, however, is that there is no pure 'Nubian' character in Northeast Africa, and merely vaguely similar artefacts elsewhere. The South Africa and Indian finds are more 'Nubian' than many 'Nubian' finds from some of the Northeast African sites which are supposedly classical.

As with diabetes, Type 1 is the more profound form, while Type 2 Nubian Levallois lies on a slippery gradient into 'normal' centripetal Levallois. Indeed, to Goder Goldberger and colleagues (2016) only Type 1 Nubian cores should be called Nubian, as Type 2 belongs to the realm of centripetal flaking. Crassard and Hilbert (2013, p. 4) likewise argued that centripetal Levallois technology graded into Nubian Levallois technology, while simultaneously arguing for the distinctiveness and uniqueness of the latter. Hilbert et al. (2016, p. 7) also suggest that the distinction between Nubian and other forms of Levallois is "problematic". While Type 1 Nubian Levallois is often defined in terms of 'distal-divergent' preparation, the reality is that most of the shaping of these cores to give lateral and distal convexity is centripetal, and often removals from the distal end are converging, parallel, or some other combination. Other scholars have long recognized the fuzzy line between Nubian and non-Nubian Levallois. Marks (1968a, p. 287) for example stated that Nubian Levallois cores are after all "no more than a Levallois point core which has a slightly modified system of preparation".

One of the clearest definitions of Nubian Levallois technology given by proponents of the Nubian Complex is that of Usik and colleagues (2013) based on a combination of different attributes, yet even here ambiguities creep in. For example, they describe different categories of steepness of the median distal ridge—the key defining feature of Nubian Levallois cores—but then state that the last of these "falls outside of Nubian Levallois, *sensu stricto*, grading into bidirectional cores or recurrent cores" (p. 250). Rose and Marks (2014) likewise emphasize the close connections between Nubian Levallois technology and non-Nubian bidirectional reduction. Usik and colleagues (2013) are to be commended for their highlighting gradation *within* the Nubian reduction method—emphasizing that the 'different organizational systems' (Type 1, Type 2, etc.) are not discrete, and were often interchangeable within a reduction sequence, as shown by refits (see also Chiotti et al. 2009). But they unfortunately fail to then take this argument to its logical conclusion, and consider how Nubian Levallois technology can flow to, and from, non-Nubian Levallois technology. Nevertheless, demonstrating the flexibility of Nubian Levallois technology suggests problems with the basic Van Peer et al. (2010) framework where different forms of Nubian Levallois technology have significant chronological significance.

Given these aspects, it is easy to imagine the repeated invention of the Nubian Levallois method, but to explain its frequency in some assemblages we clearly need a mechanism for why knappers sometimes employed this method so intensively. At sites such as Nazlet Khater and the western Dhofar sites there are hundreds of Nubian Levallois cores. Whether these reflect a trail of stone breadcrumbs or a classic case of convergent evolution, the question is the same: why were knappers sometimes so fond of this reduction method? If there are in fact pragmatic reasons that the Levallois products produced by the Nubian Levallois method were desirable, then the case for convergent evolution is perhaps strengthened. These questions have not been explored in detail by proponents of the Nubian Complex, who infer that Nubian Levallois technology is simply a stylistic choice (at least within the aim of producing 'points'), that can be studied in a culture historical framework. However, perhaps the flakes produced by Nubian Levallois reduction have particular characteristics that were desirable to their makers. Experiments could be conducted to explore what such characteristics could have been, compared to flakes produced by other methods. The current distinction between stylistic and functional aspects of lithic technology is problematic, and often based on intuition rather than detailed analysis.

In an earlier paper I considered possible pragmatic aspects driving the use and/or invention of Nubian Levallois technology (Groucutt 2014). From qualitative considerations which have not yet been formally tested, I observed that compared to Levallois points produced by unidirectional convergent methods, Nubian Levallois products tend to be straighter (i.e. less curved longitudinally). They also tend to be thicker distally. This is because the presence of the median-distal ridge on the core tends to result in the partial presence of this ridge on preferential removals, whereas with typically unidirectional convergent flaking the distal end simply feathers off. In my observation of Nubian Levallois cores from Dhofar—which I have spent months studying—I noticed that cores frequently showed that the distal termination of preferential removals has a 'feather-hinge' character. In other words, the distal termination falls somewhere between that typically described as a feather termination and that typically described as a hinge termination. I suggested that this reflecting the 'backing up' of force on the median distal ridge (Groucutt 2014). It appears to me that this is a way of making Levallois products which are both elongate and distally robust. The point here is that when we think about the 'aberrance' or otherwise of a distal termination (e.g. Dibble and Whittaker 1981), we are typically thinking from the perspective of the core, rather than the perspective of the flake removed. It is again currently a qualitative observation, but my impression is that Nubian Levallois points seem to occur in broken form far less than unidirectional-convergent Levallois points with a concorde profile.

An alternative perspective might be that there were numerous reasons why more attention would have been given to the distal end of preferential Levallois cores, converging on the idea of a median distal ridge. This may sometimes have been a way to make points, sometimes a way to produce elongated flakes, and maybe sometimes they wanted strong and straight flakes as described above. In that sense—and by observing actual rather than idealized examples of Levallois technology—the extent to which Nubian Levallois reduction is to produce 'points' can be questioned (e.g. Fig. 4.2) (see also Chiotti et al. 2009; Olszewski et al. 2010). Several scholars (e.g. Guichard and Guichard 1965; Kurashina 1978) have noted high levels of overshot preferential removals in association with Nubian Levallois reduction. This seems to be a risk associated with this reduction technique, and a risk that knappers obviously thought worth tolerating. While the scars of preferential removals on Nubian Levallois cores suggest a focus on producing pointed flakes, which were then generally removed from sites, it is interesting that at some sites very large numbers of overshot removals have been found. It is possible that these were always accidental. On the other hand, there may also be genuine 'intentional' variation in the form of preferential products produced by Nubian Levallois reduction (Fig. 4.2). For this reason, I do not favor referring to 'Nubian point cores' as is sometimes done. The extent to which Nubian Levallois reduction was specifically and exclusively to produce 'points' remains unclear (see Olszewski et al. 2010). It should be noted that the notion of 'points' is itself a complex issue (e.g. Douze et al. 2020).

Why might knappers have been particularly focused on producing relatively thick Levallois products, which perhaps had less of a tendency to break than those produced by some other Levallois methods? Here it is important to think about the locations which have produced large numbers of Nubian Levallois cores. These tend to be better watered areas on the margins of very arid regions, such as along rivers in the case of the Nile and South Africa, or reliable springs in the case of Dhofar. The significant presence of Nubian Levallois technology at K'one in Ethiopia, however, is less clearly explained by such an argument. The greater thickness of Nubian Levallois products may also have had the advantage of having greater potential for retouch than with more gracile Levallois products. Conversely though, few retouched objects are generally found at sites with Nubian Levallois technology. But this could be because these are production sites and retouched forms have been scattered elsewhere in the landscape, or it could simply be that retouch was not a prominent part of the approach typically used by the users of Nubian Levallois products. My impression is that pragmatic factors such as these are more important than trying to produce 'points'. A key argument I was trying to make in my 2014 paper (Groucutt 2014) was that 'point' subsumes a lot of technical and morphological variability (see also Douze and Delagnes 2016; Douze et al. 2020). It might be that similar morpho-technological characteristics resulted from mobility related aspects induced by either movement into arid areas (e.g. Dhofar) and/or perhaps movements away from particularly desirable raw material sources (e.g. K'one, and also Dhofar). Raw material factors are probably an important aspect of variability in the archaeological record (see also Tryon and Ranhorn 2020). This need not just be in terms of the 'quality' of the raw materials, but also factors such as their spatial distribution.

What would we expect to find if convergent evolution had been a major factor in determining the spatial and temporal distribution of Nubian Levallois technology? Surely the key point is that within broad patterns of similarity, such as in a typological sense, we should see fine scale differences, in aspects such as striking platforms, dorsal preparation, etc. In fact, as I shall outline below, there are numerous differences between Nubian Levallois reduction methods in different areas. It is rather ironic that modern proponents of the Nubian Complex are so insistent that convergent evolution is impossible, as the founding texts describing Nubian Levallois technology all emphasize that such forms occurred widely in time and space (e.g. Guichard and Guichard 1965), including in the French Neolithic. Guichard and Guichard (1965, p. 99) include a footnote in which a colleague, E. Gobert, states that "why is it that the workman, faced with different problems, led to fall back into the same solutions?". What a prescient call to consider convergent evolution! And in fact numerous specific technical and morphological features indicate high levels of variability within material described as Nubian Levallois, and that pattern of 'similarity with differences' might be more indicative of convergent evolution than purely cultural transmission. It is also of fundamental importance to consider the entire assemblage in which Nubian Levallois is (or is not) found. Proponents of the Nubian Complex have not advanced a clear theory on why it is apparently only Nubian Levallois technology which is a constant in assemblages of this technocomplex. Why were other elements of the assemblages not also transmitted?

The high levels of 'dihedral chapeau' platforms in Dhofar (Rose et al. 2011), and the absence of these elsewhere is striking. Likewise, Nubian Levallois cores from, for example, Nazlet Khater in Egypt (Vermeersch 2002) seem to be much less elongate than those from Dhofar (Rose et al. 2011). Nubian Levallois cores from K'One (Kurashina 1978), seem to be much smaller than those from places like Dhofar. All of these aspects of variability are important to consider, and need to be integrated into quantified comparative frameworks. While Nubian Levallois cores in different areas share some basic commonalities, there are also important differences which have so far been underappreciated. The act of making selective collections removes Nubian Levallois technology from its wider context and makes meaningful comparisons impossible.

Finally, in terms of the nuances of reduction methods, there are also important differences. For example, extensive refits of lithics from Egypt reveal a lot about the character of Nubian Levallois technology in a particular context. Van Peer (1992, p. 51), for example, describes a refitting sequence with a Nubian Type 1 element that shows that "the decision to create a central guiding ridge is taken in the course of a Levallois surface preparation...the design of the whole pattern is not directed towards the creation of a ridge". In this case refits show the 'opportunistic' addition of

Nubian Levallois preparation. This is not always the case, but surely from such examples, grounds for the reinvention of the Nubian Levallois method can be seen.

When pushed, proponents of the Nubian Complex argue that the defining feature is the Nubian Levallois reduction method. Yet in practice, typologically Nubian Levallois cores have been shown to be formed by various reduction sequence. Too few sites have refits—and where there are refits there are questions on how representative these are—to be really clear on what kind of reduction method is really implied by the discovery of one or two typologically Nubian Levallois cores at a site. Nubian Levallois or very closely allied cores can occur in very different settings, including: (1) as described by Guichard and Guichard (1965), (2) as at Boker Tachtit where crested blades are used in shaping, (3) in the Safahan method, where a typologically Nubian Levallois core is produced, but there is an additional step in debitage surface preparation where a removal from the distal end is the final act of preparation which removed the Levallois surface, 4) in the Taramsan method, which is halfway between Levallois and volumetric blade reduction, but phases of this can be indistinguishable from Nubian Levallois reduction.

As well as the reality that Nubian Levallois cores can occur in various reduction systems, and that there are subtle but important variations in the character of Nubian Levallois reduction, it is also important to point out that the idealized reality of Nubian Levallois reduction is often rather different from the reality. For example, the notion that lateral and distal convexity on Type 1 Nubian Levallois cores is driven almost entirely by two distal divergent removals is frequently repeated. Yet, in reality, in many cases assigned to this method the removals from the distal end are not really divergent, and there is a generally centripetal character to most Nubian Levallois debitage surfaces. Compare, for example, the stylized depiction of Type 1 Nubian Levallois reduction shown in Fig. 4.2 of Rose et al. (2011), with the reality of scar patterns of Levallois flakes produced by Nubian Levallois methods (Fig. 10 of Rose et al. 2011). Likewise, with many Type 1 Nubian Levallois cores, shown in papers such as Rose et al. (2011), removals shaping the median distal ridge are not divergent, and such cores generally have an overall centripetal character. Such aspects are important to point out, as Nubian Levallois technology is in reality a sub-type of centripetal Levallois (as in centripetally prepared and unidirectional exploited, preferential Levallois).

Recent research has demonstrated variability within the Levallois system, and the nature of distinct but broadly similar methods such as those of the Taramsan as well as the 'Howiesons Poort' cores of South Africa (Villa et al. 2010). It can be argued that both of these latter reduction methods, as well as Nubian Levallois technology, represent trajectories of change from a 'normal' Levallois base. In all three cases the key difference from classical Levallois production is that lateral and distal preparation is steeper. This took different forms, but fundamentally all can be seen as experimentation from a Levallois base towards some of the volumetric notions which later systematically picked up in Late Paleolithic systematic volumetric blade reduction.

Conclusion

There is an alluring simplicity to the idea of the Nubian Complex, with its apparently polythetic nature making falsification effectively impossible. To many, the presence of Nubian Levallois cores assigns a site to the Nubian Complex, yet the leading architect of the Nubian Complex has repeatedly argued that an assemblage does not require Nubian Levallois technology to belong to the Nubian Complex (Van Peer 1998, 2016). Those who argue for the Nubian Complex do so with remarkably certainty, and generally ignore the numerous criticisms of the concept which have been raised. To Van Peer (2016, p. 155), for example, it seems that the "only possible historic explanation" for the presence of Nubian Levallois technology in Arabia and Africa is "demic diffusion". Likewise, a few Nubian Levallois cores selected from vast assemblages of centripetal Levallois cores in central Arabia apparently "cannot" be the result of "technological convergence" (Schiettecatte et al. 2013, p. 290). Given the vast spatial and temporal scales involved, the paucity of absolute dates, and the strong indications from South Africa that convergent evolution of this form has actually clearly sometimes happened, then these ideas seem to be rather over-confident in their certainty. Occam's Razor certainly suggests the simpler null hypothesis of convergent evolution is more likely to be correct in describing the overall distribution of Nubian Levallois technology, but nothing has yet been formally tested. Within the distribution of Nubian Levallois technology there is almost certainly some cultural transmission, but the question is how much. To me the evidence indicates cultural transmission on a local scale, e.g. within southern Arabia, within the Nile Valley, etc. But cultural transmission over thousands of kilometers and tens of thousands of years has by no means yet been demonstrated, and does not seem like the simplest explanation for the overall pattern of the data.

Part of the problem seems to be that there is a rather mythical nature to the archaeological record of Northeast Africa. Proponents of the Nubian Complex imply there are large numbers of well dated MIS 5 Nubian Complex sites in this area. The reality, however, is that purportedly Nubian Complex sites are poorly dated, and occur over long time periods. It is simply not the case that the northeast African MSA is principally characterized by Nubian Levallois reduction (contra e.g. Crassard 2009, p. 163).

The extent to which convergent evolution is involved may be elucidated by wider considerations of the regional archaeological record. The two areas with high levels of Nubian Levallois technology—the Nile in southern Egypt and western Dhofar—are both located in areas with two key characteristics. Firstly, both are areas with access to water and raw materials on the edge of more arid regions. Secondly, both areas display very high levels of technical diversity through time. The Nile Valley, for instance, has all kinds of localized cultural phases, traditionally described using 'industrial' terms like Safahan, Halfan, Taramsan, etc. (e.g. Wendorf 1968; Van Peer et al. 2010). Likewise, southern Arabia has repeatedly seen the development of localized technologies (e.g. Crassard and Khalidi 2017), including highly informative examples of convergent evolution. For instance, tanged points from southern Arabia (McClure 1994) had been argued to show Aterian affinities (e.g. Beyin 2006). These, in fact, probably represent convergent evolution in the Holocene (Scerri 2012). Likewise, Dhofar and elsewhere in southern Arabia saw the production of fluted points (Charpentier et al. 2002), otherwise only known in North America. As with all such things, the devil is in the detail: whereas fluted points in the Americas are fluted from the proximal end, those in Arabia are fluted from the distal end. Of course. these findings do not prove or disprove that Nubian Levallois technologies evolved independently, but they do provide important contextual information suggesting that localized and convergent evolution is a distinct possibility. Certain geographical areas seem to have been engines for convergent evolution: places where populations could contract and perhaps become isolated, be it beside the Nile or near to springs in Dhofar.

The way to distinguish convergent evolution and cultural transmission is by comparing assemblages, rather than comparing idealized types of assemblages (As Monnier and Missal 2014, p. 78 put it). This can be done in very different ways, from quantitative comparisons of attributes (e.g. Tostevin 2012; Scerri et al. 2014) to detailed technological analyses (e.g. Crassard and Thiébaut 2011; Douze and Delagnes 2016). As long as studies analyze whole assemblages, or at least representative samples, and do so in a framework with as much chronological control as possible, then personally I think that various methods of lithic analysis will provide different sorts of useful information.

If we divide the Old World where the Middle Paleolithic/ Middle Stone Age (that could form a technological base for the independent innovation of Nubian Levallois technology) was prominent into different regions—southern Africa, eastern Africa, central Africa, western Africa, northern Africa, southwest Asia, Europe, central Asia, and southern Asia, it is notable that Nubian Levallois technology has been discovered in all of these areas except, perhaps, Europe and central Africa. These are areas that are respectively small (Europe) and barely explored by archaeologists (Central Africa), so their absence may not be particularly meaningful, and I imagine that even in areas like Europe there might be artefacts entirely consistent with Nubian Levallois technology once looked at closely, but described using different names (see e.g. Guichard and Guichard 1965, p. 99). This sporadic occurrence of Nubian Levallois technology across space offers a very strong basis for convergent evolution.

While space is hard to deny, time can be blurred. The difficulties in chronometrically dating ancient Pleistocene materials adds a level of ambiguity that can be used to proponents of a particular archaeological model. Analogously to the Nubian debate, Marks (1992), for example, has long advocated the idea of the "time transgressive early Levantine Mousterian" (see also Rose and Marks 2014). In this model a population survived in a desert for nearly two hundred thousand years, at the nexus of Africa and Eurasia, and carried on making the 'same' artefacts for all of this time. This remarkable idea is based on a tiny number of dubious age estimates and some simple comparisons of lithic technology. Faced with chronometric evidence which contracted an interpretation of lithics, Marks' student Monigal (2002, p. 9), in a thesis excellent in many other regards, asked "shall we just leave prehistoric archaeology to the geophysists?". No, we should not. But neither should we ignore them. Likewise, with the purported distinction between Lower Nile Valley Complex people and Nubian Complex people (Van Peer 1998), the fact that chronometric age estimates subsequently falsified the population dichotomy has not led to meaningful revision, but instead simply to ever looser definitions. The last few decades have seen remarkable advances in our ability to accurately date Pleistocene archaeological sites. This should be celebrated and encouraged, and the temptation to deny new discoveries simply because they contradict previous ideas should be firmly resisted. There is nothing wrong with being 'wrong', that is how science works. The problem is when people refuse to change despite findings which contradict their claims.

In terms of Nubian Levallois technology, it seems simple enough to move beyond shouting the words 'Nubian Complex' and stating that because objects look similar they imply cultural transmission. On the one hand, comparative studies of various sorts can be carried out. Secondly, experiments can be conducted to explore the functional (i.e. 'pragmatic') aspects which may have driven the use of Nubian Levallois technology (see also Shea 1995 for a call to consider the 'ultimate' rather than 'proximate' causes of variability in Levallois systems). As Shipton (2020) shows, even simple 'actualistic' experiments can be very informative. Currently in lithic analysis we tend to have a rather simple dichotomy between 'stylistic' and 'functional' aspects, yet understanding this division is perhaps fundamental in exploring convergent evolution. If we think of 'functional' reasons why convergent evolution might happen these can be both in terms of how the lithic was used—such as a good way to haft a stone tool—but also in terms of knapping procedures. I have suggested that Nubian Levallois technology is a way to produce Levallois products with some particular characteristics—straightness, thickness and robusticity. This hypothesis can easily be tested. This argument is analogous to that made about 'deliberate overshooting' in the terminal Pleistocene. A fringe view has argued that the presence of this technically difficult trait either side of the Atlantic suggests that Solutrean people crossed the Atlantic (Stanford and Bradley 2012). However, those same scholars also argue that this technique was used because it is an efficient and effective way to thin bifaces. This is surely strong grounds for convergent evolution in two populations producing small and finely shaped bifacial objects, particularly when the technique was actually rarely used (Eren et al. 2013, 2014).

In my opinion, Nubian Levallois technology represents a minor shift from 'normal' centripetally prepared preferential Levallois technology—the interesting question is whether the use of the Nubian Levallois method was driven by the same impetus (e.g. perhaps to make straight Levallois flakes), or whether it was a convergent solution to several design problems which could be addressed by installing a median distal ridge on cores. I would argue that repeated invention (or accidental production, if invention seems too teleological) is why many sites have one or two Nubian Levallois cores. An interesting area for research is in understanding why a small number of sites feature a genuine focus on this Levallois reduction method (such as some sites in Dhofar where more than 90% of the cores are Nubian Levallois cores). My reading of the available data is that the notion of a trail of stone breadcrumbs (Rose et al. 2011) breaks down when one looks in detail, and we instead face a fascinating story of spatially and temporally disconnected groups independently coming to broadly similar technological solutions. Recognizing this involves transcending the simplistic allure of culture history, with all the challenges of recognizing convergent evolution that this involves. Previous reviews (Groucutt 2015a, c) have likewise suggested that there is no simple lithic 'smoking gun' for the dispersal of *Homo sapiens* out of Africa. It should be clear how far we are from having a robust chronostratigraphic framework for the low latitudes in the Middle and Late Pleistocene.

It would be satisfying if a particular retouched tool type (e.g. Rose 2004; Beyin 2006, p. 22; Van Peer and Vermeersch 2007, p. 192) or core reduction method (e.g. Rose et al. 2011) provided strong evidence for the dispersal of our species out of Africa. However, at present convergent evolution is a simpler and better explanation for the overall pattern of observed distributions of such technologies. The meaningful elucidation of these debates is only likely to come about from detailed comparative studies that use well-dated samples and seek to test and falsify hypotheses rather than to confirm what is already thought.

Acknowledgements I thank Dr. Eleanor Scerri, in particular, for extensive discussions on the Nubian Complex. I have also discussed these themes with many other colleagues, including Michael Petraglia, Manuel Will, Alex Mackay, Katja Douze, Jeffrey Rose, James Blinkhorn, Robin Dennell, John Shea, Metin Eren, Enza Spinapolice, and Christian Tryon. I thank them all for helping improve my understanding. I also thank the other contributors to this volume for allowing me to abuse my position as editor and considerably exceed the world limit! Funding was provided by the British Academy and the Max Planck Society during the writing phase, while my study benefited from visits to study lithic collections in several countries which were funded by the Arts and Humanities Research Council and the Boise Fund (University of Oxford).

References

Al-Abri, A., Podgorna, E., Rose, J. I., Pereira, L., Mulligan, C. J., Silva, N. C., et al. (2012). Pleistocene-Holocene boundary in Southern Arabia from the perspective of human mtDNA variation. *American Journal of Physical Anthropology, 149,* 291–298.

Armitage, S. J., Jasim, A. E., Marks, A. G., Parker, A. G., Usik, V. I., & Uerpmann, H.-P. (2011). The southern route "Out of Africa": Evidence for an early expansion of modern humans into Arabia. *Science, 331,* 453–456.

Bae, C. J., Douka, K., & Petraglia, M. D. (2017). On the origin of modern humans: Asian perspectives. *Science, 358,* 1269.

Belfer-Cohen, A., & Goring-Morris, A. N. (2009). The shift from the Middle Palaeolithic to the Upper Palaeolithic: Levantine perspectives. In: M. Campes, & C. Szmidt (Eds.), *The mediterranean from 50,000 to 25,000 BP: Turning points and new directions* (pp. 87-97). Oxbow: Oxford.

Bergey, C. M., Lopez, M., Harrison, G. F., Patin, E., Cohen, J. A., Quintana-Murci, L., et al. (2018). Polygenic adaptation and convergent evolution on growth and cardiac genetic pathways in African and Asian rainforest hunter-gatherers. *Proceedings of the National Academy of Science USA, 115,* E11256–E11263.

Beshkani, A., Beuzen-Waller, T., Bonilauri, S., & Gernex, G. (2017). The first evidence of Middle Palaeolithic Nubian technology in north-central Oman. *Antiquity, 91.* https://doi.org/10.15184/aqy.2017.4.

Beverly, E. J., Driese, S. G., Peppe, D. J., Johnson, C. R., Michel, L. A., Faith, J. T., et al. (2015). Recurrent spring-fed rivers in a Middle to Late Pleistocene semi-arid grassland: Implications for environments of early humans in the Lake Pleistocene Victoria Basin, Kenya. *Sedimentology, 62,* 1611–1635.

Beyin, A. (2006). The Bab al Mandab vs the Nile-Levant: An appraisal of the two dispersal routes for early modern humans out of Africa. *African Archaeological Review, 23,* 5–30.

Blegen, N., Tryon, C. A., Faith, J. T., Peppe, D. J., Beverly, E. J., Li, B., et al. (2015). Distal tephras of the eastern Lake Victoria Basin, equatorial East Africa: Correlations, chronology and a context for early modern humans. *Quaternary Science Reviews, 122,* 89–111.

Blegen, N., Jicha, B. R., & McBrearty, S. (2018). A new tephrochonology for early diverse stone tool technologies and long-distance raw material transport in the Middle to Late

Pleistocene Kapthurin Formation, East Africa. *Journal of Human Evolution, 121,* 75–103.

Blinkhorn, J., Achyuthan, H., Petraglia, M. D., & Ditchfield, P. (2013). Middle Palaeolithic occupation in the Thar Desert during the Upper Pleistocene: The signature of a modern human exit out of Africa? *Quaternary Science Reviews, 77,* 233–238.

Blinkhorn, J., Achuthan, H., & Ajithprasad, P. (2015). Middle Palaeolithic point technologies in the Thar Desert, India. *Quaternary International, 382,* 237–249.

Boëda, E., Bourguignon, L., & Griggo, C. (1998). Activités de subsistence au Paléolithique moyen: Couche V13 b' du gisement d'Umm El Tlel (Syrie). In J. P. Brugal, L. Meignen, & M. Patou-Mathis (Eds.), *Économie préhistorique*: Les strategies de subsistence au Paléolithique. 18e Rencontres Internationales d'Histoire et d'Archéologie d'Antibes (pp. 243–258). Sophia-Antipolis: APDCA.

Boivin, N., Fuller, D. Q., Dennell, R., Allaby, R., & Petraglia, M. D. (2013). Human dispersal across diverse environments of Asia during the Upper Pleistocene. *Quaternary International, 30,* 32–47.

Bordes, F. (1961). Mousterian cultures in France. *Science, 134,* 803–810.

Bordes, F. (1980). Le débitage Levallois et ses variants. *Bulletin de la Société Préhistorique Française, 77,* 45–49.

Bouté, P., & Rosen, S. (1989). Des gisemenst moustériens dans le Néguev central: résultats prélinaires de prospections. In O. Bar-Yosef & B. Vandermeersch (Eds.), *Investigations in South Levantine Prehistory/Prehistoire de Sud-Levant* (pp. 169–185). British Archaeological Reports 497, Oxford.

Brandt, S., Fisher, E. C., Hildebrand, E. A., Vogelsang, R., Ambrose, S. H., Lesur, J., et al. (2012). Early MIS 3 occupations of Mochena Borago rockshelter, southwest Ethiopian highlands: implications for Late Pleistocene archaeology, paleoenvironments and modern human dispersals. *Quaternary International, 274,* 38–54.

Brandt, S., Hildebrand, E., Vogelsang, R., Wolfhagen, J., & Wang, H. (2017). A new MIS 3 radiocarbon chronology for Mochena Borago Rockshelter SW Ethiopia: Implications for the interpretation of Late Pleistocene chronostratigraphy and human behaviour. *Journal of Archaeological Science: Reports, 11,* 352–369.

Brown, F. H., McDougall, L., & Fleagle, J. G. (2012). Correlation of the KHS tuff of the Kibish formation to volcanic ash layers at other sites, and the age of early *Homo sapiens* (Omo I and Omo II). *Journal of Human Evolution, 63,* 577–585.

Callander, J. (2004). Dorothy Garrod's excavations in the Late Mousterian of Shukbah Cave in Palestine reconsidered. *Proceedings of the Prehistoric Society, 70,* 207–231.

Caton Thompson, G. (1952). *Kharga oasis in prehistory*. London: Athlone Press.

Charpentier, V., Inizan, M.-L., & Féblot-Augustins, J. (2002). Fluting in the old world: The Neolithic projectile points of Arabia. *Lithic Technology, 27,* 39–46.

Chavaillon, N. (1973). L'Atérien de Hassi Ouchtat dans les monts d'Ougarta (Sahara nord-occidental). *Lybica, 21,* 91–138.

Chiotti, L., Dibble, H. L., Olszewski, S. R., & Schumans, U. (2009). Middle Palaeolithic lithic technology from the western high desert, Egypt. *Journal of Field Archaeology, 34,* 307–318.

Chmielewksi, W. (1968). Early and Middle Palaeolithic sites near Arkin, Sudan. In F. Wendorf (Ed.), *The prehistory of Nubia* (pp. 110–147). Dallas: Southern Methodist University Press.

Clark, G. A., Schuldenrein, J., Donaldson, M. L., Schwarcz, H. P., Rink, W. J., & Fish, S. K. (1997). Chronostratigraphic contexts of Middle Paleolithic horizons at the 'Ain Difla rockshelter (WHS 634), West-Central Jordan. In H. G. K. Gebel, G. Kafafi, & G. O. Rollefson (Eds.), *The Prehistory of Jordan II. Perspectives from 1997* (pp. 77–100). Studies in Early Near Eastern Production, Subsistence and Environment 4. Berlin, Ex oriente.

Clark, J. D. (1954). *The prehistoric cultures of the Horn of Africa*. Cambridge: Cambridge University Press.

Clark, J. D. (1988). The Middle Stone Age of East Africa and the beginnings of regional identity. *Journal of World Prehistory, 2,* 235–305.

Clark, J. D., Schultz, D. U., Kroll, E. M., Freedman, E. E. E., Galloway, A., Batkin, A., et al. (2008). The Ateran of Adrar Bous and the Central Sahara. In D. Gifford-Gonzalez (Ed.), *Adrar Bous, Archaeology of a Central Saharan Granitic Ring Complex in Niger* (pp. 91–163). Teruven: Royal Museum of Central Africa.

Crassard, R. (2009). The Middle Palaeolithic of Arabia: The view from the Hadramawt region, Yemen. In M. D. Petraglia & J. I. Rose (Eds.), *The evolution of human populations in Araiba: Palaeoenvironments, prehistory and genetics* (pp. 151–168). Springer: Dordrecht.

Crassard, R., & Thiébaut, C. (2011). Levallois points production from eastern Yemen and some comparisons with assemblages from East-Africa, Europe and the Levant. *ERAUL* 999, Liége, 1–14.

Crassard, R., & Hilbert, Y. H. (2013). A Nubian Complex site from Central Arabia: Implications for Levallois taxonomy and human dispersals during the Upper Pleistocene. *PLoS ONEi, 8,* e69221.

Crassard, R., & Khalidi, L. (2017). Dispersals, connectivity, and indigeneity in Arabian prehistory. In N. Bovin, R. Crassard, & M. D. Petraglia (Eds.), *Human dispersal and species movement: From prehistory to present* (pp. 219–236). Cambridge: Cambridge University Press.

Crassard, R., Hilbert, Y. H., Preusser, F., Wulf, G., & Schiettecatte, J. (2018). Middle Palaeolithic occupations in central Saudi Arabia during MIS 5 and MIS 7: New insights on the origins of the peopling of Arabia. *Archaeological and Anthropological Sciences*. https://doi.org/10.1007/s12520-018-0743-2.

Dibble, H. L., Aldeias, V., Jacobs, Z., Olszewski, D. I., Rezek, Z., Lin, S., et al. (2013). On the industrial attributions of the Aterian and the Mousterian of the Maghreb. *Journal of Human Evolution, 64,* 194–210.

Dibble, H. L., & Whittaker, J. (1981). New experimental evidence on the relation between percussion flaking and flake variation. *Journal of Archaeological Science, 8,* 283–296.

Douze, K., & Delagnes, A. (2016). The pattern of emergence of a Middle Stone Age tradition at Gademotta and Kulkuletti (Ethiopia) through convergent tool and point technologies. *Journal of Human Evolution, 91,* 93–121.

Douze, K., Igreja, M., Rots, V., Cnuts, D., & Porraz, G. (2020). Technology and function of Middle Stone Age points. Insights from a combined approach at Bushman Rock Shelter, South Africa. In H. Groucutt (Ed.), *Culture history and convergent evolution: Can we detect populations in prehistory?* (pp. 1–12). Cham, Switzerland: Springer.

Eren, M. I., Patten, R. J., O'Brien, M. J., & Meltzer, D. J. (2013). Refuting the technological cornerstone of the Ice-Age Atlantic crossing hypothesis. *Journal of Archaeological Science, 40,* 2934–2941.

Eren, M. I., Patten, R. J., O'Brien, M. J., & Meltzer, D. (2014). More on the rumor of "Intentional overshot flaking" and the purported ice-age Atlantic crossing. *Lithic Technology, 39,* 55–63.

Garrod, D. A. E. (1942). Excavations at the Cave of Shukbah, Palestine, 1928. *Proceedings of the Prehistoric Society, 8,* 1–20.

Goder Goldberger, M., Gubenko, N., & Hovers, E. (2016). "Diffusion with modifications": Nubian assemblages in the central Negev highlands of Israel and their implications for Middle Paleolithic inter-regional interactions. *Quaternary International, 408,* 121–139.

Goder Goldberger, M., Ginat, H., Ragolski, G., Seri, G., & Abadi, I. (2017). Middle Palaeolithic find spots with Nubian cores from the southern Negev and the Arava, Israel. *Journal of Lithic Studies, 4*. https://doi.org/10.2218/jls.v4i1.1688.

Groucutt, H. S. (2014). Middle Palaeolithic point technology, with a focus on the site of Tor Faraj (Jordan, MIS 3). *Quaternary International, 350,* 205–226.

Groucutt, H. S., & Petraglia, M. D. (2012). The prehistory of the Arabian Peninsula: Deserts, dispersals and demography. *Evolutionary Anthropology, 21,* 113–125.

Groucutt, H. S., Petraglia, M. D., Bailey, G., Scerri, E. M. L., Parton, A., et al. (2015a). Rethinking the dispersal of *Homo sapiens* out of Africa. *Evolutionary Anthropology, 24,* 149–164.

Groucutt, H. S., White, T. S., Clark-Balzan, L., Parton, A., Crassard, R., et al. (2015b). Human occupation of the Arabian Empty Quarter during MIS 5: Evidence from Mundafan Al-Buhayrah, Saudi Arabia. *Quaternary Science Reviews, 119,* 116–135.

Groucutt, H. S., Scerri, E. M. L., Lewis, L., Clark-Balzan, L., & Blinkhorn, J. (2015c). Stone tool assemblages and models for the dispersal of *Homo sapiens* out of Africa. *Quaternary International, 382,* 8–30.

Groucutt, H. S., Breeze, P., Drake, N. A., Jennings, R., Parton, A., White, T., et al. (2016). The Middle Palaeolithic of the Nejd, Saudi Arabia. *Journal of Field Archaeology, 41,* 131–147.

Groucutt, H. S., Grün, R., Zalmout, I. S. A., Drake, N. A., Armitage, S. J., & Candy, I. (2018). *Homo sapiens* in Arabia by 85,000 years ago. *Nature Ecology and Evolution, 2,* 800–809.

Groucutt, H. S., Scerri, E. M. L., Stringer, C., & Petraglia, M. D. (2019, in press). Skhul lithic technology and the dispersal of *Homo sapiens* into Southwest Asia. *Quaternary International.* https://doi.org/10.1016/j.quaint.2017.12.027.

Grove, M., Lamb, H., Roberts, H., Davies, S., Marshall, M., et al. (2015). Climatic variability, plasticity, and dispersal: A case study from Lake Tana, Ethiopia. *Journal of Human Evolution, 87,* 32–47.

Grün, R., Stringer, C., McDermott, F., Nathan, R., Porat, N., Robertson, S., et al. (2005). U-series and ESR analyses of bone and teeth relating to the human burials from Skhul. *Journal of Human Evolution, 49,* 316–334.

Guichard, J., & Guichard, G. (1965). The Early and Middle Palaeolithic of Nubia: Preliminary results. In F. Wendorf (Ed.), *Contributions to the prehistory of Nubia* (pp. 57–116). Dallas: Southern Methodist University Press.

Guichard, J., & Guichard, G. (1968). Contributions to the study of the Early and Middle Palaeolithic of Nubia. In F. Wendorf (Ed.), *The prehistory of Nubia* (Vol. 1, pp. 148–193). Dallas: Southern Methodist University Press.

Hallinan, E., & Shaw, M. (2015). A new Middle Stone Age industry in Tanwka Karoo, Northern Cape Province, South Africa. *Antiquity Project Galley, 89,* 344.

Hawkins, A. (2012). The Aterian of the oases of the Western Desert of Egypt: Adaptation to changing climatic conditions? In J. J. Hublin & S. McPherron (Eds.), *Modern origins: A North African perspective* (pp. 157–175). Springer: Dordrecht.

Hawkins, A., Smith, J. R., Giegengack, R., McDonald, M. M. A., Kleindienst, M. R., & Schwarcz, H. P. (2001). New research on the prehistory of the escarpment in Kharga Oasis. *Nyame Akuma, 55,* 8–14.

Hilbert, Y. H., Crassard, R., Rose, J. I., Geiling, J. M., & Usik, V. I. (2016). Technological homogeneity within the Arabian Nubian Complex: Comparing chert and quartzite assemblages from central and southern Arabia. *Journal of Lithic Studies, 3.* https://doi.org/10.2218/jls.v3i2.1420.

Hilbert, Y. H., Crassard, R., Charlous, G., & Loreto, R. (2017). Nubian technology in northern Arabia: Impact on interregional variability of Middle Paleolithic industries. *Quaternary International, 435,* 77–93.

Hovers, E. (2009). *The lithic assemblages of Qafzeh Cave.* New York: Oxford University Press.

Hovers, E., Buller, A., Ekshtain, R., Oron, M., & Yeshurun, R. (2008). Ein Qashish—A new Middle Palaeolithic open-air site in northern Israel. *Journal of the Israel Prehistoric Society, 38,* 7–40.

Hussain, S. T., Richter, J., Schyle, D., Kindermann, K., Wolter, T., & Hauck, T. C. (2015). The Veiled Mousterian: Traces of Middle Palaeolithic presence in the Wadi Sabra. In D. Schyle & J. Richter (Eds.), *Pleistocene archaeology of the Petra area in Jordan* (pp. 55–85). Rahden: Marie Leidorf Verlag.

Inizan, M. L., & Ortlieb, L. (1987). Préhistoire dans la region de Shabwa au Yémen du sud (R.D.P. Yémen). *Paléorient, 13,* 5–22.

Keith, A. (1931). *New discoveries relating to the antiquity of man.* London: Williams and Norgate.

Klasen, N., Henselowsky, F., Van Peer, P., Bubezner, O., & Kindermann, K. (2018). Chronology of the Middle Stone Age sequence at Sodmein Cave (Egypt) and palaeoenvironmental implications. *Geophysical Research Abstracts, 20,* EGU2018-18167.

Kleindienst, M. R. (2006). On naming things: Behavioural changes in the Later Middle to Earlier Late Pleistocene, viewed from the eastern Sahara. In E. Hovers & S. Kuhn (Eds.), *Transitions before the transition: Evolution and stability in the Middle Paleolithic and Middle Stone Age* (pp. 13–28). New York: Springer.

Kurashina, H. (1978). *An examination of prehistoric lithic technology in east-central Ethiopia.* Ph.D. Dissertation, University of California, Berkeley.

Marks, A. E. (1968a). The Mousterian industries of Nubia. In F. Wendorf (Ed.), *The prehistory of Nubia* (Vol. 1, pp. 193–314). Dallas: Southern Methodist University Press.

Marks, A. E. (1968b). The Khormusan: An Upper Pleistocene industry in Nubia. In F. Wendorf (Ed.), *The prehistory of Nubia* (Vol. 1, pp. 315–391). Dallas: Southern Methodist University Press.

Marks, A. E. (1992). Upper Pleistocene archaeology and the origins of modern man: A view from the Levant and adjacent areas. In: T. Akazawa, K. Aoki, & T. Kimura (Eds.), *The evolution and dispersal of modern humans in Asia* (pp. 229–251). Tokyo: Hokusen-Sha, Tokyo.

Marks, A. E. (1983). The sites of Boker Tachtit and Boker Tachti: A brief introduction. In A. E. Marks (Ed.), *Prehistory and paleoenvironments in the Central Negev, Israel. Vol. 3: The Avdat/Aqev area* (pp. 15—37). Dallas; Southern Methodist University Press.

Martini, M., Sibilia, E., Zelaschi, C., Troja, S. O., Fozese, R., Gueli, A. M., et al. (1998). TL and OSL dating of fossil dune sand in the Uan Afuda and Uan Tabu sites (Tadrart Acacus, Libyan Sahara). In M. Cremaschi & S. di Lernia (Eds.), *Wadi Teshuinat. Palaeoenvironment and prehistory in Southwestern Fezzan (Libyan Sahara)* (pp. 67–72). Roma-Milano: CNR.

Masojć, M. (2010). First note on the discovery of a stratified Palaeolithic site from the Bayuda Desert (N-Sudan) within MAG concession. *Der Antike Sudan. Mitteilungen der Sudanarchäologischen Gesellschaft zu Berlin e.V.* 21, 63–70.

Masojć, M., Kusiak, J., Standzikowski, K., Paner, H., Kuc, M., Parafiniuk, M., et al. (2017). OSL/IRSL estimation for Nubian Complex Middle Stone Age settlement from Bayuda Desert in Sudan. *Journal of Archaeological Science Reports, 16,* 391–296.

Masojć, M., Nassr, A., Kim, J. Y., Krupa-Kurzynowska, J., Sohn, Y. K., Szmit, M., et al. (2019). Saharan Green corridors and Middle Pleistocene hominin dispersals across the Eastern Desert, Sudan. *Journal of Human Evolution, 130,* 141–150.

McBrearty, S., & Brooks, A. S. (2000). The revolution that wasn't: A new interpretation of the origin of modern human behaviour. *Journal of Human Evolution, 39,* 453–563.

McClure, H. A. (1994). A new Arabian stone tool assemblage and notes on the Aterian industry of North Africa. *Arabian Archaeology and Epigraphy, 5,* 1–6.

McGhee, G. (2011). *Convergent evolution: Limited forms most beautiful*. Cambridge, MA: MIT Press.

Meignen, L. (2012). Levantine perspectives on the Middle to Upper Paleolithic "Transition". *Archaeology, Ethnology and Anthropology of Eurasia, 40,* 12–21.

Mellars, P., Gori, K. C., Carr, M., Soares, P. A., & Richards, M. B. (2013). Genetic and archaeological perspectives on the initial modern human colonization of southern Asia. *Proceedings of the National Academy of Sciences USA, 110,* 10699–10704.

Mercier, N., Valladas, H., Bar-Yosef, P., Vandermeersch, B., Stringer, C., & Joron, J. L. (1993). Thermoluminescence date for the Mousterian burial site of Es Skhul, Mt. Carmel. *Journal of Archaeology Science, 30,* 169–174.

Monigal, K. (2002). *The Levantine Leptolithic: Blade production from the Lower Paleolithic to the dawn of the Upper Paleolithic*. Ph.D. Dissertation, Southern Methodist University, Dallas.

Monnier, G. F., & Missal, K. (2014). Another Mousterian debate? Bordian facies, chaîne opératoire technocomplexes, and patterns of lithic variability in the western European Middle and Upper Pleistocene. *Quaternary International, 350,* 59–83.

Montet, A. (1957). Les industries levalloisiennes d'Heliopolis ef d'Abu-Suwair (Egypt). *Bulletin de la Societé préhistorique de France, 54,* 329–339.

Morgan, L. E., & Renne, P. R. (2008). Diachronous dawn of Africa's Middle Stone Age: New ^{40}Ar/^{39}Ar ages from the Ethiopian rift. *Geology, 36,* 967–970.

Munday, F. C. (1976). Intrasite variability in the Mousterian occupation of the Avdat/Aqev area. In. A. E. Marks (Ed.), *Prehistory and paleoenvironments of the Central Negev, Israel, Vol. 1, The Avdat/Aqev area, Part 1* (pp. 57–68). Dallas: Southern Methodist University Press.

Mussi, M., Altamura, F., Macchiarelli, R., Melis, R. T., & Spinapolice, E. E. (2014). Garba III (Melka Kunture, Ethiopia): A MSA site with archaic *Homo sapiens* remains revisited. *Quaternary International, 343,* 28–39.

Nickerson, R. S. (1998). Confirmation Bias: A ubiquitous phenomenon in many guises. *Review of General Psychology, 2,* 175–200.

Nishiaki, Y. (2018). Initial Upper Paleolithic elements of the Keoue Cave, Lebanon. In: Y. Nishiaki, & T. Akazawa (Eds.), *The Middle and Upper Paleolithic archeology of the Levant and beyond* (pp. 71–86). Singapore: Springer.

O'Brien, M., Buchanan, B., & Eren, M. I. (Eds.). (2018). *Convergent evolution in stone-tool technology*. Cambridge, MA: MIT Press.

Olszewski, D. I., Dibble, H. L., McPherron, S. P., Schurmans, U. A., Chiotti, L., & Smith, J. R. (2010). Nubian Complex strategies in the Egyptian High Desert. *Journal of Human Evolution, 59,* 188–201.

Osypiński, P., & Osypiński, M. (2016). Optimal adjustment of cultural backwardness? New data on the Levallois industries in the Nile Valley. *Quaternary International, 408,* 90–105.

Pasty, J. F. (1999). *Contribution à l'étude de l'Atérien du nord mauritanien*. BAR International Series 758. Archaeopress, Oxford.

Petraglia, M. D., & Rose, J. I. (Eds.). (2009). *The evolution of human populations in Arabia: Palaeoenvironments, prehistory and genetics*. Dordrecht: Springer.

Petraglia, M. D., Alsharekh, A., Breeze, P., Clarkson, C., Crassard, R., Drake, N. A., et al. (2012). Hominin dispersal into the Nefud Desert and Middle Palaeolithic settlement along the Jubbah palaeolake, northern Arabia. *PLoS ONE, 7,* e49840.

Richter, D., Moser, J., Nami, M., Eiwanger, J., & Mikdad, A. (2010). New chronometric data from Ifri n'Ammar (Morocco) and the chronostratigraphy of the Middle Palaeolithic in the western Maghreb. *Journal of Human Evolution, 59,* 672–9.

Ronen, A. (1974). *Tirat-Carmel—A Mousterian open-air site in Israel*. Institute of Archaeology, Tel Aviv University Publication No. 3, Tel Aviv.

Rose, J. I. (2004). The question of Upper Pleistocene connections between East Africa and South Arabia. *Current Anthropology, 77,* 671–684.

Rose, J. I., & Marks, A. E. (2014). "Out of Arabia" and the Middle-Upper Palaeolithic transition in southern Levant. *Quartär, 61,* 49–85.

Rose, J. I., Usik, V. I., Marks, A. E., Hilbert, Y. H., Galetti, C. S., Parton, A., et al. (2011). The Nubian Complex of Dhofar, Oman: A Middle Stone Age industry in southern Arabia. *PLoS ONE, 6,* e28239.

Rose, J. I., Hilbert, Y. H., Marks, A. E., & Usik, V. I. (2018). *The first peoples of Oman: Palaeolithic archaeology of the Nejd plateau*. Muscat: Ministry of Heritage and Culture.

Rots, V., Van Peer, P., & Vermeersch, P. M. (2011). Aspects of tool production, use, and hafting in Palaeolithic assemblages from Northeast Africa. *Journal of Human Evolution, 60,* 637–664.

Rust, A. (1950). *Die Höhlenfunde von Yabrud*. Neumünster: Karl Wachholtz Verlag.

Samson, C. G. (1968). *The Middle Stone Age industries of the Orange River scheme area*. Bloemfontein: National Museum.

Scally, A., & Durbin, R. (2012). Revising the human mutation rate: Implications for understanding human evolution. *Nature Reviews Genetics, 13,* 745–753.

Scerri, E. M. L. (2012). A new stone tool assemblage revisited: Reconsidering the 'Aterian' in Arabia. *Proceedings of the Seminar for Arabian Studies, 42,* 357–370.

Scerri, E. M. L. (2017). The North African Middle Stone Age and its place in recent human evolution. *Evolutionary Anthropology, 26,* 119–135.

Scerri, E. M. L., Drake, N. A., Jennings, R., & Groucutt, H. S. (2014). Earliest evidence for the structure of *Homo sapiens* populations in Africa. *Quaternary Science Reviews, 101,* 207–216.

Scerri, E. M. L., Breeze, P., Parton, A., Groucutt, H. S., Stimpson, C., Clark-Balzan, L., et al. (2015). Landscape archives and Middle to Late Pleistocene human habitation in the Nefud Desert, Saudi Arabia. *Quaternary International, 382,* 200–214.

Schiettecatte, J., Al-Ghazzi, A., Charlous, G., Crassard, R., Hilbert, Y., Monchot, H., et al. (2013). The oasis of al-Kharj through time: First results of archaeological fieldwork in the province of Riyadh (Saudi Arabia). *Proceedings of the Seminar for Arabian Studies, 43,* 285–308.

Schild, R. (1998). Comment on 'The Nile Corridor and the Out of Africa Model' by P. Van Peer. *Current Anthropology, 39,* S134–S135.

Schmidt, C., Kindermann, K., Van Peer, P., & Bubenzer, O. (2015). Multi-emission luminescence dating of heated chert from the Middle Stone Age sequence at Sodmein Cave (Red Sea Mountains, Egypt). *Journal of Archaeological Science, 63,* 94–103.

Scerri, E. M. L., & Spinapolice, E. E. A. (2019). Lithics of the North African Middle Stone Age. *Journal of Anthropological Sciences, 97,* 1–36.

Seligman, C. G. (1921). The older Palaeolithic age in Egypt. *Journal of the Royal Anthropological Institute of Great Britain and Ireland, 51,* 115–153.

Shea, J. J. (1995). Behavioural factors affecting the production of Levallois points. In H. Dibbl & O. Bar Yosef (Eds.), *The definition and interpretation of Levallois technology* (pp. 279–292). Madison: Prehistory Press.

Shea, J. J. (2014). Sink the Mousterian? Named stone tool industries (NASTIES) as obstacles to investigating hominin evolutionary relationships in the Later Middle Palaeolithic Levant. *Quaternary International, 350,* 169–179.

Shipton, C. (2020). The unity of Acheulean culture. In H. Groucutt (Ed.), *Culture history and convergent evolution: Can we detect populations in prehistory?* (pp. 13–27). Cham, Switzerland: Springer.

Smith, J. R., Giegengack, R., Schwarcz, H. P., McDonald, M. M. A., Kleindienst, M. R., Hawkins, A. L., et al. (2004). A reconstruction of Quaternary pluvial environments and human occupations using stratigraphy and geochronology of fossil-spring tufas, Kharga Oasis, Egypt. *Geoarchaeology, 19,* 407–439.

Smith, J. R., Hawkins, A. L., Asmerom, Y., Polyak, V., & Giegengack, R. (2007). New age constraints on the Middle Stone Age occupations of Kharga Oasis, Western Desert, Egypt. *Journal of Human Evolution, 52,* 690–701.

Stanford, D., & Bradley, B. (2012). *Across Atlantic ice: The origin of America's Clovis culture.* Berkley: University of California Press.

Taylor, N. (2016). Across rainforests and woodlands: A systematic reappraisal of the Lupemban Middle Stone Age in Central Africa. In S. C. Jones & B. A. Stewart (Eds.), *Africa from MIS 6 to 2: Population dynamics and palaeoenvironments* (pp. 273–300). Springer.

Tostevin, G. B. (2012). *Seeing Lithics: A Middle-Range theory for testing cultural transmission in the Pleistocene.* Oxford: Oxbow Books.

Tryon, C. A., Faith, J. T., Peppe, D. J., Fox, D. L., McNulty, K. P., Jenkins, K., et al. (2012). The Pleistocene archaeology and environments of the Wasiriya Beds, Rusinga Island, Kenya. *Journal of Human Evolution, 59,* 657–671.

Tryon, C. A., & Ranhorn, K. L. (2020). Raw material and regionalization in Stone Age East Africa. In H. Groucutt (Ed.), *Culture history and convergent evolution: Can we detect populations in prehistory?* (pp. 143–156). Cham, Switzerland: Springer.

Usik, V. I., Rose, J. I., Hilbert, Y. H., Van Peer, P., & Marks, A. E. (2013). Nubian Complex reduction strategies in Dhofar in Dhofar, southern Oman. *Quaternary International, 300,* 244–266.

Van Peer, P. (1986). Présence de la technique Nubienne dans l'Atérien. *L'Anthropologie, 90,* 321–324.

Van Peer, P. (1991). New perspectives about the Nile Valley Middle Palaeolithic: Safaha method and lateralization of Levallois flakes. *Paléorient, 17,* 135–142.

Van Peer, P. (1992). *The Levallois reduction strategy.* Monographs in world prehistory, 13. Madison: Prehistory Press.

Van Peer, P. (1998). The Nile corridor and the out of Africa model: An examination of the archaeological record. *Current Anthropology, 39,* 115–140.

Van Peer, P. (2004). Did Middle Stone Age moderns of sub-Saharan African descent trigger an Upper Paleolithic revolution in the Lower Nile Valley? *Anthropologie, 42,* 215–226.

Van Peer, P. (2016). Technological systems, population dynamics, and historical process in the MSA of northern Africa. In S. C. Jones & B. A. Stewart (Eds.), *Africa from MIS 6-2: Population dynamics and paleoenvironments* (pp. 147–159). Dordrecht: Springer.

Van Peer, P., & Vermeersch, P. M. (2000). The Nubian Complex and the dispersal of modern humans in North Africa. In. L. Krzyzaniak, K. Kroeper, & M. Kobusiewicz (Eds.), *Recent research into the stone age of Northeastern Africa* (pp. 47–60). Poznan: Poznan Archaeological Museum.

Van Peer, P., & Vermeersch, P. M. (2007). The Place of Northeast Africa in the early history of modern humans: New data and interpretations of the Middle Stone Age. In P. Mellars, K. Boyle, O. Bar-Yosef, & C. Stringer (Eds.), *Rethinking the human revolution: New behavioural and biological perspectives on the origin of modern humans* (pp. 187–198). Cambridge: McDonald Institute for Archaeological Research.

Van Peer, P., Vermeersch, P., & Moeyersons, J. (1996a). Palaeolithic stratigraphy of Sodmein Cave (Red Sea Mountains, Egypt). *Geo-Eco-Trop, 20,* 61–71.

Van Peer, P., Vermeersch, P. M., Moeyersons, J., & Van Peer, W. (1996b). Palaeolithic sequence of Sodmein Cave, Red Sea Mountains, Egypt. In G. Pwiti & R. Soper (Eds.), *Aspects of African archaeology* (pp. 149–156). Harare: University of Zimbabwe publications.

Van Peer, P., Vermeersch, P. M., Paulissen, E., & Huyge, D. (2002). Palaeolithic chert quarrying at Nazlet Safaha. In P. M. Vemeersch (Ed.), *Palaeolithic chert quarrying sites in Upper and Middle Egypt* (pp. 139–210). Leuven: Leuven University Press.

Van Peer, P., Fullagar, R., Stokes, S., Bailey, R. M., Moeyersons, J., Steenhoudt, F., et al. (2003). The Early to Middle Stone Age transition and the emergence of modern human behaviour at site 8-B-11, Sai Island, Sudan. *Journal of Human Evolution, 45,* 187–193.

Van Peer, P., Rots, V., & Vroomans, J. M. (2004). A story of colourful diggers and grinders. The Sangoan and Lupemban at site 8-B-11, Sai Island. *Before Farming, 3,* 1.

Van Peer, P., Vermeersch, P. M., & Paulissen, E. (2010). *Chert quarrying, lithic technology and a modern human burial at the Palaeolihtic site of Taramsa 1, Upper Egypt.* Egyptian Prehistory Monographs, No. 5. Leuven: Leuven University Press.

Vemeersch, P. M. (Ed.). (2000). *Palaeolithic living sites in Upper and Middle Egypt.* Leuven: Leuven University Press.

Vermeersch, P. M. (2001). 'Out of Africa' from an Egyptian point of view. *Quaternary International, 75,* 103–112.

Vermeersh, P. S. (Ed.). (2002). *Palaeolithic quarrying sites in Upper and Middle Egypt.* Leuven: Leuven University Press.

Veermeesch, P. M., & Van Peer, P. (2002). Conclusions on the Palaeolithic chert extraction in Egypt. In P. M. Vermeersch (Ed.), *Palaeolithic quarrying sites in Upper and Middle Egypt* (pp. 353–360). Leuven: Leuven University Press.

Vermeersch, P. M., Paulissen, E., & Van Peer, P. (1990). Palaeolithic chert exploitation in the limestone stretch of the Egyptian Nile Valley. *African Archaeology Review, 8,* 77–102.

Vermeersch, P. M., Van Peer, P., & Paulissen, E. (2002). Middle Palaeolithic chert quarrying at Nazlet Kahter 1. In P. M. Vermeersch (Ed.), *Palaeolithic quarrying sites in Upper and Middle Egypt* (pp. 27–78). Leuven: Leuven University Press.

Vignard, M. E. (1930). Stations paléolithiques de la carrière d'Abou-el-Nour, près de Nag-Hamadi (Haute-Egypt). *Bulletin de la Société préhistorique française, 27,* 301–320.

Villa, P., Soriano, S., Teyssandier, N., & Wurz, S. (2010). The Howiesons Poort and MSA III at Klasies River main site, Cave 1A. *Journal of Archaeological Science, 37,* 630–655.

Volkman, P. (1989). *Boker Tachtit: The technological shift from the Middle to the Upper Palaeolithic in the Central Negev, Israel.* Ph. D. Dissertation. Southern Methodist University, Dallas.

Wendorf, F. (Ed.). (1968). *The prehistory of Nubia* (Vol. 1). Dallas: Southern Methodist University Press.

Wendorf, F., & Schild, R. (1974). *A Middle Stone Age sequence from the central Rift Valley, Ethiopia.* Warsaw: Polska Akademia Nauk.

Wendorf, F., Schild, R., Close, A., & Associates. (1993). *Egypt during the Last Interglacial: The Middle Palaeolithic of Bir Tarfawi and Bir Sahara East.* New York: Plenum Press.

Will, M., Mackay, A., & Philips, N. (2015). Implications of Nubian-like core reduction systems in southern Africa for the identification of early modern human dispersals. *PLoS ONE, 10,* e0131824.

Yellen, J., Brooks, A., Helgren, D., Tappen, M., Ambrose, S., Bonnefille, R., et al. (2005). The archaeology of Aduma: Middle Stone Age sites in the Awash Valley, Ethiopia. *Paleoanthropology, 10,* 25–100.

Chapter 5
Lithic Variability and Cultures in the East African Middle Stone Age

Enza Elena Spinapolice

Abstract Lithics are the most abundant archaeological evidence from the remote past, however the way they are used to reconstruct past human groups is often biased. The Middle Stone Age (MSA) is the lithic techno-complex linked to the emergence of *Homo sapiens* in Africa. However, there is no consensus in the scientific community about the significance of this lithic culture in terms of connections with particular human social groups nor its evolution. This paper focuses on the relation between lithic variability in the East African MSA and its meaning in terms of the structure of human groups, critical for interpreting the behavioral and evolutionary processes that led to *Homo sapiens* expansion within and out of Africa. Here I examine current knowledge and hypotheses and suggest some methodological advances to overcome the present difficulties.

Keywords Lithic technology • Middle Paleolithic • Africa • Paleolithic culture

Introduction

The Middle Stone Age (hereafter, MSA) has been central in debates in human evolutionary studies in recent decades, because of its connection with the emergence and spread of our species in Africa (White et al. 2003; Shea et al. 2007; Groucutt et al. 2015; Hublin et al. 2017; Stringer and Galway-Witham 2017; Brooks et al. 2018; Deino et al. 2018; Scerri et al. 2018). In fact, until now, all the fossils of early *Homo sapiens* are associated with MSA lithic industries, whose most ancient manifestation is approximately the same age as the oldest *Homo sapiens* fossils (Hublin et al. 2017; Brooks et al. 2018).

The MSA is a lithic industry spanning roughly from ~300 to ~30 thousand years ago (ka), initially conceived of as the counterpart, and sometimes used synonymously with the Middle Paleolithic (MP) of Eurasia, indicating initially, in chrono-stratigraphic terms, something following the Early Stone Age (ESA) and preceding the Later Stone Age (LSA) (Goodwin and van Riet Lowe 1929). The cultural and chronological definitions of the MSA have been the subject of much debate (for a complete review, see Douze 2011). The beginning of the MSA is generally identified by the progressive abandonment of bifaces (handaxes and cleavers) and the presence or the enhancement of the hierarchical core reduction strategies, the so-called Prepared Core Technologies (PCT, e.g. the Levallois method(s) for flake production) (Clark 1988).

It is still unclear, if the MSA is a single techno-complex or if it is the result of multiple technological traditions. As Clark (1988) noticed, in the MSA there are almost as many exceptions as conformities to the rules. Despite its importance in evolutionary terms, in fact, the study of the MSA presents serious ambiguities linked to: (i) the poor technological resolution of most studies; (ii) the large geographical and chronological span; (iii) the scarcity of well dated stratified contexts.

This general uncertainty about one of the main archaeological phases critical to our recent past, has a number of consequences affecting the quality of the models proposed to explain population dynamics (contraction, expansions, drift) in both biological and cultural terms, in this key period. Particularly, this paper focus on the relation between lithic variability in the East African MSA and its meaning in terms of human social groups, critical for interpreting the behavioral and evolutionary processes that led to *Homo sapiens* expansions within and out of Africa. In particular I analyze how and if the current knowledge of the archaeological record is able to detect meaningful social boundaries and

E. E. Spinapolice (✉)
Dipartimento di Scienze dell'Antichità, Sapienza Università di Roma, Piazzale A. Moro 5, 00185 Rome, Italy
e-mail: enzaelena.spinapolice@uniroma1.it

H. S. Groucutt (ed.), *Culture History and Convergent Evolution: Can We Detect Populations in Prehistory?*,
Vertebrate Paleobiology and Paleoanthropology, https://doi.org/10.1007/978-3-030-46126-3_5

specific human groups. In fact, it has been proposed that the MSA may correspond with the origin of regional differentiation, linked to a complex process of small-scale population fragmentation (Tryon et al. 2005). The development of regional identity is a fundamental part of the model about the MSA since the famous paper by Clark (1988), however the research about this diversity and the meaning of this identity is still lacking theoretical clarity.

One of the key open questions is whether the variability of the East African MSA is the result of different populations (defined by boundaries), or of the nature of archaeological investigation and its biases. To analyze this problem, we still have to question the anthropological meaning of the archaeologically defined "Paleolithic cultures" and test their significance in terms of human groups.

Lithics and Paleolithic Cultures

Particularly for the most ancient periods, lithics are often the only preserved data from a broader social system that produced them. One of the first questions to be addressed here is if and how lithics are expression of Paleolithic cultures.

Traditionally, archaeologists working with the Paleolithic archaeological record have relied on lithics (1) to define past "cultures", in a culture-historical perspective (i.e. Bordes 1961), and (2) to identify evolutionary trends (Foley and Lahr 1997). However, wherever ethnographic studies have been conducted on recent hunter-gatherers (e.g. Hayden 1979), they indicate that stone tools represent only a minimum portion of the technology used by the groups, and they do not necessarily reflect the complex suite of behaviors and social rules that characterize a past cultural adaption (d'Errico and Banks 2012), despite their utilization for cultural markers in the traditional culture-historical approach (Table 5.1).

Franz Boas (1938) defines culture as the totality of the relations and the activities characterizing the behavior of individuals composing a specific social group, including the products of these activities and the role they play in the life of different groups.

Archaeology relies on material culture, and material culture is an expression of a society. However, if the attribution to a group affiliation is possible for contemporaneous societies, through individual self-ascription to a group affiliation (Barth 1969), it is beyond the resolution of prehistoric archaeology (Tostevin 2012).

In general, in prehistory, we define cultural factors as elements that cannot be straightforwardly explained by practical factors, such as quantity, quality and availability of raw materials (see Tryon and Ranhorn 2020), site function, or mobility strategies. This sometimes goes under the name of "style" (Binford 1962; Binford and Binford 1968; Dunnell 1978), which includes the artefact variability not accounted for by other functional constraints (Tostevin 2012). Here, cultural traits, represent learned and shared behavior, which are acknowledged to be the landmark of a "culture".

In fact, hominins are 'culture-bearing organisms' (Foley 1985). Two factors are inherent in the "culture" concept: the capacity to transmit and receive information; and the associated aptitude to initiate, develop and change behavioral strategies, on a scale unknown in other species (Foley 1985), even if this gap between our species and the others is progressively decreasing (i.e. Whiten et al. 1999).

But how can we assess the meaning and the existence of "Paleolithic cultures"? One of the questions to be addressed is "Are we tracking cultures using the right theoretical tools?".

One of the tools we can use to build models is the capability of culture to define boundaries between one ethnic group and another (McElreath et al. 2003). Ethnic identities are given by the setting of social processes that resist a homogenizing effect, in a frame of a specific spatial structure.

Table 5.1 Short definition of the main terms used in the text

Cultural	Transmitted through social learning.
Culture	Ensemble of behaviors and ideas transmitted through social learning, identifying a social or ethnic group.
Ethnic Group	People who belong to the same social or human group and/or identify themselves as belonging to the same culture. It includes one or many social groups.
Human Group	Social or ethnic group from the past. Used here to keep it neutral with regard to ethnicity.
Social Group	Two or more people related by kinship (social bond based on common ancestry, marriage, or adoption) and/or other form of social cohesion.
Significant technological unit (STU)	Technological behavior that is culturally coded, here applied to lithics.
Techno-complex	Lithic industry belonging to a specific cultural tradition.
Tradition	Cultural phenomena transmitted within a social or ethnic group.

Wobst (1974) first assessed how style in material culture would show socially meaningful information, such as group affiliation or membership. Later, the active role of style has been questioned by Sackett (1982): he proposed an "*iso-chrestic*" (equivalent in use) model where the artisan's choices, conscious or not, regarding non-functional aspects of the artefacts, are dictated by the traditions pertaining to the social group, so the social group itself is socially bounded and consequently diagnostic of ethnicity. Wiessner (1983) has shown in ethnographic contexts that the use of a certain style is often ethic but not emic, thus unconscious: although most San artisans were not aware of making arrows whose style was indicative of their group, nevertheless they could definitely recognize their arrow among a group of arrows.

A further aspect concerns not only the meaning of the shape/morphology of the tool, but the way in which this was produced, going broadly under the term "technology" (sensu Leroi-Gourhan 1964). Technology is defined as the sequence of behaviors in the manufacture of artefacts, and it results in stylistic variation useful for culture-historical reconstruction. Technology is culturally oriented, thus two objects having the same style and the same functional properties could have been made using a different technology. The *chaîne opératoire* method allows one to regroup sets of specific gestures and relation of them to a specific "tradition", meaning by tradition a learned and established aspect of the culture (Mauss 1936; see also Maher and Macdonald 2020). In the view of *chaîne opératoire* theory, the social information of a specific society includes the knowledge necessary to perform the sequence of gestures necessary to execute a technical action. Thus, technology is the material manifestation of the society's cultural information. The gesture identified through the *chaîne opératoire* is directly connected to human social behavior. Consequently, the technology is significant as a phenomenon embedded in social action (Dobres and Hoffman 1994). The technology could fit as well in the definition of *habitus* by Bourdieu (1977), because it generates regular practices that, while not strictly determined by rules, are at the same time collectively structured. So, a different technological system could be related to a different cultural system, because the first is embedded in the second.

Finally, it has been observed that the more visible the attribute on the final artefact, the larger the inventory of possible social processes that could contribute to its variability (Tostevin 2012). This approach is often combined with the study of the "life history" of the tool (Bleed 1986, 2001; Shott 1996), connected with the "behavioral archaeology" (*sensu* Schiffer 1976), and to the "Organization of technology approach" (*sensu* Nelson 1991). Both these approaches are useful tools to detect the stylistic/cultural vs functional meaning of lithic attributes, and to investigate the characteristics of the tools that are inherent to their use and discard. Particularly important are the studies on reduction, reuse and recycling, showing that the shape in which a tool enters the archaeological record seldom reflects the shape of the same tool at the time it was made by the artisan.

In conclusion, a tool (e.g. a Gravettian point) should in general be representative of its time and place (Tostevin 2012). However, it is clear that most lithic tools that could correspond to a stylistic choice have both a chronological and geographical distribution that goes beyond any association with a specific hunter-gatherer group (e.g. Groucutt 2020). For example, the stylistic variation of arrow morphologies in a San language groups, studied by Wiessner (1983, 1984), identified groups of 1,500–2,000 persons, definitely larger than the assumed foraging band of 475 persons postulated by Wobst (1974). It must be noted, however, that the bands are fluid in their composition and their number can vary greatly, never reaching in any case the number expected by Wiessner. In general, without specific ethnographic referencing, the lithic distribution of a single tool often covers areas that are thousands of kilometers wide, impossible to superpose on the home range of any band-dimension society: tool adoption does not fit ethnic boundaries. Here a problem of time averaging also occurs because, since we cannot assert with certainty the distribution of a specific tool in a specific moment, but only in a chronological range, it is difficult to relate the geographical distribution within a narrow chronological time frame. This confusion opens the way for a certain number of simplifications that affect models for culture change and tradition in Paleolithic studies.

Lithics are then an indicative set of technical skills, knowledge and mental templates directly linked to the system that produced them, a system including social practice, symbolism and so on. We can then recognize traditions by the lithic record, and it is in tracking those specifically that maybe we can address some models for human populations.

One attempt to overcome the difficulties is to try to identify which technological and typological attributes, or set of attributes linked to specific technical behaviors, are socially meaningful. The first to relate attributes of lithics to social meaning was Carr, drawing from ethnographic data (Carr 1995). More recently, a unified (middle range) theory of artefact design was proposed by Tostevin, with the purpose to assign potential etic meanings to specific attributes of a specific class of artefacts (Tostevin 2012). The attributes should be linked with potential meanings, and with sub-attributes that are most likely to be relevant for the analyzed processes and social units.

Despite the fact that lithics are socially meaningful, however, prehistoric cultures as they are described and analyzed in the current studies, are not the expression of a single ethnic group. However, different groups of archaeological assemblages share cultural traits, and when they are not explicable by convergent evolution, are thus meaningful under the plan of culture boundaries.

Carla Sinopoli made an archaeological study of ethnographic arrows from Numic speaking groups in the American Southwest (1991): the study of 172 arrows from three different bands showed that the variables on the arrows were most distinctive between the geographically and linguistically closer groups. Eleanor Scerri and colleagues (2014), were able to combine attribute analysis on stone tools with paleoenvironmental data, showing that different population of tools were geographically connected and structured. Katja Douze (2014) positively identified the "tranchet blow" process as a meaningful chronological and cultural marker relative to the Early MSA at Gademotta.

It is only by combining the significant data from lithic attribute analysis, technological analyses, chronological data, spatial analysis and paleoenvironmental data, that it will be possible to identify meaningful social boundaries within the Paleolithic record. I will propose here to use notion of Significant Technological Units (STUs) to identify technological behaviors that can be isolated and tracked in order to relate them to specific cultural traditions.

Mechanisms of Culture Change

Traditionally, the mechanisms of culture change are identified in two main processes: "branching" and "blending" (Collard et al. 2006), or in other terms whether cultures develop by a tree-like splitting process (*phylogenesis*) or by admixture (*ethnogenesis*) (Nunn et al. 2010).

The **branching hypothesis** (phylogenesis) states that the general similarities in material culture between populations are primarily the result of within group transmission and population fissioning, in a (vertical) schema reproducing a phylogenetic tree. It has also been suggested that there are mechanisms of isolation that impede the transmission of cultural elements among contemporaneous communities by Transmission Isolating Mechanisms or TRIMS (Durham 1992).

The branching hypothesis has strong association with biological patterns, aiming to build a phylogenetic tree of related cultures: according to this hypothesis, the history of the diversity of human cultures will also be the history of human populations (Foley and Lahr 2011).

In a branching perspective, the mechanisms of culture change are described as: (1) Local adaptation; (2) Diffusion; (3) Replacement; (4) Migration; (5) Assimilation (Foley and Lahr 1997). Local adaptation can be either the result of drift or innovation.

The **blending hypothesis** (ethnogenesis) (Shennan and Collard 2005) refers to traditional "cultural diffusion" (as in Kroeber, i.e. 1949). Here cultural evolution occurs as a consequence of the borrowing of ideas and habits from contemporary societies, in a scheme of horizontal transmission. Since the beginning of the discipline, anthropology has used the concept of contact between cultures as an explanation of the cultural variation through time and space (Trigger 1996). The basis of this hypothesis is that there has always been a constant flow of ideas, goods, and cultural practices between one community to another, as much as with genes (Collard et al. 2006). This hypothesis correlates the frequency of the contact with the similar cultural patterns. Thus, different scholars state that blending is more significant than branching in human evolution (e.g. Dewar 1995; Moore 2001).

However, if this were the case, the difference within culture would be erased through time and at the present time there could be only one world culture. This is actually not the case, because the building and keeping of boundaries contributes to the big cultural diversity in *Homo sapiens*, that sharply contrasts with a relative biological uniformity, leading to the paradox of low biological diversity and high cultural diversity in modern humans (Foley and Lahr 2011).

The archaeological record itself is the proof of long enduring cultural traditions with recognizable cultural patterns lasting in space and time: the persistence of boundaries attests to social mechanisms that resist to homogenization (McElreath et al. 2003).

Furthermore, where the branching vs. blending hypotheses were tested, it was shown that the branching model is prevailing in cultural transmission (Guglielmino et al. 1995; Hewlett et al. 2002), where the blending effects are limited to trade and exchange. Archaeological inferences concerning mechanisms of cultural transmission should take into account how isolation by distance affects cultural diversity (Premo and Scholnick 2011; Scerri et al. 2014).

The greater the geographical proximity or connection of two populations the more similar two cultures are (Foley and Lahr 2011; Scerri et al. 2014) and this is likely the result of a combination of branching (direct cultural transmission) and blending (acculturation, contact, exchanges of goods and people): the way this happens is operationalized in "cultural transmission theory". Of course, neither branching nor blending alone can explain the immense variability of human cultures, and the phenomenon of convergent evolution also has to be taken into account.

Cultural Transmission Theory

Cultural transmission theory is useful for understanding the processes of transmission, modification, preservation and loss of learned behaviors, including the technical choices of artefact makers, in an evolutionary perspective (Premo and Hublin 2009; Premo and Kuhn 2010).

In cultural transmission theory, culture is defined as "information acquired by individuals from other conspecifics by teaching or imitation" (Boyd and Richerson 1988). Cultural transmission is assimilated both by mates and by people not genetically related, where the teacher is often a high-status individual. This transmission of information can thus be vertical (coming from parents), oblique (coming from other individuals in an older generation), or horizontal, from conspecifics of the same generation (Cavalli-Sforza and Feldman 1981). This generates non-adaptive cultural variants (Premo and Scholnick 2011) by innovation that can be socially fixed (i.e. transmitted), eventually by drift.

Cultural traditions are therefore the outcome of the way in which human groups reproduce themselves over generations (Foley and Lahr 2011), through social learning, defined as the transmission of all the non-genetic information from one individual to another (Galef and Laland 2005; Mesoudi 2016). Differently from genetic traits, cultural traits can be distinguished in many different ways, including their abandonment in favor of others (Foley and Lahr 2011).

Culture as a Biological Adaptation

The idea, then, that culture is a biological adaptation descends from the branching hypothesis, that has been shown to be the most effective explanation of the variation of cultural evolution and of actual human variability. Blending surely plays a role as a consequence of contacts and exchanges, but its impact over the long-term pattern of cultural evolution is limited.

In fact, there is a human selection of different cultural options, leading to cumulative cultural evolution, defined as the accumulation of beneficial modifications over successive generations (Dean et al. 2014; Mesoudi 2016); this is influenced by ecological factors, and its result is the creation and maintenance of boundaries between different communities.

To study the diversity of human cultures, over space and time, is also necessary to analyze *Homo sapiens* adaptations to different environments: in fact, our species peopled the totality of the Earth and multiplied the ways in which they adapted to environments, and the different levels of social complexity (Foley and Lahr 2011). In reconstructing past adaptations from the archaeological record, we are faced with the goal of tracking the implication of the adaptations over the material culture, thus in the archaeological record.

Particularly, when it comes to Paleolithic "cultures" we aim to understand what behavioral signatures are meaningful in terms of biological evolution:

> "*The history of the diversity of human cultures will also be the history of human populations as they have formed, moved and died out, and there will be a relationship between biological and cultural phenotypes*" (Foley and Lahr 2011).

How are biological and cultural traits connected? In Paleolithic archaeology, we have to start thinking about possible biological boundaries (i.e. different human species at the same time) associated with cultural ones, as well as significant ethnic boundaries, within *Homo sapiens*, recognizable from Paleolithic material culture.

Why is there no consistency between the biological and the archaeological records? Human populations responded to variable conditions both demographically and adaptively, engendering a complex series of changes (Lahr and Foley 2016). Different ecological circumstances promote different adaptive strategies, whether biological or cultural (Mirazon Lahr 2016). The behavioral signatures usually precede biological ones (Bateson 1988), and biological changes can be the consequences of behavioral changes (Mirazon Lahr and Foley 2001), as well as the biological changes also potentially creating behavioural change.

Transitions often are the result of the interaction between biological and cultural variation during population collapse and the subsequent loss of variation due to partial population extinction or assimilation (Mirazon Lahr 2016). Is culture merely tracking biological diversity, then?

There is another element to be taken into account, equally likely to occur in biological and cultural evolution: convergence/homoplasy. The issue of convergence is linked to independent change leading to a similar result, such as homoplasy in phylogeny; this is culturally linked to (re)invention. Convergence in cultural choices represents a common solution to limited problems, and could possibly be linked to innate mechanisms connected to brain functioning, related to the evolutionary significance of certain traits. In any case, the possibility to choose between different culturally oriented options is dominated by the primary brain functions and capabilities that are inherent to every human species, thus it has a biological signature.

Convergence is one of the big puzzling questions in the analysis of Paleolithic artefacts, and, together with branching and blending, it is one of the three hypotheses to be tested to assess similarities, contacts and descent within human groups in the Pleistocene.

Lithics and Cultures in East Africa

The multiple facets of MSA technology are currently the subject of intense investigation, and it is more and more clear that they are connected to ancestral populations likely more diverse than previously expected. First, the variability within the MSA likely includes the behavioral outcomes of multiple hominin populations and perhaps even species (Tryon and Faith 2013). The model of 'African Multiregionalism' (*sensu* Scerri 2018) helps depict a scenario that is much more complex than formerly thought, where the MSA is the result of multiple populations showing genetic and morphological differences. This model would fit with a multiple (ragged) origin of MSA, resulting in strong regional differences and a large variability overall.

Yet, most of the distinctive traits of MSA technology, such as the reliance on prepared core technology, originating as far as ~500 kya BP, are shared all over Africa. In this case, we could imagine one ancestral single population dating back to the lineage splitting from *Homo heidelbergensis* or which hominin species turns out to be ancestral to our own, leading to multiple facets and adaptations that finally were expressed into MSA.

Does the archaeological record then reflect this varied population history? Does the spatial distribution of artefacts types reflect the geographical range of specific populations? In fact, cultural change cannot be separated from its geographical and chronological dimensions (Mirazon Lahr 2016). Can we isolate human groups, in terms of populations or groups of populations, that are socially and biologically meaningful, on the basis of lithic technology?

It has been proposed that among early *Homo sapiens* populations significant behavioral novelties were associated with cognitive shifts, and thus biological evolution (Foley and Lahr 2011). The MSA origin may parallel the origin of regional differentiation, in a complex process of small-scale population fragmentation, isolation, expansion and replacement (Tryon et al. 2005; Scerri et al. 2018).

Foley and Lahr (2011) propose a model centered on East Africa, and emphasize five stages of the evolution of cultures from early *Homo sapiens* (*sensu* Bräuer 2008): (1) anatomical modernity and cultural continuity within the MSA; (2) African MSA regionalism; (3) diversification of human populations; (4) fragmentation linked to climate and environment; (5) post-Pleistocene complexity.

For the sake of this paper, I take into account the MSA context in East Africa in particular (Fig. 5.1), considering it as a chrono–cultural entity, in its original definition (Goodwin van Riet Lowe 1929; see Douze 2011 for a review), following the Acheulean and preceding the LSA. East Africa includes: South Sudan, Eritrea, Ethiopia, Djibouti, Somalia, Kenya, Uganda, Rwanda, Burundi, Tanzania.

The generalized neutral hypothesis concerning East Africa, implies that by ~200 kya BP, early *Homo sapiens* were the sole occupants of the region. This is the dominant model, mostly the outcome of current fossil and genetic evidence. However, the presence in the African continent of multiple human species at that time, should imply caution about this assumption.

The research questions regarding the MSA in East Africa involve: (i) the technological innovation developing in the archaeological record (e.g. prepared core technology, point production); (ii) the cognitive shift from the early/archaic *Homo sapiens* population to fully modern *Homo sapiens*; (iii) the expansion of the *Homo sapiens* population to eventually reach the rest of the continent and beyond.

MSA patterns can be interpreted as the gradual evolution of a variety of cultural adaptations in response to shifting regional, environmental and fluctuating demographic conditions (Kuhn 2013).

The beginning of the MSA is characterized by a number of technical innovations that follow the ESA in the archaeological sequences: (1) the (sometimes progressive, sometimes abrupt) abandonment of large cutting tools (LCT), (2) the enhanced reliance on prepared core technology (PCT), (3) blade/bladelet production; (4) the intense production and use of convergent tools. As we can see, the MSA innovations involve systems of both production (PCT, blade etc.) and use (convergent tools, microliths, etc., abandonment of handaxes) that have to be linked to a complex set of subsistence behavior. However, while those traits are incredibly stable over the early MSA, in sites often separated by thousands of kilometers and thousands of years, the modalities in which those innovations are managed and the rate of innovation and maintenance of ESA tradition change site by site. Furthermore, those innovations are not synchronous.

The differences between Eurasian MP and African MSA have been object of debate, however there are few comparative studies. Is the biological difference between *Homo sapiens* and *Homo neanderthalensis* uninfluential with regard to lithic production? Or, on the contrary are the MSA and MP more diverse then expected? After Kuhn (2013) the overall limited variability of the Middle Paleolithic is linked to the low necessity to signal identity and it is structurally different in the Eurasian MP and the African MSA. While in Europe this may be the indication of very small and dispersed groups, in the African MSA it could be the result of cumulative cultural evolution, more similar to the European Upper Paleolithic. Could this model be valuable also for the early MSA of East Africa?

It appears that in East Africa there is a persistence of some technological traits over space and time, showing no definite trend, until the explosion of what has been called the "beginning of social identity" (e.g. Wadley 2005; Scerri et al. 2014) with the large MSA variability, around MIS 4 but with different timing in the whole continent.

On another side, those peculiar traits could be stable because of an independent evolution from the ESA, leading to convergence. The phenomenon of drift and loss of peculiar technological innovation could be linked to the sparsity of populations (Kuhn 2013), and it has been

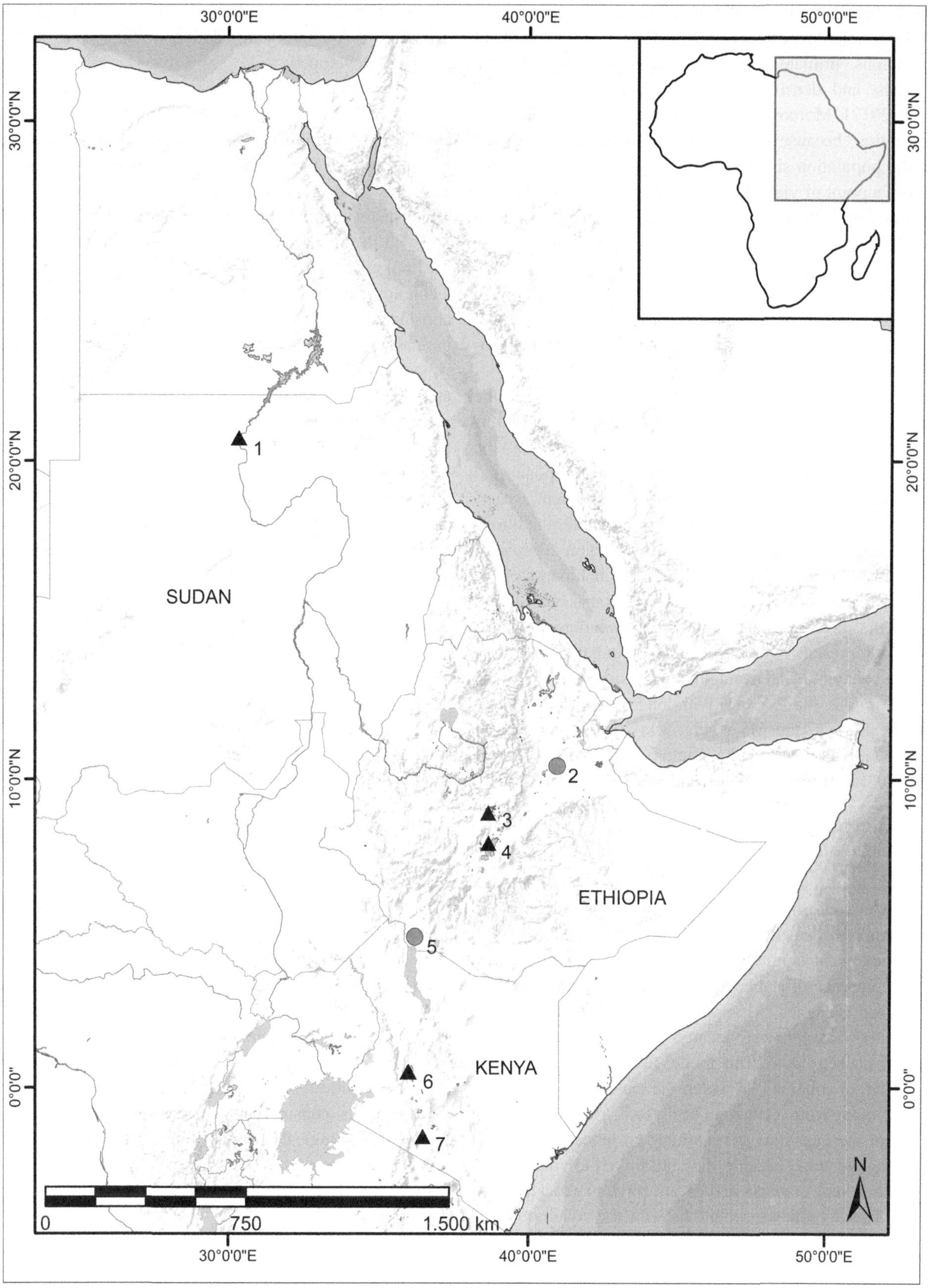

Fig. 5.1 Map of the MSA sites cited in the text. 1. Sai Island, 2. Herto, 3. Melka Kunture, 4. Gademotta, 5. Omo Kibish, 6. Kapthurin, 7. Olorgesailie

assessed that hominin population densities were low during the MSA, after an estimation based upon ethnographical comparisons, primate group sizes, environmental carrying capacities, and density of archaeological sites over time (Basell 2012). Moreover, small populations have low rates of invention, because the rate of novelty is directly linked with the population size (e.g. Kline and Boyd 2010).

From a point of view that combines cultural transmission theory and evolutionary ecology, an attempt can be made to relate lithic cultures in the East African MSA with population dynamics, expansion and isolation, within and outside East Africa. To achieve this goal, it would be necessary to build archaeologically appropriate theories to connect the research questions to the models in the (available) archaeological record (Binford 1977).

Although the MSA is often considered as a "package", the different characteristics are in fact asynchronous. It is thus important to analyze every technological aspect as the expression of a single behavior (single cultural component). It has already been shown that different technological aspects can evolve and stem independently. For example, the observations made in the Kapthurin Formation (Tryon 2006) suggest that two of the aspects considered the hallmarks of the MSA, (i) formal tools such as points (see Douze et al. 2020), and (ii) the means of flake production, including Levallois methods, represent two independent elements of hominin adaptive behavior, each having its own distinct development.

Furthermore, the few sites with long sequences spanning from the ESA to the MSA (Fig. 5.1) such as Sai Island, Gademotta, Kulkuletti, Melka Kunture, Kapthurin (Van Peer et al. 2003; McBrearty and Tryon 2006; Douze 2012; Mussi et al. 2013) in general do not show significant technological change through time and are characterized by a great variability (Clark 1988; Tryon and Faith 2013). In some sites there is a technological continuity (such as in Kapthurin); in others there are unconformities (e.g. Sai Island) that may indicate population replacement (Van Peer 2003; Tryon 2006).

Technological continuity within assemblages may indicate the presence of stable techno-cultural system over time (Douze and Delagnes 2016), while cultural diversity is more accentuated between sites located in geographically separated sites (Shea 2008). This model fits with the hypothesis of separate populations, in which the technological innovation, stemming from a common basis, took a separate course.

One of the questions to be asked is whether those traits are stable over time because they are linked to small populations that kept contacts and/or are phylogenetically connected. This hypothesis would fit with the assumption of small population sizes during the Pleistocene.

Another alternative hypothesis is that those traits are stable because they are originally linked to a single population that then split and occupied larger areas, keeping these technological traits stable. Phylogenetically the question to be asked then is whether those traits were present in the original MSA making population or they were only successively developed.

To disentangle this question, each and every single significant cultural trait shall be treated separately. It has been shown in specific cases that elements considered as the hallmark of MSA, such as methods for flake production (i.e. Levallois) and tools such the points, in reality have a different history of development (Tryon 2006).

I will in this paper introduce the definition of Significant Technological Unit (STU), as a behavioral package of technological traits identifiable in the archaeological record, considered independently of one another. STU is defined here as a technological behavior that is culturally coded, e.g. it needs to be invented, copied and reproduced, and to be archaeologically visible. Each and every STU in a specific archaeological context could be either the result of independent invention (convergent), direct transmission (branching) or cultural assimilation (blending).

I choose to examine here two different STUs, in order to discuss two technological methods: (i) the origin of Levallois technology, (ii) the origin of blade/bladelet technology; and one associated behaviour, (iii) the circulation of raw materials.

One of the "big questions" about the MSA, the production of points (see Douze et al. 2020), was intentionally left out here, for a number of reasons. Lithic point production is among the most ubiquitous cultural elements that encompass both African geography and the entire time span of the MSA. Points have been associated with MSA research since its inception; however, our knowledge about their production methods, functions and curation is still largely insufficient (Douze and Spinapolice 2016). Points are outside the aim of this paper principally because it is hard to assess how many STUs characterize point production, since sometimes they are obtained through convergent Levallois core reduction method, others from volumetric cores, and in other cases points (convergent tools) are shaped and/or retouched (Perlès 1974; Douze and Delagnes 2016). For example, Levallois point production consists at least in six interconnected steps (Leroi-Gourhan 1964) and can be considered at least a single STU. To disentangle this question would possibly require a separate work.

The STUs considered here, are, on another side, well known technological packages that are the basis for a large part of the lithic production in East African MSA: Levallois and blade technologies.

Origin of Levallois Technology

The development of Levallois methods is an aspect of lithic technological change that may provide clues to local patterns of innovation and replacement during the period of transition between ESA and MSA (Tryon et al. 2005).

The origin of Levallois is believed to be one of the main features of the MSA, as a "prepared core technology" (PCT). PCT is present in the ESA; however, the reliance of human groups on this production method in that period was poor, while in MSA contexts it becomes the most common way to produce blanks and persists all over the Late Pleistocene.

Following the definition of Boëda (1994), the Levallois method is characterized by the organization of two opposed surfaces, hierarchically patterned: the upper, dedicated to flake production, and the lower, to core preparation. The Levallois method variability includes two main forms of production: preferential and recurrent, and different flaking directions (i.e. unidirectional, convergent, centripetal).

The beginning of Levallois flaking is an event of particular importance that goes beyond lithic technology and may be an indication for the emergence of changes in hominin social, behavioral, and cognitive structures (Ambrose 2001), especially in the light of the long stasis that precede it, characterized by multiple faking systems (White and Ashton 2003).

Levallois technology is widespread in the old world, and the question on whether it comes from a single event or from a polycentric origin is a matter of debate (Rolland 1995). Among the first examples of Levallois production in Africa is the production of blanks to make cleavers (Tryon 2006), both in North Africa (Alimen and y Zuber 1978; Dauvois 1981) and in East Africa (Roche and Texier 1995).

However, there are different trajectories in the beginning and development of Levallois production strategies. Rolland (1995) identifies a dichotomy between Europe, where Levallois stems from biface production, and Africa, where it comes from successive variations of prepared cores. However, this interpretation is contradicted by some recent evidence from the European Mousterian (see for example Picin 2018).

The origin of Levallois technology has been also related to a single origin, linked to a population of archaic *Homo sapiens*, that successively spread into Eurasia (Foley and Lahr 1997). However, this single-origin hypothesis has been repeatedly challenged (see Adler et al. 2014 and references therein), and many scholars now believe in a multiple origin of Levallois technology (see Groucutt et al. 2015).

Another hypothesis states that the source of the Levallois method can be linked with handaxe production in Africa, directly evolving from existing Acheulean tradition (Biberson 1961; Dauvois 1976; Clark and Kurashina 1979). An alternative hypothesis claims that in South and East Africa the Levallois methods is possibly derived from the Victoria West cores, also called Protolevallois (Rolland 1995). However, Victoria West cores could as well be related to biface production.

The Levallois method has been classified into different sequences of production, mainly recurrent (continuous production of Levallois products) and preferential (sequence ending with the production of a preferential flake or point, Boëda 1994). It would be interesting to analyze the two methods as a separate STU, in order to identify possible trajectories of tradition and/or reinvention. Actually, there is no chronological or geographical trend in the use of recurrent vs preferential method, and both are commonly used in the same sites, often in the same assemblages, possibly to adapt to the goal of specific flake morphology, and to adapt to the shape and availability and quality of the raw materials.

In my opinion, the origin of the Levallois technology has profound cognitive and adaptive bases and consequences; however, it has to have occurred in the Middle Pleistocene, being already present in the Late Pleistocene in many sites in Africa and Eurasia, and thus has to be biologically correlated roughly with *Homo heidelbergensis* (see following paragraph).

Finally, the multiple facets linked to Levallois technology and the large variability of this method for flake production do not make this technological behavior suitable to delimit single population histories or to trace population directories within the setting of East African Middle Pleistocene. The abandonment of LCT production in favor of PCT indicates a shift in the technological strategies based on a previously acquired technology. It would be more useful, in terms of early *Homo sapiens* adaptation and behavior, to investigate the modalities and the causes for such a choice (e.g. raw material availability, environmental changes etc). The Levallois production method, being one of the hallmarks of the MSA, is therefore not suitable to answer the question: how were hominin populations structured in East Africa in the Late Pleistocene?

Origin of Blade and Bladelet Technology

Among the hierarchical core reduction strategies adopted in the MSA technological repertoire, blade and bladelet production plays an important role (e.g. for a review Bar Yosef and Kuhn 1999), because this production method has been traditionally linked to the European Upper Paleolithic "Revolution" (e.g. Mellars and Stringer 1989; Bar-Yosef 2002) and included in the hallmarks of "modern behavior". However, after the ground-breaking assumption that, from an African point of view, there was no Revolution (McBrearty and Brooks 2000), more and more evidence pushes the adoption of this strategy back in time, and it is clear now that if the systematic standardized production from prismatic cores broadly coincides with the Upper Paleolithic, the production of elongated blanks is part of the

MSA since its very beginning (Wilkins and Chazan 2012), predating the oldest currently known *Homo sapiens* fossils. After Herries (2011), the technology of blade production precedes PCT, and Levallois point production itself and these technological modifications coarsely correlate with the appearance of *Homo heidelbergensis* (Rightmire 2001). In fact, in East Africa, the earliest occurrence of non Levallois blade production is attested in the Kapthurin Formation and dated to 509 ± 9 ka (Johnson and McBrearty 2010).

The laminar technology provides evolutionary fitness, because it promotes the production of long cutting edges with a relative small technological investment (but see Eren et al. 2008). Furthermore, the rhythm of the blade production is continuous, leading to a complete reduction of the core, and the platform cores do not need a re-preparation of the surfaces as happens for Levallois cores. The continuity in the production is a characteristic shared by recurrent Levallois and blade production, while preferential Levallois requires a bigger investment of preparation and/or a discard of the core after the extraction of the preferential flake. An interpretation about the appearance of blade technology is that prior to the Upper Paleolithic, it appeared and disappeared, being linked to local adaptations and raw material availability (Wilkins and Chazan 2012).

In East Africa, the appearance of bladelets is particularly interesting. It has been proposed that the complex behavior linked to blade technology has to be shifted, in terms of efficiency, to the bladelet production, leading to the production of composite tools, and microliths (e.g. Eren et al. 2008). Bladelets in fact can be used as components of tools of greater complexity, such as composite tools, technologically more articulated than simple hafted tools (Ambrose 2001), involving a different design and an innovative set of strategies of production, use and maintenance (*sensu* Bleed 1986).

One of the most interesting aspects of bladelet technology is its relation with hafting. The evidence for hafted tools in the MSA and MP archaeological record is often discussed as a potential signature of behavioral complexity (so called "modern behavior") (Ambrose 2010; Barham 2013), involving a complex set of actions linking the tool, the joint and the haft. Hafting has also been interpreted as part of constructive memory, linked to specific cognitive abilities (Ambrose 2010; Wadley 2010). While there are some reservations on the importance on hafting in blade technology the use of bladelets and in general, microliths, is strictly linked with hafting methods. In fact, the use of adhesives appears later than the first appearance of blade technology, as in the Howiesons Poort technology in South Africa (Lombard 2006; Wadley 2010; Charrié-Duhaut et al. 2013).

One problem here is the classification of bladelets themselves, that is rather ambiguous. Bladelets are by definition smaller than blades, but their dimensional demarcation often overlaps with blades, and the quantitative definitions of blades versus bladelets differ substantially between researchers (Kaufman 1986). Quantitative descriptions of lengths and width/length ratios of artefacts can minimize the subjectivity; however, a universal definition of this boundary it rather difficult because it depends on raw material size and availability, mechanical properties, morphology of hafts and other factors (Ambrose 2002).

Despite those difficulties, a more detailed analysis of the appearance of bladelets in the archaeological record is noteworthy. Bladelets, unlike blades, appear to be a constant from the onset of the East African MSA, and could be the East African counterpart of the South African backed tools. One of the questions is whether microlithization is a mover and/or a consequence of the development of composite tool technology. Different elements contribute to considering bladelets as part of composite tools (Ambrose 2010): for example, microwear (Beyries 1988; Anderson-Gerfaud 1990), traces of mastic and red ochre (Boëda et al. 1996), and standardization of artefact size and shape (McBrearty and Brooks 2000).

The presence of bladelets (∼2–4 cm) and bladelet cores is constant in most of the assemblages from the early MSA in East Africa (contra Ambrose 2002): Gademotta (Douze 2012), Garba III (Spinapolice and Mussi in prep.), Omo Kibish (Shea 2008), Olorgesailie (Brooks et al. 2018). While these bladelets are not as standardized as their LSA/UP counterparts, still they are regular in shape and average dimensions. The question arises whether this invention is independent, thus created by convergence, or is a result of the cultural transmission of the same innovation. Is there any chronological or geographical trend in the adoption of bladelet technology in MSA? In light of recent discoveries, Olorgesailie seems to be one of the most ancient MSA sites so far discovered: the most recent report includes five localities, dating to ∼295–320 ka. Here all the characteristics of MSA are present, including prepared core technologies, and here blade and bladelet production seems to increase through time (Brooks et al. 2018). The same chronological trend has been analyzed by the author in Garba III. It is likely that the bladelets of early MSA constitute the first application of composite tools, later becoming the hallmark of the LSA, in East Africa and elsewhere (Leplongeon 2014).

Furthermore, the presence of Micro-Levallois flakes, in many of the same lithic assemblages where bladelets are present (Garba III, Gademotta, Omo Kibish), is another argument in the sense of an intentional microlithization of the assemblage, and this could be true either if the very small flakes (<2.5 cm) were the result of adaptation to raw material, or an independent technological choice (Spinapolice 2014, 2016).

Making a composite tool is a behavioral signature for planning and reliability (*sensu* Bleed 1986). It requires collecting and preparing several kinds of components and the assembling of different raw materials, which may be gathered at different times and in different places (Stout 2002). The final assembly of the functional artefact may occur much later, and some materials may be kept in reserve for maintenance and repair of composite tools. Composite-tool manufacture in the MP and MSA thus marks an increase in technological complexity compared with the single-component tools (Ambrose 2001, 2010).

The technological and cultural continuity of this tradition in East Africa is clear.

Composite-tool manufacture reflects a substantial advance in planning and hierarchical assembly of artefacts (Ambrose 2002). Bladelets have short use lives, and their use shall be coupled with a strategy for maintenance, in a system where possibly the haft is more technologically important than the tool itself. Traditionally, bladelets in UP have been associated with hunting strategies, and their presence fits well with the model of groups having complex social structure and interconnections. However, until functional analyses are applied, it cannot be excluded that bladelet were used also as simple cutting tools, as it happens for backed tools in South Africa (Igreja and Porraz 2013).

For those reasons, I believe that bladelet technology at the onset of East Africa MSA is a Significant Technological Unit that needs further investigation and has the potential to be linked to human evolution. Groucutt and colleagues (2015) argued that Levallois and blade technology evolved convergently and that there was a repeated and independent evolution of microlithic technology. However, until now, there has been no attempt for an evaluation of multiple versus single origins of bladelet technology. Further investigation and multivariate quantitative analyses could allow us to evaluate if this technological invention is suitable to test models about population contact and/or branching.

Raw Materials Transfer and Territories

The transfer of raw material over long distances has long been considered a mark of the "Upper Paleolithic" and later, of *Homo sapiens* behavior (e.g. Binford 1989, but see Spinapolice 2012). Distances from "site-to-source" (Tryon and Faith 2013) for lithic raw material provide one of the material estimates of the size of the social landscapes familiar to early hominin populations and it has long been applied for European Middle and Upper Paleolithic (Gramly 1980; Andresfky 1994; Kuhn 1995; Moncel 2004; Minichillo 2006; Féblot-Augustins 2009). Gamble (1998) considers modern humans to be associated with "extended social landscapes", defined by interaction networks that link diverse groups, occupying different areas.

The link between raw material transfer and cognitive abilities has been maintained until recent times. Ambrose (2010) considers both the passage to composite technologies and the transfer of raw material over long distance, from around 300 kya, a major shift in human cognition. Ambrose (2010), after the review of both European MP and African MSA Pleistocene hominin behavior, suggests that hominins optimize scheduling of land use developing enhanced long-term memories and understanding of seasonal environmental cues: the "culturally constructed niche".

If compared with ESA hominins, the groups making MSA artefacts in general used more frequently finer-grained rocks, particularly obsidian: there was a selection of the best raw material. The best studied lithic material is obsidian itself: from the MSA onward, obsidian is found in frequent use in almost all sites within a 50 km radius of major obsidian sources in the central Rift Valley (Merrick and Brown 1984) as well as in Ethiopia near major sources (Wendorf and Schild 1974; Muir and Hivernel 1976). Outside the immediate vicinity of the major central Rift sources, the frequency of obsidian use falls off (e.g. De Lumley et al. 2004; Tryon et al. 2005; Shea 2008); however, very small quantities of central Rift Valley obsidians are found up to 190 km from their sources (Merrick and Brown 1984; Blegen 2017; Blegen et al. 2018).

Nevertheless, despite the long tradition (Merrick and Brown 1984; Clark 1988) the geochemical characterization of the raw material sources in East Africa still covers very limited areas and focuses almost exclusively on volcanic rocks. MSA hominins regularly transported obsidian cores, flakes and tools over distances exceeding 30 km, such as in Porc Épic (Negash and Shackley 2006; Vogel et al. 2006), and sometimes exceeding 140 km, as in Songhor (McBrearty 1981), and Muguruk (McBrearty 1988) but sometimes the provisioning was mostly local, such as in Melka Kunture (Negash et al. 2006) and Gademotta/Kulkuletti (Shackley and Sahle 2017). As stated for the European MP, the difference in transported elements reflects a complex set of mobility and foraging strategies: provisioning of places vs. provisioning of individuals (sensu Kuhn 1994), or alternatively, a network of trade and exchange of tools and cores among proximity groups.

The evidence coming from recently investigated sites adds to this discussion. Recent data show that possibly the building of more complex social groups is evident since the very beginning of the MSA. Recently, the evidence from Olorgesailie pushed back in time the emergence of this behavior. According to the authors, the long-distance transport (25–50 km) of raw materials at this site suggested the existence of structured social networks among foragers at $\sim$300 Kya. In fact, exotic raw materials can indicate connections between individuals and groups occupying different territories. Raw materials can reach a site through a series of successive phases of reduction, passing hand to hand or travelling as a prepared core or tool in the hand of the same person or group. "*The distances over which exotic raw materials were obtained can be an indicator of human movement on the landscape and of inter-individual and inter-group contacts and social complexity*" (Brooks et al. 2018).

The element of raw material circulation is noteworthy because it has been suggested that the MSA is linked with an expansion into new habitats, an increased foraging range and broadened dietary basis (Tryon 2006).

Long distance raw material transport thus provides the archaeological evidence as far as ~300 Kya BP for the great extent of territories during the Pleistocene. Do these connections also imply the early structuring of *Homo sapiens* populations? The associated selection for fine grained raw material is one of the components of this behavioral package. However, it is hard to test this model. The greatest bias consists in the impossibility to test for home ranges of population territories where the sites are located in the proximity of very good raw materials sources, such as obsidian (e.g. Melka Kunture). While the long-distance raw material transfer is an indicator of large territories or of circulation of people or objects, the reliance on local raw material, especially where abundant and of good quality, is not necessarily a sign of small-scale territories or reduced social complexity. The diversity of the raw material spectrum in East Africa, the vastness of the region and the difference in biomes makes it really hard to assess anything before a specific analysis of local territories and regions. In fact, East Africa is characterized by a great variability in biomes, and it would be interesting to see the relation between the specificities of the different biomes with the different raw material transport distances. Furthermore, the variation in raw material selection and procurement can be analyzed following the changing of raw material availability over time, because climatic and/or catastrophic events can affect the procurement patterns. However, as has happens for the European MP, the analysis of raw material provisioning can add very important data to the discussion about mobility, and thus social structuring, and an attempt to further analyze this aspect in East African MSA would be very important.

Crossing the data from mobility and provisioning with the STU should be one of the goals to achieve in order to assess the structuring of different Late Pleistocene Populations in East Africa.

Towards an Understanding of MSA Human Groups

In conclusion, we can summarize two major partially complementary models for cultural transmission in the East African MSA. First the model of distinct populations/ human groups, keeping traditions stable in certain areas/regions (see interpretation for Gademotta, Douze 2012; Douze and Delagnes 2016) in the early MSA; however, for the Late Pleistocene record, hypotheses of increased interaction on larger scales have been suggested. The same technological continuity is visible in the ESA/MSA transition in Kapthurin, as a process rather than an event (Tryon et al. 2005). The second model imagines the periodic exchange of information and people from one group to another, associated with long period of separation/isolation as Scerri (Scerri 2018; Scerri et al. 2018) suggests within the model of African Multiregionalism.

Arguments against the first hypothesis are that there are not definite chronological and geographical trends linked to technological innovations, partially because at the current state of research we are not able to reconstruct phylogenetically the vast majority of significant technological units (STU). This, however, could be a derivation of the research itself, and this bias could be filled by finding more sites, and by having a more accurate chronology.

The second model seems to fit better the actual evidence, both fossil and archaeological. However, if a major contact of ideas and people occurred intermittently during the final part of the Pleistocene, one could argue that the difference in the archaeological record would be erased in a more accelerated way than we actually see in the records we have nowadays. Nevertheless, there is evidence for a 'mosaic pace' within the first half of the MSA time-scale, since typical Acheulean is still found ~200 ka (e.g. Mieso, see de la Torre et al. 2014). This could also be linked with a biological diversity within those populations. The MSA period is definitely a key period and consequently a complex one, for which convergence is probably more difficult to interpret than divergence.

The presence of specific technological behavior in specific sites, such as the *coup de tranchet* (Douze 2014), shows that a certain amount of local tradition existed and persisted in the East African MSA, as has been shown in the case of tanging/pedunculation in North Africa (Scerri et al. 2014; Scerri 2017). The more and more detailed analysis of lithic assemblages should allow the identification of other Significant Technological Units that will improve our knowledge.

It is agreed that the explosion of the MSA (post MIS 5) is characterized by the flourishing of many regional variations. However, the regions considered here are still very wide and too large to correspond to single social groups of foragers. Moreover, despite the big variability in the MSA, the technological trend is still showing a certain degree of uniformity, if considered in the basis of technological behavior.

It is possible that the whole MSA is rooted in a common lithic tradition, and this makes it difficult to identify small scale regional differences. Furthermore, a common origin could make the invention of the same technological process more likely to be a consequence of simple convergence, where the adaptive conditions in terms of ecological niches are similar.

The aim of this paper has been to enlighten the complexity of the association between lithics and cultures in the MSA, in order to open the debate through articulated models, avoiding simplistic views.

Acknowledgments This paper is dedicated to the memory of Amilcare Bietti, who taught me the concept of "culture". I wish to thank Huw Groucutt for inviting me to contribute to the volume and for his useful suggestions and comments. I thank three anonymous reviewers for their precious comments and insights, they strongly contributed to making this paper better. Thanks to Simone Ghiaroni for the anthropological clarifications. This research is part of the (H)ORIGIN Project, funded by the Italian Ministry of Research within the framework of the SIR 2014 program (Project RBSI142SRD, PI: E.E. Spinapolice). The research is also part of the RLM 2014 program.

References

Adler, D. S., Wilkinson, K., Blockley, S., Mark, D., Pinhasi, R., Schmidt Magee, B., et al. (2014). Early Levallois technology and the Lower to Middle Paleolithic transition in the Southern Caucasus. *Science, 345,* 1609–1613.

Alimen, M. H., & y Zuber, J. Z. (1978). *L'évolution de l'Acheuléen au Sahara nord-occidental: Saoura, Ougarta, Tabelbala: atlas*. Paris: Service des publications du Centre national de la recherche scientifique.

Ambrose, S. H. (2001). Paleolithic technology and human evolution. *Science, 291,* 1748–1753.

Ambrose, S. H. (2002). Small things remembered: Origins of early microlithic industries in sub-Saharan Africa. *Archeological Papers of the American Anthropological Association, 12,* 9–29.

Ambrose, S. H. (2010). Coevolution of composite-tool technology, constructive memory, and language: Implications for the evolution of modern human behavior. *Current Anthropology, 51,* S135–S147.

Anderson-Gerfaud, P. (1990). Aspects of behaviour in the Middle Paleolithic: Functional analysis of stone tools from southwest France. In P. Mellars (Ed.), *The emergence of modern humans* (pp. 389–418). Ithaca, New York: Cornell University Press.

Andrefsky, W. J. (1994). Raw material availability and the organization of technology. *American Antiquity, 59,* 21–34.

Bar Yosef, O., & Kuhn, S. (1999). The big deal about blades: Laminar technologies and human evolution. *American Anthropologist, 101,* 322–338.

Barham, L. (2013). *From hand to handle: The first industrial revolution*. Oxford: Oxford University Press.

Barth, F. (1969). *Ethnic groups and boundaries: The social organization of culture difference (Results of a Symposium Held at the University of Bergen, 23rd to 26th February 1967.)*. Oslo: Universitetsforlaget.

Bar-Yosef, O. (2002). The upper paleolithic revolution. *Annual Review of Anthropology, 31,* 363–393.

Basell, L. (2012). The Middle Stone Age of Eastern Africa. In *The Oxford handbook of African archaeology* (pp. 387–387). Oxford: Oxford University Press.

Bateson, P. (1988). The active role of behaviour in evolution. *Evolutionary Processes and Metaphors, 19,* 191–207.

Beyries, S. (1988). *Industries lithiques: tracéologie et technologie*. Oxford: British Archaeological Reports.

Biberson, P. (1961). *Le Paléolithique inférieur du Maroc Atlantique* (Vol. Fasc. 17). Rabat: Service des Antiquités du Maroc.

Binford, L. (1962). Archaeology as anthropology. *American Antiquity, 28,* 217–225.

Binford, L. (1977). *For theory building in archaeology*. San Diego: Academic Press.

Binford, L. (1989). Isolating the transition to cultural adaptations: An organizational approach. In E. Trinkaus (Ed.), *The emergence of modern humans: Biocultural adaptations in the later Pleistocene* (pp. 18–41). Cambridge: Cambridge University Press.

Binford, L., & Binford, S. R. (1968). Stone tools and human behavior. *Scientific American, 220,* 70–87.

Bleed, P. (1986). The optimal design of hunting weapons: Maintainability or reliability? *American Antiquity, 51,* 737–747.

Bleed, P. (2001). Trees or chains, links or branches: Conceptual alternatives for consideration of stone tool production and other sequential activities. *Journal of Archaeological Method and Theory, 8,* 101–127.

Blegen, N. (2017). The earliest long-distance obsidian transport: Evidence from the ~200 ka Middle Stone Age Sibilo School Road Site, Baringo, Kenya. *Journal of Human Evolution, 103,* 1–19.

Blegen, N., Jicha, B. R., & McBrearty, S. (2018). A new tephrochronology for early diverse stone tool technologies and long-distance raw material transport in the Middle to Late Pleistocene Kapthurin Formation, East Africa. *Journal of Human Evolution, 121,* 75–103.

Boas, F. (1938). *The mind of primitive man, Rev*. Oxford, England: Macmillan.

Boëda, E. (1994). *Le concept Levallois: variabilité des méthodes*. Paris: CRA-Monographies (9).

Boëda, E., Connan, J., Dessort, D., Muhesen, S., Mercier, N., Valladas, H., et al. (1996). Bitumen as hafting material on Middle Paleolithic artifacts. *Nature, 380,* 336–338.

Bordes, F. (1961). Mousterian cultures in France. *Science, 134,* 803–810.

Bourdieu, P. (1977). *Outline of a theory of practice*. Cambridge: Cambridge University Press.

Boyd, R., & Richerson, P. J. (1988). *Culture and the evolutionary process*. Chicago: University of Chicago press.

Bräuer, G. (2008). The origin of modern anatomy: By speciation or intraspecific evolution? *Evolutionary Anthropology: Issues, News, and Reviews, 17,* 22–37.

Brooks, A. S., Yellen, J. E., Potts, R., Behrensmeyer, A. K., Deino, A. L., Leslie, D. E., et al. (2018). Long-distance stone transport and pigment use in the earliest Middle Stone Age. *Science, 360,* 90–94.

Carr, C. (1995). Building a unified middle-range theory of artifact design. In C. Carr & J. Neitzel (Eds.), *Style, society, and person* (pp. 151–170). Boston: Springer.

Cavalli-Sforza, L. L., & Feldman, M. W. (1981). *Cultural transmission and evolution: A quantitative approach*. Princeton: Princeton University Press.

Charrié-Duhaut, A., Porraz, G., Cartwright, C. R., Igreja, M., Connan, J., Poggenpoel, C., et al. (2013). First molecular identification of a hafting adhesive in the late Howiesons Poort at Diepkloof Rock Shelter (Western Cape, South Africa). *Journal of Archaeological Science, 40,* 3506–3518.

Clark, J. D. (1988). The Middle Stone Age of East Africa and the beginnings of regional identity. *Journal of World Prehistory, 2,* 235–305.

Clark, J. D., & Kurashina, H. (1979). Hominid occupation of the East-Central Highlands of Ethiopia in the Plio-Pleistocene. *Nature, 282,* 33–39.

Collard, M., Shennan, S. J., & Tehrani, J. J. (2006). Branching, blending, and the evolution of cultural similarities and differences among human populations. *Evolution and Human Behavior, 27,* 169–184.

d'Errico, F., & Banks, W. E. (2012). Identifying mechanisms behind Middle Paleolithic and Middle Stone Age cultural trajectories. *Current Anthropology, 54,* S371–S387.

Dauvois, M. (1976). *Précis de dessin dynamique et structural des industries lithiques préhistoriques*. Perigueux: Fanlac.

Dauvois, M. (1981). De la simultanéité des concepts Kombewa et Levallois dans l'Acheuléen du Maghreb et du Sahara nord-occidental. *Préhistoire Africaine, 6,* 313–320.

de la Torre, I., Mora, R., Arroyo, A., & Benito-Calvo, A. (2014). Acheulean technological behaviour in the middle Pleistocene landscape of Mieso (East-Central Ethiopia). *Journal of Human Evolution, 76,* 1–25.

De Lumley, H., Beyene, Y., Bahain, J. J., Barsky, D., Cauche, D., Celiberti, V., et al. (2004). Le site de Middle Stone Age de la localité de Fejej FJ-102. In H. De Lumley & Y. Beyene (Eds.), *Les sites préhistoriques de la région de Fejej, sud-Omo, Ethiopie, dans leur contexte stratigraphique et paléontologique* (pp. 619–629). Paris: Editions Recherche sur les Civilisations.

Dean, L. G., Vale, G. L., Laland, K. N., Flynn, E., & Kendal, R. L. (2014). Human cumulative culture: A comparative perspective. *Biological Reviews, 89,* 284–301.

Deino, A. L., Behrensmeyer, A. K., Brooks, A. S., Yellen, J. E., Sharp, W. D., & Potts, R. (2018). Chronology of the Acheulean to Middle Stone Age transition in eastern Africa. *Science, 360,* 95–98.

Dewar, R. E. (1995). Of nets and trees: Untangling the reticulate and dendritic in Madagascar's prehistory. *World Archaeology, 26,* 301–318.

Dobres, M.-A., & Hoffman, C. R. (1994). Social agency and the dynamics of prehistoric technology. *Journal of Archaeological Method and Theory, 1,* 211–258.

Douze, K. (2011). L'Afrique de l'Est dans la réflexion globale sur le Middle Stone Age. *Annales d'Éthiopie, 26,* 15–51.

Douze, K. (2012). *Le Early Middle Stone Age d'Éthiopie et les changements techno-économiques à la période de l'émergence des premiers Homo sapiens*. Ph.D. Dissertation, Bordeaux.

Douze, K. (2014). A new chrono-cultural marker for the early Middle Stone Age in Ethiopia: The tranchet blow process on convergent tools from Gademotta and Kulkuletti sites. *Quaternary International, 343,* 40–52.

Douze, K., & Delagnes, A. (2016). The pattern of emergence of a Middle Stone Age tradition at Gademotta and Kulkuletti (Ethiopia) through convergent tool and point technologies. *Journal of Human Evolution, 91,* 93–121.

Douze, K., Igreja, M., Rots, V., Cnuts, D., & Porraz, G. (2020). Technology and function of Middle Stone Age points. Insights from a combined approach at Bushman Rock Shelter, South Africa. In H. Groucutt (Ed.), *Culture history and convergent evolution: Can we detect populations in prehistory?* (pp. 127–141). Cham, Switzerland: Springer.

Douze, K., & Spinapolice, E. E. (2016). "For an updated vision of Middle Stone Age point production in Africa: How does McBrearty and Brooks's (2000) map of point 'styles' look 16 years after?"—Session Presented at the 23th Biennial Meeting, SAFA. Toulouse Juin 26. July 2, 2016 (Vol. Abstract Book).

Dunnell, R. C. (1978). Style and function: A fundamental dichotomy. *American ANTIQUITY, 43,* 192–202.

Durham, W. H. (1992). Applications of evolutionary culture theory. *Annual Review of Anthropology, 21,* 331–353.

Eren, M. I., Greenspan, A., & Sampson, C. G. (2008). Are Upper Paleolithic blade cores more productive than Middle Paleolithic discoidal cores? A replication experiment. *Journal of Human Evolution, 55,* 952–961.

Féblot-Augustins, J. (2009). Revisiting European Upper Paleolithic raw material transfers: The demise of the cultural ecological paradigm? In B. Adams & B. Blades (Eds.), *Lithic materials and Paleolithic societies* (pp. 25–46). Hoboken: Wiley.

Foley, R. (1985). Optimality theory in anthropology. *Man,* 222–242.

Foley, R., & Lahr, M. M. (1997). Mode 3 technologies and the evolution of modern humans. *Cambridge Archaeological Journal, 7,* 3–36.

Foley, R. A., & Lahr, M. M. (2011). The evolution of the diversity of cultures. *Philosophical Transactions of the Royal Society B: Biological Sciences, 366,* 1080–1089.

Gamble, C. (1998). Palaeolithic society and the release from proximity: a network approach to intimate relations. *World archaeology, 29*(3), 426-449.

Galef, B. G., & Laland, K. N. (2005). Social learning in animals: empirical studies and theoretical models. *AIBS Bulletin, 55*(6), 489–499.

Goodwin, A. J. H., & Lowe, C. V. R. (1929). *The stone age cultures of South Africa.* Capetown: AMS Press.

Gramly, R. M. (1980). Raw material source, areas and curated tools assemblage. *American Antiquity, 45,* 823–833.

Groucutt, H. S. (2020). Culture and convergence: The curious case of the Nubian Complex. In H. Groucutt (Ed.), *Culture history and convergent evolution: Can we detect populations in prehistory?* (pp. 55–86). Cham, Switzerland: Springer.

Groucutt, H. S., Petraglia, M. D., Bailey, G., Scerri, E. M., Parton, A., Clark-Balzan, L., et al. (2015). Rethinking the dispersal of Homo sapiens out of Africa. *Evolutionary Anthropology: Issues, News, and Reviews, 24,* 149–164.

Guglielmino, C. R., Viganotti, C., Hewlett, B., & Cavalli-Sforza, L. L. (1995). Cultural variation in Africa: Role of mechanisms of transmission and adaptation. *Proceedings of the National Academy of Sciences, 92,* 7585–7589.

Hayden, B. (1979). *Palaeolithic reflections. Lithic technology and ethnographic excavations among Australian Aborigines.* Canberra: Humanities Press.

Herries, A. I. (2011). A chronological perspective on the Acheulian and its transition to the Middle Stone Age in southern Africa: The question of the Fauresmith. *International Journal of Evolutionary Biology, 2011.* http://dx.doi.org/10.4061/2011/961401.

Hewlett, B., DeSilvestri, A., & Guglielmino, C. R. (2002). Semes and genes in Africa. *Current Anthropology, 43,* 313–321.

Hublin, J.-J., Ben-Ncer, A., Bailey, S. E., Freidline, S. E., Neubauer, S., Skinner, M. M., et al. (2017). New fossils from Jebel Irhoud, Morocco and the pan-African origin of Homo sapiens. *Nature, 546,* 289–292.

Igreja, M., & Porraz, G. (2013). Functional insights into the innovative Early Howiesons Poort technology at Diepkloof Rock Shelter (Western Cape, South Africa). *Journal of Archaeological Science, 40,* 3475–3491.

Johnson, C. R., & McBrearty, S. (2010). 500,000 year old blades from the Kapthurin Formation, Kenya. *Journal of Human Evolution, 58,* 193–200.

Kaufman, D. (1986). A proposed method for distinguishing between blades and bladelets. *Lithic Technology, 15,* 34–40.

Kline, M. A., & Boyd, R. (2010). Population size predicts technological complexity in Oceania. *Proceedings of the Royal Society of London B: Biological Sciences, 277,* 2559–2564.

Kroeber, A. L. (1949). The concept of culture in science. *The Journal of General Education, 3,* 182–196.

Kuhn, S. L. (1994). A formal approach to the design and assembly of mobile toolkits. *American Antiquity, 59,* 426–442.

Kuhn, S. L. (1995). *Mousterian lithic technology and raw material economy: A cave study.* Princeton: Princeton University Press.

Kuhn, S. L. (2013). Roots of the middle Paleolithic in Eurasia. *Current Anthropology, 54,* S255–S268.

Leplongeon, A. (2014). Microliths in the middle and later stone age of eastern Africa: new data from Porc-Epic and Goda Buticha cave sites, Ethiopia. *Quaternary International, 343,* 100–116.

Leroi-Gourhan, A. (1964). *Le geste et la parole I, Technique et langage.* Paris: Albin Michel.

Lombard, M. (2006). Direct evidence for the use of ochre in the hafting technology of Middle Stone Age tools from Sibudu Cave. *South African Journal of Science, 18,* 57–67.

Maher, L. A., & Macdonald, D. A. (2020). Communities of interaction: Tradition and learning in stone tool production through the lens of the Epipaleolithic of Khareh IV, Jordan. In H. Groucutt (Ed.), *Culture history and convergent evolution: Can we detect populations in prehistory?* (pp. 213–243). Cham, Switzerland: Springer.

Mauss, M. (1936). Les techniques du corps. *Journal de psychologie, 32,* 271–293.

McBrearty, S. (1981). Songhor: A middle stone age site in western Kenya. *Quaternaria. Storia Naturale e Culturale del Quaternario, 23,* 171–190.

McBrearty, S. (1988). The Sangoan-Lupemban and middle stone age sequence at the Muguruk site, western Kenya. *World Archaeology, 19,* 388–420.

McBrearty, S., & Brooks, A. S. (2000). The revolution that wasn't: A new interpretation of the origin of modern human behaviour. *Journal of Human Evolution, 39,* 453–563.

McBrearty, S., & Tryon, C. (2006). From Acheulean to Middle Stone Age in the Kapthurin Formation, Kenya. In E. Hovers & S. Kuhn (Eds.), *Transitions before the transition* (pp. 257–277). New York: Springer.

McElreath, R., Boyd, R., & Richerson, P. (2003). Shared norms and the evolution of ethnic markers. *Current Anthropology, 44,* 122–130.

Mellars, P., & Stringer, C. (1989). *The human revolution. Behavioural and biological perspectives on the origins of modern humans.* Edinburgh: Edinburgh University Press.

Merrick, H. V., & Brown, F. H. (1984). Obsidian sources and patterns of source utilization in Kenya and northern Tanzania: Some initial findings. *African Archaeological Review, 2,* 129–152.

Mesoudi, A. (2016). Cultural evolution: A review of theory, findings and controversies. *Evolutionary Biology, 43,* 481–497.

Minichillo, T. (2006). Raw material use and behavioral modernity: Howiesons Poort lithic foraging strategies. *Journal of Human Evolution, 50,* 359–364.

Mirazon Lahr, M. (2016). The shaping of human diversity: Filters, boundaries and transitions. *Philosophical Transactions of the Royal Society B, 371,* 20150241.

Mirazon Lahr, M., & Foley, R. (2001). Mode 3, *Homo helmei*, and the pattern of human evolution in the Middle Pleistocene. In L. S Barham & K. A. Robson Brown (Eds.), *Human roots, African and Asia in the Middle Pleistocene* (pp. 23–39). Bristol: Western Academic & Specialist Press Ltd.

Moncel, M. H. (2004). Raw material collecting in the European Middle Palaeolithic. Stones from distant areas and local raw material: were there large and small territories? Was there an exchange of stones between neandertal groups? In O. V. Smyntyna (Ed.), *The use of living spaces in prehistory* (Vol. 1224, pp. 13–23). Oxford: Bar International Series

Moore, J. H. (2001). Ethnogenetic patterns in native North America. *Archaeology, Language, and History*, 31–56.

Muir, I., & Hivernel, F. (1976). Obsidians from the Melka-Konture prehistoric site, Ethiopia. *Journal of Archaeological Science, 3,* 211–217.

Mussi, M., Altamura, F., Macchiarelli, R., Melis, R. T., & Spinapolice, E. E. (2013). Garba III (Melka Kunture, Ethiopia): A MSA site with archaic *Homo sapiens* remains revisited. *Quaternary International, 343,* 28–39.

Negash, A., & Shackley, M. S. (2006). Geochemical provenance of obsidian artefacts from the MSA site of Porc Epic, Ethiopia. *Archaeometry, 48,* 1–12.

Negash, A., Shackley, M. S., & Alene, M. (2006). Source provenance of obsidian artifacts from the Early Stone Age (ESA) site of Melka Konture, Ethiopia. *Journal of Archaeological Science, 33,* 1647–1650.

Nelson, M. C. (1991). The study of technological organization. *Archaeological Method and Theory, 3,* 57–100.

Nunn, C. L., Arnold, C., Matthews, L., & Mulder, M. B. (2010). Simulating trait evolution for cross-cultural comparison. *Philosophical Transactions of the Royal Society B: Biological Sciences, 365,* 3807–3819.

Perlès, C. (1974). New examination of lithic implement from Porc-épic (Ethiopia) points. *Anthropologie, 78,* 529–551.

Picin, A. (2018). Technological adaptation and the emergence of Levallois in Central Europe: New insight from the Markkleeberg and Zwochau open-air sites in Germany. *Journal of Quaternary Science, 33,* 300–312.

Premo, L. S., & Hublin, J.-J. (2009). Culture, population structure, and low genetic diversity in Pleistocene hominins. *Proceedings of the National Academy of Sciences, 106,* 33–37.

Premo, L. S., & Kuhn, S. L. (2010). Modeling effects of local extinctions on culture change and diversity in the Paleolithic. *PLoS One, 5,* e15582.

Premo, L. S., & Scholnick, J. B. (2011). The spatial scale of social learning affects cultural diversity. *American Antiquity, 76,* 163–176.

Rightmire, G. P. (2001). Patterns of hominid evolution and dispersal in the Middle Pleistocene. *Quaternary International, 75,* 77–84.

Roche, H., & Texier, P. (1995). Evaluation of technical competence of Homo erectus in East Africa during the Middle Pleistocene. In J. R. F. Bower & S. Sartono (Eds.), *Human Evolution in its Evolutionary Context: Proceedings of the Pithecanthropus Centennial 1893–1993 Congress, Pithecanthropus Centenniel Foundation* (pp. 153–167). Leiden: Leiden University Press.

Rolland, N. (1995). Levallois technique emergence: Single or multiple? A review of the Euro-African record. In H. Dibble & O. Bar-Yosef (Eds.), *The definition and interpretation of Levallois technology.* Monographs in Word Archaeology *23* (pp. 333–359). Madison: Prehistory Press.

Sackett, J. R. (1982). Approaches of style in lithic archaeology. *Journal of Anthropological Archaeology, 1,* 59–112.

Scerri, E. M. (2017). The North African Middle Stone Age and its place in recent human evolution. *Evolutionary Anthropology, 26,* 119–135.

Scerri, E. M. (2018). The origin of our species. *New Scientist, 238* (3175), 34–37.

Scerri, E. M., Drake, N. A., Jennings, R., & Groucutt, H. S. (2014). Earliest evidence for the structure of Homo sapiens populations in Africa. *Quaternary Science Reviews, 101,* 207–216.

Scerri, E. M. L., Thomas, M. G., Manica, A., Gunz, P., Stock, J. T., Stringer C., et al. (2018). Did our species evolve in subdivided populations across Africa, and why does it matter? *Trends in Ecology & Evolution, 33,* 582–594.

Schiffer, M. B. (1976). *Behavioural archaeology.* New York: Academic Press.

Shackley, M. S., & Sahle, Y. (2017). Geochemical characterization of Four Quaternary obsidian sources and provenance of obsidian artifacts from the Middle Stone Age Site of Gademotta, Main Ethiopian Rift. *Geoarchaeology, 32,* 302–310.

Shea, J. J. (2008). The Middle Stone Age archaeology of the Lower Omo Valley Kibish Formation: Excavations, lithic assemblages, and inferred patterns of early Homo sapiens behavior. *Journal of Human Evolution, 55,* 448–485.

Shea, J. J., Fleagle, J. G., & Assefa, Z. (2007). Context and chronology of early *Homo sapiens* fossils from the Omo Kibish Formation, Ethiopia. In P. Mellars, K. Boyle, O. Bar-Yosef, & C. Stringer (Eds.), *Rethinking the human revolution* (pp. 153–162). Cambridge: McDonald Institute Monograph.

Shennan, S., & Collard, M. (2005). Investigating processes of cultural evolution. In R. Mace, C. J. Holden, & S. J. Shennan (Eds.), *The evolution of cultural diversity: A phylogenetic approach* (pp. 133–164). London: University College.

Shott, M. (1996). An exegesis of the curation concept. *Journal of Anthropological Research, 52,* 259–280.

Sinopoli, C. M. (1991). *Style in arrows: A study of an ethnographic collection from the western United States.* Ph.D. Dissertation, University of Michigan.

Spinapolice, E. E. (2012). Raw material economy in Salento (Apulia, Italy): New perspectives on Neanderthal mobility patterns. *Journal of archaeological science, 39,* 680–689.

Spinapolice, E. E. (2014). New Insights on the Early Middle Stone Age of Garba III, Melka Kunture, Ethiopia, Workshop "The African Roots of Hunan Behaviour", LCHES, Cambridge, May 15th 2014.

Spinapolice, E. E. (2016). East African MSA: Regionalization and Variability, SAA 81ST Annual Meeting, Orlando, Florida April 6–10, 2016.

Stout, D., Bril, B., Roux, V., DeBeaune, S., Gowlett, J. A. J., Keller, C., ... & Stout, D. (2002). Skill and cognition in stone tool production: an ethnographic case study from Irian Jaya. *Current anthropology, 43*(5), 693–722.

Stringer, C., & Galway-Witham, J. (2017). Palaeoanthropology: On the origin of our species. *Nature, 546,* 212–214.

Tostevin, G. B. (2012). *Seeing lithics: A middle-range theory for testing for cultural transmission in the Pleistocene.* American School of Prehistoric Research Monograph Series. Peabody Museum, Harvard University; and Oxbow Books.

Trigger, B. (1996). *A history of archaeological thought.* Cambridge: Cambridge University Press.

Tryon, C. A. (2006). "Early" Middle Stone Age lithic technology of the Kapthurin Formation (Kenya). *Current Anthropology, 47,* 367–375.

Tryon, C. A., & Faith, J. T. (2013). Variability in the Middle Stone Age of Eastern Africa. *Current Anthropology, 54,* S234–S254.

Tryon, C. A., McBrearty, S., & Texier, P.-J. (2005). Levallois lithic technology from the Kapthurin Formation, Kenya: Acheulian origin and Middle Stone Age Diversity. *African Archaeological Review, 22,* 199–229.

Tryon, C. A., & Ranhorn, K. L. (2020). Raw material and regionalization in Stone Age East Africa. In H. Groucutt (Ed.), *Culture history and convergent evolution: Can we detect populations in prehistory?* (pp. 143–156). Cham, Switzerland: Springer.

Van Peer, P., Fullagar, R., Stokes, S., Bailey, R. M., Moeyersons, J., Steenhoudt, F., et al. (2003). The Early to Middle Stone Age transition and the emergence of modern human behaviour at site 8-B-11, Sai Island, Sudan. *Journal of Human Evolution, 45,* 187–193.

Vogel, N., Nomade, S., Negash, A., & Renne, P. R. (2006). Forensic 40Ar/39Ar dating: A provenance study of Middle Stone Age obsidian artifacts from Ethiopia. *Journal of Archaeological Science, 33,* 1749–1765.

Wadley, L. (2005). Putting ochre to the test: Replication studies of adhesives that may have been used for hafting tools in the Middle Stone Age. *Journal of Human Evolution, 49,* 587–601.

Wadley, L. (2010). Compound-adhesive manufacture as a behavioral proxy for complex cognition in the Middle Stone Age. *Current Anthropology, 51,* S111–S119.

Wendorf, F., & Schild, R. (1974). *A Middle Stone Age sequence from the Central Rift Valley, Ethiopia.* Warsaw: Polska Akademia Nauk Instytut HistoriiKultury Materialnej.

Whiten, A., Goodall, J., McGrew, W. C., Nishida, T., Reynolds, V., Sugiyama, Y., et al. (1999). Cultures in chimpanzees. *Nature, 399* (6737), 682–685.

White, T. D., Asfaw, B., DeGusta, D., Gilbert, H., Richards, G. D., Suwa, G., et al. (2003). Pleistocene Homo sapiens from Middle Awash, Ethiopia. *Nature, 423,* 742–747.

White, M., & Ashton, N. (2003). Lower Palaeolithic core technology and the origins of the Levallois method in north-western Europe. *Current Anthropology, 44,* 598–609.

Wiessner, P. (1983). Style and social information in Kalahari San projectile points. *American Antiquity, 48,* 253–276.

Wiessner, P. (1984). Reconsidering the behavioral basis for style: A case study among the Kalahari San. *Journal of Anthropological Archaeology, 3,* 190–234.

Wilkins, J., & Chazan, M. (2012). Blade production ~500 thousand years ago at Kathu Pan 1, South Africa: Support for a multiple origins hypothesis for early Middle Pleistocene blade technologies. *Journal of Archaeological Science, 39,* 1883–1900.

Wobst, H. M. (1974). Boundary conditions for Paleolithic social systems: A simulation approach. *American Antiquity, 39,* 147–178.

Chapter 6
A Matter of Space and Time: How Frequent Is Convergence in Lithic Technology in the African Archaeological Record over the Last 300 kyr?

Manuel Will and Alex Mackay

Abstract Stone artefacts are frequently used to identify and trace human populations in the Paleolithic. Convergence in lithic technology has the potential to confound such interpretations, implying connections between unrelated groups. To further the general theoretical debate on this issue, we first delineate the concepts of independent innovation, diffusion and migration and provide archaeological expectations for each of these processes that can create similarities in material culture. As an empirical test case, we then assess how these different mechanisms play out in both space and time for lithic technology across several scales of the African Stone Age record within the last 300 thousand years (kyr). Our findings show that convergence is neither the exception nor the norm, but a scale-dependent phenomenon that occurs more often for complex artefacts than is generally acknowledged and in many different spatio-temporal contexts of the African record that can crosscut the MSA/LSA boundary. Studies using similarly-looking stone tools to recognize past populations and track human dispersals in the Stone Age thus always need to test for the potential of independent innovation and not assume migration or diffusion *a priori*.

Keywords Middle Stone Age • Later Stone Age • Human dispersal • Cultural evolution • South Africa

Introduction

Since the inception of Paleolithic archaeology as a scientific discipline, lithic technologies have been used to identify and trace prehistoric human populations across the Old and New World. Initially enshrined in the paradigm of cultural history, these research endeavors have dominated Stone Age studies for most of the 20th century and continue to be in use today, despite the rise of processual and post-processual approaches. Providing a thorough theoretical treatment and critique of cultural history in (Paleolithic) archaeology is far beyond the scope of this article (see e.g. Binford and Binford 1966; Clarke 1968; Trigger 2006: 211–313; Webster 2008; Shea 2014). Instead, we focus on the question to what extent stone artefacts can really do the job for which they have been and are still so often used for: recognizing past populations and tracking their movements across the landscape.

The study of stone artefacts takes particular importance in understanding human evolution due to their unique durability, sheer abundance, high information content and frequent spatiotemporal patterning, providing a unique long-term perspective. They represent the most tangible source of information concerning behavioral patterns of prehistoric people, as well as for reconstructing ecological adaptations, technological activities, and settlement systems (Debénath and Dibble 1994; Shott 1994; Odell 2004; Tostevin 2012). Researchers have also used the spatio-temporal distribution of stone artefacts with similar and specific morphologies for further-reaching purposes, such as identifying past human

M. Will (✉)
Department of Early Prehistory and Quaternary Ecology, University of Tübingen, Schloss Hohentübingen, 72070 Tübingen, Germany
e-mail: manuel.will@uni-tuebingen.de

Gonville and Caius College, University of Cambridge, Cambridge, UK

PAVE Research Group, Department of Archaeology and Anthropology, University of Cambridge, Cambridge, CB2 3QG, UK

A. Mackay
School of Earth and Environmental Sciences, Centre for Archaeological Science, University of Wollongong, Wollongong, Australia

Department of Archaeology, University of Cape Town, Cape Town, South Africa

H. S. Groucutt (ed.), *Culture History and Convergent Evolution: Can We Detect Populations in Prehistory?*,
Vertebrate Paleobiology and Paleoanthropology, https://doi.org/10.1007/978-3-030-46126-3_6

populations, naming cultural units and tracing dispersal events (Reynolds 2020). The phenomenon of convergence in lithic systems has the potential to confound such large-scale interpretations, implying connections between unrelated populations (e.g. Stanford and Bradley 2012; O'Brien et al. 2014; Rasmussen et al. 2014). In such cases, archaeologists reach inaccurate conclusions about the past by inferring certain processes from similarly-shaped stone tools, whose likeness resulted from other mechanisms.

Here, we are interested in how big the problem actually is: How frequent is convergence in lithic technology? If convergence is a relatively isolated phenomenon in the empirical record, we will have to scrutinize individual cases but should not throw out the baby with the bathwater. If, however, convergence is common then developing new methodical approaches that distinguish between different processes becomes essential. To answer this question, we first provide a definition of convergence, distinguish it from diffusion and migration, and discuss the archaeological expectations of all three. We then look at different scales in search for examples of convergence. Our approach is interested in how convergence plays out in both space and time across different analytical levels of the African Stone Age record throughout the last 300 kyr. We will evaluate our findings with regard to the potentials and problems in using stone tools to identify past populations, name technocomplexes and trace human dispersals in relation to the general prevalence of convergence found in this study.

The Phenomenon of Convergence: Definition, Delimitation and Archaeological Expectations

Three basic processes can produce similarities in material culture between independent populations or different areas: Migration, diffusion and convergence (Fig. 6.1). Various names have been given to these general mechanisms and their definition may differ between researchers, rendering clear demarcation of these terms necessary. Since our goal is to apply these concepts to stone tools in the African Stone Age, we also lay out the different signals that are produced by migration, diffusion and convergence in the lithic archaeological record from a theoretical perspective which can then be applied for explaining similarities observed between a given spatiotemporal context A and B.

Diffusion entails all processes of the spread of cultural information between groups without physical relocation of a population (Bellwood 2013). Diffusion is a result of cultural transmission between populations in contact with one another; it might be driven by a variety of factors including

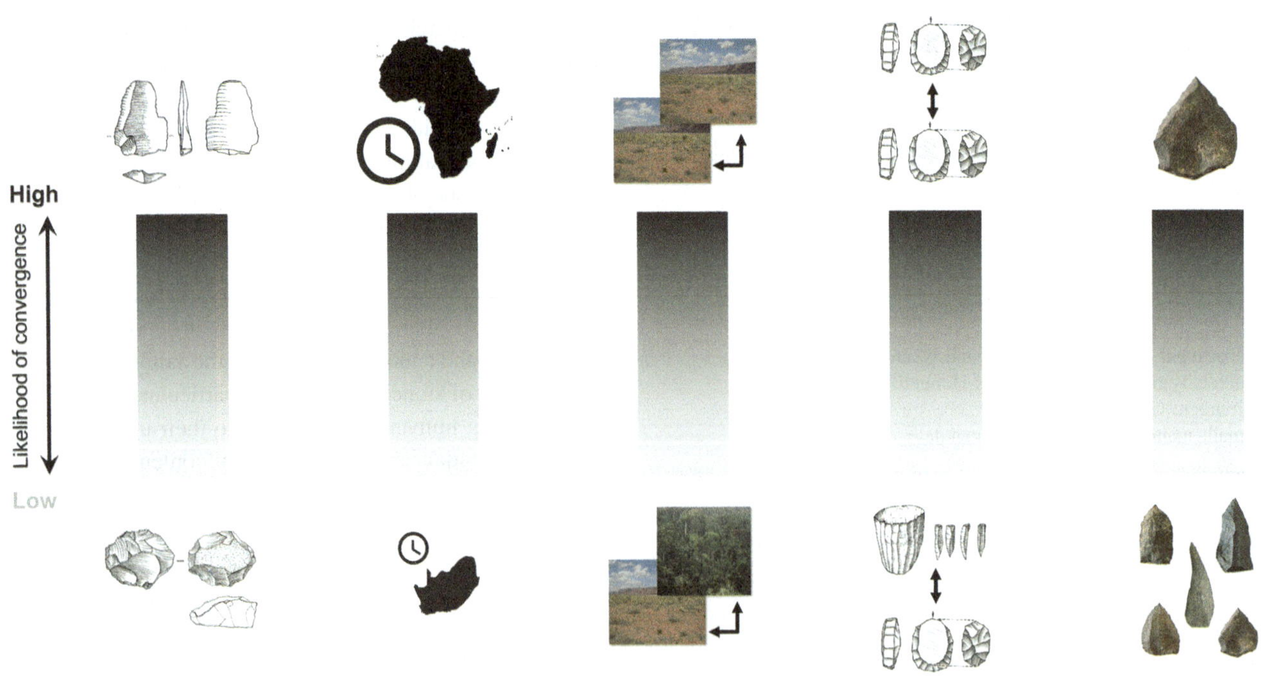

Fig. 6.1 Illustration of the likelihood of technological convergence of stone artefacts between two assemblages as a function of five variables: (1) complexity of the artefact; (2) temporal and spatial context; (3) environmental context; (4) basic technological system; (5) number of replicated elements. For detailed explanation see text

exchange, trade, warfare etc., and it may result in cultural replacement or blending between those populations (e.g. Kroeber 1940; Rouse 1986; Collard et al. 2006; Groucutt et al. 2015; Spinapolice 2020). The various modes and pathways of this complex process, both within and between populations, are the subject of much recent research (e.g. cultural transmission theory; Boyd and Richerson 1985; Henrich 2001; Eerkens and Lipo 2007). For large-scale archaeological phenomena, cultural diffusion usually denotes processes that move cultural ideas or material objects between different populations across a given time and space without concomitant migrations. Historically, the diffusionist school dominant in the 19th and early 20th century—encapsulated in the famous catchphrase "ex oriente lux" and used by eminent archaeologists such as Montelius (1903), Smith (1929) and Childe (1929)—argued that innovations typically occurred in one context and spread out from there, largely negating the possibility of independent innovation (see Trigger 2006: 217–228).

Migration denotes all processes of the physical movement of a population with the intention of inhabiting a new location on a permanent or temporary basis (Bellwood 2013). Again, multiple names with slightly different meaning and scale exist, including relocation, movement, dispersal, expansion or branching, the latter specifically denoting the common descent of a cultural variant and its subsequent dispersal (Collard et el. 2006; Groucutt et al. 2015). While migration is an essential agent of change in both biological and socio-cultural systems, throughout the mid-20th century in particular archaeologists commonly invoked this term to explain cultural similarities (i.e. migrationist theories; Trigger 2006: 217–223), albeit typically with limited critical consideration (Clark 1994).

Convergence describes the appearance of similar ideas or objects in two or more groups arising from independent innovation in each—that is, without vertical or horizontal information transmission. Note that we use the broadest sense of convergence here, and we do not distinguish specific forms of convergence such as parallel or iterative evolution (McGhee 2018). There are several reasons for this. First, 'descent' in stone tool technology is not biological: though they are shaped by historical contingency, stone tools do not descend from other stone tools. Second, parallel evolution in lithics is effectively a probability-constrained form of convergence, where pre-existing technological similarities mean that only small numbers of similar modifications are required to produce similar novel outcomes. We treat this aspect of probability in more detail below.

The term convergence is borrowed from evolutionary biology where it explains structures that are similar due to function but not ancestry (i.e. analogous). In cladistics, such a structure is also called a homoplasy and does not provide phylogenetic information on organisms. In contrast, homologous structures are similar to one another because of common ancestry (phylogenetic history) and are important in cladistics for resolving evolutionary relationships (Powell 2007; Wake et al. 2011; Futuyama 2013). Following a similar logic, convergent ideas and objects cannot be used to resolve historical relationships between populations and cultures as they arose independently due to similar functions, comparable adaptive solutions to similar environmental or social contexts, or pure chance. While being outside of the scope of this article, some archaeologists have advocated the use of more rigorous quantitative procedures for distinguishing homoplasies from homologies, borrowing approaches from phylogenetic and cladistic analysis in biology (see for overviews O'Brien et al. 2001; Lycett 2010, 2015).

Though rarely with the label applied, convergence has been an implicit mechanism in many archaeological explanations, albeit of very different characters. Convergence is central to culture evolutionist models prevalent through the 19th and early 20th century, where human technological behavior is seen to have gone through a series of universal 'stages' with similar material expression (Tylor 1871; Morgan 1877; Engels 1884), yet without any implication of contact between populations (Dunnell 1980). (These explanations could reasonably be viewed as an extreme form of parallel cultural evolution). In contrast, the adaptationalist (or selectionist) models which emerged in the late 20th century viewed stone tools as optimal solutions to problems; similar problems promoted similar solutions again without implying cultural transmission between populations (see papers in Elston and Kuhn 2002; also Hiscock et al. 2011; McCall and Thomas 2012; Clarkson et al. 2018). Problematic for the former models is the lack of a clear concept for increasing complexity in lithic technologies; problematic for the latter is the common lack both of fit between environment and technology, and of a constraining role for historical contingency (Shennan 2020). In a sense these approaches are merely the mirror of diffusionist models: technological similarities are either never the result of convergence (diffusionist) or they always are (culture evolutionist/adaptationalist). Neither brings us closer to answering our question of how common convergence really was.

Returning, then, to our three identified processes for the appearance of technological similarity—dispersal, diffusion and convergence—we expect each to produce different patterns in the material record. Because they depend on information transmission, migration and diffusion should both yield an archaeological pattern of spatio-temporal *contiguity*.[1] That is, similarity should occur in assemblages which are not separated in either space and/or time. In addition,

[1] There are exceptions, such as information spread by water craft or through tele-communications though we expect the latter to be irrelevant to the Palaeolithic.

dispersals will leave a genetic signature in the populations involved. The long-standing debate on different models for the spread of farming to Europe serves as an exemplary case for distinguishing between the two (see Ammerman and Cavalli-Sforza 1984; Pinhasi et al. 2005; Brandt et al. 2014). The archaeological object under study offers another dimension to distinguish between diffusion and dispersal. Experimental, theoretical and ethnographic studies suggest that cultural diffusion will mostly result in product copying and thus the adoption of more simple assemblage elements (e.g. blank types in lithic systems). Dispersals encompass process copying with high-fidelity transmission of more complex and multi-step systems (e.g. core reduction methods) as part of information transmission within the migrating population (e.g. Eren et al. 2011; Nigst 2012; Tostevin 2012).

Technological convergence has different archaeological expectations. First, we expect geographical and chronological gaps in the distribution of the technological element(s) under consideration (i.e. similarities are discontiguous). Here, space and time both function as measures of distance, while distance in turn functions as a proxy for likelihood of faithful transmission. Though hypothetically information could move between two separate space/time locations without leaving a signal between them, this would imply that any contact between the separated populations was not sustained, with the reasonable expectation of divergence with subsequent drift. As distance increases, the probability of faithful transmission comparably declines (Eerkens and Lipo 2005).

Second, the probability of convergence is negatively correlated with the degree of cultural similarity—the more cultural elements that are similar, the less likely this is to have occurred either by chance or similar selective pressure. Thus convergence is the more plausible process when similarities are found only in a limited subset of the technological repertoire of the assemblages (e.g. only a certain tool type). Both extent of separation and extent of similarity are necessary for distinguishing underlying processes. Similarity between assemblages in a single component could reflect convergence or diffusion (blending); similarity in many components could reflect convergence or diffusion but more likely dispersal. However, similarities in few components between spatio-temporally discrete assemblages more likely arise as a result of convergence than any other process.

Finally, we need to consider that convergence is a function of probability in terms of both analytic scale and historical contingency. If we allow, following McGhee (2018), that (a) the range of potentially functional technologies is finite, (b) the proportion of this range which has been physically realized increases with population size (itself increasing with the sampled spatio-temporal interval), and (c) the range is not explored randomly but is instead historically contingent (incremental variants on existing forms are more common than random leaps, even in modern times, e.g. Basalla 1988; Rogers 1995), a number of further expectations arises. First, leaving contingency aside, probability of convergence increases with analytic scale—the larger the spatio-temporal range we are considering the more likely convergence is to occur (e.g. the appearance of Levallois technology in Africa and Eurasia during the Middle Pleistocene and in Australia probably during the Holocene (Dortch and Bordes 1977; Tryon et al. 2006; Adler et al. 2014; Hu et al. 2019). Second, convergence is more likely to occur in populations with pre-existing technological similarities (e.g. the emergence of Levallois technology from Acheulean technology in Europe and Africa [Tryon et al. 2006; Adler et al. 2014]), either as a result of past transmission or previous convergence.

The problem of disentangling diffusion, migration and convergence arises for all elements of material culture and archaeological studies in general, but is particularly relevant for lithic technology and Paleolithic research. Due to their essentially reductive nature and the limits imposed by fracture mechanics and functional requirements, stone tools have a heavily constrained range of possible forms (McGhee 2018). Consequently, unrelated population are known to have manufactured similar artefacts (Hovers 2006). Furthermore, because cultural transmission can be both horizontal, vertical and oblique, descendent populations can make quite different artefacts from their relatively recent ancestors (Seguin-Orlando et al. 2014). Finally, even when horizontal transmission processes such as diffusion occur, variability in the properties of rocks available to the interacting populations may obscure its effects (Tryon and Ranhorn 2020).

As will be shown in the next section, researchers have tried to face this challenge by focusing on the most complex components of lithic assemblages in order to limit the confounding potential of convergence. We define complex here as those components of lithic technology which require multiple, successive, inter-dependent and hierarchical steps in a targeted flaking system (see e.g. Muller et al. 2017), such as the application of intense retouch that alters the overall shape and size of flakes (i.e. curated tools) or the long reduction chain of cores (i.e. hierarchical organization of Levallois cores). While the underlying complexity of flaking systems reduces the *a priori* probability of chance morphological similarities, for reasons noted above it is insufficient to preclude it.

The resolution of classification is an additional element requiring consideration here. While specific stone artefacts might belong to the same type (i.e. bifacial point), their morphology (e.g. lanceolate vs. tear-shaped) and/or production technique (e.g. asymmetrical vs. symmetrical shaping) might differ. When distinguishing between convergence, diffusion and migration, the empirical basis of this similarity —whether pieces resemble one another superficially or across several dimensions—should be made explicit.

We conclude that numerous factors need to be taken into account when estimating the probability of convergence, including parameters of the artefacts themselves and their context: complexity of the artefact and resolution of similarity; temporal and spatial context of the assemblages in which the artefact is found; any pre-existing similarities in technological systems and number of replicated elements for the studied assemblage. Figure 6.1 summarizes the likelihood of technological convergence of a stone artefact as a function of the individual variables. For similar looking artefacts A and B, convergence is more likely if they: (i) are less complex; (ii) are similar across fewer dimensions; (iii) are found further apart in space and time; (iv) derive from historically similar technological systems and ecological contexts; and, (v) occur in assemblages with few other similar objects. Our conception uses a relative scale with a continuous range, and is thus based on a probabilistic perspective. Artefacts can be complex, similar in numerous respects, come from the same underlying technological system and from neighboring regions, yet still be the result of independent innovation—it is just unlikely that they are. There is also no absolute figure that we can put on an artefact from which a concrete process can be logically deduced as in '5200 km and 10 kyr apart must be convergence' (e.g. Kuhn and Zwyns 2018). There is, however, an expected positive relationship between the number of different variables with matching information on the archaeological expectations of the different processes and the probability of the interpretation being correct.

What Is at Stake? Tracing and Identifying Past Populations with Stone Tools

Two broad issues are affected by distinguishing between convergence, diffusion and migration in the stone tool record of the Paleolithic. First, tracing human dispersals with stone tools, and second identifying populations or naming cultural-technological units. For both we provide a few examples from recent research in the African archaeological record of the last 300 kyr to show what is at stake. This timeframe is chosen due to the frequency of such studies in recent years, their direct relevance for human evolution, and a rich archaeological record with comparatively high spatio-temporal resolution.

Both approaches share the same basics: populations are identified based on a number and combination of specific stone artefacts, but in reality this often involves a few or single types of cores or tools. These artefacts are selected based on the perceived evidence for stylistic, idiosyncratic or historic variation: the appearance of the artefact cannot be explained by functional or economic needs but rather due to social transmission within a population according to shared ideas or to affiliate with a particular ethnic or cultural group. As such they can be seen as markers of groups that can then be identified and traced. The main difference between the two approaches is that one focuses on following the identified population as it moves from A to B, while the other develops cultural taxonomies and delineates the spatio-temporal range of cultural (and sometimes ethnic) groups on the basis of diffusion.

The identification and tracking of past populations with stone tools has been consistently employed in the interpretation of early modern human dispersals within, out of, and beyond Africa. A recent example includes the "Nubian technocomplex" in both northeast Africa and the Arabian Peninsula, defined largely on the presence of Nubian cores, which is thought to reflect the same group of people using this specific reduction technology (Rose et al. 2011; Crassard and Hilbert 2013; Usik et al. 2013; Groucutt 2020). The documentation of comparable "Nubian cores" in southern and central Arabia is seen as evidence for demographic exchange across the Red Sea, suggesting that modern humans carrying this technology entered the region before 100 ka. This would represent one of the earliest identified populations of *Homo sapiens* outside of Africa, and support arguments for the southern migration route into Asia (Crassard and Hilbert 2013). Another model is based on the archaeological site of Jebel Faya in the United Arab Emirates. Here, stone tools dating to the last interglacial are argued to show similarities to MSA lithic technology in eastern and northeastern Africa, an assessment largely based on the presence of façonnage used for the production of small hand axes and foliates (Armitage et al. 2011). These affinities between the regions are interpreted as dispersals of early modern humans from Africa across the Red Sea during times of low sea level at around $\sim$120 ka. Finally, Mellars (2006; also Mellars et al. 2013) has used the occurrence of microlithic technologies with backed segments in southern and eastern Africa (from Howiesons Poort or "HP-like" sites) and a later appearance of similar tools in India and Sri Lanka to infer population movements out of Africa along a coastal route by ca. 60–50 ka.

In all of these cases, specific tool or core types from the MSA record are used as a baseline 'African' signal for the source populations of modern humans that later dispersed to other continents. None of the approaches explicitly considers or tests the alternative hypotheses of independent innovation although they involve large spatio-temporal scales and are essentially based on single tool or core types, rendering their empirical basis susceptible to convergence. Where such tests have been undertaken, they have not been supportive (e.g. Lewis et al. 2014). Discussion on the degree to which lithic artefacts can actually help to trace early dispersals of modern humans from Africa to Eurasia is ongoing (see McBrearty 2003; Garcea 2004; Hovers 2009; Hiscock et al. 2011; White et al. 2011; Adler et al. 2014; Clarkson et al. 2018) with one recent review outright concluding that "there is no lithic 'smoking gun' for dispersal out of Africa" (Groucutt et al. 2015: 26). What unites many of these critiques is the notion that taking the phenomenon of convergence into account is a crucial step forward (Groucutt 2020).

The use of so-called type fossils to define and identify archaeological cultures within the African MSA might likewise be compromised by independent innovation. *Fossiles directeur* have a long history and played an important role in the formative years of archaeological research in Eurasia and Africa (de Mortillet 1883; Childe 1929; Goodwin and Van Riet Lowe 1929). Although there has been much criticism (e.g. Binford and Binford 1966; Clarke 1968; Renfrew 1977; Hodder 1978; Shea 2014) this has not led to the abandonment of the basic approach (apart from ethnic connotations) nor the use of archaeological cultures and named technocomplexes (see discussion in Trigger 2006: 312–313; Webster 2008; Roberts and Vander Linden 2011). Even adamant critics of cultural historical approaches such as Lewis Binford have stated that specific types of intensely worked and curated "tools" can serve as the best "ethnic markers" of groups, potentially allowing for the identification of different populations (Binford 1973: 243). This, however, is only true when specific curated forms arise from a system of shared cultural ideas and not by means of independent innovation.

In past and present studies within Africa, the presence of specific tool types or core forms is used to name technocomplexes of the MSA and LSA. Table 6.1 lists prominent examples from different regions (for illustrations see Fig. 6.2). In South Africa for example, the presence of backed tools delineates the Howiesons Poort (HP; Singer and Wymer 1982; Lombard 2005; Wadley 2008), whereas the occurrence of bifacially worked foliate or lanceolate points constitutes the typical marker of the Still Bay (SB; Goodwin and Van Riet Lowe 1929; Henshilwood et al. 2001; Wadley 2007). Some LSA technocomplexes such as the Wilton are also defined by frequent backed pieces (Deacon 1984; Lombard et al. 2012).

The recent focus on technology and variability instead of typology and normative views has resulted in much criticism towards this approach in various regions and contexts of African Stone Age archaeology. The SB of South Africa can serve as an instructive example. After the initial definition in the late 1920s in South Africa (Goodwin and Van Riet Lowe 1929) and its subsequent export to other regions such as eastern Africa (e.g. Kenyan SB; Leakey 1931; Clark 1954; Anthony 1972), critical comments on the ambiguous definition, integrity, age and status of the SB appeared in the 1950–60s (Malan 1956; Clark et al. 1966). These criticisms ultimately led to its abandonment (Sampson 1974; Deacon 1979; Volman 1981). This counter-movement has been followed by re-instatement and apparent consolidation of the concept of the SB in the 1990s and early 2000s, based on findings of bifacial points in stratified sediments from modern excavations at Hollow Rock Shelter (Evans 1994) and Blombos (Henshilwood et al. 2001), reinforced by studies on its artefactual content and temporal coherence (e.g. Wadley 2007; Jacobs et al. 2008; Henshilwood et al. 2011; Henshilwood 2012). Thus the SB appeared to be expressed as a coherent technology restricted to a finite and contiguous block of space (southernmost Africa) and time ($\sim$75–71 ka), providing a basis for assuming it to reflect a pool of information shared through transmission by diffusion or migration (Mackay et al. 2014). As a last turn of events in the colorful history of the SB, each of these bases for inferring transmission has been questioned: dates from different time periods have emerged (Tribolo et al. 2013; Conard and Porraz 2015), significant technological variability has been documented (Porraz et al. 2013; Conard et al. 2014; Archer et al. 2016; Högberg and Lombard 2016; Will and Conard 2018), and spatial gaps in the record have been identified (Archer et al. 2016).

We will scrutinize the spatio-temporal occurrence of bifacial points later, but these recent observations potentially compromise the use of bifacial artefacts (*sensu lato*) as *fossiles directeurs* or chrono-cultural markers in the MSA of southern Africa. More importantly, this detailed example echoes recent developments in other parts of Africa such as the role of tanged/pedunculated tools for the Aterian (Dibble et al. 2013; Scerri 2013) or Nubian cores for the Nubian Complex (Kleindienst 2006; Scerri et al. 2014; Groucutt et al. 2015; Will et al. 2015). Again, basic questions arise with regard to distinguishing the sources of similarities: whether certain tool and core types are part of the same cultural package (diffusion within a population) or rather adaptive solutions reached independently (convergence).

Approach and Method

The preceding examples from the last 300 kyr in Africa provide separate challenges to prevailing assumptions of cultural transmission in Stone Age research that emphasize single contexts of innovation for (complex) artefacts and feature direct equation of specific stone artefacts with past people. Seeing that we have so far intentionally picked out individual examples from a vast archaeological record, how

Table 6.1 Selected list of named MSA and LSA technocomplexes in Africa and their main typological basis of definition (see also Linstädter et al. 2012; Lombard et al. 2012; Scerri 2017)

Technocomplex	Chronology	Geography	Stone tool types	References
Sangoan	>250 ka? (pre-Lubemban)	Equatorial and southeastern Africa	Core axes, picks	McBrearty (1988), Clark (2001)
Lupemban	~300–150 ka (?); potentially until MIS 3–2 (post-Sangoan)	Central Africa (and northeastern Africa?)	Bifacial lanceolate points; Core axes	Clark (2001), Taylor (2011, 2016)
Aterian	~130–50 ka	North Africa (not Nile Valley)	Tanged tools (Bifacial foliates)	Caton-Thompson, (1946), Garcea (2004), Dibble et al. (2013), Scerri (2013, 2017)
Nubian/Afro-Arabian Nubian	~130–60 ka	Northeastern Africa and Arabian Peninsula	Nubian cores	Van Peer (1998), Van Peer and Vermeersch (2007), Rose et al. (2011), Crassard and Hilbert (2013)
Still Bay	~77–70 ka[a]	South Africa (and Namibia)	Bifacial points	Goodwin and van Riet Lowe (1929), Wadley (2007), Jacobs et al. (2008), Henshilwood (2012), Soriano et al. (2015)
Howiesons Poort	~65–60 ka	South Africa, Lesotho (and Namibia)	Backed pieces (segments)	Goodwin and van Riet Lowe (1929), Lombard (2005), Jacobs et al. (2008), Wadley (2008), Henshilwood (2012), Soriano et al. (2015)
Taramsan	~55–45 ka	Egypt	Levallois blade technology	Van Peer et al. (2010)
Nasampolai	~50–40 ka	Eastern Africa	Backed geometric microliths	Ambrose (1998, 2002)
Nasera	~40–20 ka	Eastern Africa	Small points; (backed microliths)	Mehlman (1989), Ambrose (2002)
Iberomarusian	~20–12 ka	Northwestern Africa	Backed bladelets and microliths	Barton et al. (2005), Bouzouggar et al. (2008), Olszewski et al. (2011), Linstädter et al. (2012)
Tshitolian	~15–2 ka	Central Africa	Tanged foliate points	Clark (1963), Cahen (1978), Miller (2001)
Oakhurst	12–7 ka	Southern Africa	D-shaped scrapers	Deacon (1984), Lombard et al. (2012)
Ounanian/Harifian; Ounan-Harif	<11 ka	Northeastern Africa and Sahara	Tanged points	Clark et al. (1973), Going-Morris (1991), Smith (1993)
Wilton	8–4 ka	Southern Africa	Backed microliths	Deacon (1984), Lombard et al. (2012)

[a]The age of the Still Bay in southern Africa is contested and might be considerably older (see Tribolo et al. 2013; Conard and Porraz 2015)

frequent is convergence really? Is independent innovation in stone tools a frequent phenomenon that we should expect or rather the rare exception compared to diffusion and migration? The perspective of convergence being rare in material culture has been the more dominant one in archaeology overall, but the opposite extreme is at least a tenable position in Paleolithic archaeology: considering the reductive nature of stone tools coupled with physical and functional constraints on their morphometric variability, the reduced complexity compared to later technologies (e.g. metal working), and the vast geographic spread of many basic technological systems (e.g. Mode 3), independent populations might be bound to reach similar path-dependent adaptive solutions among their stone tools frequently (see also Clarkson et al. 2018).

In order to get a grip on the evaluation of the frequency of convergence, we have previously laid out the different signals that are expected to be produced by migration, diffusion and convergence in the archaeological record from a theoretical perspective. On this basis we will identify potential cases of convergence in the African archaeological record of the last 300 kyr from the literature and provide a rough

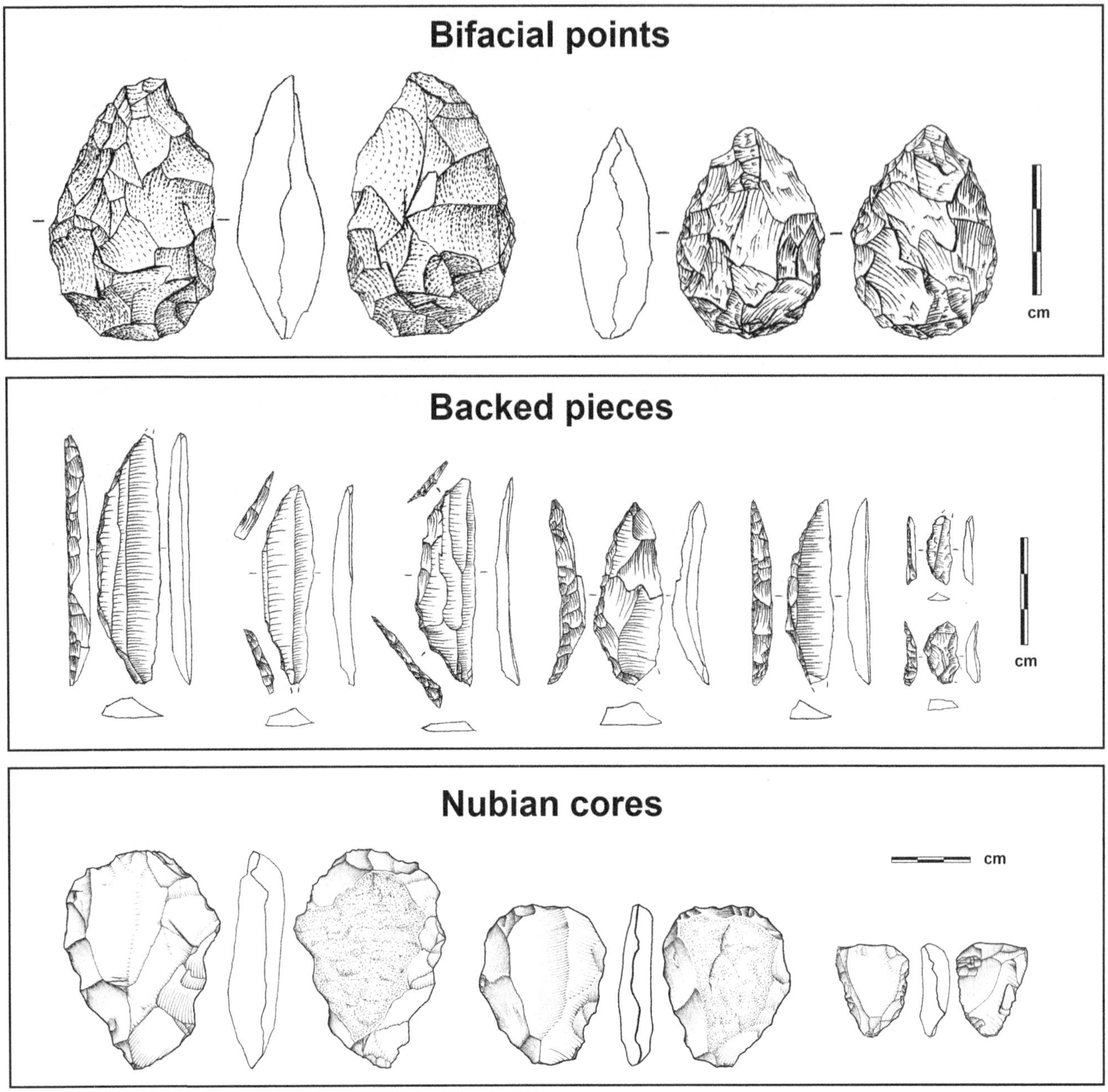

Fig. 6.2 Tool and core types used to define technocomplexes and potential candidates of convergence discussed here. Drawings by H. Würschem (bifacial points; Sibudu), A. Oechsner (backed pieces; Sibudu), and A. Sumner (Nubian cores; Uitspankraal 7)

assessment of its prevalence. Our study examines the MSA and LSA of Africa due to their rich and relatively well-resolved empirical record and chrono-cultural frameworks. As our previous discussion has identified the relevance of the analytical scale, we will make intentional use of different temporal and spatial levels of inquiry, ranging from sites to regions and continents, and from thousands to tens of thousands of years. Most of the time, archaeologists tend to view convergence as a spatial process, focusing on similar artefacts in two locations at the same time. Yet, time is also a crucial variable, particularly in distinguishing between various mechanisms. On a basic scale, we can thus keep space constant and look only at time within a single site, asking how often certain artefact classes occur in discrete periods of the sequence, before examining the spatial element and whether the (re-)emergence of a technology in an area might be a result of its importation from a contiguous pool of conserved information. We thus start by focusing on sites with long sequences in southernmost Africa to assess the frequency of recurrence in particular forms and the potential mechanisms underlying such reappearances. In a second step we broaden the scope to encompass the wider African record to explore the relationship between analytic scale and probability of convergence. While we are not under the illusion of achieving a precise quantitative assessment of convergence in the African Paleolithic record, the results help to gauge the overall tendencies on whether convergence is the rare outlier or rather a recurring feature that requires our enduring attention.

One important limitation that needs to be borne in mind is the inherently discontinuous nature of the archaeological record itself. Many of the most prominent archaeological sites in Africa have discontinuous occupational sequences, and many regions of Africa are poorly resolved. It is no more reasonable to assume that the temporal gaps reflect 'depopulation'—that is, reflect evidence of absence—than it is to assume that the spatial gaps do. The record is thus predisposed to gaps, which on a superficial reading should predispose us to inferences of convergence. Conversely, we cannot infer information transfer across those gaps, as this will predispose us to infer dispersal or diffusion. Furthermore, these gaps appear at all scales; radiocarbon ages for sequential layers might be separated by 100 or 1000 years while any given assemblage might be located 10 km or 100 km from the next nearest. At which scale is information transmission precluded? As previously noted, we can only reason that probability of convergence increases with distance in space, time and similarity.

Space and Time: Convergence in the African MSA and LSA

The Small Scale: Site Sequences in South Africa

The South African Stone Age record enjoys the benefit of providing multiple sites from caves and rockshelters with long and high-resolution stratigraphies within the last 200 kyr. There are also numerous studies on lithic assemblages and a well-developed cultural-stratigraphic system with absolute dates. Here we start with the long and well-published sequences from Sibudu, Klasies River, Diepkloof, Mertenhof, Rose Cottage Cave and Apollo 11 as test cases (see Table 6.2; Fig. 6.3). How often do specific artefact types occur in these sites in layers that are not preceded by layers containing these artefacts or at least separated from one another by large amounts of time that increase the likelihood of convergence? In accordance with recent approaches, we focus on the most complex assemblage components—retouched elements and cores—as simpler elements (blank types) are found similarly across most of the stages under study. We particularly focus on tool types that are the result of a clear design concept where retouch alters the initial form of the blank (e.g. as in the shaping of bifacial pieces) and which were often employed by scholars as (cultural) marker for identifying populations or naming technocomplexes. Finally, we need to note that for inter-site comparison our analytic resolution—the scale at which we can identify variation—is constrained by the minimum analytic scale of the available data, which is typically artefact types. From previous comments we know that this will predispose our analysis to identify superficial similarities.

Our first case study concerns backed pieces. These are blanks to which steep retouch was applied to create a blunted (80–90°) edge opposite to a sharp cutting edge. In the South African MSA and LSA, these types include a diversity of sizes and shapes (e.g. Ambrose 2002; Wadley and Mohapi 2008; Villa et al. 2010) but are most often made on blades/bladelets and encompass so-called crescents, segments or lunates (curved backing) and trapezoids (truncated backing oblique to axis of tool edge; Fig. 6.2). At Sibudu, backed pieces are present in every major MSA unit, encompassing the "pre-SB", SB, HP, "post-HP" or Sibudan, late MSA and final MSA, but not in all layers. The frequencies vary strongly between units, with only one backed piece each found in the pre-SB (Wadley 2013), four in the SB (5% of all retouched pieces; Wadley 2007), in the HP over a hundred backed pieces (~60–70% of all retouched artefacts; Wadley 2008; de la Peña 2015; M. Will own data), eight in the Sibudan but found only among four of 23 layers (1%; Will and Conard 2018), one in the late MSA (RSP; Villa et al. 2005) and 22 (~6%) in the final MSA being present in over half of the layers (Wadley 2005). There is also much morphometric variability within and between the chrono-cultural units (Wadley 2005; Wadley and Mohapi 2008; de la Peña 2015). While transmission with varying emphasis through the sequence is plausible, there are recurrent sequential gaps in the appearance of backed pieces which weaken the viability of this inference. Most notably the two phases of the record in which backed artefacts are in any way common—the HP (~62 ka) and final MSA (~38 ka)—are separated by more than a meter of sediments and an age gap of >20 kyr with continuity in few other common techno-typological elements.

Table 6.2 General information on South African sites used as case studies

Site	Rainfall zone[a]	Technocomplexes[b]	Chronology (MIS)	Key publications
Sibudu	SRZ	Pre-SB, SB, HP, "Sibudan"/post-HP, late MSA, final MSA	>77–38 ka (MIS 5–3)	Wadley (2005, 2007, 2008), Wadley and Jacobs (2006), Conard et al. (2012), Will et al. (2014)
Klasies River	YRZ	Early MSA (MSA2a, MSA2b), HP, post-HP	~110–50 ka (MIS 5d-3)	Singer and Wymer (1982),Villa et al. (2010), Wurz (2000, 2002)
Diepkloof	WRZ	Early MSA, SB, pre-HP, HP, post-HP	~120–50 ka (MIS 5d-3)	Porraz et al. (2013), Tribolo et al. (2013)
Rose Cottage Cave	SRZ	Pre-HP, HP, post-HP; final MSA, Robberg, Oakhurst, Wilton, post-Wilton	~90–2 ka (MIS 5–1)	Clark (1997a, b), Wadley (1996, 2000); Soriano et al. (2007)
Mertenhof	WRZ	Early MSA, SB, pre-HP, HP, post-HP, late MSA; Robberg	?–18 ka (MIS 5–2)	Will et al. (2015), Schmidt and Mackay (2016)
Apollo 11	WRZ	Early MSA (I, II. III), SB, pre-HP, HP, late MSA, ELSA, MicrolithicLSA, Ceramic LSA	>70–2 ka (MIS 5–1)	Wendt (1972, 1976), Vogelsang (1998), Vogelsang et al. (2010)

[a]SRZ = Summer Rainfall Zone; YRZ = Year-Round Rainfall Zone; WRZ = Winter Rainfall Zone (see Chase and Meadows 2007)
[b]HP = Howiesons Poort; SB = Still Bay; Early MSA (MIS 6–5), late MSA and final MSA (MIS 3 each) are informal designations which are sometimes further subdivided (see Lombard et al. 2012)

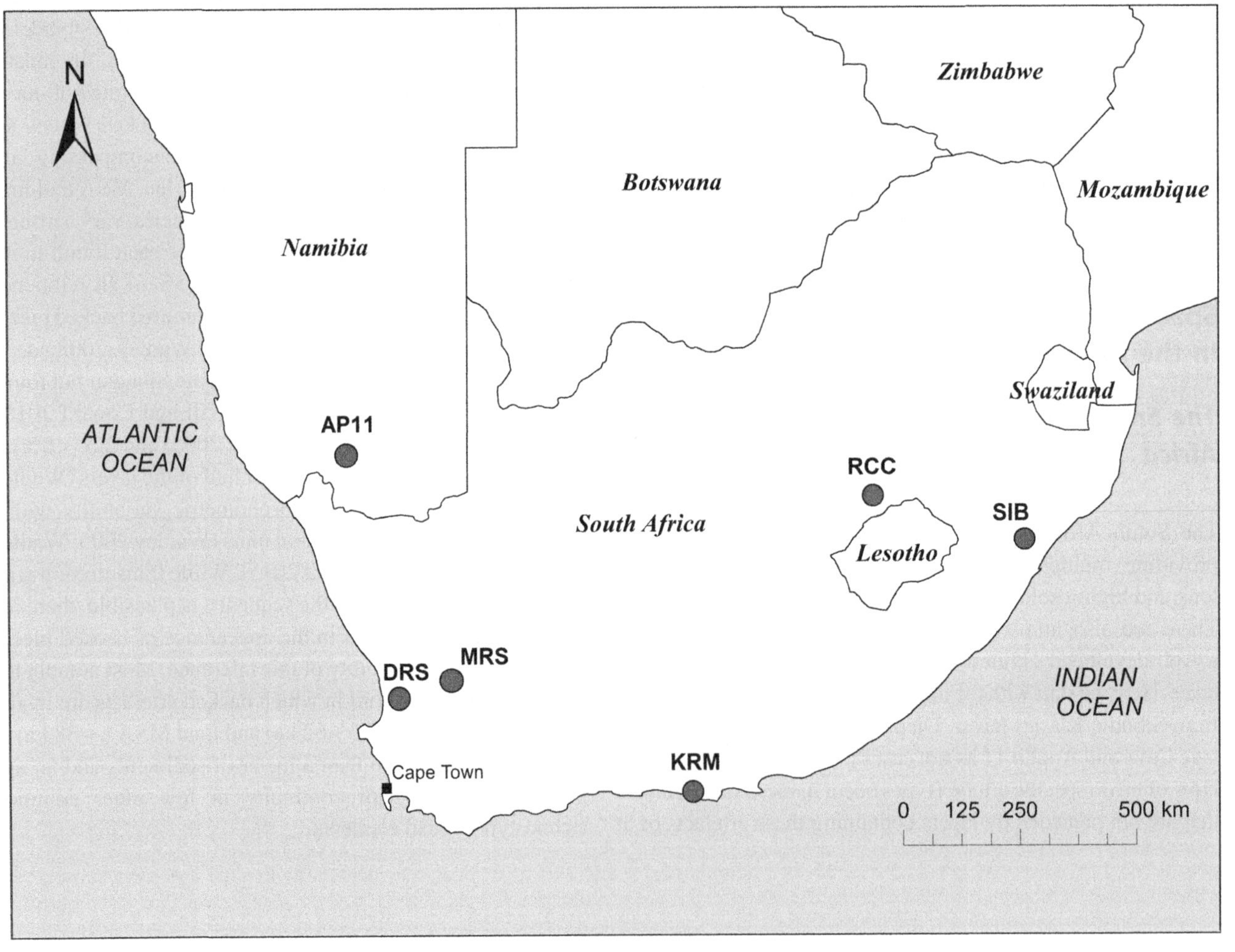

Fig. 6.3 Map of MSA and LSA sites in South Africa and Namibia discussed in text. SIB = Sibudu; KRM = Klasies River Mouth; DRS = Diepkloof; RCC = Rose Cottage Cave; MRS = Mertenhof; AP11 = Apollo 11

The famously long MSA sequence of Klasies River features rare backed pieces in the 'early MSA' or MSA II unit (n = 12) and in the post-Howiesons Poort or MSA III (n = 8). Singer and Wymer (1982) suspect that all of these pieces might be post-depositionally relocated from the intervening HP layers in which >1000 backed pieces were found. The LSA shell midden on top of the MSA deposits also produced three backed pieces (Singer and Wymer 1982: Table 9.1). In Deacon's better-resolved excavations, backed pieces make up >40% of the retouched component in the HP but are absent in the post-Howiesons Poort (Villa et al. 2010), and are not mentioned for other non-HP contexts by Wurz (2000, 2002). The only potential case for re-invention of segments at the site can be made in the LSA context, though arguably on few finds, though as this follows a substantial hiatus we cannot infer that from the Klasies sequence in isolation.

Diepkloof features backed pieces as a recurring element in the SB (2.4%), the various phases of the HP and the post-HP (8.6%). Within the HP, backed pieces vary massively in frequency for all retouched pieces (~5–46%) and also in shape (Porraz et al. 2013). The time-transgressive nature of backed pieces, and their presence through almost all sequence units and most contiguous layers most plausibly indicates information transmission (diffusion or migration), however this needs to be set against potentially massive gaps in time (~20 ka) between some of the HP sub-stages (Tribolo et al. 2013); confident assertions here are difficult given noted problems in absolute dating (Jacobs and Roberts 2015, 2017).

Our own results from the long and as yet undated Mertenhof sequence have yielded an absence of backed pieces in the earliest MSA, but their variable presence through the subsequent SB (n = 2), HP (n = 36), post-HP (n = 5), as well as the later MSA layers (n = 2; Will et al. 2015). While the HP and early post-HP backed artefacts occur through a continuous sedimentary block consistent with transmission, the rare backed pieces from the later MSA occur in a discrete sedimentary package suggestive of an independent origin.

The Namibian site of Apollo 11 with a long sequence of both MSA and LSA occupations shows presence of backed pieces in the SB (n = 2), HP (n > 100; 20.5%), and late MSA (n = 1; 4.5%), although the pieces from the SB might have been intrusive (Vogelsang 1998; Vogelsang et al. 2010). Importantly, there are also numerous small backed pieces (<20 mm; n > 300) in the microlithic LSA complex of the Wilton dated to ~9–6 ka but none in the intervening early LSA dating ~19.7–13 ka (Wendt 1972, 1974, 1976; Vogelsang et al. 2010). Thus at least two pulses with abundant backed pieces can be discerned with a distance of ~50 kyr (HP and Wilton) that crosscut underlying technological systems (MSA vs. LSA), and which are separated by considerable volumes of cultural material.

The stratigraphy at Rose Cottage Cave has yielded abundant and variable backed pieces in the HP (n = 177; 55%; Soriano et al. 2007)—but also some in the post-HP (n = 13; 5%; Soriano et al. 2007) and final MSA (n = 12; 9%; Clark 1997b)—followed much later by smaller backed pieces in all LSA layers (Clark 1997a; Wadley 2000a), but particularly frequent in the Robberg (~13 ka, n = 28; 19%; Wadley 1996), Wilton (~8–6 ka; n = 56; 17%) and post-Wilton (~2 ka; n = 101; 12%; Wadley 2000b). Given hiatuses between units, we cannot with confidence infer transmission in this case but the potential for cultural reservoir effects is clear.

The Regional Scale: Folding Space into Time in Southernmost Africa

To help resolve some of these uncertainties we turn now to the spatial element of our assessment. Here we use data both from the well-resolved sequences we have discussed and from more coarsely resolved and/or undated sites throughout southern Africa. Because the sites are widely dispersed across southernmost Africa we use the rainfall zones as an organizing principle. From east to west, South Africa can be separated into summer (SRZ), year round (YRZ) and winter rainfall zones (WRZ) based on the season during which the majority of rain falls. These zones constitute meaningful spatial units of analysis since paleoenvironmental research indicates important differences with regards to precipitation, floral and animal communities during the Late Pleistocene (Mucina and Rutherford 2006; Chase and Meadows 2007; Chevalier and Chase 2015).

Starting again with backed pieces, various forms and proportions of these artefacts are the typological hallmark of two distinct technocomplexes in South Africa: The HP in the MSA dated to ~70–60 ka (Jacobs et al. 2008; Lombard 2009; Henshilwood 2012; with the outlier of Diepkloof at ~100–50 ka; Tribolo et al. 2013) and the Wilton in the LSA dated to ~10–2 ka (Sampson 1974; Deacon 1984; Wadley 2000a, b). Both technocomplexes come from numerous sites and are geographically widespread, crosscutting ecological zones. Whereas Wilton backed pieces are generally smaller compared to HP specimens, there is large overlap in morphometrical attributes. Other than an underlying blade/bladelet basis, however, the Wilton and HP are quite distinctive in their technology, one being situated in a Levallois system and the other in prismatic blade and bipolar reduction. In between these two technocomplexes, the SRZ has provided a couple of sites with final MSA (~50–30 ka) assemblages—notably from Holley Shelter, Rose Cottage Cave, Sehonghong, Shongweni, Sibudu and Umhlatuzana (Clark 1997b; Wadley 2005; Bader et al. 2015)—that feature backed pieces as part of their retouched components. As noted in the dated Sibudu case, however, this recurrence follows an absence of >20 kyr. Backed pieces again become rare through the early LSA and Robberg, before returning to prominence in the Wilton, a further gap of >20 kyr (e.g. Deacon 1984; Wadley 1987; Mitchell 1988; summary data in Will and Mackay 2017). The separation of recurrence is even more striking in the WRZ and YRZ, where backed pieces remain rare through the late MSA and subsequent early LSA, Robberg and Oakhurst units (Low and Mackay 2016, Porraz et al. 2016, Low 2019). The time gap between the HP and Wilton pulses is in the order of 50 kyr. Viewing the WRZ/YRZ in isolation, convergence seems by far the most likely explanation for recurrence of backed pieces into the Wilton. Even if we allow for the possibility of transmission of conserved information from the adjacent SRZ, there remain aligned gaps of ~20 kyr either side of the final MSA across which transmission seems unlikely. On that basis it seems probable that backed artefacts underwent three separate instances of reinvention in southernmost Africa, crosscutting MSA/LSA technological systems.

As a next example we look at the category of bifacial points which encompasses all varieties of pieces retouched invasively on both surfaces with the aim of shaping the blank (faconnage) and with a convergent tip. As discussed above, lanceloate-shaped bifacial points in South Africa are usually seen as the *fossil directeur* of the SB (e.g. Henshilwood 2012). Starting again with Sibudu, the site currently represents a unique case for bifacial technology in southern Africa (see Will and Conard 2018) with evidence for their recurrent production in the "pre-SB", SB, HP, Sibudan, late MSA and final MSA layers. This being said, knappers made differential use of this technological option, used a variety of raw materials, employed distinct methods and produced a broad spectrum of morphometric variants during these phases (Wadley 2005, 2007; Villa and Lenoir 2006; Mohapi 2012; Conard et al. 2014; de la Peña and Wadley 2014a; de la Peña 2015; Soriano et al. 2015). While bifacial pieces from the HP and lower Sibudan show many similarities and are in direct stratigraphic connection, the bifacial technology from the "Pre-SB" (with serrated points; >80 ka; Rots et al. 2017), SB (with symmetrical leave-shaped points; ~77 ka; Wadley 2005; Soriano et al. 2015) and final MSA layers (hollow-based points; ~38 ka: Wadley 2005; see also Mohapi 2012) are distinctly different in form, opening the question of whether they reflect reinvention or drift. Rose Cottage Cave, also in the SRZ, lacks a SB phase but has yielded 5 bifacial and partly bifacial points as a minor component of the post-HP (2%)—but not in the HP (Soriano et al. 2007)—as well as two partially bifacial, stemmed points in the final MSA (~28 ka; Clark 1997b). With the exception of two small bifacial and stemmed points in the LSA of Rose Cottage Cave, one each in the ~20 ka ELSA (Clark 1997a) and ~2 ka post-Wilton (Wadley 2000b: 94), none of the LSA sequences in our sample has yielded any bifacial points. Bifacial pieces also occur in SRZ contexts at Sibudu, Umhlatuzana and Umbeli Belli in the Sibudan/post-HP, late MSA and final MSA, though with important differences in frequency, production techniques and morphometric attributes (Kaplan 1990; Wadley 2005; Villa and Lenoir 2006; Mohapi 2012, 2013; de la Peña 2015; Bader et al. 2016; Will and Conard 2018). Chronologically before the SB during MIS 7–5, the Pietersburg technocomplex (Sampson 1974) of Border Cave in northern KwaZulu-Natal features leaf-shaped bifacial points that are dated by ESR to between ~230–80 ka (Grün and Beaumont 2001; Grün et al. 2003), and similar sites with bifacials pieces such as Wonderwerk Cave, Cave of Hearths, Bushman Rock Shelter, Mwulu's Cave and Olieboompoort also likely pre-date the SB (Tobias 1949; Eloff 1969; Beaumont 1978; Volman 1984; Mason and Brain 1988; Beaumont and Vogel 2006). In addition, recent modern excavations at Sibudu have unearthed multiple layers of bifacial point production with serrated pieces dated to >77 ka, separated from the classic SB by several strata without bifacial technology (Conard et al. 2014; Conard and Porraz 2015; Rots et al. 2017).

The other MSA/LSA sites, particularly from the WRZ and YRZ, provide a different picture. Bifacial points are rare to absent in early (MIS 5) or late (MIS 3) MSA deposits of the long sequences at Klasies River (Singer and Wymer 1982; Wurz 2000, 2002), Diepkloof (Porraz et al. 2013), and Mertenhof (Will et al. 2015; own data). Klasies River has a small number (n = 5) of bifacially worked points in the upper MIS 5 deposits (MSA II) and following HP (n = 6) according to Singer and Wymer (1982)—with some additional pieces reported by Wurz (2000) from the Deacon sample, but none are similar to typical SB variants (Singer and Wymer 1982; Wurz 2002). At Diepkloof, lanceolate-shaped bifacial points characterize the SB (5.3%) and the immediately preceding "pre-SB" (0.9%), but bifacial pieces also occur in 4 out of the 24 assemblages of the "Early HP" (n = 1–8) located right above the SB (Porraz et al. 2013) and are absent thereafter in the sequence. Importantly, the few HP bifaces are "atypical relative to the SB" (Porraz et al. 2013: 3386) and might thus suggest an independent origin. This pattern is reminiscent of Apollo 11 with small numbers of fully bifacial points coming from both the SB (n = 4) and the HP (n = 3) but absent through the remaining 60 kyr of the deposit (Vogelsang et al. 2010; also Wendt 1976). At Mertenhof, bifacial points occur exclusively in the RGS stratum which is attributed to the SB, with a complete absence for the earlier and later MSA and LSA deposits (Will et al. 2015; own data).

It is possible to conceive of bifacial technology in southernmost Africa in a number of different ways. First, its persistence in the SRZ might reflect multiple instances of independent invention—something suggested by the quite different form that the technology takes when considered at higher resolution (i.e. with respect to production systems and morphology). Alternatively, it might reflect a pool of conserved information in that region, and thus diffusion through time with modification. Absent substantial gaps in their occurrence we cannot resolve these alternatives. The picture in the WRZ and YRZ is more intriguing, with occurrence restricted to a single block of time. Arguments have been made both for differences in form between the SRZ examples and those from the WRZ and YRZ signaling convergence (Archer et al. 2016), and similarities in production signaling diffusion (Hogberg and Lombard 2016). The situation here is thus unclear and requires more research.

Few core forms have been argued to be of specific significance as they occur throughout most MSA phases and tend to be flexibly used within a shared technological system of Mode 3 (e.g. prepared cores such as Levallois). The exception are so-called HP blade/bladelet cores which were first defined by Villa et al. (2010) as similar to Levallois cores but with the difference that the intersection of the debitage surface with the platform and the back of the core is not a plane but more convex. Most of our case sites (e.g. Klasies River; Diepkloof; Sibudu) feature a HP occupation with typical HP cores (Villa et al. 2010; Porraz et al. 2013; de la Peña 2015; Will, own data). No comparable cores have been attributed to other MSA or LSA phases within these localities as far as we are aware. At Sibudu, even the Sibudan that directly follows the HP and shares some other techno-typological elements lacks these cores (de la Peña and Wadley 2017; Will and Conard 2018). Overall, these observations render initial innovation and subsequent diffusion/dispersal of this blade production strategy the most likely explanation. As a limitation of our approach, the late definition of this core type makes it complicated to find it in earlier publications and its documentation in other context by newer studies will be important.

In sum, there is good evidence for convergence in backed pieces both within the MSA, between the MSA/LSA and also within the LSA in southernmost Africa. Bifacial points yield a less clear signal: they may show multiple events of independent innovation or conserved but variable information in MSA sites of the SRZ (yet not across the MSA/LSA with the exception of (rare) stemmed and party bifacially-worked points), but they occur only in one contiguous temporal block in the WRZ/YRZ, the possible connection to the SRZ samples of which is contested. 'HP cores' seem to be only present in the HP so far and we have not been able to find convincing cases of convergence in the studied sites.

The Large Scale: The African Continent

Our final scale concerns the entire continent of Africa to examine inter-regional examples of convergence in the MSA and LSA. Since the spatial scale is now much broadened and case studies might be thousands of kilometers apart from one another, the baseline probability of convergence is higher. We take up the previous case studies from smaller scales and add specific core and tool types to this discussion which have figured prominently in recent approaches to identify technocomplexes or trace human migrations.

When examining the whole continent of Africa, the temporal and spatial distribution of backed pieces grows markedly. The earliest backed tools date to about ~270–170 ka in the Lupemban of Twin Rivers (15% of all retouch), Kalambo Falls and Kabwe in Zambia, as well as ~130–100 ka at Mumbwa Caves (Barham 2000, 2002; Clark and Brown 2001). In eastern Africa, backed pieces during the MSA come from well-stratified and long sequences such as Mumba—with occasional pieces already in early MIS 5 (Mehlman 1989; Marks and Conard 2008)—but also Goda Buticha (MIS 4–3; Leplongeon et al. 2018), Mochena Borago (>50–37 ka; Brandt et al. 2012, 2017), and particularly frequent at the end of the MSA and close to the LSA transition such as at Enkapune ya Muto (Ambrose 1998, 2002). Gota Buticha features backed pieces throughout the stratigraphy running from ~63 ka in the MSA and crosscutting to the LSA until <4 ka (Leplongeon et al. 2018). Backed blades are also reported from several undated MSA sites in Tanzania (Willoughby 1996) and in Zimbabwe as part of the Bambatan and Tshangulan industries (Cooke 1963). Backed microliths have also long been seen as a hallmark of the LSA in sub-Saharan Africa (review in Ambrose 2002), the Upper Paleolithic/Epipaleolithic of northern Africa (e.g. in the Iberomarusian; Barton et al. 2005; Bouzouggar et al. 2008; Olszewski et al. 2011; Linstädter et al. 2012) and the LSA of MIS 2 in western Africa (Chevrier et al. 2018). Taking eastern Africa as one geographical example, backed blades/bladelets feature in many LSA sites in Kenya, Tanzania, Ethiopia and Somalia at the interface of an MSA/LSA transition around ~50–40 ka (e.g. Mehlman 1989; Ambrose 1998, 2002) as well as after ~15 ka (Ambrose 2002; Guthertz et al. 2014; Leplongeon et al. 2018; Shipton et al. 2018). All in all, backed pieces appear to be particularly prone to independent innovation both within MSA but particularly among LSA contexts.

On the scale of the African Stone Age, various forms of bifacial technology are present in different regions and phases of the MSA, and from its very beginning (see McBrearty and

Brooks 2000; McBrearty 2003; Tryon and Faith 2013). Most prominently, bifacial technology with large and carefully shaped points characterizes the Lupemban of central and eastern Africa at numerous sites (McBrearty 1988; Clark 2001; Taylor 2011, 2016; Tryon et al. 2012; Faith et al. 2016). Bifacially flaked points of various morphologies have also been reported from MSA assemblages in the northeastern central African rainforest (Cornelissen 2016), the early Nubian complex of north-eastern Africa (Van Peer et al. 2003; Van Peer and Vermeersch 2007; Van Peer 2016), the long MSA sequences at Mumba, Goda Buticha and other sites in eastern Africa (McBrearty and Brooks 2000; Bretzke et al. 2006; Leplongeon et al. 2018), the Aterian of North Africa (Garcea 2004; Barton et al. 2009; Dibble et al. 2013; Scerri 2013), the Bambatan of Zimbabwe (Armstrong 1931) and throughout MIS 4–2 in West Africa (Chevrier et al. 2018). Similar to our previous assessment, bifacial points are non-existent at most sites post-dating the MSA, with rare exceptions such as Ravin de la Mouche in Mali at ~9 ka (Huysecom et al. 2009), Goda Buticha at ~8–6 ka (Leplongeon et al. 2018) and small stemmed pieces of southern Africa after 2 ka (see below).

In sum, every major region of the African continent has produced various forms of bifacial points within the MSA—but not LSA—sometimes with continuous patterns of similar forms across large spans of time (Aterian: ~140–60 ka, Lupemban: ~260–30 ka) but others with multiple different morphometric types within relatively short periods (e.g. southern Africa). Much care is needed to distinguish between these different variants within the MSA, but in most cases convergence cannot be ruled out *a priori*, particularly within an MSA technological sequence.

Nubian cores are an interesting example to assess at the continental scale of Africa, as they are often considered to be of spatio-temporally constrained dispersion. Due to the elaborate and specific method of core preparation, the "Nubian Complex"—defined primarily on the presence of Nubian cores—has been equated both with an information sharing network (Van Peer 1998) and a group of people (i.e. modern humans), potentially allowing to trace early dispersals of *Homo sapiens* (Rose et al. 2011; Crassard and Hilbert 2013; Usik et al. 2013; Hilbert et al. 2017). Regarding space and time, the presence of Nubian cores in this system is thought to be limited to northeastern Africa and Arabia and dating mostly to MIS 5. Taking stock of current find spots of Nubian cores, the vast majority has been found in northeastern and eastern Africa and Arabia. This being said, examples of Nubian cores have also been found in Libya (Cremaschi et al. 1998), Algeria (Van Peer 1986) and as far west as Mauritania (Pasty 1999). Outside Africa they occur not just in the Arabian Peninsula, but also in the Negev Desert (Goder-Goldberger et al. 2016, 2017) and the Thar Desert in India (Blinkhorn et al. 2013). There is now also clear demonstration of many dozen Nubian Type 1 and 2 cores ca. 6000 km southwest and >20 kyr later in the MSA of southern Africa at a couple of sites (Hallinan and Shaw 2015; Will et al. 2015; Hallinan and Parkington 2017). Considering the spatial and temporal distance as well as the lack of other similar assemblage elements to their northeastern counterparts, the southern African Nubian cores are an exemplary case of convergence. The distribution of such cores thus cannot be simply assumed to reflect information sharing networks or distinct populations; this requires demonstration, including consideration of the space-time distribution of given samples and the presence and proportion of other techno-typological elements (e.g. Goder-Goldberger et al. 2016, 2017; Groucutt 2020).

For a long time, the presence of stemmed or tanged tools has been used to define and identify the Aterian technocomplex in the MSA of northern Africa (Caton-Thompson 1946; Cremaschi et al. 1998; Van Peer 1998). Recent research has shown that 'Aterian points' and other tanged artefacts vary in frequency and morphology, have a much less-restricted time range than previously thought dating from ~140–60 ka, and are generally insufficient markers to define the "Aterian" by itself (see Dibble et al. 2013; Scerri 2013, 2017; also Kleindienst 2006). Furthermore, tanged artefacts are often the only similarity between otherwise dissimilar 'Aterian' assemblages, while conversely sites which are otherwise technologically similar can differ in the presence/absence of such artefacts (Scerri et al. 2014). This exemplifies the problem of focussing on singular artefacts when attempting to resolve past population interaction.

This aside, while no other MSA technocomplex in Africa has yielded comparable amounts of these pieces, recent findings have also uncovered pedunculated points at ~44 ka in stratified deposits in Senegal (Niang et al. 2018) and stemmed points are included in Beaumont and Vogel's (1972) definition of the final MSA in South Africa, although they are only found in well-stratified deposits at Rose Cottage Cave at ~28 ka (Clark 1997b). Pressure-flaked barbed and tanged points are rare pieces of the LSA in southern Africa, but were found at over thirty sites in a circumscribed area of east-central interior of southern Africa, post-dating 2 ka and continuing until historic times (Humphreys 1969; Carter et al. 1988; Humphreys 1991; Mitchell 1996; Wadley 2000b). In addition, three tanged points from Nelson Bay Cave (Inskeep 1987) and two bifacial and stemmed small points in the ELSA of Rose Cottage Cave (~20 ka; Clark 1997a) predate the other finds. The LSA of Central Africa features stemmed or tanged foliate points from "Tshitolian" contexts dated to ~15–2 ka in sites such as Gombe Point (DR Congo) but often from surface contexts (Clark 1963; Cahen 1978; Miller 2001). In northeastern Africa and the Sahara, localities from the early Holocene onward (Epipaleolithic) have yielded numerous sites with tanged points including the Ounanian, Ounan-Harif or Harifian points from the Sahara and Sinai (Clark et al. 1973; Going-Morris 1991; Smith 1993).

In sum, tanging is a rare assemblage element in the MSA of Africa overall, but more frequent in different areas of the continent during the LSA. The time-transgressive nature and frequent occurrence of tanging across a coherent range of northern Africa that is not seen anywhere else in the contemporaneous MSA makes continual transmission from a conserved pool of information within this region possible. This being said, repeated innovation in similar environmental contexts and their subsequent diffusion in circumscribed areas—an interpretation reached by Scerri (2013)—cannot be excluded, particularly considering the vast geographic and temporal spread of these pieces north of the Sahara and the frequent interstratification between assemblages with and without tanged tools (Aouadi-Abdeljaouad and Belhouchet 2008; Nami and Moser 2010; Dibble et al. 2013). A combination of multiple innovation processes with accompanying diffusion and migration is likely. Considering the continental scale of the MSA, while the probability interval for convergence is large for tanged pieces, the empirical evidence for independent derivation is almost non-existent south of the Sahara. This being said, various tanged artefacts within northern Africa as well as at the MSA/LSA transition in southern Africa and in different areas of the continent in the LSA show the basic potential for its independent derivation again precluding *a priori* dismissal of this process.

How Frequent is Convergence? A Matter of Space, Time and Resolution

Convergence can be identified at various spatial and temporal scales in the African record of the last 300 kyr. Convergence is not rare but nor does it seem equally frequent for all classes of artefacts. Backed artefacts, most notably, recur within sequences and across regions in ways that are highly likely to reflect repeated independent derivation. While Clarkson et al. (2018) have discussed the many reasons for which backed artefacts would have been particularly prone to serial reinvention—and this applies at a global scale (e.g. Hiscock et al. 2011; Jochim 2018)—these arguments and observations need to be set against the tendency to consider backing technology inherently innovative (Jacobs et al. 2008; Brown et al. 2012; Ziegler et al. 2013).

The case of bifacial points seems somewhat different. They are extremely widespread across Africa, and persistent through some sequences in ways that do not always require us to invoke convergence. Equally, however, there are zones of space (e.g. WRZ/YRZ of southernmost Africa) and time (e.g., the LSA) in which they are generally very rare. This suggests that while bifacial points can be and undoubtedly were invented independently many times (O'Brien et al. 2014), and that in some areas were remarkably and persistently abundant (Shott 2020), there were also some constraints on their probability of reinvention that limits their space/time distribution. While we were unable to resolve the case of southernmost Africa, that impasse serves to reveal the importance of analytic scale. At a coarse scale, our requirement of spatio-temporal contiguity is met across the region at around 75–71 ka. Yet at a finer scale, and as others have noted, there is no documented SB near the SRZ/YRZ contact (Fisher et al. 2013; Archer et al. 2016). At what scale does a gap become meaningful? Similarly, SB assemblages resemble each other at the coarse scale—enough to suggest transmission (e.g. Mackay et al. 2014)—but at finer scales they can be considered quite different. At what scale does difference become meaningful?

Reflection on scale is also pertinent in our consideration of core types. Restricted to southernmost Africa, 'HP' Levallois cores are limited to a contiguous block of space time that almost certainly reflects transmission either through diffusion or migration. The same could potentially be said of the Nubian cores of northeast Africa and Arabia (but see Groucutt 2020). However, when we broaden our analytic scale, evidence for convergence in Nubian cores emerges. While no comparable evidence has yet appeared for HP cores, this probably reflects their lower profile as a research search image.

Perhaps the most interesting case we considered is that of tanging. First, as noted above, tanging is an artefact attribute and not an artefact type *per se*: unifacial and bifacial tools can have tangs, and tangs are found on points and scrapers. It is not necessarily conceptually or practically complex, and provides clear functional benefits. Tanging recurs repeatedly in northern Africa but rarely in southern Africa. This seems consistent with its distribution at even wider scales: tanging is noted through Eurasia, the Pacific and the Americas (e.g. Laville and Rigaud 1973; Seong 2008; Erlandson et al. 2011; Scerri 2012; Groucutt et al. 2015), though it is absent in Australia. It occurs in the Pleistocene, the Holocene, and even into recent non-lithic projectile systems. Thus, tanging does seem acutely susceptible to convergence *per se*. Viewed at the regional scale, the patchy space-time distribution of tanging in north Africa may reflect repeated reinvention (Scerri 2013). This perspective is reinforced at the global scale by its high probability of recurrence. Viewed at the meso-scale of inter-regional comparison however, and particularly through the lens of tanging's absence in the large space-time context of sub-Saharan Africa, this technique's appearance in north Africa looks more like conserved and transmitted information (Scerri 2017).

To summarize general observations from this review, even for the most derived assemblage types, such as specific tool and core forms that require multiple, interrelated steps of manufacture (e.g. bifacial points; Nubian cores), convergence occurs both within the MSA and LSA, and sometimes even across (e.g. backed pieces). We also found some hierarchical differences due to underlying technological systems such the MSA and LSA: very few bifacial points are found in LSA deposits and no Nubian cores at all. Overall, however, these observations challenge dominant models of cultural transmission in Stone Age research that often emphasize single contexts of innovation. At the same time, one should not overinflate the use of convergence since processes such as diffusion and migration did play an important role in Stone Age Africa and sometimes this might be captured in lithic artefacts, e.g. HP cores and the associated blade/bladelet technology, stemmed bifacial points in late Holocene LSA in southern Africa, and potentially Aterian tanged tools. How frequent is convergence then? While no absolute number is assignable as it is a multi-component process and hinges on the assessment of probabilities, independent innovation in Stone Age Africa is neither the exception nor the norm. It is a common process that needs to be taken seriously when explaining similarities between lithic artefacts and cannot be dismissed *a priori* or offhandedly. Good knowledge of similar types of lithic artefacts is necessary across the regional and continental scale.

Our analyses also clearly showed the importance—and complexities—of scale for resolving convergence. As Dunnell (1971: 136–137) puts it "the question of whether or not two objects share features is a direction function of the definition of the features and the scale at which they are conceived". As we change scale both in time and space, our perception of what constitutes meaningful discontinuities changes, but so too, and in the opposite direction, does the latent probability of convergence. For example at the local or site scale, we might worry about whether the 20 kyr gap ($\sim$800 generations!) between backed artefact pulses at Sibudu is enough to preclude transmission, yet at the continental scale view, the 20 kyr gap between backed artefacts at Sibudu, Mumba and Batadomba Lena appears to be less of a constraint (Mellars 2006). Similarly, the more broadly we define an artefact class the more likely we are to see it recur; the occurrence of bifacial points *sensu lato* underwrites the SB technocomplex across southernmost Africa, but that same superficial similarity was also once used to extend the SB concept from Cape Point to the Horn of Africa. At a finer resolution, variation emerges at the local scale, and broad-scale similarities are almost certain to evaporate. Given the importance of scale to the probability and perception of convergence, what is the appropriate analytic scale to take? We address this in our final section. What is clear is that future comparative analyses of similar-looking tools and cores require clear descriptions of the metrics and morphology, the modes of production and the assemblage context in which they are found to study such convergence phenomena in more detail (e.g. in shape vs. overall production).

In conclusion, convergence remains an underestimated phenomenon in the literature. Yet, some recent studies of stone tools—reflecting in particular on their reduced potential morphometric space—have taken independent innovation as a serious issue in various contexts. Examples in Africa include the Aterian tanged points (e.g. Scerri 2012, 2013), backed pieces and Nubian cores (Groucutt et al. 2015; Will et al. 2015), bifacial points in southern Africa (Archer et al. 2016) but also studies in the Middle and Upper Paleolithic of Europe (e.g. White et al. 2011; Adler et al. 2014; Jochim 2018) and global assessments (e.g. Hiscock et al. 2011; Clarkson et al. 2018). As briefly shown in the next section, however, diffusion and dispersal are still the most commonly deployed explanations.

Implications for Identifying Human Populations and Dispersals

We began this paper by asking how common convergence is. The answer to that question depends on the kinds of artefacts being considered and the analytic scale. Backed artefacts seem to recur most regularly among the artefacts we considered, while the distribution of specific core forms seems more constrained. The use of broad analytic classes increases probability of convergence, as does the use of broad spatial and temporal scales. Given these observations, we feel that global and continental-scale assessments are inevitably inappropriate, as they typically leverage the inherently patchy nature of the archaeological record to infer continuity across spatial and temporal gaps (e.g. Mellars 2006; Stanford and Bradley 2012). At this scale, the risk of convergence in lithic systems will typically be so high that it should be assumed unless extraordinary alternative (i.e. genetic) evidence is presented.

Meso-scale studies which explore continuity across adjacent regions (e.g. Armitage et al. 2011; Rose et al. 2011; Akhilesh et al. 2018; Hershkovitz et al. 2018) or in the form of technocomplexes, pose similar challenges which require scrutiny. Some of the stone artefacts utilized in these studies are particularly problematic, such as backed microliths, which are probably the weakest possible marker of transmission that we studied here given their global tendency for independent derivation (see also Clarkson et al. 2018). Studies should refrain from using single core or tool types which not only mask assemblage variability but also increase the chance of convergence as all of these forms were shown empirically to have such a tendency regardless of their complexity. The Aterian and Nubian technocomplexes present instructive examples of the plethora of problems that arise when using the presence and absence of particular *fossiles directeurs*—in this case tanged, or stemmed, artefacts and a particular sub-type of Levallois core—as the primary or single element of classification. An exclusive focus on such types hindered comparative research and cultural systematics in the northern African MSA by masking both technological variability and similarities between taxonomic units (see discussion in Dibble et al. 2013; Scerri 2013, 2017; also Kleindienst 2006), and has placed the ages and contexts of early modern human dispersals on an unsure footing. In any case, studies using similarly-looking stone tools to recognize past populations and track human dispersals always need to test for the potential of independent innovation and not assume migration or diffusion *a priori*.

Instead of using specific core and tool types to classify assemblages into one or another industry and track dispersals, other analytical scopes are more favorable to the source material. Techno-economic perspectives aided by experimental set-ups could aim to explain the variable presence and frequency of particular stone artefacts with regard to their function as part of overall adaptive solutions to variable environmental, geological and demographic conditions within the African Pleistocene and Holocene (e.g. Iovita 2011; Sisk and Shea 2011; Tomasso and Rots 2017; Clarkson et al. 2018; see Jochim 2018 for Late Pleistocene Europe). Apart from a frequent discussion about this issue in Aterian tanged tools (e.g. Scerri 2013; Tomasso and Rots 2017) and backed pieces (e.g. Clarkson et al. 2018), Nubian cores seem to often appear in similar ecological circumstances (i.e. more arid regions) and often in open-air sites (e.g. Groucutt 2020): studying the adaptive advantages and functional constraints that lie behind this convergence might be more fruitful than using them as mere tracking devices. Such studies are yet to be undertaken.

Despite the above-mentioned stumbling blocks, stone tools will continue to be used to track ancient human dispersals on various scales due to their unique durability and frequent spatiotemporal patterning. Future research should, however, refrain from using single tool or core types as tracking devices, and explicitly test between the possibilities of diffusion, dispersal and convergence for similarities observed in the archaeological record. Lithic technologies could then remain a critical guide to human population flux under favorable conditions. Research might build upon recent theoretical advancements such as models of cultural information transmission (Boyd and Richerson 1985; Henrich 2001; Eerkens and Lipo 2007; Mesoudi 2011; Shennan 2011) that can serve as bridging theory between past realities and traces of lithic technology observed by modern archaeologists. Such studies should ideally start to work within more circumscribed spatio-temporal contexts (e.g. regions) that assess multiple assemblage elements and the variability in numerous techno-typological domains within a quantitative framework and the aid of statistical methods (e.g. Tostevin 2012; Scerri et al. 2014; Groucutt et al. 2015; Goder-Goldberger et al. 2016). Use of more rigorous analytic procedures, applying phylogenetic and cladistic methods on questions of homoplasies vs. homologies (O'Brien et al. 2001; Lycett 2010, 2015), provides additional tools to test hypotheses of migration, diffusion and convergence in specific cases. Ultimately, a cross-disciplinary strategy that combines (experimental) archaeological data with fossil, genetic and paleoenvironmental information will be the most fruitful approach to study early human dispersals.

Acknowledgements We thank Huw Groucutt for inviting us to participate in this collection of papers regarding the phenomenon of convergence and the many intense discussions we had with him on this very topic. During the writing of this chapter, Manuel Will was supported by a Research Fellowship from Gonville and Caius College (Cambridge). Alex Mackay's research is funded by an Australian Research Council DECRA grant (#DE130100068).**Conflict of Interest** The authors declare that they have no conflict of interest.

References

Adler, D. S., Wilkinson, K., Blockley, S., Mark, D., Pinhasi, R., Schmidt-Magee, B., et al. (2014). Early Levallois technology and the transition from the Lower to Middle Palaeolithic in the Southern Caucasus. *Science, 345,* 1609–1613.

Akhilesh, K., Pappu, S., Rajapara, H. M., Gunnell, Y., Shukla, A. D., & Singhvi, A. K. (2018). Early Middle Palaeolithic culture in India around 385–172 ka reframes out of Africa models. *Nature, 554,* 97.

Ambrose, S. H. (1998). Chronology of the Later Stone Age and food production in East Africa. *Journal of Archaeological Sciences, 24,* 377–392.

Ambrose, S. H. (2002). Small things remembered: Origins of early microlithic industries in sub-Saharan Africa. In R. G. Elston & S. L. Kuhn (Eds.), *Thinking small: Global perspectives on microlithization* (Vol. 12, pp. 9–30). Archeological Papers of the American Anthropological Association.

Ammerman, A. J., & Cavalli-Sforza, L. L. (1984). *The Neolithic transition and the genetics of populations in Europe.* Princeton: Princeton University Press.

Anthony, B. (1972). The stillbay question. In *Congres panafricain de préhistoire*, Dakar 1967: actes de 6e session (pp. 80–82).

Aouadi-Abdeljaouad, N., & Belhouchet, L. (2008). Recent prehistoric field research in Central Tunisia: Prehistoric occupations in the Meknassy Basin. *African Archaeological Review, 25,* 75–85.

Archer, W., Pop, C. M., Gunz, P., & McPherron, S. P. (2016). What is Still Bay? Human biogeography and bifacial point variability. *Journal of Human Evolution, 97,* 58–72.

Armitage, S. J., Jasim, S. A., Marks, A. E., Parker, A. G., Usik, V. I., & Uerpmann, H.-P. (2011). The southern route out of Africa: Evidence for an early expansion of modern humans into Arabia. *Science, 331,* 453–456.

Armstrong, A. L. (1931). Rhodesian archaeological expedition (1929): Excavations in Bambata Cave and researches on prehistoric sites in Southern Rhodesia. *Journal of the Anthropological Institute of Great Britain and Ireland*, 239–276.

Bader, G. D., Will, M., & Conard, N. J. (2015). The lithic technology of Holley Shelter, KwaZulu-Natal, and its place within the MSA of Southern Africa. *South African Archaeological Bulletin, 70,* 149–165.

Bader, G. D., Cable, C., Lentfer, C., & Conard, N. J. (2016). Umbeli Belli Rock Shelter, a forgotten piece from the puzzle of the Middle Stone Age in KwaZulu-Natal, South Africa. *Journal of Archaeological Science: Reports, 9,* 608–622.

Barham, L. S. (Ed.). (2000). *The Middle Stone Age of Zambia*. Bristol: Western Academic & Specialist Press.

Barham, L. (2002). Backed tools in Middle Pleistocene of Central Africa and their evolutionary significance. *Journal of Human Evolution, 43,* 585–603.

Barton, R. N. E., Bouzouggar, J., Collcutt, S. N., Gale, R., Higham, T. F. G., Humphrey, L., et al. (2005). The Late Upper Palaeolithic occupation of the Moroccan Northwest Maghreb during the last glacial maximum. *African Archaeological Review, 22,* 77–100.

Barton, R. N. E., Bouzouggar, A., Collcutt, S. N., Schwenninger, J.-L., & Clark-Balzan, L. (2009). OSL dating of the Aterian levels at Dar es-Soltan I (Rabat, Morocco) and implications for the dispersal of modern Homo sapiens. *Quaternary Science Reviews, 28,* 1914–1931.

Basalla, G. (1988). *The evolution of technology*. Cambridge: Cambridge University Press.

Beaumont, P. B. (1978). *Border Cave*. M.A. disseration, University of Cape Town

Beaumont, P. B., & Vogel, J. C. (1972). On a new radiocarbon chronology for Africa South of the Equator. *African Studies, 31,* 155–182.

Beaumont, P. B., & Vogel, J. C. (2006). On a timescale for the past million years of human history in Central South Africa. *South African Journal of Science, 102,* 217–228.

Bellwood, P. (2013). *First migrants: Ancient migration in global perspective*. Singapore: Wiley Blackwell.

Binford, L. R. (1973). Interassemblage variability: The Mousterian and the functional argument. In C. Renfrew (Ed.), *The explanation of culture change: Models in prehistory* (pp. 227–254). Pittsburg: University of Pittsburgh Press.

Binford, L. R., & Binford, S. R. (1966). A preliminary analysis of functional variability in the Mousterian of Levallois facies. *American Anthropologist, 68,* 238–295.

Blinkhorn, J., Achyuthan, H., Petraglia, M., & Ditchfield, P. (2013). Middle Palaeolithic occupation in the Thar Desert during the Upper Pleistocene: The signature of a modern human exit out of Africa? *Quaternary Science Reviews, 77,* 233–238.

Bouzouggar, A., Barton, R. N. E., Blockley, S., Bronk-Ramsey, C., Collcutt, S. N., Gale, R., et al. (2008). Reevaluating the age of the Iberomaurusian in Morocco. *African Archaeological Review, 25,* 3–19.

Boyd, R., & Richerson, P. J. (1985). *Culture and the evolutionary process*. Chicago: University of Chicago Press.

Brandt, G., Szécsényi-Nagy, A., Roth, C., Alt, K. W., & Haak, W. (2014). Human paleogenetics of Europe: The known knowns and the known unknowns. *Journal of Human Evolution, 79,* 73–92.

Brandt, S. A., Fisher, E. C., Hildebrand, E. A., Vogelsang, R., Ambrose, S. H., Lesur, J., et al. (2012). Early MIS 3 occupation of Mochena Borago Rockshelter, Southwest Ethiopian Highlands: Implications for late Pleistocene archaeology, paleoenvironments and modern human dispersals. *Quaternary International, 274,* 38–54.

Brandt, S., Hildebrand, E., Vogelsang, R., Wolfhagen, J., & Wang, H. (2017). A new MIS 3 radiocarbon chronology for Mochena Borago Rockshelter, SW Ethiopia: Implications for the interpretation of Late Pleistocene chronostratigraphy and human behavior. *Journal of Archaeological Science: Reports, 11,* 352–369.

Bretzke, K., Marks, A. E., & Conard, N. J. (2006). Projektiltechnologie und kulturelle Evolution in Ostafrika. *Mitteilungen der Gesellschaft für Urgeschichte, 15,* 63–81.

Brown, K. S., Marean, C. W., Jacobs, Z., Schoville, B. J., Oestmo, S., Fisher, E. C., et al. (2012). An early and enduring advanced technology originating 71,000 years ago in South Africa. *Nature, 491*(7425), 590–593.

Cahen, D. (1978). New excavations at Gombe (ex-Kalina) point, Kinshasa, Zaire. *Antiquity, 52,* 51–55.

Carter, P. L., Mitchell, P. J., & Vinnicombe, P. V. (1988). *Sehonghong: The Middle and Later Stone industrial sequence at a Lesotho Rock-Shelter*. Oxford: British Archaeological Reports International Series 406.

Caton-Thompson, G. (1946). The Aterian industry: Its place and significance in the Palaeolithic world. *Journal of the Anthropological Institute of Great Britain and Ireland*, 87–130.

Chase, B. M., & Meadows, M. E. (2007). Late Quaternary dynamics of Southern Africa's winter rainfall zone. *Earth Science Reviews, 84,* 103–138.

Chevalier, M., & Chase, B. M. (2015). Southeast African records reveal a coherent shift from high-to low-latitude forcing mechanisms along the east African margin across last glacial–interglacial transition. *Quaternary Science Reviews, 125,* 117–130.

Chevrier, B., Huysecom, É., Soriano, S., Rasse, M., Lespez, L., Lebrun, B., et al. (2018). Between continuity and discontinuity: An overview of the West African Paleolithic over the last 200,000 years. *Quaternary International, 466,* 3–22.

Childe, V. G. (1929). *The Danube in prehistory*. Oxford: Clarendon Press.

Clark, G. A. (1994). Migration as an explanatory concept in Paleolithic Archaeology. *Journal of Archaeological Method and Theory, 1,* 305–343.

Clark, A. M. B. (1997a). The final Middle Stone Age at Rose Cottage Cave: A distinct industry in the Basutolian ecozone. *South African Journal of Science, 93,* 449–458.

Clark, A. M. B. (1997b). The MSA/LSA transition in southern Africa: New technological evidence from Rose Cottage Cave. *South African Archaeological Bulletin, 52,* 113–121.

Clark, J. D. (1954). A provisional correlation of prehistoric cultures north and south of the Sahara. *The South African Archaeological Bulletin, 9,* 3–17.

Clark, J. D. (1963). Prehistoric Cultures of Northeast Angola and Their Significance in Tropical Africa. In *Diamang, Publicacoes culturais* (Vol. 62). Lisbon: Museu do Dundo.

Clark, J. D. (2001). *Kalambo Falls prehistoric site volume III, the earlier Cultures: Middle and Earlier Stone Age*. Cambridge: Cambridge University Press.

Clark, J. D., & Brown, K. S. (2001). The Twin Rivers Kopje, Zambia: Stratigraphy, fauna, and artefact assemblages from the 1954 and 1956 excavations. *Journal of Archaeological Science, 28,* 305–330.

Clark, J. D., Cole, G. H., Isaac, G. L., & Kleindienst, M. R. (1966). Precision and definition in African archaeology. *The South African Archaeological Bulletin, 21,* 114–121.

Clark, J. D., Williams, M. A. J., & Smith, A. B. (1973). The geomorphology and archaeology of Adrar Bous, Central Sahara: A preliminary report. *Quaternaria, 17,* 245–297.

Clarke, D. L. (1968). *Analytical archaeology*. London: Methuen and Co LTD.

Clarkson, C., Hiscock, P., Mackay, A., & Shipton, C. (2018). Small, sharp, and standardized: Global convergence in backed-microlith technology. In B. Buchanan, M. I. Eren & M. J. O'Brien (Eds.), *Convergent evolution in stone-tool technology*. Cambridge, MA: MIT Press.

Collard, M., Shennan, S. J., & Tehrani, J. J. (2006). Branching, blending, and the evolution of cultural similarities and differences among human populations. *Evolution and Human Behavior, 27,* 169–184.

Conard, N. J., Porraz, G., & Wadley, L. (2012). What is in a name? Characterising the 'Post-Howieson's Poort' at Sibudu. South African Archaeological Bulletin, 67, 180–199.

Conard, N. J., & Porraz, G. (2015). Revising models for the cultural stratigraphic sequence of the Middle Stone Age. *South African Archaeological Bulletin, 70,* 124–130.

Conard, N. J., Bader, G. D., Schmid, V. C., & Will, M. (2014). Bringing the Middle Stone Age into clearer focus. *Mitteilungen der Gesellschaft für Urgeschichte, 23,* 121–128.

Cooke, C. K. (1963). Report on excavations at Pomongwe and Tshangula Caves, Matopo Hills, Southern Rhodesia. *South African Archaeological Bulletin, 18,* 73–151.

Cornelissen, E. (2016). The later Pleistocene in the northeastern Central African rainforest. In S. C. Jones & B. A. Stewart (Eds.), *Africa from MIS 6–2: Population dynamics and paleoenvironments* (pp. 301–319). Dordrecht: Springer.

Crassard, R., & Hilbert, Y. H. (2013). A Nubian Complex site from Central Arabia: Implications for Levallois taxonomy and human dispersals during the Upper Pleistocene. *PLoS ONE, 8,* e69221.

Cremaschi, M., Di Lernia, S., & Garcea, E. A. A. (1998). Some insights on the Aterian in the Libyan Sahara: Chronology, environment, and archaeology. *African Archaeological Review, 15,* 261–286.

de la Peña, P. (2015). The interpretation of bipolar knapping in African Stone Age studies. *Current Anthropology, 56,* 911–923.

de la Peña, P., & Wadley, L. (2014). Quartz knapping strategies in the Howiesons Poort at Sibudu (KwaZulu-Natal, South Africa). *PLoS ONE, 9,* e101534.

de la Peña, P., & Wadley, L. (2017). Technological variability at Sibudu Cave: The end of Howiesons Poort and reduced mobility strategies after 62,000 years ago. *PloS one*, 12(10).

de Mortillet, L. L. G. (1883). *Le Préhistorique antiquité de l'homme*. Paris: C. Reinwald.

Deacon, H. J. (1979). A sequence through the Upper Pleistocene and Holocene in South Africa. *World Archaeology, 10,* 241–257.

Deacon, J. (1984). *The Later Stone Age of Southernmost Africa*. Oxford: British Archaeological Reports International Series 213.

Debénath, A., & Dibble, H. L. (1994). *Handbook of Paleolithic typology. Lower and Middle Paleolithic of Europe* (Vol. I). Philadelphia: The University Museum.

Dibble, H. L., Aldeias, V., Jacobs, Z., Olszewski, D., Rezek, Z., Lin, S. C., et al. (2013). On the industrial attributions of the Aterian and Mousterian of the Maghreb. *Journal of Human Evolution, 64,* 194–210.

Dortch, C., & Bordes, F. (1977). Blade and Levallois technology in Western Australian prehistory. *Quartär, 27*(28), 1–19.

Dunnell, R. C. (1971). *Systematics in prehistory*. New York: Free Press.

Dunnell, R. C. (1980). Evolutionary theory and archaeology. *Journal of Archaeological Method and Theory, 3,* 35–99.

Eerkens, J. W., & Lipom, C. P. (2005). Cultural transmission, copying errors, and the generation of variation in material culture and the archaeological record. *Journal of Anthropological Archaeology, 24,* 316–334.

Eerkens, J. W., & Lipo, C. P. (2007). Cultural transmission theory and the archaeological record: Providing context to understanding variation and temporal changes in material culture. *Journal of Archaeological Research, 15,* 239–274.

Eloff, J. F. (1969). Bushman Rock Shelter, Eastern Transvaal: Excavations, 1967–8. *South African Archaeological Bulletin, 24,* 60.

Elston, R. G., & Kuhn, S. L. (Eds.) (2002). *Thinking small: Global perspectives on microlithization*. Archeological Papers of the American Anthropological Association 12, Airlington.

Engels, F. (1884). Der Ursprung der Familie, des Privateigentums und des Staats. Hottingen-Zürich: Verlag der Schweizerischen Volksbuchhandlung.

Eren, M. I., Lycett, S. J., Roos, C. I., & Sampson, C. G. (2011). Toolstone constraints on knapping skill: Levallois reduction with two different raw materials. *Journal of Archaeological Sciences, 38,* 2731–2739.

Erlandson, J. M., & Braje, T. J. (2011). From Asia to the Americas by boat? Paleogeography, paleoecology, and stemmed points of the Northwest Pacific. *Quaternary International, 239,* 28–37.

Evans, U. (1994). Hollow Rock Shelter, a Middle Stone Age site in the Cederberg. *South African Field Archeaology, 3,* 63–73.

Faith, J. T., Tryon, C., & Peppe, D. J. (2016). Environmental change, ungulate biogeography, and their implications for early human dispersals in equatorial East Africa. In S. C. Jones & B. A. Stewart (Eds.), *Africa from MIS 6-2: Population dynamics and paleoenvironment* (pp. 233–245). Dordrecht: Springer.

Fisher, E., Albert, R.-M., Botha, G. A., Cawthra, H., Esteban, I., Harris, J. W., et al. (2013). Archaeological reconnaissance for Middle Stone Age sites along the Pondoland Coast, South Africa. *PaleoAnthropology, 2013,* 104–137.

Futuyma, D. J. (2013). *Evolution*. Massachusetts: Sinauer Associates.

Garcea, E. A. A. (2004). Crossing deserts and avoiding seas: Aterian North African-European relations. *Journal of Anthropological Research, 60,* 27–53.

Goder-Goldberger, M., Gubenko, N., & Hovers, E. (2016). "Diffusion with modifications": Nubian assemblages in the central Negev highlands of Israel and their implications for Middle Paleolithic inter-regional interactions. *Quaternary International, 408,* 121–139.

Goring-Morris, A. N. (1991). The Harifian of the Southern Levant. *The Natufian Culture in the Levant,* 173–216.

Goldberger, M. G., Ginat, H., Ragolski, G., Seri, G., & Abadi, I. (2017). Middle Palaeolithic find spots with Nubian cores from the Southern Negev and the Arava, Israel. *Journal of Lithic Studies,* https://doi.org/10.2218/jls.v4i1.1688.

Goodwin, A. J. H., & van Riet Lowe, C. (1929). The Stone Age cultures of South Africa. *Annuals of the South African Museum, 27,* 1–289.

Groucutt, H. S. (2020). Culture and convergence: The curious case of the Nubian Complex. In H. Groucutt (Ed.), *Culture history and convergent evolution: Can we detect populations in prehistory?* (pp. 55–86). Cham, Switzerland: Springer.

Groucutt, H. S., Scerri, E. M. L., Lewis, L., Clark-Balzan, L., Blinkhorn, J., Jennings, R. P., et al. (2015). Stone tool assemblages and models for the dispersal of *Homo sapiens* out of Africa. *Quaternary International, 382,* 8–30.

Grün, R., & Beaumont, P. B. (2001). Border Cave revisited: A revised ESR chronology. *Journal of Human Evolution, 40,* 467–482.

Grün, R., Beaumont, P. B., Tobias, P. V., & Eggins, S. M. (2003). On the age of the Border Cave 5 human mandible. *Journal of Human Evolution, 45,* 155–167.

Gutherz, X., Diaz, A., Ménard, C., Bon, F., Douze, K., Léa, V., et al. (2014). The Hargeisan revisited: Lithic industries from shelter 7 of Laas Geel, Somaliland and the transition between the Middle and Late Stone Age in the Horn of Africa. *Quaternary International, 343,* 69–84.

Hallinan, E., & Shaw, M. (2015). A new Middle Stone Age industry in the Tankwa Karoo, Northern Cape Province, South Africa. *Antiquity Project Gallery.* http://antiquity.ac.uk/projgall/hallinan344.

Hallinan, E., & Parkington, J. (2017). Stone Age landscape use in the Olifants River Valley, Clanwilliam, Western Cape, South Africa. *Azania: Archaeological Research in Africa, 52,* 324–372.

Henrich, J. (2001). Cultural transmission and the diffusion of innovations: Adoption dynamics indicate that biased cultural transmission is the predominate force in behavioural change. *American Anthropologist, 103,* 992–1013.

Henshilwood, C. S. (2012). Late Pleistocene techno-traditions in southern Africa: A review of the Still Bay and Howiesons Poort, c. 75–59 ka. *Journal of World Prehistory, 25,* 205–237.

Henshilwood, C. S., Sealy, J. C., Yates, R., Cruz-Uribe, K., Goldberg, P., Grine, F. E., et al. (2001). Blombos Cave, southern Cape, South Africa: Preliminary report on the 1992–1999 excavations of the Middle Stone Age levels. *Journal of Archaeological Science, 28,* 421–448.

Henshilwood, C. S., d'Errico, F., van Niekerk, K. L., Coquinot, Y., Jacobs, Z., Lauritzen, S. E., et al. (2011). A 100,000-year-old ochre processing workshop at Blombos Cave, South Africa. *Science, 334,* 219–222.

Hershkovitz, I., Weber, G. W., Quam, R., Duval, M., Grün, R., Kinsley, L., et al. (2018). The earliest modern humans outside Africa. *Science, 359,* 456–459.

Hilbert, Y. H., Crassard, R., Charloux, G., & Loreto, R. (2017). Nubian technology in northern Arabia: Impact on interregional variability of Middle Paleolithic industries. *Quaternary International, 435,* 77–93.

Hiscock, P., Clarkson, C., & Mackay, A. (2011). Big debates over little tools: Ongoing disputes over microliths on three continents. *World Archaeology, 43,* 653–664.

Hodder, I. (Ed.). (1978). *The spatial organisation of culture.* Pittsburgh: University of Pittsburgh Press.

Högberg, A., & Lombard, M. (2016). Still Bay Point-production strategies at Hollow Rock shelter and Umhlatuzana Rock shelter and knowledge-transfer systems in Southern Africa at about 80–70 thousand years ago. *PLoS ONE, 11,* e0168012.

Hovers, E. (2006). Neandertals and modern humans in the Middle Paleolithic of the Levant: What kind of interaction? In N. J. Conard (Ed.), *When Neandertals and moderns met* (pp. 65–86). Tübingen: Kerns Verlag.

Hovers, E. (2009). *The lithic assemblages of Qafzeh Cave.* Oxford: Oxford University Press.

Hu, Y., Marwick, B., Zhang, J.-F., Rui, X., Hou, Y.-M., Yue, J.-P., et al. (2019). Late Middle Pleistocene Levallois stone-tool technology in southwest China. *Nature, 565,* 82–85.

Humphreys, A. J. B. (1969). Four bifacial tanged and barbed arrowheads from Vosburg. *South African Archaeological Bulletin, 24,* 72–74.

Humphreys, A. J. B. (1991). On the distribution and dating of bifacial tanged and barbed arrowheads in the interior of South Africa. *South African Archaeological Bulletin, 20,* 41–43.

Huysecom, E., Rasse, M., Lespez, L., Neumann, K., Fahmy, A., Ballouche, A., et al. (2009). The emergence of pottery in Africa during the tenth millennium cal BC: New evidence from Ounjougou (Mali). *Antiquity, 83,* 905–917.

Inskeep, R. R. (1987). *Nelson Bay Cave, Cape Province, South Africa: The Holocene levels.* Oxford: British Archaeological Reports International Series 357.

Iovita, R. (2011). Shape variation in Aterian tanged tools and the origins of projectile technology: A morphometric perspective on stone tool function. *PLoS ONE, 6,* e29029.

Jacobs, Z., Roberts, R. G., Galbraith, R. F., Deacon, H. J., Grün, R., Mackay, A., et al. (2008). Ages for the Middle Stone Age of southern Africa: Implications for human behaviour and dispersal. *Science, 322,* 733–735.

Jacobs, Z., & Roberts, R. G. (2015). An improved single grain OSL chronology for the sedimentary deposits from Diepkloof Rockshelter, Western Cape, South Africa. *Journal of Archaeological Science, 63,* 175–192.

Jacobs, Z., & Roberts, R. G. (2017). Single-grain OSL chronologies for the Still Bay and Howieson's Poort industries and the transition between them: Further analyses and statistical modelling. *Journal of Human Evolution, 107,* 1–13.

Jochim, M. (2018). Environmental change and technological convergence in Southern Germany. *Lithic technological organization and paleoenvironmental change* (pp. 189–202). Cham: Springer.

Kaplan, J. (1990). The Umhlatuzana rock shelter sequence: 100 000 years of Stone Age history. *South African Humanities, 2,* 1–94.

Kleindienst, M. R. (2006). On naming things: Behavioral changes in the later middle to earlier Late Pleistocene, viewed from the eastern Sahara. In E. Hovers & S. L. Kuhn (Eds.), *Transitions before the transition: Evolution and stability in the Middle Paleolithic and Middle Stone Age* (pp. 13–28). New York: Springer.

Kroeber, A. L. (1940). Stimulus diffusion. *American Anthropologist, 42,* 1–20.

Kuhn, S. L., & Zwyns, N. (2018). Convergence and continuity in the initial upper Paleolithic of Eurasia. In M. J. O'Brien, B. Buchanan & M. I. Eren (Eds.), *Convergent evolution in stone-tool technology* (pp. 131–152). Cambridge, Massachusetts: MIT Press.

Laville, H., & Rigaud, J. P. (1973). The Perigordian V industries in Perigord: Typological variations, stratigraphy and relative chronology. *World Archaeology, 4,* 330–338.

Leakey, L. S. B. (1931). *The Stone Age cultures of Kenya colony.* Cambridge: Cambridge University Press.

Leplongeon, A., Pleurdeau, D., & Hovers, E. (2018). Late Pleistocene and Holocene lithic variability at Goda Buticha (Southeastern Ethiopia): Implications for the understanding of the Middle and Late Stone Age of the horn of Africa. *Journal of African Archaeology, 15,* 1–32.

Lewis, L., Perera, N., & Petraglia, M. (2014). First technological comparison of Southern African Howiesons Poort and South Asian Microlithic industries: An exploration of inter-regional variability in microlithic assemblages. *Quaternary International, 350,* 7–25.

Linstädter, J., Eiwanger, J., Mikdad, A., & Weniger, G. C. (2012). Human occupation of Northwest Africa: A review of Middle Palaeolithic to Epipalaeolithic sites in Morocco. *Quaternary International, 274,* 158–174.

Lombard, M. (2005). The Howiesons Poort of South Africa: What we know, what we think we know, what we need to know. *South African Humanities, 17,* 33–55.

Lombard, M. (2009). The Howieson's Poort of South Africa amplified. *South African Archaeological Bulletin, 64,* 4–12.

Lombard, M., Wadley, L., Deacon, J., Wurz, S., Parsons, I., Mohapi, et al. (2012). South African and Lesotho Stone Age sequence updated. *South African Archaeological Bulletin, 67,* 123–144.

Low, M. (2019). Continuity, variability and the nature of technological change during the Late Pleistocene at Klipfonteinrand Rockshelter in the Western Cape, South Africa. *African Archaeological Review.* https://doi.org/10.1007/s10437-018-9318-y.

Lycett, S. J. (2010). Cultural transmission, genetic models and Palaeolithic variability: Integrative analytical approaches. In *New perspectives on old stones* (pp. 207–234). New York, NY: Springer.

Lycett, S. J. (2015). Cultural evolutionary approaches to artifact variation over time and space: Basis, progress, and prospects. *Journal of Archaeological Science, 56,* 21–31.

Mackay, A. (2016). Three arcs: Observations on the archaeology of the Elands Bay and Northern Cederberg landscapes. *Southern African Humanities, 29,* 1–15.

Mackay, A., Stewart, B. A., & Chase, B. M. (2014). Coalescence and fragmentation in the late Pleistocene archaeology of southernmost Africa. *Journal of Human Evolution, 72,* 26–51.

Malan, B. D. (1956). The South African Archaeological Society: Ten years of archaeology in South Africa. Presidential Address, 1955. *South African Archaeological Bulletin, 11,* 31–40.

Marks, A. E., & Conard, N. J. (2008). Technology vs. typology: The case for and against a transition from the MSA to the LSA at

Mumba Cave, Tanzania. In T. Aubry, F. Almeida, A. C. Araujo & M. Tiffagom (Eds.), *Space and time: Which diachronies, which synchronies, which scales? Typology vs technology*. 1BAR International Series, 1831 (pp. 123–131). Oxford: Archaeopress.

Mason, R. J., & Brain, C. K. (1988). *Cave of Hearths, Makapansgat*. Transvaal: Archaeological Research Unit, University of the Witswatersrand.

McBrearty, S. (1988). The Sangoan-Lupemban and Middle Stone Age sequence at the Muguruk site, Western Kenya. *World Archaeology, 19,* 388–420.

McBrearty, S. (2003). Patterns of technological change at the origin of *Homo sapiens*. *Before Farming, 3,* 1–9.

McBrearty, S., & Brooks, A. S. (2000). The revolution that wasn't: A new interpretation of the origin of modern human behaviour. *Journal of Human Evolution, 39,* 453–563.

McCall, G. S., & Thomas, J. T. (2012). Still Bay and Howiesons Poort foraging strategies: Recent research and models of culture change. *African Archaeological Review, 29,* 7–50.

McGhee, G. R. (2018). Limits on the possible forms of stoen toools: A perspective from convergent biological evolution. In M. J. O'Brien, B. Buchannan & M. I. Eren (Eds.), *Convergent evolution in stone tool technology* (pp. 23–46). Cambridge, Massachusetts: MIT Press.

Mehlman, M. J. (1989). *Late Quaternary archaeological sequences in northern Tanzania*. Ph.D. dissertation, University of Illinois, Urbana-Champaign.

Mellars, P. A. (2006). Why did modern human populations disperse from Africa ca. 60,000 years ago? A new model. *Proceedings of the National Academy of Sciences USA, 103,* 9381–9386.

Mellars, P., Gori, K. C., Carr, M., Soares, P. A., & Richards, M. B. (2013). Genetic and archaeological perspectives on the initial modern human colonization of Southern Asia. *Proceedings of the National Academy of Sciences USA, 110,* 10699–10700.

Mesoudi, A. (2011). *Cultural evolution: How Darwinian theory can explain human culture and synthesize the social sciences*. Chicago: University of Chicago Press.

Miller, S. F. (2001). Tshitolian. *Encyclopedia of prehistory* (pp. 278–286). Boiston MA: Springer.

Mitchell, P. (1988). The Late Pleistocene early microlithic assemblages of Southern Africa. *World Archaeology, 20,* 27–39.

Mitchell, P. J. (1996). Sehonghong: The Late Holocene assemblages with pottery. *South African Archaeological Bulletin, 51,* 17–25.

Mohapi, M. (2012). Point morphology and the Middle Stone Age cultural sequence of Sibudu Cave, KwaZulu-Natal, South Africa. *South African Archaeological Bulletin, 67,* 5–15.

Mohapi, M. (2013). The Middle Stone Age point assemblage from Umhlatuzana Rock Shelter: A morphometric study. *South African Humanities, 25,* 25–51.

Montelius, O. (1903). *Die älteren Kulturperioden im Orient und in Europa*. Stockholm: Asher.

Morgan, L. H. (1877). *Ancient society; or, researches in the lines of human progress from savagery, through barbarism to civilization*. H. Holt.

Mucina, L., & Rutherford, M. C. (2006). *The vegetation of South Africa, Lesotho and Swaziland*. Pretoria: South African National Biodiversity Institute.

Muller, A., Clarkson, C., & Shipton, C. (2017). Measuring behavioural and cognitive complexity in lithic technology throughout human evolution. *Journal of Anthropological Archaeology, 48,* 166–180.

Nami, M., & Moser, J. (2010). *La Grotte d'Ifri n'Ammar; tome 2: Le Paléolithique Moyen*. In Forschungen zur Archäologie Außereuropäischer Kulturen (FAAK) (Vol. 9). Wiesbaden: Reichert.

Niang, K., Blinkhorn, J., & Ndiaye, M. (2018). The oldest Stone Age occupation of coastal West Africa and its implications for modern human dispersals: New insight from Tiémassas. *Quaternary Science Reviews, 188,* 167–173.

Nigst, P. R. (2012). *The early Upper Palaeolithic of the Middle Danube region*. Leiden: Leiden University Press.

O'Brien, M. J., Darwent, J., & Lyman, R. L. (2001). Cladistics is useful for reconstructing archaeological phylogenies: Palaeoindian points from the southeastern United States. *Journal of Archaeological Sciences, 28,* 1115–1136.

O'Brien, M. J., Boulanger, M. T., Collard, M., Buchanan, B., & Tarle, L. (2014). On thin ice: Problems with Stanford and Bradley's proposed Solutrean colonisation of north America. *Antiquity, 88,* 606–613.

Odell, G. H. (2004). *Lithic analysis*. Manuals in archaeological method, theory, and technique. New York: Kluwer Academic.

Olszewski, D., Schurmans, U., & Schmidt, B. (2011). The Epipaleolithic (Iberomaurusian) from Grotte des Contrebandiers, Morocco. *African Archaeological Review, 28,* 97–123.

Pasty, J.-F. (1999). *Contribution à l'Étude de l'Atérien du Nord Mauritanien*. British Archaeological Reports International Series 758. Oxford.

Pinhasi, R., Fort, J., & Ammerman, A. J. (2005). Tracing the origin and spread of agriculture in Europe. *PLoS Biology, 3,* e410.

Porraz, G., et al. (2013). Technological successions in the Middle Stone Age sequence of Diepkloof Rock Shelter, Western Cape, South Africa. *Journal of Archaeological Science, 40,* 3376–3400.

Porraz, G., Igreja, M., Schmidt, P., & Parkington, J. E. (2016). A shape to the microlithic Robberg from Elands Bay Cave (South Africa). *Southern African Humanities, 29,* 203–247.

Powell, R. (2007). Is convergence more than an analogy? Homoplasy and its implications for macroevolutionary predictability. *Biology and Philosophy, 22,* 565–578.

Rasmussen, M., Anzick, S. L., Waters, M. R., Skoglund, P., DeGiorgio, M., et al. (2014). The genome of a Late Pleistocene human from a Clovis burial site in western Montana. *Nature, 506,* 225–229.

Renfrew, C. (1977). Alternative models for exchange and spatial distribution. In T. Earle & J, Ericson, (Eds.), *Exchange systems in prehistory* (pp. 71–90). Cambridge: Academic Press.

Reynolds, N. (2020). Threading the weft, testing the warp: Population concepts and the European Upper Palaeolithic chronocultural framework. In H. Groucutt (Ed.), *Culture history and convergent evolution: Can we detect populations in prehistory?* (pp. 187–212). Cham, Switzerland: Springer.

Roberts, B.W., & Vander Linden, M. (2011). Investigating archaeological cultures: Material culture, variability, and transmission. In B. W. Roberts & M. Vander Linden (Eds.), *Investigating archaeological cultures: Material culture, variability, and transmission* (pp. 1–21.) New York: Springer.

Rogers, E. M. (1995). *Diffusion of innovations* (4th ed.). New York: Free Press.

Rose, J. I., Usik, V. I., Marks, A. E., Hilbert, Y. H., Galetti, C. S., Parton, A., et al. (2011). The Nubian complex of Dhofar, Oman: An African Middle Stone Age industry in southern Arabia. *PLoS ONE, 6,* e28239.

Rots, V., Lentfer, C., Schmid, V. C., Porraz, G., & Conard, N. J. (2017). Pressure flaking to serrate bifacial points for the hunt during the MIS5 at Sibudu Cave (South Africa). *PLoS ONE, 12,* e0175151.

Rouse, I. (1986). *Migrations in prehistory—Inferring population movement from cultural remains*. New Haven: Yale University Press.

Sampson, C. G. (1974). *The Stone Age archaeology of southern Africa*. New York: Academic Press.

Scerri, E. M. L. (2012). A new stone tool assemblage revisited: Reconsidering the 'Aterian' in Arabia. *Proceedings of the seminar for Arabian studies, 42,* 357–369.

Scerri, E. M. L. (2013). The Aterian and its place in the North African Middle Stone Age. *Quaternary International, 300,* 111–130.

Scerri, E. M. L. (2017). The North African Middle Stone Age and its place in recent human evolution. *Evolutionary Anthropology, 26,* 119–135.

Scerri, E. M. L., Drake, N. A., Jennings, R., & Groucutt, H. S. (2014). Earliest evidence for the structure of *Homo sapiens* populations in Africa. *Quaternary Science Reviews, 101,* 207–216.

Schmidt, P., & Mackay, A. (2016). Why was silcrete heat-treated in the Middle Stone Age? An early transformative technology in the context of raw material use at Mertenhof Rock Shelter, South Africa. *PloS ONE, 11,* e0149243.

Seguin-Orlando, A., Korneliussen, T. S., Sikora, M., Malaspinas, A.-S., Manica, A., Moltke, I., et al. (2014). Genomic structure in Europeans dating back at least 36,200 years. *Science, 346,* 1113–1118.

Seong, C. (2008). Tanged points, microblades and Late Palaeolithic hunting in Korea. Antiquity, 82, 871–883.

Shea, J. J. (2014). Sink the Mousterian? Named stone tool industries (NASTIES) as obstacles to investigating hominin evolutionary relationships in the Later Middle Paleolithic Levant. *Quaternary International, 350,* 169–179.

Shennan, S. (2011). Descent with modification and the archaeological record. *Philosophical Transcation of the Royal Society B, 366,* 1070–1079.

Shennan, S. (2020). Style, function and cultural transmission. In H. Groucutt (Ed.), *Culture history and convergent evolution: Can we detect populations in prehistory?* (pp. 291–298). Cham, Switzerland: Springer.

Shipton, C., Roberts, P., Archer, W., et al. (2018). 78,000-year-old record of Middle and Later Stone Age innovation in an East African tropical forest. *Nature Communications, 9,* 1832.

Shott, M. J. (1994). Size and form in the analysis of flake debris: Review and recent approaches. *Journal of Archaeological Method and Theory, 1,* 69–110.

Shott, M. J. (2020). Toward a theory of the point. In H. Groucutt (Ed.), *Culture history and convergent evolution: Can we detect populations in prehistory?* (pp. 245–259). Cham, Switzerland: Springer.

Singer, R., & Wymer, J. (1982). *The Middle Stone Age at Klasies River Mouth in South Africa.* Chicago: University of Chicago Press.

Sisk, M. L., & Shea, J. J. (2011). The African origin of complex projectile technology: An analysis using tip cross-sectional area and perimeter. *International Journal of Evolutionary Biology.* https://doi.org/10.4061/2011/968012.

Smith, G. E. (1929). *The migrations of early culture.* Manchester: Manchester University Press.

Smith, A. B. (1993). Terminal Palaeolithic industries of Sahara: A discussion of new data. *Environmental change and human culture in the Nile Basin and Northern Africa until the second millennium BC* (pp. 69–75). Poznan Archaeological Museum: Poznan.

Soriano, S., Villa, P., & Wadley, L. (2007). Blade technology and tool forms in the Middle Stone Age of South Africa: The Howiesons Poort and post-Howiesons Poort at Rose Cottage Cave. *Journal of Archaeological Science, 34,* 681–703.

Soriano, S., Villa, P., Delagnes, A., Degano, I., Pollarolo, L., Lucejko, J. J., et al. (2015). The Still Bay and Howiesons Poort at Sibudu and Blombos: Understanding Middle Stone Age technologies. *PLoS ONE, 10,* e0131127.

Spinapolice, E. E. (2020). Lithic variability and cultures in the East African Middle Stone Age. In H. Groucutt (Ed.), *Culture history and convergent evolution: Can we detect populations in prehistory?* (pp. 87–102). Springer, Cham, Switzerland.

Stanford, D., & Bradley, B. (2012). *Across Atlantic Ice: The origin of America's Clovis culture.* Berkeley: University of California Press.

Taylor, N. (2011). The origins of hunting and gathering in the Congo basin: A perspective on the Middle Stone Age Lupemban industry. *Before Farming, 2011,* 1–20.

Taylor, N. (2016). Across rainforests and woodlands: A systematic re-appraisal of the Lupemban Middle Stone Age in Central Africa. In S. C. Jones & B. A. Stewart (Eds.), *Africa from MIS 6-2: Population dynamics and paleoenvironments* (pp. 273–299). Dordrecht: Springer.

Tobias, P. V. (1949). The excavation of Mwulu's Cave, Potgietersrus District. *South African Archaeological Bulletin, 4,* 2–13.

Tomasso, S., & Rots, V. (2017). What is the use of shaping a tang? Tool use and hafting of tanged tools in the Aterian of Northern Africa. *Archaeological and Anthropological Sciences,, 10,* 1389–1417.

Tostevin, G. B. (2012). *Seeing lithics: A middle-range theory for testing cultural transmission in the Pleistocene.* Oxford: Oxbow Books.

Tribolo, C., Mercier, N., Douville, E., Joron, J.-L., Reyss, J.-L., Rufer, D., et al. (2013). OSL and TL dating of the Middle Stone Age sequence of Diepkloof Rock Shelter (Western Cape, South Africa): A clarification. *Journal of Archaeological Science, 40,* 3401–3411.

Trigger, B. (2006). *A history of archaeological thought* (2nd ed.). Cambridge: Cambridge University Press.

Tryon, C. A., McBrearty, S., & Texier, P.-J. (2006). Levallois lithic technology from the Kapthurin formation, Kenya: Acheulian origin and Middle Stone Age diversity. *African Archaeological Review, 22,* 199–229.

Tryon, C. A., & Faith, J. T. (2013). Variability in the Middle Stone Age of Eastern Africa. *Current Anthropology, 54,* S234–S254.

Tryon, C. A., Peppe, D. J., Faith, T. J., Van Plantinga, A., Nightingale, S., Ogondo, J., et al. (2012). Late Pleistocene artefacts and fauna from Rusinga and Mfangano islands, Lake Victoria, Kenya. *Azania: Archaeological Research in Africa, 47,* 14–38.

Tryon, C. A., & Ranhorn, K. L. (2020). Raw material and regionalization in Stone Age East Africa. In H. Groucutt (Ed.), *Culture history and convergent evolution: Can we detect populations in prehistory?* (pp. 143–156). Springer, Cham, Switzerland.

Tylor, E. B. (1871). *Primitive culture: Researches into the development of mythology, philosophy, religion, art, and custom* (Vol. 2). J. Murray.

Usik, V. I., Rose, J. I., Hilbert, Y. H., Van Peer, P., & Marks, A. E. (2013). Nubian Complex reduction strategies in Dhofar, Southern Oman. *Quaternary International, 300,* 244–266.

Van Peer, P. (1986). Presence of the Nubian technique in the Aterian. *Anthropologie, 90,* 321–323.

Van Peer, P. (1998). The Nile Corridor and the Out-of-Africa Model: An examination of the archaeological record. *Current Anthropology, 39,* S115–S140.

Van Peer, P. (2016). Technological systems, population dynamics and historical process in the MSA of northern Africa. In S. C. Jones & B. A. Stewart (Eds.), *Africa from MIS 6-2: Population dynamics and paleoenvironments* (pp. 147–159). Dordrecht: Springer.

Van Peer, P., & Vermeersch, P. M. (2007). The place of Northeast Africa in the early history of modern humans: New data and interpretations on the Middle Stone Age. In P. Mellars, K. Boyle, O. Bar-Yosef, & C. B. Stringer (Eds.), *Rethinking the Human Revolution: New behavioural and biological perspectives on the origin and dispersal of modern humans* (pp. 187–198). Cambridge: McDonald Institute for Archaeological Research.

Van Peer, P., Fullagar, R., Stokes, S., Bailey, R. M., Moeyersons, J., Steenhoudt, F., et al. (2003). The Early to Middle Stone Age transition and the emergence of modern human behaviour at site 8-B-11, Sai Island, Sudan. *Journal of Human Evolution, 45,* 187–193.

Van Peer, P., Vermeersch, P. M., & Paulissen E. (2010). *Chert quarrying, lithic technology, and a modern human burial at the Palaeolithic site of Taramsa 1, Upper Egypt* (Vol. 5). Egyptian Prehistory Monographs. Leuven: Leuven University Press.

Villa, P., Delagnes, A., & Wadley, L. (2005). A late Middle Stone Age artifact assemblage from Sibudu (KwaZulu-Natal): Comparisons with the European Middle Paleolithic. *Journal of Archaeological Science, 32,* 399–422.

Villa, P., & Lenoir, M. (2006). Hunting weapons of the Middle Stone Age and the Middle Palaeolithic: Spear points from Sibudu, Rose Cottage and Bouheben. *Southern African Humanities, 18,* 89–122.

Villa, P., Soriano, S., Teyssandier, N., & Wurz, S. (2010). Howiesons Poort and MSA III at Klasies River main site, cave 1A. *Journal of Archaeological Science, 37,* 630–655.

Vogelsang, R. (1998). *Middle-Stone-Age-Fundstellen in Südwest-Namibia.* Africa Praehistorica 11. Köln: Heinrich-Barth-Institut.

Vogelsang, R., Richter, J., Jacobs, Z., Eichhorn, B., Linseele, V., & Roberts, R. G. (2010). New excavations of Middle Stone Age deposits at Apollo 11 Rockshelter, Namibia: Stratigraphy, archaeology, chronology and past environments. *Journal of African Archaeology, 8,* 185–218.

Volman, T.P. (1981). *The Middle Stone Age in the southern Cape.* Unpublished Ph.D. thesis. University of Chicago, Chicago.

Volman, T. P. (1984). Early prehistory of southern Africa. In R. G. Klein (Ed.), *Southern Africa prehistory and paleoenvironments* (pp. 169–220). Rotterdam: A.A. Balkema.

Wadley, L. (1987). *Later Stone Age hunters and gatherers of the Southern Transvaal: Social and ecological interpretations.* Oxford: British Archaeological Reports International Series 380.

Wadley, L. (1996). The Robberg Industry of Rose Cottage Cave, Eastern Free State: The technology, spatial patterns and environment. *South African Archaeological Bulletin, 51,* 64–74.

Wadley, L. (2000a). The early Holocene layers of Rose Cottage Cave, eastern Free State: Technology, spatial patterns and environment. *South African Archaeological Bulletin, 55,* 18–31.

Wadley, L. (2000b). The Wilton and pre-ceramic post-classic Wilton Industries at Rose Cottage Cave and their context in the South African sequence. *South African Archaeological Bulletin, 55,* 90–106.

Wadley, L. (2005). A typological study of the final Middle Stone Age stone tools from Sibudu Cave, KwaZulu-Natal. *South African Archaeological Bulletin, 60,* 51–63.

Wadley, L. (2007). Announcing a Still Bay industry at Sibudu Cave, South Africa. *Journal of Human Evolution, 52,* 681–689.

Wadley, L. (2008). The Howieson's Poort industry of Sibudu Cave. *South African Archaeological Bulletin, (Goodwin Series) 10,* 122–133.

Wadley, L. (2013). MIS 4 and MIS 3 occupations in Sibudu, KwaZulu-Natal, South Africa. *South African Archaeological Bulletin, 68,* 41–51.

Wadley, L., & Jacobs, Z. (2006). Sibudu Cave: Background to the excavations, stratigraphy and dating. *Southern African Humanities, 18,* 1–26.

Wadley, L., & Mohapi, M. (2008). A segment is not a monolith: Evidence from the Howiesons Poort of Sibudu, South Africa. *Journal of Archaeological Science, 35,* 2594–2605.

Wake, D. B., Wake, M. H., & Specht, C. D. (2011). Homoplasy: From detecting pattern to determining process and mechanism of evolution. *Science, 331,* 1032–1035.

Webster, G. S. (2008). Culture history: A culture-historical approach. In R. A. Bentley, H. D. Maschner, & C. Chippindale (Eds.), *Handbook of archaeological theories.* Rowman and Littlefield: Lanham.

Wendt, W. E. (1972). Preliminary report on an archaeological research programme in South West Africa. *Cimbebasia B, 2,* 1–61.

Wendt, W. E. (1974). "Art mobilier" aus der Apollo 11-Grotte in Südwest-Afrika. *Acta Praehistorica et Archaeologica, 5,* 1–42.

Wendt, W. E. (1976). "Art mobilier" from the Apollo 11 Cave, South West Africa: Africa's oldest dated works of art. *South African Archaeological Bulletin, 31,* 5–11.

White, M., Ashton, N., & Scott, R. (2011). The emergence, diversity and significance of Mode 3 (prepared core) technologies. In N. Ashton, S. Lewis, & C. Stringer (Eds.), *The ancient human occupation of Britain* (pp. 53–65). London: Elsevier Science.

Will, M., & Mackay, A. (2017). What factors govern the procurement and use of silcrete during the Stone Age of South Africa? *Journal of Archaeological Sciences: Reports, 15,* 630–645.

Will, M., & Conard, N. J. (2018). Assemblage variability and bifacial points in the lowermost Sibudan layers at Sibudu, South Africa. *Archaeological and Anthropological Sciences, 10,* 389–414.

Will, M., Mackay, A., & Phillips, N. (2015). Implications of Nubian-like core reduction systems in southern Africa for the identification of early modern human dispersals. *PLoS ONE, 10,* e0131824.

Willoughby, P. R. (1996). Middle Stone Age technology and adaptation in southwestern Tanzania. In G. Pwiti & R. Soper (Eds.), *Aspects of African Archaeology* (pp. 177–189). Harare: University of Zimbabwe Publications.

Will, M., Bader, G. D., Conard, N. J. (2014). Characterizing the late pleistocene MSA lithic technology of Sibudu, KwaZulu-Natal, South Africa. PLoS ONE 9(5), e98359.

Wurz, S. (2000). *The Middle Stone Age at Klasies River, South Africa.* Ph.D. thesis, Stellenbosch University.

Wurz, S. (2002). Variability in the Middle Stone Age lithic sequence, 115,000–60,000 years ago at Klasies River, South Africa. *Journal of Archaeological Science, 29,* 1001–1015.

Ziegler, M., Simon, M. H., Hall, I. R., Barker, S., Stringer, C., & Zahn, R. (2013). Development of Middle Stone Age innovation linked to rapid climate change. *Nature Communications, 4,* 1905.

Chapter 7
Technology and Function of Middle Stone Age Points. Insights from a Combined Approach at Bushman Rock Shelter, South Africa

Katja Douze, Marina Igreja, Veerle Rots, Dries Cnuts, and Guillaume Porraz

Abstract Edge convergence, which is typical for pointed tools, is a major morphological feature contributing to the definition of the African Middle Stone Age (MSA). The multifaceted character of points might be the key to their success and for their recurrent adoption by prehistoric populations. Whether MSA points represent a good proxy to identify populations and to discuss their interconnectedness is a question to address at several scales of observation. In this paper, we develop an approach on technological point production based on the collections from Bushman Rock Shelter (Limpopo Province, South Africa), relying on a combined study of technology and tool use. The large-scale comparison of our results with other MIS 5 occurrences in southern African show similar technological and use-wear patterns, indicating regionally-specific features. We emphasize the limits of current knowledge and the future research goals to be developed in order to better serve the interpretation of cultural contacts or convergent evolutions between ancient groups during this period of the MSA.

Keywords Point production in Africa • South African Middle Stone Age • Tool use • Residue analysis • Technological points

K. Douze (✉)
APA, Department of Genetics and Evolution, University of Geneva, Geneva, Switzerland
e-mail: Katja.Douze@unige.ch

M. Igreja
LARC-DGPC, Ministry of Culture of Portugal/ENVARCH, Cibio-Inbio, University of Porto, Porto, Portugal

V. Rots
TraceoLab/Prehistory, University of Liège, Liège, Belgium

V. Rots · D. Cnuts
FNRS, Bruxelles, Belgium

G. Porraz
CNRS, UMR 7269-LAMPEA, Aix-Marseille Université, Aix-en-Provence, France

Evolutionary Studies Institute, University of the Witwatersrand, Johannesburg, South Africa

General Overview of Middle Stone Age Point Production in Sub-Saharan Africa

Point production was one of the most widespread technologies in the Stone Age. However, as with other successful technologies such as blade technology, point production refers to multiple technical and morpho-metrical concepts. This translates into a variety of point production strategies that peak during the Middle Stone Age (MSA) across the African continent.

Although varied, MSA points conform to a morphological 'credo' which entails that the lateral edges are convergent and join in a distal part, in a geometric shape resembling a triangle. This shape offers multiple functional units, with both long cutting edges and a tip that is conducive to penetrative as well as incisive functioning. To obtain this morphology, two main concepts are dominantly used during the MSA: (1) the bifacial or unifacial shaping of undefined or non-pointed predetermined flakes into pointed forms—also called typological points; and (2) the production of triangular flakes—also called technological points—from a prepared core surface, further retouched or not. These strategies differ significantly in terms of the operational sequence to be performed and the knapping parameters to be controlled for, although both illustrate a mental projection in shape. Typological and technological points occur all over the continent during the MSA, but lithic assemblages present various combinations, from coexistence to independence.

Shaping as a technique appears during the earliest Acheulean, soon after 1.8 Ma, for the production of handaxes (e.g. Lepre et al. 2011; Beyene et al. 2013) and is implemented for the production of bifacial points in the earliest MSA. The reduction in size of bifaces into bifacial points has been considered a potential argument for identifying the beginning of the MSA (McBrearty and Tryon 2006). Shaping seems to be the result of repeated reinvention during the MSA onwards, and likely characterizes a similar

H. S. Groucutt (ed.), *Culture History and Convergent Evolution: Can We Detect Populations in Prehistory?*, Vertebrate Paleobiology and Paleoanthropology, https://doi.org/10.1007/978-3-030-46126-3_7

problem-solving solution over time. To cite only a few examples, bifacial industries that include bifacial points usually characterize the transition from the Acheulean to the MSA in different parts of Africa (e.g. McBrearty and Tryon 2006; Herries 2011) as well as the Lupemban of central Africa which spans almost the entire MSA period as at Kalambo Falls (e.g. Clark et al. 2001; Taylor 2016). Bifacial points associated with unifacial points are represented in most early and late MSA contexts in East Africa, where they usually co-occur with another point technology, mainly Levallois (e.g. Wendorf and Schild 1974; Pleurdeau 2005; Yellen et al. 2005; Tryon and Faith 2013; Ménard et al. 2014; Douze and Delagnes 2016) or Nubian Levallois (e.g. Kurashina 1978; Yellen et al. 2005; Douze 2012; see Groucutt 2020). In West Africa, the late MSA in the Falémé Valley (Senegal), as much as at Ounjougou (Mali), relies mostly on bifacial shaping of varied morphologies of points (Chevrier et al. 2016), but more diversified point production occurs in other West African locations (Scerri et al. 2016, 2017; Niang et al. 2018). Bifacial and unifacial points are also recognized in the recent phase of the so-called "Pietersburg" characterizing the MSA from the north-eastern part of South Africa (e.g. Mason 1957), including the upper MSA layers at Bushman Rock Shelter (Porraz et al. 2015). However, the most highlighted bifacial industry of the MSA is probably the Still Bay of southern Africa, dated to MIS 5–4, which is characterized by lanceolate point production (e.g. Wadley 2007; Villa et al. 2009; Högberg and Larsson 2011; Porraz et al. 2013a; Wurz 2013; Soriano et al. 2015) that additionally proves to have been occasionally heat-treated and pressure-flaked (Villa et al. 2009; Mourre et al. 2010). Unifacial point production is the common point type for the South African Post-Howiesons Poort industries (e.g. Wadley 2005; Conard et al. 2012; Bader et al. 2015; Will and Conard 2018) while bifacial points occur in a few Howiesons Poort contexts (e.g. de la Peña et al. 2013; Porraz et al. 2013b).

Technological points as desired end-products also occur for the first time within the transitional phase from the Acheulean to the MSA (Tryon et al. 2005; Douze and Delagnes 2016; Brooks et al. 2018), as well as in exceptionally early MSA collections such as Kathu Pan 1, around 500 ka (Wilkins et al. 2012). Their appearance seems to be linked to the process of diversification of core reduction methods occurring during the Acheulean to MSA transition (Tryon et al. 2005). However, as compared to the shaping technique, technological point production methods seem to be more restricted in time as they do not seem to be significant for Stone Age industries after the MSA. Their presence in MSA lithic industries is sometimes overshadowed by typological points when both types co-occur (e.g. Yellen et al. 2005), which leads to an imprecise inventory and description of triangular point production over time. By contrast, a series of southern African sites dated from MIS 6 to 5 are characterized by point predetermination as an exclusive point production strategy. This is the case for example at Klasies River Mouth (Volman 1981; Singer and Wymer 1982; Wurz 2002), Pinnacle Point 13B (Thompson et al. 2010; Schoville 2010) and Diepkloof Rock Shelter (Porraz et al. 2013b) but also in the MSA layers of Bushman Rock Shelter discussed here.

The variability in point technologies can be seen as different technical solutions to obtain tools with triangular- and alike- shapes. The adoption of one or both of these point production strategies may be explained by culturally driven factors, either by (re)invention and convergence, group interactions, or/and by adaptive behaviors to specific environments. While the demonstration of these factors and mechanisms is still largely challenging, it is established that several point styles show regional distributions over time (Clark 1988; McBrearty and Brooks 2000; Scerri et al. 2018). Therefore, knowing whether this regional mosaic of point styles reflects population structure is an important question, and needs to rely on solid data, the more so as intra-regional diversity exists and as point production varies over time (see also Shott 2020). Nonetheless, it appears that point production strategies are often insufficiently addressed to discuss these questions. Firstly, this is the result of as yet unstandardized terminologies (see Shea 2008) despite taxonomic attempts (e.g. Perlès 1974; Kurashina 1978; Yellen et al. 2005), and unequal levels of analysis and descriptions. Secondly, the fact that data on points are difficult to exploit is because the function of points is generally based on functional *a priori* (i.e. hunting armatures) rather than experimental and use-wear studies, which undermines their comprehensive study. These two factors represent a circular cause-effect relationship which places points both in the position of important tools for behavioral interpretations, while under-exploiting their informative value.

Middle Stone Age Point Functions

MSA points are still often regarded as hunting armatures, and the term "point" is sometimes used as a synonym for "projectile" (e.g. McBrearty and Brooks 2000). This leads to a general homogenization of the concept of the point, in which the purported function outweighs the technological, typological and cultural diversity of points during the MSA, as well as its other potential functions. Ethnological analogies between sub-contemporaneous hunting weaponry and Paleolithic points have further supported this alleged functional interpretation of points as hunting implements (see e.g. Tryon and Faith 2013). Furthermore, points considered projectiles acquired the status of major proxy for identifying behavioral modernity alongside pigment processing strategies, blade production and other behaviors (McBrearty and Brooks 2000). Although there is a broad scientific consensus on the fact that projectile technology existed during the entire MSA, the main challenge is still that of providing reliable evidence for pointed tool uses.

Tool Use

While analyses of tool use applied to African MSA material are still in their infancy, most studies concentrate on the identification of stone projectile technology because of its purported importance in terms of cognitive and social implications (e.g. Brooks et al. 2006; Bader et al. 2016; Wilkins et al. 2012; but see e.g. Rots and Van Peer 2006; Rots et al. 2011; Porraz et al. 2018). As it stands now, the clearest signals for projectile function are provided, in the South-African context, by Howiesons Poort backed tools (Igreja and Porraz 2013; de la Peña et al. 2018), Still Bay bifacial points (Lombard 2006) and Pre-Still Bay serrated bifacial points (Rots et al. 2017), and for elsewhere in Africa, by tanged 'Aterian' points (Tomasso and Rots 2017), as well as by technological and typological point types from northern Africa (Rots et al. 2011). Other functional studies are more debated, such as the possible projectiles from Kathu Pan 1 (Wilkins et al. 2012, 2015; Schoville et al. 2016, but see Rots and Plisson 2014), and from the Gademotta Formation (Sahle et al. 2013, 2014; Sahle and Braun 2018; but see Douze et al. 2018), or rest on limited evidence (e.g. #Gi and Aduma: Brooks et al. 2006; Cartwright's site: Waweru 2007; Sibudu: Soriano et al. 2015). Despite significant methodological descriptions on projectile identification, recent studies also highlight remaining difficulties in their detection (see Rots 2016; Hutchings 2016; Coppe and Rots 2017 and references therein).

However, these studies have also highlighted that tool types that were used as projectiles, such as MIS 5 Aterian tanged points, MIS 5/4 Still Bay points, and MIS 5–3 Howiesons Poort backed tools in particular, were also used for cutting, sawing and incising (Lombard 2006, 2007a, b; Bouzouggar et al. 2007; Igreja and Porraz 2013; Steele et al. 2016; Tomasso and Rots 2017), which indicates that tool use cannot be inferred by typology or morphology and morphometrics only. Cutting activities carried out using stone points have also been suggested for ≠Gi and Florisbad (Kuman 1989) as well as at Sodmein Cave (Rots et al. 2011). Technological points were recognized to be mainly cutting implements at Pinnacle Point based on edge damage distribution (Bird et al. 2007; Schoville 2010). At Klasies River, Milo (1998) identified embedded stone in faunal remains that he attributed to either armature use or butchery, based on comparisons with experimental work. The studies of Shea (2006) and Sisk (Sisk and Shea 2011) of Klasies River triangular flakes rather explored their penetration potential, based on morphometric values (TCSA, TCSP) indicating that, if these points were armatures, they would fall into the range of short-distance weaponry.

Use-wear results, although still scattered, have clearly changed the perception of how MSA pointed toolkits were potentially used as it is now clearly established that shape resemblance (i.e. triangular—and alike) does not equal a convergence in function.

Hafting Adhesives

The emergence of the MSA is often seen as turning point from handheld tools (i.e. bifaces) to hafted tools (i.e. points) (e.g. Clark 1988; McBrearty and Tryon 2006). While pointed forms considered hunting weapons have social implications, the identification of adhesives for hafting may have important cognitive significance (e.g. Wadley et al. 2009). Therefore, black, blackish or reddish black residues are another evidence that is looked for on pointed tools as it is assumed that these residues represent remains of adhesives used in hafting. The arguments commonly used to interpret a black residue as adhesive are its location on the assumed passive part of the tool, a viscous and fluid appearance, and a dark color (Wadley et al. 2004; Lombard 2005). Black residues interpreted as adhesives were identified on Howiesons Poort artefacts (Gibson et al. 2004; Lombard 2008; Charrié-Duhaut et al. 2013), on different types of points (Lombard 2005), and on barbs (de la Peña et al. 2018). However, methodological challenges remain as the identification of adhesives through visual observation is generally hampered by the lack of diagnostic structural elements for most glue components (e.g. resin, gum and pitch) (Cnuts et al. 2018b) and potential confusion with similar-looking residues (e.g. manganese, charcoal or fine sediment) is a significant problem. Therefore, adhesive identifications based on visual inspection only are tentative, in particular in the absence of a Scanning Electron Microscopy—Energy Dispersive X-Ray Spectroscopy (SEM-EDS) analysis that allows confirming the organic nature of the residue (Rots et al. 2017). Currently, the most reliable method for identifying the organic components of hafting adhesives is by Gas chromatography-Mass spectrometry (GC-MS) (Mills and White 1977; Regert 2004) and may lead to a botanical identification through the detection of specific biomarkers in the adhesive (Hayek et al. 1990; Evershed 2008). However, this method is destructive and requires relatively large sample sizes. Therefore, it has only been applied when adhesives were preserved in large, macroscopically visible quantities (e.g. Boëda et al. 1996; Koller et al. 2001; Hauck et al. 2013). It has been successful so far on one MSA artefact, which consists of a quartz flake bearing residue, attributed to the Late phase of the Howiesons Poort at Diepkloof Rock Shelter, and that proved to be *Podocarpus elongatus* resin (Charrié-Duhaut et al. 2013). Non-destructive techniques are currently being explored (Perrault et al. 2016; Cnuts et al. 2018a). Mineral hafting components, such as ochre, can be detected visually (Lombard 2005, 2006; Wadley et al. 2009) but again their identification requires specific analytical methods (e.g. raman spectroscopy; Helwig et al. 2014; Wojcieszak and Wadley 2018) to distinguish the exact nature of the iron oxide (e.g. hematite; e.g. Zipkin et al. 2015).

Even when chemical analyses are performed, it is essential that the analyses of potential glue remains are integrated in a broader functional study to confirm the origin of the residue, partly because a broad range of sources of residue exists, either related to the context of use (i.e. production, use, hafting) (Cnuts and Rots 2018), or to incidental or taphonomic processes (Cnuts et al. 2018b; Schmidt et al. 2015). Here, a combined approach on production strategies, use wear and residue has been attempted on assemblages from Bushman Rock Shelter, in South Africa. The results reveal the particularities of the technical organization of these MSA groups in terms of tool production strategies and use, and offer some ground to further reflect about the interpretation of residues on stone tools.

Bushman Rock Shelter MIS 5 Middle Stone Age Points

Bushman Rock Shelter (BRS) is located in the Limpopo Province of South Africa. The shelter is carved from dolomites of the Transvaal Supergroup and contains a rich LSA and MSA record of >7 m (Porraz et al. 2015). Updated geochronological dates are in progress, but place the upper MSA record of BRS within MIS 5 (Porraz et al. 2018).

The main excavations at the site were carried out by Eloff in the 1970s. He explored the lowest part of the sequence, from layer 36 to 109, as a testpit of 1×2 m^2 that remains largely unpublished (see Porraz et al. 2015; see Underhill 2012 for a study on some of the layers). For this study, we selected layers 45–56, representing 14 layers and features therein, analyzed as a combined sample as they represent a recurrent occupation of the shelter by groups bearing similar technological characteristics.

Insights from the Technological Approach

In layers 45–56, the assemblages are characterized by flakes, points, elongated flakes and blade production (Table 7.1). Points are exclusively obtained by core reduction, and the shaping process is totally absent from the production strategies used for tool production. Points represent an important feature in these assemblages, as they are numerous (15% of the lithic collection), but more importantly, because some of them seem to be introduced into the site. The most striking evidence for points being curated is the raw material representation per technological category. Points are dominantly obtained on hornfels (38%; n = 85/226), quartzite (37%; n = 84/226), and on chert (12%; n = 28/226). Quartz only represents 10% of the points while this raw material is dominant in all layers, in particular amongst cores (74%), flakes (63%) and numerous technical flakes and debris. On the contrary, raw material categories such as quartzite and chert, are almost exclusively represented by end-products (points, blades, elongated flakes), without the associated cores and waste products.

Cores are uncommon (n = 28) and small (<5 cm in max length). They show expedient knapping strategies, and more rarely discoid core reduction methods in the final phase of

Table 7.1 Assemblage composition per layer

Layer	45	46	47	48	49	50	51	52	53	54	54(i)	55	55(i)	56	Total	
General flakes	165	90	33	21	47	96	84	164	133	25	5	11	13	49	936	61%
Points (convergent flakes)	25	15	16	17	12	14	16	27	22	9	2	10	10	31	226	15%
Elongated flakes	16	11	2	10	13	13	20	14	14	6	1	2	6	19	147	10%
Blades	18	14	6	8	8	8	4	18	14	5	4	2	9	18	136	9%
Cores	4	1	4	1	0	6	0	3	7	0	0	0	0	2	28	2%
Chunks	14	5	1	0	3	16	9	10	4	1	1	0	2	5	71	5%
Total	242	136	62	57	83	153	133	236	194	46	13	25	40	124	1544	100%

flaking. They are usually on quartz (n = 21/28 cores), but also on hornfels flakes, and none of them relates to the production of triangular flakes as observed in the assemblages. Therefore, the technological assessment of points substantially rests on the characteristics of the points themselves, with the exception of the discoid reduction featured on three quartz cores and also represented by a small number of pseudo-Levallois points (5%, n = 11/226) of which five are in quartz.

Points were produced by direct hard hammer percussion, leaving prominent bulbs and large platforms. Percussion surfaces were summarily prepared by a few large removals, leaving plain (37%; n = 78) or dihedral (32%; n = 66) platforms, but the preparation by faceting also occurred (27%; n = 57). The large majority of points are the result of recurrent unidirectional convergent methods (72%; n = 163/226), mostly from a Levallois conception (n = 106/163) (e.g. Fig. 7.2A, B, E, G). The recurrence of several points being produced within the same knapping sequence rather than by a preferential method (Boëda 1994), is highlighted by the distal ends of 45% of the points that are not axially aligned with the direction of the blow (*déjeté*) (Figs. 7.1A and 7.2G). The ridges of centered symmetrical points usually meet close to the tip (Figs. 7.1A and 7.2A, B), which indicate elongation of the end-products and few blades, sometimes pointed, were likely alternately produced during the point-core exploitation. In some cases, the convergence of the guiding ridges is restored by operating a lateral removal during core exploitation (Fig. 7.1B), which also increases distal convexity of the core surface, preventing knapping accidents. Besides, points with only one central ridge are present and have been described elsewhere as *pointes accourcies* (Porraz et al. 2013b; Douze et al. 2015; Fig. 7.1A).

Point types in these assemblages also vary in size, with high standard deviations (Table 7.2), and in the morphology of their distal end (Fig. 7.1A). Distal breakage is common (33%) and often occurs as straight fractures on thick blanks (Fig. 7.2B). When points are complete, most show pointed tips, with a closed (<60°; 24%; Fig. 7.2E) or open (>60°; 13%) angle. More noticeable are the points with a distal morphology that has been named Y or T shaped (15%). They are reminiscent of a lack of convexity of the cores' exploitation surface, causing the distal overshot of the point beyond the converging guiding ridges, sometimes even resulting in blade-like products (Figs. 7.1C and 7.2F). These flakes with Y or T tips are consequently not strictly points, although that they were meant to be so in their conception is shown by the convergent guiding ridges. Retouch applied to this type of non-pointed convergent flakes did not lead to a modification of the distal ends into pointed tips, raising the question of the functional importance of the distal angle for part of the point production at BRS (Fig. 7.1C).

Retouched points usually show continuous and localized marginal semi-abrupt retouch on an edge to tip (57%, Fig. 7.2D), and to a lesser extent, single (16%) or multiple notches (7%). Only one denticulate is represented (Fig. 7.2C). A few points (n = 11) with short localized abrupt retouch on one edge of the tip stand out due to their specific location, angle and lateralization (Fig. 7.2H). This retouch creates an asymmetry in the edge angles of the tip, and could possibly indicate a lateralization of the point during use. Retouch patterns on points, i.e. continuous marginal retouch and notches, are similar to the type of retouch observed on other blanks in the collection but points are proportionally more often retouched (31%; n = 70/226) than flakes, elongated flakes or blades from which only a total of 4% are retouched (n = 51/1219).

In sum, edge convergence is a morphological characteristic which is central to the lithic system at BRS. Points are often elongated, offering long cutting edges (Table 7.2). Point production methods are not diversified, usually of Levallois conception, but there is a high diversity in point morpho-types. Long production sequences as well as a limited emphasis on the management of the core's distal convexity have led to a degree of variability in end-products' general shapes and in particular of their distal ends. The significant occurrence of Y or T shaped, broken or non-acute tips with no morphological modification by retouch, raised questions on the importance of the distal morphology of these blanks in terms of function. A series of points were selected, representative of different morpho-types and these were tested for use-wear traces.

Insights from the Use-Wear Approach

In terms of method, standard procedures were applied to examine indicators of the nature of the material worked and of the gestures performed with the points. This includes the observation of modifications visible under low magnifications in the form of scars, fractures, edge rounding, and under the microscope for polishes, striation and micro edge rounding (Semenov 1964; Keeley 1980; Plisson 1985a, b; Gonzalez-Urquijo and Ibanez-Estevez 2004; Igreja 2009; Igreja and Porraz 2013). The interpretations on the functionality of points were based on a large experimental reference collection of use-wear traces recorded on tools made of different raw materials, and based on both African Stone Age and European Paleolithic replications.

A sample of 43 points that we consider representative of the different morpho-types and raw materials described above was carefully selected for the analysis. Amongst those, 32 were rejected because their edges and surfaces presented heavy signs of natural weathering, presumably of mechanical origin, observed macroscopically. The surfaces

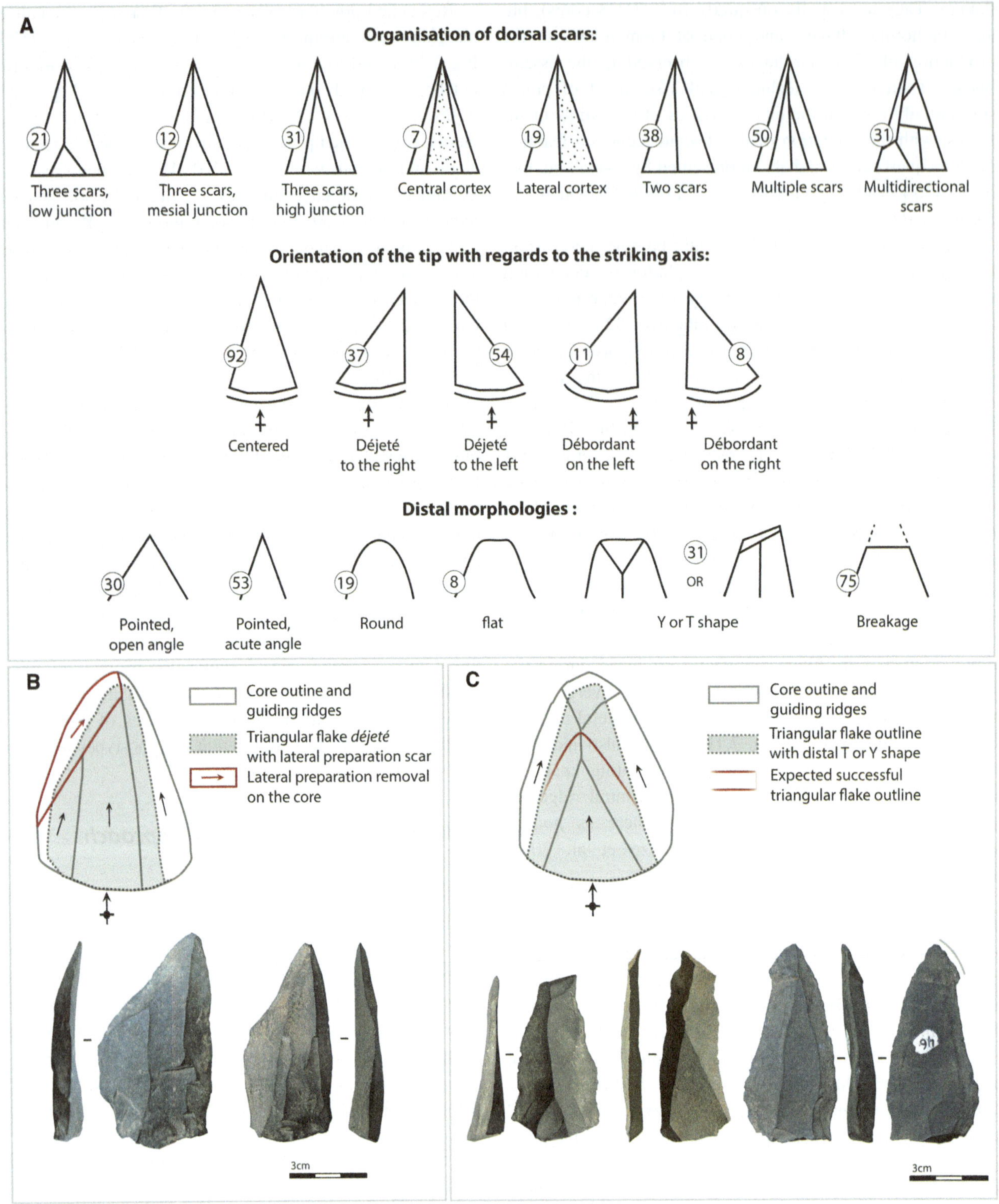

Fig. 7.1 Typological and technological characteristics of points at BRS. A General morphological features of points. Numbers in circles correspond to the number of points per category. **B** Schematic representation of lateral preparation on cores and archaeological examples. **C** Schematic representation of distally overshot points, extending beyond the triangular guiding ridges and archaeological examples. Grey line on artefact picture indicate retouched edge portions. All in hornfels

Table 7.2 Dimensions (in mm) of complete points (n = 190) from layers 45 to 56

	Length	Width	Thickness	Elongation (L/W)
Mean	52.4	31.4	8.9	1.7
Max.	103.0	62.5	16.0	5.2
Min.	24.2	11.97	3.2	0.8
St. Dev.	15.9	9.0	2.7	0.5

were marked by striae and flat bright polishes that can be distinguished from those resulting from use by their texture and random distribution on the artifacts affecting the whole microtopography of the tool's surface. However, the remaining 11 artifacts were sufficiently well preserved to allow the study in reliable analytical conditions. Among these, 10 artifacts show recognizable use-wear traces recording a total of 10 used zones (Fig. 7.2).

All traces but one are indicative of contact with raw materials with hard properties, of various natures, as documented by polishes that are compact, bright and smooth in texture, gently curved over the high points of the microtopography. The remaining one exhibits polish that has a reticular pattern associated with some rounding of the microtopography, located on both sides of the edge which is consistent with soft material cutting (Fig. 7.2G). When the nature of the hard material was determinable (n = 7/9), the traces indicated that four pieces were used on bone, in scraping (Fig. 7.2A, C, D) or cutting (Fig. 7.2E) motions. Bone polish is characterized by a highly compact texture and a mat appearance with tiny pits. It is less extensive than wood polish and more localized in a few areas of the used edge. These artifacts show well oriented polishes that enable to identify the working motion. Three artifacts were used on wood judging by the bright and smooth texture of the polish, located on both sides of the used edge, to cut (n = 2) or scrape (n = 1) (Fig. 7.2B, I, J). For two artifacts the exact nature of the worked materials could not be determined but features of the use-wear are consistent with the processing of hard materials (Fig. 7.2E, H).

The interesting results of this analysis rely on the location of the different categories of use-wear traces in relation to the working motion. The nature and development of the polishes indicate that distal edges of the points, in particular the broken ones, were exclusively worked in a transversal contact with the raw material in scraping motions, while lateral edges of points were mostly used to perform cutting motions (Fig. 7.2). The distal breakage of points does not relate with projectile functionality, and fractures related to projectile impact were not recognized in the point collection in general. Furthermore, use-wear was not recognized on retouched edges, and while the state of preservation could be the underlying cause, the hypothesis that the retouched edge portions generally play a passive functional role is considered. Finally, this small sample of tools also did not show clear use-wear evidence for hafting, but one artifact bearing on its ventral surface a potential organic macro-residue remnant of an adhesive was selected for further analysis.

Insights from Residue Analysis

Black residue could be observed with the naked eye on the bulb of a small quartz point, precisely in the area where one could imagine that hafting adhesives would be situated (Fig. 7.3A). Observations at both low and high magnification (Fig. 7.3B) show a black residue with an amorphous texture, similar to previously observed hafting adhesives from Diepkloof Rock shelter (Charrié-Duhaut et al. 2013: Fig. 5). The results of the microscopic analysis in combination with the location of the residue on the passive part of the point provided ample arguments to assess the black residue as potential hafting adhesive and to decide upon a more detailed analysis.

The molecular analysis of the Volatile Organic Components (VOC) with comprehensive two-dimensional gas chromatography-high-resolution time-of-flight mass spectrometry (HS-SPME GC×GC–HRTOFMS) (Cnuts et al. 2018a) did not yield any diagnostic results. A subsequent elemental analysis with a SEM-EDS (JEOL IT300) demonstrated that the matrix of the residue consisted of authigenic amorphous silica (opal) (Fig. 7.3D, E) with inclusions of phytoliths (Fig. 7.3C). The dense association of opal minerals and phytoliths in archaeological sediments has been argued to derive from vegetal tissues that were used as combustion fuel and that, given their incombustible nature, can be found in high concentrations in ashes (Schiegl et al. 1994, 1996; Karkanas et al. 1999; Delhon 2010). In less favorable preservation conditions, the charcoal and other carbonate ash components will dissolve and the silica minerals (i.e. phytoliths and opal) will be the only remnants of the ash. It can thus be concluded that the blackish residue is not linked with hafting, but is taphonomic in nature. The existence of hearths within the living area is a real issue and has a significant impact on what can be expected in terms of residues on stone tools (Schmidt et al. 2015; Cnuts et al.

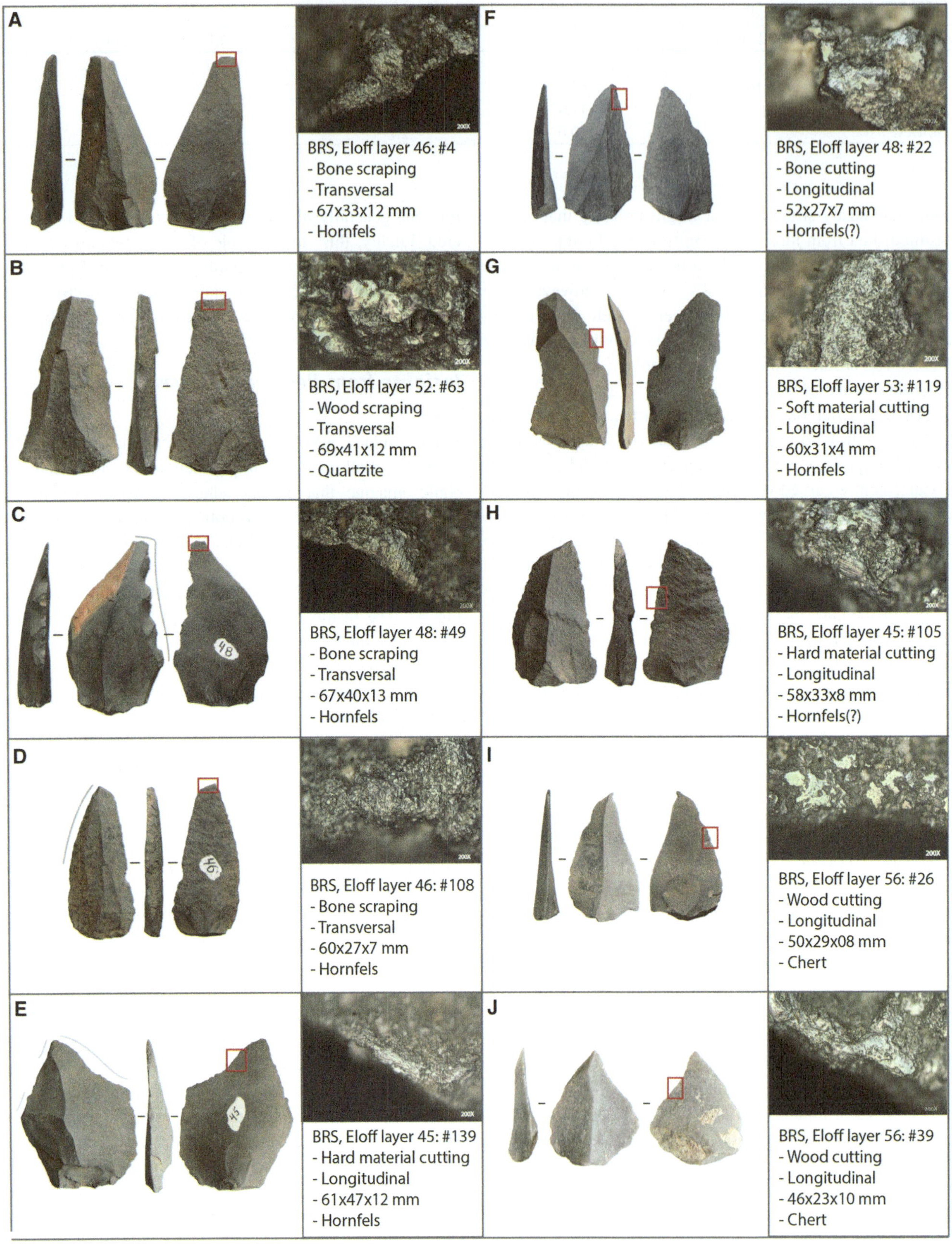

Fig. 7.2 Synthesis of use-wear results on points from BRS. Photomicrographs were taken with a digital camera Canon EOS 600D (see Igreja 2009; Igreja and Porraz 2013 for details on methods). Artifacts were examined using a binocular microscope (Olympus, magnifications up to 100×) and a metallurgical incident light microscope equipped with Differential Interference Contrast (DIC) (Olympus, magnification up to 200×). All artefacts are represented at the same scale. Grey lines on artefact pictures indicate retouched edge portions

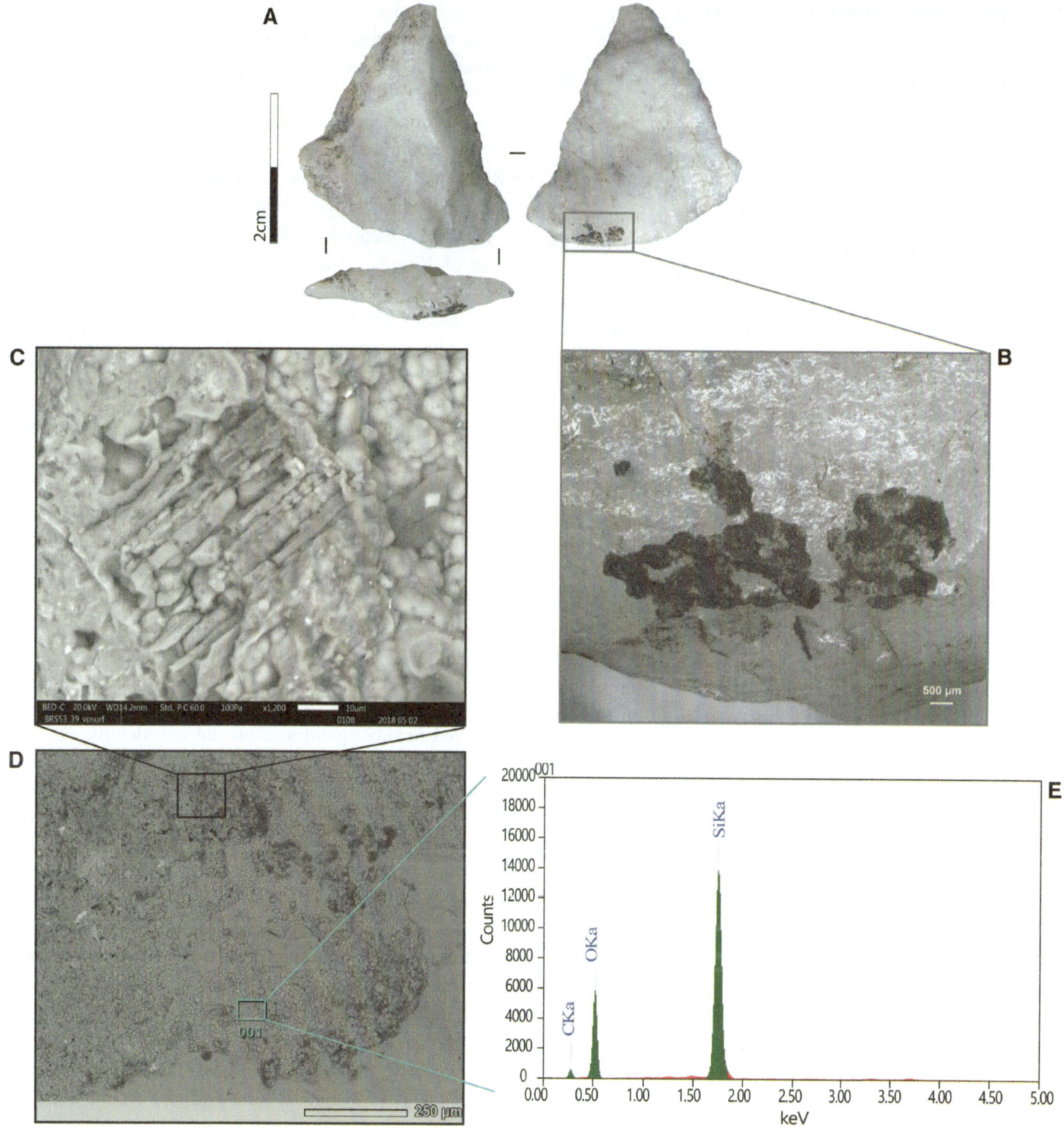

Fig. 7.3 Analysis of black residue on a quartz point from BRS. A Small quartz point with adhering black residue on the bulb; **B** Detail of black residue x22.5; **C** SEM—backscattered electron image of a phytolith inclusion x1200; **D** SEM—backscattered electron image of the black residue showing amorphous silica minerals x180; **E** SEM-EDS spectrum of the residue showing the presence of Carbon (C), Oxygen (O) and Silica (S)

2018b). Indeed, resinous residues can be incidentally deposited on tools that are discarded close to a fire or near fire woods without the residue having any link with use or hafting. It emphasizes the importance of an integrated functional and chemical study before linking a blackish residue with hafting (see Rots et al. 2016). Nevertheless, the exceptional preservation of the residue indicates that a more extensive analysis of the surfaces of the artefacts from the test-pit is promising and may reveal other types of residues in the future.

Bushman Rock Shelter in Context

The nature of the production of points at BRS reflects the implementation of a distinct technical knowledge, predominantly the unidirectional convergent Levallois method, which is not a response to a single functional need. In terms of the operating scheme, the knappers oriented their strategy towards an overall production of flakes with convergent edges that, within the same knapping sequence, provided for axially symmetrical and morphologically balanced points, as well as a majority of points showing a high degree of variability in terms of shapes and metrics. Besides, a discoidal reduction sequence has been applied for some points. It can be argued that the production of points as main lithic strategy was, at BRS, a suitable and economic technical solution to supply for a variety of cutting and scraping edges, and possibly for hunting armatures, although the latter function could not be ascertained.

The lithic system of BRS shows similarities with broadly contemporaneous contexts, which opens avenues for reflecting on the existence of convergent evolution or on interacting populations during MIS 5 in South Africa. Although scarce comparative data are available for this region, sites of north-eastern South Africa have historically been considered part of a regionally-specific complex, yet loosely defined, named the Pietersburg, based on the collections from the Cave of Hearths (COH), located some 150 km north-west of BRS. Technological points are notably documented in Beds 4–8 of COH attributed to the Early and Middle Pietersburg by Mason (1957; see also Sinclair 2009) but also at Mwulu's Cave (de la Peña et al. 2019; Feathers et al. 2020) and at Border Cave in members 4WA and 5BS (Backwell et al. 2018), although the study of the assemblages of both sites is still ongoing.

However, most close comparisons are provided by sites located much farther away, in the Cape regions of South Africa, more than ca. 1300 km southwest of BRS. Similarities had already been noticed by Singer and Wymer (1982) between a phase alternately named MSA II, MSA 2b and Mossel Bay (Volman 1981; Singer and Wymer 1982; Wurz 2002) at Klasies River Mouth (KRM) in the Eastern Cape Province, and Beds 6–8 at COH, a comparison further ascertained by Volman (1981). This phase at KRM is characterized by an abundant production of technological points and the use of hard hammer percussion, and largely complies with the MIS 5 lithic system at BRS presented here. The core reduction operational scheme shows differences, probably in direct relation with the use of different raw material volumes and morphologies (see also Tryon and Ranhorn 2020). Contrary to BRS, point production at KRM is almost exclusively performed on local quartzite cobbles abundantly available locally, and largely knapped on site (Wurz 2000). At BRS, because points are often curated and represent more diverse raw materials, it can be assumed that their production requires a larger range of knapping approaches in the first stages of the *chaîne opératoire*, adapted to the larger diversity in naturally available raw material volumes. Nevertheless, the produced points show strong similarities between KRM and BRS, and a notable production of blades characterizes the assemblages of both sites.

The MIS 5 locations of Pinnacle Point 13 (PP13), the M3 phase of Blombos Cave (BBC) and the "MSA Mike" at Diepkloof Rockshelter (DRS) are other examples of occupation phases that left industries similar to BRS, with local specificities occurring, in particular in the amounts of notched tools (see Douze et al. 2015 for synthesis). Technological points are significant while the shaping process never occurs. In addition, evidence for point functionality has been investigated at PP13B. Several studies have shown that points served for defleshing tasks (Schoville et al. 2016; Bird et al. 2007; Schoville 2010) but evidence for points used as projectiles was also proposed through the identification of stone fragments embedded in size 3 mammal bones (Thompson 2008; O'Driscoll and Thompson 2014). For the same time period, only points from the M3 phase at BBC have been reported as bearing high amounts of "Diagnostic Impact Fractures" (21% of 180 points) by Lombard (2007b), although our analysis (KD) of 11 layers of the same phase excavated in subsequent seasons did not identify such features (Douze et al. 2015; see Coppe and Rots 2017 for problems of defining "DIFs"). At PP13B, it has been suggested that the "little evidence for spear-points based on impact fractures" (Schoville et al. 2016: 23) in the MIS 5 technological point production could be linked with a discard of broken tools on the landscape during foraging strategies. This hypothesis could also be applied to BRS that shows multiple functions and morphologies of points which could be expected for a living site, while evidence for hunting armatures are deficient yet.

While similarities in technological systems are recorded on comparative bases between these localities, the patterns of interaction between these human groups are more difficult to discuss as evidence is so far apart. No archaeological record is currently known—or investigated—for this time period in the area between the north-eastern part of South Africa, where BRS, COH, BC and Mwulu's Cave are located and the sites from the Cape region. The direct consequence has been to encourage researchers to study these regions independently, eluding possible similarities between COH and BRS on one hand, with the Mossel Bay phase of KRM on the other hand. The current state of research perpetuates the dilemma of whether similarities between remote sites should be interpreted as the result of evolutionary convergences in isolated areas or if they demonstrate a degree of interaction between populations to a point that allowed behavioral influences; an issue that is often less discussed when scales of comparison are more reduced (see Will and Mackay 2020, Chap. 6 in this volume). A consensual point of view would suggest an alternative scenario in which the southern African territory, during MIS 5, was intermittently a place for population networks and regionally restrained cultural developments leading to loose—yet existent—cultural ties that allowed parallel developments of behaviors adapted to specific environments. However, this explanation would only be a hypothesis amongst others, often very general ones, that does not provide tangible elements to approach the archaeological reality any further.

Point Production as a Way to Approach Population Patterning

There is a hierarchy in the information provided by lithic industries to approach human cultural groups that could open new perspectives for a better understanding of population dynamics (Gardin 1980; Gallay 1986; Tostevin 2012). Here, convergence in behaviors regarding point production strategies and point functional variability have been seen by comparing assemblages containing technological points, yet it has not been possible to clearly state if these convergences between the different sites of South Africa are due to their occupation by groups sharing social and cultural norms. While distance is one argument provided to question this hypothesis (including external adaptive factors such as raw material availability), it may also relate to a methodological weakness in lithic data exploitation, in particular the behavioral implications of the interwoven criteria of typology, technology and tool use.

The typology of points may provide evidence on human cultural ties, since style is influenced by the cultural norms of a group (e.g. Sackett 1977). Technologies in turn, allow approaching the way points are made, the gestures and technical choices opted for by a group of knappers (i.e. raw material, percussors, core reduction methods, shaping *chaîne opératoire*, etc.) that are culturally influenced by inter-generational transmission and/or by group interaction (Lemonnier 2012; see also Shennan 2020). However, technological options to obtain pointed forms by core reduction or by shaping are not infinite (see Tixier 1978), as they are constrained by the physical properties of conchoidal fracture of hard rocks, which is a factor leading to convergent solutions over time to produce these forms. Similar points can be obtained following different reduction sequences and different functions can be performed with a single point type. To optimize the categorization, the comparison and the interpretation of points, we urgently need better chronological and spatial control. Also the reconstitution of tool uses certainly remains a key element for reducing the versatility of the interpretative frames on human cultural and spatial patterning.

Functional needs do not change drastically over time and space for hunter-gatherer groups, but the implements do. One decisive factor to be taken into account when considering point production within the lithic system is the possible complementary functional role of points with regards to other tool types in the toolkit or, on the contrary, the central functional role of points within a given system as it is the case at BRS. Most other MSA industries provide a large number of tool types and end-products beside points, which constitute varied toolkits composed of different tool morpho-types. Indeed, while assemblages like that of BRS or from the Still Bay industry show that points are produced numerously within a poorly-diversified technical system, presumably for performing different types of actions, other MSA sites show a moderate production of points, in diversified technical systems, for less varied or even specific functionalities, dominantly for their use as projectiles in some cases (e.g. Rots et al. 2017) or for other specific tasks (Rots et al. 2016). The relative representativeness of points —and their function—within different MSA assemblages may be an important asset for approaching contrasting social and economic organization of MSA groups, as well as for a better outline of behavioral regularities in given regions. Documenting open-air or more ephemeral occurrences of MIS 5 occupations would provide a better grasp on site functions (e.g. hunting localities *versus* residential sites) and should therefore constitute another priority for future research in South Africa.

Concluding on Points as Equivocal Tools

The overwhelming use of pointed tool morphology during the MSA could be linked to a recurrent dissemination of populations through the continent, in diversified patterns of progression, retreats and interactions (see also Scerri et al. 2018). The fragmentary archaeological evidence fundamentally restrains our grasp of these complex dynamics, which likely include multi-regional origins for this successful tool morphology.

As much as the term *point* is polysemic in the context of MSA lithic studies, it also designates an equivocal tool. Besides its convergent morphology, this tool encapsulates a large variation of conceptions in terms of technology, style and function. Our study encompasses Groucutt's (2014) and others (e.g. Kandel et al. 2016) views that advocate for the description of point variability as a key method to better approach the cultural load of point manufacture but we also advocate for a systematic investigation of their function. In recent years, use-wear analyses in Africa have progressed towards a larger scope on points' functions, that slowly fills the scientific gap with research that has shown functional dualities for points in other regions, in Europe (e.g. Rots 2013, 2015), in the Levant (e.g. Plisson and Beyries 1998; Bonilauri 2010), and in North Africa (Rots et al. 2011; Tomasso and Rots 2017).

Our study has shown the potential of studying point production from different angles as well as some methodological limitations and research avenues to be developed with the purpose of better approaching past behavioral patterns. Nevertheless, regularities are observed between examples of MIS 6–5 point production in South Africa, and even if the underlying population and cultural dynamics cannot be demonstrated at this stage, the growing bulk of precise data on lithic assemblages in context, supported by adequate experimental reference collections to interpret ancient tool-use, provides optimistic perspectives for future research in this direction.

Acknowledgements The Bushman excavation is funded by the French Ministry of Foreign affairs (excavation permit #2475). We also acknowledge the South African Heritage Resources Agency for providing the temporary export permit for residue analysis (permit ID 2404). The research of V. Rots and D. Cnuts has been supported by the European Research Council under the European Union's Seventh Framework Programme (FP/2007-2013)/ERC Grant Agreement n. 312283 (V. Rots) and the Fund for Scientific Research (FNRS-FRS).

References

Backwell, L. R., d'Errico, F., Banks, W. E., de la Peña, P., Sievers, C., Stratford, D., ... & Wadley, L. (2018). New excavations at Border Cave, KwaZulu-Natal, South Africa. *Journal of Field Archaeology, 43*(6), 417-436.

Bader, G. D., Cable, C., Lentfer, C., & Conard, N. J. (2016). Umbeli Belli Rock Shelter, a forgotten piece from the puzzle of the Middle Stone Age in KwaZulu-Natal, South Africa. *Journal of Archaeological Science: Reports, 9,* 608–622.

Bader, G. D., Will, M., & Conard, N. J. (2015). The lithic technology of Holley Shelter, KwaZulu-Natal, and its place within the MSA of southern Africa. *The South African Archaeological Bulletin, 70,* 149–165.

Beyene, Y., Katoh, S., WoldeGabriel, G., Hart, W. K., Uto, K., Sudo, M., et al. (2013). The characteristics and chronology of the earliest Acheulean at Konso, Ethiopia. *Proceedings of the National Academy of Sciences, 110,* 1584–1591.

Bird, C., Minichillo, T., & Marean, C. W. (2007). Edge damage distribution at the assemblage level on Middle Stone Age lithics: An image-based GIS approach. *Journal of Archaeological Science, 34,* 771–780.

Boëda, E. (1994). *Le concept Levallois: variabilité des méthodes.* Archéo éditions.

Boëda, E., Connan, J., Dessort, D., Muhesen, S., Mercier, N., Valladas, H., et al. (1996). Bitumen as a hafting material on Middle Palaeolithic artefacts. *Nature, 380,* 336–338.

Bonilauri, S. (2010). *Les outils du Paléolithique moyen, une mémoire technique oubliée? Approche techno-fonctionnelle appliquée à un assemblage lithique de conception Levallois provenant du site d'Umm el Tlel (Syrie centrale).* Ph.D. Dissertation, University of Nanterre, Paris 10.

Bouzouggar, A., Barton, R. N. E., & De Araujo Igreja, M. (2007). *A brief overview of recent research into the Aterian and Upper Palaeolithic of northern and eastern Morocco* (Vol. 12, pp. 473–488). Scienze dell'Antichità, Edizioni Quasar.

Brooks, A. S., Nevell, L., Yellen, J. E., & Hartman, G. (2006). Projectile technologies of the African MSA. In E. Hovers & S. L. Kuhn (Eds.), *Transitions before the transition* (pp. 233–255). Boston, MA: Springer.

Brooks, A. S., Yellen, J. E., Potts, R., Behrensmeyer, A. K., Deino, A. L., Leslie, D. E., et al. (2018). Long-distance stone transport and pigment use in the earliest Middle Stone Age. *Science, 360,* 90–94.

Charrié-Duhaut, A., Porraz, G., Cartwright, C. R., Igreja, M., Connan, J., Poggenpoel, C., et al. (2013). First molecular identification of a hafting adhesive in the late Howiesons Poort at Diepkloof Rock Shelter (Western Cape, South Africa). *Journal of Archaeological Science, 40,* 3506–3518.

Chevrier, B., Rasse, M., Lespez, L., Tribolo, C., Hajdas, I., Fígols, M. G., et al. (2016). West African Palaeolithic history: New archaeological and chronostratigraphic data from the Falémé valley, eastern Senegal. *Quaternary International, 408,* 33–52.

Clark, J. D. (1988). The Middle Stone Age of East Africa and the beginnings of regional identity. *Journal of World Prehistory, 2,* 235–305.

Clark, J. D., Cormack, J., & Chin, S. (2001). *Kalambo Falls prehistoric site: Volume 3, The earlier Cultures: Middle and Earlier Stone Age* (Vol. 3). Cambridge University Press.

Cnuts, D., Perrault, K. A., Stefanuto, P. H., Dubois, L. M., Focant, J. F., & Rots, V. (2018a). Fingerprinting glues using HS-SPME GC× GC–HRTOFMS: A new powerful method allows tracking glues back in time. *Archaeometry, 60,* 1362–1376.

Cnuts, D., & Rots, V. (2018). Extracting residues from stone tools for optical analysis: Towards an experiment-based protocol. *Archaeological and Anthropological Sciences, 10,* 1717–1736.

Cnuts, D., Tomasso, S., & Rots, V. (2018b). The role of fire in the life of an adhesive. *Journal of Archaeological Method and Theory, 25,* 839–862.

Conard, N. J., Porraz, G., & Wadley, L. (2012). What is in a name? Characterising the 'Post-Howieson's Poort' at Sibudu. *The South African Archaeological Bulletin, 67,* 180–199.

Coppe, J., & Rots, V. (2017). Focus on the target. The importance of a transparent fracture terminology for understanding projectile points and projecting modes. *Journal of Archaeological Science: Reports, 12,* 109–123.

de la Peña, P., Taipale, N., Wadley, L., & Rots, V. (2018). A techno-functional perspective on quartz micro-notches in Sibudu's Howiesons Poort indicates the use of barbs in hunting technology. *Journal of Archaeological Science, 93,* 166–195.

de la Peña, P., Val, A., Stratford, D. J., Colino, F., Esteban, I., Fitchett, J. M., ... & Moll, R. (2019). Revisiting Mwulu's Cave: New insights into the Middle Stone Age in the southern African savanna biome. *Archaeological and Anthropological Sciences, 11*(7), 3239–3266.

de la Peña, P., Wadley, L., & Lombard, M. (2013). Quartz bifacial points in the Howiesons Poort of Sibudu. *The South African Archaeological Bulletin, 68,* 119–136.

Delhon, C. (2010). Phytolithes et taphonomie, apport de l'expérimentation à la quantification des phytolithes dans les cendres de bois. In I. Théry-Parisot, L. Chabal, & S. Costamagno (Eds.), *Taphononomie des résidus organiques brûlés et des structures de combustion en milieu archéologique. Actes de la table ronde, Valbonne, 27–29 mai 2008* (pp. 97–107). Toulouse: Palethnologie.

Douze, K. (2012). *Le Early Middle Stone Age d'Éthiopie et les changements techno-économiques à la période de l'émergence des premiers Homo sapiens*. Ph.D. Dissertation, University of Bordeaux.

Douze, K., & Delagnes, A. (2016). The pattern of emergence of a Middle Stone Age tradition at Gademotta and Kulkuletti (Ethiopia) through convergent tool and point technologies. *Journal of Human Evolution, 91,* 93–121.

Douze, K., Delagnes, A., Rots, V., & Gravina, B. (2018). A reply to Sahle and Braun's reply to 'The pattern of emergence of a Middle Stone Age tradition at Gademotta and Kulkuletti (Ethiopia) through convergent tool and point technologies' [J. Hum. Evol. 91 (2016) 93–121]. *Journal of Human Evolution, 125,* 207–214.

Douze, K., Wurz, S., & Henshilwood, C. S. (2015). Techno-cultural characterization of the MIS 5 (c. 105–90 Ka) Lithic Industries at Blombos Cave, Southern Cape, South Africa. *PLoS ONE, 10,* e0142151.

Evershed, R. P. (2008). Organic residue analysis in archaeology: The archaeological biomarker revolution. *Archaeometry, 50,* 895–924.

Feathers, J. K., Evans, M., Stratford, D. J., & de la Peña, P. (2020). Exploring complexity in luminescence dating of quartz and feldspars at the Middle Stone Age site of Mwulu's cave (Limpopo, South Africa). *Quaternary Geochronology*, 101092.

Gallay, A. (1986). *L'archéologie demain*. Paris: Belfond Sciences.

Gardin, J. C. (1980). *Archaeological constructs: An aspect of theoretical archaeology*. Cambridge, Paris: Cambridge University Press and Éditions de la Maison des Sciences de l'Homme.

Gibson, N. E., Wadley, L., & Williamson, B. S. (2004). Microscopic residues as evidence of hafting on backed tools from the 60 000 to 68 000 year-old Howiesons Poort layers of Rose Cottage Cave, South Africa. *Southern African Humanities, 16,* 1–11.

Gonzalez-Urquijo, J. E., & Ibanez-Estevez, J. J. (2004). *Metodologia de analisis funcional de instrumentos tallados en silex*. Bilbao: Universidad de Deusto.

Groucutt, H. S. (2014). Middle Palaeolithic point technology, with a focus on the site of Tor Faraj (Jordan, MIS 3). *Quaternary International, 350,* 205–226.

Groucutt, H. S. (2020). Culture and convergence: The curious case of the Nubian Complex. In H. Groucutt (Ed.), *Culture history and convergent evolution: Can we detect populations in prehistory?* (pp. 55–86). Cham, Switzerland: Springer.

Hauck, T. C., Connan, J., Charrié-Duhaut, A., Le Tensorer, J.-M., & Sakhel, H. A. (2013). Molecular evidence of bitumen in the Mousterian lithic assemblage of Hummal (central Syria). *Journal of Archaeological Science, 40,* 3252–3262.

Hayek, E. W., Krenmayr, P., Lohninger, H., Jordis, U., Moche, W., & Sauter, F. (1990). Identification of archaeological and recent wood tar pitches using gas chromatography/mass spectrometry and pattern recognition. *Analytical Chemistry, 62,* 2038–2043.

Helwig, K., Monahan, V., Poulin, J., & Andrews, T. D. (2014). Ancient projectile weapons from ice patches in northwestern Canada: Identification of resin and compound resin-ochre hafting adhesives. *Journal of Archaeological Science, 41,* 655–665.

Herries, A. I. (2011). A chronological perspective on the Acheulian and its transition to the Middle Stone Age in southern Africa: The question of the Fauresmith. *International Journal of Evolutionary Biology*. https://doi.org/10.4061/2011/961401.

Högberg, A., & Larsson, L. (2011). Lithic technology and behavioural modernity: New results from the Still Bay site, Hollow Rock Shelter, Western Cape Province, South Africa. *Journal of Human Evolution, 61,* 133–155.

Hutchings, W. K. (2016). When is a point a projectile? Morphology, impact fractures, scientific rigor, and the limits of inference. In R. Iovita & K. Sano (Eds.), *Multidisciplinary approaches to the study of Stone Age weaponry* (pp. 3–12). Dordrecht: Springer.

Igreja, M. (2009). Use-wear analysis of non-flint stone tools using DIC microscopy and resin casts: A simple and effective technique. In Recent functional studies on non flint stone tools: Methodological improvements and archaeological inferences, 23–25 May 2008, *LISBOA—Proceedings of the Workshop [CD-ROM]*. Lisboa (Padrão dos Descobrimentos).

Igreja, M., & Porraz, G. (2013). Functional insights into the innovative Early Howiesons Poort technology at Diepkloof Rock Shelter (Western Cape, South Africa). *Journal of Archaeological Science, 40,* 3475–3491.

Kandel, A. W., Bolus, M., Bretzke, K., Bruch, A. A., Haidle, M. N., Hertler, C., et al. (2016). Increasing behavioral flexibility? An integrative macro-scale approach to understanding the Middle Stone Age of southern Africa. *Journal of Archaeological Method and Theory, 23,* 623–668.

Karkanas, P., Kyparissi-Apostolika, N., Bar-Yosef, O., & Weiner, S. (1999). Mineral assemblages in Theopetra, Greece: A framework for understanding diagenesis in a prehistoric cave. *Journal of Archaeological Science, 26,* 1171–1180.

Keeley, L. H. (1980). *Experimental determination of stone tool uses: A microwear analysis*. University of Chicago Press.

Koller, J., Baumer, U., & Mania, D. (2001). High-tech in the Middle Palaeolithic: Neandertal-manufactured pitch identified. *European Journal of Archaeology, 4,* 385–397.

Kuman, K. A. (1989). Florisbad and # Gi: The contribution of open-air sites to study of the Middle Stone Age in southern Africa. Ph.D. Dissertation, University of Pennsylvania.

Kurashina, H. (1978). An examination of prehistoric lithic technology in east-central Ethiopia. Ph.D. Dissertation, University of California.

Lemonnier, P. (2012). Des objets pour penser l'indicible. La nécessaire convergence des théories de la culture matérielle. In N. Schlanger (Ed.), *La préhistoire des autres: Perspectives archéologiques et anthropologiques* (pp. 275–289). Paris: La Découverte.

Lepre, C. J., Roche, H., Kent, D. V., Harmand, S., Quinn, R. L., Brugal, J. P., et al. (2011). An earlier origin for the Acheulian. *Nature, 477,* 82.

Lombard, M. (2005). Evidence of hunting and hafting during the Middle Stone Age at Sibidu Cave, KwaZulu-Natal, South Africa: A multianalytical approach. *Journal of Human Evolution, 48,* 279–300.

Lombard, M. (2006). First impressions of the functions and hafting technology of Still Bay pointed artefacts from Sibudu Cave. *Southern African Humanities, 18,* 27–41.

Lombard, M. (2007a). The gripping nature of ochre: The association of ochre with Howiesons Poort adhesives and Later Stone Age mastics from South Africa. *Journal of Human Evolution, 53,* 406–419.

Lombard, M. (2007b). Evidence for change in Middle Stone Age hunting behaviour at Blombos Cave: Results of a macrofracture analysis. *South African Archaeological Bulletin, 62,* 62–67.

Lombard, M. (2008). Finding resolution for the Howiesons Poort through the microscope: Micro-residue analysis of segments from Sibudu Cave, South Africa. *Journal of Archaeological Science, 35,* 26–41.

Mason, R. J. (1957). The Transvaal Middle Stone Age and statistical analysis. *The South African Archaeological Bulletin, 12,* 119–137.

McBrearty, S., & Brooks, A. S. (2000). The revolution that wasn't: A new interpretation of the origin of modern human behavior. *Journal of human evolution, 39,* 453–563.

McBrearty, S., & Tryon, C. (2006). From Acheulean to middle stone age in the Kapthurin formation, Kenya. In E. Hovers & S. L. Kuhn (Eds.), *Transitions before the Transition* (pp. 257–277). Boston: Springer.

Ménard, C., Bon, F., Dessie, A., Bruxelles, L., Douze, K., Fauvelle, F. X., et al. (2014). Late Stone Age variability in the Main Ethiopian

Rift: New data from the Bulbula River, Ziway-Shala basin. *Quaternary International, 343,* 53–68.

Mills, J. S., & White, R. (1977). Natural resins of art and archaeology their sources, chemistry, and identification. *Studies in Conservation, 22,* 12–31.

Milo, R. G. (1998). Evidence for hominid predation at Klasies River Mouth, South Africa, and its implications for the behaviour of early modern humans. *Journal of Archaeological Science, 25,* 99–133.

Mourre, V., Villa, P., & Henshilwood, C. S. (2010). Early use of pressure flaking on lithic artifacts at Blombos Cave, South Africa. *Science, 330,* 659–662.

Niang, K., Blinkhorn, J., & Ndiaye, M. (2018). The oldest Stone Age occupation of coastal West Africa and its implications for modern human dispersals: New insight from Tiémassas. *Quaternary Science Reviews, 188,* 167–173.

O'Driscoll, C. A., & Thompson, J. C. (2014). Experimental projectile impact marks on bone: Implications for identifying the origins of projectile technology. *Journal of Archaeological Science, 49,* 398–413.

Perlès, C. (1974). Réexamen typologique de l'industrie du Porc-Épic (Éthiopie): Les pointes et pièces pointues. *L'Anthropologie, 78,* 529–552.

Perrault, K. A., Stefanuto, P. H., Dubois, L., Cnuts, D., Rots, V., & Focant, J. F. (2016). A new approach for the characterization of organic residues from stone tools using GC× GC-TOFMS. *Separations, 3,* 16.

Pleurdeau, D. (2005). Human technical behavior in the African middle stone age: The lithic assemblage of Porc-Epic cave (Dire Dawa, Ethiopia). *African Archaeological Review, 22,* 177–197.

Plisson, H. (1985a). *Étude fonctionnelle d'outillages lithiques préhistoriques par l'analyse des micro-usures: Recherche méthodologique et archéologique* (p. 357). PhD dissertation, University of Paris I, Paris.

Plisson, H. (1985b). Contribution de la tracéologie a la localisation des aires d'activité et d'occupation. *Anthropologie (L'), 89,* 473–478.

Plisson, H., & Beyries, S. (1998). Pointes ou outils triangulaires? Données fonctionnelles dans le Moustérien levantin [with Commentry of Shea, J., Marks, A., & Geneste, J. M.]. *Paléorient, 24,* 5–24.

Porraz, G., Parkington, J. E., Rigaud, J. P., Miller, C. E., Poggenpoel, C., Tribolo, C., et al. (2013a). The MSA sequence of Diepkloof and the history of southern African Late Pleistocene populations. *Journal of Archaeological Science, 40,* 3542–3552.

Porraz, G., Texier, P. J., Archer, W., Piboule, M., Rigaud, J. P., & Tribolo, C. (2013b). Technological successions in the Middle Stone Age sequence of Diepkloof Rock Shelter, Western Cape, South Africa. *Journal of Archaeological Science, 40,* 3376–3400.

Porraz, G., Val, A., Dayet, L., De La Pena, P., Douze, K., Miller, C. E., et al. (2015). Bushman Rock Shelter (Limpopo, South Africa): A perspective from the edge of the Highveld. *The South African Archaeological Bulletin, 70,* 166–179.

Porraz, G., Val, A., Tribolo, C., Mercier, N., de la Peña, P., Haaland, M. M., et al. (2018). The MIS5 Pietersburg at '28'Bushman Rock Shelter, Limpopo Province, South Africa. *PLoS ONE, 13,* e0202853.

Regert, M. (2004). Investigating the history of prehistoric glues by gas chromatography–mass spectrometry. *Journal of Separation Science, 27,* 244–254.

Rots, V. (2013). Insights into early Middle Palaeolithic tool use and hafting in Western Europe. The functional analysis of level IIa of the early Middle Palaeolithic site of Biache-Saint-Vaast (France). *Journal of Archaeological Science, 40,* 497–506.

Rots, V. (2015). Hafting and the interpretation of site function in the European Middle Palaeolithic. In N. Conard & A. Delagnes (Eds.), *Settlement dynamics of the Middle Paleolithic and Middle Stone Age* (pp. 383–410). Tübingen: Kerns Verlag.

Rots, V. (2016). Projectiles and hafting technology. In R. Iovita & K. Sano (Eds.), *Multidisciplinary approaches to the study of Stone Age weaponry* (pp. 167–185). Dordrecht: Springer.

Rots, V., Lentfer, C., Schmid, V. C., Porraz, G., & Conard, N. J (2016). There is more to life than subsistence: Use-wear and residue analyses on pre-Still Bay stone tools at Sibudu, *ESHE Conference. Abstracts*, Madrid, 210.

Rots, V., Lentfer, C., Schmid, V. C., Porraz, G., & Conard, N. J. (2017). Pressure flaking to serrate bifacial points for the hunt during the MIS5 at Sibudu Cave (South Africa). *PLoS ONE, 12,* e0175151.

Rots, V., & Plisson, H. (2014). Projectiles and the abuse of the use-wear method in a search for impact. *Journal of Archaeological Science, 48,* 154–165.

Rots, V., & Van Peer, P. (2006). Early evidence of complexity in lithic economy: Core-axe production, hafting and use at Late Middle Pleistocene site 8-B-11, Sai Island (Sudan). *Journal of Archaeological Science, 33,* 360–371.

Rots, V., Van Peer, P., & Vermeersch, P. M. (2011). Aspects of tool production, use, and hafting in Palaeolithic assemblages from Northeast Africa. *Journal of Human Evolution, 60,* 637–664.

Sackett, J. A. (1977). The meaning of style in archaeology: A general model. *American Antiquity, 42,* 369–380.

Sahle, Y., & Braun, D. R. (2018). A reply to Douze and Delagnes's 'The pattern of emergence of a Middle Stone Age tradition at Gademotta and Kulkuletti (Ethiopia) through convergent tool and point technologies' [J. Hum. Evol. 91 (2016) 93–121]. *Journal of Human Evolution, 125,* 201–206.

Sahle, Y., Hutchings, W. K., Braun, D. R., Sealy, J. C., Morgan, L. E., Negash, A., et al. (2013). Earliest stone-tipped projectiles from the Ethiopian Rift date to >279,000 years ago. *PLoS ONE, 8,* e78092.

Sahle, Y., Morgan, L. E., Braun, D. R., Atnafu, B., & Hutchings, W. K. (2014). Chronological and behavioral contexts of the earliest Middle Stone Age in the Gademotta formation, main Ethiopian rift. *Quaternary International, 331,* 6–19.

Scerri, E. M., Blinkhorn, J., Groucutt, H. S., & Niang, K. (2016). The Middle Stone Age archaeology of the Senegal river valley. *Quaternary International, 408,* 16–32.

Scerri, E. M., Blinkhorn, J., Niang, K., Bateman, M. D., & Groucutt, H. S. (2017). Persistence of Middle Stone Age technology to the Pleistocene/Holocene transition supports a complex hominin evolutionary scenario in West Africa. *Journal of Archaeological Science: Reports, 11,* 639–646.

Scerri, E. M., Thomas, M. G., Manica, A., Gunz, P., Stock, J. T., Stringer, C., et al. (2018). Did our species evolve in subdivided populations across Africa, and why does it matter? *Trends in Ecology & Evolution, 33,* 582–594.

Schiegl, S., Goldberg, P., Bar-Yosef, O., & Weiner, S. (1996). Ash deposits in Hayonim and Kebara caves, Israel: Macroscopic, microscopic and mineralogical observations, and their archaeological implications. *Journal of Archaeological Science, 23,* 763–781.

Schiegl, S., Lev-Yadun, S., Bar-Yosef, O., El Goresy, A., & Weiner, S. (1994). Siliceous aggregates from prehistoric wood ash: A major component of sediments in Kebara and Hayonim caves (Israel). *Israel Journal of Earth Sciences, 43,* 257–278.

Schmidt, P., Porraz, G., Bellot-Gurlet, L., February, E., Ligouis, B., Paris, C., et al. (2015). A previously undescribed organic residue sheds light on heat treatment in the Middle Stone Age. *Journal of Human Evolution, 85,* 22–34.

Schoville, B. J. (2010). Frequency and distribution of edge damage on Middle Stone Age lithic points, Pinnacle Point 13B, South Africa. *Journal of Human Evolution, 59,* 378–391.

Schoville, B. J., Brown, K. S., Harris, J. A., & Wilkins, J. (2016). New experiments and a model-driven approach for interpreting Middle

Stone Age lithic point function using the edge damage distribution method. *PLoS ONE, 11,* e0164088.
Semenov, S. A. (1964). *An experimental study of the oldest tools and artefacts from traces of manufacture and wear.* New Jersey: Barnes & Noble Books.
Shea, J. J. (2006). The origins of lithic projectile point technology: Evidence from Africa, the Levant, and Europe. *Journal of Archaeological Science, 33,* 823–846.
Shea, J. J. (2008). The Middle Stone Age archaeology of the Lower Omo Valley Kibish Formation: Excavations, lithic assemblages, and inferred patterns of early Homo sapiens behavior. *Journal of Human Evolution, 55,* 448–485.
Shennan, S. (2020). Style, function and cultural transmission. In H. Groucutt (Ed.), *Culture history and convergent evolution: Can we detect populations in prehistory?* (pp. 291–298). Cham, Switzerland: Springer.
Shott, M. J. (2020). Toward a theory of the point. In H. Groucutt (Ed.), *Culture history and convergent evolution: Can we detect populations in prehistory?* (pp. 245–259). Cham, Switzerland: Springer.
Sinclair, A. (2009). The MSA Stone Tool Assemblage from the Cave of Hearths, Beds 4-9. In J. McNabb & A. Sinclair (Eds.), *The Cave of Hearths: Makapan Middle Pleistocene Research Project: Field research by Anthony Sinclair and Patrick Quinney, 1996–2001* (pp. 105–137). Oxford, GB: BAR Archaeopress.
Singer, R., & Wymer, J. (1982). *The Middle Stone Age at Klasies River Mouth in South Africa.* University of Chicago Press.
Sisk, M. L., & Shea, J. J. (2011). The African origin of complex projectile technology: An analysis using tip cross-sectional area and perimeter. *International Journal of Evolutionary Biology.* https://doi.org/10.4061/2011/968012.
Soriano, S., Villa, P., Delagnes, A., Degano, I., Pollarolo, L., Lucejko, J. J., et al. (2015). The Still Bay and Howiesons Poort at Sibudu and Blombos: Understanding Middle Stone Age technologies. *PLoS ONE, 10,* e0131127.
Steele, T. E., Mackay, A., Fitzsimmons, K., Igreja, M., Marwick, B., Orton, J., et al. (2016). Varsche Rivier 003: A Middle Stone Age site with Still Bay and Howiesons Poort assemblages in southern Namaqualand, South Africa. *PaleoAnthropology, 2016,* 100–163.
Taylor, N. (2016). Across rainforests and woodlands: A systematic reappraisal of the Lupemban Middle Stone Age in Central Africa. In S. C. Jones & B. A. Stewart (Eds.), *Africa from MIS 6-2* (pp. 273–299). Dordrecht: Springer.
Thompson, J. C. (2008). *Zooarchaeological tests for modern human behavior at Blombos Cave and Pinnacle Point Cave 13B, southwestern Cape, South Africa.* Dissertation, Arizona State University.
Thompson, E., Williams, H. M., & Minichillo, T. (2010). Middle and late Pleistocene Middle Stone Age lithic technology from Pinnacle point 13B (Mossel Bay, western Cape province, South Africa). *Journal of Human Evolution, 59,* 358–377.
Tixier, J. (1978). Méthode pour l'étude des outillages lithiques. Ph.D. Dissertation, University of Nanterre.
Tomasso, S., & Rots, V. (2017). What is the use of shaping a tang? Tool use and hafting of tanged tools in the Aterian of Northern Africa. *Archaeological and Anthropological Sciences, 10,* 1389–1417.
Tostevin, G. B. (2012). *Seeing lithics: A middle-range theory for testing for cultural transmission in the Pleistocene.* American School of Prehistoric Research Monograph Series. Peabody Museum, Harvard University, and Oxbow Books.
Tryon, C. A., & Faith, J. T. (2013). Variability in the Middle Stone Age of Eastern Africa. *Current Anthropology, 54,* S234–S254.
Tryon, C. A., McBrearty, S., & Texier, P. J. (2005). Levallois lithic technology from the Kapthurin formation, Kenya: Acheulian origin and Middle Stone Age diversity. *African Archaeological Review, 22,* 199–229.
Tryon, C. A., & Ranhorn, K. L. (2020). Raw material and regionalization in Stone Age East Africa. In H. Groucutt (Ed.), *Culture history and convergent evolution: Can we detect populations in prehistory?* (pp. 143–156). Cham, Switzerland: Springer.
Underhill, D. (2012). The Earliest Middle Stone Age of northern South Africa: The Cave of Hearths and Bushman Rock Shelter. In K. Ruebens, I. Romanowska, & R. Bynoe (Eds.), *Unravelling the Palaeolithic.* (pp. 13–26). Oxford: BAR International Series 2400.
Villa, P., Soressi, M., Henshilwood, C. S., & Mourre, V. (2009). The Still Bay points of Blombos Cave (South Africa). *Journal of Archaeological Science, 36,* 441–460.
Volman, T. P. (1981). *The Middle Stone Age in the southern Cape.* Ph. D. Dissertation, University of Chicago.
Wadley, L. (2005). A typological study of the final Middle Stone Age stone tools from Sibudu Cave, KwaZulu-Natal. *The South African Archaeological Bulletin, 60,* 51–63.
Wadley, L. (2007). Announcing a Still Bay industry at Sibudu Cave, South Africa. *Journal of Human Evolution, 52,* 681–689.
Wadley, L., Hodgskiss, T., & Grant, M. (2009). Implications for complex cognition from the hafting of tools with compound adhesives in the Middle Stone Age, South Africa. *Proceedings of the National Academy of Sciences, 106,* 9590–9594.
Wadley, L., Williamson, B., & Lombard, M. (2004). Ochre in hafting in Middle Stone Age southern Africa: A practical role. *Antiquity, 78,* 661–675.
Waweru, V. N. (2007). Middle Stone Age of technology at Cartwright's site, Kenya. Ph.D. Dissertation, University of Connecticut.
Wendorf, F., & Schild, R. (1974). *A Middle Stone Age Sequence from the Central Rift Valley, Ethiopia.* Wroclaw: Zakład Narodowy im. Ossolińskich.
Wilkins, J., Schoville, B. J., Brown, K. S., & Chazan, M. (2012). Evidence for early hafted hunting technology. *Science, 338,* 942–946.
Wilkins, J., Schoville, B. J., Brown, K. S., & Chazan, M. (2015). Kathu Pan 1 points and the assemblage-scale, probabilistic approach: A response to Rots and Plisson, "Projectiles and the abuse of the use-wear method in a search for impact". *Journal of Archaeological Science, 54,* 294–299.
Will, M., & Conard, N. J. (2018). Assemblage variability and bifacial points in the lowermost Sibudan layers at Sibudu, South Africa. *Archaeological and Anthropological Sciences, 10,* 389–414.
Wojcieszak, M., & Wadley, L. (2018). Raman spectroscopy and scanning electron microscopy confirm ochre residues on 71 000-year-old bifacial tools from Sibudu, South Africa. *Archaeometry, 60,* 1062–1076.
Wurz, S. (2002). Variability in the Middle Stone Age lithic sequence, 115,000–60,000 years ago at Klasies river, South Africa. *Journal of Archaeological Science, 29,* 1001–1015.
Wurz, S. (2013). Technological trends in the Middle Stone Age of South Africa between MIS 7 and MIS 3. *Current Anthropology, 54,* S305–S319.
Wurz, S. J. D. (2000). *The Middle Stone Age at Klasies River, South Africa* (Doctoral dissertation, University of Stellenbosch).
Yellen, J., Brooks, A., Helgren, D., Tappen, M., Ambrose, S., Bonnefille, R., et al. (2005). The archaeology of Aduma Middle Stone Age sites in the Awash Valley, Ethiopia. *PaleoAnthropology, 10,* e100.
Zipkin, A. M., Hanchar, J. M., Brooks, A. S., Grabowski, M. W., Thompson, J. C., & Gomani-Chindebvu, E. (2015). Ochre fingerprints: Distinguishing among Malawian mineral pigment sources with Homogenized Ochre Chip LA–ICPMS. *Archaeometry, 57,* 297–317.

Chapter 8
Raw Material and Regionalization in Stone Age Eastern Africa

Christian A. Tryon and Kathryn L. Ranhorn

Abstract Stone tools are the dominant artifact type at Paleolithic sites, and the kinds of stone tools used and their methods of manufacture form some of the richest datasets to assess temporal and geographic patterning in hominin behavior. Using these datasets to compare different lithic assemblages requires comprehensive analytical frameworks that be applied across multiple sites, but this is complicated by the varied nature of the different rock types used in the past. The bedrock lithology of eastern Africa is particularly varied, and we show for a range of Early Pleistocene-to-Holocene-aged archaeological sites that the type and frequency of raw material used, particularly quartz, has significant impacts on a number of typological, technological, and metric variables used to measure variation across time and space, severely weakening our abilities to assess the extent to which past geographic variation in the archaeological record in particular can be attributed to hominin behavior or bedrock geology. Convergence (homoplasy) in particular may be difficult to discern, as even similar behaviors resulting from shared cultural traditions (homology) may result in very different looking artifact types because of the nature of the rock types used.

Keywords Lithic analysis • Inter-assemblage variability • Toolstone • Data comparability

C. A. Tryon (✉)
Department of Anthropology, University of Connecticut, 354 Mansfield Rd., Storrs, CT 06269, USA

K. L. Ranhorn
Institute of Human Origins, School of Human Evolution and Social Change, Arizona State University, PO Box 872402, Tempe, AZ, 85287, USA

Introduction

Recognizing regional 'identities' is a persistent feature of analyses of Paleolithic variability and the processes that produced it, whether these identities are defined as industries, industrial complexes, facies, populations, or some other taxonomic term (cf. Will et al. 2014; Shea 2014; see also discussion in Reynolds 2020; O'Brien and Bentley 2020). For decades, particular efforts have been made to define regional variants among Middle Stone Age (MSA) sites in Africa, where broadly defined industrial complexes such as the Aterian and Lupemban seem to spatially co-vary with major biomes such as the Sahara Desert and the Central African Rainforest, respectively (e.g. Clark 1988, 1993; McBrearty and Brooks 2000; Scerri et al. 2014; Jones and Stewart 2016; Scerri 2017). Inter-regional variation for MSA sites appears to be greater than seen in preceding Acheulian sites, although variation in some 'terminal' Acheulian assemblages may anticipate later MSA patterns (Tryon et al. 2005; Potts et al. 2018). Because MSA sites in Africa appear to be associated with *Homo sapiens*, there is a particular interest in linking the developing regionalization in the archaeological record (i.e. geographically distinct behavioral variability) to other lines of evidence that suggest extensive population structure (i.e. geographically distinct biological variability) among Middle and Late Pleistocene *H. sapiens* (Mackay et al. 2014; Scerri et al. 2018; see also discussion in Groucutt 2020; Spinapolice 2020).

Despite the widespread recognition of a number of large scale regional MSA variants, few are formally defined (see Scerri 2017), particularly in eastern Africa, where in some cases particular industries are well defined, but are reported from one or two sites at most, contributing to a general sense of regional heterogeneity rather than providing a useful comparative tool (reviewed in Tryon and Faith 2013; see also Ranhorn and Tryon 2018). We believe that carefully defining regional variation is an important research

H. S. Groucutt (ed.), *Culture History and Convergent Evolution: Can We Detect Populations in Prehistory?*,
Vertebrate Paleobiology and Paleoanthropology, https://doi.org/10.1007/978-3-030-46126-3_8

objective. As a prelude to this, our goal here is to outline the impacts of lithic raw material variability on attempts to do so. Certainly, there are a number of instances where the selection of specific kinds of raw material may be a defining feature of an artifact industry and presumably the populations that made them. Well-known examples include the preference for finer-grained rocks seen among Howiesons Poort MSA assemblages in southern Africa (Ambrose and Lorenz 1990; Brown 2011; Oestmo 2017), and the association of particular obsidian quarries with Savanna Pastoral Neolithic and Elmenteitan pastoralists in the late Holocene of eastern Africa (Ambrose 2012; Goldstein and Munyiri 2017). These patterns reflect clear choices made by past human groups, choices likely caused in part by various properties of the rocks chosen for use, such as predictable fracture qualities, edge durability, color, or other properties (e.g. Braun et al. 2009; Pargeter and Hampson 2019).

However, rather than focusing on the role of human choice in using particular kinds of rocks, we emphasize the extent to which the properties of the rocks themselves influence the types of variables archaeologists use to define regional variability. Specifically, we highlight some of the ways in which the properties of the available stone raw material used for tool production may create archaeologically detectable variability that masks behavioral similarity, in effect making cases of behavioral convergence difficult to identify. While some studies in northern Africa (Scerri et al. 2014) suggest that in some cases, geographic patterning in hominin lithic reduction strategies can transcend local differences in raw material, this issue remains under-explored in eastern Africa.

Lithic Raw Material Variability in Eastern Africa

Hominins in eastern Africa used a wide range of different types of rocks for the manufacture of stone tools, with the diversity of the available types of rocks a reflection of the region's complex geological history (Fig. 8.1). Sedimentary rocks include chert, often formed in past and present saline lakes, with well-known outcrops found at Lake Natron and Olduvai Gorge in Tanzania and Lake Magadi in Kenya (Stiles et al. 1974; Hay 1968, 1976), and shale at Mtongwe on the Kenyan coast (Omi 1984). Igneous rocks include lavas that vary in texture and composition, from coarse-grained basalts at Koobi Fora and Olorgesailie, Kenya (Noll 2000; Isaac et al. 1997), trachytes from West Turkana (Harmand 2007), phonolites from Rusinga Island (Tryon et al. 2014), and obsidians, particularly from the Central Rift Valley in Kenya (Brown et al. 2013). Quartz is available as part of the pre-Cambrian basement rocks found throughout the region, best exposed outside of the Rift Valley, found in abundance at places such as Lukenya Hill in Kenya (Gramly 1976), Nasera in Tanzania (Mehlman 1977), and throughout neighboring portions of central Africa, as at Matupi Cave, D.R.C. (Van Noten 1977; Muya wa Bitanko 1985–1986). Metamorphic rocks include various metasomatized volcanic rocks from Kanjera (Braun et al. 2008), mylonitized lavas from Isimila, Tanzania (Howell et al. 1962), and quartzites found at sites such as Nsongezi, Uganda (Cole 1967). Fossil wood was used at some sites on the eastern shores of Lake Turkana in Kenya (Kelly 1996).

These examples of rock types are but a small sample of the kinds of lithic raw material used for tool production at Stone Age sites in eastern Africa, and this diversity of rock types and their geological sources has been useful in petrographic and geochemical studies of artifact provenance (Merrick and Brown 1984; Feblot-Augustins 1990; Merrick et al. 1994; Noll 2000; Harmand 2007; Braun et al. 2008). However, these different kinds of rocks also vary substantially in terms of their hardness, durability, grain size, texture, and fracture mechanics. These properties affect how stone tools were made and how they were used, and as we argue below through a series of examples drawn from Early Pleistocene- to Holocene-aged sites, have important impacts on some of the qualitative and quantitative variables potentially useful in the construction of regionally specific ways of tool manufacture.

Handaxe Variability at Olduvai Gorge, Tanzania

In a remarkable series of papers, Jones (1979, 1980, 1994) outlined the results of the experimental replication and use of various artifact types to aid in the analysis of Acheulian and other Early Stone Age assemblages at Olduvai Gorge (Fig. 8.2). Several results are worth reiterating here, as they have broad implications for understanding more general patterning in the archaeological record. First, the form in which lithic raw material is initially available can impact the nature of the finished tool (see also Andrefsky 1994; White 1998). Specifically, bifaces made on cobbles will generally have thicker cross-sections rather than those made on slabs (or large flakes). At Olduvai Gorge, this contrast is seen most clearly between thicker basalt bifaces (made on cobbles derived from local streams; Fig. 8.2A) and thinner phonolite (Fig. 8.2B) and quartzite bifaces (Fig. 8.2C), the latter typically made on slabs or spalls found near outcrops.

Second, hardness, grain-size and texture influence the amount of retouch on an artifact. Jones experimented in making cleavers and handaxes made of various rock types available at Olduvai Gorge, and used them to butcher animal carcasses. This work showed that quartzite tools in particular could be used for long periods with little retouch or

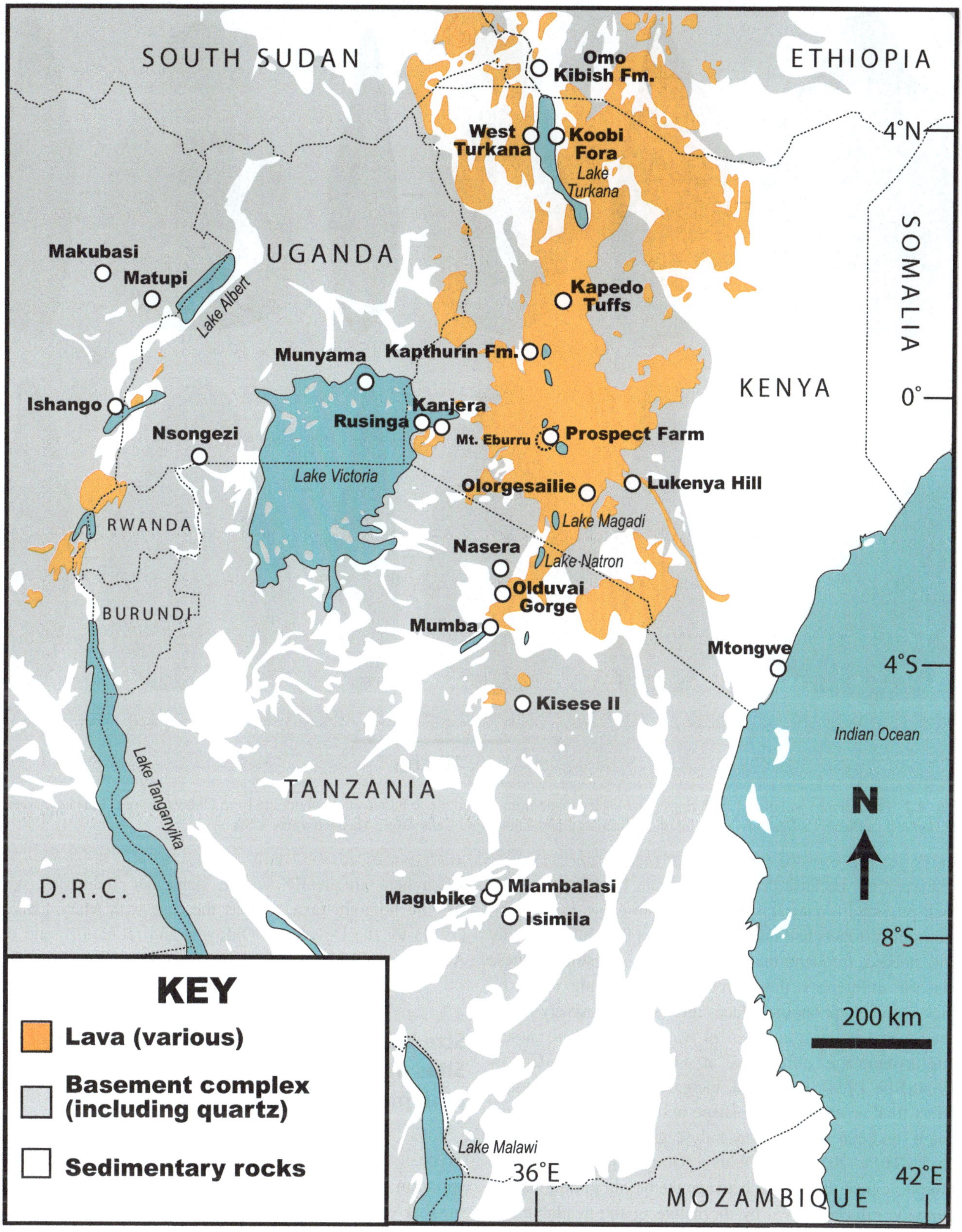

Fig. 8.1 Simplified geological map of eastern and central Africa, showing outcrops of lava, pre-Cambrian igneous and metamorphic basement rocks (including quartz), and sedimentary deposits, as well as key archaeological sites and geographic features mentioned in the text. Base map redrawn from Choubert and Faure-Muret (1985)

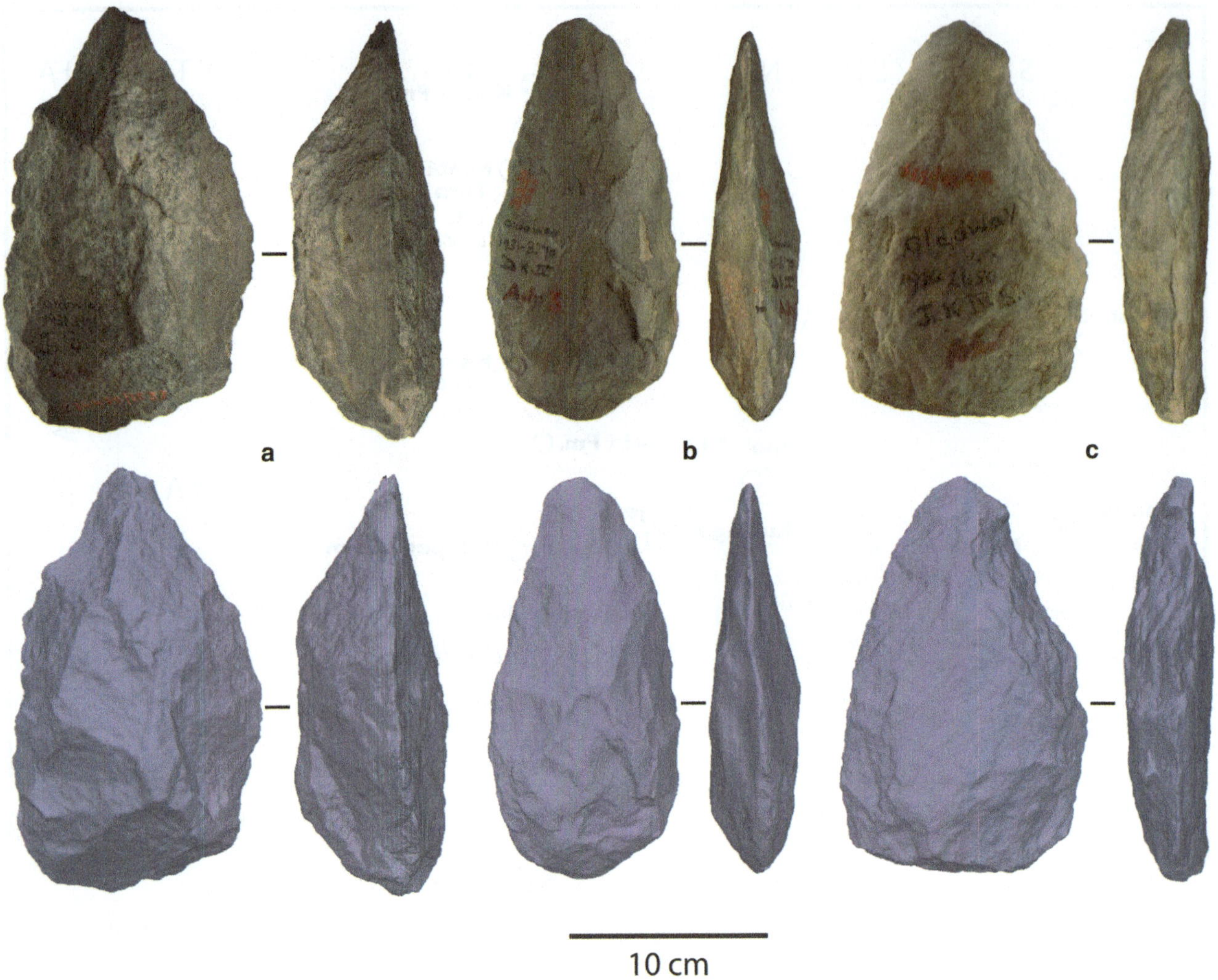

Fig. 8.2 Photographs and digital renderings of handaxes made from **a** quartzite, **b** basalt, and **c** phonolite from Olduvai Gorge, Tanzania, housed at Harvard University's Peabody Museum of Archaeology and Ethnology, Cambridge, Massachusetts, USA

resharpening, because the irregular grain boundaries in the rock provided a natural sort of saw-tooth edge. In contrast, phonolite artifacts had rather brittle edges that dulled rapidly and needed frequent re-sharpening. As a result of these material differences, the degree of re-sharpening varies by rock type (i.e. phonolite bifaces are more intensively retouched), impacting a number of variables used to assess inter-assemblage differences at Olduvai Gorge (Callow 1994; Roe 1994). In general, increased amounts of retouch affect final artifact size and shape at discard as well as scar count, and can reduce inter-analyst agreement (Proffitt and de la Torre 2014). At Olduvai Gorge in particular, the morphological differences between the basalt, phonolite, and quartzite bifaces caused by the nature of the available raw material are the sorts of variables that had previously been used to distinguish between different artifact industries, in this case, the Developed Oldowan and the Acheulian. The significance of these industrial classifications goes beyond issues of archaeological nomenclature, however, as researchers frequently equate particular industries with specific hominin taxa, as was the case with Mary Leakey (1971) for the Developed Oldowan and *H. habilis*, and the Acheulian and *H. erectus*.

Size and Retouch Intensity Among MSA Sites in Northern Kenya and Southern Ethiopia

Tryon et al. (2008) explored the extent to which raw material variation explained differences among MSA sites found in the Rift Valley of northern Kenya (the Kapthurin Formation, the Kapedo Tuffs, and the Lake Turkana basin) and southern Ethiopia (Omo Kibish). They focused on two variables similar to those originally examined by Jones at Olduvai Gorge: The size of naturally occurring raw material packages and the influence of rock type on retouch intensity.

They demonstrated that distance from the Rift margin is a good predictor of average artifact size, a fact driven by (1) hominin use of cobbles and (2) progressive declines in cobble size with increased transport distance by water along rivers and streams. In this model, the actual location of the raw material sources used is unknown, but the assumption is that transport of clasts begins at erosional nickpoints at the Rift margin (Fig. 8.3A) and continues towards the axis of the Rift. As shown in Fig. 8.3A, the pattern of size decline with distance is true both for lava and chert cores, with the further finding that for any given site, chert cores were on average smaller than lava ones. Eren et al. (2013) report the same pattern further south from MSA sites at Olduvai Gorge. This general size effect exists because the lava flows from which the cobbles ultimately derive are vertically thick and spatially extensive, whereas chert beds or nodules are thin and/or spatially constrained. The initial form of the raw material impacted the final size of the artifact, which varied by raw material type (lava vs. chert). Because of this, artifact size alone is a difficult variable to use in analyzing geographic patterns of behavioral variability.

The second finding reported by Tryon et al. (2008) was that retouch intensity was consistently higher on chert compared to lava artifacts (Fig. 8.3B) for these northern Kenyan and southern Ethiopian sites. The reasons for this are unclear. It may relate to the durability of the raw material, such that the edges of the more brittle chert artifacts required more frequent re-sharpening, as suggested above for the phonolite handaxes at Olduvai Gorge, or that hominins preferentially selected chert for artifacts needing retouch. It may also be that retouch is simply more visible to the archaeologist on chert rather than on lava artifacts, a problem more often explicitly recognized for quartz artifacts (David et al. 1981; Mehlman 1989; Bisson 1990). Whatever the reason, the fact that retouch intensity varies by raw material type causes a number of problems for comparing across assemblages made up of different rock types, particularly when employing classical typologies that rely on retouched tools such as that developed by Bordes (1961). Assemblages made by the same group of hominins in different geological zones could look very different simply because of the kinds of lithic raw material available, a phenomenon also long recognized in assessments of variability among similarly-aged western European Mousterian sites (Rolland and Dibble 1990).

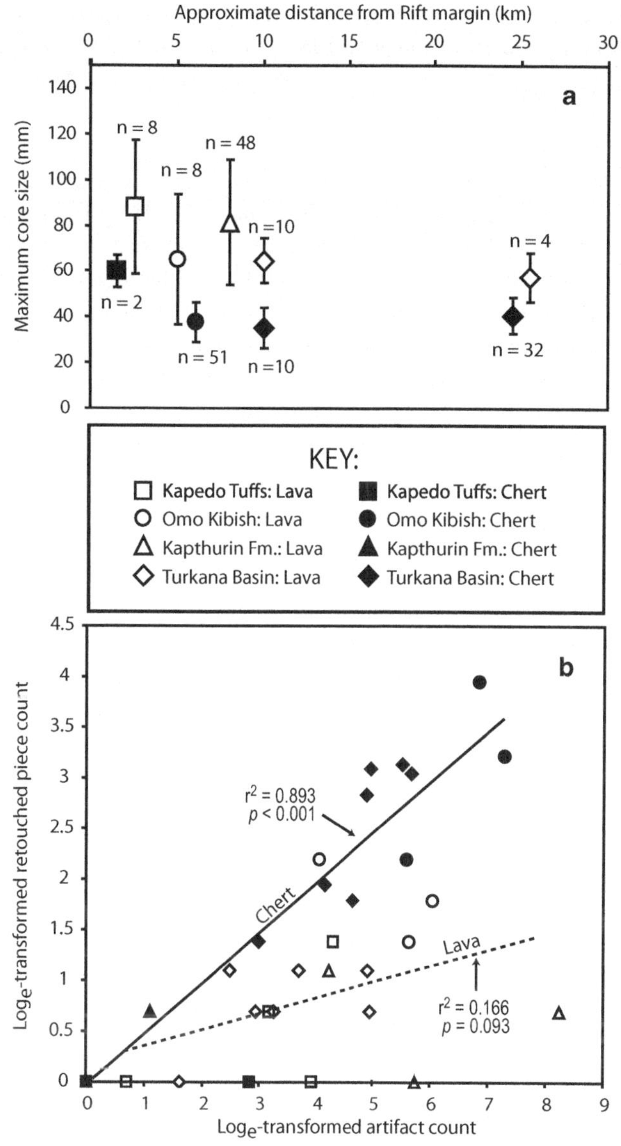

Fig. 8.3 Bivariate plots showing the impact of different raw material types (lava and chert) on core size and retouched piece frequency from Middle Stone Age sites in northern Kenya and southern Ethiopia, adapted from Tryon et al. (2008). **A** Shown are mean and standard deviation of core maximum size in mm; horizontal position of each raw material type slightly offset to accommodate size range. Core size diminishes with increased transport distance (from Rift Valley axis), but lava cores are always larger than chert ones, in part because of larger initial sizes at outcrop form. **B** Plot of retouched piece count by raw material type, showing that retouched pieces are always significantly more common among chert artifacts than those of lava. Values are log-transformed to accommodate variance in sample size

Quantifying Quartz Variability at Nasera

The use of systematic comparative analytical frameworks is central to our understanding of technological regional variability, but our work at Nasera rockshelter in northern Tanzania has emphasized in particular some of the problems involved when applying analytical strategies developed using rocks such as flint, chert, or silcrete such as those developed for Pinnacle Point 5-6 (Wilkins et al. 2017) to artifact assemblages made of quartz. Nasera rockshelter is on the margin of the Serengeti Plains, and is an important reference site for much of the Late Pleistocene in eastern Africa, containing quartz-dominated MSA, 'transitional,' and Later Stone Age (LSA) lithic assemblages including the Mumba industry, one of the regional variants considered in narratives of eastern Africa as a central region for the origin and dispersal of modern humans (Mehlman 1989; McBrearty and Brooks 2000; Mellars 2006; Mellars et al. 2013). Temporal changes at Nasera have been argued to represent in part the impact of demographic shifts leading to larger or denser human populations (Tryon and Faith 2016) based on evidence for changes in the local environment and increased occupation intensity. Ranhorn (2017) focused on understanding the nature of lithic technology that occurred with these shifts in environmental or demographic variables, specifically investigating aspects of lithic technology demonstrated to be related to aspects of flintknapper learning and copying, using analytical approaches initially developed by Tostevin (2012).

Tostevin (2012) divided comparative analyses of lithic artifacts into four technological domains related to choices made during the process of tool manufacture, including core modification, platform maintenance, direction of core exploitation, and dorsal surface convexity (Table 8.1). A fifth category was also used in his inter-assemblage comparisons, toolkit morphology, which includes how retouched pieces were further selected and modified for use. Tostevin's system was developed specifically for comparing sites in areas rich in flint in both the Mediterranean basin and eastern Europe, and in those contexts, we believe that the system works remarkably well. Scerri et al. (2014) also applied Tostevin's analytical approaches to assemblages in northern Africa, and with the addition of multivariate analyses such as principal component analysis (PCA), demonstrated regional patterns in lithic technology.

Ranhorn's (2017) application of the Tostevin analytical approach to Nasera, and to other quartz-dominated assemblages in eastern Africa, however, revealed a number of interesting complications, which can be distilled down to two important and distinct issues. The first of these involves the basic reading of the artifacts themselves, and the second involves the broader analytical framework, specifically the arrangement of attributes and their associated technological domains.

Firstly, and as noted above, the ways in which quartz fractures as well as the optical properties of the rock make it difficult to reliably recognize many of the lithic attributes seen on other rock types, a widely recognized problem with the material in eastern Africa and central Africa (David et al. 1981; Mehlman 1989; Bisson 1990; Cornelissen 2003; Diez-Martin et al. 2009) and elsewhere (e.g. Driscoll 2011). Ranhorn applied the Tostevin framework to Nasera, as well as other sites in Kenya with artifacts made up of other raw material types. At the Kenyan sites of Prospect Farm, Prolonged Drift, and multiple localities in Koobi Fora, where the majority of the artifacts were made on obsidian or various types of chert, Tostevin's comparative system worked well. At Nasera, many of the same measured attributes were not easily measured in a replicable way, and therefore removed from the comparative analyses. For example, patterns of flake scar directionality were obscure and difficult to confidently 'read,' and this lack of confidence in these data precluded analyses relating to early- and late-stage core reduction methods.

Table 8.1 Summary of the comparative approach developed by Tostevin (2012) and estimates of our confidence in its application to quartz-based artifact assemblages

Technological domain	Flintknapping step/Attribute	Estimated confidence of measurement in quartz
Core modification	Core orientation	Low
	Core management	Low
Platform maintenance	Platform treatment	Low
	External platform angle (degrees)	High
	Platform thickness	High
Direction of core exploitation	Direction of early exploitation	Low
	Direction of late exploitation	Low
	Percentage cortex	Medium
Dorsal surface convexity	Elongation of the longitudinal convexity: length/width ratio	High
	Shape of convexity: debitage lateral edges	Medium
	Curvature of convexity: profile	High
	Lateral convexity: cross-section	High
	Vertical convexity: width/thickness ratio	High
Toolkit morphology	Tool laminarity	High
	Tool vertical convexity	High
	Shape of tool cutting edges	Medium
	Shape of distal terminus	High
	Curvature of cutting edge	Medium
	Application of unique retouch	Medium
	Location of tool retouch	Medium

Secondly, the arrangement of attributes and their associated technological domains may differ in quartz-based technologies than in those for which Tostevin's (2012) approach was devised. In bipolar technologies specifically, the knapper transfers energy such that it travels from two opposing sides of the cobble or pebble, and one can vary this method by striking obliquely, vertically, or transversely towards the anvil, thus creating lateral, split, or transverse flake fragments (see Callahan 1987 and Jones 2006 for detailed descriptions). One resulting and recurring phenomenon involves the formation of forms such as "splinter pieces", wedges, and *pieces esquillees,* or scaled pieces, abundant in the Nasera assemblage, which are difficult to classify as either a core or flake (Villa et al. 2012), an issue encountered elsewhere, such as in the case of Karari scrapers (Harri Isaac et al. 1976) and other "core tools" (cf. McPherron 2009). In our Nasera analysis, this ambiguity had downstream effects for which attributes we measured on the piece and how, and importantly for our analysis, to which technological domain their associated measurements belonged. All of these issues further complicate the comparability of quartz datasets to those from elsewhere in Africa where quartz technologies are less common.

Table 8.1 summarizes our estimates of the utility of Tostevin's (2012) system as currently applied to quartz-based technologies. Fundamentally, attribute-based analyses require inter-analyst and intra-analyst reproducibility. Our ability to consistently measure these attributes varied along an ordinal scale which we divided into "low", "medium", and "high". Attributes with "low" utility indicate that a re-evaluation of the domain itself may be warranted; in effect this rating is nearly equivalent to a "not applicable" score. Attributes with a "medium" value indicate ambiguities that tend to be associated with the readability or measurability of the attribute in various quartz grades. Finally, a "high" value indicates an attribute that can be systematically measured utilizing commonly used measurement techniques such as caliper measurements. Refitting studies similar to those outlined by Scerri et al. (2016), and specifically attuned to the various forms of quartz grades, may help refine these estimates or better isolate useful variables in the future.

Quartz and the Abundance of Typical Later Stone Age (LSA) Tools

As a further test of the impact of raw material on the typological composition of archaeological assemblages, particularly quartz, we consider here the role of retouch and raw material on the presence of artifact types considered diagnostic of the Later Stone Age, in particular backed microlithic crescents. The presence of backed microlithic crescents is frequently used as means to assess the onset of the Middle/Later Stone Age transition, the timing of which appears to vary regionally across eastern Africa (reviewed in Tryon et al. 2018). Here, we begin to explore the extent to which raw material may play a role in this apparent temporal heterogeneity in the adoption of new artifact forms. Specifically, we examine the relationship between the frequency of the use of quartz as a raw material and the abundance of backed microlithic crescents. Our sample includes only crescent-shaped microliths, which we define as generally small (≤ 30 mm) elongated flakes or flake fragments with deliberate, abrupt retouch (backing) on one of the long edges that provides a curved or crescent shape to the piece (see Leplongeon 2014). We do not include naturally backed or otherwise unmodified pieces that may well have served the same function as the deliberately backed ones.

In order to reduce conflating the impacts of temporal, spatial, and raw material variability, we draw on sites that generally date to the Last Glacial Maximum, ~18–26 ka (Clark et al. 2009) and include sites not only from across a range of longitudes and depositional contexts (from rainforest to savanna, open air sites and rockshelters) but also a number of different sites from the same geographic locale, summarized in Table 8.2, with site locations shown in Fig. 8.1. Our comparative sample includes 10 sites from four countries. From the DRC, we include Matupi Cave, 90–170 cm below surface (Van Noten 1977, 1982; Muya wa Bitanko 1985–1986), Ishango 11 NFP, SJ, and NT levels (Mercader and Brooks 2001), Makubasi SE Level 1 (Mercader and Brooks 2001), and Matangai Turu Level 1 (Mercader and Brooks 2001). The Ugandan sample is limited to Munyama Cave levels 80–120 cm (Van Noten 1971; Valcke 1974), with five sites from Kenya, all from the Lukenya Hill area, including GvJm22, Occurrence E (Gramly 1976), GvJm62 units B/C (Marean 1992; Barut 1997), GvJm46 LSA level (Barut 1997), GvJm16, 98.50–97.90 cm below datum (Merrick 1975) and GvJm19, 115–150 cm below surface (Barut 1997). The Lemuta industry at Nasera (levels 4–5) was formerly the primary Tanzanian lithic assemblage believed to date to the LGM that has abundant data on the abundance of backed microlithic crescents made on different raw material types (Mehlman 1989). However, more recent dating of the site suggests that the Lemuta industry substantially pre-dates the LGM (Ranhorn and Tryon 2018) and thus we excluded these data from our analyses.

Within our sample, there is a strong, significant negative relationship ($r^2 = 0.679$, $p = 0.003$) between the abundance of quartz in an assemblage and the frequency of backed microlithic crescents (Fig. 8.4). Because all sites generally date to the LGM, this relationship is unlikely due to temporal differences among sites in the sample. And the fact that this observed patterns holds for sites across ca. 900 km as well as among those less than a km apart indicates that geographic distance was not a factor. Therefore, we conclude that

Table 8.2 Tabulation of assemblage size, quartz abundance, retouched tool count, and backed microlithic crescent frequency from 10 Equatorial African sites dating to approximately the Last Glacial Maximum

Site	Artifact sample size (n)	% quartz (of total)	Retouched tool count	% backed microlithic crescents
Matupi Cave (90–170 cm)	1,376	96.0	69	1.4
Ishango 11 (NFP, SJ, NT)	1,678	98.2	26	7.7
Matangai Turu NW (Levels 1–2)	727	100.0	12	0.0
Makubasi SE (Level 1)	240	99.6	10	0.0
Munyama Cave (80–120 cm)	54,945	98.7	1522	30.6
Lukenya Hill, GvJm22 (Occurrence E)	40,757	65.2	936	73.2
Lukenya Hill, GvJm62, units B/C	19,893	76.6	400	13.5
Lukenya Hill, GvJm46 (LSA)	14,418	91.7	262	13.7
Lukenya Hill, GvJm16 (98.50–97.90 cm)	7,612	45.8	298	60.1
Lukenya Hill, GvJm19 (115–150 cm)	13,081	84.9	344	20.1

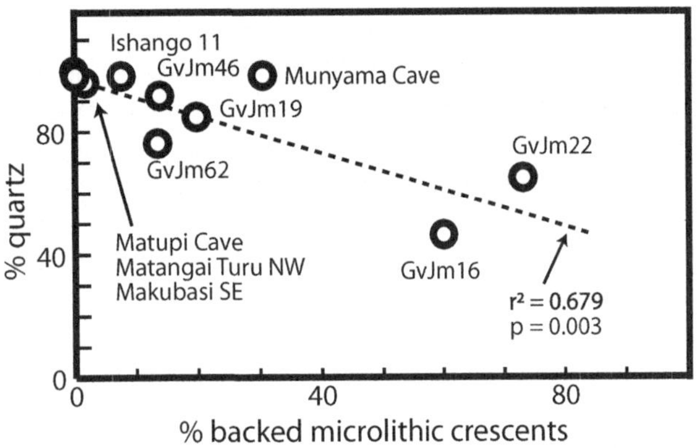

Fig. 8.4 Bivariate plot showing the negative relationship between the abundance of quartz in an archaeological assemblage and the frequency of backed microlithic crescents, a typical tool defining the African Later Stone Age. All 10 sites in this sample approximately date to the Last Glacial Maximum, and include assemblages from across Equatorial Africa. Data listed in Table 8.2

backed microlithic crescents are less frequent in quartz-based assemblages in eastern (and central) Africa. More broadly, this means that the frequency of at least some of those elements used to define regional variants are strongly dependent on the type of raw material used. In terms of defining the early LSA, at least part of the reason that backed microlithic crescents are so rare in some levels at sites such as Mumba (Mehlman 1989; Diez-Martin et al. 2009) and Kisese II (Tryon et al. 2018) might be the local abundance of quartz at those sites.

Backed Pieces and the Later Stone Age Eburran in Kenya

The Eburran is one of the better studied LSA industries in eastern Africa, consisting of a developmental sequence of blade-based industries beginning in the early Holocene and persisting after the introduction of groups of dedicated 'Neolithic' pastoralists in the region ~5 ka. Five different Eburran phases are recognized on the basis of radiometric dates, stratigraphic superposition, and changes in the size and shape of blades and backed pieces seen among sites near the eponymous Mt. Eburru (Fig. 8.1) (Ambrose 1984; Wilshaw 2016), the geographic center for Eburran sites and the primary obsidian source for artifacts found at them (Frahm and Tryon 2018). Wilshaw (2012, 2016) has recently presented a detailed consideration of the definition of the Eburran, based on a number of qualitative and quantitative lithic analyses. Recognizing the impacts of raw material variation on studies of inter-assemblage variation, he restricted his study to assemblages dominated by obsidian (Wilshaw 2012: 65). While methodologically sound, this approach is ultimately restrictive in terms of defining regional entities, as it makes it difficult to directly compare contemporary, geographically adjacent archaeological entities found in areas where obsidian is absent, such as the quartz-dominated Kansyore and other late Holocene forager sites in the Lake Victoria basin (Seitsonen 2010). The presence of obsidian from Mt. Eburru at Lake Victoria sites (Merrick and Brown 1984; Frahm et al. 2017) implies some sort of connection between contemporary groups in the two regions, raising questions as to why the lithic artifacts they made are classified differently. For example, are these different cultural traditions or a by-product of archaeologist's analytical procedures?

Some of the difficulties in classifying Holocene LSA assemblages that are characterized by raw material variability can been seen at the site of GvJm22 at Lukenya Hill (Fig. 8.1), adjacent to but outside of the Rift Valley, where obsidian, chert, and quartz are locally available for tool manufacture (Gramly 1976; Merrick and Brown 1984; Merrick et al. 1994; Tryon et al. 2015). Because backed microlithic crescent size is one of the variables used to define different phases of the Eburran, we use the GvJm22 sample as a test of the impacts of different raw material type on

Table 8.3 Basic descriptive data for microlithic backed crescents from Holocene LSA Occurrence D and Late Pleistocene LSA Occurrence E from site GvJm22 at Lukenya Hill, Kenya, subdivided by lithic raw material type, including artifact count and maximum length in mm. Additional data from GvJm22 Occurrence C and from GvJm16 from Merrick (1975); dates from Merrick (1975), Gramly (1976), and Tryon et al. (2015). PN = Pastoral Neolithic

Site	Level	Archaeology	Age (ka)	Raw material	Count (n)	Average length (mm)	Range length (mm)
GvJm22	C	PN	2.2	Obsidian	40	18.4	12.0–32.0
GvJm22	C	PN	2.2	Chert	7	18.0	12.0–26.0
GvJm22	C	PN	2.2	Quartz	0	NA	NA
GvJm22	D	LSA	6.7	Obsidian	6	12.2±2.5	8.8–15.1
GvJm22	D	LSA	6.7	Chert	9	16.6±4.0	9.5–23.0
GvJm22	D	LSA	6.7	Quartz	2	12.2±4.1	9.4–15.2
GvJm22	E	LSA	37–15	Obsidian	21	23.0±7.2	13.0–43.6
GvJm22	E	LSA	37–15	Chert	98	21.2±4.0	12.6–34.8
GvJm22	E	LSA	37–15	Quartz	9	17.9±2.5	14.8–23.1
GvJm16	C	PN	2.2	Obsidian	13	16.7	12.0–23.0
GvJm16	C	PN	2.2	Chert	19	20.4	12.0–27.0
GvJm16	C	PN	2.2	Quartz	0	NA	NA
GvJm16	B	LSA	20–15	Obsidian	13	17.2	11.0–25.0
GvJm16	B	LSA	20–15	Chert	18	23.8	15.0–43.0
GvJm16	B	LSA	20–15	Quartz	0	NA	NA

backed microlithic crescent size from a single site. Summary data are reported in Table 8.3. Crescents, and backed microliths in general, are an appropriate artifact type for this comparison, as they tend to be replaced rather than re-sharpened with extensive use (Hiscock 2006), and what re-sharpening occurs affects lateral edge shape but has little impact on tool length. Therefore, a decline in size from repeated use as might be expected for artifacts from distant sources (e.g. Newman 1994) is unlikely to explain any observed differences within this artifact sample.

We first draw on our own measured sample of backed microlithic crescents drawn from the 1970s excavations at the site. For the Late Pleistocene LSA assemblages from GvJm22 Occurrence E (Gramly 1976; Tryon et al. 2015), a Kruskal-Wallis test of sample medians ($H = 8.868$, $p = 0.012$) indicates variation within our sample in terms of the length of quartz, chert, and obsidian crescents, with quartz crescents significantly smaller than those of either chert or obsidian (Mann-Whitney post-hoc test, Bonferroni corrected p values of 0.013 and 0.005, respectively). Occurrence D at GvJm22 overlaps in age with Eburran sites in the Central Rift, but the sample size of crescents ($n = 17$) and particularly those of quartz ($n = 2$) is lower than for Occurrence E because of a smaller excavated volume (Gramly 1975). Comparisons within Occurrence D suggest no significant differences between sample medians ($H = 5.359$, $p = 0.069$), although this is driven at least in part by the very small and highly variable quartz sample. Excluding quartz, an unequal variance t-test indicates that mean obsidian crescent length is significantly smaller than mean chert crescent length from Occurrence D at GvJm22 ($t = 2.400$, $p = 0.020$).

Comparisons *within* raw material types shows that at GvJm22, Holocene backed microlithic crescents are significantly smaller than Late Pleistocene ones for those made of obsidian ($p < 0.001$) and chert ($p = 0.008$), but not for quartz ($p = 0.313$).

Our measured sampled of backed microlithic crescents from GvJm22 is very small, limiting the strength of any inferences drawn from our statistical tests. However, published data from site GvJm16, ~500 m north of GvJm22 with a similar Late Pleistocene-Holocene archaeological sequence (Table 8.3), show patterns similar to those at GvJm22. At GvJm16, chert backed microlithic crescents are generally larger than those of obsidian, average backed microlithic crescent size declines over time, and quartz backed microlithic crescents are comparatively rare (Merrick 1975).

The presence of significant size differences of the same artifact class made on different raw materials from both Late Pleistocene and Holocene contexts strongly suggests a general pattern noted by Wilshaw (2016) specifically for the Eburran. That is, that raw material type can affect artifact size, one of several metric variables commonly used to characterize archaeological patterns.

Discussion and Conclusions

Taken in broader context, our results demonstrate that raw material must be considered when discerning Stone Age patterns of convergence (homoplasy) from shared cultural entities (homology), as similar behaviors may result in very different looking artifacts because of the nature of the rock types used. eastern Africa has a highly variable bedrock geology, with a number of different kinds of rocks available, many showing wide variance in a number of properties relevant to an archaeological analysis of stone tools. One outcome of this variability is the difficulty of devising straightforward criteria for the recognition of regional variation, particularly approaches that rely on metric criteria alone. As our examples drawn from throughout the Pleistocene and Holocene suggest, variation in the type of raw material used can drive variation in artifact size, retouch intensity, and recognition of a range of criteria used to construct geographically or temporally distinctive patterns of artifact manufacture.

We can have confidence in studies that rely on homology only when similarities that result from convergence, or homoplasy, can be reliably identified. In the context of this volume, we note that studying convergent evolution is useful for demonstrating that populations with distinctively different histories and developmental trajectories can produce remarkably similar things or evolve comparable forms. We have tried to show here using a number of examples from Early Pleistocene to Holocene archaeological sites in eastern Africa that the converse may also be true. That is, in the case of stone tools, sometimes even similar behaviors (i.e. ways of making tools), histories, or evolutionary trajectories can produce a very different seeming archaeological record, simply because of the types of rocks available and their mechanical properties. Of course, we can consider an alternative, that is, does use of similar rock types or raw material packages (e.g. cobbles, slabs, etc.) lead to similar technologies of raw material reduction and tool production? This may be the case in eastern Africa.

Clearly, understanding the underlying lithic raw material variability is a requisite step in explaining regional and inter-regional archaeological variation. One recent synthesis of eastern African lithic variability among MSA and early LSA sites (Faith et al. 2015) detected geographic differences in the presence and absence of particular artifact types between sites north and south of Equator, interpreted as possible evidence for the presence of spatially-defined boundaries in artifact production methods. As shown in Fig. 8.1, the Tanzanian record of sites such as Mumba, Nasera, Kisese II, Magubike, and Mlambalasi are all quartz dominated, while those in Kenya (with the exception of Lukenya Hill) are in areas rich in various types of lava. It is possible, therefore, that the apparently distinctive Middle and Late Stone Age technologies of Tanzania (e.g. the Mumba and Nasera industries) simply reflect properties related to the local availability of quartz. Whether or not the differences seen north and south of the Equator represent different behavioral traditions in the kinds of stone tools made or their methods of manufacture, or are simply a by-product of geology reflect hypotheses that remain to be tested.

We have sought to emphasize the importance of raw material for the analysis of eastern African Stone Age assemblages. This is of course nothing new, as lithic analysts have recognized for a long time that different rock types have very different properties that affect artifact form. Our point, however, is that approaches that seek to understand artifact variation at large temporal and geographic scales across Africa need to systematically take these differences into account (see Will and Mackay 2020 for a similar discussion). Otherwise excellent analytical approaches developed for isotropic rocks such as chert and flint are not easily adopted to rocks such as quartz, and developing different analytical protocols for different rock types leads to incomparable datasets, a problem similar to that caused by comparisons between Middle and Upper Paleolithic sites made using fundamentally different stone tool typologies (Grayson and Cole 1998). Ongoing work by the Comparative Analysis of Middle Stone Age Artefacts (CoMSAfrica) project (Will et al. 2019) may resolve some of these issues.

Statistical approaches such as the use of multiple regression and multivariate analyses (e.g. Scerri et al 2014) can deal with some of the impacts of raw material type (especially when comparing artifact frequencies), but the more pervasive problem is the role of raw material type on metric attributes, which are less easy to tease out with post hoc numerical tests. Digitization efforts might help reduce some of the ambiguity caused in the analysis of different raw materials, but as emphasized by Magnani (2014) and visible in Fig. 8.2, quartzite, quartz, and similar materials consistently cause a problem when using these approaches. Certainly, additional experimental approaches that elucidate the nuances of flaking mechanics in different raw materials are needed. Experimental replication of artifacts in quartz and other rock types suggests one way forward (e.g. Jones 2006; Gurtov et al. 2015; Pargeter and de la Peña 2017), as might efforts to better quantify rock texture and its impact on flaking mechanics (Brantingham et al. 2000; Noll 2000). Controlled experiments that use varied geological or synthetic materials in a series of standardized tests such as those devised by Pelcin (1997a, b, c) using glass represent another way forward. Substantial work remains to be done, but the promise of the development of a more accurate approach to geographic variation in lithic technology is one that makes these efforts worthwhile.

Acknowledgements We thank Huw Groucutt for the opportunity to contribute to this volume, Jess McNeil for the crescent measurements from Occurrence D from the R. M. Gramly excavations at site GvJm22, Lukenya Hill, Tyler Faith, who helped clarify some of our thinking about the relationship between quartz and backed microlithic crescents, and Manuel Will for some of the inspiration to write it. We thank the reviewers' comments for improving the quality of this manuscript. Research that contributed to this paper was conducted in Kenya by Tryon under research permits MOEST/13/001/30C229, NCST/5/002/R/576 and by Ranhorn in Tanzania under research permits COSTECH 2015-120-NA-2015-24, Antiquities 03/2015/2016 ERV3896941, Ngorongoro Conservation Area Authority NCAA/D/157/Vol. V/101. Funding for the field and laboratory research presented here was provided by the Leakey Foundation, the National Geographic Society (7994-06 and 8762-10), the US National Science Foundation (BCS-0841530 and BCS-0852609), the American School of Prehistoric Research, and by Harvard University.

References

Ambrose, S. H. (1984). Holocene environments and human adaptations in the Central Rift Valley, Kenya. Ph.D. thesis, University of California, Berkeley.

Ambrose, S. H. (2012). Obsidian dating and source exploitation studies in Africa: Implications for the evolution of human behavior. In I. Liritzis & C. M. Stevenson (Eds.), *Obsidian and ancient manufactured glasses* (pp. 56–72). Albuquerque, NM: University of New Mexico Press.

Ambrose, S. H., & Lorenz, K. G. (1990). Social and ecological models for the Middle Stone Age in southern Africa. In P. Mellars (Ed.), *The emergence of modern humans: An archaeological perspective* (pp. 3–33). Edinburgh: Edinburgh University Press.

Andrefsky, W., Jr. (1994). Raw material availability and the organization of technology. *American Antiquity, 59,* 21–34.

Barut, S. (1997). Later Stone Age lithic raw material use at Lukenya Hill, Kenya. Ph.D. dissertation, University of Illinois.

Bisson, M. S. (1990). Lithic reduction sequences as an aid to the analysis of Late Stone Age quartz assemblages from Luano Spring, Chingola, Zambia. *African Archaeological Review, 8,* 103–138.

Bordes, F. (1961). *Typologie du Paléolithique Ancien et Moyen.* Bordeaux: Publication de l'Institut de Préhistoire de l'Univesité de Bordeaux Mémoire 1.

Brantingham, P. J., Olsen, J. W., Rech, J. A., & Krivoshapkin, A. I. (2000). Raw material quality and prepared core technologies in Northeast Asia. *Journal of Archaeological Science, 27,* 255–271.

Braun, D. R., Plummer, T., Ditchfield, P., Ferraro, J. V., Maina, D., Bishop, L. C., et al. (2008). Oldowan behavior and raw material transport: Perspectives from the Kanjera Formation. *Journal of Archaeological Science, 35,* 2329–2345.

Braun, D. R., Plummer, T., Ferraro, J. V., Ditchfield, P., & Bishop, L. C. (2009). Raw material quality and Oldowan hominin toolstone preferences: Evidence from Kanjera South, Kenya. *Journal of Archaeological Science, 36,* 1605–1614.

Brown, K. S. (2011). The Sword in the Stone: Lithic raw material exploitation in the Middle Stone Age at Pinnacle Point Site 5–6, Southern Cape, South Africa. Ph.D. thesis, University of Cape Town.

Brown, F. H., Nash, B. P., Fernandez, D. P., Merrick, H. V., & Thomas, R. J. (2013). Geochemical composition of source obsidians from Kenya. *Journal of Archaeological Science, 40,* 3233–3251.

Callahan, E. (1987). *An evaluation of the lithic technology in Middle Sweden during the Mesolithic and Neolithic.* Uppsala: Societas Archaeologica Upsaliensis.

Callow, P. (1994). The Olduvai bifaces: Technology and raw materials. In M. D. Leakey & D. A. Roe (Eds.), *Olduvai Gorge, Volume 5: Excavations in Beds III, IV, and the Masek Beds, 1968–1971* (pp. 235–253). Cambridge: Cambridge University Press.

Choubert, G., & Faure-Muret, A. (1985). *1:5,000,000 international geological map of Africa.* Paris: Commission for the Geological Map of the World/UNESCO.

Clark, J. D. (1988). The Middle Stone Age of East Africa and the beginnings of regional identity. *Journal of World Prehistory, 2,* 235–305.

Clark, J. D. (1993). African and Asian perspectives on the origins of modern humans. In M. J. Aitken, C. B. Stringer, & P. A. Mellars (Eds.), *The origin of modern humans and the impact of chronometric dating* (pp. 148–178). Princeton: Princeton University Press.

Clark, P. U., Dyke, A. S., Shakun, J. D., Carlson, A. E., Clark, J., Wohlfarth, B., et al. (2009). The last glacial maximum. *Science, 325,* 710–714.

Cole, G. H. (1967). The Later Acheulian and Sangoan of southern Uganda. In W. W. Bishop & J. D. Clark (Eds.), *Background to evolution in Africa* (pp. 481–528). Chicago: Chicago University Press.

Cornelissen, E. (2003). On microlithic quartz industries at the end of the Pleistocene in Central Africa: The evidence from Shum Laka (NW Cameroon). *African Archaeological Review, 20,* 1–24.

David, N., Harvey, P., & Goudie, C. J. (1981). Excavations in the southern Sudan 1979. *Azania, 16,* 7–54.

Diez-Martin, F., Domínguez-Rodrigo, M., Sanchez, P., Mabulla, A. Z. P., Tarriño, A., Barba, R., et al. (2009). The Middle to Later Stone Age technological transition in East Africa. New data from Mumba rockshelter Bed V (Tanzania) and their implications for the origin of modern human behavior. *Journal of African Archaeology, 7,* 147–173.

Driscoll, K. (2011). Identifying and classifying vein quartz artefacts: An experiment conducted at the World Archaeological Congress, 2008. *Archaeometry, 53,* 1280–1296.

Eren, M. I., Durant, A. J., Prendergast, M., & Mabulla, A. Z. P. (2013). Middle Stone Age archaeology at Olduvai Gorge, Tanzania. *Quaternary International, 322–323,* 292–313.

Faith, J. T., Tryon, C. A., Peppe, D. J., Beverly, E. J., Blegen, N., Blumenthal, S., et al. (2015). Paleoenvironmental context of the Middle Stone Age record from Karungu, Lake Victoria Basin, Kenya, and its implications for human and faunal dispersals in East Africa. *Journal of Human Evolution, 83,* 28–45.

Feblot-Augustins, J. (1990). Exploitation des matières premières dans l'Acheuléen d'Afrique: Perspectives comportementales. *Paléo, 2,* 27–42.

Frahm, E., Goldstein, S. T., & Tryon, C. A. (2017). Late Holocene forager-fisher and pastoralist interactions along the Lake Victoria shores, Kenya: Perspectives from portable XRF of obsidian artifacts. *Journal of Archaeological Science Reports, 11,* 717–742.

Frahm, E., & Tryon, C. A. (2018). Later Stone Age toolstone acquisition in the Central Rift Valley of Kenya: Portable XRF of Eburran obsidian artifacts from Leakey's excavations at Gamble's Cave II. *Journal of Archaeological Science: Reports, 18,* 475–486.

Goldstein, S. T., & Munyiri, J. M. (2017). The Elmenteitan obsidian quarry (GsJj50): New perspectives on obsidian access and exchange during the Pastoral Neolithic in Southern Kenya. *African Archaeological Review, 34,* 43–73.

Gramly, R. M. (1975). Pastoralists and hunters: Recent prehistory in Southern Kenya and Northern Tanzania. Ph.D. thesis, Harvard University.

Gramly, R. M. (1976). Upper Pleistocene archaeological occurrences at site GvJm/22, Lukenya Hill, Kenya. *Man, 11,* 319–344.

Grayson, D. K., & Cole, S. C. (1998). Stone tool assemblage richness during the Middle and Early Upper Paleolithic in France. *Journal of Archaeological Science, 25,* 927–938.

Groucutt, H. S. (2020). Culture and convergence: The curious case of the Nubian Complex. In H. Groucutt (Ed.), *Culture history and convergent evolution: Can we detect populations in prehistory?* (pp. 55–86). Cham, Switzerland: Springer.

Gurtov, A. N., Buchanan, B., & Eren, M. I. (2015). Dissecting quartzite and basalt bipolar flake shape: A morphometric comparison of experimental replications from Olduvai Gorge, Tanzania. *Lithic Technology, 40,* 332–341.

Harmand, S. (2007). Economic behaviors and cognitive capacities of early hominins between 2.34 Ma and 0.7 Ma in West Turkana, Kenya. *Mitteilungen der Gesellschaft für Urgeschichte, 16,* 11–23.

Harris, J. W. K., & Isaac, G. (1976). The Karari Industry: Early Pleistocene archaeological evidence from the terrain east of Lake Turkana, Kenya. *Nature, 262,* 102–107.

Hay, R. L. (1968). Chert and its sodium-silicate precursors in sodium-carbonate lakes of East Africa. *Contributions to Mineralogy and Petrology, 17,* 255–274.

Hay, R. L. (1976). *Geology of the Olduvai Gorge.* Berkeley: University of California Press.

Hiscock, P. (2006). Blunt and to the point: Changing technological strategies in Holocene Australia. In I. Lilley (Ed.), *Archaeology of Oceania: Australia and the Pacific Islands* (pp. 69–95). Malden, MA: Blackwell Publishing.

Howell, F. C., Cole, G. H., & Kleindienst, M. R. (1962). Isimila: An Acheulian occupation site in the Iringa highlands, Southern Highlands Province, Tanganyika. In G. Mortlemans & J. Nenquin (Eds.), *Actes du IVe Congrès Panafricain de Préhistoire et de l'Étude du Quarternaire* (pp. 45–60). Tervuren: Annales Musée Royal de l'Afrique Central, Serie in 8, Sciences Humaines 40.

Isaac, G. L. (Ed.). (1997). *Koobi Fora Research Project Volume 5: Plio-Pleistocene Archaeology.* Cambridge: Cambridge University Press.

Isaac, G. L., Harris, J. W. K., & Kroll, E. M. (1997). The stone artefact assemblages: A comparative study. In G. Isaac & B. Isaac (Eds.), *The Koobi Fora Research Project, Volume: 5: Plio-Pleistocene Archaeology* (pp. 262–306). Oxford: Clarendon Press.

Jones, P. R. (1979). Effects of raw materials on biface manufacture. *Science, 204,* 835–836.

Jones, P. R. (1980). Experimental butchery with modern stone tools and its relevance for Palaeolithic archaeology. *World Archaeology, 12,* 153–165.

Jones, P. R. (1994). Results of experimental work in relation to the stone industries of Olduvai Gorge. In M. D. Leakey & D. A. Roe (Eds.), *Olduvai Gorge, Volume 5: Excavations in Beds III, IV and the Masek Beds, 1968–1971* (pp. 254–296). Cambridge: Cambridge University Press.

Jones, S. (2006). Quartz tool technology in the Northeast Georgia Piedmont. *The Society for Georgia Archaeology, 34,* 27–88.

Jones, S. C., & Stewart, B. A. (Eds.). (2016). *Africa from MIS 6-2: Population dynamics and paleoenvironments.* New York: Springer.

Kelly, A. J. (1996). Intra-regional and Inter-regional Variability in the East Turkana (Kenya) and Kenyan Middle Stone Age. Ph.D. thesis, Rutgers University.

Leakey, M. D. (1971). *Olduvai Gorge Volume 3: Excavations in Beds I and II, 1960–1963.* Cambridge: Cambridge University Press.

Leplongeon, A. (2014). Microliths in the Middle and Later Stone Age of eastern Africa: New data from Porc-Epic and Goda Buticha cave sites, Ethiopia. *Quaternary International, 343,* 100–116.

Mackay, A., Stewart, B. A., & Chase, B. M. (2014). Coalescence and fragmentation in the late Pleistocene archaeology of southernmost Africa. *Journal of Human Evolution, 72,* 26–51.

Magnani, M. (2014). Three-dimensional alternatives to lithic illustration. *Advances in Archaeological Practice, 2,* 285–297.

Marean, C. W. (1992). Implications of late Quaternary mammalian fauna from Lukenya Hill (south-central Kenya) for paleoenvironmental change and faunal extinctions. *Quaternary Research, 37,* 239–255.

McBrearty, S., & Brooks, A. S. (2000). The revolution that wasn't: A new interpretation of the origin of modern human behavior. *Journal of Human Evolution, 39,* 453–563.

McPherron, S. P. (2009). *Tools versus cores: Alternative approaches to stone tool analysis.* Cambridge: Cambridge Scholars Publishing.

Mehlman, M. J. (1977). Excavations at Nasera Rock, Tanzania. *Azania, 12,* 111–118.

Mehlman, M. J. (1989). Late quaternary archaeological sequences in Northern Tanzania. Ph.D. thesis, University of Illinois.

Mellars, P. (2006). Why did modern human populations disperse from Africa *ca.* 60,000 years ago? A new model. *Proceedings of the National Academy of Sciences USA, 103,* 9381–9386.

Mellars, P., Gori, K. C., Carr, M., Soares, P. A., & Richards, M. B. (2013). Genetic and archaeological perspectives on the initial modern human colonization of southern Asia. *Proceedings of the National Academy of Sciences, 110,* 10699–10704.

Mercader, J., & Brooks, A. S. (2001). Across forests and savannas: Later Stone Age assemblages from Ituri and Semliki, Democratic Republic of Congo. *Journal of Anthropological Research, 57,* 197–217.

Merrick, H. V. (1975). *Change in Later Pleistocene Lithic Industries in Eastern Africa.* Ph.D. dissertation, University of California.

Merrick, H. V., & Brown, F. H. (1984). Obsidian sources and patterns of source utilization in Kenya and northern Tanzania: Some initial findings. *African Archaeological Review, 2,* 129–152.

Merrick, H. V., Brown, F. H., & Nash, W. P. (1994). Use and movement of obsidian in the Early and Middle Stone Ages of Kenya and northern Tanzania. In S. T. Childs (Ed.), *Society, culture, and technology in Africa* (pp. 29–44). Philadelphia: MASCA.

Muya wa Bitanko, K. (1985–1986). *Préhistoire du Zaïre Oriental: Essai de synthèse des âges de la pierre taillé.* Ph.D. dissertation, Catholic University of Louvain.

Newman, J. R. (1994). The effects of distance on lithic material reduction technology. *Journal of Field Archaeology, 21,* 491–501.

Noll, M. P. (2000). *Components of Acheulian lithic assemblage variability at Olorgesailie, Kenya.* Ph.D. dissertation, University of Illinois.

O'Brien, M. J., & Bentley, R. A. (2020). Learning strategies and population dynamics during the Pleistocene colonization of North America. In H. Groucutt (Ed.), *Culture history and convergent evolution: Can we detect populations in prehistory?* (pp. 261–281). Cham, Switzerland: Springer.

Oestmo, S. (2017). *A formal modeling approach to understanding stone tool raw material selection in the African Middle Stone Age: A case study from Pinnacle Point, South Africa.* Ph.D., Arizona State University

Omi, G. (Ed.). (1984). *Mtongwe: An Interim Report of the East and Northeast African Prehistory Project 1982.* Matsumoto, Japan: Shinshu University.

Pargeter, J., & de la Peña, P. (2017). Milky quartz bipolar reduction and lithic miniaturization: Experimental results and archaeological implications. *Journal of Field Archaeology, 42,* 551–565.

Pargeter, J., & Hampson, J. (2019). Quartz crystal materiality in Terminal Pleistocene Lesotho. *Antiquity, 367,* 11–27.

Pelcin, A. W. (1997a). The effect of core surface morphology on flake attributes: Evidence from a controlled experiment. *Journal of Archaeological Science, 24,* 749–756.

Pelcin, A. W. (1997b). The effect of indentor type on flake attributes: Evidence from a controlled experiment. *Journal of Archaeological Science, 24,* 613–621.

Pelcin, A. W. (1997c). The formation of flakes: The role of platform thickness and exterior platform angle in the production of flake initiations and terminations. *Journal of Archaeological Science, 24,* 1107–1113.

Potts, R., Behrensmeyer, A. K., Faith, J. T., Tryon, C. A., Brooks, A. S., Yellen, J., et al. (2018). Environmental dynamics during the onset of the Middle Stone Age in eastern Africa. *Science, 360,* 86–90.

Proffitt, T., & de la Torre, I. (2014). The effect of raw material on inter-analyst variation and analyst accuracy for lithic analysis: A case study from Olduvai Gorge. *Journal of Archaeological Science, 45,* 270–283.

Ranhorn, K. L. (2017). *Cultural transmission and lithic technology in Middle Stone Age Eastern Africa.* Ph.D., The George Washington University.

Ranhorn, K., & Tryon, C. A. (2018). New radiocarbon dates from Nasera Rockshelter (Tanzania): Implications for studying spatial patterns in Late Pleistocene technology. *Journal of African Archaeology, 16,* 211–222.

Reynolds, N. (2020). Threading the weft, testing the warp: Population concepts and the European Upper Palaeolithic chronocultural framework. In H. Groucutt (Ed.), *Culture history and convergent evolution: Can we detect populations in prehistory?* (pp. 187–212). Cham, Switzerland: Springer.

Roe, D. A. (1994). A metrical analysis of selected sets of handaxes and cleavers from Olduvai Gorge. In M. D. Leakey & D. A. Roe (Eds.), *Olduvai Gorge, Volume 5: Excavations in Beds III, IV and the Masek Beds, 1968–1971* (pp. 146–234). Cambridge: Cambridge University Press.

Rolland, N., & Dibble, H. L. (1990). A new synthesis of Middle Paleolithic variability. *American Antiquity, 55,* 480–499.

Scerri, E. M. L. (2017). The North African Middle Stone Age and its place in recent human evolution. *Evolutionary Anthropology, 26,* 119–135.

Scerri, E. M. L., Blinkhorn, J., Gravina, B., & Delagnes, A. (2016). Can lithic attribute analyses identify discrete reduction trajectories? A quantitative study using refitted lithic constellations. *Journal of Archaeological Method and Theory, 23,* 669–691.

Scerri, E. M. L., Drake, N. A., Jennings, R., & Groucutt, H. S. (2014). Earliest evidence for the structure of *Homo sapiens* populations in Africa. *Quaternary Science Reviews, 101,* 207–216.

Scerri, E. M. L., Thomas, M. G., Manica, A., Gunz, P., Stock, J. T., Stringer, C., et al. (2018). Did our species evolve in subdivided populations across Africa, and why does it matter? *Trends in Ecology & Evolution, 33,* 582–594.

Seitsonen, O. (2010). Lithics use at Kansyore sites in East Africa: Technological organization at four recently excavated sites in Nyanza Province, Kenya. *Azania: Archaeological Research in Africa, 45,* 49–82.

Shea, J. J. (2014). Sink the Mousterian? Named stone tool industries (NASTIES) as obstacles to investigating hominin evolutionary relationships in the Later Middle Paleolithic Levant. *Quaternary International, 350,* 169–179.

Spinapolice, E. E. (2020). Lithic variability and cultures in the East African Middle Stone Age. In H. Groucutt (Ed.), *Culture history and convergent evolution: Can we detect populations in prehistory?* (pp. 87–102).Cham, Switzerland: Springer.

Stiles, D. N., Hay, R. L., & O'Neil, J. R. (1974). The MNK Chert Factory Site, Olduvai Gorge, Tanzania. *World Archaeology, 5,* 285–308.

Tostevin, G. (2012). *Seeing lithics: A middle-range theory for testing for cultural transmission in the Pleistocene.* Cambridge: American School of Prehistoric Research Monograph Series, Peabody Museum, Harvard University, & Oxbow Books.

Tryon, C. A., Crevecoeur, I., Faith, J. T., Ekshtain, R., Nivens, J., Patterson, D., et al. (2015). Late Pleistocene age and archaeological context for the hominin calvaria from GvJm-22 (Lukenya Hill, Kenya). *Proceedings of the National Academy of Sciences, 112,* 2682–2687.

Tryon, C. A., & Faith, J. T. (2013). Variability in the Middle Stone Age of Eastern Africa. *Current Anthropology, 54,* S234–S254.

Tryon, C. A., & Faith, J. T. (2016). A demographic perspective on the Middle to Later Stone Age transition from Nasera rockshelter, Tanzania. *Philosphical Transactions of the Royal Society B, 371,* 20150238.

Tryon, C. A., Faith, J. T., Peppe, D. J., Keegan, W. F., Keegan, K. N., Jenkins, K. H., et al. (2014). Sites on the landscape: Paleoenvironmental context of late Pleistocene archaeological sites from the Lake Victoria basin, equatorial East Africa. *Quaternary International, 331,* 20–30.

Tryon, C. A., Lewis, J. E., Ranhorn, K. L., Kwekason, A., Alex, B., Laird, M. F., et al. (2018). Middle and Later Stone Age chronology of Kisese II rockshelter (UNESCO World Heritage Kondoa Rock-Art Sites), Tanzania. *PLoS ONE, 13,* e0192029.

Tryon, C. A., McBrearty, S., & Texier, P.-J. (2005). Levallois lithic technology from the Kapthurin Formation, Kenya: Acheulian origin and Middle Stone Age diversity. *African Archaeological Review, 22,* 199–229.

Tryon, C. A., Roach, N. T., & Logan, M. A. V. (2008). The Middle Stone Age of the northern Kenyan Rift: Age and context of new archaeological sites from the Kapedo Tuffs. *Journal of Human Evolution, 55,* 652–664.

Valcke, J. (1974). *De Late Steentijd van de Munyamagrot op het eiland Buvuma in het Victorianyanzameer (Uganda).* Licentiaatsverhandeling: University of Ghent.

Van Noten, F. (1971). Excavations at Munyama Cave. *Antiquity, 45,* 56–58.

Van Noten, F. (1977). Excavations at Matupi Cave. *Antiquity, 51,* 35–40.

Van Noten, F. (1982). *The archaeology of Central Africa.* Graz, Austria: Akademische Druck-u. Verlagsanstalt.

Villa, P., Soriano, S., Tsanova, T., Degano, I., Higham, T. F. G., d'Errico, F., et al. (2012). Border Cave and the beginning of the Later Stone Age in South Africa. *Proceedings of the National Academy of Sciences of the United States of America*, 13208–13213.

White, M. J. (1998). On the significance of Acheulean biface variability in southern Britain. *Proceedings of the Prehistoric Society*, 15–44.

Wilkins, J., Brown, K. S., Oestmo, S., Pereira, T., Ranhorn, K. L., Schoville, B. J., et al. (2017). Lithic technological responses to Late Pleistocene glacial cycling at Pinnacle Point Site 5–6, South Africa. *PLoS ONE, 12,* e0174051.

Will, M., Bader, G. D., & Conard, N. J. (2014). Characterizing the Late Pleistocene MSA lithic technology of Sibudu, KwaZulu-Natal, South Africa. *PLoS ONE, 9,* e98359.

Will, M., & Mackay, A. (2020). A matter of space and time: How frequent is convergence in lithic technology in the African archaeological record over the last 300 kyr? In H. Groucutt (Ed.), *Culture history and convergent evolution: Can we detect populations in prehistory?* (pp. 103–126). Cham, Switzerland: Springer.

Will, M., Tryon, C. A., Shaw, M., Scerri, E. M. L., Ranhorn, K. L., Pargeter, J., et al. (2019). Comparative analysis of Middle Stone Age artifacts (CoMSAfrica). *Evolutionary Anthropology, 28,* 57–59.

Wilshaw, A. (2012). *An investigation into the LSA of the Nakuru-Naivasha Basin and Surround, Central Rift Valley, Kenya: Technological classifications and population considerations.* Ph.D. dissertation, Cambridge University.

Wilshaw, A. (2016). The current status of the Kenya Capsian. *African Archaeological Review, 33,* 13–27.

Chapter 9
The Middle-Upper Paleolithic Transition: A Long-Term Biocultural Effect of Anatomically Modern Human Dispersal

Aaron Jonas Stutz

Abstract Neanderthals and anatomically modern humans made the Middle-Upper Paleolithic technological transition together. Perhaps more accurately stated, the technological innovations reflected in Early Upper Paleolithic archaeological assemblages were developed, adopted, and spread by a western Eurasian metapopulation that encompassed variable admixture histories. This is the unavoidable implication of robust analyses of ancient human, omnivorous prey, and microbial genomes, which document long-term—if sporadic—interaction and successful family formation between geographically expanding anatomically modern humans and indigenous Neanderthals. This social and population interaction occurred within a broad time-frame, likely ca. 120–40 ka, involving complex, multi-scalar niche construction and biocultural evolutionary dynamics. This chapter reconsiders theoretical, methodological, and empirical issues surrounding the study of lithic assemblages that define the Middle-Upper Paleolithic transition, considering how we can better answer a key question. If the Middle-Upper Paleolithic transition was an indirect consequence of anatomically modern human dispersal and interaction with Neanderthals, then what, if anything, did technological change have to do with Neanderthal extinction, ca. 40 ka?

Keywords The last Neanderthals • Anatomically modern humans • Niche construction dynamics • Biocultural evolution • Technological innovation • Enchronic cultural discourse • Cultural transmission and change

A. J. Stutz (✉)
Department of Anthropology, Emory University, 1557 Dickey Dr. Atlanta, Atlanta, GA 30322, USA
e-mail: aaronjstutz@gmail.com

Introduction

Explanatory narratives about anatomically modern human (AMH) range expansion and Neanderthal extinction necessarily incorporate stone tools as key characters. And with good reason. It is not just about the hominin populations themselves. Human biological evolution has played out through deeply inextricable relationships among cultural systems, their constituent populations and wider ecological conditions. Niche construction dynamics have generated a cascade of archaeological and environmental traces, complementing the hominin fossil and paleo-molecular record, especially from the Late Pleistocene onward (Boivin et al. 2016). This chapter examines and expands on Tostevin's (2003, 2007, 2011, 2013) theoretical approach to the study of stone tools, intimate-scale social interaction, cultural transmission, and population migration waves (see also Tryon and Ranhorn 2020). I emphasize that Tostevin identifies necessary components of a widely applicable conceptual framework for explaining long-term—that is, millennial-scale—conservatism in social reproduction. I argue, though, that his approach is not entirely theoretically sufficient for investigating how technological change in the Middle-Upper Paleolithic transition was systemically related to Neanderthal-AMH population turnover. I highlight a fundamental, challenging, even potentially confounding point. In Paleolithic foraging societies the very same intimate, cooperative contexts and core social networks could have contributed alternatively to technological continuity or to innovation-adoption and change.

Indeed, intimate cooperative settings involve laughter and encouragement, judgment and social levelling, and management of complex relationship-networks (Boehm 2012), all in the context of family formation and transfers to offspring (Hill et al. 2009; Kaplan et al. 2010). Such core social contexts may more often favor unreflected or culturally rationalized conformity-biased transmission. They can also support

H. S. Groucutt (ed.), *Culture History and Convergent Evolution: Can We Detect Populations in Prehistory?*,
Vertebrate Paleobiology and Paleoanthropology, https://doi.org/10.1007/978-3-030-46126-3_9

mobilization of collective action, sometimes involving decision-making about adopting innovations. In this chapter I explore how—in light of extensive evidence for long-term, if sporadic interaction and admixture between Neanderthals and AMH populations—we may better explain and contextualize those variable elements of persistence and change in late Middle Paleolithic (LMP) and early Upper Paleolithic (EUP) technological practices, transmitted through complex, recurrently negotiated cooperative social networks.

Background: Continuity and Change in Populations, Technologies and Social Networks

Recent fossil morphological and paleogenomic results reveal new details about the population-biological turnover involving AMH demic expansion and Neanderthal extinction in western Eurasia (Hublin et al. 2012; Fu et al. 2014, 2015, 2016; Pimenoff et al. 2017; Prüfer et al. 2017; Weyrich et al. 2017; Groucutt et al. 2018; Hajdinjak et al. 2018; Petr et al. 2019). In light of these new data, Tostevin's theoretical framework—alongside related methodological efforts at quantitatively characterizing multivariate similarity among lithic assemblages from the later Middle and Upper Pleistocene (Hovers and Raveh 2000; Hovers 2009; Shea 2011; Scerri et al. 2014a, b; 2016; Tryon and Ranhorn 2020)—provides a still-pertinent reminder that "stone tools are not people." As reviewed in more detail below, when we look at the western Eurasian archaeological record, we cannot plausibly claim that variability in LMP and EUP lithic assemblages closely maps onto Neanderthal extinction and AMH population expansion. Tostevin, in particular, has theoretically deepened constructive critiques and applications of lithic assemblage analyses (see Dibble 1995; Hovers and Raveh 2000; Bar-Yosef and Van Peer 2009; Hovers 2009; Teyssandier et al. 2010; Shea 2011; Tsanova et al. 2012; Kuhn 2013; Douze et al. 2020; Groucutt 2020; Tryon and Ranhorhn 2020). This chapter extends the theoretical consideration of the intimate cooperative contexts of social learning and cultural reproduction, which Tostevin brings clearly to the forefront, considering in depth its interrelated social and embodied practical dimensions.

Expanding the theoretical treatment of lithic assemblages in reconstructing Paleolithic culture history, Tostevin (2011, 2013) delves into Christopher Carr's "multi-level theoretical bridging" approach, which aims to reconstruct those socio-cultural processes that drove chronological and spatial variation—or relative lack of it—among archaeological assemblages (Carr and Neitzel 1995). This bridging argument is necessary to link archaeological assemblage variability to small-scale, intercorporeal cultural reproduction practices. In studying the MP-UP transition as a critical period of prehistoric technological change, unfolding systemically—albeit nonlinearly—with AMH-Neanderthal population-biological turnover, Tostevin connects bodily learning and practical technological activity to intimate social interaction contexts (2003, 2007, 2011, 2013). Such spatially close settings structure and are structured by bodily practices that are also social, fundamental for teaching and learning, cooperative bonding, social judgment and norm enforcement. Among prehistoric human foragers, socially intimate practices are embedded—in turn—within complex fission-fusion social dynamics and landscape mobility (Hill et al. 2011, 2014), which involved more widely dispersed meetings, occurring during visits to a camp's edge, along paths, and across landscape-scale vistas (Tostevin 2013). Thus, Tostevin argues compellingly that, among Late Pleistocene foragers, effective embodied interaction with percussors, stone cores, and detached blanks would have unfolded in intimate, cooperative social settings. These activity sites (that is, processing sites, logistical camps, or multi/extended-family residential camps) tended to facilitate or constrain the cultural transmission of knowledge in local Late Pleistocene social networks, whether they were constituted by Neanderthal, AMH, or admixed populations.

In developing a finely detailed picture of cultural transmission processes, Tostevin makes an underappreciated—and as noted above, arguably logically necessary—contribution to biocultural evolutionary inquiry into niche construction and adaptation dynamics in human evolution (Table 9.1). His theoretical and methodological framework leads our inquiry into an even deeper, murkier question in paleoanthropology—one that puts Neanderthal-AMH population-biological turnover in long-term, million-year-scale hominin evolutionary perspective. How did human complex-skill teaching and learning co-evolve with other remarkable, derived traits that comprise an embodied cognitive-behavioral and life-history adaptive complex underpinning cumulative culture and recurrently disruptive eco-social niche construction dynamics (Tomasello 2008; Hill et al. 2009; Stout and Hecht 2017; Stutz 2019; Stout et al. 2019)? These traits include:

- extended bouts of embodied selective attention, often involving complex, fine-motor brachial-manual learned skills (Stutz 2014a);
- frequent cooperative solicitation of joint attention (Tomasello 2008);
- endurance ambulatory and other gross-motor activities (Langdon 2005; Kaplan et al. 2010);
- cross-domain, highly associative mental hierarchical representation and abstraction processes (Fischmeister et al. 2017);

- cooperative, extended goal-oriented bouts of selective attention during resource search, acquisition, and transport (Goren-Inbar 2011);
- social judgment and norm enforcement (Boyd and Richerson 2005; Hill et al. 2009);
- teaching of sequence-dependent, goal oriented, skill-intensive practices, as key to intergenerational and horizontal cooperative bonding (Gärdenfors and Högberg 2017; Stout et al. 2019);
- and intergenerationally transferred social network ties embedded in a multi-scalar fission-fusion social system, in which delayed reciprocity becomes a fundamental, emergent feature (Grove 2012; Hill et al. 2014).

In this chapter I revisit Tostevin's (2013) careful consideration of embodied cultural transmission processes, which variably unfold at different scales of socio-spatial intimacy, activity, selective attention, and joint attention. I take into account more recent paleogenomic and radiometric dating results, which robustly document sporadic, yet significant, Neanderthal-AMH biocultural interaction, having spanned a substantial portion of the Late Pleistocene, ca. 120–40 ka (Prüfer et al. 2017; Hajdinjak et al. 2018). Along with complementary statistical analyses of later Middle Stone Age (MSA)/Middle Paleolithic (MP) technological variability in North Africa and the Arabian Peninsula (Scerri et al. 2014a, b; Groucutt et al. 2015; Scerri 2017), the ancient genomic data direct our attention away from a longstanding model for the MP-UP transition, toward a more complex—if still-poorly-understood alternative. The MP-UP transition can no longer be seen as a direct reflection of AMH demographic range expansion into western Eurasia. We are left to grapple with AMH dispersal as one contributing causal factor, helping to drive—while also being impacted by—the technological changes that define the archaeological transition. At the same time, Tostevin's theoretical framework—with its thorough connection to lithic analysis and assemblage-comparison methodology—emphasizes explanations for cultural conservatism, critically including contexts involving demic diffusion. I argue, however, that it does not sufficiently address decision-making and change, which would have been made in the very same intimate-scale cooperative situations that more often reproduced cultural practices (Reynolds 1986).

By focusing mainly on theoretical issues of technological continuity and change—as I extend Tostevin's work in studying traces of cultural dynamics—this chapter dives deeply into the problem of the western Eurasian MP-UP transition and its systemic connection to AMH range expansion and Neanderthal extinction. As the problem stands now, we cannot reasonably pin major population biological changes on particular archaeological technocomplexes. How, then, do we explain the MP-UP transition as part of complex, long-term metapopulation dynamics—on the one hand—and biocultural evolution and multi-scalar niche construction—on the other?

Table 9.1 Key theoretical concepts

Biocultural evolution
Biocultural evolution is a phrase most often used in the fields of bioarchaeology, human biology, and medical anthropology, with the concept frequently defined in introductory textbooks in biological anthropology. Here, in a typical example, *biocultural evolution* is seen as "[t]he mutual, interactive evolution of human biology and culture; the concept that biology makes culture possible and that developing culture further influences the direction of biological evolution; a basic concept in understanding the unique components of human evolution" (Jurmain et al. 2017, p. 7). Use of the term "biocultural" appears to have been independently adopted by Eugene Ruyle (who built the conceptual foundation of dual-inheritance theory) and Dennis van Gerven and colleagues (who helped to develop a continuous tradition of studying the feedbacks between culture and biology, as key for explaining human biological variation, microevolution, and health outcomes) (Ruyle 1973; van Gerven et al. 1973). Today, dual-inheritance models (Boyd and Richerson 2005) and biocultural research in human biology and medical anthropology (Zuckerman and Martin 2016) have become—in no small part—quite separated research fields. Still, the term *biocultural* remains heuristically valuable for linking these areas of anthropological inquiry to research on human evolution (Stutz 2013a, b).
Niche construction
Niche construction was introduced by Odling-Smee and colleagues, building on a key conceptual argument that Lewontin introduced concerning the emergence of adaptation (Lewontin 1983; Odling-Smee et al. 2003). Lewontin formalized the idea that adaptation and niche are coupled in a complex feedback system. A population's current phenotypes change as a function of the prevailing environment, but that environment—or instantaneous niche—changes as a function of the population's current phenotypes. A key theoretical advance with the niche-adaptation co-evolutionary model is that systems ecology may be more effectively integrated with evolutionary biology. Recent high-profile debates about human niche construction have focused on priorities: documenting the chronology and scope of direct food-web impacts and physical system-alterations—driven by human omnivory, demography, and extractive technologies—or modeling and explaining the population-niche feedback dynamics that provide comparative and theoretical insights into the nature of niche construction as an evolutionary process (Boivin et al. 2016; Ellis et al. 2016; Erlandson et al. 2016). These are not mutually exclusive goals. Both documenting and explaining niche construction in the MP-UP transition will be important for better understanding AMH-Neanderthal population turnover as biocultural evolution!

The Big Picture: Biocultural Evolution in the MP-UP Transition

The broader conceptual issue at the core of this chapter—how to investigate and understand the MP-UP transition as integral to complex, nonlinear biocultural and niche construction dynamics—remains deeply challenging. I emphasize that it has already been cogently posed and preliminarily explored by Brantingham and colleagues (2004b; see also Kuhn 2013). In a comprehensive edited volume—taking on a Eurasian geographical scope—they and their contributors present extensive support for the hypothesis that, during the first millennia of the Middle-Upper Paleolithic (MP-UP) transition, local or regional hunter-gatherer networks developed and adopted variations on "canonical" EUP technologies—sometimes involving convergent innovation—encompassing heterogeneous volumetric knapping approaches to blade(let) production, usually alongside manufacture of endscrapers, burins, and various retouched knives, microlithic barbs, and points on blade(let)s (Brantingham et al. 2004a). I note that there remains theoretical need for integrating a multi-scalar complex dynamical systems approach (Kuhn 2013; Stiner and Kuhn 2016) with observations about Upper Paleolithic emergence. Indeed, arguments for a complex, mosaic MP-UP scenario—with variable, indirect ties to Neanderthal extinction—have mainly been based on detailed empirical observations. Here, researchers have focused on characterizing lithic technology and tool morphology from particular sites and regions, described in the context of available stratigraphic information and radiometric dating. Perhaps it is more accurate to say that evaluations of lithic data—mainly involving qualitative techno-typological comparisons across a supra-continental scale—have argued against straightforward biocultural expansion and replacement hypotheses, leaving a mosaic cultural transition process as the plausible alternative (d'Errico et al. 1998; Brantingham et al. 2004a; Straus 2005; Teyssandier 2006; Tsanova et al. 2012).

Fossil finds, ancient human genomes, and microbial genomes show modern human presence and admixture with Neanderthal populations in western Eurasia, well prior to the MP-UP transition (Sankararaman et al. 2012; Fu et al. 2014, 2015, 2016; Hershkovitz et al. 2015; Kuhlwilm et al. 2016; Lazaridis et al. 2016; Pimenoff et al. 2017; Prüfer et al. 2017; Weyrich et al. 2017; Hajdinjak et al. 2018). The mosaic-transition perspective has been preliminarily considered in light of complete Neanderthal genome data (d'Errico and Stringer 2011; Zilhão 2011), but it has been caught up in debates over details concerning stratigraphic documentation, dating methods, paleoclimatic correlations, and interpretation of taphonomic data, especially concerning how particular regional technological traditions may—or may not—be associated with morphologically or molecularly diagnostic Neanderthals (Teyssandier 2006, 2008; Ofer Bar-Yosef and Bordes 2010; Riel-Salvatore 2010; Teyssandier et al. 2010; Hublin et al. 2012; Zilhão 2013; Zilhão et al. 2015; Benazzi et al. 2015; Devièse et al. 2017; Staubwasser et al. 2018; Teyssandier and Zilhão 2018; see Reynolds 2020, for a discussion of current debates over artifacts and populations in the later Upper Paleolithic of Europe). Still, the recent paleogenomic and fossil data are very clearly non-concordant with a straightforward hypothesis about one hominin species expanding demographically, replacing a closely related one across western Eurasia (Fig. 9.1) (Dediu and Levinson 2013, 2018).

This carries enormous evolutionary-theoretical implications for explaining the MP-UP transition. No longer can one easily assert that the appearance of certain EUP technologies directly marks Neanderthal extinction or the arrival of AMH groups, as some researchers continue to assume (Benazzi et al. 2015; Staubwasser et al. 2018). Rather, *the MP-UP transition must be explained as part and parcel of a broader evolutionary change—one which had already been shaped by long-term AMH-Neanderthal biocultural interaction across much of the Middle Paleolithic period in western Eurasia*. Considering the critical population-biological turnover that resulted in Neanderthal extinction, ca. 40 ka, the stone tools would remain characters in the story, but they would now be players in a biocultural evolutionary process that unfolded before, during and after AMH populations successfully dispersed substantially north of the Saharo-Arabian belt (Groucutt et al. 2015a, b, 2018; Scerri 2017; Scerri et al. 2014b; Stutz 2019; Groucutt 2020).

From Big Picture to Intimate-Scale Biocultural Dynamics

Sometimes the simplest explanation does not suffice. As reviewed above, recent archaeological, hominin fossil, and paleogenomic data arguably falsify the more parsimonious explanation for Neanderthal extinction and AMH dispersal into western Eurasia. Contrary to previous, quite convincing arguments about then-available observations (Stringer and Andrews 1988; Klein 1995; Bar-Yosef 2000; Hublin 2000; Bar-Yosef 2002; Mellars 2006a, b; Klein 2008; Shea 2008; Müller et al. 2011), East African *Homo sapiens* populations *did not* spread entirely singularly or decisively into southwestern Asia at a certain point after 50 ka.

Moreover, it is not entirely clear if a well-defined, short-lived AMH—or even an admixed AMH-Neanderthal—demographic expansion-wave drove the adoption of EUP technological traditions. Ancient DNA recovered from the Oase 1 mandible provides striking evidence for recent

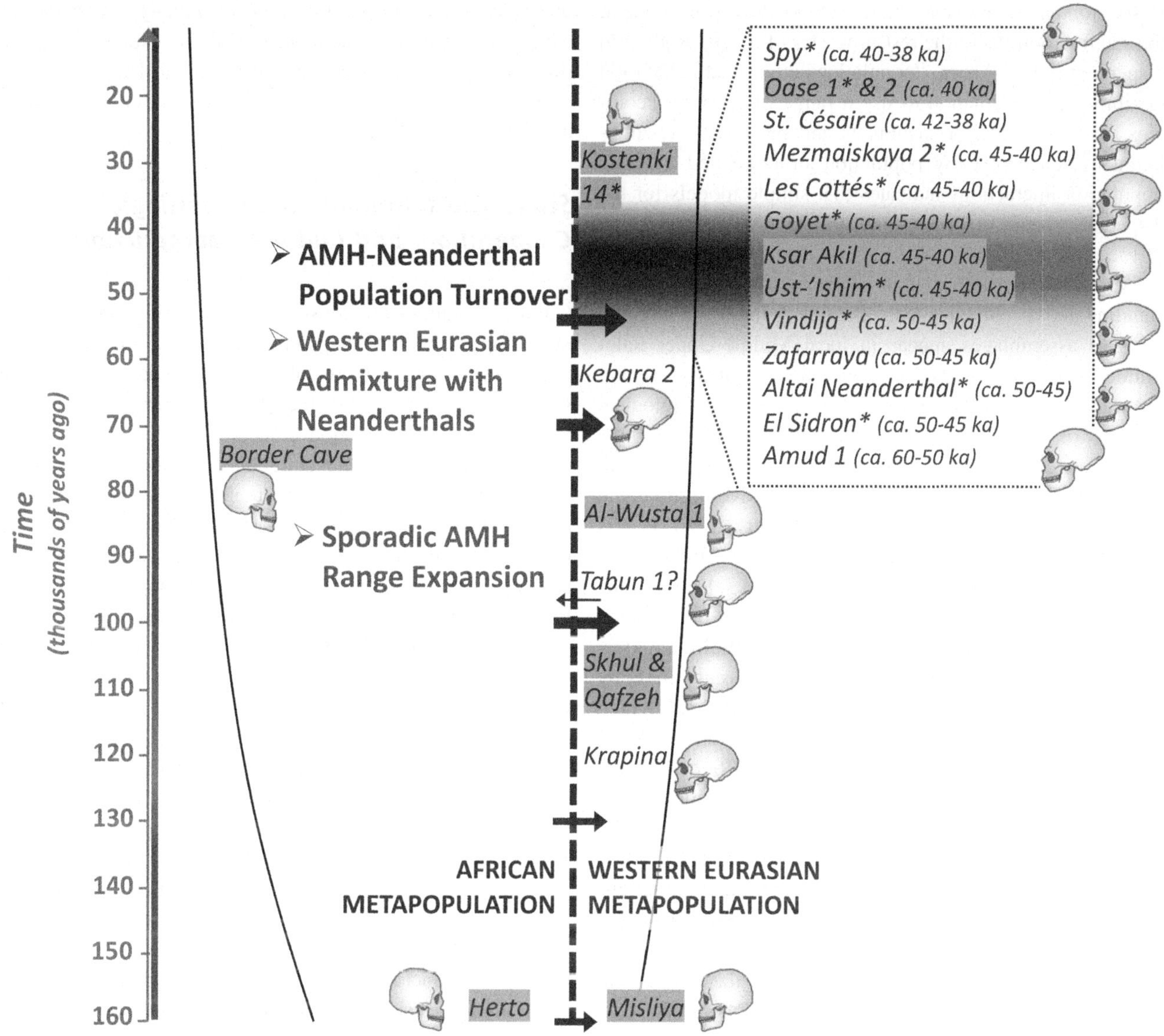

Fig. 9.1 Schematic illustration of anatomically modern human (AMH) sporadic range expansion into western Eurasia, long-term admixture with Neanderthals, and AMH-Neanderthal population turnover. Key fossil hominin specimens are labelled according to site name and whether their cranio-dental or body-proportion morphology is closer to AMH or Neanderthal samples, respectively. Specimens with * indicate that genomic data has been recovered for these individuals. Specimens classified as Neanderthals are labelled with unhighlighted text. Those classified as AMH are labelled with gray highlighted text

Neanderthal ancestry in an individual with anatomically modern features, ca. 40 ka (Fu et al. 2015). Admixture with Neanderthals was part of AMH range expansion, from Africa and into western Eurasia, both prior to and during the critical EUP interval 45–40 ka. There may have been several successive demographic waves, with variable admixture and separate or partially overlapping geographic ranges (Tostevin 2013; Lazaridis et al. 2014, 2016; Fu et al. 2016; Hajdinjak 2018; Pimenoff et al. 2017; Weyrich et al. 2017). Perhaps more significantly, phenotypically AMH populations did not carry with them a signature Later Stone Age (LSA) or Early Upper Paleolithic (EUP) tool kit, tracing their path toward expansion. This latter point is not without controversy (Benazzi et al. 2015; Zilhão et al. 2015). However, recent work on the Late Pleistocene archaeology of North Africa and the Arabian Peninsula indicates that AMH demographic range expansion developed out of well-established regional forager social networks, separated by limited geographical connections across riverine corridors and wetland waypoints. These regional networks—with restricted connections to each other—transmitted local technological practices over many generations (Scerri et al. 2014a, b; Groucutt et al. 2015b). This documented lithic technological variability does not directly map how African

Homo sapiens demes may have expanded into the Levant, the Zagros foothills or the rest of western Eurasia. Rather, in dispersing out of Africa, "populations employed behavioral flexibility and adaptation to use a range of different ecologies, including interior savannahs and the coast. Accumulating evidence of early population structure and multiple population interactions indicates that simple models for the dispersal process are no longer sufficient" (Groucutt et al. 2015a, 161).

An underlying aim of this chapter is to highlight long-standing and recent evidence, alike, falsifying the claim that, as EUP assemblages appear in local Middle-Upper Paleolithic (MP-UP) sequences, we are basically seeing the arrival of AMH groups. The population-biological process of AMH-Neanderthal turnover unfolded significantly between 50 and 40 ka (Devièse et al. 2017; Hajdinjak et al. 2018). It began, quite likely, at least several millennia before 50 ka (Richter et al. 2008; Fu et al. 2014). As emphasized above, it happened only after a very long period of sporadic interaction, admixture, drift, and selection, during which an AMH-Neanderthal reproductive barrier may have been favored but remained permeable (Wolf and Akey 2018) (see Fig. 9.1). Although Neanderthals went extinct by ca. 40 ka (Pinhasi et al. 2011; Hublin et al. 2012; Higham et al. 2014; Talamo et al. 2016; Devièse et al. 2017), AMH populations did not just outcompete them. Rather, demic diffusion, variable admixture, and selection favoring a range of AMH traits drove the population-biological turnover (Fu et al. 2014, 2015; Sankararaman et al. 2016; Wolf and Akey 2018). And persistent stone-tools-equal-people assumptions to the contrary (Staubwasser et al. 2018), nor did they simply fill the already-abandoned territories once occupied by widespread Neanderthal metapopulations, the constituent groups of which had long utilized Middle Paleolithic technologies (Lowe et al. 2012).

With this evolutionary background, archaeological evidence takes on particular importance. Only archaeological traces and their depositional contexts can clarify the western Eurasian mosaic of change in climatic conditions, forager behavior, and human impacts on the prevailing ecosystems during the Late Pleistocene. The joint population-biological and archaeological questions can no longer simply be stated, "Where and when did AMH groups replace Neanderthal ones, and what material culture assemblages did the respective taxa produce, use, and discard?" Straightforward replacement or turnover may have occurred in some regions, but it cannot be asserted across the board (Fu et al. 2014, 2015; Hajdinjak et al. 2018).

It is thus more challenging than ever to explain how the western Eurasian MP-UP transition, as an archaeological phenomenon, was systemically intertwined with AMH expansion, long-term (albeit occasional) AMH-Neanderthal interaction and admixture, and eventually, Neanderthal extinction. Among other theoretical and methodological difficulties, it is necessary to isolate human behavioral patterns related to cultural transmission, cooperative discourse, and innovation adoption.

Stone Tools, Intimate Social Settings, Cooperation, and Cultural Reproduction

Cultural dynamics are constituted by teaching and learning, normative technological practice and social judgment, and the development and social spread of innovations. The latter—which may include the adoption of new technologies, along with communication strategies, socio-political organizational principles, and culturally constituted institutions—can spread via forms of cultural stimulus diffusion and demic diffusion (Tostevin 2013). Still, cultural transmission of embodied technological competence is highly local and socially intimate, critically shaping and shaped by interaction in small-group cooperative settings. Here, enchronic—that is, socio-temporally intertwined, dynamically affectively attuned—intercorporeal interactions reflect and constitute network ties among close kin and allies (Enfield 2013; Sinha 2015a, b). It is from this perspective that Tostevin (2013) makes this salient point: cultural transmission practices in such dyadic or small-group cooperative settings can structure archaeological markers of demographic expansion waves, which unfold over multiple generations.

In a series of publications—including his comprehensive monograph (2013)—Tostevin argues that Paleolithic research on cultural transmission must focus on those stoneworking traces and artifact forms associated with sequence-dependent technological gesture systems that could only be observed, discriminated, and learned in such intimate social settings. In contrast, an observant visitor standing at a residential or logistical camp's edge could glean only grosser characteristics—including overall core and blank outline, blank and tool distal form, and proximal shape (e.g. Levallois versus punctiform). She or he would not necessarily comprehend how gesture-sequences may hierarchically tie specific actions to socially relevant technological goals, such as achieving particular blank quantities or shapes (Tostevin 2003, 2007, 2011, 2013). Archaeological artifact attributes key to intimate cultural-transmission settings include dorsal scar-patterns on blanks and tools, core platform preparation, and light retouch patterns, all of which can be linked to intergenerationally structuring or persistent cultural processes: teaching, learning, and conformity-biased technological performance. Such culturally reproduced gesture systems could follow expanding populations, potentially over centennial or millennial timeframes (Tostevin 2013).

In aiming to explain the MP-UP transition over much of western Eurasia, Tostevin's main claim is that EUP cultural traditions, discourse strategies, and embodied know-how would have been taught and socially reproduced in intimate contexts, primarily among the occupants of residential camps, potentially reinforced on raw-material provisioning and logistical foraging trips, through social judgment, negotiation, and emblemic, assertive signaling. He has methodologically integrated his theoretically robust "behavioral approach to cultural transmission," or BACT (Monnier and Missal 2014), with continuous metric and categorical attributes describing core and blank size and shape, core preparation and management patterns, and blank morphology on formal tools. Assemblage-level statistical patterns in these features describe learned, highly complex, sequence-dependent behaviors. Careful, attribute-based quantitative comparison of technological learning in any two assemblages can provide a baseline for investigating similarities and differences in local, culturally transmitted stoneworking practices (Tostevin 2013). In developing a high-resolution statistical approach, Tostevin has nuanced the picture of stoneworking variability—especially in terms of core management and blank production—across several Early Upper Paleolithic sites, from the southern Levant to Central Europe (Tostevin 2003, 2007; Tostevin and Skrdla 2006).

I would suggest that Tostevin's BACT studies have yielded substantial results, underpinned by compelling biocultural and archaeological theory. His work critically complements recent qualitative comparisons among western Eurasian LMP and EUP assemblages (Goring-Morris and Belfer-Cohen 2003, 2006; Svoboda and Bar-Yosef 2003; Brantingham et al. 2004a; Conard 2006; Hovers and Belfer-Cohen 2006, 2013; Teyssandier 2006; Bar-Yosef and Belfer-Cohen 2010; Bordes and Teyssandier 2011; Hoffecker 2011; Meignen 2012; Nigst 2012; Tsanova et al. 2012; Kuhn and Zwyns 2014; Teyssandier and Zilhão 2018). As I discuss below, Tostevin's pairwise assemblage-comparison methodology also provides granular resolution that, in the future, may be constructively integrated with multivariate approaches for elucidating similarity and difference among large sets of assemblages (Hovers and Raveh 2000; Scerri et al. 2014a, 2016). Here, I take up the pressing theoretical issue of broadening the BACT model. I underscore that future research must continue the integration of theoretical and methodological approaches, but more immediately, I argue that the BACT framework may be extended to encompassing how enchronic discourse and practice shaped innovation-adoption, influencing long-term change in lithic technology practices.

Expanding the BACT Framework: From Teaching and Learning to Discourse and Innovation

Tostevin zooms in on core management and blank production as an intricate, sequence-dependent cultural practice—that is, a cultural domain requiring intense, prolonged teaching and learning in spatially proximate, socially cooperative activity settings (Gärdenfors and Högberg 2017; Stout and Hecht 2017; Stout et al. 2019). Along with language, free-hand percussion stoneworking is an example *par excellence* of the unique derived hominin cognitive-behavioral capacity to learn and deploy general, hierarchical models of goal-oriented, complex embodied interaction with one's surroundings, based on episodic experiences in socially and physically heterogeneous situations (Hauser et al. 2002; Fitch 2011; Fischmeister et al. 2017). As a key theoretical foundation for understanding how complex, learned behaviors may be strongly conservatively transmitted, BACT constitutes a more satisfying, thorough explanatory bridge between past forager behavior and lithic assemblage patterning than does the *chaîne-opératoire* research tradition, which has remained strongly influenced by Leroi-Ghouran's notion that an abstract, structuring concept (analogous to Saussure's *langue*) must lie behind technological gesture traces in the archaeological record (see, e.g. Tsanova et al. 2012; see critiques in Bar-Yosef and van Peer 2009; Tostevin 2013).

In part because of its focus on stratigraphically securely provenienced lithic assemblages (Tsanova 2006; Teyssandier 2008; Bordes and Teyssandier 2011; Tostevin 2013; Tsanova et al. 2012), and in part because of its theoretical necessity (Tostevin 2011), BACT provides further, robust support for falsifying one key hypothesis: that indigenous Neanderthal groups could have adopted EUP technologies by acculturation that involved gift exchange or imitation of migrating AMH communities (d'Errico et al. 1998; Teyssandier 2006; Tsanova et al. 2012). A Neanderthal group practicing Mousterian stoneworking traditions—approaching the outskirts of a newly arrived AMH camp, the occupants of which worked stone following more recently adopted technological practices—would not have been able to glimpse the nuances necessary to imitate an entire Early Upper Paleolithic reduction sequence.

Just as important, Tostevin's theoretically grounded results also highlight that the technological and formal differences between late Mousterian and EUP lithic assemblages are substantial—at least in his southern Levantine and Central

European case studies—suggesting that the MP-UP transition involved the adoption of significantly different stoneworking practices. This makes it difficult to identify a cultural ancestor-descendant link between any particular LMP or Middle Stone Age (MSA) archaeological culture and the succeeding EUP technocomplex in a given region (Bordes and Teyssandier 2011). Indeed, the differences among some EUP assemblages—especially those belonging to different named stone tool industries, which encompass Initial Upper Paleolithic, Early Ahmarian, and Levantine Aurignacian in Tostevin's assemblage samples—also emerge as potentially significant (for a more qualitative perspective on arguably gradual, culturally continuous patterns of MP-UP technological change and diversity in the Levant, see Belfer-Cohen and Goring-Morris (2017) and Goring-Morris and Belfer-Cohen (2018).

Table 9.2 Reliably dated *first appearance* of EUP technocomplexes in western Eurasia, >40 ka.[a]

Technocomplex	Core management strategy	Key blank forms	Key point, microlith, and tool forms	Regional occurrences	Date of first appearance[b]	Associated hominin fossils[c]
Early Ahmarian	Single and opposed platform blade(let) removals on narrow or broad-fronted cores	Slightly curved or straight blade (let)s	el-Wad Points	Central Levant	47–44 ka	
Initial Upper Paleolithic (IUP)	Bidirectional elongated Levallois cores with lateral prismatic blade removals; hard hammer percussion	Elongated, relatively thick Levallois blanks; straight or slightly curved blades	Emireh Points, el-Wad Points, Umm el-Tlel Points; chamfered pieces	Northern Levant	45–44 ka	Likely AMH maxilla fragment from Ksar Akil Rockshelter Layer XXV
			Emireh Points, el-Wad Points	Southern Levant	>40 ka	
			Retouched Levallois points; Foliate points	Eastern Balkans	>40 ka	
				Central Europe (Bohunician)	52–45 ka	
Early Aurignacian	Broad-fronted single-platform blade cores and carinated bladelet cores	Relatively thick blades and curved, sometimes twisted bladelets	Aurignacian blades and Dufour bladelets	Swabian Jura	ca. 42–40 ka	
Kostenki-Borchova EUP ("Eastern Proto-Aurignacian")	Uni- and bipolar blade(let) cores on narrow or broad-fronted cores	Blades and bladelets	Endscrapers on thick blades	Eastern European Plain (southern Russia)	>40 ka	
Uluzzian	Levallois and non-Levallois flake cores and bipolar cores on flakes (splintered pieces)	Bladelets	Abruptly backed lunates	Italian Peninsula and possibly Greece	ca. 43–41 ka	
Chatelperronian	Primarily unidirectional blade removal from broad-fronted cores	Blades of variable width and thickness	Chatelperron curved-backed points; bilaterally retouched endscrapers	France & N. Iberia	>40 ka	St. Césaire Neanderthal burial, directly dated 42.5–39.5 ka

(continued)

Table 9.2 (continued)

Technocomplex	Core management strategy	Key blank forms	Key point, microlith, and tool forms	Regional occurrences	Date of first appearance[b]	Associated hominin fossils[c]
Lincombian-Ranisian-Jerzmanowician (LRJ)	Bipolar blade removal from broad-fronted cores, often with soft-hammer percussion	Relatively wide, long, thin blades	Bifacially retouched blades, utilized as elongated—sometimes pointed—core-tools	Central Europe to southern England	>40 ka	Spy fragmentary Neanderthal remains, directly dated ca. 40–38 ka

[a]The description of technocomplexes here focuses only on lithic assemblages (Conard and Bolus 2003; Goring-Morris and Davidzon 2006; Teyssandier 2008; Stiner et al. 2010; Flas 2011; Meignen 2012; Tsanova et al. 2012; Moroni et al. 2013; Tostevin 2013; Kuhn and Zwyns 2014; Peresani et al. 2016; Alex et al. 2017; Goring-Morris and Belfer-Cohen 2018; Teyssandier and Zilhão 2018; Shea et al. in press). There is heated debate over the stratigraphic association of many bone, antler, ivory, and shell artifacts with EUP lithic material (Teyssandier 2008; Riel-Salvatore 2010; Hublin et al. 2012; Zilhão 2013; Benazzi et al. 2015; Zilhão et al. 2015). The definition of "*first-appearance* EUP technocomplex" in a given region is—at present—necessarily arbitrary. Earlier or penecontemporaneous assemblages in Eastern and Central Europe—representing Szeletian or Bohunician (IUP) technocomplexes—would only add to the mosaic of stoneworking practices adopted in the 45–40 ka interval (Tostevin and Skrdla 2006; Tsanova et al. 2012). Those technocomplexes included here have evidence of representative assemblages being stratigraphically superposed over Mousterian assemblages in many sites

[b]Dates of first appearance remain necessarily based on imprecise observations. Here, I have focused on accuracy, requiring one or more of the following criteria: assemblages with techno-typological characteristics of a given technocomplex must be (a) dated by thermoluminescence (TL) on securely associated heated flint artifacts (Richter et al. 2008); (b) have associated high-quality (bone ultra-filtration or careful charcoal pretreatment) radiocarbon assays; or (c) stratigraphically underlie independently dated soil or tephra horizons (Higham et al. 2009; Kuhn et al. 2009; Semal et al. 2009; Stiner et al. 2010; Hoffecker 2011; Rebollo et al. 2011; Higham et al. 2012; Hublin et al. 2012; Talamo et al. 2012; Haesaerts et al. 2013; Tostevin 2013; Bosch et al. 2015; Stutz et al. 2015, Alex et al. 2017)

[c]Associated human remains are very rare from EUP contexts >40 ka, and often, debate persists over stratigraphic details and the sufficiency of sample pretreatment and measurement in radiometric dating. I have focused, where possible, on directly dated, well-preserved hominin specimens, pretreated with the collagen ultrafiltration protocol (Semal et al. 2009; Hublin et al. 2012). The depositional context of the "Ethelruda" maxilla specimen from Ksar Akil Rockshelter, Lebanon, is discussed in Douka et al. (2013), and the age is approximated based on Bosch et al. (2015)

As broadly summarized in Table 9.2 and Fig. 9.2, the techno-typological attribute states that characterize regional groups of western Eurasian EUP assemblages, ca. 45–40 ka, suggest geographically mosaic differences in semiotic and practical transmission—via teaching, discourse, judgment and negotiation—in cooperative intimate social settings. The variability in core management strategies, blank shapes, and point, knife, or microlith forms among and within Initial Upper Paleolithic (IUP), Early Ahmarian, Chatelperronian, Early Aurignacian, Uluzzian, or Lincombian-Ranisian-erzmanowician (LRJ) assemblage clusters—considered in light of the BACT theoretical framework—falsifies another, broader hypothesis. There is no ostensible support for the claim that any one of these technocomplexes were models for imitation and acculturation among those forager societies who came to produce any of the other EUP technocomplexes listed in Table 9.2 (see also d'Errico 1998; Tsanova et al. 2012). Here, I use "acculturation" in the sense emphasized in debates over purported Neanderthal acquisition of AMH technological practices, where a clear *biocultural* boundary to intergroup social intimacy is suggested (d'Errico et al. 1998; Flas 2011; Tostevin 2013). As Tsanova and colleagues (2012, 471) have already argued, evidence for a mosaic western European MP-UP transition favors "separating the evolution of material culture from that of biological morphologies." From Tostevin's theoretical point of departure, it is important to point out that—regardless of her recent African, Neanderthal, or admixed genealogy—an EUP forager with learned, embodied fluency in producing Early Aurignacian carinated endscrapers and twisted bladelets would not have been able to visit the edge of a strange group's camp, picking up the details of LRJ strategies for producing thick blades and managing them as elongated core-tools.

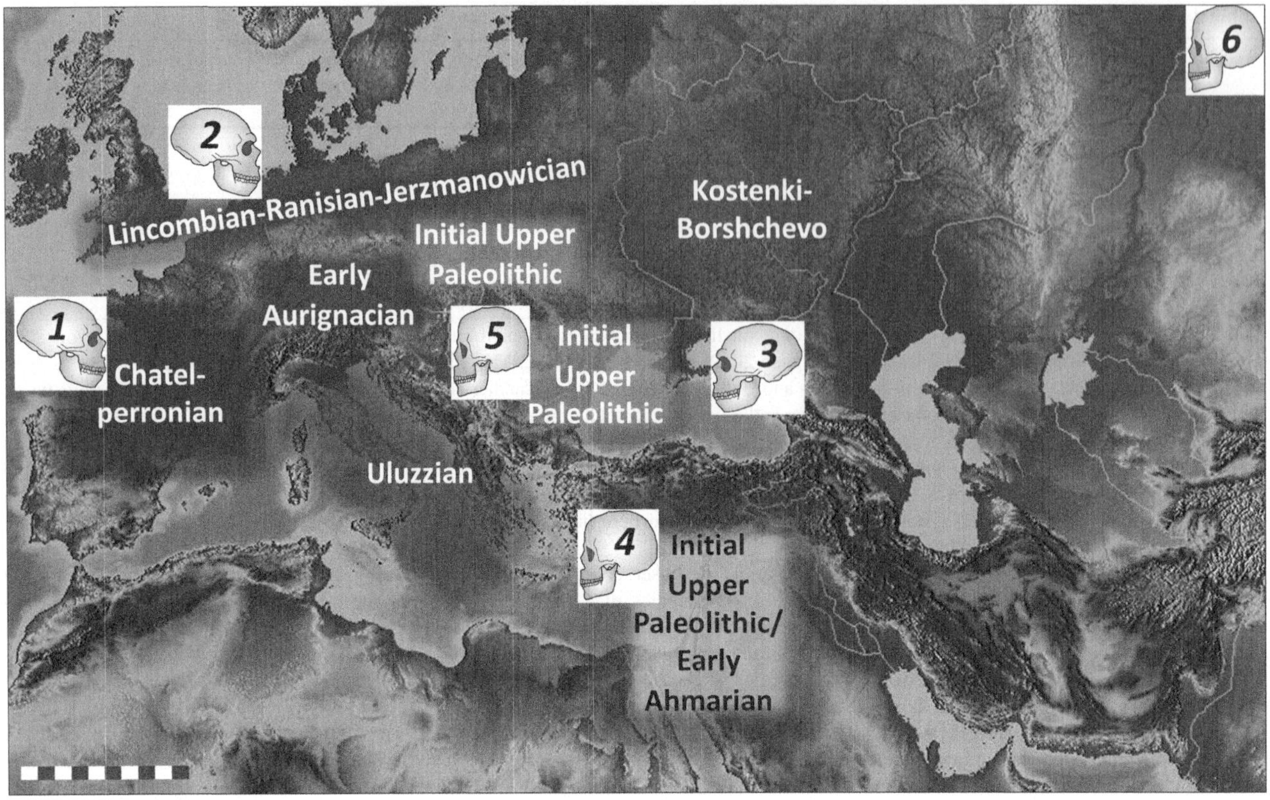

Fig. 9.2 Geographic distribution of technologically distinct early Upper Paleolithic (EUP) assemblage clusters dating to the 50–40 ka time interval. Key, dated fossil specimens show a mosaic of specimens with Neanderthal or anatomically modern human (AMH) traits: (1) St. Césaire Neanderthal, ca. 43–39 ka; (2) Spy Neanderthals, ca. 40–38 ka; (3) Mezmaiskaya 2 subadult Neanderthal burial, ca. 45–40 ka; (4) Initial Upper Paleolithic/Early Ahamarian assemblages, likely associated with AMH populations; (5) Oase 1 and 2 AMH cranium and mandible, ca. 40 ka; (6) Ust-'Ishim isolated AMH femur, ca. 45–40 ka. Basemap from Open Street Map contributors, licensed under a Creative Commons Attribution-ShareAlike 2.0 (CC-SA) license

I suggest that it is precisely here that the decision-making and innovation-adoption in cooperative hunter-gatherer contexts comes into focus. It is here that we can develop a highly relevant case study for considering expansion of the BACT framework, to encompass behavioral change and its socio-ecological context, as well as transmission and cultural reproduction. The chronological overlap of at least six—likely more—EUP technocomplexes strongly suggests that, across the ca. 45–40 ka interval, permeable and shifting hunter-gatherer social networks maintained different discourses over innovation and conservation of stoneworking, tool use, curation, and discard strategies (summarized in Table 9.2).

It is possible that some of the technotypological similarities among geographically disparate assemblages—for example, Chatelperronian and Early Ahmarian blade(let) production, dated to ca. 45–40 ka—reflect convergent innovation, perhaps in response to similar changes in land tenure, social network structure, on-site task activities, and mobility practices (Shea 2008). Yet, other factors—such as the politico-ideological development of long-distance alliances, shaping ritualized visitation and intermarriage practices—could alter social network boundaries and favor the spread of innovations, especially in response to imbalances between demographic rates (mouths to feed and hands to contribute labor), on the one hand, and energy/nutrient extraction and distribution rates (social technologies and wider ecological conditions), on the other.

In short, a "behavioral approach to cultural transmission"—when considered in light of the remarkable geographical mosaic of EUP technological change, mainly in the 45–40 ka timeframe, unfolding during a still-complex population-biological shifting balance between AMH and Neanderthal alleles and phenotypic traits—may be modified, specifically to investigate the intimate social context of discourse, decision-making and cooperation surrounding invention and innovation adoption.

In taking on intimate-scale cooperation and innovation—as well as transmission—the approach would also have to stretch theoretically, covering behavior's ecological and social settings across multiple scales. I address this issue below. For now, I argue that we can begin to formulate a wider theoretical approach, encompassing the behavioral dynamics and ecological conditions of cultural change and transmission. The question about what happened in western Eurasia between ca. 45 and 40 ka may now be posed as follows. Given that AMH range expansion, Neanderthal resilience, and complex admixture history would have already persistently perturbed hunter-gatherer biocultural niches in western Eurasia, can we clarify and explain variation in local and regional eco-social contexts that may have shaped population turnover in some regions, but local innovation in others?

From Intimate and Embodied Contexts to Metapopulation Dynamics and Neanderthal-AMH Admixture

As illustrated in Fig. 9.3, human forager mobility, cooperative social-network, and fission-fusion behaviors—which are all critical to shaping daily energy flux and extracting sufficient omnivorous food resources to maintain metabolic energy balance—span a spatially extensive domain, relative to ecological scaling patterns, which in turn encompass biocultural adaptation and niche partitioning, biological microevolution, and long-term macroevolution (involving major niche construction trends or equilibria, the emergence of adaptive complexes, and speciation and extinction). The relatively high complexity of hunter-gatherer biocultural systems, from daily to intergenerational temporal scales, would have played a key role in the technologically heterogeneous MP-UP transition across western Eurasia.

The recent paleogenomic evidence robustly shows that the pattern of Neanderthal and AMH biocultural interaction was intricate, unfolding over many millennia. When Neanderthals and anatomically modern humans met, it led—at least occasionally—to cooperation and social network formation, involving family formation and significant admixture, well before and during the MP-UP transition (Fu et al. 2014, 2015, 2016; Lazaridis et al. 2014; Kuhlwilm et al. 2016; Sankararaman et al. 2016; Pimenoff et al. 2017; Prüfer et al. 2017; Weyrich et al. 2017; Hajdinjak et al. 2018). The basis for this characterization is straightforward. As the Middle Pleistocene unfolded, European Neanderthal metapopu-

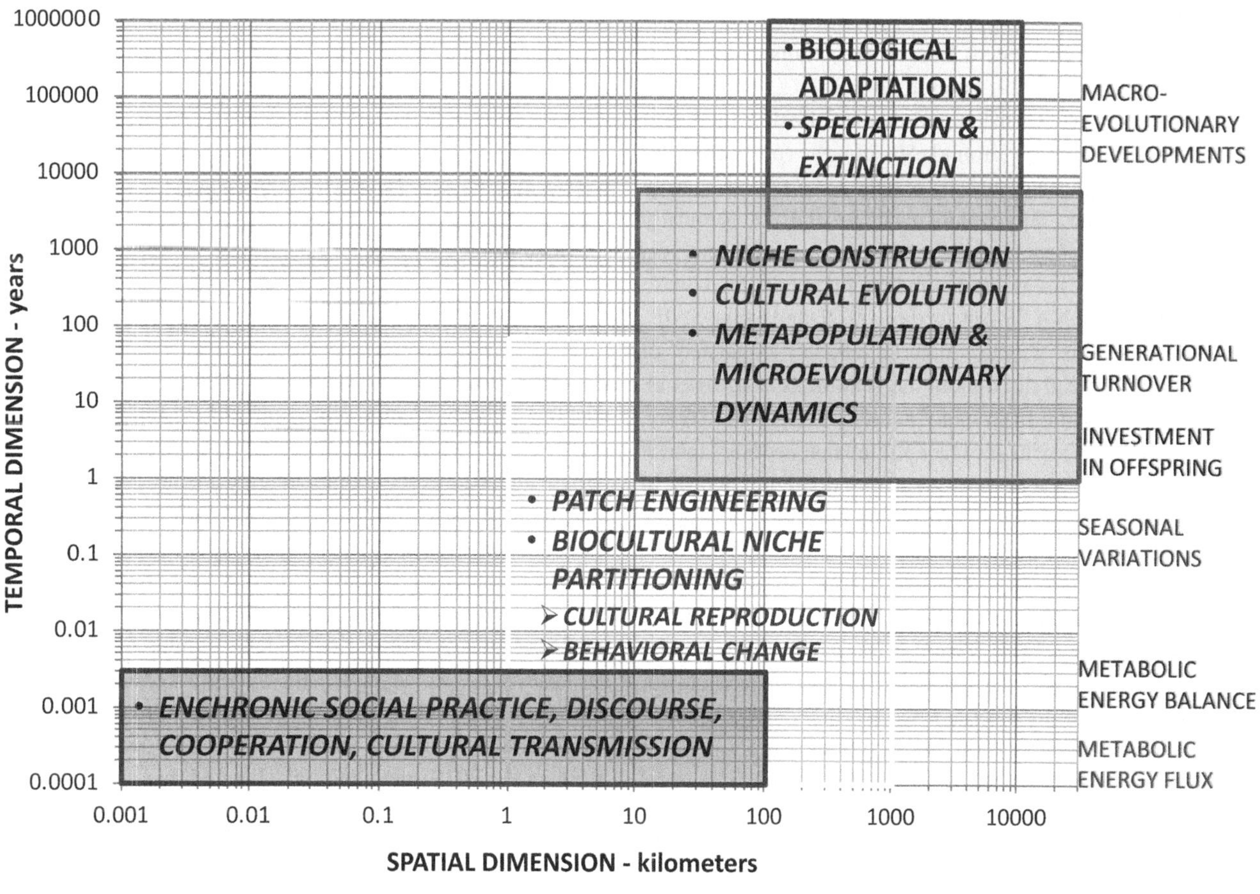

Fig. 9.3 Ecological spatio-temporal scaling diagram, modeling how short-term (that is, intra-daily) human behavior and environments are "tail-heavy," encompassing surprisingly wide spatial scales, in which metabolic energy flux and energy balance shape and are shaped by complex shifts from cooperative in-camp tasks, play, ritual and social discourse to long-distance logistical and residential mobility trips

lations emerged and evolved in relative isolation—likely due to distance and restricted migration across the Saharo-Arabian belt (Groucutt et al. 2015a)—and thus, they accumulated numerous derived single nucleotide polymorphisms (SNPs). A well-defined metapopulation waxed and waned in geographic extent, but in the face of later AMH range expansion and demic diffusion, Neanderthals succumbed to population turnover by ca. 40 ka. Still, many Neanderthal derived genetic markers have been inherited by several thousand generations of AMH populations outside of Africa, up to the present (Sankararaman et al. 2016; Wolf and Akey 2018; Petr et al. 2019). Among EUP anatomically modern human bones yielding well-preserved nuclear DNA, the Neanderthal derived SNP markers are spread across unlinked loci, throughout every chromosome (Sankararaman et al. 2016; Vernot et al. 2016). It is implausible that the broad array of unlinked Neanderthal SNPs came from a very few archaic Eurasian ancestors. After many millennia, in the face of AMH range expansion, such limited Neanderthal ancestry would almost inevitably have disappeared through drift. It is dramatically more likely that sporadic—yet recurring—cooperation, family formation and ethnogenesis contributed to significantly admixed populations, whose Neanderthal ancestry was later diluted by up to several waves of AMH demic diffusion into western Eurasia (Lazaridis et al. 2014; Sankararaman et al. 2014; Fu et al. 2016).

As AMH range expansion and demic diffusion increased in pace—very roughly around 50 ka—the proportion of African to Neanderthal ancestry would have tipped sharply toward the former. This would result in EUP anatomically modern humans each inheriting between a great-great (ca. 6%) or great-great-great-great grandparent's (ca. 1.5%) worth of recombinant DNA linked to derived Neanderthal SNPs. Yet, most of this genetic material already consisted of unlinked loci, spread throughout the genome (Lazaridis et al. 2014). Thus, it came from many admixed ancestors. Today, while many people with very recent Eurasian ancestry inherit only a great-great-great-great-great grandparent's worth of Neanderthal recombinant DNA—again sprinkled across the nuclear genome—*at least* 20% of the Neanderthal reference genome can be found in a sufficiently large sample of extant humans (Sankararaman et al. 2014; Vernot and Akey 2014; Vattathil and Akey 2015; Vernot et al. 2016). Substantially admixed ancestry was the metapopulation-scale vehicle for how EUP populations inherited Neanderthal derived genetic markers. Given the long-term, high expense that cooperative hominin parenting entails—with calorically rich resources and cumulative, complex-learning-based social capital transfers to offspring—the AMH-Neanderthal admixture history cannot be explained exclusively by sporadic sexual trists or pregnancies from sexual violence. Socially embedded family formation, involving biparental care and resource transfers—a core feature of the human adaptive system (Hill et al. 2009; Hrdy 2009; Kaplan et al. 2010; Stutz 2014a)—likely facilitated the bulk of the gene flow.

This genomic picture of Neanderthal-AMH biocultural interaction supports a possibility that Tostevin (2007, 2011, 2013) preliminarily explored, as he analyzed MP and EUP assemblages from central Europe (see also Tostevin and Skrdla 2006). Because Middle Paleolithic Neanderthal-AMH interaction was sporadically constituted by family formation and admixture in various parts of western Eurasia, it necessarily follows—considering the prolonged, cooperative transfers of nutrients, calories, social capital, and cultural embodied knowledge that structure human life history adaptations and their systemic connection to our intensely social niche (Kaplan et al. 2009, 2010; Stutz 2009)—that strangers were sometimes invited into residential camps, becoming new allies or kin, learning local stoneworking and tool-use practices.

It is now especially necessary to broaden application of Tostevin's theoretical framework, expanding beyond the behavioral basis for cultural transmission and conservatism. The mobile foraging contexts in which intimate-scale cooperative interaction is negotiated can shape not only the limits, but even the possibilities of making decisions that alter technological norms. Such settings provide the ground for small-group discourse—among kin and allies—over considering and implementing innovations.

Intimate-Scale Decision-Making and Cooperative Innovation: From the Oldowan to the Upper Paleolithic

Evidence for intimate-scale, enchronic interaction and adaptive cultural change—in contrast to comformity-biased, conservative, socio-politically driven cultural transmission (Boyd and Richerson 2005; Whiten et al. 2005; Tostevin 2013)—in stoneworking and tool-use traditions comes from recent research on the earliest Oldowan. Focusing on a series of sites in the Gona locality, Ethiopia, ca. 2.6 ma, Stout and colleagues (2019) have developed a theoretically and methodologically holistic approach—similar in many ways to Tostevin's—as they study the very emergence of hominin cultural-technological traditions for integrating raw material provisioning, complex stoneworking, and stone tool-use. Stout et al. (2019) may certainly be studying archaeological traces from hominins who mainly communicated with indexical, deictic and iconic—as opposed to habitually symbolic and recursively organized—gestural and vocal signs, employed to cooperate and manage joint attention at intimate social scales (Tomasello 2008; Fitch 2011; Stutz 2014a). However, diachronic change in earliest Oldowan stoneworking practices suggests that discourse—whether mediated by language, protolanguage, or limited gestural enchronic management of self-control, joint attention, and collective action—led to significant shifts in learned core exploitation patterns.

Here, the methodological approach is similar to the lithic analysis protocols for the Late Pleistocene samples discussed here (Kuhn 1995, 2013; Hovers 2009; Tostevin 2013; Scerri et al. 2014a), in that it utilizes experimental modeling and highly detailed core and debitage attribute analyses (Stout et al. 2019). The findings also provide foundational support for carrying Tostevin's theoretical work further. At the very beginning of the Earlier Stone Age, complex social learning processes already interacted with—and were likely evolutionarily shaped by—eco-cultural niche construction, specifically driving the cultural emergence of bi- and multi-facial flaking as a normative strategy for more efficiently and intensely obtaining whole flakes from available cobbles. Over millennia of hominin forager activity—with small-group stone-working and food processing embedded in a much wider fission-fusion terrestrial social system—unifacial flaking fell by the wayside (Stout et al. 2019).

The long-term result was arguably an Earlier Stone Age eco-cultural niche construction dynamic, which strengthened cooperative bonding and social network formation, while further expanding the landscape contexts in which cooperative hominin groups provisioned places with raw materials, utilizing efficient, learned technological practices to capture and extract calorie-dense omnivorous food resources (Hill et al. 2009; Kaplan et al. 2009, 2010; Stout and Hecht 2017; Stout et al. 2019). As Stout and colleagues emphasize, such adaptive discourse and transmission of resulting practices certainly contributed to eco-cultural niche construction, shaping the terrestrial, omnivorous, extractive and socially intense ("TOES") dimensions of the conspicuously non-equilibrium human niche (Stutz 2014a). Construction in the TOES niche is fundamentally related to the emergence of cumulative culture, intensifying Baldwinian evolutionary processes, recursively favoring selection for language, an extended juvenile growth period, and prolongued post-reproductive survival (Hill et al. 2009; Stout and Hecht 2017).

Of course, neither Neanderthals nor AMH populations—nor admixed AMH-Neandertal groups, for that matter—may be confused with the hominin authors of the earliest Oldowan traditions. Genomic and fossil-anatomical evidence confirms that, whatever the variation in AMH and Neanderthal evolutionary anatomy, there was a shared neuromotor capacity for language (Sankararaman et al. 2016; Dediu and Levinson 2018). I underscore that biocultural evolutionary trends—including cumulative culture, language evolution, and stabilization of regional metapopulation equilibria—would have jointly, gradually contributed to larger modal social network sizes over the Early and Middle Pleistocene. In turn, language-mediated social judgment, self-discipline, gossip, and norm-enforcement would have strengthened conformity-biased processes, driven by mundane practice, ritualization and the production of identities, memory, ideology, and the reinforcement of in-group/out-group distinctions. At the same time, the potential for adopting innovations remained. Moreover, the niche construction impact of such cultural innovations—involving increases in local, sustainable resource extraction rates, fertility, and survival—would have risen in larger, more culturally resilient cooperative social networks (Stiner and Kuhn 2016; Blinkhorn and Grove 2018; Malinsky-Buller and Hovers 2019).

Technological development in the EUP remains incompletely understood. Invention, innovation adoption, demic diffusion, cultural stimulus-diffusion patterns, and even loss of technological knowledge all contributed to the MP-UP transition's complex biocultural underpinnings (Hovers and Belfer-Cohen 2006, 2013; Belfer-Cohen and Hovers 2010). Yet, there are common themes among the mosaic archaeological technocomplexes dating to ca. 45–40 ka (see Table 9.2). These include more frequent: use of volumetric approaches for detaching large numbers of thin flakes and blade(let)s; retouch modification of blade(let)s into normative point, knife, and barb forms; and production and hafting of processing tools. Overall, it appears that in many regions, EUP practices involved compound-tool innovations employed in two key areas: hunting and butchery in a wider range of environmental settings (Shea 2008; Shea and Sisk 2010; Teyssandier et al. 2010; Tostevin 2013), and more diverse in-camp task activities, suggesting marginally longer residential camp stays, with normative culturally structured task roles (Kuhn and Stiner 2006; Stutz et al. 2015; Stutz and Nilsson Stutz 2017). Taking into account the derived hominin potential for small-group, intimate social-scale discourse and decision-making over invention and innovation, I argue that the mosaic range of technological change in EUP assemblage variability reflects at least several instances of local innovation-adoption (see Fig. 9.2). Testing this suggestion will involve substantial collaborative research; it requires further resolving chronological, ecological, and behavioral details in the MP-UP transition.

Considering Innovation-Adoption in the MP-UP Transition: The Initial Upper Paleolithic (IUP) as a Case Study

AMH geographic spread and Neanderthal extinction involved supra-regional, long-range population expansions and metapopulation dynamics, spanning western Eurasia in the Late Pleistocene. As Kuhn (2013) has recently underscored, we cannot assume that such regional and supra-regional dispersal and migration behaviors would leave obvious Paleolithic archaeological signatures, at least in individual sites or regional assemblage samples. In the previous section, I discussed how the BACT socio-behavioral framework (Tostevin 2013; Monnier and Missal 2014) can be theoretically expanded, explaining how small-group learning, cultural reproduction, and cooperative discourse processes can alternatively shape conservatism and innovation among lithic assemblages. Enchronic, intimate social interactions are embedded in daily activities, decisions, and intragroup heterogeneity in metabolic dynamics (which, in turn, can be intricately affected by myriad genetic, life history, ecological, sex-linked and biosocial factors). How, then, might the subtleties of cultural transmission and innovation adoption have been systemically tied to the broader-scale biocultural evolutionary trends that shaped the MP-UP transition and AMH-Neanderthal turnover? It may be surprisingly difficult to distinguish conformity-biased transmission patterns—structured in humans by mundane embodied practice and *habitus*, social judgment, intra-group norm enforcement, and emblemic stylistic signaling (Bourdieu 1977; Wiessner 1983; Bourdieu 1990; Boyd and Richerson 1992; Hill et al. 2009; Tostevin 2013)—from enchronic bouts of discourse and decision-making that contribute to shifting equilibria between foraging adaptations and socio-ecological niche construction. We may measure and analyze as many technological attributes as we can conceive, but we still need to consider how our similarities among lithic assemblages may be explained as temporal and chronological variability in traces of cultural practices that were adopted and socially reproduced in wider, complex behavioral and ecological settings. In studying archaeological lithic assemblages associated with the western Eurasian MP-UP transition, in particular, our challenge is even greater. We still seek to discriminate traces of population turnover—in which one culturally conservative group largely replaced another—from a situation of population continuity or admixture, in which technological changes occurred due to intimate-scale discourse and innovation-adoption. Moreover, we need to recognize that innovations adopted in one region could have spread more widely by ongoing discourse, ideology production, and intricate patterns of conformity-biased adoption. In this section I examine lithic technology and chronological evidence associated with Initial Upper Paleolithic occupations in the Levant and Central Europe, in order to consider possible alternative explanations for innovation adoption and cultural transmission that shaped archaeological traces of this key EUP technocomplex.

Initial Upper Paleolithic (IUP) technologies were developed and adopted sometime between 50 and 45 ka (Richter et al. 2008; Tostevin 2013; Kuhn and Zwyns 2014; Bosch et al. 2015; Stutz et al. 2015). Tostevin (2013) convincingly argues that the technological-attribute-state similarities between Levantine (sometimes referred to as "Emiran") and Central European (Bohunician) IUP assemblages are so thorough—from initiating core reduction to obtaining and modifying blanks of a particular morphology—that we can rule out independent innovation in the ca. 50–45 ka time-frame. In this scenario, Levantine groups are hypothesized to have grown and spread demographically, transmitting what would have become traditional IUP technologies from generation to generation, during a centennial or millennial-scale process of demic diffusion into Central Europe. However, available high-quality radiometric dates from the Brno-Bohunice site are also consistent with IUP technology being adopted first in Central Europe, ca. 50 ka—very possibly by local, admixed Neanderthal-AMH groups (Tostevin and Skrdla 2006; Richter et al. 2008; Tostevin 2013). Thus, IUP traditions may have then spread to the Levant slightly later, ca. 45 ka (Kuhn et al. 2009; Boëda et al. 2015; Bosch et al. 2015). In this alternative scenario, occasional, influential long-distance kinship alliances could have driven—and been constituted by—rare visits and small-group discourse, leading to innovation-diffusion, from Central Europe to southeastern Europe, across Anatolia, and into the northern Levant (see, e.g., Tostevin and Skrdla 2006; Tsanova 2006; Tostevin 2013; Kuhn and Zwyns 2014; Boëda et al. 2015).

We may now outline two hypotheses for the development, adoption, and geographic spread of IUP stoneworking practices. The first hypothesis—involving southern Levantine technological development and adoption, followed by demic diffusion into southeastern and Central Europe—is more in line with long-held expectations about stone tools (at least partly) reflecting AMH out-of-Africa range expansion during the MP-UP transition. The second hypothesis is about IUP technological development and adoption reflecting mainly inter-group sociopolitical, relationship-building responses to the larger-scale—albeit indirect—systemic effects of AMH demic diffusion. This would have unfolded with a geographically mosaic pattern of AMH-Neanderthal social interaction and admixture. The hypothetical spread of IUP technological practices—occurring against the tide of AMH population expansion—is admittedly counterintuitive, but it could plausibly have been driven by influential individuals or small kin-groups, migrating from Central Europe to the Southeast, as they negotiated long-distance marriage and exchange alliances, in response to sporadic imbalances among local human demographic systems, cultural institutions for economic production and distribution, and ecological patch productivity and predictability.

The latter hypothesis is theoretically reasonable. It is interesting, in that it helps us to consider an alternative proposal in which AMH demic diffusion—whether with or without Neanderthal admixture—was *indirectly* connected to technological change. Here, innovative practices for initiating and managing cores, in order to produce a range of elongated blanks, would have developed in Central Europe, far from the likely Near Eastern source of AMH population expansion. The new technology would have subsequently spread through renegotiation of long-distance alliances and social boundaries. Generations of experienced knappers would have moved fitfully—but repeatedly—across a widespread forager network, from Central Europe to the Levant, modeling, teaching and utilizing new approaches to blank and formal tool production.

On its face, this alternative hypothesis is more consistent with recently available high-quality radiometric dates (Hublin et al. 2020; Richter et al. 2008). However, the chronometric measurement samples (TL and ^{14}C dates) are still limited in scope. Other things being equal, then, there is currently no strong reason to expect that one hypothesis is substantially more likely than the other. Additional radiometric dates will allow us to refine the chronology of IUP occupations in western Eurasia. These alternative hypotheses may then be tested more comprehensively by measuring multivariate inter-assemblage similarities—across technological attributes in blanks, cores and formal tool—and their associations with independent data on foraging, onsite task activities, and prevailing ecological conditions, in a more comprehensive site sample from the Levant, southeastern and Central Europe.

From Theory to Method: Evaluating Innovation and Transmission Patterns in a Mosaic MP-UP Transition

Such a study would integrate the theoretical expansion of Tostevin's BACT framework—as discussed in this chapter to focus on the archaeologically visible effects of intimate-scale cultural dynamics in hunter-gatherer societies —with a corresponding, comprehensive methodological approach that addresses the mosaic MP-UP transition as reflecting complex biocultural evolution and niche construction dynamics. The methodology must yield reproducible observations that are predicted by alternative propositions about forager innovation and conservatism in social and ecological context. Hunter-gatherer cultural dynamics were driven by intimate-scale teaching and learning, performance and judgment, and—occasionally—collective decision-making to adopt novel technological practices. Yet, they were systemically influenced by social networks, demographic conditions, and broader environmental systems. Thus, as illustrated in the previous section, relevant hypotheses may take on a challenging intricacy. From a behavioral-ecological standpoint, the main biocultural pressures favoring technological conservatism would have been the maintenance of cooperative social relationships—which constituted and were constituted by intimate-scale discourse, in-camp task division, and mobilization of logistical provisioning trips, and decision-making about group fission-fusion changes under visits, camp stays, and residential moves. Other things being roughly equal—even in the face of occasional kin-structured dispersal, group-fissioning, the unpredictable demands or needs of long-distance allies, and migration driven by AMH range expansion or climatic fluctuations—cultural conservatism would be a safe strategy for holding onto critical social capital. Yet, the archaeological record of the MP-UP transition makes clear that innovative technologies were repeatedly adopted during Marine Isotope Stage 3 (ca. 60–30 ka), in assocation with AMH-Neanderthal admixture and population turnover. In the MP-UP transition the feedbacks among cultural dynamics, social networks, demography, population biology and wider niche construction followed a long-term non-equilibrium trajectory. In this section, I review and comment on a methodology for tackling the biocultural and ecological intricacy of the mosaic MP-UP transition across western Eurasia. I address the two most straightforward aspects of such a methodology—those involving multivariate measures of technological similarity among lithic assemblages (Scerri et al. 2014a, 2016), on the one hand, and improved chronological resolution, on the other—before turning to an approach for evaluating the systemic ecological and behavioral context of technological conservatism and innovation.

Multivariate Statistical Approaches to Conservatism and Innovation in Late Pleistocene Hunter-Gatherer Social Networks

Previous efforts at characterizing Middle Paleolithic (MP) and Middle Stone Age (MSA) interassemblage variability point toward promising approaches, building on Tostevin's pair-wise assemblage-comparison methods (2003, 2007,

2013). Focusing on the statistical methodology itself, Hovers and Raveh (2000) have analyzed MP assemblages from successive stratigraphic layers in Qafzeh Cave, Israel, demonstrating how dissimilarity plots—generated through multidimensional scaling—can also include arrays (shown as arrows), whose orientation and length from the centroid quantify how particular attributes correlate with the sample assemblages (see also Hovers 2009). Focusing on attribute analyses of North African MSA assemblages, Scerri et al. (2014a) have taken a geographically wider, spatially explicit approach, illustrating how multivariate comparisons among the sampled assemblages can be integrated with paleoenvironmental data, in order to test and refine hypotheses about hunter-gatherer mobility, social networks, and restricted paths to interaction. These analytical approaches are entirely applicable to studying the IUP and other EUP technocomplexes. They can render the bridge between theory and method more robust. A more comprehensive multivariate analysis of assemblage similarity, within and between currently defined technocomplexes, has the potential to explain more fully the mosaic MP-UP transition in western Eurasia. Ongoing radiometric dating work will continue to refine site chronologies. With better resolved mapping of technological variability in time and space, it should be possible to distinguish local, gradual technological change from abrupt innovation-adoption—that is, archaeological first-appearance of technologies that are relatively dissimilar from those in preceding occupations and neighboring areas. With a sufficiently large assemblage sample at a western Eurasian geographic scale, it should be possible to trace the spatial spread of innovative technological practices, reflected in particular attributes or correlated attribute sets, encompassing not only core initiation, management, and discard, but also blank modification and tool-use. Such an analysis could adopt "wombling" tests, which involve Monte Carlo simulation to randomize geographic distances, revealing significant transition zones. (Such methods have been primarily developed in spatial population genetic and epidemiological studies [Barbujani and Sokal 1990; Oden et al. 1993; Lu and Carlin 2005; Liang et al. 2009].)

Of course, in order to evaluate gradual local change, abrupt innovation-adoption and geographic spread of novel technologies that fall within the IUP technocomplex, it would even be desirable to expand the assemblage sample to represent the range of LMP and EUP technological traditions in western Eurasia in the ca. 55–35 ka timeframe. Because of the complexity likely involved in the mosaic MP-UP transition, zones of significant chronological or spatial variability could then be considered in paleoenvironmental and behavioral-ecological context, in order to investigate possible pressures shaping innovation or conservatism.

Methodological and Sampling Challenges to Chronology-Building

As noted above, this effort would require ongoing work to expand and refine chronological databases. Table 9.2 (see above) summarizes the well-dated EUP technocomplexes from the Levant and Europe, including those relying on ^{14}C and thermoluminescence (TL) results from studies specifically focused on sample pretreatment and measurement challenges near the older limits of accurate radiocarbon dating. It remains vital to continue improving the database of high-quality radiometric dates. The field-sampling and laboratory challenges for obtaining accurate ^{14}C dates on bone, charcoal, and shell have been recently discussed (Douka et al. 2010; Rebollo et al. 2011; Higham et al. 2014; Bosch et al. 2015; Stutz et al. 2015; Talamo et al. 2016; Alex et al. 2017; Devièse et al. 2017; see also Reynolds 2020). To give one example, I did not add to Table 9.2 recent charcoal dates associated with Early Baradostian archaeological assemblages from the Zagros foothills and Persian Plateau. Observations on the lithic assemblages from the Zagros foothills and Persian Plateau provide further—albeit qualified—support for a mosaic, complex MP-UP transition scenario, but it is not yet strongly established that Early Baradostian and related occupations significantly predate 40 ka. Assemblages that may broadly be assigned to the Early Baradostian technocomplex overlie Mousterian deposits at Shanidar Cave and Warwasi Rockshelter (Olszewski and Dibble 1994; Olszewski 1999; Conard and Ghasidian 2011; Olszewski 2009; Otte et al. 2011; Tsanova et al. 2012; Tsanova 2013). Recently published AMS ^{14}C dates on charcoal samples, although based on careful sampling and pretreatment protocols, reveal persisting challenges over preservation (Becerra-Valdivia et al. 2017). Without a sufficient sample of high-quality radiometric dates, we cannot yet robustly warrant placing EUP assemblages from the Zagros and Persian Plateau region in an accurate, detailed chronological framework. This is only to say that compelling sampling and methodological considerations remain for further improving the radiometric database for MP-UP transition contexts.

Measuring Economic and Socio-political Factors in Innovation Adoption

Perhaps my most substantial, broader methodological concern is to operationalize the theoretical expansion of Tostevin's BACT framework, via a more thorough, multi-scalar consideration of wider environmental and behavioral contexts in which blanks were produced, exchanged, modified, curated, and discarded. What were the mobility, foraging and task diversity, group-size and fission-fusion patterns, and normative social judgment practices that jointly influenced stoneworking and tool-use behaviors? Figure 9.4 provides a graphic, *hypothetical* model—and here, I underscore that it is a schematic model for developing testable hypotheses, based on general distillations of technological variability inferred in a wide range of MP and EUP assemblages—that largely relies on BACT's focus on core management, blank production, and retouch intensity (Tostevin 2013), while adding more explicit focus on retouched tool and point diversity. Here, I outline the three main dimensions in this graphical model, discussing how MP and EUP technocomplexes appear to vary.

The "Core Management Steps" axis highlights variation in the sometimes-arbitrary intricacy of a core-initiation, management, and overall exploitation practice. In general, initiating, orienting, and managing a blade(let) core discarded with one striking platform and single-fronted removal surface—that is, a core-exploitation pattern so important in Early Ahmarian assemblages in the Levant—requires fewer steps than an opposed-platform Nubian or IUP core (Goring-Morris and Davidzon 2006; Meignen 2012; Shea 2011, 2013a; Tostevin 2013; Kuhn and Zwyns 2014; Goder-Goldberger et al. 2016; Belfer-Cohen and Goring-Morris 2017; Goring-Morris and Belfer-Cohen 2018). Other MP and MSA traditions vary in complexity. Centripetal Levallois or radial flaking management strategies—which dominate core exploitation patterns commonly employed in the MP occupation of Qafzeh Cave, the earlier Mousterian in Italy, and many typical Mousterian or Charentian sites—involve fewer steps than do bifacial or foliate leaf point traditions (Kuhn 1995; Hovers 2009; Delagnes and Rendu 2011; Faivre et al. 2014, 2017).

The "Blank Provisioning" axis accounts for assemblage-level variability in blank thickness, core-flake ratios, and retouch intensity, reflecting tradeoffs in provisioning mobile individuals versus encamped groups with utilizable flakes (Kuhn 1992, 1995; Stiner and Kuhn 1992; Hovers 2009). In general, centripetal Levallois and radial flaking approaches yield fewer, thicker flakes, which may be

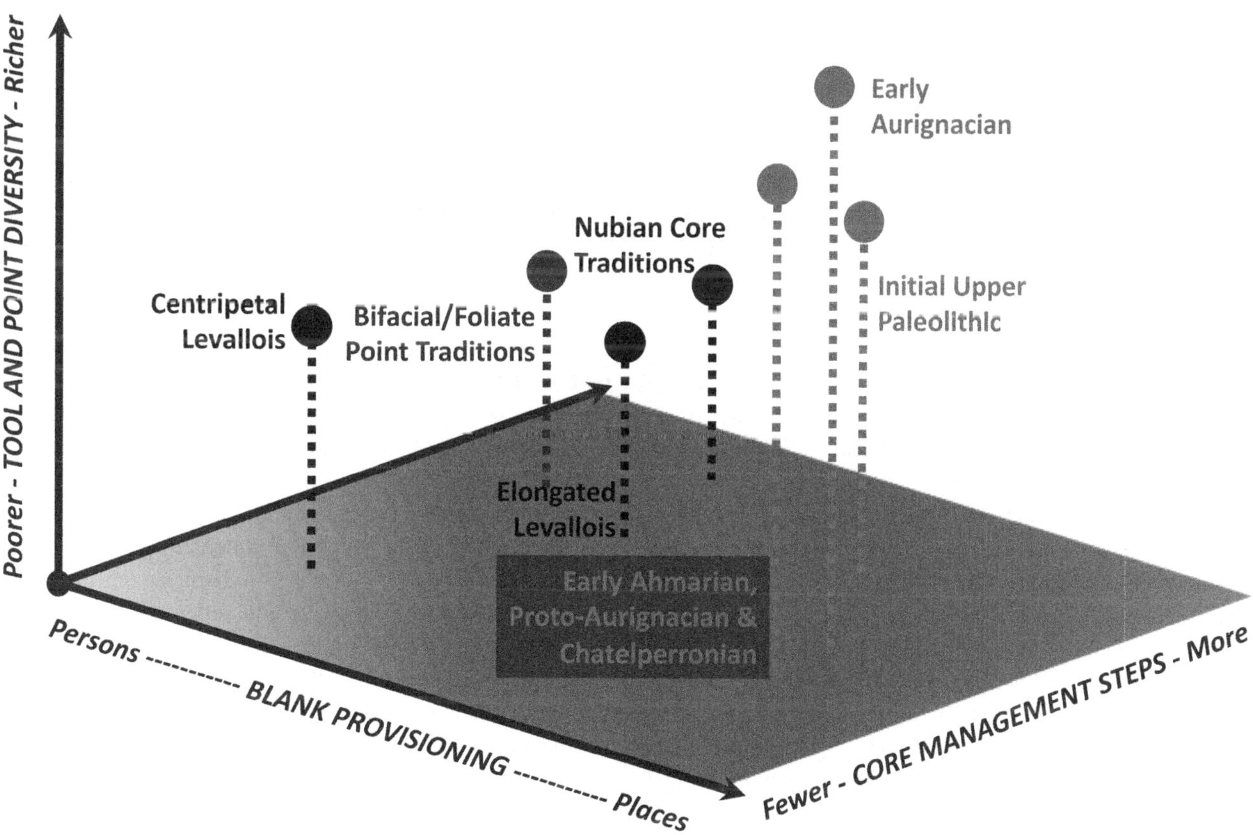

Fig. 9.4 Graphical schematic model of technological variability in Middle Paleolithic and early Upper Paleolithic technocomplexes. In general, EUP technologies exhibit a greater emphasis on producing thinner, longer blanks to provision mobile camp occupants; an increase in formally retouched implement diversity; and a tendency toward marginally reducing the complexity of core reduction practices

efficiently carried and rejuvenated prior to discard (Kuhn 1995). Prismatic blade production, laminar and convergent Levallois strategies, and bidirectional flaking can yield a larger number of thinner flakes, more efficiently generating sharp edges and points (Kuhn 1995; Shea 2011). In a range of publications, Kuhn has emphasized how this can give us insight into human fission-fusion patterns and land-use strategies, especially in Paleolithic contexts where social-network scales and residential mobility strategies varied well outside the bounds of familiar ethnographic cases (1991, 1995; see also Stiner and Kuhn 1992). It is clear that western Eurasian MP assemblages already exhibited complex variability in blank-provisioning patterns—over time and space, from ca. 200 to 40 ka (Bar-Yosef and Kuhn 1999; Delagnes and Rendu 2011; Shea 2011; Kuhn and Zwyns 2014; Faivre et al. 2017). Moreover, some technocomplexes included hybrid, hierarchically organized strategies—including the use of bifaces as elongated, pointed core-tools—flexibly provisioning individuals with portable, resilient cutting edge, while also having the capacity to produce thin, sharp blanks (Conard and Bolus 2003; Teyssandier 2008; Flas 2011; Meignen 2012; Tostevin 2013). Others encompassed shifting strategies. In such instances, as the core shrinks in size, a focus on detaching thinner elongated blanks—whether obtained through a uni- or bidirectional Levallois technique or prismatic blade production—can give way to producing fewer, thicker flakes (Dibble 1995; Hovers 2009; Tostevin 2013; Shea 2016). Many MP and UP technocomplexes variably add bipolar percussion on flakes, carinated burins, and similar strategies to provision places expediently with sharp bladelets, regardless of the predominant free-hand core-management and blank-production practices (Tsanova 2006; Stutz et al. 2015; Kadowaki 2018; Shea et al. in press). It is important to point out that the initially geographically patchy EUP abandonment of bifacial and Levallois core-management strategies—broadly alongside the adoption of diverse prismatic blade(let)-core approaches (narrow or broad-fronted, carinated, multiplatform, and pyramidal)—tended to go hand-in-hand with a substantial shift toward provisioning places with thin, sharp-edged or pointed blanks (Shea 2013b).

Finally, the "Tool and Point Diversity" axis reflects the relative abundance and techno-morphological richness of retouched artifacts. This gets at one of the most conspicuous technological shifts observed in the western European MP-UP transition, with formal endscrapers, burins, and point/knife forms becoming much more common in EUP assemblages. These forms come to complement truncations, denticulated and notched flakes, and flakes with continuous lateral or transverse retouch—that is, those formal tools classified by Bordes as sidescrapers and transverse scrapers, which are common in many western Eurasian MP assemblages. Normative EUP retouched tool and point production was likely related to an increased emphasis on provisioning places with more complex gear, including compound implements that could be distributed and utilized in small cooperative groups, mobilized for logistical provisioning trips or in-camp task activities (Kuhn and Stiner 2006). In turn, the up-front investment in task-specific gear—including wooden spear shafts and foreshafts, endscraper handles, and various free-hand tools, from bone awls and points, to burins, knives and denticulates—would have been favored in socio-ecological contexts involving marginally longer mobile-camp stays, a wider range of co-residential group sizes, higher richness in logistical provisioning activities, and more diverse in-camp task patterns. Kuhn and Stiner (2006) have argued that these EUP developments would have co-evolved bioculturally with practically structured social roles and identities, likely with gender shaping division of labor in hunting, gathering, raw-material provisioning, and on-site tasks (e.g. tool-making and material processing, food preparation and distribution, shelter construction, etc.). Such normative structuration in co-residential task and logistical trip practices appears to have begun in certain regions in the later Middle Paleolithic ca. after 65–55 ka (Stiner and Kuhn 1992; Madella et al. 2002; Henry 2003; Alperson-Afil and Hovers 2005; Speth et al. 2012; Speth 2013; Hartman et al. 2015; Estalrrich et al. 2017), increasing gradually in organizational resilience and complexity over many millennia in the EUP, ca. 45–30 ka (Kuhn et al. 2009; Kuhn 2013).

While Fig. 9.4 only estimates the relative positions of assemblages usually assigned to diverse MP and EUP technocomplexes, the work of Tostevin, Eren, Dibble, Kuhn, Hovers, Scerri and others demonstrates that the variability mapped therein can be defined and measured in a logically consistent, more detailed, rigorous, and reproducible way (Dibble 1995; Eren et al. 2005; Hovers 2009; Kuhn 2013; Tostevin 2013; Scerri et al. 2014a, 2016; Groucutt 2020; see Douze et al. 2020, for additional methodological approaches). Here, this graphical approach to hypothesis formation does suggest some new lines of argument and inquiry. Depending on cultural and environmental context, variation in core-management complexity—to focus on one dimension of technological variability—could reflect heterogeneity in blank demand, but it could also reflect socio-politically imposed, ritualized, symbolic costs on learning. Initiating, preparing, managing, and reducing a Nubian Levallois core, for example (Goder-Goldberger et al. 2016; Groucutt 2020; Will and Mackay 2020), could have been an important performative activity in small-group mobile camp contexts, shaping the knapper's identity within her or his social network, while also influencing the network's constituent relationships, via aesthetic judgment and exchange of blade and convergent Levallois point products. In a late MSA, LMP and EUP biocultural niche that must have generally involved sufficient fertility and adult survivorship rates to support long-term demic diffusion, complex core-management practices can be hypothesized as one strategy to mediate land-tenure claims and long-distance alliance-building.

Back to the Initial Upper Paleolithic: Reconsidering Economic and Sociopolitical Factors in the Adoption or Spread of IUP Technologies

The graphic visualization approach in Fig. 9.4 suggests a spectrum of alternative hypotheses concerning those environmental and behavioral contexts that shaped and were shaped by IUP technological development, adoption, and spread. Regardless of which region may lay claim to Initial Upper Paleolithic first adoption, recent radiometric dates and associated archaeological data help to clarify the behavioral and environmental context that may have favored or constrained the use of IUP technologies in certain parts of the Levant. We might see the IUP as a genuinely transitional blank-provisioning strategy, bimodally emphasizing the production of portable, thicker Levallois points and blades, alongside thinner, narrower, straight-profiled blades and bladelets (Meignen 2012; Kuhn and Zwyns 2014). Yet, we might see the same technocomplex as reflecting an arbitrarily complex, sequence-dependent, conservatively reproduced norm, in which—once adopted in a social network—IUP (and occasionally, at least in the southern Levant, Nubian) core reduction was performed and socially judged through culturally reproduced, ritualized practices, in small-group camp contexts. Both factors—intensification of blank provisioning for on-site tasks and logistical mobility, on the one hand, and ritualized elaboration of sequence-dependent practices, on the other—may have played out in the Levant, ca. 45 ka, in a broader biocultural context of long-term, sporadic demic diffusion, admixture, and marginally higher human population densities across western Eurasia.

As discussed above, IUP technologies in the Levant may actually have been adopted via long-distance social ties with forager networks in Central Europe. The IUP assemblages from Brno Bohunice have been dated by thermoluminescence assays to ca. 52–47 ka (Richter et al. 2008). Other geoarchaeological data are consistent with this early age for the Central European MP-UP transition, although IUP technologies appear to have been culturally transmitted and exploited for many millennia; some Bohunician assemblages also likely date to the 45–40 ka interval (Svoboda and Bar-Yosef 2003; Tostevin and Skrdla 2006; Tostevin 2013). Technologically and formally similar lithic assemblages from Ksar Akil Units XXV–XXI and Üçağızlı Cave Layer I have been dated to ca. 45–40 ka (Kuhn et al. 2009; Bosch et al. 2015; Stutz et al. 2015). If long-distance alliance networks mediated local responses to demic diffusion, admixture, and marginally rising population, we might expect to see an arbitrarily complicated core exploitation and management strategy that—in cultural practice—constituted a hard-to-fake signal of cooperative learning, participation, and familiarity (indeed, a kind of performed metaphorical kinship) in intimate social contexts.

It must also be observed that the absolutely oldest calibrated dates for any EUP context in the Levant come from layers associated with Early Ahmarian assemblages: those from the southern Levantine Mediterranean-zone sites of Kebara and Manot Caves, dating to ca. 47–45 ka (Rebollo et al. 2011; Alex et al. 2017). In this setting, relative to Early Ahmarian camp settings (including Mughr el-Hamamah in the Jordan Valley, ca. 45–39 ka [Stutz et al. 2015; Stutz and Nilsson Stutz 2017; Shea et al. in press]), IUP technologies would have been adopted in Levantine foraging territories with more rugged terrain or lower rainfall—that is, in a broad arc around the southern Levantine Mediterranean vegetation zone and the warm, well-watered Jordan Valley. The Levantine IUP adoption zone would have ranged from the northern coastal Levant to the semi-arid el-Kowm Basin, Syria, the southern Transjordanian Plateau, and the Negev Desert (Fig. 9.5). In these regions with lower biomass productivity or higher mobility and resource-transport costs, logistical provisioning of thicker elongated blanks would have more economically complemented blade(let) production in residential camps.

IUP technologies do not appear to have been systematically exploited for blank production in Mediterranean vegetation zone sites (Stutz et al. 2015; Stutz and Nilsson Stutz 2017). At Kebara and Manot Caves, Early Ahmarian contexts—dated to ca. 45 ka—are stratigraphically overlain by Levantine Aurignacian deposits (Bar-Yosef et al. 1996; Tostevin 2003; Rebollo et al. 2011; Tostevin 2013; Alex et al. 2017). In contrast, IUP-associated layers from the northern Levant and southernmost semi-arid zones are most often succeeded by Early Ahmarian occupations (see Fig. 9.5) (Marks 1977a, 1983; Bergman 1988; Monigal 2003; Fox and Coinman 2004; Kuhn et al. 2009; Boëda et al. 2015; Bosch et al. 2015). According to the best-dated stratified contexts—those from Ksar Akil Rockshelter and Üçağızlı Cave (Kuhn et al. 2009; Bosch et al. 2015)—IUP core management strategies gave way to Early Ahmarian ones ca. 42–39 ka. Thus, available high-quality, calibrated ^{14}C dates indicate that Mediterranean vegetation zone Early Ahmarian and surrounding IUP sites were broadly contemporaneous, ca. 45 ka. A vital theoretical point emerges here. Seemingly well-defined, penecontemporaneous, neighboring technocomplexes—in this case, the Early Ahmarian and the

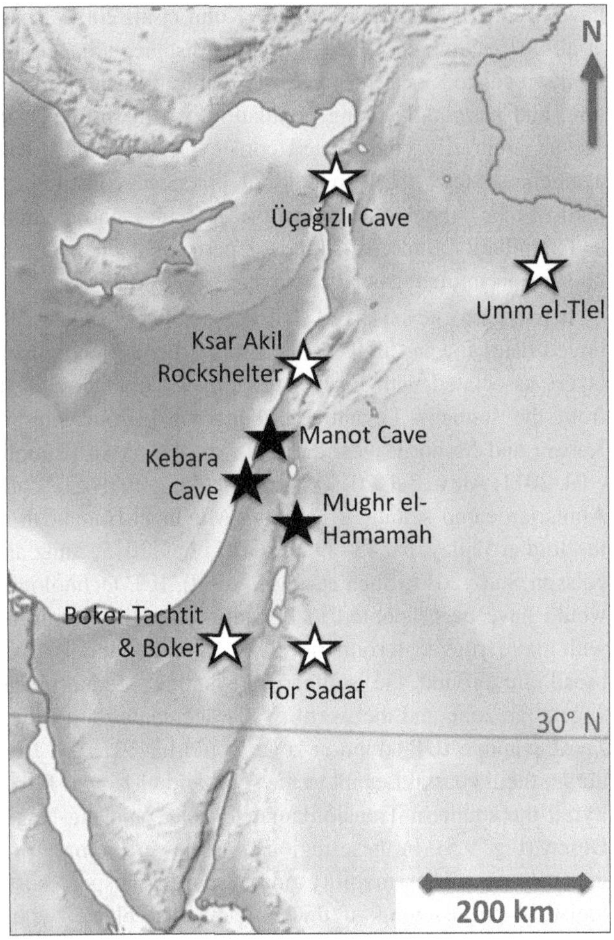

Fig. 9.5 Key early Upper Paleolithic sites in the Levant with stratigraphic and radiometric data relevant to the first appearance of EUP technologies and their long-term development, adoption, and spread. The geographic distribution of penecontemporaneous Early Ahmarian (black stars) and Initial Upper Paleolithic (IUP) technocomplexes (white stars), ca. 45–40 ka, requires us to reconsider the environmental, socio-political, cultural, and technological changes that constituted the Levantine Middle-Upper Paleolithic (MP-UP) transition. Base map from Wikimedia Commons user Fulvio314, licensed under a Creative Commons 3.0 unported license (CC-BY)

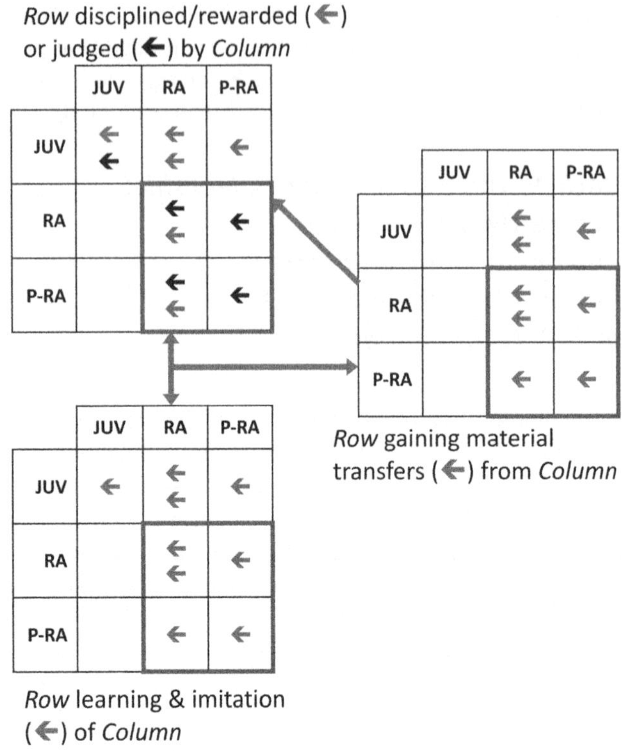

Fig. 9.6 Schematic diagram of how social judgment, learning, and material and nutritional transfers structure and are structured by the human life-history strategy, in which the juvenile (JUV) stage is unusually long; the reproductive adult (RA) stage is dependent on cooperation or altruism from other members of one's social network; and the unusually well-defined, extended post-reproductive adult (PR-A) stage, which contributes to the mobilization of collective action, the cultural reproduction of normative practices, and the social distribution and cultural resilience of memory

Initial Upper Paleolithic in the Levant—could plausibly reflect two alternative stoneworking and tool-use technologies adopted by the same forager social networks, utilized across key ecological productivity, population density, and residential mobility gradients (see also Tryon and Ranhorn 2020, for a discussion of the role of raw material variability).

There is an instructive comparison with Bohunician IUP assemblage variability. As Tostevin (2013) highlights, some—but hardly all—Central European deposits associated with IUP technologies, broadly dated to ca. 50–40 ka, incorporate bifaces with flaking patterns similar to late Middle Paleolithic Micoquian bifaces or Szeletian leaf points (Svoboda and Bar-Yosef 2003; Tostevin and Skrdla 2006; Tostevin 2013). It is possible that the penecontemporeneity of named Szeletian and Bohunician technocomplexes also reveals that a regional, mobile forager network employed alternative stoneworking and blank-use strategies across environmental gradients. Under this scenario, IUP core exploitation and management strategies—on the one hand—and biface knapping, curation, and resharpening strategies—on the other—would still be culturally reproduced in small-group camp settings. Yet, as structured and structuring gesture systems that could have jointly constituted—and been constituted by—a widespread, mobile forager network, "Bohunician" IUP core reduction and "Szeletian" biface production and curation would have formed grounded semiotic media for cooperative bond-formation, ritualized performance, and social judgment. However, the co-adoption of these strategies in a single Eastern-Central European mobile forager network would involve a kind of cultural intensification of social network management and land-tenure, requiring environmentally context-dependent code-switching. Such socio-political, alliance-mediated innovations could also build cultural-institutional resilience in the face of those demographic fluctuations, whether driven by climatic oscillations (Staubwasser et al. 2018) or sporadic AMH range expansion and admixture with local populations of mainly Neanderthal ancestry (Fu et al. 2016).

Conclusion: Why So Much Change Between 50 and 40 ka?

Material culture variability is shaped not only by teaching and learning (d'Errico and Banks 2015; Gärdenfors and Högberg 2017). It also reflects enchronic discourse among socially networked adults—often involving judgment, gossip, self-discipline, and display of bodily cultural competence (Stutz 2009, 2014a; Enfield 2013; Sinha 2015a, b). Among mobile Late Pleistocene foragers, diverse small-group social settings would have constituted the ground for cultural transmission and learning, influencing opportunities for dependent juveniles and more tenuously socially integrated adults (e.g. recent immigrants to a region) to master local knowledge, generating embodied and social capital distributed across networks of vertical (adult-juvenile) and horizontal (adult-adult) bonds. This would also have been the setting in which core cooperative group members, distant or fictive kin, and prestigious visitors could have introduced, advocated for, and taught innovative technologies.

As schematically illustrated in Fig. 9.6, the intimate-scale social interactions so important for cultural transmission and innovation adoption involve complex feedbacks among: social judgment, reward, and disciplining; learning and imitation; and material and energy transfers. Moreover, these enchronic social dynamics are structured by the derived human life history strategy, which is mainly constituted by symbolic and material transfers from reproductive-age adults (*R-A*) and post-reproductive adults (*PR-A*) to juveniles (*JUV*) who grow up over an extended developmental period (Kaplan and Robson 2002, 2009; Kaplan et al. 2009, 2010; Stutz 2009; Stutz 2014b). In this chapter, an overarching theoretical concern—in delving into the dynamics of cultural reproduction and change that shaped late Middle and early Upper Paleolithic archaeological assemblages—has been to tie these intricate, intimate-scale, temporally short-term social dynamics to much larger-scale niche construction and biocultural evolutionary processes (see Fig. 9.2). My point of departure has been set squarely at a long-term evolutionary level of resolution. As reviewed here, recent paleogenomic and radiometric data confirm a long history of AMH-Neanderthal biocultural interaction—occasionally involving successful family formation, leading to substantial admixture—prior to the final phase of population-biological turnover, ca. 50–40 ka. It was at this point, ca. 40 ka, that Neanderthal skeletal phenotypes and alleles became quite rare, or even disappeared. Ancient genomic data track a long-term, likely drift-driven decline in derived Neanderthal alleles in the descendants of admixed AMH-Neanderthal late Middle Paleolithic and EUP groups (Petr et al. 2019) (Fig. 9.7). In other words, this was when Neanderthals—as a plausibly defined western Eurasian metapopulation, possibly in the process of evolving a reproductive barrier with geographically expanding AMH groups—went extinct (Fu et al. 2014; Hublin 2017).

In tying the broader population-biological scale to the intimate social scale of cultural reproduction, I have stressed that the key paleoanthropological questions should not be posed as macroevolutionary ones, dealing with specific-level range expansion, competitive exclusion or extinction. Rather, the focus must be on niche construction and biocultural evolutionary processes in complex dynamical systems, critically constituted by varied and open hominin metapopulations (Brantingham et al. 2004b; Kuhn 2013; Stutz 2019). During the Late Pleistocene, how did variable dispersal, migration, and admixture behaviors interact with local niche construction, extinction and biocultural adaptation? More to the point, what were the ecological and socio-behavioral conditions favoring Neanderthal morphologies and activity patterns across most of Western Eurasia, even in the face of gene flow with AMH groups, until ca. 50 ka? And after this juncture, why did the situation appear to have flipped toward favoring AMH morphologies, likely with at least a marginal reduction in mobility and activity levels (Snodgrass and Leonard 2009; Froehle et al. 2013; Stutz et al. 2015; Goldfield et al. 2018; Stutz 2019)?

One potential answer is that Neanderthal and AMH admixture occurred in a biocultural setting in which hunter-gatherer social networks—and the ecological clines their territories covered—were gradually transformed by cultural and economic intensification. The role of economic intensification in the MP-UP transition has been preliminarily explored (O'Connell 2006; Morin 2008), but I suggest that we have barely begun to elucidate the nature of intensification and productivity-raising innovation in highly mobile hunter-gatherer social networks (Stutz 2009, 2012). In this chapter I have built on Tostevin's BACT framework to derive new hypotheses for the geogaphically mosaic pattern of EUP innovation, which emerged mainly in the 45–40 ka time interval. Critical to my argument—centered on the theme that diverse EUP technologies were developed and adopted in a joint process of cultural and economic

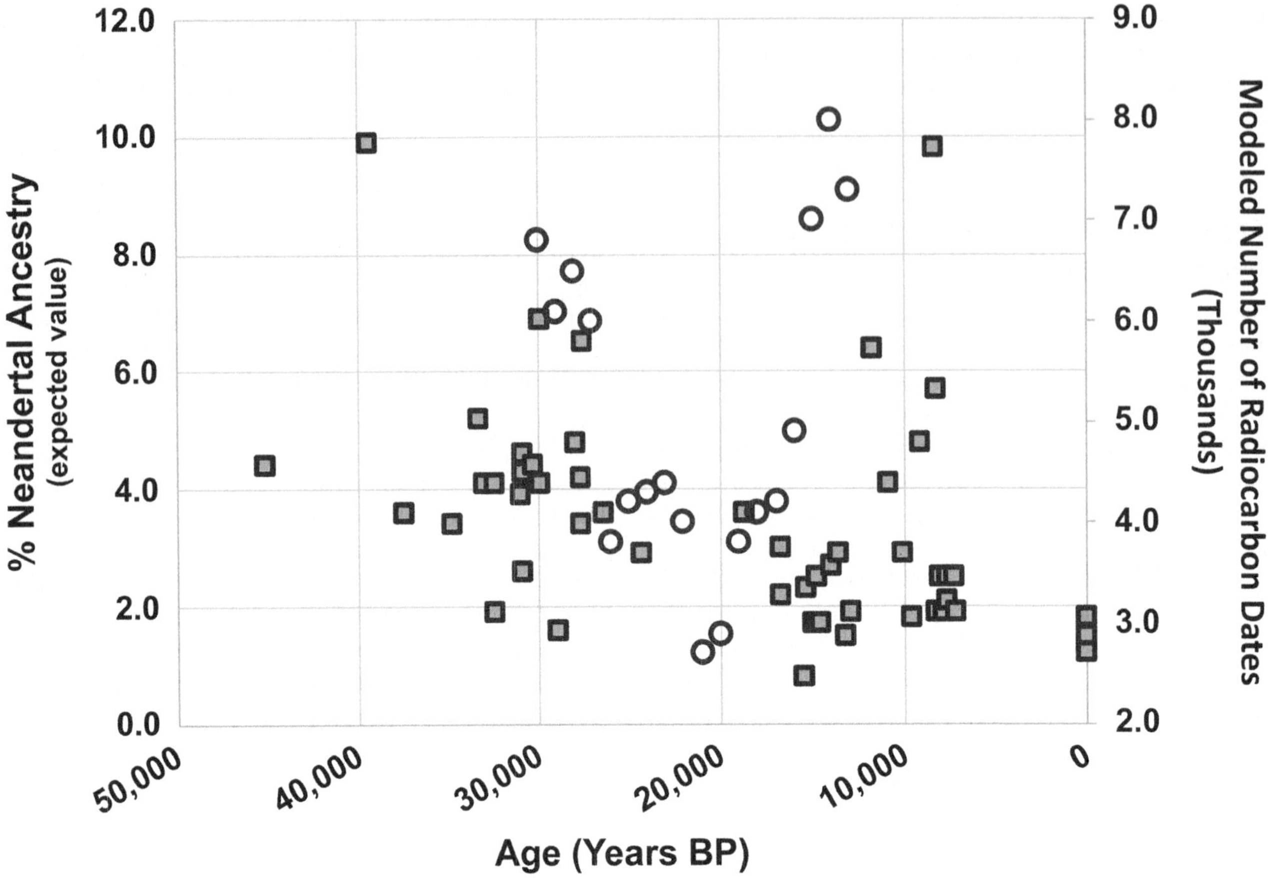

Fig. 9.7 The level of admixed Neanderthal ancestry in anatomically modern human (AMH) individuals (squares) gradually declined in Europe, as population—measured by radiocarbon date proxy modelling (circles)—also declined, from ca. 45 ka to the Last Glacial Maximum (LGM), ca. 23–18 ka. Sporadic but significant interaction and family formation occurred before and during the MP-UP transition. The apparent rebound in Neanderthal admixture levels in the Tardiglacial and early Holocene periods, ca. 15–7 ka, was likely driven by drift and migration, as hunter-gatherer populations recovered after the LGM and early farming populations began to expand (see Lazaridis et al. 2016). Paleogenomic and demographic-proxy radiocarbon model data from published sources (Tallavaara et al. 2015; Fu et al. 2016)

intensification—is the fact that, in most regional or local contexts, *admixed Neanderthal-AMH metapopulations constituted and negotiated the forager social networks that made the MP-UP transition.*

More specifically, I have suggested the hypothesis that at least some archaeologically defined EUP technocomplexes do not directly correspond to the territories that persistent forager social networks held onto. Rather, distinct technocomplexes—often discussed as chronologically succeeding or geographically neighboring one another, such as the IUP and Early Ahmarian in the Levant or the Szeletian and Bohunician in Central Europe—are suggested to have been constituted by technologies culturally reproduced *within* the same regional forager networks. Lithic attributes defining these respective technocomplexes are found in different sites, but within the same regions, in the ca. 50–40 ka timeframe. I have proposed that these technological alternatives were used by the same highly mobile hunter-gatherer networks, as they exploited ecological gradients via alternative mobility and blank-production and curation strategies. While currently only a hypothesis, cultural intensification/techno-ecological code-switching may be a variation on a theme that also explains regional technological diversification in the late Middle Paleolithic of southwestern Europe (Monnier and Missal 2014; Ruebens and Wragg Sykes 2016; Faivre et al. 2017). Moreover, most or all of the earliest western Eurasian technocomplexes with diagnostic EUP features—volumetric blade production, retouched blade(let)s, and tool forms such as burins and endscrapers—may have been part of longer-term LMP trend, in which networked populations constituted substantially by Neanderthal ancestry maintained their highest densities in ecologically richer refugia, while extended-kin and allied subgroups regularly exploited lower-productivity zones, stretching over several hundred kilometers or more. In the face of AMH demic diffusion and mosaic admixture with Neanderthals, the MP-UP archaeological transition may significantly reflect renegotiated social ties and land-tenure practices, leading to the fragmentation of such Neanderthal super-territories that extended over major ecological productivity clines, for instance between low-elevation Mediterranean zones and higher elevation or more open, cold habitats.

This social-network/land-tenure renegotiation hypothesis has the theoretical advantage of accounting for why the MP-UP transition, ca. 45–41 ka, followed a regional mosaic pattern, while subsequent millennia, ca. 41–38 ka, are marked by the geographically extensive, successive adoption of retouched forms (el-Wad/Font-Yves points and Dufour bladelets of the Dufour subtype) associated with Early Ahmarian, Proto-Aurignacian, and—quite likely—Early Baradostian technocomplexes (Mellars 2006a; Olszewski 2009; Bar-Yosef and Belfer-Cohen 2010; Hoffecker 2011; Banks et al. 2013a; Stutz 2019). The roughly contemporaneous production, use, and discard of inversely retouched Dufour bladelets at Boker (Negev Desert) and Les Cottés (western France), respectively associated with Early Ahmarian and Proto-Aurignacian core management technologies (Marks 1977b; Monigal 2003; Teyssandier 2006), indicate that long-distance alliances, maintained stepwise across a matrix of the relatively reduced mobility territories held by regional social networks, may have been especially important around 40 ka. By the time that Proto-Aurignacian technologies were adopted across most of western and southern Europe—likely in the centuries prior to 40 ka (Banks et al. 2013a, b; Schmidt et al. 2013)—it may be predicted that mobile territory sizes had already shrunk, base-camp stays had become marginally longer, and in-camp task and out-of-camp logistical trips were more varied. Adults and juveniles alike would have experienced lower average daily calorie expenditures, while more often interacting with larger co-residential groups. This constructed eco-cultural niche would have favored AMH body proportions and marginally higher fecundity (Stutz et al. 2015; Stutz 2019). It may have also favored neuromotor and social cognition capacities more frequent in African populations. Regardless of how cold pulses restricted exploitation of colder habitats, the biocultural feedbacks would have been significant. Natural selection against Neanderthal body proportions and cranio-dental morphology would have become established, while demic diffusion would have accelerated, further raising African ancestry in EUP populations—at the expense of Neanderthal genealogical background—ca. 40–35 ka (see Fig. 9.7).

Testing this proposal requires integration of many lines of evidence across western Eurasia, involving a comprehensive effort to develop research questions, assemblage and radiometric-dating sampling schemes, refined attribute definitions, and statistical approaches that can trace and consider fine-scale local and regional changes in logistical task provisioning, on-site task diversification, regional social network management (via long-distance raw material and blank transport), and intimate-scale teaching and learning, ritualized performance, and norm enforcement. A spatially and chronologically explicit multivariate statistical methodology (e.g. Scerri et al. 2014a, 2016) can effectively build on Tostevin's BACT framework, applied to a much larger sample of LMP and EUP lithic assemblages, with the study augmented by a further refined focus on diversity in retouched tool technology—as formal points, knives, barbs, hafted endscrapers, burins, and truncations appear to be important foraging and in-camp task-related EUP innovations, across the geographically mosaic technocomplexes that define the MP-UP transition in western Eurasia. Rather than pushing us toward particularistic explanations, this work would provide sufficient resolution to identify biocultural evolutionary and niche construction dynamics among human demography, ecological conditions, foraging institutions, and social strategies for alliance management and intimate-scale cooperation. Such inquiry can clarify the evolutionary conditions and trends that alternately favored Neanderthal and AMH life history strategies, body proportions, cranio-facial features, and—quite likely—related neuroanatomical, activity-related and metabolic adaptations.

Acknowledgments I express my sincere gratitude to Huw Groucutt for his editorial encouragement and critical but constructive comments. Three anonymous reviewers provided critical feedback, and I have sought to improve the original manuscript by considering their comments as thoroughly as possible. I owe a substantial intellectual debt to Gil Tostevin, whose theoretical work and methodological care have inspired me to try standing on his shoulders. With all of the risk that entails. The Mughr el-Hamamah project team—supported by the funding we have received from the National Science Foundation, the Wenner-Gren Foundation for Anthropological Research, the LSB Leakey Foundation, Emory University, the Gregory-Rackley Faculty Development Award, and our crowdfunding backers (http://experiment.com/paleoplants)—has helped me integrate the ideas and data considered in this chapter. With deepest respect and warmth, I dedicate this chapter to to the memory of Ofer Bar-Yosef, a most extraordinary mentor.

References

Alex, B., Barzilai, O., Hershkovitz, I., Marder, O., Berna, F., Caracuta, V., et al. (2017). Radiocarbon chronology of Manot Cave, Israel and Upper Paleolithic dispersals. *Science Advances, 3*(11), e1701450. https://doi.org/10.1126/sciadv.1701450.

Alperson-Afil, N., & Hovers, E. (2005). Differential use of space in the Neandertal site of Amud Cave, Israel. *Eurasian Prehistory, 3*, 3–22.

Banks, W. E., d'Errico, F., & Zilhão, J. (2013a). Human–climate interaction during the Early Upper Paleolithic: Testing the hypothesis of an adaptive shift between the Proto-Aurignacian and the Early Aurignacian. *Journal of Human Evolution, 64*(1), 39–55. https://doi.org/10.1016/j.jhevol.2012.10.001.

Banks, W. E., d'Errico, F., & Zilhão, J. (2013b). Revisiting the chronology of the Proto-Aurignacian and the Early Aurignacian in Europe: A reply to Higham et al.'s comments on. *Journal of Human Evolution, 65*(6), 810–817. https://doi.org/10.1016/j.jhevol.2013.08.004.

Barbujani, G., & Sokal, R. R. (1990). Zones of sharp genetic change in Europe are also linguistic boundaries. *Proceedings of the National Academy of Sciences, 87*(5), 1816–1819.

Bar-Yosef, O. (2000). The impact of radiocarbon dating on old world archaeology: Past achievements and future expectations. *Radiocarbon, 42*(1), 23–39.

Bar-Yosef, O., Arnold, M., Mercier, N., Belfer-Cohen, A., Goldberg, P., Housley, R., et al. (1996). The dating of the upper Paleolithic layers in Kebara Cave, Mt Carmel. *Journal of Archaeological Science, 23*(2), 297–306. https://doi.org/10.1006/jasc.1996.0028.

Bar-Yosef, O., & Van Peer, P. (2009). The Chaîne Opératoire approach in Middle Paleolithic archaeology. *Current Anthropology, 50*(1), 103–131. https://doi.org/10.1086/592234.

Bar-Yosef, O. (2002). The upper Paleolithic revolution. *Annual Review of Anthropology, 31*(1), 363–393. https://doi.org/10.1146/annurev.anthro.31.040402.085416.

Bar-Yosef, O., & Belfer-Cohen, A. (2010). The Levantine Upper Palaeolithic and Epipaleolithic. In E. A. A. Garcia (Ed.), *South-Eastern Mediterranean peoples between 130,000 and 10,000 years ago* (pp. 144–167). Oxford: Oxbow.

Bar-Yosef, O., & Bordes, J.-G. (2010). Who were the makers of the Châtelperronian culture? *Journal of Human Evolution, 59*(5), 586–593. https://doi.org/10.1016/j.jhevol.2010.06.009.

Bar-Yosef, O., & Kuhn, S. L. (1999). The big deal about Blades: Laminar technologies and human evolution. *American Anthropologist, 101*(2), 322–338.

Becerra-Valdivia, L., Douka, K., Comeskey, D., Bazgir, B., Conard, N. J., Marean, C. W., et al. (2017). Chronometric investigations of the Middle to Upper Paleolithic transition in the Zagros Mountains using AMS radiocarbon dating and Bayesian age modelling. *Journal of Human Evolution, 109,* 57–69. https://doi.org/10.1016/j.jhevol.2017.05.011.

Belfer-Cohen, A., & Goring-Morris, A. N. (2017). The Upper Palaeolithic in Cisjordan. In Y. Enzel & O. Bar-Yosef (Eds.), *Quaternary environments, climate change and humans in the Levant* (pp. 627–637). New York: Cambridge University Press.

Belfer-Cohen, A., & Hovers, E. (2010). Modernity, enhanced working memory, and the Middle to Upper Paleolithic record in the Levant. *Current Anthropology, 51*(S1), S167–S175. https://doi.org/10.1086/649883.

Benazzi, S., Slon, V., Talamo, S., Negrino, F., Peresani, M., Bailey, S. E., et al. (2015). The makers of the Protoaurignacian and implications for Neandertal extinction. *Science*, aaa2773. https://doi.org/10.1126/science.aaa2773.

Bergman, C. A. (1988). Ksar Akil and the Upper Palaeolithic of the Levant. *Paléorient, 14*(2), 201–210. https://doi.org/10.3406/paleo.1988.4468.

Blinkhorn, J., & Grove, M. (2018). The structure of the Middle Stone Age of eastern Africa. *Quaternary Science Reviews, 195,* 1–20. https://doi.org/10.1016/j.quascirev.2018.07.011.

Boëda, É., Bonilauri, S., Kaltnecker, E., Valladas, H., & Al-Sakhel, H. (2015). Un débitage lamellaire au Proche-Orient vers 40 000 ans cal BP. Le site d'Umm el Tlel, Syrie centrale. *L'Anthropologie, 119*(2), 141–169. https://doi.org/10.1016/j.anthro.2015.04.001.

Boehm, C. (2012). Ancestral hierarchy and conflict. *Science, 336* (6083), 844–847. https://doi.org/10.1126/science.1219961.

Boivin, N. L., Zeder, M. A., Fuller, D. Q., Crowther, A., Larson, G., Erlandson, J. M., et al. (2016). Ecological consequences of human niche construction: Examining long-term anthropogenic shaping of global species distributions. *Proceedings of the National Academy of Sciences, 113*(23), 6388–6396. https://doi.org/10.1073/pnas.1525200113.

Bordes, J.-G., & Teyssandier, N. (2011). The Upper Paleolithic nature of the Châtelperronian in South-Western France: Archeostratigraphic and lithic evidence. *Quaternary International, 246*(1–2), 382–388. https://doi.org/10.1016/j.quaint.2011.08.001.

Bosch, M. D., Mannino, M. A., Prendergast, A. L., O'Connell, T. C., Demarchi, B., Taylor, S. M., et al. (2015). New chronology for Ksâr 'Akil (Lebanon) supports Levantine route of modern human dispersal into Europe. *Proceedings of the National Academy of Sciences, 112*(25), 7683–7688. https://doi.org/10.1073/pnas.1501529112.

Bourdieu, P. (1977). *Outline of a Theory of Practice*. (R. Needham, Trans.). New York: Cambridge University Press.

Bourdieu, P. (1990). *The logic of practice*. Palo Alto: Stanford University Press.

Boyd, R., & Richerson, P. J. (1992). Punishment allows the evolution of cooperation (or anything else) in sizable groups. *Ethology and Sociobiology, 13*(3), 171–195. https://doi.org/10.1016/0162-3095(92)90032-Y.

Boyd, R., & Richerson, P. J. (2005). *The origin and evolution of cultures*. Oxford: Oxford University Press.

Brantingham, P. J., Kuhn, S. L., & Kerry, K. W. (2004a). *The early Upper Paleolithic beyond Western Europe*. Berkeley: University of California Press.

Brantingham, P. J., Kuhn, S. L., & Kerry, K. W. (2004b). On the difficulty of Middle-Upper Paleolithic transitions. *The early Upper Paleolithic beyond Western Europe* (pp. 1–13). Berkeley: University of California Press.

Carr, C., & Neitzel, J. E. (1995). *Style, society, and person: Archaeological and ethnological perspectives*. New York: Springer Science & Business Media.

Conard, N., & Ghasidian, E. (2011). The Rostamian cultural group and the taxonomy of the Iranian Upper Paleolithic. In N. Conard, P. Drechsler, & A. Morales (Eds.), *Between sand and sea: The archaeology and human ecology of Southwestern Asia* (pp. 33–52). Tübingen: Kerns Verlag.

Conard, N. J., & Bolus, M. (2003). Radiocarbon dating the appearance of modern humans and timing of cultural innovations in Europe: New results and new challenges. *Journal of Human Evolution, 44* (3), 331–371. https://doi.org/10.1016/S0047-2484(02)00202-6.

Conard, N. J. (2006). *When Neanderthals and modern humans met*. Tübingen: Kerns Verlag.

d'Errico, F., & Banks, W. E. (2015). The archaeology of teaching: A conceptual framework. *Cambridge Archaeological Journal, 25*(4), 859–866. https://doi.org/10.1017/S0959774315000384.

d'Errico, F., & Stringer, C. B. (2011). Evolution, revolution or saltation scenario for the emergence of modern cultures? *Philosophical Transactions of the Royal Society of London. Series B, Biological Sciences, 366*(1567), 1060–1069. https://doi.org/10.1098/rstb.2010.0340.

d'Errico, F., Zilhão, J., Julien, M., Baffier, D., & Pelegrin, J. (1998). Neanderthal acculturation in Western Europe? A critical review of the evidence and its interpretation. *Current Anthropology, 39*(S1), S1–S44. https://doi.org/10.1086/204689.

Dediu, D., & Levinson, S. C. (2013). On the antiquity of language: The reinterpretation of Neandertal linguistic capacities and its consequences. *Language Sciences, 4,* 397. https://doi.org/10.3389/fpsyg.2013.00397.

Dediu, D., & Levinson, S. C. (2018). Neanderthal language revisited: Not only us. *Current Opinion in Behavioral Sciences, 21,* 49–55. https://doi.org/10.1016/j.cobeha.2018.01.001.

Delagnes, A., & Rendu, W. (2011). Shifts in Neandertal mobility, technology and subsistence strategies in western France. *Journal of Archaeological Science, 38*(8), 1771–1783. https://doi.org/10.1016/j.jas.2011.04.007.

Devièse, T., Karavanić, I., Comeskey, D., Kubiak, C., Korlević, P., Hajdinjak, M., et al. (2017). Direct dating of Neanderthal remains from the site of Vindija Cave and implications for the Middle to Upper Paleolithic transition. *Proceedings of the National Academy of Sciences, 114*(40), 10606–10611. https://doi.org/10.1073/pnas.1709235114.

Dibble, H. L. (1995). Biache Saint-Vaast, level IIA: A comparison of analytical approaches. In H. L. Dibble & O. Bar-Yosef (Eds.), *The definition and Interpretation of Levallois variability* (pp. 93–116). Madison: Prehistory Press.

Douka, K., Hedges, R. M., & Higham, T. G. (2010). Improved AMS 14C dating of shell carbonates using high-precision X-ray diffraction and a novel density separation protocol (CarDS). *Radiocarbon, 52*(2), 735–751. https://doi.org/10.2458/azu_js_rc.v.3707.

Douka, K., Bergman, C. A., Hedges, R. E. M., Wesselingh, F. P., & Higham, T. F. G. (2013). Chronology of Ksar Akil (Lebanon) and Implications for the Colonization of Europe by Anatomically Modern Humans. *PLoS ONE, 8*(9), e72931. https://doi.org/10.1371/journal.pone.0072931.

Douze, K., Igreja, M., Rots, V., Cnuts, D., & Porraz, G. (2020). Technology and function of Middle Stone Age points. Insights from a combined approach at Bushman Rock Shelter, South Africa. In H. Groucutt (Ed.), *Culture history and convergent evolution: Can we detect populations in prehistory?* (pp. 127–141). Cham, Switzerland: Springer.

Ellis, E. C., Richerson, P. J., Mesoudi, A., Svenning, J.-C., Odling-Smee, J., & Burnside, W. R. (2016). Evolving the human niche. *Proceedings of the National Academy of Sciences, 113*(31), E4436–E4436. https://doi.org/10.1073/pnas.1609425113.

Enfield, N. J. (2013). *Relationship thinking: Agency, enchrony, and human sociality: Agency, enchrony, and human sociality*. Oxford: Oxford University Press.

Eren, M. I., Dominguez-Rodrigo, M., Kuhn, S. L., Adler, D. S., Le, I., & Bar-Yosef, O. (2005). Defining and measuring reduction in unifacial stone tools. *Journal of Archaeological Science, 32*(8), 1190–1201. https://doi.org/10.1016/j.jas.2005.03.003.

Erlandson, J. M., Zeder, M. A., Boivin, N. L., Crowther, A., Denham, T., Fuller, D. Q., et al. (2016). Reply to Ellis et al.: Human niche construction and evolutionary theory. *Proceedings of the National Academy of Sciences, 113*(31), E4437–E4438. https://doi.org/10.1073/pnas.1609617113.

Estalrrich, A., El Zaatari, S., & Rosas, A. (2017). Dietary reconstruction of the El Sidrón Neandertal familial group (Spain) in the context of other Neandertal and modern hunter-gatherer groups. A molar microwear texture analysis. *Journal of Human Evolution, 104*, 13–22. https://doi.org/10.1016/j.jhevol.2016.12.003.

Faivre, J.-P., Discamps, E., Gravina, B., Turq, A., Guadelli, J.-L., & Lenoir, M. (2014). The contribution of lithic production systems to the interpretation of Mousterian industrial variability in south-western France: The example of Combe-Grenal (Dordogne, France). *Quaternary International, 350*, 227–240. https://doi.org/10.1016/j.quaint.2014.05.048.

Faivre, J.-P., Gravina, B., Bourguignon, L., Discamps, E., & Turq, A. (2017). Late Middle Palaeolithic lithic technocomplexes (MIS 5–3) in the northeastern Aquitaine Basin: Advances and challenges. *Quaternary International, 433*, 116–131. https://doi.org/10.1016/j.quaint.2016.02.060.

Fischmeister, F. P., Martins, M. J. D., Beisteiner, R., & Fitch, W. T. (2017). Self-similarity and recursion as default modes in human cognition. *Cortex, 97*, 183–201. https://doi.org/10.1016/j.cortex.2016.08.016.

Fitch, W. T. (2011). The evolution of syntax: An exaptationist perspective. *Frontiers in Evolutionary Neuroscience, 3*. https://doi.org/10.3389/fnevo.2011.00009.

Flas, D. (2011). The Middle to Upper Paleolithic transition in Northern Europe: The Lincombian-Ranisian-Jerzmanowician and the issue of acculturation of the last Neanderthals. *World Archaeology, 43*(4), 605–627. https://doi.org/10.1080/00438243.2011.624725.

Fox, J. R., & Coinman, N. R. (2004). Emergence of the Levantine Upper Paleolithic: Evidence from the Wadi al-Hasa. In P. J. Brantingham, S. L. Kuhn, & K. W. Kerry (Eds.), *The early Upper Paleolithic beyond Western Europe* (pp. 97–112). Berkeley: University of California Press.

Froehle, A. W., Yokley, T. R., & Churchill, S. E. (2013). Energetics and the origin of modern humans. In F. H. Smith & J. C. M. Ahern (Eds.), *The origins of modern humans: Biology reconsidered* (2nd ed., pp. 285–320). New York: Wiley.

Fu, Q., Hajdinjak, M., Moldovan, O. T., Constantin, S., Mallick, S., Skoglund, P., et al. (2015). An early modern human from Romania with a recent Neanderthal ancestor. *Nature, 524*(7564), 216–219. https://doi.org/10.1038/nature14558.

Fu, Q., Li, H., Moorjani, P., Jay, F., Slepchenko, S. M., Bondarev, A. A., et al. (2014). Genome sequence of a 45,000-year-old modern human from western Siberia. *Nature, 514*(7523), 445–449. https://doi.org/10.1038/nature13810.

Fu, Q., Posth, C., Hajdinjak, M., Petr, M., Mallick, S., Fernandes, D., et al. (2016). The genetic history of Ice Age Europe. *Nature, 534* (7606), 200–205. https://doi.org/10.1038/nature17993.

Gärdenfors, P., & Högberg, A. (2017). The archaeology of teaching and the evolution of homo docens. *Current Anthropology, 58*(2), 188–208. https://doi.org/10.1086/691178.

Goder-Goldberger, M., Gubenko, N., & Hovers, E. (2016). "Diffusion with modifications": Nubian assemblages in the central Negev highlands of Israel and their implications for Middle Paleolithic inter-regional interactions. *Quaternary International, 408*(Part B), 121–139. https://doi.org/10.1016/j.quaint.2016.02.008.

Goldfield, A. E., Booton, R., & Marston, J. M. (2018). Modeling the role of fire and cooking in the competitive exclusion of Neanderthals. *Journal of Human Evolution, 124*, 91–104. https://doi.org/10.1016/j.jhevol.2018.07.006.

Goren-Inbar, N. (2011). Culture and cognition in the Acheulian industry: A case study from Gesher Benot Yaʿaqov. *Philosophical Transactions of the Royal Society B: Biological Sciences, 366* (1567), 1038–1049. https://doi.org/10.1098/rstb.2010.0365.

Goring-Morris, A. N., & Belfer-Cohen, A. (2003). *More than meets the eye: Studies on Upper Palaeolithic diversity in the Near East*. Oxford: Oxbow.

Goring-Morris, A. N., & Belfer-Cohen, A. (2006). A hard look at the "Levantine Aurignacian": How real is the taxon? In Ofer Bar-Yosef & J. Zilhão (Eds.), *Towards a Definition of the Aurignacian: Proceedings of the Symposium Held in Lisbon, Portugal, June 25–30, 2002* (pp. 297–314). Lisbon: Instituto Português de Arqueologia.

Goring-Morris, N., & Belfer-Cohen, A. (2018). The Ahmarian in the context of the earlier Upper Palaeolithic in the Near East. In Y. Nishiaki & T. Akazawa (Eds.), *The Middle and Upper Paleolithic Archeology of the Levant and Beyond* (pp. 87–104). Singapore: Springer Singapore. https://doi.org/10.1007/978-981-10-6826-3_7.

Goring-Morris, N., & Davidzon, A. (2006). Straight to the point: Upper paleolithic Ahmarian lithic technology in the levant. *Anthropologie, 44*(1), 93–111.

Groucutt, H. S. (2020). Culture and convergence: The curious case of the Nubian Complex. In H. Groucutt (Ed.), *Culture history and convergent evolution: Can we detect populations in prehistory?* (pp. 55–86). Cham, Switzerland: Springer.

Groucutt, H. S., Grün, R., Zalmout, I. A. S., Drake, N. A., Armitage, S. J., Candy, I., et al. (2018). Homo sapiens in Arabia by 85,000 years ago. *Nature Ecology & Evolution*, 1. https://doi.org/10.1038/s41559-018-0518-2.

Groucutt, H. S., Petraglia, M. D., Bailey, G., Scerri, E. M. L., Parton, A., Clark-Balzan, L., et al. (2015a). Rethinking the dispersal of Homo sapiens out of Africa. *Evolutionary Anthropology: Issues, News, and Reviews, 24*(4), 149–164. https://doi.org/10.1002/evan.21455.

Groucutt, H. S., Scerri, E. M. L., Lewis, L., Clark-Balzan, L., Blinkhorn, J., Jennings, R. P., et al. (2015b). Stone tool assemblages and models for the dispersal of Homo sapiens out of Africa. *Quaternary International, 382,* 8–30. https://doi.org/10.1016/j.quaint.2015.01.039.

Grove, M. (2012). Space, time, and group size: A model of constraints on primate social foraging. *Animal Behaviour, 83*(2), 411–419. https://doi.org/10.1016/j.anbehav.2011.11.011.

Haesaerts, P., Damblon, F., Nigst, P., & Hublin, J.-J. (2013). ABA and ABOx radiocarbon cross-dating on Charcoal from Middle Pleniglacial Loess deposits in Austria, Moravia, and Western Ukraine. *Radiocarbon, 55*(2), 641–647. https://doi.org/10.1017/S0033822200057799.

Hajdinjak, M., Fu, Q., Hübner, A., Petr, M., Mafessoni, F., Grote, S., et al. (2018). Reconstructing the genetic history of late Neanderthals. *Nature, 555*(7698), 652–656. https://doi.org/10.1038/nature26151.

Hartman, G., Hovers, E., Hublin, J.-J., & Richards, M. (2015). Isotopic evidence for Last Glacial climatic impacts on Neanderthal gazelle hunting territories at Amud Cave, Israel. *Journal of Human Evolution, 84,* 71–82. https://doi.org/10.1016/j.jhevol.2015.03.008.

Hauser, M. D., Chomsky, N., & Fitch, W. T. (2002). The faculty of language: What is it, who has it, and how did it evolve? *Science, 298*(5598), 1569–1579. https://doi.org/10.1126/science.298.5598.1569.

Henry, D. O. (2003). *Neanderthals in the Levant: Behavioural organization and the beginnings of human modernity*. New York: Continuum International Publishing Group.

Hershkovitz, I., Marder, O., Ayalon, A., Bar-Matthews, M., Yasur, G., Boaretto, E., et al. (2015). Levantine cranium from Manot Cave (Israel) foreshadows the first European modern humans. *Nature, 520*(7546), 216. https://doi.org/10.1038/nature14134.

Higham, T., Brock, F., Peresani, M., Broglio, A., Wood, R., & Douka, K. (2009). Problems with radiocarbon dating the Middle to Upper Palaeolithic transition in Italy. *Quaternary Science Reviews, 28*(13–14), 1257–1267.

Higham, T., Basell, L., Jacobi, R., Wood, R., Ramsey, C. B., & Conard, N. J. (2012). Testing models for the beginnings of the Aurignacian and the advent of figurative art and music: The radiocarbon chronology of Geißenklösterle. *Journal of Human Evolution, 62*(6), 664–676. https://doi.org/10.1016/j.jhevol.2012.03.003.

Higham, T., Douka, K., Wood, R., Ramsey, C. B., Brock, F., Basell, L., et al. (2014). The timing and spatiotemporal patterning of Neanderthal disappearance. *Nature, 512*(7514), 306–309. https://doi.org/10.1038/nature13621.

Hill, K., Barton, M., & Hurtado, A. M. (2009). The emergence of human uniqueness: Characters underlying behavioral modernity. *Evolutionary Anthropology: Issues, News, and Reviews, 18*(5), 187–200. https://doi.org/10.1002/evan.20224.

Hill, K. R., Walker, R. S., Božičević, M., Eder, J., Headland, T., Hewlett, B., et al. (2011). Co-residence patterns in Hunter-Gatherer societies show unique human social structure. *Science, 331*(6022), 1286–1289. https://doi.org/10.1126/science.1199071.

Hill, K. R., Wood, B. M., Baggio, J., Hurtado, A. M., & Boyd, R. T. (2014). Hunter-Gatherer inter-band interaction rates: Implications for cumulative culture. *PLoS ONE, 9*(7), e102806. https://doi.org/10.1371/journal.pone.0102806.

Hoffecker, J. F. (2011). The early upper Paleolithic of Eastern Europe reconsidered. *Evolutionary Anthropology: Issues, News, and Reviews, 20*(1), 24–39. https://doi.org/10.1002/evan.20284.

Hovers, E. (2009). *The lithic assemblages of Qafzeh cave*. Oxford University Press.

Hovers, E., & Belfer-Cohen, A. (2006). "Now you see it, now you don't"—Modern human behavior in the Middle Paleolithic. In E. Hovers & S. L. Kuhn (Eds.), *Transitions before the transition: Evolution and stability in the Middle Paleolithic* (pp. 295–304). New York: Springer. Accessed 8 July 2013.

Hovers, E., & Belfer-Cohen, A. (2013). On variability and complexity: Lessons from the Levantine Middle Paleolithic record. *Current Anthropology, 54*(S8), S337–S357. https://doi.org/10.1086/673880.

Hovers, E., & Raveh, A. (2000). The use of a multivariate graphic display technique as an exploratory tool in the analysis of inter-assemblage lithic variability: A case study from Qafzeh cave, Israel. *Journal of Archaeological Science, 27*(11), 1023–1038. https://doi.org/10.1006/jasc.1999.0511.

Hrdy, S. B. (2009). *Mothers and others: The evolutionary origins of mutual understanding*. Cambridge, MA: Harvard University Press.

Hublin, J.-J. (2000). Modern-nonmodern hominid interactions: a Mediterranean perspective. In Ofer Bar-Yosef & D. Pilbeam (Eds.), *The geography of Neandertals and modern humans in Europe and the Greater Mediterranean* (pp. 157–182). Cambridge, MA: Peabody Museum of Archaeology and Ethnology, Harvard University.

Hublin, J.-J. (2017). The last Neanderthal. *Proceedings of the National Academy of Sciences, 114*(40), 10520–10522. https://doi.org/10.1073/pnas.1714533114.

Hublin, J.-J., Talamo, S., Julien, M., David, F., Connet, N., Bodu, P., et al. (2012). Radiocarbon dates from the Grotte du Renne and Saint-Césaire support a Neandertal origin for the Châtelperronian. *Proceedings of the National Academy of Sciences, 109*(46), 18743–18748. https://doi.org/10.1073/pnas.1212924109.

Hublin, J.-J., Sirakov, N., Aldeias, V., Bailey, S., Bard, E., Delvigne, V., et al. (2020). Initial upper palaeolithic homo sapiens from bacho kiro cave, Bulgaria. *Nature*, 1–4. https://doi.org/10.1038/s41586-020-2259-z.

Jurmain, R., Kilgore, L., Trevathan, W., Ciochon, R. L., & Bartelink, E. (2017). *Introduction to physical anthropology* (Fifteenth ed.). Boston: Cengage Learning.

Kadowaki, S. (2018). Ahmarian or Levantine Aurignacian? Wadi Kharar 16R and new insights into the Upper Palaeolithic lithic technology in the Northeastern Levant. In Y. Nishiaki & T. Akazawa (Eds.), *The Middle and Upper Paleolithic archeology of the Levant and Beyond* (pp. 105–116). Singapore: Springer Nature.

Kaplan, H. S., Gurven, M., Winking, J., Hooper, P. L., & Stieglitz, J. (2010). Learning, menopause, and the human adaptive complex. *Annals of the New York Academy of Sciences, 1204*(1), 30–42. https://doi.org/10.1111/j.1749-6632.2010.05528.x.

Kaplan, H. S., Hooper, P. L., & Gurven, M. (2009). The evolutionary and ecological roots of human social organization. *Philosophical Transactions of the Royal Society B: Biological Sciences, 364*(1533), 3289–3299. https://doi.org/10.1098/rstb.2009.0115.

Kaplan, H. S., & Robson, A. J. (2002). The emergence of humans: The coevolution of intelligence and longevity with intergenerational transfers. *Proceedings of the National Academy of Sciences, 99*(15), 10221–10226. https://doi.org/10.1073/pnas.152502899.

Kaplan, H. S., & Robson, A. J. (2009). We age because we grow. *Proceedings of the Royal Society B: Biological Sciences, 276*(1663), 1837–1844. https://doi.org/10.1098/rspb.2008.1831.

Klein, R. G. (1995). Anatomy, behavior, and modern human origins. *Journal of World Prehistory, 9*(2), 167–198. https://doi.org/10.1007/BF02221838.

Klein, R. G. (2008). Out of Africa and the evolution of human behavior. *Evolutionary Anthropology: Issues, News, and Reviews, 17*(6), 267–281. https://doi.org/10.1002/evan.20181.

Kuhlwilm, M., Gronau, I., Hubisz, M. J., de Filippo, C., Prado-Martinez, J., Kircher, M., et al. (2016). Ancient gene flow from early modern humans into Eastern Neanderthals. *Nature, 530* (7591), 429–433. https://doi.org/10.1038/nature16544.

Kuhn, S. L. (1991). "Unpacking" reduction: Lithic raw material economy in the mousterian of west-central Italy. *Journal of Anthropological Archaeology, 10*(1), 76–106. https://doi.org/10.1016/0278-4165(91)90022-P.

Kuhn, S. L. (1992). On planning and curated technologies in the Middle Paleolithic. *Journal of Anthropological Research, 48*(3), 185–214.

Kuhn, S. L. (1995). *Mousterian lithic technology: An ecological perspective*. Princeton: Princeton University Press.

Kuhn, S. L. (2013). Questions of complexity and scale in explanations for cultural transitions in the Pleistocene: A case study from the early Upper Paleolithic. *Journal of Archaeological Method and Theory, 20*(2), 194–211. https://doi.org/10.1007/s10816-012-9146-3.

Kuhn, S. L., & Stiner, M. C. (2006). What's a mother to do? The division of labor among Neandertals and modern humans in Eurasia. *Current Anthropology, 47*(6), 953–981. https://doi.org/10.1086/507197.

Kuhn, S. L., Stiner, M. C., Güleç, E., Özer, I., Yılmaz, H., Baykara, I., et al. (2009). The early Upper Paleolithic occupations at Üçağızlı Cave (Hatay, Turkey). *Journal of Human Evolution, 56*(2), 87–113. https://doi.org/10.1016/j.jhevol.2008.07.014.

Kuhn, S. L., & Zwyns, N. (2014). Rethinking the initial Upper Paleolithic. *Quaternary International, 347*(Supplement C), 29–38. https://doi.org/10.1016/j.quaint.2014.05.040.

Langdon, J. H. (2005). *The human strategy: An evolutionary perspective on human anatomy*. Oxford: Oxford University Press.

Lazaridis, I., Nadel, D., Rollefson, G., Merrett, D. C., Rohland, N., Mallick, S., et al. (2016). Genomic insights into the origin of farming in the ancient Near East. *Nature, 536*(7617), 419–424. https://doi.org/10.1038/nature19310.

Lazaridis, I., Patterson, N., Mittnik, A., Renaud, G., Mallick, S., Kirsanow, K., et al. (2014). Ancient human genomes suggest three ancestral populations for present-day Europeans. *Nature, 513* (7518), 409–413. https://doi.org/10.1038/nature13673.

Lewontin, R. C. (1983). Gene, organism and environment. In D. S. Bendall (Ed.), *Evolution from molecules to men* (pp. 273–285). Cambridge: Cambridge University Press.

Liang, S., Banerjee, S., & Carlin, B. P. (2009). Bayesian wombling for spatial point processes. *Biometrics, 65*(4), 1243–1253. https://doi.org/10.1111/j.1541-0420.2009.01203.x.

Lowe, J., Barton, N., Blockley, S., Ramsey, C. B., Cullen, V. L., Davies, W., et al. (2012). Volcanic ash layers illuminate the resilience of Neanderthals and early modern humans to natural hazards. *Proceedings of the National Academy of Sciences*. https://doi.org/10.1073/pnas.1204579109.

Lu, H., & Carlin, B. P. (2005). Bayesian areal wombling for geographical boundary analysis. *Geographical Analysis, 37*(3), 265–285. https://doi.org/10.1111/j.1538-4632.2005.00624.x.

Madella, M., Jones, M. K., Goldberg, P., Goren, Y., & Hovers, E. (2002). The exploitation of plant resources by Neanderthals in Amud cave (Israel): The evidence from phytolith studies. *Journal of Archaeological Science, 29*(7), 703–719. https://doi.org/10.1006/jasc.2001.0743.

Malinsky-Buller, A., & Hovers, E. (2019). One size does not fit all: Group size and the late middle Pleistocene prehistoric archive. *Journal of Human Evolution, 127,* 118–132. https://doi.org/10.1016/j.jhevol.2018.11.002.

Marks, A. E. (1977a). *Prehistory and Paleoenvironments in the Central Negev, Israel: The Avdat*. Dallas: Southern Methodist University Press.

Marks, A. E. (1977b). The Upper Paleolithic sites of Boker Tachtit and Boker: A preliminary report. *Prehistory and Paleoenvironments in the Central Negev, Israel, Volume II: The Avdat/Aqev Area, Part 2, and the Har Harif* (pp. 61–79). Dallas: Southern Methodist University Press.

Marks, A. E. (1983). *Prehistory and Paleoenvironment in the Central Negev, Israel* (Vol. 3). Dallas: Southern Methodist University Press.

Meignen, L. (2012). Levantine Perspectives on the Middle to Upper Paleolithic "Transition". *Archaeology, Ethnology and Anthropology of Eurasia, 40*(3), 12–21. https://doi.org/10.1016/j.aeae.2012.11.003.

Mellars, P. (2006a). Archeology and the dispersal of modern humans in Europe: Deconstructing the "Aurignacian". *Evolutionary Anthropology: Issues, News, and Reviews, 15*(5), 167–182. https://doi.org/10.1002/evan.20103.

Mellars, P. (2006b). Why did modern human populations disperse from Africa ca. 60,000 years ago? A new model. *Proceedings of the National Academy of Sciences, 103*(25), 9381–9386. https://doi.org/10.1073/pnas.0510792103.

Monigal, K. (2003). Monigal, K. (2003). Technology, economy, and mobility at the beginning of the Levantine Upper Paleolithic. In A. N. Goring-Morris & A. Belfer-Cohen (Eds.), *More than meets the eye: Studies on Upper Palaeolithic diversity in the Near East* (pp. 118–133). Oxford: Oxbow Books.

Monnier, G. F., & Missal, K. (2014). Another Mousterian Debate? Bordian facies, chaîne opératoire technocomplexes, and patterns of lithic variability in the western European Middle and Upper Pleistocene. *Quaternary International, 350,* 59–83. https://doi.org/10.1016/j.quaint.2014.06.053.

Morin, E. (2008). Evidence for declines in human population densities during the early Upper Paleolithic in western Europe. *Proceedings of the National Academy of Sciences, 105*(1), 48–53. https://doi.org/10.1073/pnas.0709372104.

Moroni, A., Boscato, P., & Ronchitelli, A. (2013). What roots for the Uluzzian? Modern behaviour in Central-Southern Italy and hypotheses on AMH dispersal routes. *Quaternary International, 316,* 27–44. https://doi.org/10.1016/j.quaint.2012.10.051.

Müller, U. C., Pross, J., Tzedakis, P. C., Gamble, C., Kotthoff, U., Schmiedl, G., et al. (2011). The role of climate in the spread of modern humans into Europe. *Quaternary Science Reviews, 30*(3–4), 273–279. https://doi.org/10.1016/j.quascirev.2010.11.016.

Nigst, P. R. (2012). *The early Upper Palaeolithic of the Middle Danube region*. Amsterdam: Amsterdam University Press.

O'Connell, J. F. (2006). How did modern humans replace Neanderthals? Insights from hunter-gatherer ethnography and archaeology. In N. J. Conard (Ed.), *When Neanderthals and modern humans met* (pp. 43–64). Tübingen: Kerns Verlag.

Oden, N. L., Sokal, R. R., Fortin, M.-J., & Goebl, H. (1993). Categorical wombling: Detecting regions of significant change in spatially located categorical variables. *Geographical Analysis, 25* (4), 315–336. https://doi.org/10.1111/j.1538-4632.1993.tb00301.x.

Odling-Smee, F. J., Feldman, M. W., & Laland, K. N. (2003). *Niche construction: The neglected process in evolution*. Princeton: Princeton University Press.

Olszewski, D. I. (1999). The early Upper Palaeolithic in the Zagros Mountains. In W. Davies & R. Charles (Eds.), *Dorothy Garrod and the progress of the Palaeolithic: Studies in the Prehistoric archaeology of the Near East and Europe* (pp. 167–180). Oxford: Oxbow Books.

Olszewski, D. I. (2009). Whither the Aurignacian in the Middle East? Assessing the Zagros Upper Paleolithic. In M. Otte, F. Biglari, & J. Jaubert (Eds.), *Iran Palaeolithic/Le Paléolithique d'Iran* (pp. 39–45). Oxford: Archaeopress.

Olszewski, D. I., & Dibble, H. L. (1994). The Zagros Aurignacian. *Current Anthropology, 35*(1), 68–75. https://doi.org/10.1086/204242.

Otte, M., Shidrang, S., Zwyns, N., & Flas, D. (2011). New radiocarbon dates for the Zagros Aurignacian from Yafteh cave, Iran. . *Journal of Human Evolution, 61*(3), 340–346. https://doi.org/10.1016/j.jhevol.2011.05.011.

Peresani, M., Cristiani, E., & Romandini, M. (2016). The Uluzzian technology of Grotta di Fumane and its implication for reconstructing cultural dynamics in the Middle-Upper Palaeolithic transition of Western Eurasia. *Journal of Human Evolution, 91,* 36–56. https://doi.org/10.1016/j.jhevol.2015.10.012.

Petr, M., Pääbo, S., Kelso, J., & Vernot, B. (2019). Limits of long-term selection against Neandertal introgression. *Proceedings of the National Academy of Sciences, 116*(5), 1639–1644. https://doi.org/10.1073/pnas.1814338116.

Pimenoff, V. N., Oliveira, D., Mendes, C., & Bravo, I. G. (2017). Transmission between Archaic and Modern human ancestors during the evolution of the oncogenic human papillomavirus 16. *Molecular Biology and Evolution, 34*(1), 4–19. https://doi.org/10.1093/molbev/msw214.

Pinhasi, R., Higham, T. F. G., Golovanova, L. V., & Doronichev, V. B. (2011). Revised age of late Neanderthal occupation and the end of the Middle Paleolithic in the northern Caucasus. *Proceedings of the National Academy of Sciences, 108*(21), 8611–8616. https://doi.org/10.1073/pnas.1018938108.

Prüfer, K., de Filippo, C., Grote, S., Mafessoni, F., Korlević, P., Hajdinjak, M., et al. (2017). A high-coverage Neandertal genome from Vindija Cave in Croatia. *Science, 358*(6363), 655–658. https://doi.org/10.1126/science.aao1887.

Rebollo, N. R., Weiner, S., Brock, F., Meignen, L., Goldberg, P., Belfer-Cohen, A., et al. (2011). New radiocarbon dating of the transition from the Middle to the Upper Paleolithic in Kebara Cave, Israel. *Journal of Archaeological Science, 38*(9), 2424–2433. https://doi.org/10.1016/j.jas.2011.05.010.

Reynolds, N. (2020). Threading the weft, testing the warp: Population concepts and the European Upper Palaeolithic chronocultural framework. In H. Groucutt (Ed.), *Culture history and convergent evolution: Can we detect populations in prehistory?* (pp. 187–212). Cham, Switzerland: Springer.

Reynolds, R. G. (1986). An adaptive computer model for the evolution of plant collecting and early agriculture in the Eastern Valley of Oaxaca. In K. V. Flannery (Ed.), *Guila Naquitz: Archaic foraging and early agriculture in Oaxaca, Mexico* (pp. 439–500). New York: Academic Press.

Richter, D., Tostevin, G., & Škrdla, P. (2008). Bohunician technology and thermoluminescence dating of the type locality of Brno-Bohunice (Czech Republic). *Journal of Human Evolution, 55*(5), 871–885. https://doi.org/10.1016/j.jhevol.2008.04.008.

Riel-Salvatore, J. (2010). A niche construction perspective on the Middle-Upper Paleolithic transition in Italy. *Journal of Archaeological Method and Theory, 17*(4), 323–355. https://doi.org/10.1007/s10816-010-9093-9.

Ruebens, K., & Wragg Sykes, R. M. (2016). Spatio-temporal variation in late Middle Palaeolithic Neanderthal behaviour: British bout coupé handaxes as a case study. *Quaternary International, 411,* 305–326. https://doi.org/10.1016/j.quaint.2015.04.037.

Ruyle, E. E. (1973). Genetic and cultural pools: Some suggestions for a unified theory of biocultural evolution. *Human Ecology, 1*(3), 201–215.

Sankararaman, S., Mallick, S., Dannemann, M., Prüfer, K., Kelso, J., Pääbo, S., et al. (2014). The genomic landscape of Neanderthal ancestry in present-day humans. *Nature, 507*(7492), 354–357. https://doi.org/10.1038/nature12961.

Sankararaman, S., Mallick, S., Patterson, N., & Reich, D. (2016). The combined landscape of Denisovan and Neanderthal ancestry in present-day humans. *Current Biology, 26*(9), 1241–1247. https://doi.org/10.1016/j.cub.2016.03.037.

Sankararaman, S., Patterson, N., Li, H., Pääbo, S., & Reich, D. (2012). The date of interbreeding between Neandertals and modern humans. *PLoS Genetics, 8*(10), e1002947. https://doi.org/10.1371/journal.pgen.1002947.

Scerri, E. M. L. (2017). The North African Middle Stone Age and its place in recent human evolution. *Evolutionary Anthropology: Issues, News, and Reviews, 26*(3), 119–135. https://doi.org/10.1002/evan.21527.

Scerri, E. M. L., Drake, N. A., Jennings, R., & Groucutt, H. S. (2014a). Earliest evidence for the structure of Homo sapiens populations in Africa. *Quaternary Science Reviews, 101,* 207–216. https://doi.org/10.1016/j.quascirev.2014.07.019.

Scerri, E. M. L., Groucutt, H. S., Jennings, R. P., & Petraglia, M. D. (2014b). Unexpected technological heterogeneity in northern Arabia indicates complex Late Pleistocene demography at the gateway to Asia. *Journal of Human Evolution, 75,* 125–142. https://doi.org/10.1016/j.jhevol.2014.07.002.

Scerri, E. M. L., Gravina, B., Blinkhorn, J., & Delagnes, A. (2016). Can lithic attribute analyses identify discrete reduction trajectories? A Quantitative study using refitted lithic sets. *Journal of Archaeological Method and Theory, 23*(2), 669–691. https://doi.org/10.1007/s10816-015-9255-x.

Schmidt, C., Sitlivy, V., Anghelinu, M., Chabai, V., Kels, H., Uthmeier, T., et al. (2013). First chronometric dates (TL and OSL) for the Aurignacian open-air site of Româneşti-Dumbrăviţa I, Romania. *Journal of Archaeological Science, 40*(10), 3740–3753. https://doi.org/10.1016/j.jas.2013.04.003.

Semal, P., Rougier, H., Crevecoeur, I., Jungels, C., Flas, D., Hauzeur, A., et al. (2009). New data on the late Neandertals: Direct dating of the Belgian Spy fossils. *American Journal of Physical Anthropology, 138*(4), 421–428. https://doi.org/10.1002/ajpa.20954.

Shea, J. J. (2008). Transitions or turnovers? Climatically-forced extinctions of Homo sapiens and Neanderthals in the east Mediterranean Levant. *Quaternary Science Reviews, 27*(23–24), 2253–2270. https://doi.org/10.1016/j.quascirev.2008.08.015.

Shea, J. J. (2011). Homo sapiens Is as Homo sapiens Was. *Current Anthropology, 52*(1), 1–35. https://doi.org/10.1086/658067.

Shea, J. J. (2013a). *Stone tools in the Paleolithic and Neolithic of the Near East: A guide*. New York: Cambridge University Press. Accessed 10 June 2013.

Shea, J. J. (2013b). Lithic modes A-I: A new framework for describing global-scale variation in stone tool technology illustrated with evidence from the East Mediterranean Levant. *Journal of Archaeological Method and Theory, 20*(1), 151–186. https://doi.org/10.1007/s10816-012-9128-5.

Shea, J. J. (2016). The lithic assemblage of Mughr El-Hamamah (Jordan) 39-45 ka BP. *Presented at the Paleoanthropology Society Annual Meeting*, Atlanta.

Shea, J. J., & Sisk, M. L. (2010). Complex projectile technology and Homo sapiens dispersal into Western Eurasia. *PaleoAnthropology, 2010,* 100–122. https://doi.org/10.4207/PA.2010.ART36.

Shea, J. J., Stutz, A. J., & Nilsson Stutz, L. (in press). An early Upper Paleolithic stone tool assemblage from Mughr El-Hamamah, Jordan. *Journal of Field Archaeology*.

Sinha, C. (2015a). Language and other artifacts: Socio-cultural dynamics of niche construction. *Frontiers in Psychology*, 1601. https://doi.org/10.3389/fpsyg.2015.01601.

Sinha, C. (2015b). Ontogenesis, semiosis and the epigenetic dynamics of biocultural niche construction. *Cognitive Development, 36,* 202–209. https://doi.org/10.1016/j.cogdev.2015.09.006.

Snodgrass, J., & Leonard, W. R. (2009). Neandertal energetics revisited: Insights into population dynamics and life history evolution. *PaleoAnthropology, 2009,* 220–237.

Speth, J. D. (2013). Middle Paleolithic Large-Mammal Hunting in the Southern Levant. In J. D. Speth & J. L. Clark (Eds.), *Zooarchaeology and modern human origins: Human hunting behavior during The Later Pleistocene*. New York: Springer.

Speth, J. D., Meignen, L., Bar-Yosef, O., & Goldberg, P. (2012). Spatial organization of Middle Paleolithic occupation X in Kebara Cave (Israel): Concentrations of animal bones. *Quaternary International, 247,* 85–102. https://doi.org/10.1016/j.quaint.2011.03.001 .

Staubwasser, M., Drăgușin, V., Onac, B. P., Assonov, S., Ersek, V., Hoffmann, D. L., et al. (2018). Impact of climate change on the transition of Neanderthals to modern humans in Europe. *Proceedings of the National Academy of Sciences, 115*(37), 9116–9121. https://doi.org/10.1073/pnas.1808647115.

Stiner, M. C., & Kuhn, S. L. (1992). Subsistence, technology, and adaptive variation in Middle Paleolithic Italy. *American Anthropologist, 94*(2), 306–339. https://doi.org/10.1525/aa.1992.94.2.02a00030.

Stiner, M. C., & Kuhn, S. L. (2016). Are we missing the "sweet spot" between optimality theory and niche construction theory in archaeology? *Journal of Anthropological Archaeology, Part B, 44,* 177–184. https://doi.org/10.1016/j.jaa.2016.07.006.

Stiner, M. C., Kozlowski, J. K., Kuhn, S. L., Karkanas, P., & Koumouzelis, M. (2010). Klissoura Cave 1 and the Upper Paleolithic of southern Greece in cultural and ecological context. *Eurasian Prehistory, 7,* 309–321.

Stout, D., & Hecht, E. E. (2017). Evolutionary neuroscience of cumulative culture. *Proceedings of the National Academy of Sciences, 114*(30), 7861–7868. https://doi.org/10.1073/pnas.1620738114.

Stout, D., Rogers, M. J., Jaeggi, A. V., & Semaw, S. (2019). Archaeology and the origins of human cumulative culture: A case study from the Earliest Oldowan at Gona, Ethiopia. *Current Anthropology, 60*(3), 309–340. https://doi.org/10.1086/703173.

Straus, L. G. (2005). A mosaic of change: The Middle-Upper Paleolithic transition as viewed from New Mexico and Iberia. *Quaternary International, 137*(1), 47–67. https://doi.org/10.1016/j.quaint.2004.11.019.

Stringer, C. B., & Andrews, P. (1988). Genetic and fossil evidence for the origin of modern humans. *Science, 239*(4845), 1263–1268. https://doi.org/10.1126/science.3125610.

Stutz, A. J. (2009). The "Nature of Transitions" in the Stone Age: A comparative perspective. In J. J. Shea & D. E. Lieberman (Eds.), *Transitions in prehistory: Papers in honor of Ofer Bar-Yosef* (pp. 477–498). Cambridge, MA, USA: Harvard University American School of Prehistoric Research.

Stutz, A. J. (2012). Culture and politics, behavior and biology: Seeking synthesis among fragmentary anthropological perspectives on Hunter-Gatherers. *Reviews in Anthropology, 41*(1), 23–69. https://doi.org/10.1080/00938157.2012.649183.

Stutz, A. J. (2013a, May 22). Biocultural evolution–An overview. *The Biocultural Evolution Blog*. Retrieved December 11, 2018, from https://bioculturalevolution.net/biocultural-overview/.

Stutz, A. J. (2013b, June 8). More on the origins of "Biocultural." *The Biocultural Evolution Blog*. Retrieved December 12, 2018, from https://bioculturalevolution.net/2013/06/08/more-on-the-origins-of-biocultural/.

Stutz, A. J. (2014a). Embodied niche construction in the hominin lineage: semiotic structure and sustained attention in human embodied cognition. *Frontiers in Psychology, 5,* Article 834. https://doi.org/10.3389/fpsyg.2014.00834.

Stutz, A. J. (2014b). Modeling the pre-industrial roots of modern super-exponential population growth. *PLoS ONE, 9*(8), e105291. https://doi.org/10.1371/journal.pone.0105291.

Stutz, A. J. (2019). Near East (Including Anatolia): Geographic description and general chronology of the Paleolithic and Neolithic. In *Encyclopedia of Global Archaeology* (2nd Edition.). New York: Springer Nature Switzerland AG. https://doi.org/10.1007/978-3-319-51726-1_1910-3.

Stutz, A. J., & Nilsson Stutz, L. (2017). Mughr el-Hamamah: An early Upper Paleolithic cave site on the Eastern Jordan Valley Flanks. In Y. Enzel & O. Bar-Yosef (Eds.), *Quaternary environments, climate change and humans in the Levant* (pp. 285–290). New York: Cambridge University Press.

Stutz, A. J., Shea, J. J., Rech, J. A., Pigati, J. S., Wilson, J., Belmaker, M., et al. (2015). Early Upper Paleolithic chronology in the Levant: New ABOx-SC accelerator mass spectrometry results from the Mughr el-Hamamah Site, Jordan. *Journal of Human Evolution, 85,* 157–173. https://doi.org/10.1016/j.jhevol.2015.04.008.

Svoboda, J., & Bar-Yosef, O. (Eds.). (2003). *Stranska Skala: Origins of the Upper Paleolithic in the Brno Basin, Moravia, Czech Republic.* Cambridge, MA, USA: Peabody Museum of Archaeology and Ethnology, Harvard University. Retrieved July 8, 2013, from http://www.alibris.com/Stranska-Skala-Origins-of-the-Upper-Paleolithic-in-the-Brno-Basin-Moravia-Czech-Republic-J-Svoboda/book/19801243.

Talamo, S., Hajdinjak, M., Mannino, M. A., Fasani, L., Welker, F., Martini, F., et al. (2016). Direct radiocarbon dating and genetic analyses on the purported Neanderthal mandible from the Monti Lessini (Italy). *Scientific Reports, 6,* 29144. https://doi.org/10.1038/srep29144.

Talamo, S., Soressi, M., Roussel, M., Richards, M., & Hublin, J.-J. (2012). A radiocarbon chronology for the complete Middle to Upper Palaeolithic transitional sequence of Les Cottés (France). *Journal of Archaeological Science, 39*(1), 175–183. https://doi.org/10.1016/j.jas.2011.09.019.

Tallavaara, M., Luoto, M., Korhonen, N., Järvinen, H., & Seppä, H. (2015). Human population dynamics in Europe over the Last Glacial maximum. *Proceedings of the National Academy of Sciences, 112*(27), 8232–8237. https://doi.org/10.1073/pnas.1503784112.

Teyssandier, N. (2006). Questioning the first Aurignacian: Mono or multi cultural phenomenon during the formation of the Upper Paleolithic in Central Europe and the Balkans. *Anthropologie Brno, XLIV*(1), 9–29.

Teyssandier, N. (2008). Revolution or evolution: The emergence of the Upper Paleolithic in Europe. *World Archaeology, 40*(4), 493–519. https://doi.org/10.1080/00438240802452676.

Teyssandier, N., Bon, F., & Bordes, J.-G. (2010). Within projectile range: Some thoughts on the appearance of the Aurignacian in Europe. *Journal of Anthropological Research, 66*(2), 209–229.

Teyssandier, N., & Zilhão, J. (2018). On the entity and antiquity of the Aurignacian at Willendorf (Austria): Implications for modern human emergence in Europe. *Journal of Paleolithic Archaeology, 1*(2), 107–138. https://doi.org/10.1007/s41982-017-0004-4.

Tomasello, M. (2008). *Origins of human communication*. Cambridge, MA, USA: MIT Press.

Tostevin, G. B. (2003). A quest for antecedents: A comparison of the terminal Middle Palaeolithic and Earuly Upper Paleolithic of the Levant. In A. N. Goring-Morris & A. Belfer-Cohen (Eds.), *More than meets the eye: Studies on Upper Palaeolithic diversity in the Near East* (pp. 54–67). Oxford: Oxbow.

Tostevin, G. B. (2007). Social intimacy, artefact visibility, and acculturation models of Neanderthal-modern human interaction. In P. Mellars, K. V. Boyle, O. Bar-Yosef, & C. B. Stringer (Eds.), *Rethinking the*

Human Revolution: New behavioural and biological perspectives on the origins and dispersal of modern humans (pp. 341–357). Cambridge: MacDonald Institute Research Monograph Series.

Tostevin, G. B. (2011). Levels of theory and social practice in the reduction sequence and Chaîne Opératoire Methods of Lithic Analysis. *PaleoAnthropology, 2011* (Special Issue: Reduction Sequence, Chaîne Opératoire, and Other Methods: The Epistemologies of Different Approaches to Lithic Analysis), 351–375. https://doi.org/10.4207/pa.2011.art64.

Tostevin, G. B. (2013). *Seeing lithics: A middle-range theory for testing for cultural transmission in the Pleistocene*. Oxford: Oxbow.

Tostevin, G. B., & Skrdla, P. (2006). New excavations at bohunice and the question of the uniqueness of the type-site for the Bohunician industrial type. *Anthropologie, 44*(1), 31–48.

Tryon, C. A., & Ranhorn, K. L. (2020). Raw material and regionalization in Stone Age East Africa. In H. Groucutt (Ed.), *Culture history and convergent evolution: Can we detect populations in prehistory?* (pp. 143–156). Cham, Switzerland: Springer.

Tsanova, T. (2006, December 12). *Les débuts du Paléolithique supérieur dans l'Est des Balkans. Réflexion à partir de l'étude taphonomique et techno-économique des ensembles lithiques des sites de Bacho Kiro (couche 11), Temnata (couches VI et 4) et Kozarnika (niveau VII)* (phdthesis). Sciences du Vivant, Géoscience, Sciences de l'environnement, Université Bordeaux 1. Retrieved from https://tel.archives-ouvertes.fr/tel-00541845/document.

Tsanova, T. (2013). The beginning of the Upper Paleolithic in the Iranian Zagros. A taphonomic approach and techno-economic comparison of Early Baradostian assemblages from Warwasi and Yafteh (Iran). *Journal of Human Evolution, 65*(1), 39–64. https://doi.org/10.1016/j.jhevol.2013.04.005.

Tsanova, T., Zwyns, N., Eizenberg, L., Teyssandier, N., Le Brun-Ricalens, F., & Otte, M. (2012). Le plus petit dénominateur commun: réflexion sur la variabilité des ensembles lamellaires du Paléolithique supérieur ancien d'Eurasie. Un bilan autour des exemples de Kozarnika (Est des Balkans) et Yafteh (Zagros central). *L'Anthropologie, 116*(4), 469–509. https://doi.org/10.1016/j.anthro.2011.10.005.

van Gerven, D. P., Carlson, D. S., & Armelagos, G. J. (1973). Racial history and bio-cultural adaptation of Nubian archaeological populations. *The Journal of African History, 14*(4), 555–564.

Vattathil, S., & Akey, J. M. (2015). Small amounts of archaic admixture provide big insights into human history. *Cell, 163*(2), 281–284. https://doi.org/10.1016/j.cell.2015.09.042.

Vernot, B., & Akey, J. M. (2014). Resurrecting surviving Neandertal lineages from modern human genomes. *Science, 343*(6174), 1017–1021. https://doi.org/10.1126/science.1245938.

Vernot, B., Tucci, S., Kelso, J., Schraiber, J. G., Wolf, A. B., Gittelman, R. M., et al. (2016). Excavating Neandertal and Denisovan DNA from the genomes of Melanesian individuals. *Science*, aad9416. https://doi.org/10.1126/science.aad9416.

Weyrich, L. S., Duchene, S., Soubrier, J., Arriola, L., Llamas, B., Breen, J., et al. (2017). Neanderthal behaviour, diet, and disease inferred from ancient DNA in dental calculus. *Nature, 544*(7650), 357–361. https://doi.org/10.1038/nature21674.

Wiessner, P. (1983). Style and social information in Kalahari San projectile points. *American Antiquity, 48*(2), 253–276. https://doi.org/10.2307/280450.

Whiten, A., Horner, V., & Waal, F. B. M. de. (2005). Conformity to cultural norms of tool use in chimpanzees. *Nature, 437*(7059), 737–740. https://doi.org/10.1038/nature04047.

Will, M., & Mackay, A. (2020). A matter of space and time: How frequent is convergence in lithic technology in the African archaeological record over the last 300 kyr? In H. Groucutt (Ed.), *Culture history and convergent evolution: Can we detect populations in prehistory?* (pp. 103–126). Cham, Switzerland: Springer.

Wolf, A. B., & Akey, J. M. (2018). Outstanding questions in the study of archaic hominin admixture. *PLOS Genetics, 14*(5), e1007349. https://doi.org/10.1371/journal.pgen.1007349.

Zilhão, J. (2011). Aliens from outer time? Why the "Human Revolution" is wrong, and where do we go from here? In S. Condemi & G.-C. Weniger (Eds.), *Continuity and discontinuity in the peopling of Europe* (pp. 331–366). Dordrecht: Springer Netherlands. Retrieved August 28, 2013, from http://link.springer.com/chapter/10.1007/978-94-007-0492-3_25.

Zilhão, J. (2013). Neandertal-modern human contact in western Eurasia: Issues of Dating, Taxonomy, and Cultural Associations. In T. Akazawa, Y. Nishiaki, & K. Aoki (Eds.), *Dynamics of learning in Neanderthals and modern humans volume 1: Cultural perspectives* (pp. 21–57). New York: Springer.

Zilhão, J., Banks, W. E., d'Errico, F., & Gioia, P. (2015). Analysis of site formation and assemblage integrity does not support attribution of the Uluzzian to modern humans at Grotta del Cavallo. *PLoS ONE, 10*(7), e0131181. https://doi.org/10.1371/journal.pone.0131181.

Zuckerman, M. K., & Martin, D. L. (2016). *New directions in biocultural anthropology*. Hoboken, New Jersey: Wiley.

Chapter 10
Threading the Weft, Testing the Warp: Population Concepts and the European Upper Paleolithic Chronocultural Framework

Natasha Reynolds

Abstract Interpretations of the European Upper Paleolithic archaeological record have long relied on concepts of past populations. In particular, cultural taxonomic units—which are used as a framework for describing the archaeological record—are commonly equated with past populations. However, our cultural taxonomy is highly historically contingent, and does not necessarily accurately reflect variation in the archaeological record. Furthermore, we lack a secure theoretical basis for the description of past human populations based on taxonomic units. In order to move past these problems and satisfactorily address questions of Upper Paleolithic populations, we need to entirely revise our approach to chronocultural framework building. Here, I outline a specific way of describing the archaeological record that deliberately avoids the use of cultural taxonomic units and instead concentrates on individual features of material culture. This approach may provide a more appropriate basis for the archaeological study of Upper Paleolithic populations and for comparisons with genetic data.

Keywords Cultural taxonomy • Genetics • Aurignacian • Gravettian • Magdalenian

Introduction

The European Upper Paleolithic represents a special case in the study of past populations within the Paleolithic archaeological record. Leaving aside questions concerning the authorship of the "transitional" industries (Hublin 2015), the European Upper Paleolithic relates, as far as we know, to a single hominin taxon: *Homo sapiens*. This is in contrast with many other parts of the Paleolithic archaeological record, where multiple taxonomically distinct hominin groups need to be considered. Furthermore, the Upper Paleolithic archaeological record of Europe is abundant and relatively well-studied, and we have extensive associated data on ancient human genomes in comparison with other parts of the world.

However, the Upper Paleolithic populations of Europe remain poorly understood archaeologically. There is little agreement on what archaeology can tell us about Upper Paleolithic populations: we lack consensus or even much explicit discussion concerning the definition of populations, an epistemological framework, the formulation of research questions, and the methods and theoretical approaches we might employ.

Population concepts are often used in studies of the Middle to Upper Paleolithic transition, where the Middle Paleolithic is associated with a Neanderthal population and the Upper Paleolithic with an anatomically modern human population: here, usage of the population concept is generally quite clear and in line with biological understandings of the term. There has also been substantial research into Upper Paleolithic demography, again usually demonstrating a clear understanding of population concepts in the biological sense (e.g. Bocquet-Appel and Demars 2000; Gamble et al. 2005; French 2015; Tallavaara et al. 2015).

However, population concepts are also widely invoked as explanations for variation within the Upper Paleolithic archaeological record. For example, as we shall see below, the differences between lithic assemblages in two regions might be explained by the idea that different populations were present in each area at some point in the past. In these cases, the word seems to be used without a formal definition, meaning, essentially, "a group of people". However, it usually appears to refer to a group of people posited to have been linked by common cultural traditions as well as perhaps common ancestry and/or identity.

N. Reynolds (✉)
UMR 5199 PACEA, Université de Bordeaux, Bâtiment B8, Allée Geoffroy Saint Hilaire, CS 50023, 33615 Pessac Cedex, France
e-mail: natasha.reynolds@u-bordeaux.fr

H. S. Groucutt (ed.), *Culture History and Convergent Evolution: Can We Detect Populations in Prehistory?*, Vertebrate Paleobiology and Paleoanthropology, https://doi.org/10.1007/978-3-030-46126-3_10

Population concepts are closely linked with one of our most important analytical approaches to the Upper Palaeolithic: cultural taxonomy. Cultural taxonomy concerns the definition and description of archaeological taxonomic units (e.g. Aurignacian, Badegoulian, Ahrensburgian). These might be called "technocomplexes", "archaeological cultures", etc, and archaeologists vary strongly in how they use these concepts (e.g. Clarke 1968; Dunnell 1971; Gamble et al. 2005; Clark and Riel-Salvatore 2006; Roberts and Vander Linden 2011; Sørensen 2014; Hermon and Niccolucci 2017; Reynolds accepted manuscript). Taxonomic units, established based on the study of material culture and chronology, are frequently postulated to have been associated with particular past "populations". Depending on the example, this population may be more or less explicitly defined, more or less discrete, and more or less persistent. Populations are often described based on the existence of taxonomic units, and are often named after them: hence e.g. "the Gravettians", "the Solutreans", etc.

For better or worse, the study of the Upper Paleolithic never had a backlash against ideas of "stone tools equal people" to the degree that the study of later prehistory in the West took a turn against ideas that "pots equal people" (Kramer 1977; Van Oyen 2017). Many archaeologists appear comfortable with, for example, thinking of a group of people called "the Aurignacians", distinct in their traditions, ancestry and identity, who made and deposited the archaeological assemblages that we now call Aurignacian. Furthermore, archaeologists might think of these people as clearly different from "the Gravettians" who apparently succeeded them. These ideas may be explicitly stated and meant literally, or they may be hidden assumptions or used as heuristic tools.

In this chapter I discuss several aspects of the continuing importance of population concepts in the study of the Upper Paleolithic and how they manifest themselves as part of the chain of reasoning that leads us from collections of excavated artefacts to the re-creation of social and cultural processes during the Late Pleistocene. I begin with a discussion of some explicit uses of the population concept in Upper Paleolithic archaeological interpretation. I then discuss the present cultural taxonomic system and some of its shortcomings, in order to argue that taxonomic units should not be naïvely correlated with past populations. In an attempt to create a better basis for the archaeological understanding of late Pleistocene populations, I devote the middle part of this chapter to advocating a specific way of building and revising the Upper Paleolithic chronocultural framework, based on an emphasis on coherence and a dialectical consideration of chronological and material culture data. I also outline a specific way of conceptualizing this framework, which focuses not on the construction of abstract taxonomic units, but rather on describing multiple links between assemblages based on the co-occurrence of index fossils or other well-defined features. I then discuss our prospects for establishing a robust archaeological approach to populations by comparing the chronocultural framework against the results obtained from paleogenetic studies. Although at present we are far from being able to make reliable inferences about Upper Paleolithic populations from the archaeological record, there is much potential for future progress.

Populations in the European Upper Paleolithic

The use of population concepts in the study of the Upper Paleolithic, particularly as expressed in references to "the Aurignacians", "the Gravettians", etc., has a long history. The early twentieth century history of these concepts also demonstrates—although a full treatment of this subject is beyond the scope of the current paper—their development in a context of essentialist and often racist approaches to populations and ethnic groups (see e.g. MacCurdy 1914, 1915; Macalister 1921: 385; Hřdlicka 1927; Collie 1928; Burkitt and Childe 1932; cf. McNabb 2020). The prejudices that shaped archaeological concepts during this time may have had more influence than we would like to think on modern archaeological ideas of Paleolithic populations and may go some way towards explaining their deficiencies.

To gain an impression of some ideas that were in circulation and without repeating here the more odious racist comparisons, we can consider the following quotation from Macalister (1921, pp. 580–2): "One of the most difficult problems of the Upper Paleolithic Term is the relation of the Solutreans to the Aurignacians which preceded them, and to the Magdalenians which followed them. … That the Solutrean culture is associated with a people of different racial affinities from the Aurignacian is indicated by the bones from Předmost and Brünn. … Some circumstances drove the Solutreans back from central and eastern Europe along the way by which their ancestors had come. They crowded back on the Aurignacians and for a time kept them suppressed." Similarly, Burkitt and Childe (1932: 192) state that "The Solutreans invaded parts of Western Europe and dominated the Aurignacians." From these quotations we can see that archaeological cultures were seen as being the product of groups of people named for them, and these groups of people were seen as discrete populations or ethnic groups whose histories of migration and development could be reconstructed. The descriptions of postulated interactions between past populations using a vocabulary of invasion, suppression or domination, and the notions of essential "racial" difference between them, now seem clearly of their time. However, ideas concerning the existence of "Solutreans", "Aurignacians" etc. have been passed down to us in the present day and continue to live on in archaeological discussions. Although the language used has typically been toned down to more neutral terminology of "population replacement" etc, we shall see that notions of essential differences between "Solutreans", "Magdalenians" etc still permeate much archaeological interpretation in the present day, despite the lack of convincing archaeological or genetic evidence for such discrete populations.

Although not all modern archaeologists refer to populations in the Upper Paleolithic with respect to cultural groupings, many do so explicitly. Reference to "Aurignacians", "Gravettians" etc. is still fairly common in modern archaeological practice (e.g. Bodu 1998; Finlayson and Carrión 2007; Otte 2010, 2013; Ronchitelli et al. 2015; Svoboda 2015; Tejero 2016), even if the intended meaning of these terms varies between authors. Some go further, and link changes in the archaeological record with putative population extinctions and movements in the past (e.g. Gamble et al. 2005; Banks et al. 2008; Schmidt et al. 2012; Kozłowski 2015; Djindjian 2016). For example, it has been explicitly argued that the population that created Aurignacian assemblages went extinct and was replaced by another population that created Gravettian assemblages (Finlayson and Carrión 2007; Bradtmöller et al. 2012), and, on a different scale, that the appearance of Badegoulian assemblages in France represents a population incursion from Central Europe (Gamble et al. 2005; cf. Banks et al. 2008).

Of course, the idea that movements of populations are responsible for changes in the archaeological record is itself logically dependent on the idea that distinct populations co-existed during the Upper Paleolithic. One modern example of this is the idea that the Epigravettian and Magdalenian, or Epigravettian and Solutrean, technocomplexes are evidence for distinct contemporary populations during the Late Upper Paleolithic (Banks et al. 2008; Bradtmöller et al. 2012). However, in many other cases the idea of the co-existence of separate populations is not discussed directly, especially where work is focused on diachronic change within small regions. Rather, the idea that discrete populations co-existed during the Upper Paleolithic (either within Europe or in a larger geographic area) is an assumption implicit within the argument for the replacement of one population by another.

Upper Paleolithic Cultural Taxonomy

As we have already seen, explicit discussions of populations are usually framed around named cultural taxonomic units (or "technocomplexes", "archaeological cultures", etc), i.e. taxonomic units are seen to correspond to past populations. But what are these taxonomic units and how robust is the inference of populations from them? Understandings of cultural taxonomy among Upper Paleolithic archaeologists are highly diverse (e.g. Djindjian et al. 1999; Gamble et al. 2005; Clark and Riel-Salvatore 2006; Riede 2011; Reynolds and Riede 2019; Reynolds accepted manuscript), and the strength of the theoretical and empirical foundations of these understandings similarly differs strongly. In practice, these units can be treated as time periods, as sets of assemblages, as past populations, as traditions or sets of traditions, as geographical distributions, as combinations of all these things, or as different things at different times (Reynolds accepted manuscript). Archaeologists usually do not think of all these units in the same way, and might think about "the Gravettian" differently than "the Badegoulian", and "the Ahrensburgian" differently than "the Aurignacian", if only because their temporal and geographical scales differ. Nevertheless, most archaeologists work with the assumption that these units are to some extent meaningful and useful in describing the structure of the archaeological record of Upper Paleolithic Europe.

The following is a brief and partisan summary of the status of these major taxonomic units as they are currently used. The earliest "transitional" Upper Paleolithic industries in Europe remain enigmatic and heavily disputed regarding their association with Neanderthals and/or anatomically modern humans (Hublin 2015). "Proto-Aurignacian" assemblages appear to relate to a distinct chronological phase, earlier than "Aurignacian" assemblages sensu stricto (Le Brun-Ricalens et al. 2009; Teyssandier et al. 2010; Bordes et al. 2011; Banks et al. 2013a, b). Numerous chronologically restricted types of Aurignacian assemblages can be identified in various parts of Europe based on lithic and osseous evidence (especially the presence/absence of index fossils such as *burins busqués*); examples include the Early Aurignacian and Evolved Aurignacian groups in Western Europe (e.g. Noiret 2009; Michel 2010; Sinitsyn 2010; Bordes et al. 2011; Anghelinu and Niţă 2014; Chu et al. 2018). The situation concerning Gravettian assemblages is rather similar, in that numerous Gravettian *faciès* are described for different time periods and areas based on assemblage contents, particularly the presence of particular index fossils: our knowledge of these across Europe is perhaps better than for Aurignacian assemblages, and examples include the Rayssian, Noaillian and Kostënki-Avdeevo Culture groups (e.g. Klaric 2007; Noiret 2009, 2013; de la Peña and Vega Toscana 2013; Pesesse 2013; Reynolds 2014; Lengyel 2016). For later periods, the situation becomes more complicated. In parts of Western Europe, Solutrean and Badegoulian assemblages post-date Gravettian assemblages, which in turn are post-dated by Magdalenian assemblages (Straus 2000; Ducasse and Langlais 2007; Renard 2011; Ducasse 2012; Langlais et al. 2016). However, in much of Eastern, Central and Southern Europe, where Solutrean assemblages have not been identified, Late Upper Paleolithic assemblages post-dating Gravettian assemblages may be described as Epigravettian, Magdalenian or Epiaurignacian (Burdukiewicz 2001; Svoboda and Novák 2004; Verpoorte 2009; Maier 2015). The latest Upper Paleolithic assemblages in Europe have been attributed to a multiplicity of taxonomic units including Azilian, Hamburgian, and Swiderian (e.g. Grimm and Weber 2008; Burdukiewicz 2011; Fat Cheung et al. 2014; Sauer and Riede 2019). However, the validity of the distinctions between many of the Late and Final Upper Paleolithic taxonomic units is in fact rather questionable (Svoboda and Novák 2004; Maier 2015: 236–237, Naudinot et al. 2017; Sobkowiak-Tabaka and Winkler 2017; Sauer and Riede 2019). Finally, some geographically restricted taxonomic units, especially in Eastern Europe (e.g. Streletskian, Gorodtsovian) have resisted inclusion into the main European chronocultural framework and their significance remains difficult to understand (Sinitsyn 2010, 2015).

Much archaeological research continues to be carried out based on an assumption of the robusticity and essentiality of these units and the differences between them. However, our taxonomic units are not mutually equivalent in their salience, their temporal and geographical scope, or the amount of material culture variation they incorporate. The major taxonomic units (Aurignacian, Gravettian, Magdalenian, Epigravettian) relate to many thousands of years and huge geographical areas, subsuming a significant amount of variation in material culture, subsistence practices, mobility patterns, and so on. Although each of these groups are, in principle, united by certain aspects of their material culture, and relate to coherent periods of time and contiguous geographical areas, as outlined above the fact of internal variation and phasing within each of these taxonomic units is extremely well-established. On the other hand, the distinctions between many taxonomic units are questionable, and there may be significant continuity in material culture variability between groups of assemblages conventionally attributed to different units. This includes similarities between units that are separated chronologically (e.g. Gravettian and Epigravettian: Mihailovic and Mihailovic 2007; Anghelinu et al. 2018) and between those that are separated geographically (e.g. the numerous Late Upper Paleolithic industries of Central Europe: Sobkowiak-Tabaka and Winkler 2017; Sauer and Riede 2019).

As is widely understood, the existing system of units has developed historically and is far from systematically constructed. Certain regions (especially, of course, Aquitaine) have been far more important for the construction of units than others, and the taxonomic units originally defined based on Aquitanian material have been subsequently applied across Europe (e.g. Otte 1981; Noiret 2009; Sinitsyn 2015). Political factors and nationalist frameworks have heavily shaped the system of taxonomic units that we use (Tomášková 2003; Vander Linden and Roberts 2011; Sauer and Riede 2019). Quite aside from the complex history of development of the taxonomic framework, the nature of the archaeological record itself does not always lend itself easily to the systematic definition of equivalent units. Some parts of the Upper Paleolithic are more obviously distinctive in their surviving material culture than others, which may or may not reflect past cultural distinctiveness. Furthermore, the heterogeneous geology of Europe has created great variation in depositional contexts. Short-term open-air sites in Eastern Europe present very different challenges and opportunities for defining taxonomic units than do dense cave sequences from further west. Finally, of course, even if we can obtain a full understanding of the archaeological record and its history of interpretation, the definition of taxonomic units from first principles is by no means a settled matter (e.g. Clarke 1968; Dunnell 1971; Gamble et al. 2005; Clark and Riel-Salvatore 2006; Riede 2011; Shea 2014).

Given the known problems with the cultural taxonomic framework as it currently exists, it is clearly inappropriate to equate cultural taxonomic units with past populations. In some cases, there may have been population continuity between chronologically or geographically distinct taxonomic units; in others, taxonomic units may subsume multiple distinct prehistoric populations. Cultural taxonomic units, at whatever scale, should not be treated as representing discrete, monolithic cultural phases; nor should they be correlated with discrete, distinctive past populations.

However, this critique of the cultural taxonomic framework should *not* be taken to question the existence of clear patterning within the Upper Paleolithic archaeological record. Similarities and differences between sites and assemblages do often reflect past sociocultural relationships, and these can be used to examine questions of population dynamics in the Upper Paleolithic. But in order to start addressing questions of population dynamics more accurately, we need to find a better approach to building and conceptualizing our chronocultural framework. In the following sections I want to explicitly outline one particular approach to chronology and material culture comparison that can be used across the European Upper Paleolithic record. Most of this is not new, and my version of this approach is built on the work of numerous other researchers (e.g. Garrod 1938; Rogachëv 1957; de Sonneville-Bordes 1966; Demars and Laurent 1992; Grigor'ev 1993; Bordes 2006; Le Brun-Ricalens et al. 2009; Klaric et al. 2009; Noiret 2009; Teyssandier and Zilhão 2018). However, since this kind of approach is not universally used or understood, I think it is worth describing it explicitly.

The European Upper Paleolithic Chronocultural Framework: Warp and Weft

A large part of the history of progress in European Upper Paleolithic studies is a history of improved understanding both of the chronology and sequencing of assemblages (what I refer to in this chapter as the "warp") and of intra- and inter-regional comparisons based on material culture (the "weft") (Fig. 10.1). Together this knowledge can be combined to form what we can call the chronocultural framework of Upper Paleolithic Europe: an overview of the material variability of the archaeological record in its chronological and geographical framework. I have deliberately chosen this warp and weft analogy because it helps to illustrate a fundamental point: in the approach I am outlining here, there is an assumption that there is a certain underlying regularity to the archaeological record that can and should be used to help us synthesize our understanding, and that both chronological and material culture evidence should be used dialectically. Where artefact or assemblage types are well-described, they often cluster in time and space, even if the scale of the clusters varies depending on the aspect of material culture we are examining. Some aspects of material culture were highly persistent, lasting for thousands of years; others were far more ephemeral. Some material culture features have been found across Europe; others have only been found in limited regions. The temporal and geographical restriction of certain features is what allows us to construct a useful chronocultural framework for Upper Paleolithic Europe.

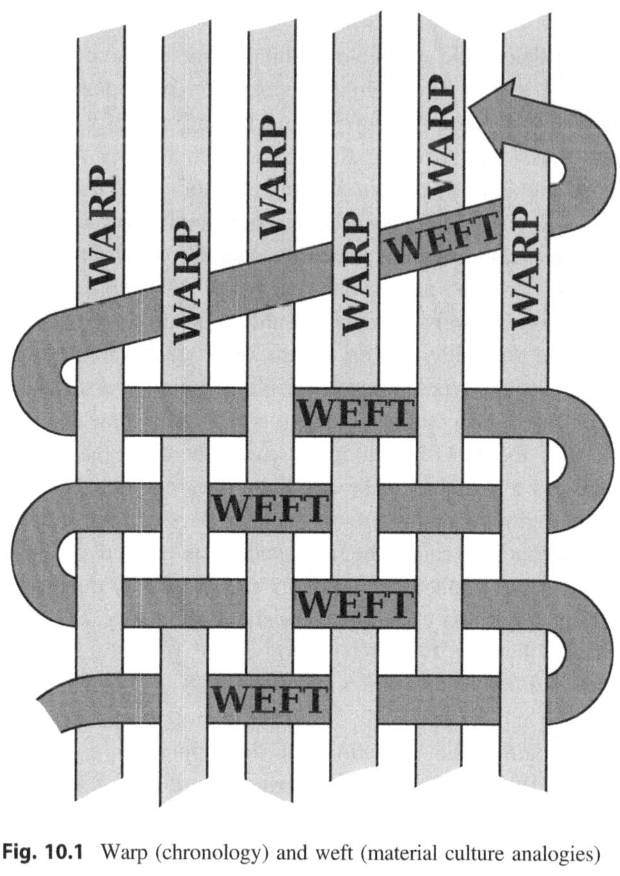

Fig. 10.1 Warp (chronology) and weft (material culture analogies)

This type of work remains utterly essential to archaeological practice: we have no hope of understanding complex processes such as population dynamics, the spread of technologies, or responses to environmental changes, without comprehensive knowledge of what material culture variability looks like. It should be noted from the outset that I count the construction of taxonomic units as entirely secondary to the identification of similarities and differences in the archaeological record. Upper Paleolithic taxonomic units, if employed, should be treated as heuristic, revisable concepts, useful largely for summarizing variability rather than as analytic units (Reynolds accepted manuscript). In other parts of the archaeological record traditional taxonomic units have also been the subject of critical attention and their usefulness for describing variability or as analytical units is in many cases questionable (e.g. Scerri et al. 2014; Shea 2014; Groucutt 2020). I hold that the description of variability is best done from the bottom up, with an explicit focus on specific features of material culture and other aspects of the archaeological record, and that it is not in fact necessary to attempt to place assemblages into discrete taxonomic units. For example, when evaluating the chronocultural framework of Gravettian sites, I consider it far more important to consider the differing distributions of the numerous Gravettian lithic index fossils (Gravette points, *éléments bitronqués*, shouldered points, Font-Robert points, Raysse burins, etc.), female figures and other features than to try and place sites into discrete taxonomic units or groupings.

An up-to-date synthesis of the chronocultural framework for the whole of the European Upper Paleolithic remains, at present, far from reach. Many good partial summaries of the archaeological record of particular regions or time periods are available but overall we have yet to find a way of integrating all the available information together in a way that formally describes our uncertainties and is useful as an interpretive model. To continue with our textile analogy, we would like a complete, smoothly woven canvas to work with, but although we have some good, strong threads in the right place, in both the warp and weft, there are also numerous fibers that need to be tied together, not to mention large holes to be filled and various mistakes to be undone and rewoven. Nevertheless, real incremental progress is being made in our understanding of the chronocultural framework of Upper Paleolithic Europe. In the following sections I describe the approach that is allowing this progress to be made. The approach I outline is not universally endorsed, and later in the chapter I will discuss some of its detractors' arguments.

There are two principal aspects to the construction and ongoing revision of the chronocultural framework for Upper Paleolithic Europe: material culture comparison and chronology. Both are necessary, and both have their strengths and weaknesses. In the next pages I discuss how we can use each of them to describe and revise our chronocultural framework.

Threading the Weft: Comparative Material Culture Study

Upper Paleolithic material culture from approximately contemporary sites across Europe often shows profound similarities. For example, there are apparent strong similarities between Aurignacian bladelets found across Europe (Le Brun-Ricalens et al. 2009; Tsanova et al. 2012; Dinnis et al. 2019), early Gravettian microgravette assemblages in Italy, Russia and elsewhere (Sinitsyn 2007, 2013; Moreau 2010; Wierer 2013; Reynolds 2014), and between female figures ("Venus figurines") found across Europe in late Gravettian assemblages (Mussi et al. 2000; Soffer et al. 2000; Paris et al. 2017; Khlopachev et al. 2018). The strength of these similarities varies from case to case, and during some time periods there appears to have been more regionalisation in material culture than during others. Most archaeologists recognize the reality of these similarities, and furthermore assume that the similarities in material culture dating to the same time are because there were similarities between what people were doing at more than one location at the same time and that this is due to sociocultural connections between them, either direct or historical. In theoretical terms, this is based on ideas, in all their great variety, of the critical importance of social and cultural factors in shaping material culture and technological practice (e.g. Leroi-Gourhan 1964–65; Sackett 1982; Pigeot 1990; Dobres 1999; Pelegrin 2007; Mesoudi and O'Brien 2009; Knappett 2011; Jordan 2015; O'Brien and Bentley 2020).

There are numerous aspects of material culture that can be examined from a comparative perspective. Lithic assemblages are the main basis for the description of variation, but other aspects of material culture (e.g. personal ornaments, osseous assemblages) and, indeed, evidence beyond the strict definition of "material culture", such as the remains of dwelling structures, faunal assemblages and site distribution with respect to landscape, can also be used to tell us something about past similarities and differences (e.g. Iakovleva 2003; Vanhaeren and d'Errico 2006; Svoboda 2007; Goutas 2013; Perlès 2013; Gaudzinski-Windheuser and Jöris 2015; Wojtal et al. 2018).

However, lithic techno-typology remains a key aspect of material culture comparison for the European Upper Paleolithic. Typology—when done well—is a powerful archaeological tool that is highly relevant to contemporary archaeological practice. Modern lithic artefact typology usually takes into account technological information, and the term "techno-typology", from the French "*techno-typologie*" is increasingly encountered in English-language literature, emphasizing that lithic technology needs to be studied in combination with lithic typology: from the point of view of cultural taxonomy in particular, the two are inseparable. Modern day techno-typology studies the morphology *and* technology of lithic artefacts with a view to understanding the "types", either emic or etic, *and* the technological practices underlying artefact variation. Much modern work of this kind is highly revisionist, and applies a critical approach to previously established artefact types (e.g. Soriano 1998; Hays and Lucas 2000; Pesesse and Michel 2006; Le Brun-Ricalens et al. 2009; Klaric et al. 2009, 2015; Pesesse 2009–2010; Lev et al. 2011).

Techno-typology is vital to the definition and usage of index fossils (*fossiles directeurs*): chronologically and geographically restricted artefact types that are key to the comparison of assemblages. (The term "type fossils", often encountered in English-language archaeological literature, is a somewhat misleading usage, since in biology this term refers to the "type specimen" or "holotype" used as a reference for the formal definition of a species or population.) In Upper Paleolithic archaeology, lithic index fossils continue to be key to the definition and correlation of archaeological deposits. Because they have been a major focus of work over the years, and because they have been heavily used for inter-site comparison, our existing chronocultural framework and taxonomic units have largely been built using them.

The relationship between a defined, ideal "type" or "class" and an actual physical group of archaeological artefacts is rarely straightforward. Questions of how to manage variability within groups of artefacts, how best to define formal types, whether to split or lump, and so on, are part of archaeological techno-typological practice and debate: the fact that in many cases there are no "right" answers to many of these questions does not mean that the whole enterprise is worthless (neither does it mean that our understandings cannot be improved). It is perfectly possible to carry out a pragmatic typology of artefacts by treating all of our types and classes as heuristic, preliminary, etic categories that are nonetheless potentially reflective of past sociocultural relationships (Hayden 1984; Dunnell 1986; Adams and Adams 2009: 282–284) (although it is important to recognize the limitations of this approach; e.g. Odell 1981). What this means is that we can use archaeologically recognizable, defined types and classes for comparative purposes, regardless of our level of confidence that they were purposefully created by past people or that they were used, for example, to self-consciously demonstrate group affiliation (Wobst 1977; Sackett 1982, 1985; Wiessner 1983, 1985). Unconscious technological habits are just as important as conscious efforts in creating the traits and patterns that we see in past material culture, and are also subject to the forces of cultural inheritance, transmission and drift (e.g. Barton 1997; Hurt and Rakita 2001; Lyman and O'Brien 2004; Collard et al. 2009).

It needs to be noted that Upper Paleolithic index fossils are best established with a view not only to any inherent techno-typological distinctiveness but also to their distribution in the record. By this I mean, as has long been established, that useful index fossils are clearly restricted chronologically and often also geographically in the record (de Sonneville-Bordes 1966; Demars and Laurent 1992). Endscrapers, burins *sensu lato* and other very common tool types are not suitable index fossils for defining the chrono-cultural framework of the European Upper Paleolithic. A good index fossil is one that can be clearly and explicitly defined using technological and morphological criteria, and that is clearly restricted within the archaeological record.

The question of the technological relationships between index fossils, and particularly whether separate index fossils reflect stages in the reduction of a single tool type, is also important. It is clear that the majority of Upper Paleolithic index fossils cannot feasibly have been converted from one type into another, in contrast to e.g. Middle Paleolithic scraper "types" (Dibble 1995). The risk of misidentifying index fossils as incomplete or modified versions of other index fossils, e.g. fragments of incompletely backed bladelets as shouldered points (Reynolds 2014; Polanská and Hromadová 2015; Wilczyński 2015), is widely understood among lithic specialists. Some of the variation that we do see within particular groups of artefacts may well be a reflection of the application of additional retouch to modify given tool types for use, as argued by Neeley and Barton (1994) for some Levantine Epipaleolithic tools. However, this is not a problem unless it causes the inappropriate definition of multiple index fossils (rather than the description of some artefacts as "atypical" examples): in any case, if one index fossil turns out to be simply an ad hoc modification of another tool type, then their geographical and chronological distributions should coincide.

We have come a long way since the formative studies of Upper Paleolithic lithic typology by de Sonneville Bordes and Perrot (1954, 1955, 1956a, b), and even since the useful updates to this work by Demars (1990) and Demars and Laurent (1992). Recent work has focused closely on individual artefact types, their formal definition, the technology of their creation (often informed by a *chaîne opératoire* approach; see also Maher and Macdonald 2020), and consideration of ariability within the groups of artefacts attributed to each type. This has led to the definition of new index fossils (e.g. *éléments bitronqués*/Late Gravettian rectangles in Central Europe; Polanská and Hromadová 2015; Wilczyński et al. 2015), the correction of previous misclassifications of artefacts (e.g. Kostënki knives in Western and Central Europe; Lev et al. 2011; Klaric et al. 2015); critical analysis of the coherence of particular types (e.g. northern European Final Paleolithic tanged points: Serwatka and Riede 2016), and systematic formal comparison and reclassification of traditional types (e.g. Early Upper Paleolithic bladelets from across Europe: Le Brun-Ricalens et al. 2009). This work is being carried out all over Europe and is making real, if necessarily piecemeal, improvements to our understanding of material culture variability.

The study of lithic technology *sensu stricto*—the full process of production of stone tools, from the first blows to a nodule to the final stages of retouch or resharpening of an artifact—can also be used to compare assemblages. There are abundant possibilities for this type of comparison: for example, blank production strategies (e.g. specific features of blade and bladelet production in Proto-Aurignacian assemblages: Le Brun-Ricalens et al. 2009; Teyssandier et al. 2010; Bordes et al. 2011), retouch characteristics (e.g. lateralization of backing in Gravettian assemblages; Harrold 1993; Reynolds 2014), and the use of different types of percussion (e.g. varying usage of soft stone and organic hammers throughout the Upper Paleolithic in Western Europe; Pelegrin 2012).

Studies of lithic artefacts are informative at various scales. The production of backed bladelets and general pervasiveness of the use of abrupt backing for many thousands of years all over Europe, as seen in Gravettian and later assemblages, tells us something about the persistence of particular traditions on a large time-scale. On the other hand, the chronological and geographical restrictedness of certain distinctive index fossils and technological habits (e.g. bladelet production using various types of carinated artefacts; Bordes 2006) provides insights of a different kind.

Other types of material culture can also be used to explore the differences and similarities between sites, and, fascinatingly, often give a different picture of variation than lithic assemblages do. Osseous artefacts are in some cases already recognised as at least as important to inter-site comparisons as lithic artefacts, as is the case for Aurignacian osseous points and Gorodtsovian bone "paddles" (Sinitsyn 2010; Doyon 2019). Upper Paleolithic personal ornaments show complex patterns of variation that do not always map straightforwardly onto patterns seen in other aspects of material culture (Vanhaeren and d'Errico 2006; Perlès 2013; Rigaud et al. 2014). Female figures ("Venus figurines") have been found in late Gravettian assemblages across Europe dating to approximately the same time, but the lithic assemblages with which they are associated show clear typological differences (Efimenko 1958; Otte 1981; Gvozdover 1998; Lev 2009; Simonet 2012; Paris et al. 2017).

Comparisons between different aspects of material culture —including different aspects of lithic assemblages as well as of non-lithic assemblages—sometimes mirror each other and sometimes contradict each other. This should not be seen as a problem. These variances can tell us something about the complexity of social and population processes during the Upper Paleolithic (Vanhaeren and d'Errico 2006; Hromadova 2012; Perlès 2013; Goutas 2016). As further discussed below, the key to managing and understanding this complexity within our chronocultural framework is to consider each aspect of material culture separately and to treat them all as potentially informative of past sociocultural processes.

The approach advocated in this chapter focuses on the presence/absence of particular features in assemblages. The high degree of variability of the Upper Paleolithic record makes it well-suited to this type of approach. Although it does not provide a full picture of the similarities and differences between sites it is an excellent way to build a comprehensive preliminary bottom-up framework that does not rely on traditional top-down cultural taxonomy. It should be noted however that statistical comparisons of the technological or morphological attributes of assemblages have also proved useful for evaluating material culture variability and testing traditional taxonomic units (e.g. Scerri et al. 2014; Serwatka and Riede 2016; Doyon 2019). Here, given that the area of interest is the European Upper Paleolithic record as a whole, I have deliberately chosen an approach that facilitates the rapid comparison of a large number of assemblages and provides a very clear basis for comparisons. The incorporation of data on e.g. relative abundances of artefact types, or the results of multivariate statistical analyses, could in principle be combined with presence/absence data as part of the same framework, but this would require careful planning and would add greatly to the complexity of the functional, raw material and other factors that need to be considered in order to enable valid comparisons. A bottom-up chronocultural framework based on the presence/absence of particular features already provides many advantages over the traditional cultural taxonomic framework and, importantly, can feasibly be constructed for the entire European Upper Paleolithic record.

Testing the Warp: The Importance of Chronology

The second main axis of our chronocultural framework is chronology, unquestionably key to the study of the Upper Paleolithic. Examining change through time on the site, regional or continental level requires understanding of both relative and absolute chronologies of assemblages. However, chronology is not a value-neutral field. Different archaeologists and archaeological scientists place varying emphasis on each aspect of chronology building, and these differences in emphases help to explain many of the most heated debates in Upper Paleolithic archaeology in recent years. Researchers have different ideas of what is best practice and differ in how the relationship between chronology and material culture comparisons should be managed. Here, I discuss stratigraphy and absolute dating in turn.

Stratigraphy

Stratigraphy has been a key aspect of archaeological chronology building since the earliest days of the discipline and remains so. Its most basic principle—that archaeological material was physically deposited in chronological order—is simple and inarguable. In practice, of course, there are many nuances that need to be taken into account and that become increasingly important as we build chronologies in greater detail.

There are numerous recurrent problems in the study of stratigraphy. The lack of reliable and detailed stratigraphic information for many key excavations, especially early excavations, causes frequent difficulties (e.g. Gravina et al. 2018; Teyssandier and Zilhão 2018). Even where stratigraphic recording is impeccable, the complexity of formation processes can pose serious problems for interpretation. Stratigraphic units are also often treated as individually uniform despite the fact that we know that they do not necessarily relate to discrete collections of archaeological material and the divisions between them are often subject to error (e.g. Discamps et al. 2015). This issue comes into sharpest focus when we consider the problem of "mixing" between stratigraphic units. Refitting studies at numerous sites (e.g. Hahn 1988; Morin et al. 2005; Discamps et al. 2015; Gravina et al. 2018) have shown that contemporary material may be found in separate stratigraphic units, either due to taphonomic processes or to misinterpretations of stratigraphy during excavation. Problems with stratigraphy underlie many of the most intractable problems that we have in understanding the relationships between certain Upper Paleolithic assemblages.

Radiocarbon Chronology

For the Upper Paleolithic, the most important method of absolute dating remains radiocarbon dating, although luminescence dating is of increasing importance, especially as precision and reliability improve (e.g. Lomax et al. 2014;

Frouin et al. 2017). Radiocarbon dating is key to chronological comparison of assemblages, particularly from single-layer sites, but also for refining the chronology of long sequences.

Unfortunately, radiocarbon dating remains far from infallible. The fact that only tiny amounts of contamination containing modern carbon can skew results for Upper Paleolithic samples by thousands of years has rendered many published dates highly misleading (Higham 2011). Certain labs appear to be more reliable than others in producing accurate dates. There is considerable variation in methods used, and presumably in adherence to protocols during sample pretreatment and measurement. Methods have also changed over the years, with some clear improvements in reliability, at least at certain labs.

Methods for the AMS measurement of isotopes have already reached an extremely high level of accuracy (Bronk Ramsey et al. 2004). The part of the radiocarbon dating process that is more potentially problematic in the present day is the pretreatment of samples ahead of AMS measurement. Pretreatment generally involves the attempted isolation of a particular part of a sample. For charcoal, available pretreatment methods appear to be largely reliable for producing accurate results (Brock and Higham 2009; Haesaerts et al. 2013). However, the isolation of collagen from bone samples for dating is more methodologically challenging than sometimes understood: standard methods such as ABA (acid-base-acid washes), with or without ultrafiltration, cannot be said to reliably remove all non-collagen material from a sample, and the removal of conservation materials, even with the use of additional solvent washes, appears particularly problematic (Brock et al. 2013, 2018; Marom et al. 2013).

One of the most interesting and promising recent developments in radiocarbon dating has been the application of single amino acid (hydroxyproline) dating to bone samples. This method very effectively ensures the isolation of collagen material only, by using high-performance liquid chromatography (HPLC) to isolate the amino acid hydroxyproline, found almost uniquely in collagen. The isolation of hydroxyproline means that almost all contaminants (apart from collagen-based glues and preservatives) can be excluded, in principle leading to far more accurate dates than before. The method is expensive and labor-intensive, and continues to be subject to methodological improvements (McCullagh et al. 2010; Marom et al. 2013; Nalawade-Chavan et al. 2014; Devièse et al. 2018). Nevertheless, it has produced some outstandingly interesting results, including the first convincing, consistent dates for the burials from Sungir', Russia (Marom et al. 2012), which have been difficult to date due to the heavy contamination of the human remains and other archaeological material with preservatives.

Recent results obtained using this method, however, should also focus attention on the potential shortcomings of more established radiocarbon dating methods. Although the problems with contaminated material are well known, new results of single amino acid dating suggest that even material with a pristine curatorial history may be difficult to date accurately. In one study, two bone samples from recent excavations at Abri Blanchard, France were dated using both the established ABA/ultrafiltration method and the single amino acid method: the latter produced results that were several thousand years older (Bourrillon et al. 2018). The authors of the study suggested that the site's geochemistry may have something to do with the discrepancy in dating, as humic acids deriving from groundwater may have become cross-linked with collagen molecules, causing the results obtained from conventional methods to be incorrect. In a further example, bones and personal ornaments from Kostënki 17/II, Russia, dated using both ABA/ultrafiltration methods and the single amino acid dating method produced dates where, again, the results from the latter method were several thousand years older than those from the former (Dinnis et al. 2019). Although the dated bones were from twentieth-century excavations and their curatorial history is incompletely known, they were not visibly treated and they were washed with solvents at the beginning of sample pretreatment, in an attempt to remove any invisible glues or preservatives (Brock et al. 2010).

In both these studies the bone samples were very similar to many others that have been assumed to be entirely suitable for standard ABA/ultrafiltration dating. The only reason that we know that in these cases the dates produced using ABA/ultrafiltration are inaccurate is because we also have results obtained using the hydroxyproline method. These archaeological examples echo the results of experiments where a ^{14}C-depleted bone, ca. 60–70 thousand years old, was soaked in hot tea for one hour to mimic the effects of humic and fulvic acids on buried archaeological samples (Marom et al. 2013). Despite applying pretreatment methods including ultrafiltration, the radiocarbon date subsequently produced from the treated sample was ca. 22 kya ^{14}C BP, showing that ca. 6% of the dated carbon derived from the modern tea. Single amino acid methods, however, successfully produced an infinite radiocarbon date for the treated sample indistinguishable from that obtained from control samples.

Unfortunately, we have no routine way at present of determining whether an archaeological sample has been affected by, for example, contact with humic and fulvic acids in soil and groundwater. This is a significant blow to efforts to create detailed chronologies based on radiocarbon dates from bone samples: it means that any date not produced using the single amino acid technique, especially for the earlier part of the Upper Paleolithic where contamination causes more acute problems, must be treated as questionable until these processes are better understood. Furthermore, where dates are wrong, they will likely appear to be younger than they should: this means that this factor does not introduce random statistical noise, but in fact causes bias in one direction. Therefore, Bayesian statistical methods as currently employed in radiocarbon chronology building are not appropriate to counteract this source of error. The effects of geochemistry on radiocarbon dating samples must be treated as a priority area for research; so too must the development of statistical modelling methods for compensating bias in radiocarbon dates using stratigraphic and archaeological information.

The known problems with stratigraphy and radiocarbon dating are a principal reason for my support of a dialectical approach to the construction of the Upper Paleolithic chronocultural framework. To obtain a strong chronological framework, it is not enough to uncritically accept the results of absolute dating of particular artefacts or stratigraphic units: we need to use material culture comparisons to inform our chronological inferences. In the next section, I outline how this works in practice.

Coherence and Convergence

The preceding sections described both principal axes of the Upper Paleolithic chronocultural framework: material culture comparison and chronology. Because neither chronology nor material culture comparison are infallible and neither alone can describe the chronocultural framework, we need to combine them dialectically, carefully weighing evidence from both sides. In order to do this, we need a theoretical position. The position advocated here is to assume coherence in the archaeological record: i.e. to assume that similarities in archaeological material cluster geographically and temporally. This is based on the assumption that similarities that we see in the archaeological record are the result of similarities in behavior between people in the past, and that people who were closer in time and space tended to be more similar in behavior. This approach sees variation in material culture as having been shaped by historically situated activity within a social context: in other words, that many of the similarities and differences we see are the result of relationships between people, either through contemporary interaction or through relationships of inheritance from a common ancestor (e.g. Sackett 1982; Dobres 1999; Mesoudi and O'Brien 2009; Knappett 2011; Tixier 2012; Jordan 2015; O'Brien and Bentley 2020). Furthermore, it assumes that the mobile hunter-gatherers of the Upper Paleolithic were highly socially connected across long distances, and that cultural changes spread quickly by diffusion. Therefore, we should not expect to see the static, geographically restricted existence of particular traditions over many thousands of years in a small area.

In practice, and based on the examples where we have good understandings of both chronology and material variation, what does the archaeological record of Upper Paleolithic Europe look like? It can be envisaged as a three-dimensional model, with time in the vertical dimension and space in the two horizontal dimensions (Fig. 10.2), although for the sake of illustration we can also envisage it as a two-dimensional model, with time in the vertical dimension and space in the single horizontal dimension (Fig. 10.3). But in any case, when we focus on the most chronologically and geographically restricted index fossils and other features, we can use them to link series of assemblages across space. This leads back to the "warp and weft" analogy used in this chapter: we can connect assemblages according to their temporally most specific aspects, in which case they cluster closely in the vertical, time dimension, and are dispersed to a greater or lesser extent in the horizontal, space dimension, just like colored threads on a loom. However, we can also use less temporally specific aspects of assemblages (e.g. backed lithic technology in Gravettian and later assemblages) to link large groups of sites over long periods of time. Artefact categories with very little geographical or temporal specificity, including non-specific burins, endscrapers, retouched blades, etc. are not useful for this exercise.

The assumption of coherence creates certain expectations, with consequences for how we evaluate archaeological information. For example, if a particular well-defined index fossil is found at eight sites within a region, and six of these sites are radiocarbon dated to within two thousand years of each other, but two sites are dated to six and ten thousand years younger than the other sites, then at the final two sites both the identification of the index fossil and the dating of the assemblage should be questioned (Fig. 10.4). This extends to the occasional claims for extremely precocious appearances of certain types of assemblage, further discussed below. In another example, if the same well-defined and rare technological feature appears in a number of sites in two different regions, dated to approximately the same time, it is fair to ask whether there was some kind of connection between them even if there appears to be a geographical discontinuity in their distribution. In both cases, there are ways of further investigating the situation: in the first, by re-examining the lithics and re-dating the assemblages, in the second, by searching in collections from geographically intermediate sites to see if the same technological feature can be identified. Conversely, where similar archaeological features are found in assemblages that are securely dated to different periods and are perhaps geographically distant,

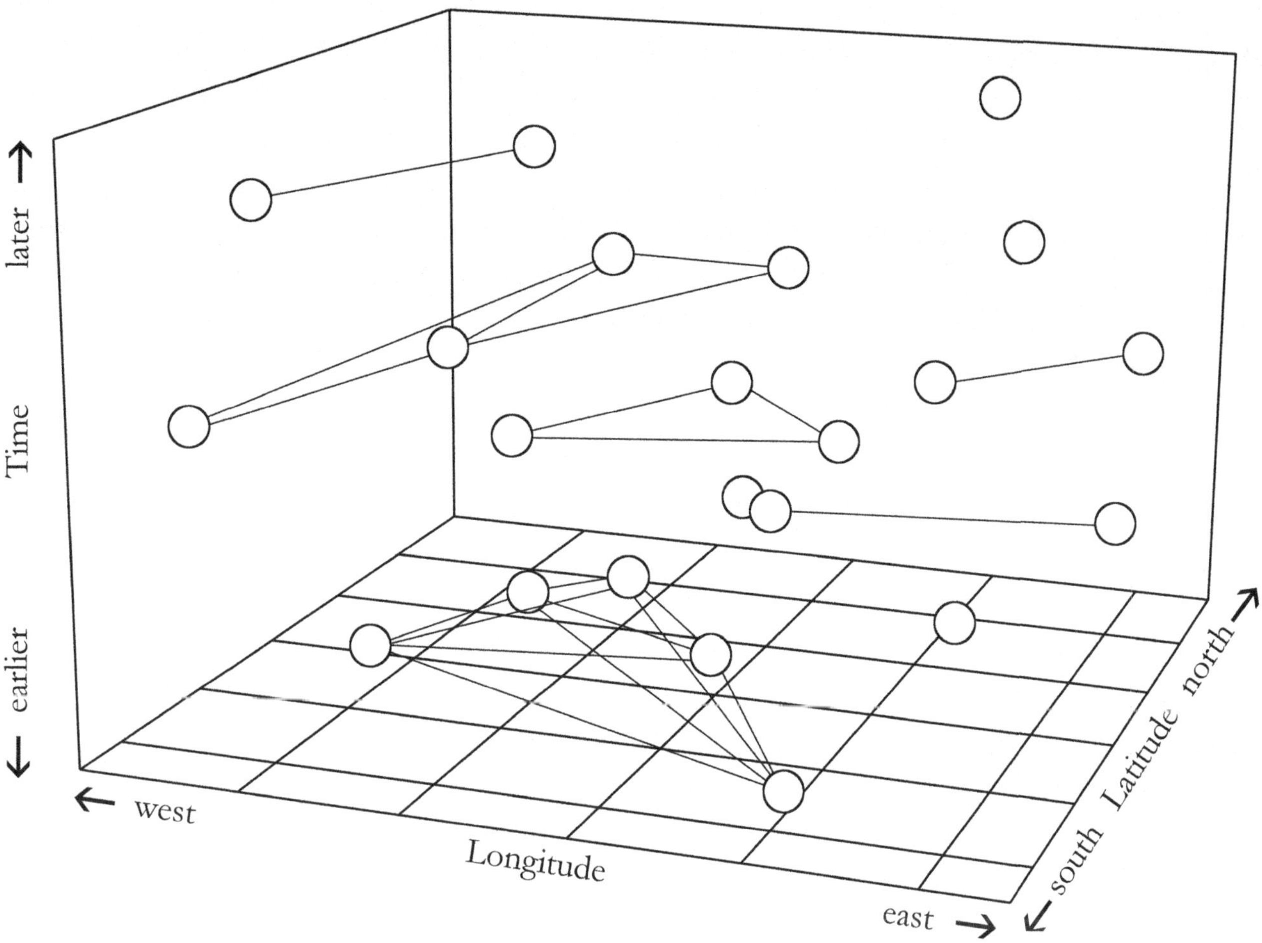

Fig. 10.2 Threading the weft: a simplified schematic diagram representing one way that we can visualize the Upper Paleolithic archaeological record. Spheres represent assemblages, and are placed in a chronospatial framework according to their estimated age and geographical location. The spheres are connected by lines where assemblages share a temporally restricted, well-defined material culture feature (e.g. an index fossil, a particular technology). This way of conceptualizing the archaeological record is the basis of the approach advocated in this chapter

without any evidence for the same features in intermediate part of the record, then we may have identified a case of convergence.

The question of convergence in Upper Paleolithic industries is more easily addressed than sometimes claimed (e.g. Clark and Riel-Salvatore 2006). Examples of convergence in artefact form and technological features can be identified within the Upper Paleolithic archaeological record. For example, Anosovka points, found in late Gravettian assemblages in Russia and Ukraine are similar to backed points found in Late Upper Paleolithic assemblages in northern and western Europe, including Federmesser points (Schwabedissen 1954; Baales et al. 2001; Beliaeva 2002; Sinitsyn 2007, 2014; Sobkowiak-Tabaka 2017; Reynolds et al. 2019). The assumption of coherence in the archaeological record greatly facilitates the evaluation of convergence. In order to assess possible convergence, we identify the assemblages where a particular feature is present, and consider whether they are geographically and chronologically contiguous. Gravettian sites with Anosovka points and Late Upper Paleolithic sites with Federmesser and other backed points each form a geographically and chronologically coherent group, but these two groups are independent both geographically and chronologically. We can assume that the finds of Anosovka points at several sites in eastern Ukraine and western Russia at the end of the Mid Upper Paleolithic were *not* the result of convergence. However, their similarities with much later, and geographically distant, Late Upper Paleolithic backed points *are* the result of convergence. The separation of geographically and chronologically distinct groups of assemblages allows us, if necessary, to use similar or even identical material culture criteria for defining more than one group of assemblages.

It gets more difficult to identify true cases of convergence as features become more frequent in the archaeological record (see also Will and Mackay 2020): is the common, but not universal, appearance of simple truncated backed bladelets in Gravettian assemblages the result of convergence,

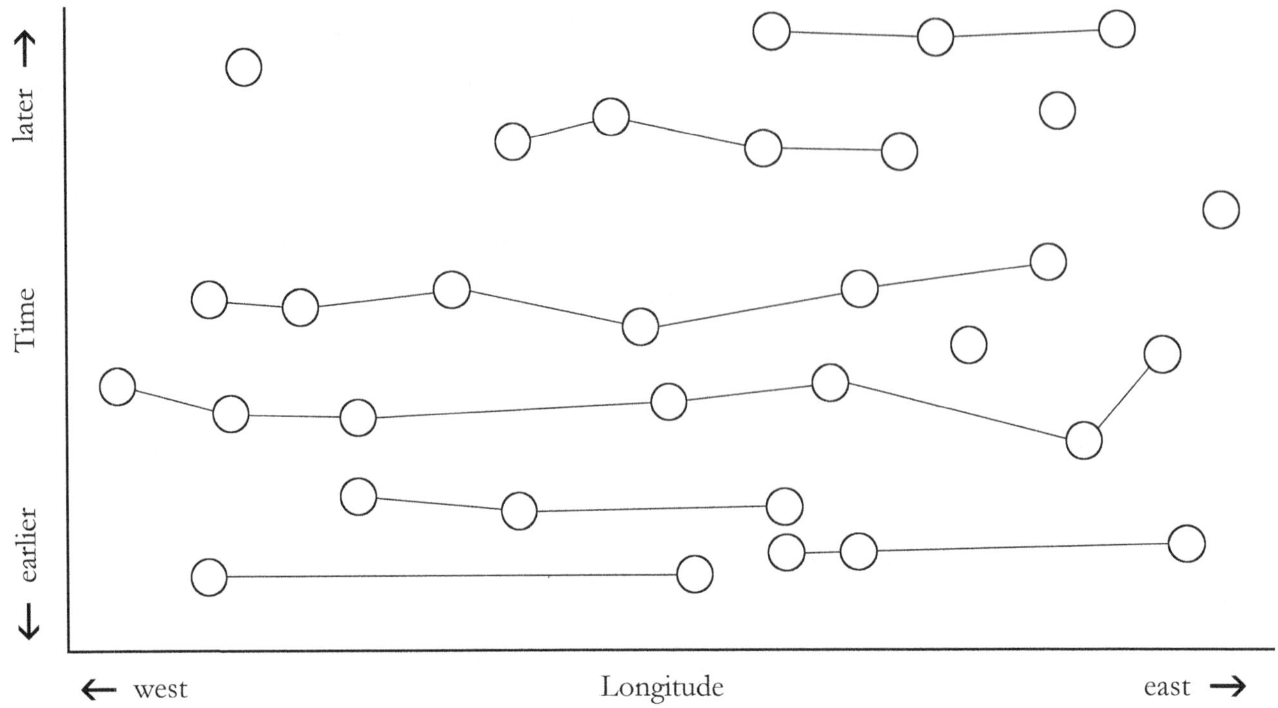

Fig. 10.3 An example of a simplified 2-dimensional depiction of the relationships between sites, similar to that in Fig. 10.2 but omitting the third axis (Latitude). This type of figure will be used in the rest of this chapter for the sake of simplicity

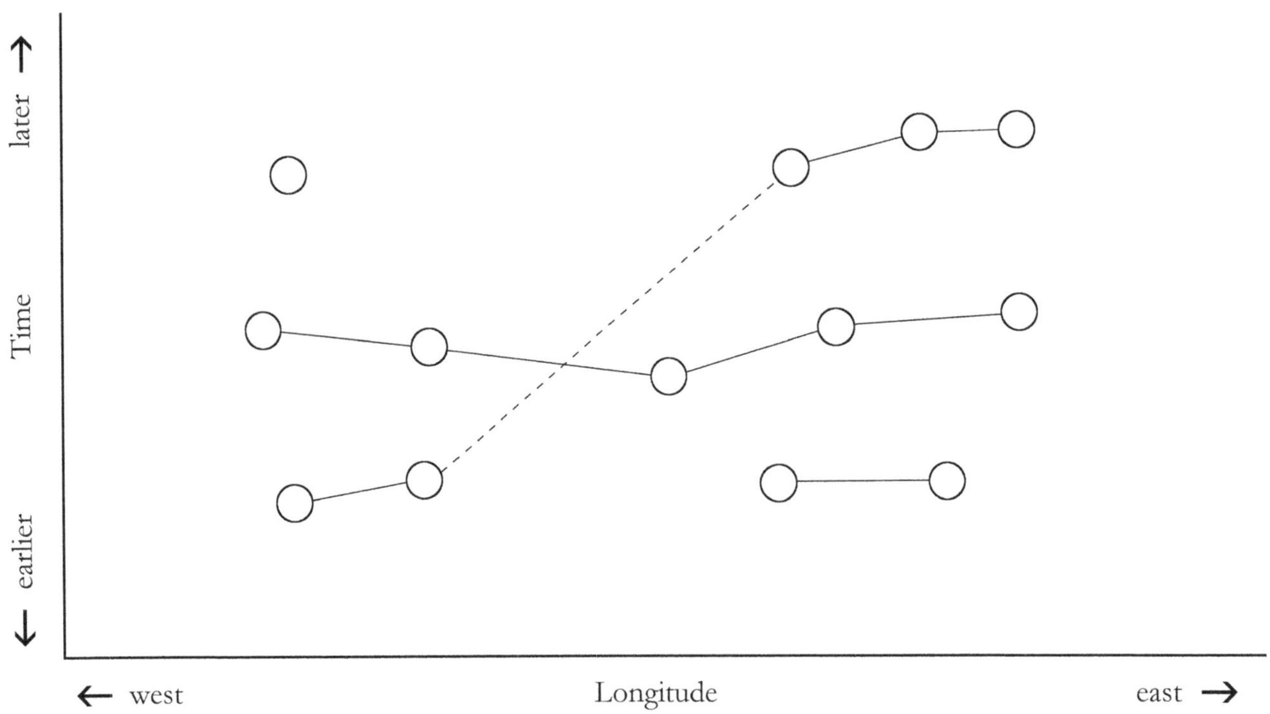

Fig. 10.4 Testing the warp: the identification of problems with the data or of convergence. Here, the dashed lines join assemblages that are distant in time but apparently share a material culture feature. Where this kind of result is obtained, both the accuracy of the material culture comparison and of the chronology of the assemblages should be questioned. If both are judged to be robust, and if there are no chronologically intermediate assemblages with similar material culture, and if there is a significant period of time between the younger and older assemblages (perhaps more than several thousand years) then the similarities between the assemblages are best treated as an example of convergence and no cultural link inferred or described between them

or can we use it to establish links between assemblages? There are several possible responses to this problem. First, we can simply avoid using a particular feature as the basis of establishing material culture similarity where such doubts are present. Second, we can examine carefully the exact distribution of the artefact type in the record, to see whether there is more patterning to its distribution than we previously thought. Finally, we can try to refine the criteria used, in order to see if distinctions can be made. For example, truncated backed bladelets *sensu lato* may not be useful for establishing connections between assemblages, but the specific type of truncated backed bladelet known as *éléments bitronqués*/Late Gravettian rectangles probably are: they appear to form a geographically coherent, temporally restricted group in Central and Eastern Europe in assemblages dating to the late Mid Upper Paleolithic (Rogachëv and Anikovich 1982; Reynolds 2014; Polanská and Hromadová 2015; Wilczyński et al. 2015; Lisitsyn 2015).

Alternative Perspectives

Not all researchers subscribe to the approach set out above. The theoretical conflict between approaches that do and do not prioritize the "big picture" of chronocultural coherence has been the cause of some acrimonious recent debates in Paleolithic archaeology. This especially relates to claims for precocious evidence of the appearance of particular types of assemblage in particular small regions, often several thousand years before they appear elsewhere in Europe. These claims are typically based on the dating of individual assemblages. If we adhere to the concepts of coherence outlined above, we will likely reject the possibility of such early, localized appearances.

Some prominent recent examples of this include claims for extremely early Aurignacian assemblages at Geissenklösterle and Willendorf II (Conard and Bolus 2003; Higham et al. 2012, 2013; Nigst et al. 2014), and extremely early Gravettian assemblages at Buran-Kaya III (Yanevich 2014). These ideas contravene the theoretical approach outlined above, and have been criticized on such grounds. For example, the recent criticism by Teyssandier and Zilhão (2018) of the extremely early dating of an Early Aurignacian assemblage at Willendorf II was in part initiated by the observation that the dating of an Early Aurignacian assemblage to several thousand years earlier than any other Early Aurignacian assemblage violated principles of coherence similar to those described above. ("Early Aurignacian" assemblages form a distinct sub-group of Aurignacian assemblages, characterized by a series of technological and typological features, and the term does not simply refer to Aurignacian assemblages with early dates). The authors confirmed the validity of their critique by examining, and finding significant flaws in, the stratigraphic association between the dated samples and the Early Aurignacian assemblage at Willendorf II. However, the critique was provoked by a theoretical observation. The claim for an extremely early Aurignacian assemblage at Geissenklösterle has similarly been criticized based on a detailed examination of the stratigraphy of the site, a critique which was again provoked by the observation that the claim violated a theoretical model of the appearance of Aurignacian assemblages across Europe (Zilhão and d'Errico 2003; Banks 2015). Likewise, the claim for an extremely early Gravettian assemblage at Buran-Kaya III has been rejected on the grounds of its chronological difference from all other Gravettian assemblages (Hublin 2015; Reynolds and Green 2019). Critiques of this kind have an important part to play in strengthening the chronocultural framework for Upper Paleolithic Europe as a whole, although they need to be backed up with empirical evidence to be truly convincing.

The sociocultural significance of material culture variation within the Upper Paleolithic record has also been challenged from various perspectives, often resulting in some degree of dissent from the approach set out above. For example, it has been argued that in the vast majority of cases it is not possible to reconstruct the intentions of the manufacturers and users of Paleolithic stone tools, and that this has a bearing on the typological study of assemblages (e.g. Marks et al. 2001; Dibble et al. 2016). However, such arguments do not undermine the approach advocated in this paper. First, in many cases the recognition of broad-scale patterning in Paleolithic material culture can be achieved regardless of whether we have fully analyzed the material and attempted to reconstruct the intentions of its creators: certain lithic index fossils and other features are very clearly restricted to certain parts of the archaeological record even if they are not yet satisfactorily understood from a technological and functional perspective. Second, I would argue (following many others, including Mellars 1989; Pelegrin 1991; Tixier 2012) that in many cases from the Upper Paleolithic we *can* reconstruct past intentions of creation to some degree. In assemblages where, for example, there are many hundreds of examples of a particular, technologically and morphologically homogeneous and meticulously created stone tool, dominating the retouched assemblage (e.g. microgravettes at Kostënki 8/II, Noailles burins at level IV of Isturitz; Sinitsyn 2007; Lacarrière et al. 2011; Reynolds 2014), it seems perfectly reasonable to assume that these artefact types were deliberately and systematically created. The idea of *systematic* creation may in fact be particularly important to the interpretation of material culture variation: a single artefact may be intrusive or an ad hoc creation; several hundred highly similar artefacts probably are not. A long debate played out during the late twentieth century concerning the difference between emic and etic typological categories (e.g. Hayden 1984; Dunnell 1986; Read 1989; Lyman and O'Brien 2004; Adams and Adams 2009; Van Oyen 2015); however, for the purposes of chronocultural framework building, it is perfectly acceptable to assume that all our typological categories are etic in nature. As long as our index fossils and other features are well-defined and restricted in the record, we can use them for chronocultural framework building, whether or not we think that the people who created and used them would have identified them as constituting a single category.

Some authors have also doubted the degree of cultural significance that should be attributed to Paleolithic lithic artefacts, seeing variability rather as largely the result of functional and mobility factors (e.g. Riel-Salvatore and Clark 2001; Clark and Riel-Salvatore 2006). The same authors also doubt that the resolution of the archaeological record is great enough to allow us to discern any cultural component that might exist in lithic variability. This sort of criticism perhaps fails to take into account the numerous examples where we do have excellent evidence for geographically and temporally restricted artefact variation. Although many assemblages are certainly palimpsests of multiple phases of occupation, this does not necessarily prevent us from defining variation in the record, much of which is best understood on a long-term scale in any case. For most archaeologists, at least some of this variation is best explained by cultural factors.

Finally, an obvious criticism to be levelled at the approach described here concerns the possibility of "leads" and "lags" in the distribution of particular material culture types: in other words, in identifying the spread of particular traditions. If we assume general contemporaneity between assemblages with similar material culture, and if we preferentially question radiocarbon dates and stratigraphic information that contradicts this assumption of contemporaneity, then it could be argued that we are excluding the possibility of identifying the earliest (or latest) occurrence of a particular type of material culture.

There are both theoretical and methodological responses to this. In theoretical terms, it must be remembered that we are dealing with the scanty material traces of mobile hunter-gatherers, and so the particular geographical "origin" of a given material cultural trait may be extremely difficult to define (Teyssandier and Zilhão 2018). The earliest appearance of a given trait in the archaeological record may post-date the diffusion of the trait across hundreds or thousands of kilometers. Furthermore, for most of the Upper Paleolithic our radiocarbon chronology is insufficient in resolution to identify leads or lags of less than a millennium, even though most transitions probably took place across Europe faster than this (d'Errico and Banks 2015; Reynolds and Green 2019). In this context, the assumption of near-contemporaneity between materially similar assemblages is acceptable, at least when building the first approximation of this framework. Further refinements, including the identification of possible leads and lags, become easier as the framework is established.

A Brief Case Study: Mid Upper Paleolithic Russia

Perhaps the best way to clarify the approach outlined in this paper is to present a case study of how it works in practice. Here, I discuss the Mid Upper Paleolithic (MUP; ca. 30,000–22,000 ^{14}C BP or ca. 34,000–26,000 cal BP) Gravettian record of European Russia. This is to show how a dialectic approach to radiocarbon chronology and assemblage comparison can be used to develop a working hypothesis of a chronocultural framework. Particular assemblages can then be targeted for further work, allowing us to strengthen and refine the framework.

There is only one Gravettian site in Russia dating to the early MUP: Kostënki 8/II, with a rich assemblage of microgravettes, dating to around 28--27,000 ^{14}C BP (Reynolds et al. 2015). There are no clear analogies in Eastern Europe for this site: comparisons have, however, been made with assemblages of approximately the same age containing microgravettes from sites across Europe such as Grotta Paglicci, Grotta della Cala, Geiβenklösterle and Abri Pataud (Sinitsyn 2007, 2013; Moreau 2010; Wierer 2013; Reynolds 2014).

Two Gravettian sites in Russia have now been directly dated to ca. 25,000 ^{14}C BP: Kostënki 4 and Borshchëvo 5 (Reynolds et al. 2015). Although it is very difficult to find strong contemporary analogies for the site of Kostënki 4 (Reynolds 2014; Zheltova 2015), the Borshchëvo 5 assemblage does find clear similarities in that from Kostënki 9 (Sinitsyn 2007, 2015; Lisitsyn 2015), due to the shared presence of *éléments bitronqués*. There are no radiocarbon dates yet available for Kostënki 9, but it seems reasonable to assume that Kostënki 9 dates to approximately the same time as Borshchëvo 5 based on their assemblage similarities. If radiocarbon dates can be obtained for Kostënki 9, this will help to test and refine this proposition.

A relatively large group of Gravettian sites is attributed to the Kostënki-Avdeevo Culture, including Kostënki 1/I, 13, 14/I and 18, Avdeevo, and Zaraisk, probably dating to ca.

24,000–22,500 ^{14}C BP (Sinitsyn et al. 1997; Amirkhanov 2000; Abramova et al. 2001; Haesaerts et al. 2017; Reynolds et al. 2017). Shouldered points were found at all of these sites, and they have other lithic techno-typological features in common. Female figures were found at Kostёnki 1/I, 13, Avdeevo and Zaraisk (Abramova 1995; Amirkhanov and Lev 2008). Long lines of hearths associated with pits were also found at Kostёnki 1/I, Avdeevo and Zaraisk (Efimenko 1958; Bulochnikova 2008; Amirkhanov 2009).

The relationship of the site of Gagarino to this group has long been debated. Female figures and small shouldered points were found there but it lacks the large shouldered points found at the other sites, while its available radiocarbon dates are relatively young, and suggest that the site post-dates ca. 22,000 ^{14}C BP (Tarassov 1971; Tarasov 1979; Sinitsyn et al. 1997; White 1997; Bulochnikova 1998; Sinitsyn 2007; Reynolds et al. 2019). However, techno-typological study of the lithic assemblage and of the shouldered points suggests that the absence of large shouldered points at the site may be due to raw material factors (Es'kova 2015; Reynolds et al. 2019). Gagarino may well be earlier in age than its radiocarbon dates suggest, and closer to the age of Kostёnki 1/I, Avdeevo and Zaraisk. Again, further radiocarbon dating may help to test this proposition.

The site of Khotylёvo 2 presents more difficult challenges. This site is dated to ca. 23,000 ^{14}C BP (Gavrilov et al. 2015), and, like Gagarino, its relationship to the Kostёnki-Avdeevo Culture sites has been the subject of debate. Female figures were found at the site (Abramova 1995; Gavrilov et al. 2015) but the artefacts previously identified as shouldered points are in fact better described as variants of Gravette points (Reynolds 2014). Here, we can use the female figures and some lithic types (Kostёnki knives, backed bladelets) to link the site with Kostёnki 1/I etc; the question of whether the site should be described as belonging to the Kostёnki-Avdeevo Culture or not can be left aside under the approach followed here.

Finally, we can link a series of late Gravettian sites in Russia and Ukraine where Anosovka points have been identified: Kostёnki 21/III (North), Kostёnki 11/II, Pushkari I, and Klyusy (Praslov and Ivanova 1982; Rogachёv and Popov 1982; Ivanova 1985; Beliaeva 2002; Sinitsyn 2007, 2014, 2015; Gavrilov 2016; Reynolds et al. 2019). Although there is some uncertainty over the dating of these sites (e.g. there are no radiocarbon dates available for Klyusy), it seems safe to assume their approximate contemporaneity as a working hypothesis.

These examples allow us to build up a basic chronocultural framework for the Gravettian record in Russia based on the presence/absence of particular assemblage features (Fig. 10.5). Backed lithic technology is present at all sites mentioned. Systematic production of microgravettes is attested at the earliest site; full-sized Gravette points and *éléments bitronqués* appear later; shouldered points appear for the first time about a thousand years after that, usually in association with female figures although one site has female figures and no shouldered points; finally, Anosovka points appear. This is a highly simplified schema: to these index fossils we could add further lithic techno-typological features, specific aspects of personal ornament and osseous assemblages, and details of the remains of dwelling structures (e.g. Efimenko 1958; Hromadova 2012; Goutas 2013).

It should be emphasized that although I do acknowledge pre-existing taxonomic units (e.g. Kostёnki-Avdeevo Culture), I do not attempt to construct further taxonomic units based on this record (which might be called Tel'manskian, Alexandrovskian, Borshchevskian, Anosovskian) because in my view it is not essential to understanding or analyzing variation in the record. What is perhaps more interesting is to consider the different geographical distributions of each of the discussed index fossils: from the distribution of microgravettes (and Gravette points) across Europe (Sinitsyn 2007, 2013; Moreau 2010; Wierer 2013), to the more restricted distributions of *éléments bitronqués* and shouldered points to Eastern and Central Europe (Grigor'ev 1993; Lisitsyn 2015; Polanská and Hromadová 2015; Wilczyński et al. 2015), to the much smaller distribution of Anosovka points in a small area of southwestern Russia and eastern Ukraine (Reynolds et al. 2019). Whether the distributions of these index fossils map onto past populations, or whether they do in all cases, remains impossible at present to answer, but they provide a far better dataset to address such questions than the traditional cultural taxonomic framework.

Can We Infer the Existence of Past Populations from the Archaeological Record?

In previous sections of this paper, I have briefly outlined how population concepts are commonly used in the modern study of the Upper Paleolithic, especially their frequent correlation with cultural taxonomic units (Aurignacian, Magdalenian, etc). I have also shown that the conceptualization of discrete populations associated with such taxonomic units (Aurignacians, Magdalenians, etc) has a long history, and outlined some of the many, widely acknowledged problems with the Upper Paleolithic cultural framework as it stands. We cannot assume that every taxonomic

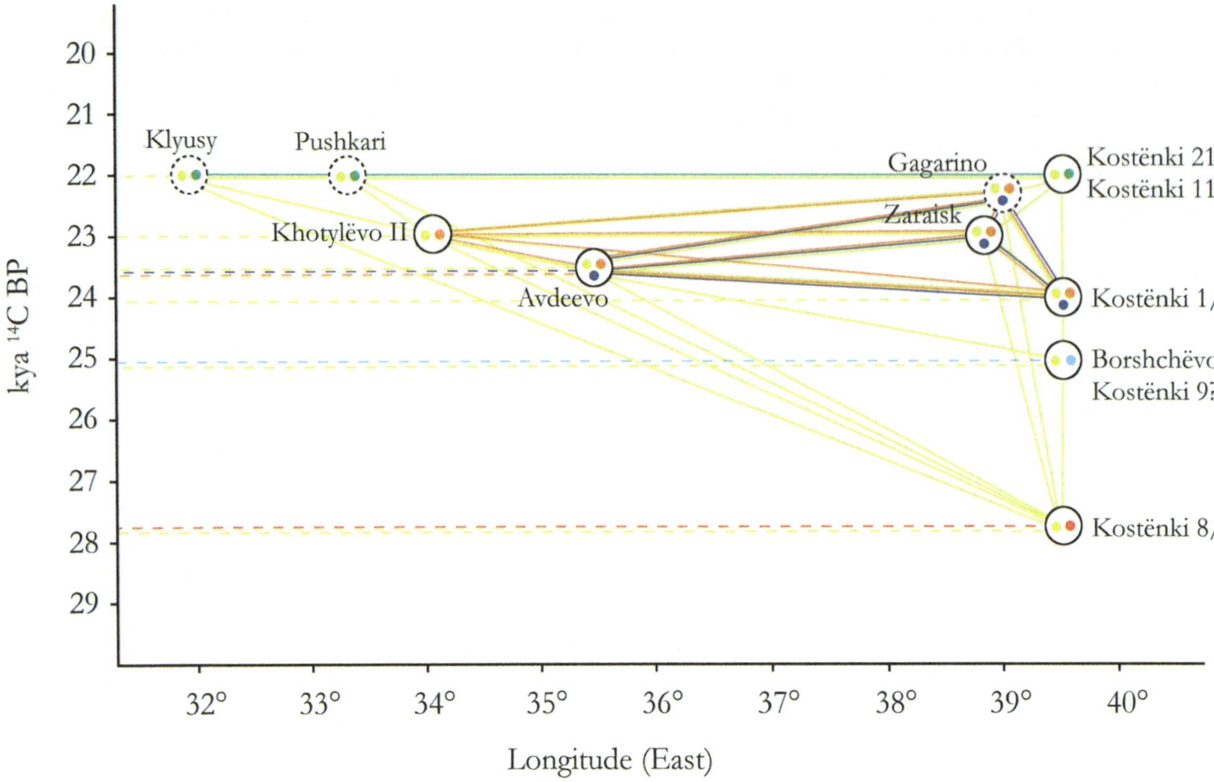

Fig. 10.5 A provisional chronocultural framework for Gravettian sites in Russia and eastern Ukraine. Key: Solid circles: well-dated sites. Dashed circles: sites whose dating is postulated based on material culture comparisons. Dots in circles: indicate presence of an index fossil or other feature used for comparison. Solid lines: indicate co-presence of index fossils or other features. Dashed lines: indicate co-presence at sites further west (exact chronological relationships not defined here). Colors of dots and lines indicate which material culture features are present: yellow—backed lithic technology; green—Anosovka points; orange—female figures; dark blue—Kostënki-Avdeevo-type shouldered points; light blue—*éléments bitronqués*; red—systematic microgravette production

unit is equivalent in terms of its robusticity, its discreteness, or the amount of variation it subsumes. Without even going into some of the more sophisticated possible theoretical objections, this fact alone means that it is inappropriate to equate taxonomic units with past populations.

In an attempt to go beyond these problems, I then described an approach to the study of the Upper Paleolithic whereby the chronocultural framework of this period is established based on an assumption of geographical and temporal coherence in the material culture record. This kind of approach is widely used in the study of the European Upper Paleolithic, and holds out the possibility of significant further progress in developing and revising our chronocultural framework despite the known problems with chronology and the current incompleteness of our knowledge of material culture variation.

The reason for going into so much detail regarding this approach is that I think it is necessary to fundamentally reconsider our entire cultural taxonomic framework if we are to successfully engage with questions concerning populations during the Upper Paleolithic. I also think that the approach outlined here is the best available method for doing so. One of the bases of this approach is the assumption that the relatively short-lived, often geographically restricted groupings of sites that we can establish based on the presence of particular carefully defined index fossils or other features do reflect past social connections between people. As a result, they have a direct bearing on questions of populations.

There is wide agreement within our discipline on the importance of synthesizing material culture and chronological data and of the identification of patterning and coherent groupings in the archaeological record. The distinctiveness of the approach put forward here, if any, lies in its insistence on a firmly bottom-up rather than top-down description of the archaeological record. Since the approach treats the definition of taxonomic units as an additional, optional step to working out the similarities and contrasts between sites, it does not require the definition of taxonomic units such that all sites can be placed into discrete taxonomic units (Fig. 10.6). Furthermore, it does not attempt to place units into a hierarchical system, unlike most of the current cultural taxonomic system as described above, Clarke's approach set

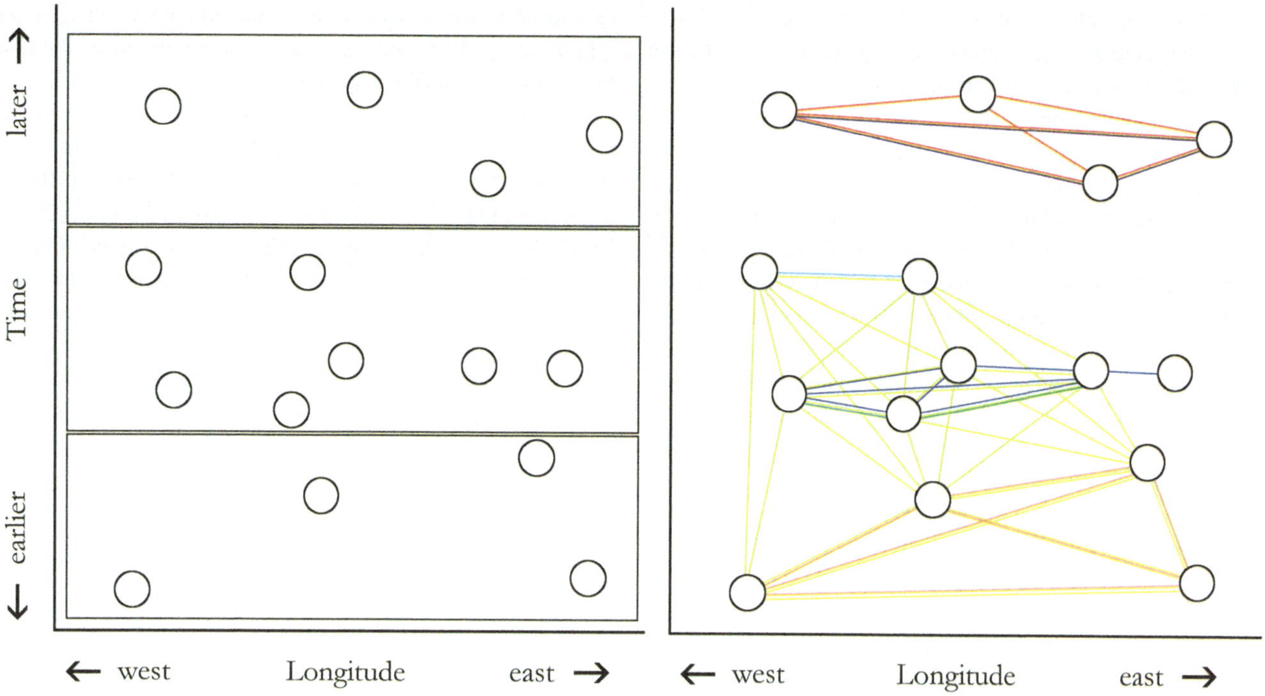

Fig. 10.6 One way of visualizing the differences between top-down (left) and bottom-up (right) approaches. Traditional approaches to cultural taxonomy (left) focus on placing assemblages into cultural taxonomic units, which often fail to reflect the complexity of the differences and similarities between sites. The approach advocated in this chapter (right) focuses on defining individual links between sites based on the co-presence of index fossils and other features

out in *Analytical Archaeology* (1968), or some of the evolutionary archaeological approaches advocated by other researchers (e.g. Riede 2011). Instead, we can define connections between sites based on many different kinds of evidence, and at many different scales. Some of these connections will overlap and echo each other, while others will diverge. So, for example, we can describe a geographically and chronologically coherent group of sites dating to the Mid and Late Upper Paleolithic all over Europe showing evidence for the systematic production of backed lithic artefacts; another coherent group of sites where backed bladelets are found; a coherent group or groups of sites where Gravette points are found, or where female figures are found and so on. (In some cases particular types of artefact or other features will appear to be important within a single site but will not have any clear analogies in chronologically and geographically proximate sites. The existence of *sui generis* features is also important and should also be used to add to our picture of variation within the record as a whole.)

The approach set out here aims to avoid essentialist conceptions of cultural taxonomy by making the use of taxonomic units an entirely optional add-on to the chrono-cultural framework itself. Nevertheless, by relying on index fossils and other "index features" it does, arguably, continue to take an essentialist view of material culture variation itself, rather than one that is based in population thinking (meant in the philosophical sense, rather than for any relationship with the human populations that are the subject of this chapter; Sober 1980; Leonard and Jones 1987; O'Brien and Holland 1990; Riede 2011). However, it should be noted that the usage of index fossils and features is advocated only as a heuristic tool to link assemblages, rather than as providing anything approaching a full picture of material culture variation. By using many different definitions of index fossils, which are not mutually exclusive, and which vary in their strictness (so that a single artefact can be defined as e.g. both a backed bladelet and as an *élément bitronqué*), it is hoped that a nuanced, population thinking approach to material culture variation can in fact be approximated, even if strictly speaking the approach advocated here has a different epistemological and methodological basis.

A framework such as the one I describe, which is constructed in a bottom-up fashion directly using links between sites, is ideal for addressing questions of past populations. This framework (which can be characterized as a presence/absence matrix of site and assemblage features, with associated geographical and chronological information) can be analyzed using numerous network analysis methods (e.g. Knappett 2011; Brughmans 2013; Collar et al. 2015) and other statistical and modeling approaches (e.g. Baxter 2009; Shennan et al. 2015; Rigaud et al. 2018).

We can analyse this data to look for break-points where many different types of material culture changed simultaneously, for slow change over time, for cases where certain types of material culture changed but others didn't. All of these examples have different implications in terms of the cultural and demographic processes that may have caused them. However, material culture can and does change for numerous reasons, many of which are quite independent from population changes. The question of the inference of population structure from material culture variation is, therefore, one of the hard problems of prehistoric archaeology in general.

Although there are numerous theoretical approaches that could be applied to this, especially based on ethnographic analogy, for the European Upper Paleolithic we are dealing with very large timescales in the context of significant environmental changes. This makes it very difficult to draw robust analogies from anything that we can observe in the present. In fact, the best way to improve our inference of population structure during the Upper Paleolithic may be to compare the archaeological record with the results now being obtained from a fundamentally different perspective on prehistory: ancient DNA studies.

Comparing Archaeological and Paleogenetic Evidence

In recent years several major papers have been published on human genetic diversity during the Upper Paleolithic (e.g. Fu et al. 2016; Posth et al. 2016; Sikora et al. 2017). The results of this work permit direct testing—and improvement—of archaeological inferences regarding Upper Paleolithic population structure.

Some of the results from ancient DNA studies have profound, widespread implications. For example, according to Fu et al. (2016), all analyzed individuals in Europe from between ca. 37,000 cal BP and 14,000 cal BP (or ca. 33,000 ^{14}C BP and 12,000 ^{14}C BP) "seem to derive from a single ancestral population with no evidence of substantial genetic influx from elsewhere". The spread of Gravettian traditions does appear to have been associated with at least some population movements, as attested by the distribution of the "Věstonice Cluster" that they identify. The dearth of human remains associated with Aurignacian assemblages makes the task of understanding population processes for this part of the archaeological record more difficult. However, an Early Upper Paleolithic individual from Goyet Cave belonged to a population that did not disappear with the appearance of Gravettian assemblages, but whose descendants became widespread again during the Late Upper Paleolithic (Fu et al. 2016).

The most significant identified turnover in European populations during the Upper Paleolithic in fact occurred during the Late Upper Paleolithic, 14,500–14,000 cal BP (ca. 12,500–12,000 ^{14}C BP) (Posth et al. 2016; Fu et al. 2016). This does not correlate with a clear and major pan-European transition in archaeological taxonomic units. If this turnover in populations continues to be supported by further research, it provides a good example of why the current cultural taxonomic framework should not be seen as providing a straightforward reflection of past population dynamics.

However, the results from ancient DNA studies do suggest that in some cases, cultural taxonomic transitions were indeed associated with population changes—for example, the Aurignacian-Gravettian transition (albeit not, apparently, associated with a full demographic replacement: Fu et al. 2016; Sikora et al. 2017). Other observations also coincide with archaeological interpretations. For example, Layer I of Kostёnki 12 has, despite its Mid Upper Paleolithic age, been consistently described as Gorodtsovian rather than Gravettian due to the composition of its assemblage, which does not contain backed lithics (Sinitsyn 2010, 2015). The attribution of the human remains found at Kostenki 12 to a population that was distinct from the "Věstonice cluster" associated with Gravettian assemblages (Fu et al. 2016; Sikora et al. 2017), suggests that this distinction between Gravettian and Gorodtsovian may be a reflection of past population differences, although a DNA study of remains found in association with Gravettian assemblages in Eastern Europe would be interesting to further explore this.

Genetic data provides an independent line of evidence for comparison against archaeological interpretations (as also argued by Shennan 2020). In order to strengthen the archaeological understanding of past populations, we can compare our chronocultural frameworks against the population histories determined using genetic studies. From such comparisons we can gain an understanding of what types of archaeological evidence and argument can be used for discerning past population structure and dynamics, and how reliable they are. We can then use those same types of evidence and argument in parts of the archaeological record where we have less genetic evidence. Systematic comparison against the results of genetic studies can greatly enhance the archaeological study of Upper Paleolithic populations, and provides an opportunity to move past ad hoc and intuitive reasoning.

None of this is to imply that archaeologists can or should cede the study of the Upper Paleolithic to geneticists. Culture change is fundamentally different from biological genetic change, and we should not expect variation in the archaeological record ever to exactly follow the picture given by genetic data. In fact, it is perhaps in the areas where the results from archaeology and genetics diverge that some of the most interesting future studies will be focused. The purpose of archaeology is not just to describe past populations: it is far more than that. Genetics can give us valuable insights into past populations, but it is archaeology that can make sense of the processes underlying past population dynamics, and the great diversity of associated social and cultural outcomes.

Conclusions

Population concepts are profoundly important in Upper Paleolithic archaeology, underlying many of our most basic interpretations. However, the present archaeological understanding of Upper Paleolithic populations is far from satisfactory.

We are working within a cultural taxonomic system that has many arbitrary elements and does not provide an accurate overall picture of variation in the Upper Paleolithic archaeological record. In particular, our taxonomic units are not all equivalent in their salience or the amount of variation that they encompass. One of the underlying assumptions underlying our material culture comparisons and taxonomic unit construction is that they reflect something about sociocultural and population processes. However, an uncritical reading of the conventional Upper Paleolithic taxonomic framework cannot be used to infer the existence of past populations.

Although many of our current taxonomic units do have definite descriptive value, we must treat them as heuristic and revisable, or abandon them altogether. Substantial progress has been made on understanding the full, detailed picture of variation in the archaeological record, although this by necessity tends to be done on relatively small scales. Further comparative work is needed, perhaps especially for the Late Upper Paleolithic record.

The best approach to improving our chronocultural framework considers both material culture comparison and chronological evidence, the warp and the weft, within a paradigm that expects coherence in the archaeological record itself. A chronocultural framework does not need to consist of abstract, top-down taxonomic units but can instead exist as a formal bottom-up systematization of the individual similarities and differences between sites, as expressed in the presence/absence of particular index features—not only lithic index fossils, but also technological features, personal ornament types, dwelling structure types, and so on. This can be visualized as a network in a chronospatial framework but can also be expressed as matrices recording the presence and absence of particular features at different sites, allowing many different types of analyses to be carried out.

Paleogenomics, which in recent years has begun to provide highly interesting results concerning European Upper Paleolithic populations, offers an opportunity to establish some basic principles for the inference of past population structure and dynamics from archaeological data. A chronocultural framework based on a bottom-up examination of archaeological similarities and differences between sites, as proposed here, is ideal for direct comparison against the results of genetic studies, which are similarly based on a bottom-up treatment of the similarities and differences between individual genomes. Where archaeology and genetics truly give different pictures of the past (i.e. where this is not just the result of naïve interpretation of cultural taxonomic units) this provides us with an opportunity to gain a fuller understanding of the complexity of cultural change in the past and its differences from biological population change.

At present we are not able to properly evaluate the existence and nature of Upper Paleolithic populations using archaeological evidence, although the occasions where genetic evidence agrees with archaeological data offer intriguing hints that this may, in principle, be possible. However, if we can work at large scales to gain a fuller, more consistent picture of the chronocultural framework for Upper Paleolithic Europe, and if we can systematically compare this framework with the results now being provided by paleogenomic studies, then we have an excellent opportunity to finally establish a solid epistemological basis for the archaeological study of Upper Paleolithic populations. This in turn will greatly enrich our understanding of cultural processes during the Upper Paleolithic and open up new avenues of archaeological interpretation.

The European Upper Paleolithic remains a special case within the Paleolithic as a whole. Not all of the observations made in this chapter are necessarily extensible to the rest of the Paleolithic. Nevertheless, if we can gain a stronger understanding of populations within this small part of the archaeological record, we may perhaps gain insights that can be transferred to the study of other areas and time periods. The study of past populations is at the heart of many of the questions that we ask about the Paleolithic. There are significant theoretical and methodological challenges to be overcome in order to make progress in this area, and correspondingly large gains to be made in our understanding of the human past.

Acknowledgements I am grateful to Will Banks, Julia Kotthaus and Felix Riede for conversations and correspondence that helped me to clarify many of the arguments put forward in this chapter, and to Huw Groucutt for the invitation to contribute to this volume. I also thank the anonymous reviewers and HG for their comments, which were very helpful in improving the manuscript. This work was supported by the European Union's Horizon 2020 research and innovation programme under Marie Skłodowska-Curie grant agreement No. 747400. All figures in this chapter were created by NR and are used under a CC-BY 4.0 licence; original figures are available at https://doi.org/10.6084/m9.figshare.7770986.

References

Abramova, Z. A. (1995). *L'art paléolithique d'Europe orientale et de Sibérie*. Grenoble: Jérôme Millon.

Abramova, Z. A., Grigorieva, G. V., & Zaitseva, G. I. (2001). The age of Upper Paleolithic sites in the Middle Dnieper River basin of Eastern Europe. *Radiocarbon, 42,* 1077–1084.

Adams, W. Y., & Adams, E. W. (2009). *Archaeological typology and practical reality: A dialectical approach to artifact classification and sorting*. Cambridge: Cambridge University Press.

Amirkhanov, Kh. A. (2000). *Zaraiskaia stoianka*. Moscow: Nauchnyi mir.

Amirkhanov, Kh. A. (Ed.). (2009). *Issledovaniia paleolita v Zaraiske 1999–2005*. Moscow: Paleograf.

Amirkhanov, Kh. A., & Lev, S. Iu. (2008). New finds of Paleolithic art from the Zaraisk site. *Anthropology & Archeology of Eurasia, 46,* 24–45.

Anghelinu, M., & Niţă, L. (2014). What's in a name: The Aurignacian in Romania. *Quaternary International, 351,* 172–192.

Anghelinu, M., Niţă, L., & Murătoreanu, G. (2018). Le Gravettien et l'Épigravettien de l'Est de la Roumanie: une réévaluation. *L'anthropologie, 122,* 183–219.

Baales, M., Grimm, S., & Jöris, O. (2001). Hunters of the "Golden Mile": The late Allerød *Federmessergruppen* Site at Bad Breisig, Central Rhineland, Germany. *Notae Praehistoricae, 21,* 67–72.

Banks, W. E. (2015). Constructing chronologies for the late Middle Paleolithic and Upper Paleolithic: Limitations and means to overcome them. *World Archaeology, 47,* 585–600.

Banks, W. E., d'Errico, F., Townsend Peterson, A., Vanhaeren, M., Kageyama, M., Sepulchre, P., et al. (2008). Human ecological niches and ranges during the LGM in Europe derived from an application of eco-cultural niche modeling. *Journal of Archaeological Science, 35,* 481–491.

Banks, W. E., d'Errico, F., & Zilhão, J. (2013a). Human–climate interaction during the Early Upper Paleolithic: Testing the hypothesis of an adaptive shift between the Proto-Aurignacian and the Early Aurignacian. *Journal of Human Evolution, 64,* 39–55.

Banks, W. E., d'Errico, F., & Zilhão, J. (2013b). Revisiting the chronology of the Proto-Aurignacian and the Early Aurignacian in Europe: A reply to Higham et al.'s comments on Banks et al. (2013). *Journal of Human Evolution, 65,* 810–817.

Barton, C. M. (1997). Stone tools, style, and social identity: An evolutionary perspective on the archaeological record. In C. M. Barton & G. Clark (Eds.), *Rediscovering Darwin: Evolutionary theory and archaeological explanation* (pp. 141–156). Washington, DC: Archeological Papers of the American Anthropological Association, No. 7.

Baxter, M. J. (2009). Archaeological data analysis and fuzzy clustering. *Archaeometry, 51,* 1035–1054.

Beliaeva, V. I. (2002). Kremnëvaia industriia Pushkarei 1. In N. D. Praslov (Ed.), *Verkhnii paleolit – verkhnii pleistotsen: dinamika prirodnykh sobytii i periodizatsiia arkheologicheskikh kul'tur. Materialy mezhdunarodnoi konferentsii, posviashchennoi 90-letiiu so dnia rozhdeniia A. N. Rogachëva* (pp. 132–137). Saint Petersburg: Nauka.

Bocquet-Appel, J. P., & Demars, P. Y. (2000). Population kinetics in the Upper Palaeolithic in western Europe. *Journal of Archaeological Science, 27,* 551–570.

Bodu, P. (1998). Magdalenians-Early Azilians in the centre of the Paris Basin: A filiation? The example of Le Closeau (Rueil-Malmaison, France). In S. Milliken (Ed.), *The organization of lithic technology in Late Glacial and Early Postglacial Europe (BAR International Series 700)* (pp. 131–147). Oxford: Oxbow.

Bordes, J.-G. (2006). News from the West: A reevaluation of the classical Aurignacian sequence of the Périgord. In O. Bar-Yosef & J. Zilhão (Eds.), *Towards a definition of the Aurignacian, Proceedings of the Symposium held in Lisbon, Portugal, June 25–30, 2002 (Trabalhos de Arqueologia 45)* (pp. 147–171). Lisbon: Instituto Português de Arqueologia.

Bordes, J.-G., Bachellerie, F., le Brun-Ricalens, F., & Michel, A. (2011). Towards a new "transition": New data concerning the lithic industries from the beginning of the Upper Palaeolithic in Southwestern France. In A. P. Derevianko & M. Shunkov (Eds.), *Characteristic features of the Middle to Upper Palaeolithic Transition in Eurasia, Proceedings of the International Symposium, "Characteristic features of the Middle to Upper Palaeolithic Transition in Eurasia: Development of Culture and Evolution of Homo Genus" (July 4–10, Denisova Cave, Altai)* (pp. 116–129). Novosibirsk: Department of the Institute of Archaeology and Ethnography SB RAS.

Bourrillon, R., White, R., Tartar, E., Chiotti, L., Mensan, R., Clark, A., et al. (2018). A new Aurignacian engraving from Abri Blanchard, France: Implications for understanding Aurignacian graphic expression in Western and Central Europe. *Quaternary International, 491,* 46–64.

Bradtmöller, M., Pastoors, A., Weninger, B., & Weniger, G.-C. (2012). The repeated replacement model—Rapid climate change and population dynamics in Late Pleistocene Europe. *Quaternary International, 247,* 38–49.

Brock, F., & Higham, T. F. G. (2009). AMS radiocarbon dating of Paleolithic-aged charcoal from Europe and the Mediterranean rim using ABOx-SC. *Radiocarbon, 51,* 839–846.

Brock, F., Higham, T., Ditchfield, P., & Bronk Ramsey, C. (2010). Current pretreatment methods for AMS radiocarbon dating at the Oxford Radiocarbon Accelerator Unit (ORAU). *Radiocarbon, 52,* 103–112.

Brock, F., Geoghegan, V., Thomas, B., Jurkschat, K., & Higham, T. F. G. (2013). Analysis of bone "collagen" extraction products for radiocarbon dating. *Radiocarbon, 55,* 445–463.

Brock, F., Dee, M., Hughes, A., Snoeck, C., Staff, R., & Bronk Ramsey, C. (2018). Testing the effectiveness of protocols for removal of common conservation treatments for radiocarbon dating. *Radiocarbon, 60,* 35–50.

Bronk Ramsey, C., Higham, T., & Leach, P. (2004). Towards high-precision AMS: Progress and limitations. *Radiocarbon, 46,* 17–24.

Brughmans, T. (2013). Thinking through networks: A review of formal network methods in archaeology. *Journal of Archaeological Method and Theory, 20,* 623–662.

Bulochnikova, E. V. (1998). Vchera i segodnia poniatiia "vostochnyi gravet'en". In Kh. A. Amirkhanov (Ed.), *Vostochnoi gravett* (pp. 67–72). Moscow: Nauchnyi mir.

Bulochnikova, E. V. (2008). Khronologiia verkhnepaleoliticheskikh stoianok v sisteme arkheologicheskikh i estestvennonauchnykh dannykh. In G. A. Khlopachëv (Ed.), *Khronologiia, periodizatsiia i krosskul'turnye sviazi v kamennom veke: Zamiatninskii sbornik. Vypusk 1* (pp. 63–73). Saint Petersburg: Nauka.

Burdukiewicz, J. M. (2001). The last Ice Age and settlement break in the northern past of Central Europe. *Fontes Archaeologici Posnanienses, 39,* 15–29.

Burdukiewicz, J. M. (2011). Late Glacial hunter-gatherer reactions to the Younger Dryas cooling event in the southern and eastern Baltic regions of Europe. *Quaternary International, 242,* 302–312.

Burkitt, M., & Childe, V. G. (1932). A chronological table of prehistory. *Antiquity, 6,* 185–201.

Chu, W., Lengyel, G., Zeeden, C., Péntek, A., Kaminská, L., & Mester, Z. (2018). Early Upper Paleolithic surface collections from loess-like sediments in the northern Carpathian Basin. *Quaternary International, 485,* 167–182.

Clark, G. A., & Riel-Salvatore, J. (2006). Observations on systematics in Paleolithic archaeology. In E. Hovers & S. L. Kuhn (Eds.), *Transitions before the transition* (pp. 29–56). Boston, MA: Springer.

Clarke, D. L. (1968). *Analytical archaeology* (2nd ed.). London: Methuen.

Collar, A., Coward, F., Brughmans, T., & Mills, B. J. (2015). Networks in archaeology: Phenomena, abstraction, representation. *Journal of Archaeological Method and Theory, 22,* 1–32.

Collard, M., Shennan, S. J., Buchanan, B., & Bentley, R. A. (2009). Evolutionary biological methods and cultural data. In R. A. Bentley, H. D. G. Maschner, & C. Chippindale (Eds.), *Handbook of archaeological theories* (pp. 203–223). Lanham: Altamira Press.

Collie, G. L. (1928). *The Aurignacians and their culture.* Beloit: Logan Museum.

Conard, N. J., & Bolus, M. (2003). Radiocarbon dating the appearance of modern humans and timing of cultural innovations in Europe: New results and new challenges. *Journal of Human Evolution, 44,* 331–371.

de la Peña, P., & Vega Toscano, G. (2013). The Early Upper Palaeolithic puzzle in Mediterranean Iberia. *Quartär, 60,* 85–106.

Demars, P.-Y. (1990). Proposition pour une nouvelle liste typologique des outillages lithiques du Paléolithique supérieur. *Paléo, 2,* 191–201.

Demars, P.-Y., & Laurent, P. (1992). *Types d'outils lithiques du Paléolithique supérieur en Europe.* Paris: Presses du CNRS.

d'Errico, F., & Banks, W. E. (2015). Tephra studies and the reconstruction of Middle-to-Upper Paleolithic cultural trajectories. *Quaternary Science Reviews, 118,* 182–193.

de Sonneville-Bordes, D. (1966). L'évolution du Paléolithique supérieur en Europe occidentale et sa signification. *Bulletin de la Société préhistorique française, 63,* 3–34.

de Sonneville Bordes, D., & Perrot, J. (1954). Lexique typologique du Paléolithique supérieur. *Bulletin de la Société préhistorique française, 51,* 327–335.

de Sonneville Bordes, D., & Perrot, J. (1955). Lexique typologique du Paléolithique supérieur. *Bulletin de la Société préhistorique française, 52,* 76–79.

de Sonneville Bordes, D., & Perrot, J. (1956a). Lexique typologique du Paléolithique supérieur. *Bulletin de la Société préhistorique française, 53,* 408–412.

de Sonneville Bordes, D., & Perrot, J. (1956b). Lexique typologique du Paléolithique supérieur. *Bulletin de la Société préhistorique française, 53,* 547–559.

Deviese, T., Comeskey, D., McCullagh, J., Bronk Ramsey, C., & Higham, T. (2018). New protocol for compound-specific radiocarbon analysis of archaeological bones. *Rapid Communications in Mass Spectrometry, 32,* 373–379.

Dibble, H. L. (1995). Middle Paleolithic scraper reduction: Background, clarification, and review of the evidence to date. *Journal of Archaeological Method and Theory, 2,* 299–368.

Dibble, H. L., Holdaway, S. J., Lin, S. C., Braun, D. R., Douglass, M. J., Iovita, R., et al. (2016). Major fallacies surrounding stone artifacts and assemblages. *Journal of Archaeological Method and Theory, 24,* 813–851.

Dinnis, R., Bessudnov, A., Reynolds, N., Devièse, T., Pate, A., Sablin, M., et al. (2019). New data for the Early Upper Paleolithic of Kostenki (Russia). *Journal of Human Evolution, 127,* 21–40.

Discamps, E., Gravina, B., & Teyssandier, N. (2015). In the eye of the beholder: Contextual issues for Bayesian modelling at the Middle-to-Upper Palaeolithic transition. *World Archaeology, 47,* 601–621.

Djindjian, F. (2016). Territories and economies of hunter–gatherer groups during the last glacial maximum in Europe. *Quaternary International, 412,* 37–43.

Djindjian, F., Kozlowski, J., & Otte, M. (1999). *Le Paléolithique supérieur en Europe.* Paris: Armand Colin.

Dobres, M.-A. (1999). Technology's links and chaînes: The processual unfolding of technique and technician. In M.-A. Dobres & C. R. Hoffman (Eds.), *The social dynamics of technology: Practice, politics, and world views* (pp. 124–146). Washington, D.C.: Smithsonian Institution Press.

Doyon, L. (2019). On the shape of things: A geometric morphometrics approach to investigate Aurignacian group membership. *Journal of Archaeological Science, 101,* 99–114.

Ducasse, S. (2012). What is left of the Badegoulian "interlude"? New data on cultural evolution in southern France between 23,500 and 20,500 cal. BP. *Quaternary International, 272–273,* 150–165.

Ducasse, S., & Langlais, M. (2007). Entre Badegoulien et Magdalénien, nos coeurs balancent… Approche critique des industries lithiques du Sud de la France et du Nord-Est espagnol entre 19000 et 16500 BP. *Bulletin de la Société préhistorique française, 104,* 771–785.

Dunnell, R. C. (1971). *Systematics in prehistory.* New York: Free Press.

Dunnell, R. C. (1986). Methodological issues in Americanist artifact classification. *Advances in Archaeological Method and Theory, 9,* 149–207.

Efimenko, P. P. (1958). *Kostënki I.* Moscow: Izdatel'stvo Akademii Nauk SSSR.

Es'kova, D. K. (2015). Tekhnologiia rasshchepleniia kamnia na stoiankakh vostochnogo gravetta Russkoi ravniny. Ph.D Dissertation *avtoreferat,* Institut Arkheologii RAN, Moscow.

Fat Cheung, C., Chevallier, A., Bonnet-Jacquement, P., Langlais, M., Ferrié, J.-G., Costamagno, S., et al. (2014). Comparaison des séquences aziliennes entre Dordogne et Pyrénées: état des travaux en cours. In M. Langlais, N. Naudinot, & M. Peresani (Eds.), *Les groupes culturels de la transition Pléistocène-Holocène entre Atlantique et Adriatique. Actes de la séance de la Société préhistorique française de Bordeaux, 24–25 mai 2012* (pp. 17–44). Paris: Société préhistorique française.

Finlayson, C., & Carrión, J. S. (2007). Rapid ecological turnover and its impact on Neanderthal and other human populations. *Trends in Ecology and Evolution, 22,* 213–222.

French, J. C. (2015). The demography of the Upper Palaeolithic hunter–gatherers of Southwestern France: A multi-proxy approach using archaeological data. *Journal of Anthropological Archaeology, 39,* 193–209.

Frouin, M., Guérin, G., Lahaye, C., Mercier, N., Huot, S., Aldeias, V., et al. (2017). New luminescence dating results based on polymineral fine grains from the Middle and Upper Palaeolithic site of La Ferrassie (Dordogne, SW France). *Quaternary Geochronology, 39,* 131–141.

Fu, Q., Posth, C., Hajdinjak, M., Petr, M., Mallick, S., Fernandes, D., et al. (2016). The genetic history of Ice Age Europe. *Nature, 534,* 200–205.

Gamble, C., Davies, W., Pettitt, P., Hazelwood, L., & Richards, M. (2005). The archaeological and genetic foundations of the European population during the Late Glacial: Implications for 'agricultural thinking'. *Cambridge Archaeological Journal, 15,* 193–223.

Garrod, D. A. E. (1938). The Upper Palaeolithic in the light of recent discovery. *Proceedings of the Prehistoric Society, 4,* 1–26.

Gaudzinski-Windheuser, S., & Jöris, O. (2015). Contextualising the female image—Symbols for common ideas and communal identity in Upper Palaeolithic societies. In F. Coward, R. Hosfield, M. Pope, & F. Wenban-Smith (Eds.), *Settlement, society and cognition in human evolution: Landscapes in mind* (pp. 288–314). Cambridge: Cambridge University Press.

Gavrilov, K. N. (2016). *Verkhnii paleolit basseina Desny: preemstvennost' i variabel'nost' v razvitii material'noi kul'tury.* Moscow: Nestor-Istoriia.

Gavrilov, K. N., Voskresenskaya, E. V., Maschenko, E. N., & Douka, K. (2015). East Gravettian Khotylevo 2 site: Stratigraphy, archeozoology, and spatial organization of the cultural layer at the newly explored area of the site. *Quaternary International, 359–360,* 335–346.

Goutas, N. (2013). De Brassempouy À Kostienki: l'exploitation technique des ressources animales dans l'Europe gravettienne. In M. Otte (Ed.), *Les Gravettiens* (pp. 105–160). Paris: Éditions Errance.

Goutas, N. (2016). Gravettian projectile points: Considerations about the evolution of osseous hunting weapons in France. In M. Langley (Ed.), *Osseous Projectile Weaponry* (pp. 89–107). Dordrecht: Springer.

Gravina, B., Bachellerie, F., Caux, S., Discamps, E., Faivre, J.-P., Galland, A., et al. (2018). No reliable evidence for a Neanderthal-Châtelperronian association at La Roche-à-Pierrot. *Saint-Césaire. Scientific Reports, 8,* 15134.

Grigor'ev, G. P. (1993). The Kostenki-Avdeevo archaeological culture and the Willendorf–Pavlov–Kostenki–Avdeevo cultural unity. In O. Soffer & N. D. Praslov (Eds.), *From Kostenki to Clovis: Upper Paleolithic—Paleo-Indian adaptations* (pp. 51–65). New York: Plenum Press.

Grimm, S. B., & Weber, M.-J. (2008). The chronological framework of the Hamburgian in the light of old and new ^{14}C dates. *Quartär, 55,* 17–40.

Groucutt, H. S. (2020). Culture and convergence: The curious case of the Nubian Complex. In H. Groucutt (Ed.), *Culture history and convergent evolution: Can we detect populations in prehistory?* (pp. 55–86). Cham, Switzerland: Springer.

Gvozdover, M. V. (1998). Kremnëvyi inventar' Avdeevskoi verkhnepaleoliticheskoi stoianki. In K. A. Amirkhanov (Ed.), *Vostochnoi gravett* (pp. 234–278). Moscow: Nauchnyi mir.

Haesaerts, P., Damblon, F., Nigst, P., & Hublin, J.-J. (2013). ABA and ABOx radiocarbon cross-dating on charcoal from Middle Pleniglacial loess deposits in Austria, Moravia, and western Ukraine. *Radiocarbon, 55,* 641–647.

Haesaerts, P., Damblon, F., van der Plicht, J., Otte, M., & Nigst, P. R. (2017). Additional data to the stratigraphy and the chronology of the Kostenki 1 (Poliakov) sequence, Voronezh, Russia. In S. Vasilyev, A. Sinitsyn, & M. Otte (Eds.), *Le Sungirien, Saint-Pétersbourg 2016* (pp. 135–142). Liège: ÉRAUL 147.

Hahn, J. (1988). *Die Geißenklösterle-Höhle im Achtal bei Blaubeuren I. Fundhorizontbildung und Besiedlung im Mittelpalaölithikum und im Aurignacien.* Stuttgart: Konrad Theiss Verlag.

Harrold, F. B. (1993). Variability and function among Gravette points from southwestern France. *Archeological Papers of the American Anthropological Association, 4,* 69–81.

Hayden, B. (1984). Are emic types relevant to archaeology? *Ethnohistory, 31,* 79–92.

Hays, M. A., & Lucas, G. (2000). A technological and functional analysis of carinates from Le Flageolet I, Dordogne, France. *Journal of Field Archaeology, 27,* 455–465.

Hermon, S., & Niccolucci, F. (2017). Formally defining the time-space-archaeological culture relation: Problems and prospects. *Archeologia e Calcolatori, 28,* 93–108.

Higham, T. F. G. (2011). European Middle and Upper Palaeolithic radiocarbon dates are often older than they look: Problems with previous dates and some remedies. *Antiquity, 85,* 235–249.

Higham, T., Basell, L., Jacobi, R., Wood, R., Bronk Ramsey, C., & Conard, N. J. (2012). Testing models for the beginnings of the Aurignacian and the advent of figurative art and music: The radiocarbon chronology of Geißenklösterle. *Journal of Human Evolution, 62,* 664–676.

Higham, T., Wood, R., Moreau, L., Conard, N., & Bronk Ramsey, C. (2013). Comments on 'Human–climate interaction during the early Upper Paleolithic: Testing the hypothesis of an adaptive shift between the Proto-Aurignacian and the Early Aurignacian' by Banks et al. *Journal of Human Evolution, 65,* 806–809.

Hřdlicka, A. (1927). The Neanderthal phase of man. *The Journal of the Royal Anthropological Institute of Great Britain and Ireland, 57,* 249–274.

Hromadova, B. (2012). *Ispol'zovanie syr'ia iz kosti, bivnia i roga na stoiankakh kostënkovsko-avdeevskoi kul'tury (vostochnoi gravett).* Ph.D. Dissertation *avtoreferat*, MGU, Moscow.

Hublin, J. J. (2015). The modern human colonization of western Eurasia: when and where? *Quaternary Science Reviews, 118,* 194–210.

Hurt, T. D., & Rakita, G. F. M. (Eds.). (2001). *Style and function: Conceptual issues in evolutionary archaeology.* Westport, CT: Bergin & Garvey.

Iakovleva, L. (2003). Les habitats en os de mammouths du Paléolithique supérieur d'Europe orientale: les données et leur interprétation. In S. A. Vasiliev, O. Soffer, & J. K. Kozłowski (Eds.), *Perceived landscapes and built environments: the cultural geography of late palaeolithic Eurasia. Actes du XIV° Congrès UISPP, Université de Liège, Belgique, 2–8 septembre 2001: colloques 6.2 & 6.5 (BAR International Series 1122)* (pp. 47–57). Oxford: Oxbow Books.

Ivanova, M. A. (1985). Struktura Gmelinskogo paleoliticheskogo poselniia (po rezul'tatam planigraficheskogo i tipologicheskogo analiza kremnëvogo inventaria). Ph.D Dissertation *avtoreferat*, LOIA RAN, Leningrad.

Jordan, P. (2015). *Technology as human social tradition: Cultural transmission among hunter-gatherers.* Oakland: University of California Press.

Khlopachev, G., Vercoutère, C., & Wolf, S. (2018). Les statuettes féminines en ivoire des faciès gravettiens et post-gravettiens en Europe centrale et orientale: modes de fabrication et de représentation. *L'anthropologie, 122,* 492–521.

Klaric, L. (2007). Regional groups in the European Middle Gravettian: A reconsideration of the Rayssian technology. *Antiquity, 81,* 176–190.

Klaric, L., Guillermin, P., & Aubry, T. (2009). Des armatures variées et des modes de productions variables. *Gallia Préhistoire, 51,* 113–154.

Klaric, L., Lev, S., Giria, Y., & Polanská, M. (2015). 'Couteaux de Kostienki et lames aménagées par technique de Kostienki'. Retour sur un malentendu historique. *Bulletin de la Société préhistorique française, 112,* 421–474.

Knappett, C. (2011). *An archaeology of interaction: Network perspectives on material culture and society.* Oxford: Oxford University Press.

Kozłowski, J. K. (2015). The origin of the Gravettian. *Quaternary International, 359,* 3–18.

Kramer, C. (1977). Pots and people. In L. D. Levine & T. C. Young (Eds.), *Mountains and lowlands: Essays in the archaeology of greater Mesopotamia* (pp. 91–112). Malibu: Undena Publications.

Lacarrière, J., Goutas, N., Normand, C., Simonet, A., & Schwab, C. (2011). Vers une redéfinition des occupations gravettiennes de la grotte d'Isturitz (Pyrénées-Atlantiques, France): révision critique des collections «anciennes» par l'approche intégrée des données lithiques, fauniques et de l'industrie osseuse. In N. Goutas, L. Klaric, D. Pesesse, & P. Guillermin (Eds.), *À la recherche des identités gravettiennes: actualités, questionnements et perspectives: actes de la table ronde sur le Gravettien en France et dans les pays limitrophes, Aix-en-Provence, 6–8 octobre 2008 (Mémoire SPF 52)* (pp. 67–83). Paris: Société préhistorique française.

Langlais, M., Sécher, A., Caux, S., Delvigne, V., Gourc, L., Normand, C., et al. (2016). Lithic tool kits: A metronome of the evolution of the Magdalenian in southwest France (19,000–14,000 cal BP). *Quaternary International, 414,* 92–107.

Leonard, R. D., & Jones, G. T. (1987). Elements of an inclusive evolutionary model for archaeology. *Journal of Anthropological Archaeology, 6,* 199–219.

Le Brun-Ricalens, F., Bordes, J.-G., & Eizenberg, L. (2009). A crossed-glance between southern European and Middle-Near Eastern early Upper Palaeolithic lithic technocomplexes. Existing models, new perspectives. In M. Camps & C. Szmidt (Eds.), *The Medterranean from 50 000 to 25 000 BP: Turning points and new directions* (pp. 11–33). Oxford: Oxbow Books.

Lengyel, G. (2016). Reassessing the Middle and Late Upper Palaeolithic in Hungary. *Acta Archaeologica Carpathica, 51,* 47–66.

Leroi-Gourhan A. (1964–1965). *Le geste et la parole, vols. 1 and 2.* Paris: A. Michel.

Lev, S. Y. (2009). Kamennyi inventar' Zaraiskoi stoianki: Tipologicheskii aspekt. In K. A. Amirkhanov (Ed.), *Issledovaniia paleolita v Zaraiske 1999–2005* (pp. 37–185). Moscow: Paleograf.

Lev, S. Y., Klarik, L., & Giria, E. Y. (2011). Nozhi kostёnkovskogo tipa i plastiny s podrabotkoi kontsa: fenomen konvergentnogo razvitiia ili rodstvo tekhnologii? In K. N. Gavrilov (Ed.), *Paleolit i mezolit vostochnoi Evropy* (pp. 235–279). Moscow: Institut arkheologii RAN/Taus.

Lisitsyn, S. N. (2015). The late Gravettian of Borshevo 5 in the context of the Kostenki-Borshevo sites (Don basin, Russia). *Quaternary International, 359–60,* 372–383.

Lomax, J., Fuchs, M., Preusser, F., & Fiebig, M. (2014). Luminescence based loess chronostratigraphy of the Upper Palaeolithic site Krems-Wachtberg, Austria. *Quaternary International, 351,* 88–97.

Lyman, R. L., & O'Brien, M. J. (2004). A history of normative theory in Americanist archaeology. *Journal of Archaeological Method and Theory, 11,* 369–396.

Macalister, R. A. S. (1921). *A text-book of European archaeology. Vol. I: The Palaeolithic period.* Cambridge: Cambridge University Press.

MacCurdy, G. G. (1914). La Combe, a Paleolithic cave in the Dordogne. *American Anthropologist, 16,* 157–184.

MacCurdy, G. G. (1915). [Book Review] Ancient hunters and their modern representatives. By WJ Sollas, D. Sc., FRS, Professor of Geology and Palaeontology in the University of Oxford. Second Edition. *The American Historical Review, 21,* 135–137.

Maher, L. A., & Macdonald, D. A. (2020). Communities of interaction: Tradition and learning in stone tool production through the lens of the Epipaleolithic of Khareh IV, Jordan. In H. Groucutt (Ed.), *Culture history and convergent evolution: Can we detect populations in prehistory?* (pp. 213–243). Cham, Switzerland: Springer.

Maier, A. (2015). *The central European Magdalenian: Regional diversity and internal variability.* Dordrecht: Springer.

Marks, A. E., Hietala, H. J., & Williams, J. K. (2001). Tool standardization in the Middle and Upper Palaeolithic: A closer look. *Cambridge Archaeological Journal, 11,* 17–44.

Marom, A., McCullagh, J. S. O., Higham, T. F. G., Sinitsyn, A. A., & Hedges, R. E. M. (2012). Single amino acid radiocarbon dating of Upper Paleolithic modern humans. *Proceedings of the National Academy of Sciences of the USA, 109,* 6878–6881.

Marom, A., McCullagh, J. S. O., Higham, T. F. G., & Hedges, R. E. M. (2013). Hydroxyproline dating: Experiments on the ^{14}C analysis of contaminated and low-collagen bones. *Radiocarbon, 55,* 698–708.

McCullagh, J. S. O., Marom, A., & Hedges, R. E. M. (2010). Radiocarbon dating of individual amino acids from archaeological bone collagen. *Radiocarbon, 52,* 620–634.

McNabb, J. (2020). Problems and pitfalls in understanding the Clactonian. In H. Groucutt (Ed.), *Culture history and convergent evolution: Can we detect populations in prehistory?* (pp. 29–53). Cham, Switzerland: Springer.

Mellars, P. (1989). Major issues in the emergence of modern humans. *Current Anthropology, 30,* 349–385.

Mesoudi, A., & O'Brien, M. J. (2009). Placing archaeology within a unified science of cultural evolution. In S. Shennan (Ed.), *Pattern and process in cultural evolution* (pp. 21–32). Berkeley: University of California Press.

Michel, A. (2010). *L'Aurignacien récent (post-ancien) dans le Sud-Ouest de la France: variabilité des productions lithiques. Révision taphonomique et techno-économique des sites de Caminade-Est, abri Pataud, Roc-de-Combe, Le Flageolet I, La Ferrassie et Combemenue.* Ph.D. Dissertation, Université de Bordeaux.

Mihailovic, D., & Mihailovic, B. (2007). Considérations sur le Gravettien et l'Épigravettien ancien des Balkans de l'ouest. *Paléo, 19,* 115–128.

Moreau, L. (2010). Geißenklösterle. The Swabian Gravettian in its European context. *Quartär, 57,* 79–93.

Morin, E., Tsanova, T., Sirakov, N., Rendu, W., Mallye, J.-B., & Lévêque, F. (2005). Bone refits in stratified deposits: Testing the chronological grain at Saint-Césaire. *Journal of Archaeological Science, 32,* 1083–1098.

Mussi, M., Cinq Mars, J., & Bolduc, P. (2000). Echoes from the mammoth steppe: The case of the Balzi Rossi. In W. Roebroeks, M. Mussi, J. Svoboda, & K. Fennema (Eds.), *Hunters of the Golden Age* (pp. 105–124). Leiden: University of Leiden Press.

Nalawade-Chavan, S., McCullagh, J., & Hedges, R. (2014). New hydroxyproline radiocarbon dates from Sungir', Russia, confirm early Mid Upper Palaeolithic burials in Eurasia. *PLoS ONE, 9,* e76896.

Naudinot, N., Tomasso, A., Messager, E., Finsinger, W., Ruffaldi, P., & Langlais, M. (2017). Between Atlantic and Mediterranean: Changes in technology during the Late Glacial in Western Europe and the climate hypothesis. *Quaternary International, 428* (B), 33–49.

Neeley, M. P., & Barton, C. M. (1994). A new approach to interpreting late Pleistocene microlith industries in southwest Asia. *Antiquity, 68,* 275–288.

Nigst, P. R., Haesaerts, P., Damblon, F., Frank-Fellner, C., Mallol, C., Viola, B., et al. (2014). Early modern human settlement of Europe north of the Alps occurred 43,500 years ago in a cold steppe-type environment. *PNAS, 111,* 14394–14399.

Noiret, P. (2009). *Le Paléolithique supérieur de Moldavie. Essai de synthèse d'une évolution multi-culturelle.* Liège: ÉRAUL 121.

Noiret, P. (2013). De quoi Gravettien est-il le nom ? In M. Otte (Ed.), *Les Gravettiens* (pp. 29–64). Paris: Éditions Errance.

O'Brien, M., & Holland, T. D. (1990). Variation, selection, and the archaeological record. *Archaeological Method and Theory, 2,* 31–79.

O'Brien, M. J., & Bentley, R. A. (2020). Learning strategies and population dynamics during the Pleistocene colonization of North

America. In H. Groucutt (Ed.), *Culture history and convergent evolution: Can we detect populations in prehistory?* (pp. 261–281). Cham, Switzerland: Springer.

Odell, G. (1981). The morphological express at function junction: Searching for meaning in lithic tool types. *Journal of Anthropological Research, 37,* 319–342.

Otte, M. (1981). *Le Gravettien en Europe Centrale*. Bruges: De Tempel (Dissertationes Archaeologicae Gandenses, vol. XX).

Otte, M. (Ed.). (2010). *Les Aurignaciens*. Paris: Éditions Errance.

Otte, M. (Ed.). (2013). *Les Gravettiens*. Paris: Éditions Errance.

Paris, C., Deneuve, E., Fagnart, J.-P., Coudret, P., Antoine, P., Peschaux, C., et al. (2017). Premières observations sur le gisement gravettien à statuettes féminines d'Amiens-Renancourt 1 (Somme). *Bulletin de la Société préhistorique française, 114,* 423–444.

Pelegrin, J. (1991). Les savoir-faire: une très longue histoire. *Terrain, 16,* 106–113.

Pelegrin, J. (2007). Réflexions sur la notion de spécialiste dans la taille de la pierre au Paléolithique. In R. Desbrosse & A. Thévenin (Eds.), *Arts et cultures de la Préhistoire: hommages à Henri Delporte* (pp. 315–318). Paris: Éditions du Comité des travaux historiques et scientifiques.

Pelegrin, J. (2012). Sur les débitages laminaires du Paléolithique supérieur. In F. Delpech & J. Jaubert (Eds.), *François Bordes et la Préhistoire: colloque international François Bordes, Bordeaux, 22–24 avril 2009* (pp. 141–152). Paris: Éditions du Comité des Travaux historiques et scientifiques.

Perlès, C. (2013). Tempi of change: When soloists don't play together. Arrhythmia in 'continuous' change. *Journal of Archaeological Method and Theory, 20,* 281–299.

Pesesse, D. (2013). Le Gravettien existe-t-il ? Le prisme du système technique lithique. In M. Otte (Ed.), *Les Gravettiens* (pp. 67–104). Paris: Éditions Errance.

Pesesse, D. (2009–2010). La pointe de Font-Yves et les productions lithiques des derniers Aurignaciens. *Paléo, 21,* 203–222.

Pesesse, D., & Michel, A. (2006). Le burin des Vachons: apports d'une relecture technologique à la compréhension de l'Aurignacien final. *Paléo, 18,* 143–160.

Pigeot, N. (1990). Technical and social actors in Prehistory. Flintknapping specialists and apprentices at Magdalenian Étiolles. *Archaeological Review from Cambridge, 9,* 126–141.

Polanská, M., & Hromadová, B. (2015). Réflexion autour des industries gravettiennes «post-pavloviennes» de Slovaquie occidentale et de Moravie (25,500/24,500–22,000 BP non calibré). In S. Sázelová, M. Novák, & A. Mizerová (Eds.), *Forgotten times and spaces: New perspectives in paleoanthropological, paleoetnological and archeological studies* (pp. 132–154). Brno: Institute of Archeology of the Czech Academy of Sciences and Masaryk University.

Posth, C., Renaud, G., Mittnik, A., Drucker, D. G., Rougier, H., Cupillard, C., et al. (2016). Pleistocene mitochondrial genomes suggest a single major dispersal of non-Africans and a Late Glacial population turnover in Europe. *Current Biology, 26,* 827–833.

Praslov, N. D., & Ivanova, M. A. (1982). Kostënki 21 (Gmelinskaia stoianka). In N. D. Praslov & A. N. Rogachëv (Eds.), *Paleolit kostënkovsko-borshchëvskogo raiona na Donu 1879–1979: Nekotorye itogi polevykh issledovanii* (pp. 198–210). Leningrad: Nauka.

Read, D. W. (1989). Intuitive typology and automatic classification: Divergence or full circle? *Journal of Anthropological Archaeology, 8,* 158–188.

Renard, C. (2011). Continuity or discontinuity in the Late Glacial Maximum of south-western Europe: The formation of the Solutrean in France. *World Archaeology, 43,* 726–743.

Reynolds, N. (2014). *The Mid Upper Palaeolithic of European Russia: Chronology, culture history and context. A study of five Gravettian backed lithic assemblages*. DPhil Dissertation, University of Oxford.

Reynolds, N. (Accepted manuscript). The Gravettian is dead: Against equivocation and reification in chronocultural studies of the Upper Palaeolithic. In O. Touzé, N. Goutas, H. Salomon, & P. Noiret (Eds.), *North-Western Europe During the Gravettian. Contributions of recent research to the understanding of the societies and their environments. Proceedings of the International congress of Liège (12–13 April 2018)*. Liège and Brussels: Presses universitaires de Liège and Royal Belgian Society of Anthropology and Prehistory.

Reynolds, N., & Green, C. (2019). Spatiotemporal modelling of radiocarbon dates using linear regression does not indicate a vector of demic dispersal associated with the earliest Gravettian assemblages in Europe. *Journal of Archaeological Science*: Reports, 27, 101958.

Reynolds, N., & Riede, F. (2019). House of cards: Cultural taxonomy and the study of the European Upper Palaeolithic. *Antiquity, 93,* 1350–1358.

Reynolds, N., Lisitsyn, S. N., Sablin, M. V., Barton, N., & Higham, T. (2015). Chronology of the European Russian Gravettian: New radiocarbon dating results and interpretation. *Quartär, 62,* 121–132.

Reynolds, N., Dinnis, R., Bessudnov, A. A., Devièse, T., & Higham, T. F. G. (2017). The Kostënki 18 child burial and the cultural and funerary landscape of Mid Upper Palaeolithic European Russia. *Antiquity, 91,* 1435–1450.

Reynolds, N., Germonpré, M., Bessudnov, A. A., & Sablin, M. V. (2019). The Late Gravettian site of Kostënki 21 Layer III, Russia: A chronocultural reassessment based on a new interpretation of the significance of intra-site spatial patterning. *Journal of Paleolithic Archaeology, 2,* 160–210.

Riede, F. (2011). Steps towards operationalizing an evolutionary archaeological definition of culture. In B. W. Roberts & M. Vander Linden (Eds.), *Investigating archaeological cultures: Material culture, variability, and transmission* (pp. 245–270). New York: Springer.

Riel-Salvatore, J., & Clark, G. A. (2001). Grave markers: Middle and Early Upper Paleolithic burials and the use of chronotypology in contemporary Paleolithic research. *Current Anthropology, 42,* 449–479.

Rigaud, S., Roussel, M., Rendu, W., Primault, J., Renou, S., Hublin, J.-J., et al. (2014). Les pratiques ornementales à l'Aurignacien ancien dans le Centre-Ouest de la France. L'apport des fouilles récentes aux Cottés (Vienne). *Bulletin de la Société préhistorique française, 111,* 19–38.

Rigaud, S., Manen, C., & García-Martínez de Lagrán, I. (2018). Symbols in motion: Flexible cultural boundaries and the fast spread of the Neolithic in the western Mediterranean. *PLoS ONE, 13,* e0196488.

Roberts, B. W., & Vander Linden, M. (2011). Investigating archaeological cultures: Material culture, variability, and transmission. In B. W. Roberts & M. Vander Linden (Eds.), *Investigating archaeological cultures: Material culture, variability, and transmission* (pp. 1–21). New York: Springer.

Rogachëv, A. N. (1957). Mnogosloinye stoianki kostënkovsko-borshëvskogo raiona na Donu i problema razvitiia kul'tury v epokhu verkhnego paleolita na Russkoi ravnine. *Materialy i issledovaniia po arkheologii SSSR, 59,* 9–134.

Rogachëv, A. N., & Anikovich, M. V. (1982). Kostënki 9 (Biriuch'ii Log). In N. D. Praslov & A. N. Rogachëv (Eds.), *Paleolit kostënkovsko-borshchëvskogo raiona na Donu 1879–1979: Nekotorye itogi polevykh issledovanii* (pp. 109–113). Leningrad: Nauka.

Rogachëv, A. N., & Popov, V. V. (1982). Kostënki 11 (Anosovka 2). In N. D. Praslov & A. N. Rogachëv (Eds.), *Paleolit kostënkovsko--borshchëvskogo raiona na Donu 1879–1979: Nekotorye itogi polevykh issledovanii* (pp. 116–132). Leningrad: Nauka.

Ronchitelli, A., Mugnaini, S., Arrighi, S., Atrei, A., Capecchi, G., Giamello, M., et al. (2015). When technology joins symbolic behaviour: The Gravettian burials at Grotta Paglicci (Rignano Garganico–Foggia–Southern Italy). *Quaternary International, 359–360,* 423–441.

Sackett, J. R. (1982). Approaches to style in lithic archaeology. *Journal of Anthropological Archaeology, 1,* 59–112.

Sackett, J. R. (1985). Style and ethnicity in the Kalahari: A reply to Wiessner. *American Antiquity, 50,* 154–159.

Sauer, F., & Riede, F. (2019). A critical reassessment of cultural taxonomies in the Central European Late Palaeolithic. *Journal of Archaeological Method and Theory, 26,* 155–184.

Scerri, E. M. L., Drake, N. A., Jennings, R., & Groucutt, H. S. (2014). Earliest evidence for the structure of *Homo sapiens* populations in Africa. *Quaternary Science Reviews, 101,* 207–216.

Schmidt, I., Bradtmöller, M., Kehl, M., Pastoors, A., Tafelmaier, Y., Weninger, B., et al. (2012). Rapid climate change and variability of settlement patterns in Iberia during the Late Pleistocene. *Quaternary International, 274,* 179–204.

Schwabedissen, H. (1954). *Die Federmesser-Gruppen des nordwesteuropäischen Flachlandes: Zur Ausbreitung des Spät-Magdalénien (Offa-Bücher 9).* Neumünster: Karl Wachholtz Verlag.

Serwatka, K., & Riede, F. (2016). 2D geometric morphometric analysis casts doubt on the validity of large tanged points as cultural markers in the European Final Palaeolithic. *Journal of Archaeological Science: Reports, 9,* 150–159.

Shea, J. J. (2014). Sink the Mousterian? Named stone tool industries (NASTIES) as obstacles to investigating hominin evolutionary relationships in the Later Middle Paleolithic Levant. *Quaternary International, 350,* 169–179.

Shennan, S. (2020). Style, function and cultural transmission. In H. Groucutt (Ed.), *Culture history and convergent evolution: Can we detect populations in prehistory?* (pp. 291–298). Cham, Switzerland: Springer.

Shennan, S. J., Crema, E. R., & Kerig, T. (2015). Isolation-by-distance, homophily, and "core" vs. "package" cultural evolution models in Neolithic Europe. *Evolution and Human Behavior, 36,* 103–109.

Sikora, M., Seguin-Orlando, A., Sousa, V. C., Albrechtsen, A., Korneliussen, T., Ko, A., et al. (2017). Ancient genomes show social and reproductive behavior of early Upper Paleolithic foragers. *Science, 358,* 659–662.

Simonet, A. (2012). *Brassempouy (Landes, France) ou la matrice gravettienne de l'Europe.* Liège: ÉRAUL 133.

Sinitsyn, A. A. (2007). Variabilité du Gravettien de Kostienki (bassin moyen du Don) et des territoires associés. *Paléo, 19,* 181–202.

Sinitsyn, A. A. (2010). The Early Upper Palaeolithic of Kostenki: Chronology, taxonomy, and cultural affiliation. In C. Neugebauer-Maresch & L. R. Owen (Eds.), *New aspects of the Central and Eastern European Upper Palaeolithic—Methods, chronology, technology and subsistence (Symposium by the Prehistoric Commission of the Austrian Academy of Sciences, Vienna, November 9–11, 2005)* (pp. 27–48). Vienna: Österreichischen Akademie der Wissenschaften.

Sinitsyn, A. A. (2013). Gravett Kostënok v kontekste gravetta Vostochnoi Evropy. In G. V. Sinitsyna (Ed.), *Problemy zaseleniia severo-zapada Vostochnoi Evropy v verkhnem i final'nom paleolite (kul'turno-istoricheskie protsessy)* (pp. 4–32). Saint Petersburg: IIMK RAN.

Sinitsyn, A. A. (2014). K probleme kul'turnoi prinadlezhnosti Pushkarei 1. In D. G. Savinov & A. I. Murashkin (Eds.), *Problemy arkheologii epokhi kamnia: k 70-letiiu Valentiny Ivanovny Beliaevoi: Sbornik nauchnykh statei (Trudy istoricheskogo fakul'teta SPbGU, Vol. 18)* (pp. 234–244). Saint Petersburg: Izdatel'stvo SPbGU.

Sinitsyn, A. A. (2015). Perspectives on the Palaeolithic of Eurasia: Kostënki and related sites. In N. Sanz (Ed.), *Human origin sites and the World Heritage Convention in Eurasia, vol. 1 (HEADS 4)* (pp. 163–189). Paris & Mexico City: UNESCO.

Sinitsyn, A. A., Praslov, N. D., Svezhentsev, Iu. S., & Sulerzhitskii, L. D. (1997). Radiouglerodnaia khronologiia verkhnego paleolita Vostochnoi Evropy. In A. A. Sinitsyn & N. D. Praslov (Eds.), *Radiouglerodnaia khronologiia paleolita Vostochnoi Evropy i Severnoi Azii. Problemy i perspektivy* (pp. 21–66). Sankt-Peterburg: IIMK RAN.

Sober, E. (1980). Evolution, population thinking, and essentialism. *Philosophy of Science, 47,* 350–383.

Sobkowiak-Tabaka, I. (2017). *Rozwój społeczności Federmesser na Nizinie Środkowoeuropejskiej.* Poznań: Wydawnictwo Instytutu Archeologii i Etnologii Polskiej Akademii Nauk.

Sobkowiak-Tabaka, I., & Winkler, Katja. (2017). The Ahrensburgian and the Swiderian in the area around the middle Oder River: Reflections on similarities and differences. *Quartär, 64,* 217–240.

Soffer, O., Adovasio, J. M., & Hyland, D. C. (2000). The "Venus" figurines: Textiles, basketry, gender, and status in the Upper Paleolithic. *Current Anthropology, 41,* 511–537.

Sørensen, M. L. S. (2014). The archaeological culture concept: Hot or cold understandings. In H. Alexandersson, A. Andreeff, & A. Bünz (Eds.), *Med Hjärta Och Hjärna: En Vänbok till Professor Elisabeth Arwill-Nordbladh* (pp. 247–258). Gothenburg: Department of Historical Studies, University of Gothenburg.

Soriano, S. (1998). Les microgravettes du Périgordien de Rabier à Lanquais (Dordogne): analyse technologique fonctionnelle. *Gallia préhistoire, 40,* 75–94.

Straus, L. G. (2000). A quarter-century of research on the Solutrean of Vasco-Cantabria, Iberia and beyond. *Journal of Anthropological Research, 56,* 39–58.

Svoboda, J. A. (2007). The Gravettian on the Middle Danube. *Paléo, 19,* 203–220.

Svoboda, J. (2015). Perspectives on the Upper Palaeolithic in Eurasia: The Case of the Dolní Vestonice-Pavlov Sites. In N. Sanz (Ed.), *Human origin sites and the World Heritage Convention in Eurasia* (Vol. 1, pp. 190–204)., *HEADS 4* Paris & Mexico City: UNESCO.

Svoboda, J. A., & Novák, M. (2004). Eastern Central Europe after the Upper Pleniglacial: Changing points of observation. *Archäologisches Korrespondenzblatt, 34,* 463–477.

Tallavaara, M., Luoto, M., Korhonen, N., Järvinen, H., & Seppä, H. (2015). Human population dynamics in Europe over the Last Glacial Maximum. *Proceedings of the National Academy of Sciences, 112,* 8232–8237.

Tarasov, L. M. (1979). *Gagarinskaia stoianka i eë mesto v paleolite Evropy.* Leningrad: Nauka.

Tarassov [Tarasov], L. M. (1971). La double statuette paléolithique de Gagarino. *Quartär, 22,* 157–163.

Tejero, J.-M. (2016). Spanish Aurignacian projectile points: An example of the first European Paleolithic hunting weapons in osseous materials. In M. Langley (Ed.), *Osseous projectile weaponry* (pp. 55–69). Dordrecht: Springer.

Teyssandier, N., & Zilhão, J. (2018). On the entity and antiquity of the Aurignacian at Willendorf (Austria): Implications for modern human emergence in Europe. *Journal of Paleolithic Archaeology, 1,* 107–138.

Teyssandier, N., Bon, F., & Bordes, J.-G. (2010). Within projectile range: Some thoughts on the appearance of the Aurignacian in Europe. *Journal of Anthropological Research, 66,* 209–229.

Tixier, J. (2012). *A method for the study of stone tools: Guidelines based on the work of J. Tixier = Méthode pour l'étude des outillages lithiques: notice sur les travaux scientifiques de J. Tixier*. Luxembourg: Musée National d'Histoire et d'Art/Centre National de Recherche Archeologique.

Tomášková, S. (2003). Nationalism, local histories and the making of data in archaeology. *Journal of the Royal Anthropological Institute, 9,* 485–507.

Tsanova, T., Zwyns, N., Eizenberg, L., Teyssandier, N., Le Brun-Ricalens, F., & Otte, M. (2012). Le plus petit dénominateur commun: réflexion sur la variabilité des ensembles lamellaires du Paléolithique supérieur ancien d'Eurasie. Un bilan autour des exemples de Kozarnika (Est des Balkans) et Yafteh (Zagros central). *L'anthropologie, 116,* 469–509.

Vander Linden, M., & Roberts, B. W. (2011). A tale of two countries: Contrasting archaeological culture history in British and French archaeology. In B. W. Roberts & M. Vander Linden (Eds.), *Investigating archaeological cultures: Material culture, variability, and transmission* (pp. 23–40). New York: Springer.

Vanhaeren, M., & d'Errico, F. (2006). Aurignacian ethno-linguistic geography of Europe revealed by personal ornaments. *Journal of Archaeological Science, 33,* 1105–1128.

Van Oyen, A. (2015). Actor-Network Theory's take on archaeological types: Becoming, material agency and historical explanation. *Cambridge Archaeological Journal, 25,* 63–78.

Van Oyen, A. (2017). Material culture and mobility: A brief history of archaeological thought. In C. Heitz & R. Stapfer (Eds.), *Mobility and pottery production: Archaeological & anthropological perspectives* (pp. 53–65). Leiden: Sidestone Press.

Verpoorte, A. (2009). Limiting factors on early modern human dispersals: The human biogeography of late Pleniglacial Europe. *Quaternary International, 201,* 77–85.

White, R. (1997). Perspectives technologiques sur la «double statuette» de Gagarino. *Techne, 5,* 6–16.

Wierer, U. (2013). Variability and standardization: The early Gravettian lithic complex of Grotta Paglicci, Southern Italy. *Quaternary International, 288,* 215–238.

Wiessner, P. (1983). Style and social information in Kalahari San projectile points. *American Antiquity, 48,* 253–276.

Wiessner, P. (1985). Style or isochrestic variation? A reply to Sackett. *American Antiquity, 50,* 160–166.

Wilczyński, J. (2015). Lithic artefacts. In J. Wilczyński (Ed.), *A Gravettian site in southern Poland: Jaksice II* (pp. 33–62). Krakow: ISEA PAS.

Wilczyński, J., Wojtal, P., Łanczont, M., Mroczek, P., Sobieraj, D., & Fedorowicz, S. (2015). Loess, flints and bones: Multidisciplinary research at Jaksice II Gravettian site (southern Poland). *Quaternary International, 359–60,* 114–130.

Will, M., & Mackay, A. (2020). A matter of space and time: How frequent is convergence in lithic technology in the African archaeological record over the last 300 kyr? In H. Groucutt (Ed.), *Culture history and convergent evolution: Can we detect populations in prehistory?* (pp. 103–126). Cham, Switzerland: Springer.

Wobst, H. M. (1977). Stylistic behavior and information exchange. In C. E. Cleland (Ed.), *For the director: Research essays in honor of James B. Griffin (Anthropological Papers 61)* (pp. 317–342). Ann Arbor: University of Michigan Museum of Anthropology.

Wojtal, P., Wilczyński, J., Wertz, K., & Svoboda, J. A. (2018). The scene of a spectacular feast (part II): Animal remains from Dolní Věstonice II, the Czech Republic. *Quaternary International, 466* B, 194–211.

Yanevich, A. (2014). Les occupations gravettiennes de Buran-Kaya III (Crimée): contexte archéologique. *L'anthropologie, 118,* 554–566.

Zheltova, M. N. (2015). Kostenki 4: Gravettian of the east—Not Eastern Gravettian. *Quaternary International, 359–360,* 362–371.

Zilhão, J., & d'Errico, F. (2003). The chronology of the Aurignacian and Transitional technocomplexes. Where do we stand? In J. Zilhão & F. d'Errico (Eds.), *The chronology of the Aurignacian and of the transitional technocomplexes: Dating, stratigraphies, cultural implications. (Trabalhos de Arqueologia 33)* (pp. 313–349). Lisbon: Instituto Português de Arqueologia.

Chapter 11
Communities of Interaction: Tradition and Learning in Stone Tool Production Through the Lens of the Epipaleolithic of Kharaneh IV, Jordan

Lisa A. Maher and Danielle A. Macdonald

Abstract Between 23 and 11.5 ka Epipaleolithic groups of Southwest Asia initiated and experienced dramatic changes —on a previously unprecedented scale—in economy and settlement, with the appearance of semi-sedentary villages and intensified interdependent relationships with each other and specific plants and animals. These events provide a rare opportunity to study the long-term development of social processes in the region and the increasingly obvious fact that social, economic and technological changes were manifested as complex, entangled and non-linear developments. Most recent attempts to explain change in the material culture record typically highlight the earliest evidence for plant management or cultivation, ritual funerary practices, and dwelling and architecture. While these are important contributions that serve as the foundation for challenging our traditional notions of hunter-gatherer to farmer transitions, they center on changes in the economic or symbolic realms of prehistoric life, arguably downplaying the role of technology. This paper attempts to explore the role of technology in our reconstructions of the lifeways of hunter-gatherers by examining the social role of technology, the centrality of the technological process to everyday practice, and the transmission of technological knowledge (and, thus, culture) through communities of practice. We use chipped stone tools and their associated debris from the site of Kharaneh IV, eastern Jordan, as an illustrative case study of how we currently study chipped stone tools in this region. Using a *chaîne opératoire* approach to the study of EP assemblages, we consider how different groups of knappers at the EP site of Kharaneh IV, and beyond, interacted in fluid and ever-changing interactions to share knowledge or reinforce existing social traditions.

Keywords Social practice • Technology • Communities of practice • Situated learning • Skill • Lithics • *Chaîne opératoire* • Hunter-gatherers • Aggregation site

Introduction

Between 23 and 11.5 ka, Epipaleolithic (EP) groups of the Southern Levant initiated and experienced dramatic changes —on a previously unprecedented scale—in economy and settlement, with the appearance of semi-sedentary villages and intensified interdependent relationships with specific plants and animals. Most recent attempts to explain the associated changes in the material culture record are typically 'origins'-focused (e.g., Gamble 2007), highlighting, for example, the earliest evidence for plant management or cultivation (Snir et al. 2015), ritual funerary practices (Grosman et al. 2008; Munro and Grosman 2010; Maher et al. 2011; Nadel et al. 2013), and dwelling and architecture (Nadel 2000; Nadel et al. 2004; Nadel et al. 2011; Maher et al. 2012). While these are important contributions that serve as the foundation for challenging our traditional notions of the hunter-gatherer to farmer transition (see also Finlayson and Warren 2010; Finlayson and Makarewicz 2013), they center on changes in the economic or symbolic realms of prehistoric life, arguably and inadvertently downplaying the role of technology.

This paper attempts to explore the role of technology in our reconstructions of the lifeways of EP hunter-gatherers by examining the social role of technology, the centrality of the technological process to everyday practice, and the transmission of technological knowledge through communities of

L. A. Maher (✉)
Department of Anthropology, University of California, Berkeley, 232 Kroeber Hall, Berkeley, CA 94720-3710, USA
e-mail: maher@berkeley.edu

D. A. Macdonald
Department of Anthropology, The University of Tulsa, Harwell Hall, 800 S. Tucker Drive, Tulsa, OK 74104, USA
e-mail: danielle-macdonald@utulsa.edu

H. S. Groucutt (ed.), *Culture History and Convergent Evolution: Can We Detect Populations in Prehistory?*,
Vertebrate Paleobiology and Paleoanthropology, https://doi.org/10.1007/978-3-030-46126-3_11

practice. We use chipped stone tools and their associated debris from the Epipaleolithic (EP) site of Kharaneh IV, eastern Jordan, as an illustrative case study for the investigation of chipped stone tool production in this region. We suggest here that a social approach to chipped stone technologies provides a rich context within which to think about hunter-gatherers actively engaged in change through socialized landscapes (Langley 2013) and landscape learning (Rockman 2013; Hussain and Floss 2016) during this period of prehistory that archaeologists typically characterize as a transitional part of the process of 'becoming Neolithic' (Belfer-Cohen and Goring-Morris 2011; Goring-Morris and Belfer-Cohen 2011; Watkins 2013). We focus on how individual stone tool producers, or flintknappers, made stone tools as knowledgeable agents in a material world (Dobres 2000), using a *chaîne opératoire* approach to the study of archaeological assemblages. We also consider how different groups of knappers at the EP site of Kharaneh IV, and beyond, interacted in fluid and ever-changing networks of shared knowledge that may have served to reinforce existing social traditions and resulted in particular patterns of variation within the archaeological record across time and space.

In this light, we suggest that stone tool production during this period of so-called 'transition' is better thought of as a technological experience that shaped many aspects of these hunter-gatherer lifeways. This demonstrates that the world of EP hunter-gatherers was already a socially-complex landscape and that the so-called hunter-gatherer-to-farmer transition was not a unilineal trajectory from hunter to farmer or mobile to sedentary; at best, it was a bumpy and winding series of paths that were not straightforward or inevitable (Belfer-Cohen and Bar-Yosef 2000; Valentin 2008). Rather than viewing the EP as an important period of prehistory because it was a time of dramatic cultural change related to the origins or emergence of the Neolithic—a pre-cursor to Neolithic society where some threshold was overcome to tip the balance towards agriculture—we instead focus on the evidence for complex social interactions in a dynamic EP landscape. The evidence for aggregation and far-reaching interaction networks of exchange of material objects (Richter et al. 2011; Maher 2016; Maher et al. 2016) and stone tool technological knowledge at Kharaneh IV is but one example of where we can trace these evidences.

The Epipaleolithic Period in Southwest Asia

The EP period covers just over 10,000 years of prehistory in Southwest Asia where hunter-gatherer communities actively shaped the world around them and engaged in social networks of interactions over increasingly large scales. The EP archaeological record is notable for exhibiting features related to the gradual emergence of sedentism, food production, and otherwise 'Neolithic' lifeways. These include the appearance of 'permanent' stone-built houses (although see Boyd 2006), agglomerated into villages and occupied over long periods of time (Cucchi et al. 2005; Weissbrod et al. 2017), changes in plant management strategies (Snir et al. 2015; Ramsey et al. 2016) and intensified use of cereals (Asouti and Fuller 2012; Asouti 2013), evidence for bread-making (Arranz-Otaegui et al. 2018) and, potentially, beer-brewing (Hayden et al. 2012; Liu et al. 2018), as well as a wide variety of complex symbolic practices related to changing worldviews (Goring-Morris and Belfer-Cohen 2002; Goring-Morris and Belfer-Cohen 2010; Watkins 2010; Watkins 2011). Recently, scholars have described the latest phases of the EP as being within the throes of Neolithization, or 'becoming Neolithic' (Belfer-Cohen and Goring-Morris 2011; Goring-Morris and Belfer-Cohen 2011; see also Maher In press). In addition, researchers recognize that within this vast time span, like elsewhere, hunter-gatherer communities were far from static; the respective changes in lifeways were complicated, not directional or isolated, or easily correlated with changes in environment or other aspects of social life (Finlayson and Warren 2010; Finlayson and Warren 2017).

The EP is unified temporally and behaviorally through technological similarities, manifested as hunter-gatherers whose chipped stone tool technology focused on the production of microliths.[1] Despite overall continuity in material culture throughout the EP period, it is conventionally divided into three main phases—Early, Middle and Late EP—each of which can be further subdivided into many geographically-bounded industries, facies, and entities (Fig. 11.1). Traditionally, the basis for internal EP subdivisions rests on variations in the relative proportions of microlith types and are interpreted to represent different cultural groups (Pirie 2004). For example, Early EP groups such as the Kebaran are archaeologically-recognizable by a chipped stone tool assemblage dominated by gracile micropoints and obliquely truncated (and usually backed) bladelets formed from narrow-faced cores and made without the use of the microburin technique (versus Nebekian groups who did use the microburin technique to produce non-geometric microliths); Middle EP groups such as the Geometric Kebaran are best known for producing geometric microliths, particularly trapeze/rectangles, also without consistent use of the microburin technique, on broad-faced cores (versus Mushabian groups who used the microburin technique to make Mushabian points); Late EP Natufians are best-known

[1]As defined in Southwest Asia by Bar-Yosef (1970), Goring-Morris (1987) and (Tixier 1963, Tixier et al. 1980).

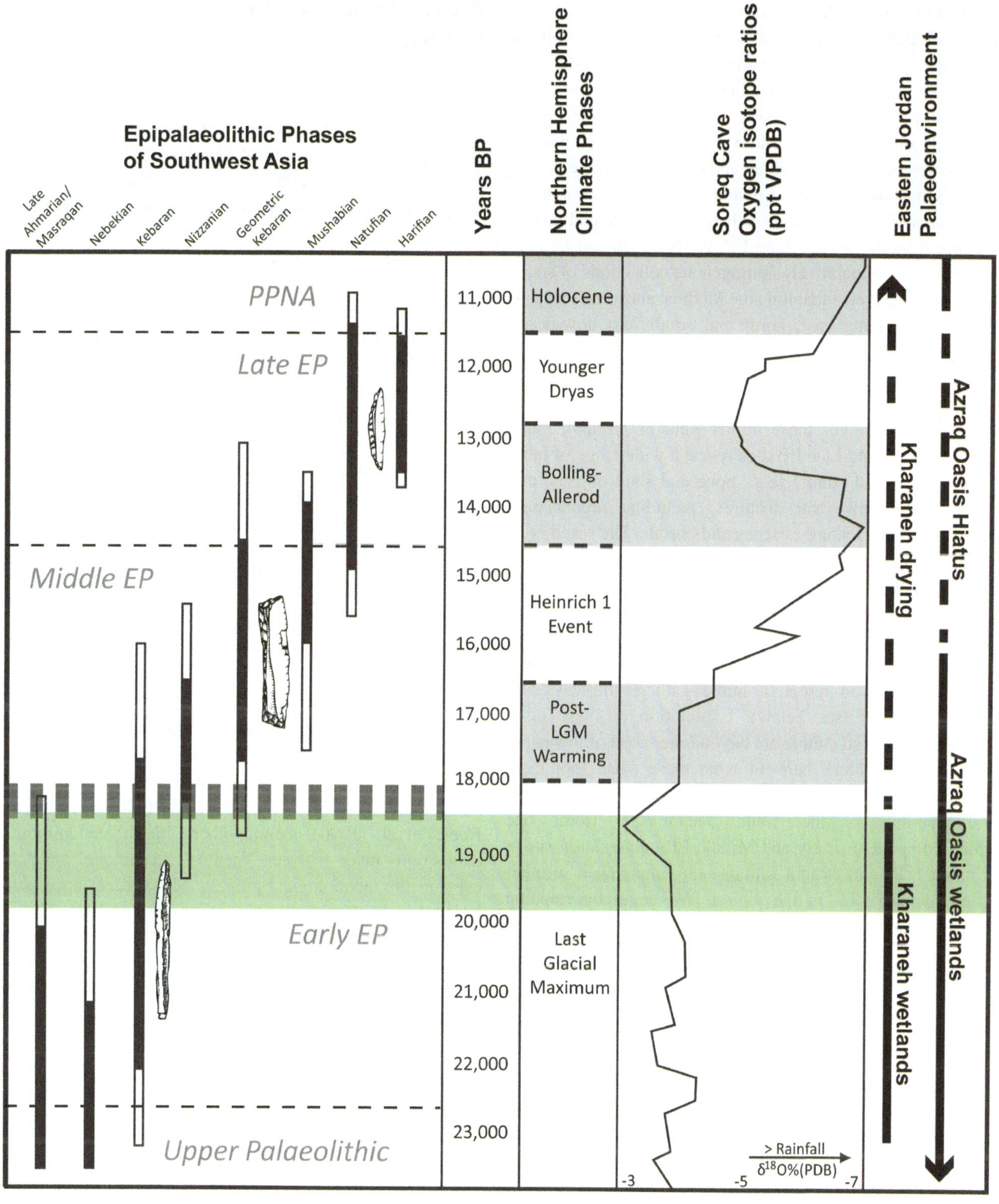

Fig. 11.1 A simplified cultural-chronological framework for the EP period (c. 23–11.5 ka), noting individual cultural entities within the Early, Middle and Late EP (the large dashed lines for the onset of the Middle EP reflect the early dates from Azraq sites), major northern hemisphere paleoenvironmental phases alongside regional oxygen isotope records from Soreq Cave, and paleoenvironment reconstructions from the Azraq Basin (modified from Maher, In press). The grey marks global periods of warming. The green marks the timing and duration of occupation of Kharaneh IV

for making lunate-shaped geometric microliths from a wide range of small flakes and blades, with use of the microburin technique, and from highly variable cores (versus Harifians who also make the Harifian point). Yet, the differences in material culture between Early, Middle, and Late phases generally outweigh those within these phases, such that the differences between Mushabian and Geometric Kebaran sites are less than those between Kebaran, Geometric Kebaran and Natufian sites. Over time, these heterogeneous social groups disappear so that by the Late EP, the Natufian can be considered as a comparatively homogenous collections of social groups[2] intensively inhabiting the Mediterranean core area, as well as extending east, north and south into now-desert regions.

Aside from considerations of stone tool technologies (although the best-preserved, they are certainly not the only line of evidence for these hunter-gatherer groups), many Early, Middle and Late EP sites reveal a wide range of other types of material culture (e.g., bone and shell objects) and highly informative site features, including architecture, hearths, storage features, caches and burials. The result is an overly complex picture of EP groups over space and time (Maher and Richter 2011), where variability in microlith technology does not necessarily exhibit a one-to-one correlation with variations in other aspects of material culture, prompting questions about what chipped stone variability actually means in terms of hunter-gatherer lifeways and group identities (see below). Exploration of these other aspects of material culture are beyond the scope of this paper and, thus, we focus here on what these stone tool technologies can tell us about the nature of variability through examination of the lithic technology from one multi-component (Early and Middle EP) aggregation site—Kharaneh IV—in eastern Jordan and comparing it to other contemporary sites in the region. We begin by stepping back, outside of the EP of Southwest Asia, to provide a framework for understanding the role of stone tool technology in shedding light on EP hunter-gatherer lifeways.

[2]We emphasize here that this homogeneity is relative to what is exhibited earlier in the EP—the Natufian is not the 'same' across the entire region, but, rather, these Late EP sites share a number of features that make their ascription to the Natufian accepted by most researchers. Notably, the most widely-recognized of these features is a chipped stone tool technology focused on the production of small geometric microliths, namely crescents or lunates. These sites are found from the Sinai Peninsula north to southern Anatolia across to the Iranian Plateau and, possibly, west to the island of Cyprus.

Archaeological Approaches to Technology

Technology, as both a process and a practice, is a fundamental theme in anthropology (e.g., Hodges 1989; Lemonnier 1992; Pfaffenberger 1992; Hegmon 1998; Schiffer 2001; Crown 2007; Miller 2007; Lemonnier 2013). The engagement with material objects to answer questions about past societies has always been a central part of the discipline (Mauss 1936; Forbes 1964; Leroi-Gourhan 1964; Lemonnier 1976; Lemonnier 1986; Haudricourt 1987; Lave and Wenger 1991; Lemonnier 1992; Dietler and Herbich 1998; Schiffer 2004). Within the early history of archaeology, with its focus on function and key 'technological' milestones that defined cultural-chronological periods (i.e., the Stone Age, Copper Age, Bronze Age, Iron Age, even the Industrial Age; Franklin 1999; Suchman 2001), rarely did researchers consider technology as something to be studied beyond its role in economy and social organization as a cultural-historical indicator (see, for example, summaries in Hodges 1970; Basalla 1988), highlighted also in Binford's consideration of *technomic* artifacts (Binford 1962). These early approaches to technology focused on relationships between form and function where tool inventions and innovations developed to fulfil a specific need, or on unilineal progressions of technological development that related technological stages to human evolution. However, considerations of the role of style in technology (Lechtman and Merrill 1977) and a focus on the *techniques* of material culture production (Lemonnier 1976) in the 1970s highlighted other aspects of technology. For example, Tixier (1979), Perlés and Phillips (1991), and Boëda et al. (1990), among others, highlighted important distinctions made between 'method', 'technique' and 'concept' in the technological process. More recent approaches have taken on themes of design theory (Bleed 1986; Eerkens 1991; Kingery 2001), technical choice (Stark 1998; Sillar and Tite 2000; Bleed 2001), performance and process (Schiffer 2001; Schiffer et al. 2001; Schiffer 2004; Skibo and Schiffer 2008) and ecological or evolutionary considerations (Kuhn 2004; McClure 2007; Surovell 2012; Shennan 2020). Critical of these economic and functionalist approaches to technology, called the 'Standard View' of technology by Pfaffenberger (Pfaffenberger 1992) and 'black box' approaches by Dobres (Dobres 2000), these researchers take cues from the *chaîne opératoire* approach to technology (Mauss 1936; Leroi-Gourhan 1964; Pelegrin et al. 1988; Karlin et al. 1991; Lemonnier 1992; Sellet 1993; Schlanger 1994; Chazan 2009; Soressi and Geneste 2011; Audouze et al. 2017; Delage 2017). They advocate an approach that focuses on the social behaviors decipherable through interactions between a maker and their materials and environment, as well as the social interactions between people (various makers and users), including the performances and gestures involved in making and using things (see also Tostevin 2007; Tostevin 2013). These social approaches explore a range of issues related to standardization, specialization, innovation and invention, social identity and boundaries, status, skill, learning and gender (e.g., Gero 1991; Costin 2001; Ingold 2001; Bamforth and Finlay 2008; Wendrich 2013a; Gibbs 2015).

The Study of Technology in the EP of Southwest Asia

In Southwest Asia, excavations at prehistoric sites have produced an extensive and impressive range of material objects that include a wealth of chipped stone tools and pottery, elaborately decorated stone vessels, textiles and basketry, plaster statues, beads made from bone and shell, and intricately cast copper objects. However, here, as in many parts of the world, such objects have most often been approached typologically (see Pirie 2004), with the aim of creating descriptive classificatory schemes that can be used to date sites and define cultural groups. These studies focus more on later prehistory and the rise of state-level societies and implicitly suggest that technology is a passive reflection of cultural norms, and innovations were made primarily (although, certainly not solely) to meet economic or functional needs (e.g., Moorey 1999; Bourriau and Phillips 2004; Baker 2018). While there is no doubt that these were important concerns to prehistoric practitioners and motivating factors in the development (and limitations) of certain technologies, we emphasize technology here as an active social phenomenon in prehistory and examine the decisions made and actions performed (the *chaîne opératoire*) during chipped stone tool production, use and discard within EP technological traditions in order to investigate the reasons for these choices and actions, how they varied across space and time, and how they influenced (created, maintained and transformed) social relationships in EP lifeways (see Gibbs 2015 for an example of this for Near Eastern pottery technologies).

The social role of technology is of great interest to archaeologists, as demonstrated by a number of publications (Pfaffenberger 1988; Gero 1989; Gero 1991; Dobres and Hoffman 1992; Lemonnier 1992; Pfaffenberger 1992; Schlanger 1994; Hegmon 1998; Franklin 1999; Dobres 2000; Sillar and Tite 2000; Pfaffenberger 2001; Lemonnier 2002; Killick 2004; Hurcombe 2007; Miller 2007; Dobres 2010; Schiffer 2011; Lemonnier 2013). These works have been extremely valuable in promoting theoretical issues related to technology as a social phenomenon, and in demonstrating the importance of examining the choices made by prehistoric craftspeople. Building on these works, this paper attempts to apply this to the study of EP technology in Southwest Asia through an overview of long-term technological developments at the Early and Middle EP aggregation site of Kharaneh IV, Jordan, as a case study tracing the relationships between technology, social interaction and culture change. While we recognize that technologies never exist in isolation as the products of one technological process often form the tools for, or components of, another technological system, only chipped stone is addressed here. In addition, we focus on particular aspects of technology, including the *chaîne opératoire* and learning, identity and standardization, while acknowledging the importance of other aspects not addressed directly, including gender, specialization and skill. The concepts presented here could easily be argued to apply to a wide range of technological traditions and materials in the EP.

Lithic Technology as Social Practice

Lithic technology studies took off with the work of Leroi-Gourhan (Leroi-Gourhan and Brézillon 1973) and Haudricourt (1964; 1987), both of whom considered the study of technology as the science of human activities (cf. Soressi and Geneste 2011). Continuing this technology-focused approach to lithic analysis, J. Tixier and colleagues (Tixier 1979; Tixier et al. 1980; Pelegrin 1990) focused on the social significance of techniques used by prehistoric peoples. Attention shifted from the study of people through their tools to the study of societies through their techniques (Soressi and Geneste 2011). "From this perspective, a technique is understood as a social product, as well as a founding element of the society, which constitutes the technique, conditions it, reproduces it and shapes it" (Soressi and Geneste 2011:336).

Studies of technology in archaeology focus on how material culture was made, as well as frameworks for seeing technology through a lens of social relations (i.e., that it plays an active role in creating, negotiating and maintaining social relationships). Viewing technology primarily as a social phenomenon, M.-A. Dobres (2000:1; *our emphasis*) advocates for a focus on "understanding past *social relationships* [and mindful communities of practice] and how they were forged, mediated, and made meaningful during the everyday practice of material culture production and use". Dobres (2000:1–2) argues for "the necessity of understanding the intertwined social and material constitution of technological practice in prehistory, and how these simultaneously tangible and intangible dynamics contributed, dialectically, to long-term cultural stability and change". Thus, if understanding technology allows one to reconstruct social relationships, these social relationships, which define a society or culture (and differences between them), allow one to examine culture(s) from a long-term and comparative perspective. Practice theory and the concept of *habitus* (Bourdieu 1977; Bourdieu 1990; Ortner 2006) focus our attention to the daily or routine aspects of making and using an object, acknowledging the importance of technology as a (social) *process*, emphasizing gestures, actions and interactions, whether the outcome is a material object, performance, or takes other forms (Leroi-Gourhan 1964). But, when the outcome does have a material manifestation (i.e., results in the production of a physical product), this process can be (partially) traced and reconstructed. Spatial and temporal patterns in these material manifestations within and between sites allow us to scale up practices to the group and inter-group levels.

Linking together studies of technological process and social practice, a communities of practice approach (see below) focuses on communities of embodied practitioners (e.g., knappers, potters, farmers, etc.) and how they understand both materiality and the world in which they live (Dobres 2000; Sassaman and Rudolphi 2001). In this approach, focus is placed not just on identifying what knappers, for example, were making when they sat down to produce stone tools, or even how they went about making them, but also on the socially-mediated processes of learning and knowledge transmission—how a knapper learned how to make what they make, who they learned it from, who they knapped with (and who they didn't). In other words, we attempt to rediscover how the technological process of making and using stone tools creates situated learning contexts and mutable communities composed of people with different levels of engagement with stone tools and, thus, engaged in different types of social relationships. With this framework, we attempt to move beyond describing, measuring and interpreting chipped stone tools and debitage to attempt to understand how that tool (its production, use, maintenance, and discard) reflects social interaction and performance between people at the varying scales of individuals and groups.

We acknowledge that the practices involved in the technical processes of stone tool production are both learned and habitual, conscious and unconscious, spontaneous, and deeply entrenched in a social context. Since the "knowledge, understandings, and awareness that derive from one's encounters with their material world are neither neutral or 'merely' practical, [but are created, transformed and maintained through social encounters]; they also reaffirm one's understanding of the world and how it should be worked" (Dobres 2000:5). In this way, chipped stone technologies cannot be understood in isolation from other aspects of society—the social and the material are integrated through practice. In other words, raw material acquisition relates to movements and networks of exchange across the landscape, decisions about design and functionality relate to daily tasks at hand and style to expressions of identity and belonging (or even something else altogether). The challenge, then, with lithic technologies becomes how to interpret past social relationships interwoven into material remains (including the body)? Would we, for example, be able to identify a specialist or master knapper in either the level of quality of their products (or lack of mistakes in debris) or their tell-tale bodily expressions (i.e., pathologies related to decades of habitual behavior)? Can we use lithic assemblages to explore social relationships at intra- or inter-group (and intra- or inter-site) levels? Although the intertwining of social phenomena and technology has been extensively applied in European prehistoric archaeology, there has been considerably less research highlighting the social nature of Levantine Epipalaeolithic assemblages. If we see technology as a social practice, then the *chaîne opératoire* approach to material culture provides a useful hermeneutic tool to access these links.

Approaches to Lithic Technology: *Chaîne Opératoire*, Refitting and Experimental Archaeology

There are several recent and excellent summaries of the history of the *chaîne opératoire* concept in lithic studies (Geneste 1991; Julien 1992; Chazan 2009; Soressi and Geneste 2011; Texier and Meignen 2012; Audouze et al. 2017; Delage 2017), and so we only summarize some key concepts here. Studying 10,000–20,000 year-old hunter-gatherer base camps from France, anthropologist André Leroi-Gourhan used the distribution of different stone tool categories to reconstruct spatial patterns in their production and thus the organization of different on-site activities (Leroi-Gourhan 1964; Leroi-Gourhan and Brézillon 1973; Leroi-Gourhan 1993). Since this highly influential work, a variety of 'schools' with varying approaches to stone tool production have operated within Europe and North America. Although simplistic distinctions between them sometimes persist where the former are characterized as adhering to the concepts of the *chaîne opératoire* as first described by Leroi-Gourhan (see also the excellent review by Audouze et al. 2017), and all that it has come to encompass (Dobres 2010; Soressi and Geneste 2011), and the latter seemingly focused on reconstructing behavioral or reduction sequences (Bleed 2001; Odell 2004; Andrefsky 2005), we note that practitioners of each recognize fundamental differences in approaches to method and theory.[3]

A *chaîne opératoire* approach in the study of chipped stone technologies has at its core the notion of a mental concept or template that guides action, and is inclusive of the planning, preparations, and bodily performances of making, using and discarding an object, as well as the social contexts within which these things take place. A knapper has an intended goal in mind and a plan to carry it out; however, this plan is flexible and fluid throughout the process, changing situationally. This approach incorporates the physical actions of the knapper (i.e., gestures) with his or her knowledge of how to produce a tool (*connaissance*) and skill to perform these actions (*savoir faire*) (Pelegrin 1990). It includes the 'rules' for production as a series of spatial relationships between knowledge, skill, action and materials (method), and consideration of the physical means of transmitting energy in the knapping process, or technique (Leroi-Gourhan 1993; Chazan 2009). It includes a multitude of 'steps', both physical and cognitive, visible and invisible, that occur simultaneously and in a non-linear fashion, and may involve the physical and mental activities of one or multiple individuals (Chazan 2009), as well as the dynamics of making and correcting mistakes, or going 'back-to-the-drawing-board'. Thus, understanding stone tool production as a linear sequence of events is much too simplistic.

The technological process described by any *chaîne opératoire* includes all aspects of 'becoming a material object' and is embedded within a social context. The *chaîne opératoire* attempts to elucidate the (social) experience of being a transformative part of the material world (sensu stricto Dobres 2001). It provides detailed qualitative and quantitative data on artifact 'life histories' and shared technical processes (Pelegrin, Karlin and Bodu 1988). It allows one to integrate the intersecting social and material aspects of people making things (Dobres 2001; Ingold 2001; Pfaffenberger 2001; Dobres 2010).

The best ways that we have to detect these complexities in thought, action and practice (beyond context-specific, albeit highly useful, flow charts) are through refitting studies and experimental archaeology. Refitting involves working backwards from a collection of individual tools and by-products, essentially putting the pieces back together like a jigsaw in order to parse out, or reconstruct and understand, how they originally were removed (the actions and materials used) and in what order (Pigeot 1990; Hofman and Enloe 1992; Close 2000; Laughlin and Kelly 2010). Equifinality is always an issue in stone tool production as there is more than one way to produce almost any type of tool. However, people tend to rely on one of these ways over others depending on how they were taught to make particular tools; that is, their socially-constituted and situated traditions. So, refitting studies can provide insights into a knapper's original production plan[4]—an intended final product and a mental template of how to get there—as well as changes to this plan through the production process. Alongside detailed spatial mapping of debitage during excavation, refitting can potentially indicate when more than one knapper was involved in the manufacturing of tools; where one knapper began a core and one or more others continued to knap it. This could occur in participatory learning contexts, situationally by picking up where someone left off (or selecting bladelets from a knapped set of bladelets), or in the context of an organized, multi-person production sequence, such as that suggested for some Middle EP sites in the northern Negev (Goring-Morris 1987). Since idiosyncrasies in knapping are often reflective of individual styles and can be represented on more than one type of debitage, refitting provides one of the best ways to identify the signatures of individual versus multiple knappers.

Since both the mental plans and more situational activities result from learned social knowledge that is transmitted in social contexts—knappers learn how to make tools from others (masters) who thus enculturate their students (apprentices) into a particular tradition—socially-constituted practices are imparted into the production of the tools. A tradition of stone tool production can be understood through refitting and, thus, we can examine social learning processes and social identity through understanding the technological process.

Similarly, the experimental replication of stone tool technologies has proven invaluable for connecting artifact types to particular methods and techniques. Continuing with the early experimental work of Bordes, Crabtree, and Tixier (Crabtree and Davis 1968; Bordes and Crabtree 1969; Crabtree 1970; Crabtree 1972; Tixier 1974; Tixier, Inizan and Roche 1980), for example, Dibble and colleagues (Dibble and Bernard 1980; Dibble 1997; Dibble and Rezek 2009; Dibble et al. 2017) have had great success in demonstrating the cause-and-effect of specific techniques (i.e., hard vs. soft hammer and platform lipping) on stone that provide us with a known range of conditions under which particular features and debitage are produced. Since

[3]We should note that whether you adhere to one school or the other results from your training, not your nationality. Thus, many North American scholars trained in Europe follow a *chaîne opératoire* approach. Additionally, these boundaries, such as they are, are increasingly blurred today with the predominance of collaborative research projects.

[4]Even in so-called opportunistic knapping events, the knapper still begins the process with a plan.

techniques leave distinct material traces, reconstructing technological processes through the study of archaeological collections can be verified by artifact refitting and experimental reproduction of these processes (Soressi and Geneste 2011). Integrating these allows one to move beyond classifying types of tools to identify prehistoric cultures to understanding motivation, learning, gestures, actions, and material products as related to social practices (Dobres 2000). People learn and make choices about how to make things in socially-meaningful ways. This combined approach also means that tools cannot be seen in isolation, but instead become part of larger technological systems that include both other materials (i.e., recognition that stone tools are often used to make other tools) and link sites to their larger physical and social landscape (i.e., production may include more than one archaeological site, group of people, or place in the landscape).

The *Chaîne Opératoire* and Communities of Practice

One of the major criticisms of the *chaîne opératoire* approach is that it attempts to get at the unknowable—such as what was in the minds of prehistoric people (e.g., Bar-Yosef and Van Peer 2009; Tostevin 2013). We would argue that while this approach certainly recognizes that a major part of the technological process occurs within individual minds and, while it is true that we will never be able to decipher the minds of prehistoric individuals, approaching the *chaîne opératoire* within the framework of communities of practice can help us to reconstruct the much more relevant and accessible social relationships of the prehistoric groups we seek to understand. It also serves to broaden our perspective from individual mindsets and actions to emphasize social interactions between people, and between people and the material world through the study of knowledge transmission and social group interaction reflected in lithic artifacts.

Incorporating communities of practice and concepts of situated learning (Lave and Wenger 1991; Wenger 1998) into lithic studies provides a framework to understand the larger social environments of prehistoric knappers (and the communities they participated within) and the way these social environs may have influenced their choices and decisions. And, while the *chaîne opératoire* approach has analytical limitations, such as resolving issues of co-occurrence, representation and completeness (Soressi and Geneste 2011), it has still proven useful for taking the archaeologist from a pile of broken rocks to a meaningful interpretation of past actions, economies, and social relationships. Indeed, our understanding of the latter of these—social relationships—has benefitted from recent analytical frameworks that explore knowledge transmission and situated learning (Lave and Wenger 1991; Wenger 1998; Minar and Crown 2001; Sassaman and Rudolphi 2001; Wallaert-Pêtre 2001; Crown 2007), allowing the archaeologist to put social practice into practice.

The EP of Southwest Asia

In the EP, chipped stone is usually the most abundant artifact class and forms the foundation for our reconstructions of past behavior (Fig. 11.1). Comprehensive histories of the analysis of lithic typologies and technologies from the EP have been covered by many scholars elsewhere (e.g., Olszewski 2004; Pirie 2004; Goring-Morris et al. 2009; Richter and Maher 2013b), and so we do not review these here. Traditional studies use the occurrence of particular tool types (e.g., microlith types) to identify different social (sometimes referred to as "ethnic") groups, labelled by the predominance of specific 'tool kits' or suites of microlith forms, and use changes in these groups to mark culture change over space and time (Bar-Yosef 1970; Byrd 1987; Goring-Morris 1987; Fellner 1990; Bar-Yosef 1991; Henry 1995; Henry 1996; Garrard and Byrd 2013). While there is no doubt that this recognition of different industries and facies has proven useful, it has also produced a somewhat confusing picture of the relationships between EP assemblages and sites across the region (Maher 2010).

Work over the last decade or so has provided a great deal of new data with which to understand these assemblages, allowing researchers to employ greater nuance to the picture of stone tool production throughout the EP and better understand patterns of temporal and spatial variability in stone tool assemblages. More recent technological studies and *chaîne opératoire* approaches have improved our understanding of the decisions and actions involved in the production and use of EP stone tools (Olszewski 2001; Richter 2007; Yaroshevich et al. 2010; Richter 2011; Macdonald 2013; Maher and Macdonald 2013; Yaroshevich et al. 2013; Richter 2014; Olszewski and al-Nahar 2016; Macdonald et al. 2018; Macdonald and Maher In press), but also leave us with many questions regarding our interpretations of the meaning of microlith variability (Olszewski 2006; Maher 2010; Maher and Richter 2011; Olzsewski 2011; Richter and Maher 2013b). Building on these studies, we attempt here to apply an integrative approach to chipped stone technology that enacts concepts of *chaîne opératoire* and communities of practice to explore why specific technological traditions came to dominate in specific times and places in the EP of Southwest Asia, concentrating on data from the Early and Middle EP site of Kharaneh IV, eastern Jordan, and its possible connections to other contemporary sites.

In particular, we focus on questions related to understanding the nature of variability in microlith technologies: namely, what does variability in final tool form mean? This is directly relevant to a number of related questions surrounding microlith production and use: how does this variability relate to microlith style and/or function; what were they used for; why are there differences over time in core shape and core trimming elements if it is the final tool form that matters; how does their production relate to the apparent differing trajectories of blade and bladelet production? In order to address issues of microlith variability, we employ a *chaîne opératoire* microlith approach to stone tool analysis within the framework of the social role of technology. We focus here on understanding flintknapping communities of practice at Kharaneh IV by exploring technological differences in assemblages within and between occupation phases and their potential role in negotiating relationships between social groups aggregating at Kharaneh IV and in interactions across a broader EP social landscape.

Kharaneh IV, Eastern Jordan

Kharaneh IV is an exceptionally large EP site located in the Azraq Basin of eastern Jordan (Figs. 11.2, 11.3). Previous work in the basin provides the archaeological and paleoenvironment foundation upon which this work is based, and includes earlier work at the site itself in the 1980s (Garrard and Stanley-Price 1975; Muheisen 1983; Copeland and Hours 1989; Garrard and Byrd 1992; Garrard et al. 1994a; Garrard et al. 1994b; Betts 1998; Cordova et al. 2013; Garrard and Byrd 2013). Covering more than 21,000 m^2 and with over 2 m of dense archaeological deposits, it is one of the largest EP sites in Southwest Asia. Test soundings across the site suggest that while some of the site's marginal horizontal extent is surficial and results from deflation and erosion,[5] outside of an approximately 1 m perimeter, dense and *in situ* EP deposits are found subsurface, at increasing depths (2.4 m to-date, without yet finding sterile deposits) towards the center of the site (Maher 2017). The first systematic excavations at the site, led by M. Muheisen in the 1980s (Muheisen 1983; Muheisen 1988a; Muheisen 1988b), uncovered an incredible density of highly-stratified, *in situ* EP occupations in two main areas of the site (Areas A and B; Fig. 11.3), which included hearth features, postholes, floors, pits and burials—all within less than 15 m^2 of excavated area.

In 2008, the Epipalaeolithic Foragers in Azraq Project (EFAP) began work at the site, with eight excavation seasons and two study seasons completed to-date. EFAP expanded greatly the excavated areas in Areas A and B in order to better expose and document features noted by Muheisen, and opened several entirely new areas in the northern, southern and eastern parts of the site, altogether totaling over 120 m^2 in excavated area. Here EFAP documented a greater extent for many of the features noted by Muheisen (esp. hearths and postholes), discovered several Early EP hut structures (Maher et al. 2012), a human burial, and communal Middle EP food-processing features (Spyrou In review), linked together aspects of site formation to changes in the local landscape (Jones et al. 2016b) and demonstrated clear stratigraphic relationships between the Early and Middle EP occupations at the site (Macdonald, Allentuck and Maher 2018). Currently, excavations are focused on Early EP site organization and use of space surrounding the above-mentioned hut structures and associated human remains. Results of the excavations to-date are published in numerous other venues (see below) and are thus not detailed here. Instead, we present a brief summary of the excavations in Areas A and B (those areas under discussion here) and aspects of the analysis of material culture from these occupations insofar as they provide context for the following discussion of the chipped stone material. Below, we focus on the results of our ongoing analyses of the chipped stone tool assemblages from the site and what they can elucidate about EP technology and social interactions.

In its initial phases, EFAP posed two key questions regarding Kharaneh IV: (1) why here? and (2) what kinds of on-site activities resulted in such an immense site with such high artifact densities? Addressing the first question, paleoenvironmental reconstructions based on geoarchaeological, zooarchaeological and archaeobotanical datasets from the on-site deposits and surrounding landscape have allowed us to trace landscape change over the last 25,000 years and articulate it with initial occupation of the site and changes in site use over time, as well as its abandonment by ~18.5 ka (Jones et al. 2016a; Jones et al. 2016b; Maher 2017). The site has evidence for both grassland- and wetland-dependent animal species, (Martin et al. 2010; Martin et al. 2016) and plant remains (macrobotanicals and charcoal, phytoliths; Asouti et al. 2015; Ramsey et al. 2016; Ramsey and Rosen 2016; Ramsey et al. 2018), as well as the presence of freshwater lake and wetland sediments in the surrounding terraces and basal occupation deposits on-site (even containing freshwater ostracods; Jones et al. 2016b). These provide multiple lines of evidence to corroborate the presence of substantial and persistent bodies of water within the vicinity of, and sometimes inundating, the small terrace upon

[5]Despite this exposure, only EP material culture is noted from the surface and subsurface deposits.

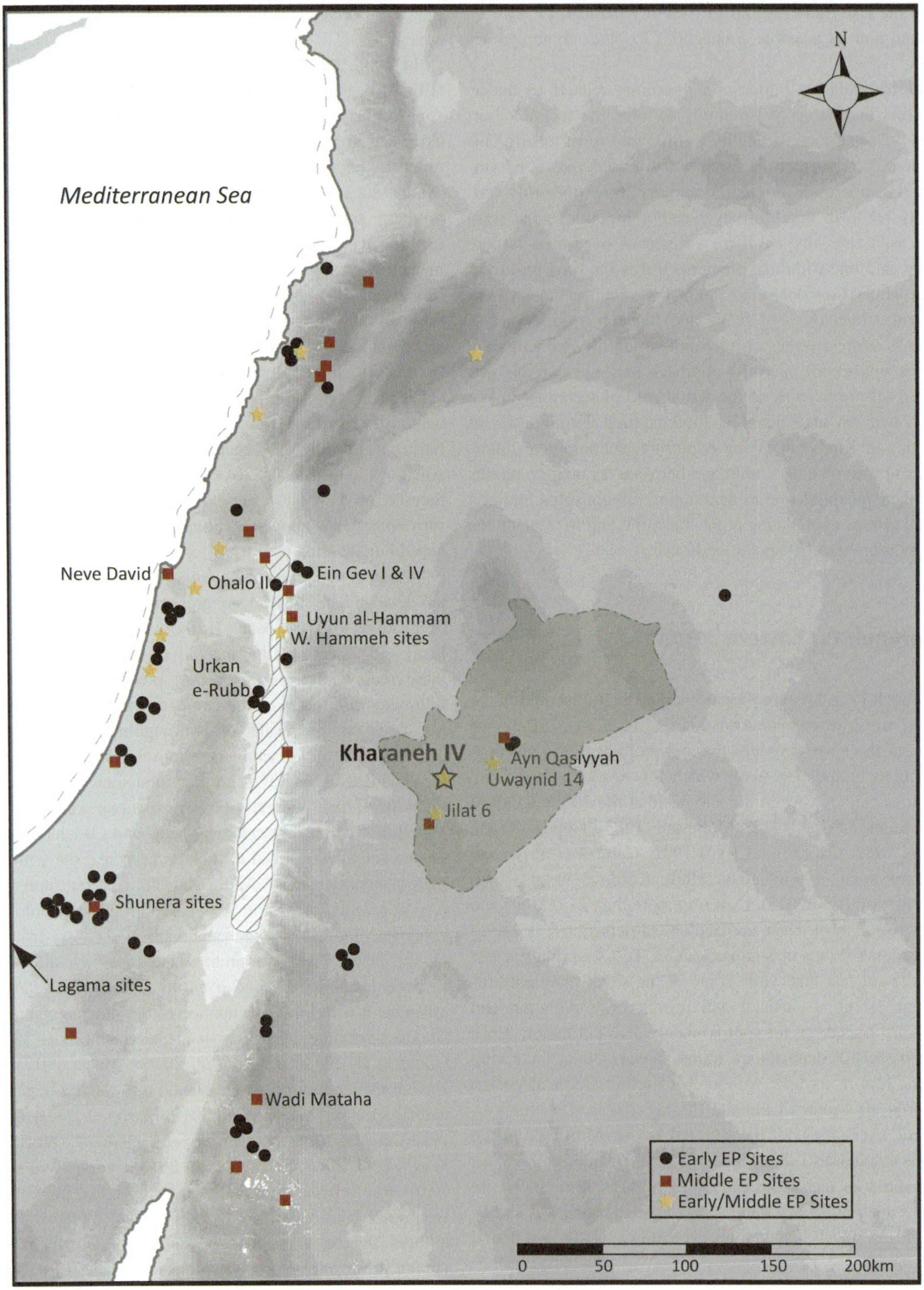

Fig. 11.2 Map of the southern Levant showing the location of Kharaneh IV with respect to other major Early and Middle EP sites in the region (modified from Maher, In press). The Azraq Basin, located in eastern Jordan, is highlighted in dark green

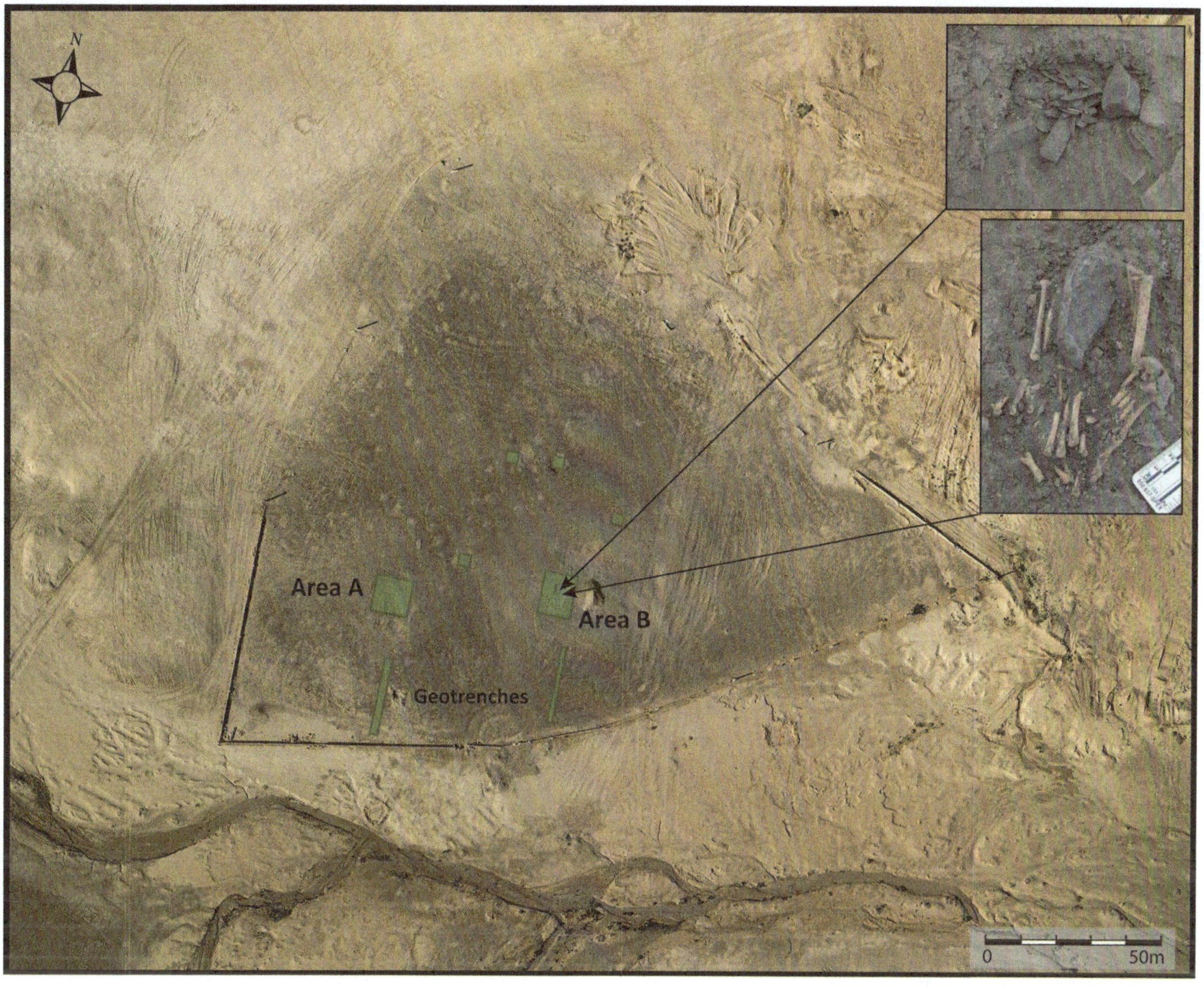

Fig. 11.3 Composite aerial photograph of Kharaneh IV showing the locations of excavated areas. Areas A and B are discussed in detail in the text. Inset are two images from the Early EP deposits showing a lithic cache outside Hut Structure 1 (upper) and bladelet core stashed in the remains of a fox pelt bag (lower) (Aerial photo courtesy of Fragmented Heritage, University of Bradford, inset photos from EFAP archive)

which the site was founded. Prior to and during all phases of occupation of the site, the landscape was a comparatively well-watered, lush and resource-rich habitat. This abundance of plants and animals, likely also including migratory animals, from a range of ecological niches (wetlands, grasslands and parklands—the latter attested to by small quantities of arboreal macrobotanics and phytoliths) was probably a major draw for EP peoples during the height of the Last Glacial Maximum, where otherwise cool and dry conditions are documented elsewhere (Maher 2017). Indeed, Kharaneh IV appears to have been a particularly favorable locale during the Early and Middle EP, a pattern similarly noted elsewhere in the Azraq Basin throughout prehistory (Garrard, Baird and Byrd 1994a; Jones and Richter 2011; Cordova et al. 2013; Garrard and Byrd 2013; Richter and Maher 2013a; Richter 2014).

Initial occupation of the site by mobile Early EP groups was likely episodic, and extended (probably discontinuously) across much of the site. Early EP lithics and fauna are not seen in the same densities as in later occupations and, with the exception of postholes in a northern sounding, the earliest deposits lack evidence of clear occupational features (although only small 1 × 1 m trenches have reached these depths). These earliest occupations are found within carbonate- and ostracod-rich lake deposits that likely represent a wetland or playa lake context; periods of local lowered water tables dried out the terrace and EP groups occupied the shores of the receded wetland, while periods of higher water levels submerged the terrace and EP groups lived elsewhere. Once the area dried permanently (still within the Early EP phases), occupation of the site was notably larger in scope, with repeated and persistent use for c. 1200 years. Carbonate development within the wetland marls on terraces surrounding the site indicates water sources began drying up sometime within the latter phases of the Early EP and continued through the Middle EP occupations. The latest Middle EP phases of occupation date to c. 18.5 ka, when the wetlands disappeared and there is no evidence for subsequent reoccupation of the site. The lack of Late EP and early Neolithic remains suggests that even the area surrounding the site does not appear to have been reoccupied again until later in the Holocene when this area was already a semi-arid to arid steppe.

Excellent preservation of charcoal from virtually every context provides an abundance of datable material. To-date, EFAP has obtained 21 radiocarbon dates from well-stratified contexts covering the entire span of excavated deposits; 16 dates from Early EP contexts and five dates from Middle EP contexts. These dates place occupation of the site to between 19,830 and 18,600 cal BP, with Area A dating from 18,600 to 18,850 cal BP and Area B from 18,750 to 19,830 cal BP (Richter et al. 2013). In addition, OSL dates from wetland deposits on- and off-site bracket the earliest occupations to sometime after 21,000 cal BP (Jones et al. 2016b). There is no evidence of a substantial depositional hiatus, suggesting intensive occupation for the 1200 years the site was occupied. This intensive occupation was likely characterized by repeated periods of multi-seasonal site habitation, some of which may have been prolonged. Recent analysis of the lithic assemblages from a deep sounding in Area A (AS42), covering 2.4 m in depth of stratified Middle and Early EP contexts corroborates that the distinctions between Early and Middle EP lithic technologies are more detectable than those within individual Early and Middle EP contexts (Macdonald, Allentuck and Maher 2018). Therefore, while further analyses remain ongoing, we ascribe the assemblages to three main phases of occupation, two within the Early EP equivalent to a series of early and later Kebaran industries and the Middle EP to a series of highly variable Geometric Kebaran industries.[6]

In Area B (Early EP) we have excavated a number of hearth and pit features as well as extensive midden deposits. Of particular note, we have fully excavated two adjacent hut structures (Structures 1 and 2), and discovered the remains of at least two more structures nearby (Maher et al. 2012; Maher and Conkey 2019). Both structures exhibit several superimposed floors and caches, as well as intentional destruction through burning, with Structure 2 also containing the remains of an adult female buried on the hut floor prior to its destruction (Maher et al., forthcoming). The same sequence of events between these adjacent structures and similarities in their construction and maintenance indicate a degree of relative contemporaneity within the Early EP and very specific uses of space within this time frame of occupation, perhaps related to shared traditions or memory. The spatially bounded and distinct nature of the structures and associated hearths, caches, middens and other features throughout this phase suggest persistence in site organization, even if the locations of these features changed with different occupations. In contrast, the Middle EP deposits are characterized by several superimposed, horizontally-extensive compact earthen surfaces, each associated with several hearth features, often ringed by postholes, and extensive middens. The Early EP deposits show a complicated sequence of numerous thin, discrete contexts—multiple hut structures, knapping areas, hearths, compact surfaces, ash dumps, dense, discrete middens, caches, and burials. The Middle EP deposits are less numerous, thicker and horizontally-extensive—large and thick compact earthen surfaces with no clear boundaries, large midden and fill deposits, large, overlapping hearths, postholes, few discrete knapping areas, and no caches or hut structures. The very different character of deposits between these phases of occupation appear to relate to changes in the use of space, with a shift from 'private' or discrete activities in the Early EP to more communal living in the Middle EP (Maher 2018; Maher and Conkey 2019).

Analyses of the faunal remains from the site indicate exploitation of a notably wide range of species, but with a clear dominance of gazelle (especially goitered gazelle) throughout the occupation deposits (>80% in the Early EP and reaching >90% in some Middle EP contexts) (Martin et al. 2010; Martin et al. 2016; Macdonald, Allentuck and Maher 2018; Allentuck In prep). Differences between contexts within the Early and Middle EP phases are specific to features, rather than any notable changes over time. For example, within the Early EP, distinctions are evident in species (and animal parts) represented inside the hut structures, between the hut structures, and with the 'outside' contexts, such that 'inside' the huts, only five species—gazelle, fox, hare, tortoise and wild ass—are found, while outside assemblages are much more diverse. There are no notable differences between occupations within the Early and Middle EP phases; instead, differences here seem to be largely between the Early and Middle EP. In particular, various analyses of the gazelle remains suggest multi-season or year-round occupation of the site (esp. in the Middle EP) (Jones 2012; Henton et al. 2017) and a shift towards communal hunting, large-scale processing for meat drying and

[6]See Macdonald, Allentuck and Maher (2018) and Maher (2018) for details on these stratigraphic relationships and phasing of the site.

meat-sharing and storage in the Middle EP (Martin et al. 2010; Spyrou In review).

The use of various plant resources by the site's occupants suggest the knowledgeable use of a variety of local wetland, grassland and parkland plants for foods, construction, bedding, and fuel (Ramsey et al. 2016; Ramsey et al. 2018). Other aspects of material culture, including worked bone, ochre, groundstone and other features are found in both Early and Middle EP phases of occupation, but possible worked bone notation devices are more common to the Middle EP, perhaps related to a need or desire to keep records between aggregating groups? The abundance of marine shell (n =>2000), some 200–300 km from its nearest possible sources in the Mediterranean and Red Seas, is of particular note for our discussion here about long-distance networks of social interaction. Microscopic analyses of traces of manufacture and use indicate that much of the shell is intentionally pierced or modified, shows evidence of being used (strung together and worn or sewn onto something), and shows intentional ochre staining (Allcock 2009; Richter et al. 2011). Shell appears in all contexts in the Early and Middle EP, but in notably higher frequencies in the Middle EP, reinforcing the idea of groups coming from the coast, or interacting with groups from the coast in networks involving the movement of people and objects (Maher 2016).

Analyzing the Kharaneh IV Assemblage

Our conservative estimates indicate that, to-date, EFAP has excavated well over 3 million lithics, and almost equal numbers of faunal remains. While approximately half of these counts come from deflated surface cleanings (0–2 cm), uppermost disturbed layers (top ~2–5 cm), rodent burrows and other small subsurface disturbances, the remainder derive from well-preserved, intact subsurface features. Obviously, only a fraction of these have hitherto been analyzed. We present here an overview of the results of this work within the framework described above that attempts to place chipped stone tool production as a socially-situated technology and as part of a larger EP technological world. With a clear technological focus on the production of microliths as easily-replaceable parts of larger composite tools (possibly with wood or bone handles), they were made and used in conjunction with other materials and are, thus, best understood when considered alongside other Kharaneh IV datasets.

EFAP's approach to analyzing the lithic assemblage is a techno-typological classification scheme modified from that developed by Wilke and Quintero for naviform core blade reduction and with particular attention to a diversity of technologically-diagnostic core trimming elements (Wilke and Quintero 1994) and defined and detailed elsewhere (Maher and Macdonald 2013; Macdonald, Allentuck and Maher 2018).[7] An assemblage is divided into categories of retouched pieces (including microliths and microburins[8]), debitage (chips, shatter, complete and fragmentary flakes and blades, including primary pieces, platform isolation elements and edge preparation), core-trimming elements (CTEs) and cores. Core-trimming elements (CTEs) were divided into two categories: those related to initial core preparation and those related to non-initial and ongoing core maintenance. Core preparation elements reflect the design used to shape the core in preparation for subsequent removals and include initial platform spalls that initiate and prepare a platform, lateral core trimming pieces that remove cortex and crested blades that prepare the core face for bladelet removals. Core maintenance elements are involved in the ongoing shaping of the core for continued bladelet removals, and include pieces that fix mistakes, change platform angles and renew platform and core faces. They include angle correction elements and core tablets that relate to platform maintenance, as well as core face rejuvenation elements, profile correction blades and partially crested blades that ensure the successful extraction of target bladelet blanks. Cores are classified by shape (often determined by the extent of utilized core face) and the nature of the targeted removals (e.g., blade, bladelet, and/or flake).

The retouched tools, including microliths, were classified according to the conventional and generally accepted EP typologies of Bar-Yosef (1970) and Goring-Morris (1987). While variations on these typologies have been produced by others (e.g., Muheisen 1988b; Byrd 1989; Henry 1995; Garrard and Byrd 2013), they are specific to research projects investigating a small number of sites within geographically-localized areas. Thus, the former ones remain the most widely used and applicable classification schemes, especially for interregional comparisons. As an exception, the typological list created by Muheisen (Muheisen 1988b) specific to the geometric microliths of Kharaneh IV was consulted. Of the retouched pieces, analysis focused on the microliths as diagnostic indicators of specific EP entities and, thus, cultural and chronological affiliation (Figs. 1.1, 11.4). Microliths were divided into geometric, non-geometric, and fragmentary microliths. Geometric microliths at Kharaneh IV are defined as bladelets retouched into a geometric shape, usually trapezoidal or rectangular in form, although lunate-like pieces are also noted.

[7]Following Wilke and Quintero, we pay particular attention to technologically-diagnostic core trimming pieces. A publication that details and illustrates our analytical approach is forthcoming.

[8]While microburins generally represent unmodified and technically debitage, they are categorized alongside tools because of their highly distinctive appearance and value as a diagnostic cultural-chronological feature, like microliths, of specific EP entities.

Fig. 11.4 **a** Gracile, non-geometric microliths (arched pieces and obliquely truncated and backed bladelets) and narrow-faced cores from the Early EP levels in Area B. **b** Geometric microliths variants (trapezes, rectangles, unbacked trapezes) and broad-faced cores from the Middle E contexts in Area A

Non-geometric microliths are bladelets unobtrusively retouched into straight, pointed, arched, curved, oblique, or other irregular forms.

Fragmentary pieces are retouched bladelets or microliths with one or both ends broken. However, this seemingly straightforward distinction has quite complex implications for interpretation that highlight the importance of considering both the end product *and* the prevailing concepts of blank production. First, the gross distinction between geometric and non-geometric microliths and, thus, Early and Middle/Late EP phases necessitates having complete pieces that allows one to distinguish between, for example, an Early EP obliquely truncated and backed bladelet and a Middle EP trapeze (the difference being whether or not the proximal end was retouched to another oblique angle). Fragmentary pieces can mask these distinctions and, when occurring in large numbers, affect distinctions between microlith assemblages as being predominantly geometric or non-geometric. Second, it is often impossible to tell whether these pieces are broken intentionally (snapped) or unintentionally during manufacture or through use, also obscuring distinctions between geometrics 'in-progress' and utilized non-geometrics. In current typological schemes fragmentary pieces are often lumped with non-geometric microliths (for example, as broken backed bladelets (Bar-Yosef 1970) or retouched/backed bladelet fragments (Goring-Morris 1987), inflating the proportions of non-geometric tools. In order to avoid this, especially given the large numbers of fragmentary pieces in the Kharaneh IV assemblages, we employ the category 'fragmentary microlith' to remove the potential bias of misclassifying these as either non-geometrics or geometrics. Thus, all backed bladelets with two broken ends (medial fragments) were put in this category. Where possible, backed pieces with one broken end were identified to a specific microlith type on the basis of morphological similarities on remaining features; they were classified as fragmentary if the original tool type was unclear. Within the non-geometric and geometric classes, microliths were further subdivided into types based on the typologies developed by Bar-Yosef (1970) and Goring-Morris (1987).

Stone Tool Production at EP Kharaneh IV: The Nature of Occupation Over Space and Time

We have analyzed just over 300,000 lithics from the site (Tables 11.1, 11.2), predominantly from Areas A and B, and uncovered some important differences in technological strategies between these main excavation areas (Fig. 11.4). As noted by Muheisen (Muheisen 1988a), broad distinctions can be made between the strictly Early EP occupations, with assemblages overwhelmingly dominated by non-geometric microliths excavated from Area B, and the predominantly Middle EP occupations, with assemblages overwhelmingly dominated by geometric microliths excavated from Area A. However, recent analysis of a deep trench (AS42) in Area A demonstrates a) the presence of stratified Early and Middle EP occupations here and b) meaningful distinctions between upper and lower Early EP contexts that might represent at least two phases of Early EP occupation (Macdonald, Allentuck and Maher 2018). These latter distinctions may also play out in Area B as we focus on further analyzing the retouched assemblages from this area.

Area B and the Early EP

With over 1.5 m of well-preserved, fine-grained stratified deposits, only clearly Early EP contexts have been excavated from Area B (Figs. 11.3, 11.4). These Early EP assemblages are not highly diverse and suggest a focus on relatively standardized production of particular non-geometric microliths forms (Table 11.1). They are overwhelmingly dominated by narrow-faced, single platform cores (>60%) with smaller numbers of multi-directional and opposed platforms cores, usually made from small, eroded cobbles of a local, grey-brown flint (see below). Given the small size of most cores (>15 cm max. length) and the high proportion of cores with substantial cortex remaining, it seems as though these small cores were primarily for the production of non-geometric microliths; the larger blade tool (endscrapers and burins dominate) component of the assemblages were produced from notably larger cores in a separate reduction sequence and these cores not commonly found remaining on-site.

Both flakes (n = 18,400) and blades/bladelets (n = 14,392) are found in relatively equal numbers. Flakes generally represent the initial stages of core preparation before consistent blade/bladelet removals and, thus, microliths-focused production sequences often produce large numbers of flakes. Here, a flake-to-blade ratio of 1.28 suggests emphasis was placed on early core shaping that would have produced a large number of flakes in relation to blades/bladelets. These counts include primary pieces, but not lateral core trimming pieces (flat to medially curved expanding flakes, often with cortex on the distal end, removed to flatten platforms and/or shape the base and sides of a core during core preparation or maintenance). These are flake removals usually containing cortex on their distal dorsal surfaces removed primarily (although not exclusively) during early stages of core shaping. These are by far the most common type of CTE here (>30%), alongside initial and faceted platform spalls and crested blades. These core shaping CTEs form 40% of all CTEs in the Early EP occupations, whereas in the Middle EP occupations they form only 22% of all CTEs (see below) (Fig. 11.5). Bladelet debitage is generally narrow and gracile in shape, and highly standardized in overall form.

Analysis of the retouched assemblage from the Early EP deposits in Area B indicates that over 80% of the tools are microliths, and >50% are identified as non-geometric microliths (Maher and Macdonald 2013; Macdonald, Allentuck and Maher 2018) (Fig. 11.6). These non-geometric forms are predominantly gracile obliquely-truncated and backed bladelets, finely backed bladelets and, in lower levels, microgravettes. The microburin technique is present, but in low frequencies, with a restricted microburin index of 2.4 (Bar-Yosef 1970), suggesting its inconsistent use. However, changes in the use of the microburin technique within the Early EP occupations were noted within the lower levels of the deep sounding in AS42 (Area A), suggesting changes in technological practices for manufacturing microliths. This, alongside an emphasis on microgravettes in these lower contexts, suggests there may be at least two phases of Early EP occupation of the site (Macdonald, Allentuck and Maher 2018).

Area a and the Middle EP

While the excavated deposits from Area A are not as vertically extensive as those in Area B, individual strata are more substantial and show more horizontal continuity, thus exhibiting simpler stratigraphic relationships over space and with depth. The chipped stone assemblage exhibits a wide range of tools, debitage, CTEs and cores, and this diversity within lithic classes is noted throughout all contexts (Table 11.2). In other words, in comparison to Area B

Table 11.1 Area B lithic assemblage totals by analytical category and locus. Note the loci are listed from stratigraphically latest on the left to earliest on the right. **a** Early EP debitage categories by locus. **b** Early EP tool classes by locus

Debitage Category

Locus Number	Blades/Bladelets	Secondary Blades	Flakes	Secondary Flakes	Primary Pieces	Platform Isolation Elements	Edge Preparation Elements	Chips	Shatter	Burnt Shatter	Triangular Pieces	*All Unretouched Debitage*	First Burin Spall (all)	Sharpening Burin Spall	Plunging Burin Spall	Hinging Burin Spall	Twisted Burin Spall	*All Burin Spalls*	Non-Initial Spontaneous Core Tablets	Non-Initial Corrective Core Tablets	Initial Core Tablet	Initial Faceting Platform Spalls	Profile Correction Blades	Core Face Rejuvenation	Partial Ridged Blades	Lateral Core Trimming Piece	Crested Blade	Angle Correction Element	Bottom Partial Ridged Blade	Varia	*All Core Trimming Pieces*	Total
000	10	-	1	-	-	-	-	-	-	-	-	*11*	-	-	-	-	-	*0*	-	-	-	-	-	-	-	-	4	-	-	-	*4*	**15**
001	648	-	757	-	-	-	-	28	78	-	-	*1511*	2	-	-	-	-	*2*	-	41	-	-	-	-	-	-	38	-	-	8	*87*	**1600**
002	1447	-	498	-	-	-	-	2510	69	-	-	*4524*	-	-	-	-	-	*0*	-	12	-	-	-	-	-	-	10	-	-	2	*24*	**4548**
003	829	-	720	-	-	7	-	2952	49	-	2	*4559*	-	-	-	-	-	*0*	-	5	1	-	1	3	-	5	5	-	-	2	*22*	**4581**
004	128	-	533	-	16	13	-	1095	172	-	2	*1959*	2	-	-	-	-	*2*	3	-	-	-	-	41	-	3	-	-	-	-	*47*	**2006**
009	24	-	44	-	-	4	-	305	10	-	-	*387*	1	-	-	-	-	*1*	-	-	-	-	-	9	-	6	1	-	-	-	*16*	**404**
027	160	-	142	-	2	12	-	1275	18	-	-	*1609*	-	-	-	-	-	*0*	-	-	-	-	-	22	2	8	-	-	-	-	*32*	**1641**
032	1	-	0	-	-	-	-	-	-	-	-	*1*	-	-	-	-	-	*0*	-	-	-	-	-	-	-	-	-	-	-	-	*0*	**1**
033	2	-	0	-	-	-	-	-	-	-	-	*2*	-	-	-	-	-	*0*	-	-	-	-	-	-	-	-	-	-	-	-	*0*	**2**
040	1478	-	1076	-	99	26	53	2227	50	107	-	*5116*	13	7	1	-	2	*23*	12	28	5	-	4	63	48	139	7	-	-	134	*440*	**5579**
043	14	-	3	-	-	-	-	-	-	-	-	*17*	-	-	-	-	-	*0*	-	-	-	-	-	-	-	-	-	-	-	-	*0*	**17**
060	135	-	80	-	5	1	-	880	74	-	-	*1175*	-	-	-	-	-	*0*	2	-	-	-	2	24	3	2	-	2	-	-	*35*	**1210**
084	373	12	272	16	47	12	10	1420	31	52	-	*2245*	2	-	-	-	-	*2*	4	6	-	1	7	7	15	30	2	-	-	-	*72*	**2319**
088	3427	109	3724	259	115	27	5	5554	106	749	3	*14078*	8	4	-	2	-	*14*	17	50	23	20	48	55	87	210	-	30	1	-	*541*	**14633**
095	4	-	3	-	3	1	3	-	1	-	-	*15*	-	-	-	-	-	*0*	4	-	-	-	-	5	-	6	-	2	-	-	*17*	**32**
100	-	-	-	-	-	-	-	-	-	-	-	*0*	-	-	-	-	-	*0*	-	-	-	-	-	-	-	-	-	-	-	-	*0*	**0**
128	4	-	2	-	2	-	1	-	-	-	-	*9*	-	-	-	-	-	*0*	-	-	-	-	-	1	1	-	-	-	-	-	*2*	**11**
158	-	-	-	-	-	-	-	-	-	-	-	*0*	-	-	-	-	-	*0*	-	-	-	-	-	-	-	-	-	-	-	-	*0*	**0**
172	-	-	-	-	-	-	-	-	-	-	-	*0*	-	-	-	-	-	*0*	1	-	-	-	-	-	-	-	-	-	-	-	*1*	**1**
176	3	-	-	-	-	-	-	-	-	-	-	*3*	-	-	-	-	-	*0*	-	-	-	-	-	-	-	-	-	-	-	-	*0*	**3**
200	316	11	561	64	37	3	2	1230	22	462	-	*2708*	1	2	-	-	-	*3*	-	11	2	7	22	3	25	82	5	4	-	-	*161*	**2872**
203	-	-	-	-	-	-	-	-	-	-	-	*0*	-	-	-	-	-	*0*	-	-	-	-	-	-	-	-	-	-	-	-	*0*	**0**
208	240	7	423	3	10	-	-	896	6	159	-	*1744*	-	1	-	-	-	*1*	-	2	1	-	-	4	2	10	1	-	-	-	*20*	**1765**
209	301	14	344	17	6	-	-	1180	15	284	1	*2162*	-	-	-	-	-	*0*	1	4	-	-	3	3	7	8	1	3	-	-	*30*	**2192**
210	660	19	881	65	46	5	1	3128	20	557	2	*5384*	1	2	-	-	1	*4*	3	8	2	3	1	15	16	14	1	9	1	-	*73*	**5461**
212	25	-	9	1	1	-	-	-	-	-	-	*36*	-	-	-	-	-	*0*	-	-	-	-	-	-	2	-	-	-	-	-	*2*	**38**
213	9	-	4	-	-	-	-	-	-	-	-	*13*	-	-	-	-	-	*0*	-	-	-	-	-	-	1	-	-	-	-	-	*1*	**14**
214	352	9	269	38	20	4	-	1596	29	89	3	*2409*	5	2	-	-	-	*7*	-	4	1	2	7	11	3	7	-	5	-	-	*40*	**2456**
215	476	7	594	26	12	2	1	1638	14	386	-	*3156*	2	-	-	1	-	*3*	2	7	1	1	5	-	4	15	1	3	1	-	*40*	**3199**
217	374	21	515	36	24	3	1	997	15	24	-	*2010*	2	1	-	-	-	*3*	-	4	2	1	5	4	13	5	-	6	-	-	*40*	**2053**
218	1	-	1	-	-	-	-	-	-	-	-	*2*	-	-	-	-	-	*0*	-	-	-	-	-	-	-	-	-	-	-	-	*0*	**2**
220	1	-	-	-	-	-	-	-	-	100	-	*101*	-	-	-	-	-	*0*	-	-	-	-	-	-	-	-	-	-	-	-	*0*	**101**
222	1187	31	2004	72	57	-	9	2664	1	2763	-	*8788*	1	1	-	-	-	*2*	1	19	7	2	17	4	25	39	1	11	-	-	*126*	**8916**
232	-	-	-	-	-	-	-	-	-	100	-	*100*	-	-	-	-	-	*0*	-	-	-	-	-	-	-	-	-	-	-	-	*0*	**100**
234	1	-	-	-	-	-	-	-	-	-	-	*1*	-	-	-	-	-	*0*	-	1	-	-	-	-	-	-	-	-	-	-	*1*	**2**
242	934	18	2054	33	69	1	4	3798	26	318	-	*7255*	2	-	-	-	-	*2*	-	11	3	7	24	14	15	41	-	11	-	-	*126*	**7383**
243	1	-	3	-	1	-	-	-	-	153	-	*158*	-	-	-	-	-	*0*	-	-	-	-	-	1	-	-	-	-	-	-	*1*	**159**
258	416	20	1155	74	21	2	4	880	46	112	4	*2734*	5	-	-	-	-	*5*	-	9	1	4	29	26	3	24	-	5	-	-	*101*	**2840**
270	124	9	410	13	8	1	5	432	8	44	3	*1057*	2	-	-	-	-	*2*	-	1	1	-	4	3	-	9	-	2	-	-	*20*	**1079**
Total	**14105**	**287**	**17082**	**717**	**601**	**124**	**99**	**36685**	**860**	**6459**	**20**	*77039*	**49**	**20**	**1**	**3**	**3**	*74*	**50**	**223**	**50**	**48**	**179**	**318**	**272**	**663**	**77**	**93**	**3**	**146**	*2122*	**79235**

(A) Early EP debitage categories by locus.

Table 11.1 (continued)

Locus Number	Scrapers	Multiple Tools	Burins	Retouched Burin Spalls	Retouched Pieces	Backed Blades	Truncations	Points	Non-geometric microliths	Geometric microliths	Fragmentary microliths	Utilized microliths	Perforators	Notches and Denticulates	Heavy Duty Tools	Utilized Pieces	Pieces Esquilles	Varia	Microburins	Totals
	Tool Class																			
000	12	4	-	-	2	-	-	-	1	-	-	-	-	-	3	-	-	-	-	**22**
001	14	4	12	-	79	4	5	-	11	2	2	-	1	15	1	-	-	-	-	**150**
002	2	-	5	-	14	3	2	-	68	-	50	-	-	2	-	-	-	-	1	**147**
003	13	5	1	-	6	12	-	-	57	-	104	-	1	1	1	-	-	-	2	**203**
004	15	1	8	-	20	2	-	-	47	-	-	-	-	5	-	-	-	-	12	**110**
009	2	-	-	-	-	1	-	-	7	-	-	-	-	-	-	-	-	-	-	**10**
027	2	-	2	-	1	-	-	-	14	-	6	-	3	-	-	-	-	-	7	**35**
032	-	-	-	-	-	-	-	-	-	-	-	-	-	-	-	-	-	-	-	**0**
033	-	-	-	-	-	-	-	-	-	-	-	-	-	-	-	-	-	-	-	**0**
040	5	1	3	-	5	3	2	-	137	-	-	-	-	1	-	1	3	-	-	**161**
043	-	-	-	-	-	-	-	-	-	-	-	-	-	-	-	-	-	-	-	**0**
060	4	-	3	-	5	-	-	-	11	-	-	-	-	3	-	-	-	-	26	**52**
084	-	-	-	-	-	5	-	-	63	-	7	1	-	-	-	1	-	-	-	**77**
088	14	-	5	-	14	12	6	-	241	-	124	11	-	13	-	11	-	-	-	**451**
095	-	-	-	-	-	-	-	-	-	-	-	-	-	-	-	-	-	-	-	**0**
100	-	-	-	-	-	-	-	-	-	-	-	-	-	-	-	-	-	-	-	**0**
128	1	-	1	-	-	1	-	-	-	-	-	-	-	-	-	-	-	-	-	**3**
158	-	-	-	-	-	-	-	-	-	-	-	-	-	-	-	-	-	-	-	**0**
172	-	-	-	-	-	-	-	-	-	-	-	-	-	2	-	-	-	-	-	**2**
176	2	1	-	-	1	-	-	-	-	-	-	-	-	-	-	1	-	-	-	**5**
200	2	1	-	-	4	-	-	-	39	-	40	-	1	-	-	-	-	-	1	**88**
203	-	-	-	-	-	-	-	-	-	-	-	-	-	-	-	-	-	-	-	**0**
208	-	-	-	-	2	-	-	-	46	-	7	1	-	-	-	2	-	-	-	**58**
209	1	1	-	-	2	3	-	-	47	-	24	1	-	-	-	4	-	-	1	**84**
210	1	2	2	-	7	1	-	-	61	-	57	-	-	3	-	10	-	-	-	**144**
212	-	-	-	-	-	-	-	-	-	-	-	-	-	-	-	-	-	-	-	**0**
213	-	-	-	-	-	-	-	-	1	-	-	-	-	-	-	-	-	-	-	**1**
214	1	-	2	-	7	1	1	-	34	-	50	-	-	4	-	3	-	-	-	**103**
215	-	-	-	-	3	1	-	-	34	-	48	-	-	1	-	6	-	-	1	**94**
217	-	-	-	-	1	3	-	-	44	-	7	-	-	-	-	2	-	-	-	**57**
218	-	-	-	-	-	-	-	-	-	-	-	-	-	-	-	-	-	-	-	**0**
220	-	-	-	-	-	-	-	-	-	-	-	-	-	-	-	-	-	-	-	**0**
222	6	-	1	-	9	6	-	-	148	-	87	8	-	-	-	11	-	-	-	**276**
232	-	-	-	-	-	-	-	-	-	-	-	-	-	-	-	-	-	-	-	**0**
234	-	2	-	-	1	-	-	-	-	-	-	-	-	-	-	-	-	-	-	**3**
242	4	4	-	1	3	6	2	-	127	1	86	-	-	2	-	-	-	-	-	**236**
243	1	-	-	-	-	-	-	-	-	-	-	-	-	-	-	-	-	-	-	**1**
258	14	-	1	-	14	3	-	-	65	-	15	-	1	2	-	-	-	-	1	**116**
270	7	-	1	-	5	-	-	-	24	-	7	-	-	1	1	1	-	-	-	**47**
Totals	**123**	**26**	**47**	**1**	**205**	**67**	**18**	**0**	**1327**	**3**	**721**	**22**	**7**	**55**	**6**	**53**	**3**	**0**	**52**	**2736**

(B) Early EP tool classes by locus.

Table 11.2 Area A lithic assemblage totals by analytical category and locus. Note the loci are listed from stratigraphically latest on the left to earliest on the right. **a** Middle EP debitage categories by locus. **b** Middle EP tool classes by locus

Locus Number	Debitage Category: Blades/Bladelets	Secondary Blades	Flakes	Secondary Flakes	Primary Pieces (100%)	Platform Isolation Elements	Edge Preparation Elements	Chips	Shatter	Burnt Shatter	Triangular Pieces	All Unretouched Debitage	First Burin Spall (all)	Sharpening Burin Spall	Plunging Burin Spall	Hinging Burin Spall	Twisted Burin Spall	All Burin Spalls	Non-Initial Spontaneous Core Tablets	Non-Initial Corrective Core Tablets	Initial Core Tablet	Initial Faceting Platform Spalls	Profile Correction Blades	Core Face Rejuvenation	Partial Ridged Blades	Lateral Core Trimming Piece	Crested Blade	Angle Correction Element	Bottom Partial Ridged Blade	Varia	All Core Trimming Pieces	Total
001	1168	-	786	-	-	11	5	5	102	17	-	2094	4	1	1	-	4	10	2	12	1	-	4	21	10	1	28	-	-	8	87	**2191**
002	1645	-	1263	-	3	-	2	1782	104	-	-	4799	1	-	1	-	-	2	5	5	1	-	-	14	5	8	14	-	-	6	58	**4859**
003	3	-	11	-	2	-	-	-	-	-	-	16	-	-	-	-	-	0	-	-	-	-	-	4	2	1	-	-	-	-	7	**23**
004	1	-	1	-	-	-	-	-	-	-	-	2	-	-	-	-	-	0	-	2	-	-	-	-	1	-	-	1	-	-	4	**6**
008	18641	421	21956	1482	967	623	285	54611	2886	0	18	101890	116	1	-	-	-	117	47	127	9	34	456	1226	124	418	124	134	84	-	2783	**104790**
010	4	-	-	-	-	-	-	-	-	-	-	4	-	-	-	-	-	0	-	1	-	-	-	-	-	-	-	-	-	-	1	**5**
023	4	-	6	-	-	-	-	64	-	32	-	106	-	-	-	-	-	0	-	-	-	-	-	-	-	-	-	-	-	-	0	**106**
034	1768	34	2852	67	46	45	18	1986	75	141	1	7033	6	2	1	-	5	14	1	25	2	8	53	25	30	31	10	14	-	-	199	**7246**
035	557	17	755	7	1	3	4	999	35	53	-	2431	2	1	-	-	-	3	-	7	-	1	11	18	6	12	4	4	-	-	63	**2497**
041	1106	3	222	13	7	1	3	228	52	93	-	1728	-	-	-	-	-	0	0	2	-	-	9	-	-	2	1	2	-	-	16	**1744**
047	197	9	349	17	13	2	5	148	2	58	-	800	1	-	-	-	-	1	-	-	-	-	5	1	1	1	-	3	-	-	11	**812**
057	443	5	648	12	16	7	3	1005	92	45	1	2277	-	-	-	-	-	0	1	3	1	-	7	43	1	18	12	1	-	-	87	**2364**
065	57	1	49	3	1	-	-	82	2	11	-	206	-	-	-	-	-	0	-	-	-	-	-	1	2	-	-	-	-	-	3	**209**
067	119	1	113	1	1	-	1	95	3	15	-	349	1	-	-	-	-	1	-	1	1	-	1	1	1	1	-	2	-	-	8	**358**
076	597	-	626	-	1	21	-	2036	69	-	-	3350	-	-	-	-	-	0	-	-	-	-	2	52	-	11	7	-	-	-	72	**3422**
080	347	11	640	48	31	4	5	109	10	16	1	1222	-	-	-	-	-	0	-	8	-	-	19	2	9	13	6	8	-	-	65	**1287**
092	25	-	16	-	1	-	-	35	-	5	-	82	-	-	-	-	-	0	-	-	-	-	-	-	-	-	-	-	-	-	0	**82**
099	3480	104	5617	128	134	26	36	3179	35	125	4	12868	11	3		2	1	17	4	46	5	15	145	68	45	58	2	67	-	-	455	**13340**
100	10672	344	15559	664	428	143	76	15457	431	759	21	44554	31	8	2	2	10	53	7	137	18	22	201	163	172	117	24	138	7	-	1006	**45613**
107	2690	85	3721	143	100	20	23	4589	36	642	3	12052	7	5	2	-	1	15	4	25	3	10	71	23	37	45	11	36	-	-	265	**12332**
114	-	-	-	-	-	-	-	-	-	-	-	-	-	-	-	-	-	0	-	-	-	-	-	-	-	-	-	-	-	-	0	**0**
115	3690	92	5269	232	124	13	41	3909	200	243	16	13829	6	2	3	1	-	12	4	52	9	12	109	53	31	61	23	48	-	-	402	**14243**
124	147	3	433	7	5	-	-	108	7	32	1	743	-	-	-	-	-	0	-	1	-	-	5	1	-	2	1	-	-	-	10	**753**
139	-	-	-	-	-	-	-	-	-	16	-	16	-	-	-	-	-	0	-	-	-	-	-	-	-	-	-	-	-	-	0	**16**
179	-	-	-	-	-	-	-	-	-	-	-	0	-	-	-	-	-	0	-	-	1	-	-	-	-	-	-	-	-	-	1	**1**
Total	**47361**	**1130**	**60892**	**2824**	**1881**	**919**	**507**	**90427**	**4141**	**2303**	**66**	**212451**	**186**	**23**	**10**	**5**	**21**	**245**	**75**	**454**	**51**	**102**	**1098**	**1716**	**477**	**800**	**267**	**458**	**91**	**14**	**5603**	**218299**

A) Middle EP debitage categories by locus.

Table 11.2 (continued)

Locus Number	Tool Class																			Total
	Scrapers	Multiple Tools	Burins	Retouched Burin Spalls	Retouched Pieces	Backed Blades	Truncations	Points	Non-geometric microliths	Geometric microliths	Fragmentary microliths	Utilized microliths	Perforators	Notches and Denticulates	Heavy Duty Tools	Utilized Pieces	Pieces Esquilles	Varia	Microburins	
001	19	5	6	-	69	2	1	-	46	56	25	-	-	9	-	25	-	-	1	264
002	11	12	2	-	26	8	1	-	46	55	79	-	-	8	-	4	-	-	1	253
003	2	2	-	-	2	1	-	-	-	1	1	-	-	1	-	-	-	-	-	10
004	-	-	-	-	-	-	-	-	-	-	-	-	-	-	1	-	-	-	-	1
008	36	52	39	16	418	59	23	-	128	710	718	-	3	51	5	129	-	68	12	2467
010	-	-	-	-	-	-	-	-	-	-	-	-	-	-	-	-	-	-	-	0
023	-	-	-	-	-	-	-	-	-	-	-	-	-	-	-	-	-	-	-	0
034	7	6	-	1	15	12	3	-	28	28	79	11	-	1	-	5	-	-	-	196
035	-	2	1	-	3	4	-	-	4	23	24	2	-	1	-	5	-	-	-	69
041	-	1	-	-	-	-	-	-	4	5	27	-	-	-	-	-	-	-	1	38
047	2	-	-	-	4	2	-	-	4	10	38	-	-	-	-	3	-	-	-	63
057	1	2	2	-	7	5	1	-	8	23	40	-	1	-	-	1	-	-	1	92
065	-	3	-	-	2	1	1	-	1	2	9	1	-	-	-	1	-	-	-	21
067	1	-	-	-	2	2	-	-	2	3	10	-	-	-	-	2	-	-	1	23
076	4	-	1	-	-	1	1	-	-	31	33	-	-	-	-	-	-	-	-	71
080	1	2	1	-	7	6	-	-	7	12	46	-	-	-	-	-	-	-	-	82
092	-	-	-	-	-	-	-	-	1	2	1	-	-	-	-	1	-	-	-	5
099	10	8	8	1	20	6	2	-	25	241	291	2	-	2	-	19	-	-	3	638
100	39	28	19	7	69	28	5	-	268	202	521	117	2	14	-	63	-	-	3	1385
107	4	6	1	1	12	12	3	-	42	121	164	18	1	7	-	6	-	-	-	398
114	-	1	-	-	-	-	-	-	-	-	-	-	-	-	-	-	-	-	-	1
115	43	26	11	1	50	26	-	1	69	107	231	45	-	15	3	19	-	-	18	665
124	2	-	-	-	-	1	-	-	2	13	35	-	-	-	-	1	-	-		54
139	-	-	-	-	-	-	-	-	-	-	-	-	-	-	-	-	-	-	-	0
179	-	-	-	-	-	-	-	-	-	-	-	-	-	-	-	-	-	-	-	0
Total	**182**	**156**	**91**	**27**	**706**	**176**	**41**	**1**	**685**	**1645**	**2372**	**196**	**7**	**109**	**9**	**284**	**0**	**68**	**41**	**6796**

B) Middle EP tool classes by locus.

(characterized by production standardization), these contexts are more variable within individual contexts, but this variability is documented consistently between contexts. Chipped stone tool production appears very flexible and fluid, characterized by a lack of standardization.

This variability is also borne out in the wide variety of core types evidenced here. Although narrow-faced cores persist (likely a result of continued use of particular local raw material sources characterized by small, flattish cobbles and narrow tabular flint outcrops (see below), broad-faced cores now comprise over 40% of the represented core types. Multi-directional cores are also more common. Raw material types are also notably more diverse in the Middle EP. With less constrained use of the highly-localized brown-grey flint, probably related to changes in the distances and scheduling of movements throughout the Azraq Basin, it seems knappers took advantage of nodules of more variable sizes and shapes.

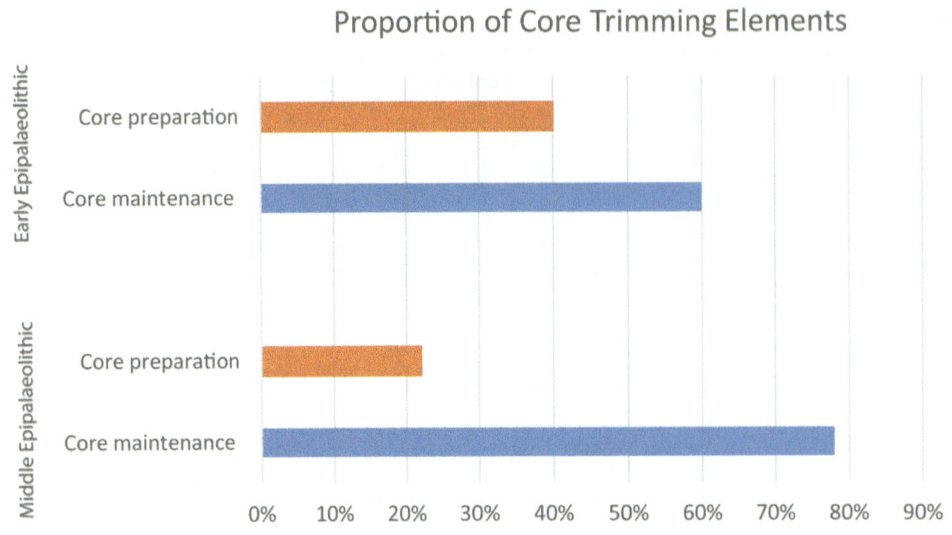

Fig. 11.5 Proportion of core preparation elements to core maintenance elements in the Early and Middle Epipaleolithic deposits at Kharaneh IV

Here the proportion of flakes (n = 65,600) to blades (n = 48,500) is slightly greater (flake-to-blade ratio of 1.35), but with fewer primary pieces, so it seems that there is less effort placed on producing specifically-shaped bladelet blanks; flakes and blades were (intentionally?) produced throughout all stages of core reduction, rather than initial flake removals and subsequent bladelet removals. Indeed, flakes were sometimes used as microlith blanks. This flexibility in flake versus blade blanks is also documented in the CTEs, with notably greater emphasis placed on core maintenance or 'fixing and shaping as they go'. The CTEs are overwhelmingly dominated by core face rejuvenation flakes and profile correction blades. Altogether, core maintenance pieces represent 78% of all CTEs (n = 5603) (Fig. 11.5). Indeed, distinctions between flakes and blades are more difficult in these assemblages as bladelet blanks are highly variable in overall size and shape—standardization was simply not important in blank production (although it may have continued to be important in final tool form).

Of the retouched tools, 70% are microliths. Larger blade tools are dominated by variously retouched pieces, end-scrapers, backed blades, and multiple tools. Of the microliths, 35% are geometrics (trapeze/rectangle variants), while 53% of these are fragmentary microliths; however, this is a result of very conservative identifications of broken microliths (discussed above) (Fig. 11.6). Given the diagnostic features remaining on these fragments, it is likely that most of the fragmentary microliths are actually broken geometrics and this would bring the proportion of geometric microliths in the retouched tool assemblage to ca. 60%. The geometric microliths are highly variable, but trapeze-rectangles predominate in the form of backed and unbacked trapezes, and other variants (Muheisen and Wada 1995; Macdonald 2013; Maher and Macdonald 2013), as well as asymmetrical trapezoids, lunates, triangles, and a variety of other geometric forms. Many of these forms are found at other sites throughout the region, hinting at the possibility for multiple aggregating groups coming together with their own lithic traditions (see below). Many of these pieces are heavily retouched and are manufactured on flake blanks, reinforcing the idea that blank shape was highly flexible and retouch was used to make these pieces into a desired geometric shape.

There are several possible explanations for the diversity of microliths noted in the Middle EP occupations; they could reflect the diversity of social groups at the site, a diversity of tasks engaged in at the site (i.e., different and more variable than the tasks performed by Early EP groups), mixing and time-averaging of occupations (al-Nahar and Olszewski 2016), or diversity could result from the lack of standardization in blank production. The extremely well-preserved nature of the deposits at the site (detected through both fine-grained excavation strategy and program of micromorphology) and program of radiocarbon dating demonstrate the integrity of these *in situ* deposits. This diversity is clear within individual contexts, and these contexts do not represent substantial time-averaged deposits.[9] Use-wear analyses of microliths from both Early and Middle EP contexts does not indicate any difference in the types of activities performed with these tools (see below), and geometric morphometrics of microliths shows no correlation between geometric form and blank type (Macdonald 2013). The only other Middle EP site with such a high diversity microliths is Jilat 6, also located in the Azraq Basin and interpreted as a substantial aggregation site (Garrard and Byrd 1992; Garrard

[9]Of course, some degree of time-averaging cannot be ruled out; however, the time gaps between adjacent contexts is very small.

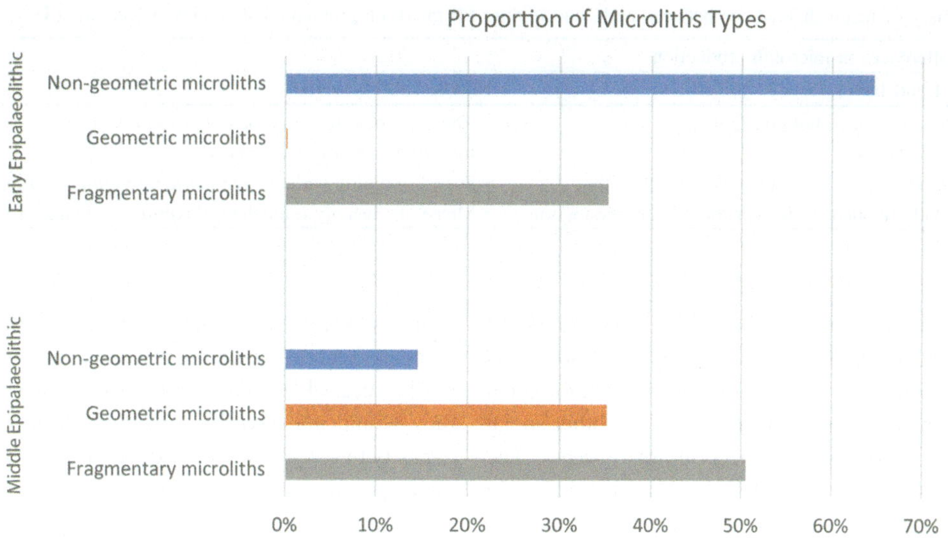

Fig. 11.6 Proportion of non-geometric, geometric, and fragmentary microliths in the Early and Middle Epipaleolithic deposits at Kharaneh IV

and Byrd 2013). In contrast, other, smaller Middle EP sites in the Azraq Basin and elsewhere do not show this degree of variability; instead, one or just a few geometric types dominate these assemblages.[10] These sites also tend to be geographically-clustered so that sites with similar types of microlith assemblage tend to be close to each other, leading to the above-mentioned, spatially-bounded industries. Furthermore, the 'types' of microliths documented at Kharaneh IV do not represent new types; they are found at other Middle EP sites, just not in combination in such substantial numbers. Thus, we suggest that the diversity reflects socially-constituted choice and the implications of this are discussed below.

Reconstructing Phases of Occupation

Analysis of lithics throughout the stratigraphic sequence indicate that there were at least three distinct phases of occupation at the site, the first two characteristic of the Early EP and the final phase characteristic of the Middle EP (Macdonald, Allentuck and Maher 2018) (Table 11.3). The earliest phase is characterized by microgravettes, the microburin technique, and bipolar backing.[11] The second phase, also Early EP, lacks the microburin technique and contains a large percent of obliquely truncated and backed bladelets. The microlith assemblage of the final phase is dominated by geometric microliths, placing this typological style within the Middle EP. There is no break in the stratigraphic record, just fluctuations in artifact density, suggesting that occupation at Kharaneh IV was virtually continuous. As discussed elsewhere, we do not necessarily imply that there were no breaks in occupation, just that they were relatively temporary breaks; the site was repeatedly occupied for prolonged periods during its 1200-year span (Maher and Conkey 2019). In addition, the final phase of occupation sees an increase in the diversity of microlith types, perhaps a reflection of an increasing diversity in group membership at the site over time and increasing complexity in terms of 'who' was at the site and 'when' (Macdonald, Allentuck and Maher 2018).

Stone Tool Production at EP Kharaneh IV: What to Do with Variability?

A knapping community is a flexible, fluid community (perhaps even an imagined one, Anderson 2006) that can exist within and between prehistoric groups. It is situational and can include members from the same household, next door, or hundreds of miles away who periodically interact. It consists of participants of various skill levels, from master to apprentice, and the traces of these skills may be detected in the products and by-products of knapping (Pigeot 1990). Change in the structure of these communities is reflected in the processes of knowledge transmission in stone tool production, including the appearance of craft specialization,

[10]We recognize that these sites do exhibit a degree of variability in geometric (and non-geometric) microlith forms; however, only a few of these types usually make up a large proportion of the microlith assemblage.

[11]Although backing is not discussed in detail here, in general, different backing styles (bipolar, abrupt, fine, inverse, alternating, etc.) are used as attributes to define different microlithic tools forms and, thus, industries or cultural groups.

Table 11.3 Summary of major distinctions between early and middle EP microlith production strategies at Kharaneh IV

Broad temporal differences in microlith production	
Early EP (Phases I and II)	Middle EP (Phase III)
• Raw material choice is local, but constrained	• Raw material choice is local, but more flexible
• Investment in core shaping	• Investment in core maintenance
• Focus on standardized, gracile, non-geometric microliths	• Focus on highly variable geometric microliths resulting in many 'types'
• Microliths minimally retouched; investment in blank production	• Microliths heavily retouched; flexibility in blank production

industrial-scale production and large-scale networks of exchange for raw materials (e.g., Carter et al. 2008). This approach emphasizes the social aspects of technological learning, which include gender (Gero 1991) and other social relations, as well as the role of objects in this process. The lithic assemblages at Kharaneh IV suggest a change in the nature of these knapping communities over time, with differences in technological choices during knapping, as well as differences in formal tools. These changes can be linked to changing populations at the site as Kharaneh IV becomes a locale for more aggregating groups, reflected in increasing diversity of microlith tool form through time.

Raw Material Choice

A geological survey within a 30 km radius of the site by C. Delage and X. Mangado in 2010 focused on identifying flint raw material sources documented on-site in order to assess whether the range of materials utilized by EP knappers, especially in the Middle EP, were accessible within the local vicinity of the site. This survey was paired with detailed comparative petrographic analyses of the flint collected from various geological outcrops and the archaeological materials from all phases of occupation. The raw material survey and petrographic analysis confirmed that all of the examined lithics, regardless of quality, color, or other features, were available locally—within less than 20 km, or a day's walk—from the site (Delage and Mandago, pers. comm. 2010). A wide variety of nodule size, shape, color and quality are available for all local sources, providing little constraint on knapping choices, such as on considerations of size or standardization.

While almost all of the knapped flint recovered from the excavations is of good quality, analysis of the archaeological assemblages noted a distinction in raw material use between the Early and Middle EP occupations. Early EP knappers primarily worked a brown-grey flint while in the Middle EP material used by knappers exhibits a very wide range of raw material types, including a notable preference for very fine, pink flints. This distinction has two important implications for resource acquisitioning and movements within the landscape. The first is that the wide range of flint sources used by Middle EP groups did not include, at least in all the samples analyzed to-date, so-called 'exotic' or non-local sources. Second, since all of these geological outcrops are easily accessible today, and we have no reason to believe they would not have been throughout the EP,[12] the differences between Early and Middle EP raw material use reflect choice—a preference for particular sources over others. The brown-grey flint favored by Early EP knappers is found eroding out of the limestone immediately surrounding the site. Middle EP knappers certainly continued to take advantage of this readily-available material, but also clearly preferred to go somewhat farther afield and sought out flints of various quality and color, some of which are quite spectacular and may have been favored materials because of their aesthetic qualities.

The final point here regarding Middle EP movements and preferences in raw material has implications for our reconstructions of large-scale social interaction and communities of knappers. Like at Gönnersdorf (Jöris et al. 2011), we argue for aggregation of hunter-gatherer groups from different geographic areas; however, these movements are traced through microliths 'traditions' rather than through raw material sources. It is clear that the raw material on which these highly variable geometric microliths are being made is local and the debris from their manufacturing is found throughout the Middle EP deposits in clear, in situ knapping contexts and caches. Thus, it seems that the making of this wide diversity in geometric microlith types is taking place on-site, probably in various socially-mediated knapping contexts. One might envision a situation where members of

[12]Geomorphological work in the Azraq Basin, especially around Kharaneh IV, indicates that rates of surface deposition and erosion are slow, with deflation being a major cause of landscape change (Fuchs, M., Dietze, M., Al-Qudah, K., & Lomax, J. (2015). Dating desert pavements–First results from a challenging environmental archive. *Quaternary Geochronology*, 30, 342–349.) This deflation disproportionately affects unconsolidated Quaternary sediments more so than consolidated limestone bedrock from which the flint erodes. In any case, over time this erosion would only ensure an ongoing supply of highly local flint and not explain the increase in diversity in the Middle EP, or explain why Early EP groups, whose sites are documented throughout the basin, would not have had access to other (nearby!) sources.

different 'groups' sat down together and each made these tools according to their own particular local tradition, or even shared this knowledge in situated communities of practice. We are still exploring the spatial distributions of these knapping contexts to explore these possibilities; however, as we have argued elsewhere it does seem clear that the ideas about making these various geometric microliths were highly mobile, this knowledge and preference for particular 'styles' moved with people to and from the site, rather than the raw material or finished tools themselves (Maher and Macdonald 2013).

Knowledge, Skill and Learning

Ongoing work on the Kharaneh IV lithic assemblage is now focused on whether we can a) detect spatial patterns in the chipped stone and other materials that might indicate households or groups, and their interactions, and b) detect knapping communities that might be reflected in expressions of social traditions in microliths production or exhibit varying levels of skill. For example, were there specialists, apprentices, or children involved in learning these technologies (Pigeot 1990; Bamforth and Finlay 2008)? Indeed, the concepts of skill and learning are invaluable to our reconstructions of past technological practice and notions of 'doing' and making (see Ingold 2001 for a useful discussion of skill in technological studies), as well as our assessments of the creation of social relationships between makers (and makers) and users (Wendrich 2013b).

At Kharaneh IV, it seems that 'mistakes', characterized as hinged terminations[13] or failed flake removals on cores, as plunging, stepped or hinged flakes and blades, and hinged dorsal scars on flakes and blades are noted throughout all analyzed contexts. However, while these mistakes are ubiquitous, they are not overwhelmingly frequent. Of course, the mark of a skilled knapper is their ability to fix these mistakes; to 'keep calm and carry on'. Thus, we are beginning to try and map out varying levels of skill, measured as proclivities to make mistakes and abilities to fix mistakes in various contexts. However, there are some complicating factors in this assumption. For example, we must wonder whether these so-called 'mistakes' in the Middle EP are easily recognizable as such. Mistakes, alongside a highly variable microliths assemblage, where emphasis was placed not on core shaping to produce standard blanks but on core maintenance where blank shape was flexible and even flakes were considered suitable, might suggest that the Middle EP knappers exhibited less 'skill' in knapping in comparison to Early EP knappers whose skill is reflected in highly uniform products. These technical mistakes can be very illuminating when understood within the overall character of a core reduction sequence, similar to the methods used by Pigeot (1990); perhaps they were apprentices, novices, or children learners? We are beginning to explore these latter possibilities (e.g., Finlay 1997; Milne and Wendrich 2013; Wendrich 2013c). However, there may be a couple of other contributing factors: First, perhaps with lots of locally-available material, there was little need to fix mistakes. Knappers could make-do with highly unstandardized pieces so production was more flexible and ideas of what was a mistake were more relaxed. Second, if our ideas about communities of practice and fluid and communal knapping groups is correct then so-called mistakes may result from 'trying out' another groups methods of microliths production – they may result from otherwise expert knappers in intra-group acts of knowledge transmission or exchange.

That Middle EP lithic production was notably less standardized than in the Early EP phases at Kharaneh IV is also a feature well-documented in Late EP (Natufian) assemblages (Goring-Morris, Hovers and Belfer-Cohen 2009) where microliths production similarities across a broad area of the region are taken to reflect highly interconnected social groups or spheres of interaction. Little effort seems placed on maintaining or conserving cores, blanks could be bladelets or small flakes and size and shape were unimportant as they were heavily retouched to the desired dimensions. Technological trajectories were highly divergent and mutable within assemblages, and these highly variable assemblages—sharing primarily the production and use of lunates, regardless of how a knapper 'got here'—may contribute to the heterogeneity documented at Natufian sites throughout the region. The establishment of wide-reaching social interaction spheres, perhaps beginning in the EP, might have cultivated the sharing of knowledge, ideas and techniques, resulting in increased variability within assemblages and the breakdown of highly localized stone tool traditions. In this situation, aggregation sites like Kharaneh IV would play a key role in the creation and maintenance of these networks, with diverse smaller bands of hunter-gatherers aggregating at the site. Processes of aggregation lead to knowledge exchange and the transmission of knapping techniques across group boundaries, creating permeable communities of practice.

Style and Function

Archaeologists have approached material culture variability in numerous ways. This variability often relates to the material properties of the artifact that are highly visible and easily compared. As archaeologists, we often equate this

[13]Although, of course, not all hinged terminations are necessarily mistakes (Tixier 1979).

variability with 'style'. This paradigm has greatly influenced EP archaeology, with the 'style' of microliths delineating chronological and regional cultures. Broader discussions about whether material culture variability is a reflection of learning communities can only be had after 'functional' interpretations are tested (Richter 2007; Yaroshevich et al. 2010; Macdonald 2013; Yaroshevich, Nadel and Tsatskin 2013). Focusing on the use of technology, functional studies of variability can test whether morphological variability is a reflection of function, rather than the result of cultural practices (e.g., Kingery 2001; Schiffer et al. 2001; Schiffer 2003). Microwear studies of Middle EP microliths from Kharaneh IV show that there is a high frequency of used microliths (48.5%). The majority of these were projectiles (27.5%), followed by cutting tools (10.5%). Interestingly, there is no correlation between microlith form and function at Kharaneh IV, with distinct microlith types being used for several difference functions (Macdonald 2013). For example, both trapeze-rectangles and unbacked trapezes functioned as projectile inserts, cutting tools, and scraping tools. Diagnostic impact fractures occur as parallel fractures (14.3%) and oblique/perpendicular fractures (85.7%). The majority of the geometric microliths have the latter type of fractures indicating that they were not used as tips for arming projectiles. Non-projectile functions include cutting meat and hide, butchering activities with bone contact, and working hard materials. This suggests that the form of the microlith did not dictate the function of the object, thus form is more the result of cultural/social, rather than functional, processes.

Changes in material culture style can reflect a wide diversity of social processes unrelated to functional constraints. Visible elements of material culture can communicate social boundaries, as in Wiessner's 'emblemic style' (Wiessner 1984), or they can be easily transmitted between groups through learning communities. We suggest that microlith production at Kharaneh IV, at least, was malleable and widespread, and the variability documented within the Middle EP was a result of technological diffusion—shared ideas and knapping experiences—as different social groups from within the Azraq Basin and beyond engaged in long-distance movements and maintained social networks of interaction and exchange (of material culture and technological knowledge).

Discussion: What's the Point?

The social lives of people at Kharaneh IV did not begin or end with their flintknapping. The complicated deposits found in both the Early and Middle EP deposits on the site suggest that daily practice was filled with interactions and negotiations. However, the archaeological record at Kharaneh IV suggests a shift in social life between the Early and Middle EP occupations. The Early EP deposits are characterized by complicated stratigraphy of numerous, thin, discrete deposits. These include multiple hut structures, knapping areas, hearths, compact surfaces, ash dumps, dense, discrete middens, caches, intentional destruction deposits, and human burials. These defined deposits, with clear boundaries and walls (in the case of the huts) suggests delineation and maintenance of space, dividing tasks and activities into different locales, potentially performed by different people. In contrast, there are few Middle EP occupational surfaces, and each one is characterized by a broad horizontal extent. Deposits from this final phase are large and thick compact earthen surfaces with no clear boundaries, large middens and fill deposits, large, overlapping hearths, usually ringed by many postholes. In contrast to the Early EP occupational deposits, there are no discrete knapping areas, caches, huts or burials. The differences in occupation suggests that people in the Early and Middle EP were interacting with each other in different ways and using space differently. Patterns during the Early EP suggest boundaries between activities, while during the Middle EP activities are less bounded, perhaps reflecting a shift towards more communal lifeways.

This increase in evidence for communal living is witnessed in multiple aspects of the Kharaneh IV Middle EP assemblage. Evidence suggests that the people occupying Kharaneh IV during the Middle EP participated in communal hunting and processed large amounts of meat. Postholes surrounding hearths hint at evidence for meat drying racks and gazelle carcass processing indicates that meat might have been stored for later use by the community. Finally, an increase in the diversity of the microlith assemblage may reflect an increase in the diversity of people aggregating at Kharaneh IV during the Middle EP. Changes from inwardly-focused practices during the Early EP to community-focused practices in the Middle EP suggest a shift and expansion of communities of practice over time. These changes may also reflect a key change in the population at the site, with either larger groups or increasing numbers of groups aggregating at the site during the Middle EP, and changing patterns of site re-use over time leading to complicated reconstructions of individual 'occupations' (see also al-Nahar and Olszewski 2016 for a discussion of issues related to time-averaging). These increasing populations at the site suggest that knowledge of knapping techniques and microlith form was widely shared during the Middle EP. People with these shared traditions, skills, and knowledge would have brought these ideas with them as they travelled within the landscape—from near and far—sharing and exchanging ideas as they moved. Thus, we argue that the widespread pattern of shared geometric microlith form, such as the trapeze-rectangle, seen throughout the Levant is not necessarily the result of convergent evolution, but rather the result of aggregation and interaction of knapping communities at sites such as Kharaneh IV. Within the context of multiple, overlapping networks of EP social groups, interacting with varying degrees of intimacy (Gamble 1998) across time and space, lithic technologies and traditions could be shared in the form of both knowledge and material objects over long distances. We suggest that knowledge and skills necessary for the creation of lithic technologies was, in part, shared between different groups through specific 'communities of practice' created at aggregation sites. Thus, widespread forms of microliths found throughout the Levant are the result of shared traditions and knowledge—the transmission of information—rather than independent developments and convergent evolution. This, of course, does not answer the question of why similar microlith forms, geometric or otherwise, emerged outside of the Levant in earlier and later times; in other words, we still grapple with the prevalence of convergent evolution in the *emergence* of microlith technologies and the role convergent evolution may play at different scales (Clarkson et al. 2018; Shipton 2020; Shott 2020; Will and Mackay 2020). It is possible that explanations related to knowledge exchange, broad cultural changes in subsistence economy, or functional constraints on stone tools can operate at different scales.

Given the density of archaeological deposits at Kharaneh IV, even after a decade of research at the site our work remains ongoing. Several ways forward including pairing ongoing techno-typological analyses of the archaeological assemblages from Kharaneh IV with further functional analysis such as microwear and residue analysis. The integration of 'form' with 'function' elucidates the complicated relationship between tool morphology and how material objects were integrated into daily life through action (Macdonald 2013). Experimental flintknapping is being conducted to understand the contexts of experience/skill and situated learning during the technological processes of EP chipped stone tool production by recreating it in the present. Coupled with experimental flintknapping, our refitting studies aid in the reconstruction of the *chaîne opératoire* for microlith production. Finally, placing Kharaneh IV in context with other contemporaneous sites is shedding valuable light onto larger community networks during the EP.

Conclusions: Implications for Understanding Prehistoric Technology Today

Situated within the bigger picture of hunter-gatherer studies, exploring the nature of material culture at the intra-site and inter-site levels allows us to address issues of technological change in relation to larger questions of material culture variability and culture change. A focus on technology, innovation and associated learning practices are now at the forefront of material culture studies (e.g., Lave and Wenger 1991; Lemonnier 1992; Pfaffenberger 1992; Hegmon 1998; Crown 2007; Miller 2007; Lemonnier 2013). This has great relevance for our understanding of patterns and variability on-site at Kharaneh IV and within the larger field of Levantine prehistory. For example, stone tool production occurs within a social context and is often integrative to other technologies (Ambrose 2001; Bamforth and Finlay 2008; Lombard and Haidle 2012), thus unraveling the social and learning aspects of lithic technology can provide valuable insights into a range of other related activities and technologies, such as wood, textiles, bone and shell. Considering Kharaneh IV as a persistent place of aggregation and interaction in a highly social EP landscape highlights aspects of landscape learning in changes in hunting practices, raw material exploitation, and movement as these groups travelled within the Azraq Basin, and beyond (Maher 2018). It also allows us to reconsider some of our assumptions about the distances, frequencies and durations of hunter-gatherer movements within a landscape and how EP and early Neolithic hunter-gatherers created and transformed their physical and social worlds.

A central theme of this chapter is that chipped stone technologies are seen as learned sets of activities enacted within social contexts by communities of knappers. These contexts reflect social interactions between individuals and groups and the type, degree, and distances of these interactions can be explored through inter- and intra-site variability. Beyond simple 'tool-making', knappers engage in knowledgeable interactions with the environment, materials, other people, and the traditions and worldviews that shape identity. Taken in this framework, the study of technology from an archaeological perspective is little different from the study of technology today and, indeed, can draw heavily on contemporary thought and literature (Lechtman 1977; Franklin 1999; Suchman et al. 1999; Suchman 2001). Archaeology, in this sense, has a lot to offer on contextualizing how people interact with technology, placing global issues of technological change into a long-term perspective that considers how humans have constructed and engaged with a material world that structures and shapes culture and culture change.

References

al-Nahar, M., & Olszewski, D. I. (2016). Early Epipaleolithic lithics, time-averaging, and site interpretations: Wadi al-Hasa region, Western Highlands of Jordan. *Quaternary International, 396,* 40–51.

Allcock, S. (2009). *Beyond trade and subsistence: the use of shell ornaments to infer social interaction and increasing complexity during the early and middle Epipalaeolithic.* Jordan: University College London.

Allentuck, A. (In prep). Taphonomy, subsistence and deposition of faunal remains in the early Epipalaeolithic. In L. Maher & T. Richter (Eds.), *A prehistoric oasis in the azraq basin, Jordan: The Epipalaeolithic foragers in Azraq project 2005–2015.*

Ambrose, S. (2001). Paleolithic technology and human evolution. *Science, 291,* 1748–1753.

Anderson, B. (2006). *Imagined communities: Reflections on the origin and spread of nationalism.* London: Verso Books.

Andrefsky, W. (2005). *Lithics: Macroscopic approaches to analysis. Cambridge manuals in archeology.* Cambridge; New York: Cambridge University Press.

Arranz-Otaegui, A., Carretero, L. G., Ramsey, M. N., Fuller, D. Q., & Richter, T. (2018). Archaeobotanical evidence reveals the origins of bread 14,400 years ago in northeastern Jordan. *Proceedings of the National Academy of Sciences, 115,* 7925–7930.

Asouti, E. (2013). Evolution, history and the origin of agriculture: Rethinking the Neolithic (plant) economies of South-west Asia. *Levant, 45,* 210–218.

Asouti, E., Ceren Kabukcu, C., White, C. E., Kuijt, I., Finlayson, B., & Makarewicz, C. (2015). Early Holocene woodland vegetation and human impacts in the arid zone of the southern Levant. *The Holocene, 25,* 1565–1580.

Asouti, E., & Fuller, D. Q. (2012). From foraging to farming in the southern Levant: The development of Epipalaeolithic and Pre-pottery Neolithic plant management strategies. *Vegetation History and Archaeobotany, 21,* 149–162.

Audouze, F., Bodu, P., Karlin, C., Julien, M., Pelegrin, J., & Perlès, C. (2017). Leroi-Gourhan and the chaîne opératoire: A response to Delage. *World Archaeology, 49,* 718–723.

Baker, J. (2018). *Technology of the Ancient Near East: From the Neolithic to the Early Roman Period*: Routledge.

Bamforth, D. B., & Finlay, N. (2008). Archaeological approaches to lithic production skill and craft learning. *Journal of Archaeological Method and Theory, 15,* 1–27.

Bar-Yosef, O. (1970). The Epipalaeolithic cultures of Palestine, Hebrew University.

Bar-Yosef, O. (1991). The search for lithic variability among Levantine Epi-Palaeolithic industries. In (Eds.), *25 Ans d'Etudes Technologiques en Préhistoire* (pp. 319–336). Juan-les-Pins: Editions APDCA.

Bar-Yosef, O., & Van Peer, P. (2009). The Chaine operatoire approach in middle paleolithic archaeology. *Current Anthropology, 50,* 103–131.

Basalla, G. (1988). *The evolution of technology.* Cambridge University Press.

Belfer-Cohen, A., & Bar-Yosef, O. (2000). Early Sedentism in the Near East: A bumpy ride to village life. In I. Kuijt (Ed.), *Life in Neolithic farming communities: Social organization, Identity, and Differentiation, Fundamental issues in archaeology* (pp. 19–37). New York: Kluwer Academic/Plenum Publishers.

Belfer-Cohen, A., & Goring-Morris, N. (2011). Becoming farmers: The inside story. *Current Anthropology, 52,* S209–S220.

Betts, A. V. G. (1998). *The Harra and the Hamad: Excavations and Surveys in Eastern Jordan* (Vol. 1). Sheffield: Sheffield Academic Press.

Binford, L. R. (1962). Archaeology as anthropology. *American Antiquity, 28,* 217–225.

Bleed, P. (1986). The optimal design of hunting weapons: Mantainability or reliability. *American Antiquity, 51,* 737–747.

Bleed, P. (2001). Trees or Chains, links or branches: Conceptual alternatives for consideration of stone tool production and other sequential activities. *Journal of Archaeological Method and Theory, 8,* 101–127.

Boëda, E., Geneste, J.-M., & Meignen, L. (1990). Identification de chaîne opératoires lithiques du Paléolithique ancien et moyen. *Paléo, 2,* 43–79.

Bordes, F., & Crabtree, D. E. (1969). The Corbiac blade technique and other experiments. *Tebiwa, 12,* 1–21.

Bourdieu, P. (1977). *Outline of a theory of practice.* Cambridge: Cambridge University Press.

Bourdieu, P. (1990). *The logic of practice.* Stanford: Stanford University Press.

Bourriau, J., & Phillips, J. (2004). *Invention and innovation: The social context of technological change 2: Egypt, the Aegean and the Near East, 1650–1150 BC.* Oxbow Books.

Boyd, B. (2006). On sedentism in the Later Epipalaeolithic (Natufian) Levant. *World Prehistory, 38,* 164–178.

Byrd, B. (1987). *Beidha and the Natufian: Variability in Levantine settlement and subsistence,* University of Arizona.

Byrd, B. (1989). *The Natufian Encampment at Beidha: Late Pleistocene Adaptation in the Southern Levant; Excavations at Beidha 1. Jutland Archaeological Society Publications XXIII:1.* Aarhus: Aarhus University Press.

Carter, T., Dubernet, S., King, R., Le Bourdonnec, F.-X., Mili, M., Poupeau, G., et al. (2008). Eastern Anatolian obsidians at Çatalhöyük and the reconfiguration of regional interaction in the Early Ceramic Neolithic. *Antiquity, 82,* 900–909.

Chazan, M. (2009). Pattern and technology: Why the Chaine operatoire matters. In J. Shea & D. Lieberman (Eds.), *Transitions in prehistory: Essays in honor of ofer bar-yosef, American school of oriental research* (pp. 467–498). Cambridge: Oxbow.

Clarkson, C., Hiscock, P., Mackay, A., & Shipton, C. (2018). Small, sharp, and standardized: Global convergence in backed-microlith technology. In M. O'Brien, B. Buchanan, & M. Eren (Eds.), *Convergent evolution in stone-tool technology* (pp. 175–200). Cambridge: MIT Press.

Close, A. E. (2000). Reconstructing movement in prehistory. *Journal of Archaeological Method and Theory, 7,* 49–75.

Copeland, L., & Hours, F. (1989). *The hammer on the rock: Studies in the early Palaeolithic of Azraq, Jordan Part I.* Lyon: C.N.R.S. Universite Lumiere.

Cordova, C. E., Nowell, A., Bisson, M., Ames, C. J., Pokines, J., Chang, M., et al. (2013). Interglacial and glacial desert refugia and the middle Paleolithic of the Azraq Oasis, Jordan. *Quaternary International, 300,* 94–110.

Costin, C. L. (2001). Craft Production Systems. In G. M. Feinman & T. D. Price (Eds.), *Archaeology at the millennium: A sourcebook* (pp. 273–327). New York: Kluwer Academic/Plenum Publishers.

Crabtree, D. E. (1970). Flaking stone with wooden implements. *Science, 169,* 146.

Crabtree, D.E. (1972). *An introduction to flintworking. Occasional papers of the Idaho State University Museum, no. 28.* Pocatello: Idaho State University Museum.

Crabtree, D. E., & Davis, E. L. (1968). Experimental manufacture of wooden implements with tools of flaked stone. *Science, 159,* 426–428.

Crown, P. L. (2007). Learning about learning. In J. M. Skibo, M. W. Graves, & M. T. Stark (Eds.), *Archaeological anthropology: Perspectives on method and theory* (pp. 198–217). Tuscon: University of Arizona Press.

Cucchi, T., Vigne, J.-D., & Auffray, J.-C. (2005). First occurrence of the house mouse (Mus musculus domesticus Schwarz & Schwarz, 1943) in the Western Mediterranean: A zooarchaeological revision of subfossil occurrences. *Biological Journal of the Linnean Society, 84,* 429–445.

Delage, C. (2017). Once upon a time… the (hi) story of the concept of the chaîne opératoire in French prehistory. *World Archaeology, 49,* 158–173.

Dibble, H. L. (1997). Platform variability and flake morphology: A comparison of experimental and archaeological data and implications for interpreting prehistoric lithic technological strategies. *Lithic Technology, 22,* 150–170.

Dibble, H. L., & Bernard, M. C. (1980). A comparitive study of basic edge Angle measurement techniques. *American Antiquity, 45,* 837–865.

Dibble, H. L., Holdaway, S. J., Lin, S. C., Braun, D. R., Douglass, M. J., Iovita, R., et al. (2017). Major fallacies surrounding stone artifacts and assemblages. *Journal of Archaeological Method and Theory, 24,* 813–851.

Dibble, H. L., & Rezek, Z. (2009). Introducing a new experimental design for controlled studies of flake formation: Results for exterior platform angle, platform depth, angle of blow, velocity, and force. *Journal of Archaeological Science, 36,* 1945–1954.

Dietler, M., & Herbich, I. (1998). Habitus, techniques, style: An integrated approach to the social understanding of material culture and boundaries. In M. Stark (Ed.), *The archaeology of social boundaries* (pp. 232–263). Washington D.C.: Smithsonian Institution Press.

Dobres, M. A. (2000). *Technology and social agency: Outlining a practice framework for archaeology.* Oxford: Blackwell.

Dobres, M.A. (2001). Meaning in the making: Agency and the social embodiment of technology and art. *Anthropological perspectives on technology,* pp. 47–76.

Dobres, M. A. (2010). The phenomenal promise of Chaîne Opératoire: Mindfully engaged bodies and the manufacture of personhood in a regional perspective. In R. Barndon, A. Engevik, & I. Øye (Eds.), *The archaeology of regional technologies: Case studies from the paleolithic to the age of the vikings* (pp. 51–67). Lampeter: Edwin Mellon Press.

Dobres, M. A., & Hoffman, C. R. (1992). Social agency and the dynamics of prehistoric technology. *Journal of Archaeological Method and Theory, 1,* 211–258.

Eerkens, J. W. (1991). Reliable and maintainable technologies: Artifact standardization and the early to later mesolithic transition in northern England. *Lithic Technology, 23,* 42–53.

Fellner, R. (1990). The problems and prospects of cultural evolution. *Papers from the Institute of Archaeology, 1,* 45–55.

Finlay, N. (1997). Kid knapping: the missing children in lithic analysis. In J. Moore & E. Scott (Eds.), *Invisible people and processes: Writing gender and childhood into European archaeology* (pp. 203–221). New York: Leichester University.

Finlayson, B., & Makarewicz, C. (2013). Neolithic stereotypes: Has South-west Asian archaeology outlived the Neolithic? *Levant, 45,* 119.

Finlayson, B., & Warren, G. (2010). *Changing natures: Hunter-Gatherers, first farmers and the modern world. Duckworth debates in archaeology.* Bristol: Bristol Classical Press.

Finlayson, B., & Warren, G. (Eds.). (2017). *Diversity of Hunter-Gatherer pasts.* Oxford: Oxbow Books.

Forbes, R. J. (1964). *Studies in ancient technology.* Leiden: Brill.

Franklin, U. (1999). *The real world of technology.* Toronto: House of Anansi.

Fuchs, M., Dietze, M., Al-Qudah, K., & Lomax, J. (2015). Dating desert pavements–First results from a challenging environmental archive. *Quaternary Geochronology, 30,* 342–349.

Gamble, C. (1998). Palaeolithic society and the release from proximity: A network approach to intimate relations. *World Archaeology, 29,* 426–449.

Gamble, C. (2007). *Origins and revolutions: Human identity in earliest prehistory.* Cambridge University Press.

Garrard, A., Baird, D., & Byrd, B. (1994a). The chronological basis and significance of the Late Palaeolithic and Neolithic sequence in the Azraq Basin, Jordan. In O. Bar-Yosef & R. Kra (Eds.), *Late Quaternary Chronology and palaeoclimates of the Eastern Mediterranean* (pp. 177–199). Radiocarbon.

Garrard, A., Baird, D., Colledge, S., Martin, L., & Wright, K. (1994b). Prehistoric environment and settlement in the Azraq Basin: An interim report on the 1987 and 1988 excavation seasons. *Levant, 26,* 73–109.

Garrard, A., & Byrd, B. (1992). New dimensions to the epipalaeolithic of the Wadi el-Jilat in Central Jordan. *Paléorient, 18,* 47 62.

Garrard, A., & Byrd, B. (2013). *Beyond the fertile crescent: Late Palaeolithic and Neolithic Communities of the Jordanian Steppe. Volume 1: Project background and the Late Palaeolithic-Geological context and technology. CBRL Levant Supplementary series.* Oxford: Oxbow Books.

Garrard, A., & Stanley-Price, N. (1975). A survey of prehistoric sites in the Azraq Basin, Eastern Jordan. *Paléorient, 3,* 109–126.

Geneste, J.-M. (1991). Systèmes techniques de production lithique. Variations techno-économiques dans les processus de réalisation des outillages paléolithiques. *Techniques & Culture, 17–18,* 1–35.

Gero, J. M. (1989). Assessing social information in material objects: How well do lithics measure up? In R. Torrence (Ed.), *Time, energy, and stone tools* (pp. 67–77). Cambridge: Cambridge University Press.

Gero, J. M. (1991). Genderlithics: Women's roles in stone tool production. In J. M. Gero & M. W. Conkey (Eds.), *Engendering aechaeology* (pp. 163–193). Cambridge: Cambridge University Press.

Gibbs, K. (2015). Pottery invention and innovation in East Asia and the Near East. *Cambridge Archaeological Journal, 25,* 339–351.

Goring-Morris, A.N. (1987). *At the Edge. Terminal Pleistocene Hunter-Gatherers in the Negev and Sinai. BAR international series 361.* Oxford: British Archaeological Reports.

Goring-Morris, A. N., & Belfer-Cohen, A. (2002). Symbolic behaviour from the Epipalaeolithic and Early Neolithic of the Near East: Preliminary observations on continuity and change. In H. G. K. Gebel, G. B. D. Hermansen, & V. C. Hoffman-Jensen (Eds.), *Magic practices and ritual in the near eastern Neolithic, studies in early near eastern production, subsistence, and environment* (Vol. 8, pp. 67–79). Berlin: Ex oriente.

Goring-Morris, A. N., & Belfer-Cohen, A. (2010). Different ways of being, different ways of seeing … changing worldviews in the near East. In W. Finlayson & G. Warren (Eds.), *Landscapes in transition: Understanding hunter-gatherer and farming landscapes on the early Holocene of Europe and the Levant* (pp. 9–22). London: CBRL Monographs.

Goring-Morris, A. N., & Belfer-Cohen, A. (2011). Neolithization processes in the levant: The outer envelope. *Current Anthropology, 52,* S195–S208.

Goring-Morris, A. N., Hovers, E., & Belfer-Cohen, A. (2009). The dynamics of pleistocene and early holocene settlement patterns and human adaptations in the levant: An overview. In J. Shea & D. Lieberman (Eds.), *Transitions in prehistory: Essays in honor of ofer bar-yosef, American school of prehistoric research monographs* (pp. 185–252). Oxford: Oxbow Books.

Grosman, L., Munro, N. D., & Belfer-Cohen, A. (2008). A 12,000-year-old Shaman burial from the southern Levant (Israel). *Proceedings of the National Academy of Sciences of the United States of America, 105,* 17665–17669.

Haudricourt, A.-G. (1964). La technologie, science humaine. *La Pensée, 115,* 28–35.

Haudricourt, A.-G. (1987). *La technologie, science humaine*. Paris: Recherche d'histoire et d'ethnologie des Techniques.

Hayden, B., Canuel, N., & Shanse, J. (2012). What was brewing in the Natufian? An archaeological assessment of brewing technology in the Epipaleolithic. *Journal of Archaeological Method and Theory, 20,* 102–150.

Hegmon, M. (1998). Technology, style, and social practices: Archaeological approaches. In M. T. Stark (Ed.), *The archaeology of social boundaries*. Washington, D.C.: Smithsonian Institution Press.

Henry, D. O. (1995). *Prehistoric cultural ecology and evolution: Insights from southern Jordan*. New York: Plenum Press.

Henry, D. O. (1996). Functional minimalism versus ethnicity in explaining lithic patterns in the Levantine Epipalaeolithic. *Antiquity, 70,* 135–136.

Henton, E., Martin, L., Garrard, A., Jourdan, A.-L., Thirlwall, M., & Boles, O. (2017). Gazelle seasonal mobility in the Jordanian steppe: The use of dental isotopes and microwear as environmental markers, applied to Epipalaeolithic Kharaneh IV. *Journal of Archaeological Science: Reports, 11,* 147–158.

Hodges, H. (1970). *Technology in the ancient world*. London: Allen Lane The Penguin Press.

Hodges, H. (1989). *Artifacts: An introduction to early materials and technology*. London: Duckworth.

Hofman, J.L., & Enloe, J.G. (1992). *Piecing together the past: Applications of refitting studies in archaeology*. Vol. 578. *BAR international series*. Oxford: BAR.

Hurcombe, L. (2007). *Archaeology, artefacts and material culture*. London: Routledge.

Hussain, S. T., & Floss, H. (2016). Streams as entanglement of nature and culture: European Upper Paleolithic river systems and their role as features of spatial organization. *Journal of Archaeological Method and Theory, 23,* 1162–1218.

Ingold, T. (2001). Beyond art and technology: The anthropology of skill. In M.B. Schafer (ed.), *Anthropological perspectives on technology* (pp. 17–31). Alburqueque: University of New Mexico Press.

Jones, J. (2012). Using gazelle dental cementum studies to explore seasonality and mobility patterns of the Early-Middle Epipalaeolithic Azraq Basin, Jordan. *Quaternary International, 252,* 195–201.

Jones, M., Maher, L., Richter, T., Macdonald, D., & Martin, L. (2016a). Human-environment interactions through the Epipalaeolithic of Eastern Jordan. In D. Contreras (Ed.), *Correlation is not enough: Building better arguments in the archaeology of human-environment interactions* (pp. 121–140). New York: Routledge.

Jones, M., & Richter, T. (2011). Palaeoclimatic and archaeological implications of Pleistocene and Holocene environments in Azraq, Jordan. *Quaternary Research, 76,* 363–372.

Jones, M. D., Maher, L. A., Macdonald, D. A., Ryan, C., Rambeau, C., Black, S., et al. (2016b). The environmental setting of Epipalaeolithic aggregation site Kharaneh IV. *Quaternary International, 396,* 95–104.

Jöris, O., Street, M., & Turner, E. (2011). Spatial analysis at the Magdalenian site of Gönnersdorf (Central Rhineland, Germany). An Introduction. In S. Gaudizinski-Windheuser, O. Joris, M. Sensburg, M. Street & E. Turner (Eds.), *Site-internal spatial organization of hunter-gatherer societies: Case studies from the European Palaeolithic and mesolithic* (pp. 53–80). Mainz: Verlag des Römisch-Germanischen Zentralmuseums.

Julien, M. (1992). Du fossile directeur à la chaîne opératoire. In J. Garanger (Ed.), *La Préhistoire dans le monde* (Vol. 1992, pp. 163–193). Paris: PUF.

Karlin, C., Bodu, P., & Pelegrin, J. (1991). Processus techniques et chaînes opératoires: Comment les préhistoriens s'approprient un concept élaboré par les ethnologues. In H. Balfet (Ed.), *Observer l'Action technique: Des Chaînes Opératoires, pour quoi faire?* (pp. 101–117). Paris: Éditions du CNRS.

Killick, D. (2004). Social constructionist approaches to the study of technology. *World Archaeology, 36,* 571–578.

Kingery, W. D. (2001). The design process as a critical component of the anthropology of technology. In M. Schiffer (Ed.), *Anthropological perspectives on technology* (pp. 123–138). Dragoon, AZ: Amerind Foundation.

Kuhn, S. (2004). Evolutionary perspectives on technology and technological change. *World Archaeology, 36,* 561–570.

Langley, M. C. (2013). Storied landscapes makes us (Modern) Human: Landscape socialisation in the Palaeolithic and consequences for the archaeological record. *Journal of Anthropological Archaeology, 32,* 614–629.

Laughlin, J. P., & Kelly, R. L. (2010). Experimental analysis of the practical limits of lithic refitting. *Journal of Archaeological Science, 37,* 427–433.

Lave, J., & Wenger, E. (1991). *Situated learning: Legitimate peripheral participation*. Cambridge: Cambridge University Press.

Lechtman, H. (1977). Style in technology-some early thoughts. In H.R. Lechtman & J.M. Merrill (Eds.), *Material culture, organization and dynamics of technology, proceedings of the American ethnological society* (pp. 3–20). St. Paul: West Publishing Co.

Lechtman, H., & Merrill, R.S. (1977). *Material culture: styles, organization, and dynamics of technology*. West Publishing Company.

Lemonnier, P. (1976). La Description des Chaines operatoires: Contribution a l'analyse des systemes techniques. *Techniques et Cultures,* 1.

Lemonnier, P. (1986). The study of material culture today: toward an anthropology of technical systems. *Journal of Anthropological Archaeology, 5,* 147–186.

Lemonnier, P. (1992). *Elements for an anthropology of technology*. Ann Arbor: University of Michgan.

Lemonnier, P. (2002). *Technological choices: Transformation in material culture since the neolithic. material cultures: Interdisciplinary studies in the material construction of social worlds*. New York: Routledge.

Lemonnier, P. (2013). *Mundane objects. Materiality and non-verbal communication*. Walnut Creek: Left Coast Press.

Leroi-Gourhan, A. (1964). *Le Geste et la Parole. I, technique et langage. Sciences d'Aujourd'hui*. Paris: Albin Michel.

Leroi-Gourhan, A. (1993). *Gesture and speech. An october book*. Cambridge: Massachusetts Insitute of Technology.

Leroi-Gourhan, A., & Brézillon, M.N. (1973). *Fouilles de Pincevent: essai d'analyse ethnographique d'un habitat magdalénien.(La section 36)*. Centre national de la Recherche Scientifique.

Liu, L., Wang, J., Rosenberg, D., Zhao, H., Lengyel, G., & Nadel, D. (2018). Fermented beverage and food storage in 13,000 y-old stone mortars at raqefet cave, Israel: Investigating Natufian ritual feasting. *Journal of Archaeological Science: Reports, 21,* 783–793.

Lombard, M., & Haidle, M. (2012). Thinking a bow-and-arrow set: Cognitive implications of middle stone age bow and stone-tipped arrow technology. *Cambridge Archaeological Journal, 22,* 237–264.

Macdonald, D. (2013). Interpreting variability through multiple methodologies: The interplay of form and function in Epipalaeolithic microliths. University of Toronto.

Macdonald, D., & Maher, L. (In press). Domestic tasks at Kharaneh IV: Understanding the Epipalaeolithic toolkit through microwear. In J. Gibaja, I. Clemente, N. Mazzucco & J. Marreiros (Eds.), *Hunter-Gatherers Tool Kit: a functional perspective*. Newcastle: Cambridge Scholars Press.

Macdonald, D. A., Allentuck, A., & Maher, L. A. (2018). Technological change and economy in the Epipalaeolithic: Assessing the shift from early to middle Epipalaeolithic at Kharaneh IV. *Journal of Field Archaeology, 43,* 437–456.

Maher, L. A. (2010). People and their places at the end of the Pleistocene: Evaluating perspectives on physical and cultural landscape change. In G. Warren & B. Finlayson (Eds.), *Landscapes in transition: Understanding hunter-gatherer and farming landscapes on the early Holocene of Europe and the levant* (pp. 34–44). London: CBRL Monographs.

Maher, L.A (2016). A Road Well-Travelled? Exploring Terminal Pleistocene Hunter-Gatherer Activities, Networks and Mobility in Eastern Jordan. In M. Chazan & K. Lillios (Eds.), *Fresh fields and pastures new: Papers presented in honor of Andrew M.T. Moore* (pp. 49–75). Leiden: Sidestone Press.

Maher, L.A. (2017). Late quaternary refugia: Aggregations and Palaeoenvironments in the Azraq Basin. In O. Bar Yosef & Y. Enzel (Eds.), *Quaternary environments, climate change and humans in the levant* (pp. 679–689). Cambridge: Cambridge University Press.

Maher, L. A. (2018). Persistent place-making in prehistory: The creation, maintenance and transformation of an Epipalaeolithic landscape. *Journal of Archaeological Method and Theory, 25,* 1–86.

Maher, L.A. (In press). Hunter-gatherer home-making? building landscape and community in the Epipalaeolithic. In I. Hodder (Ed.), *Consciousness and creative at the dawn of settled life: The test-case of Çatalhöyük*. Cambridge: Cambridge University Press.

Maher, L. A., & Conkey, M. (2019). Homes for hunters? Exploring the concept of home at hunter-gatherer sites in upper palaeolithic Europe and Epipalaeolithic Southwest Asia. *Current Anthropology, 60,* 91–137.

Maher, L.A., & Macdonald, D. (2013). Assessing typo-technological variability in Epipalaeolithic assemblages: Preliminary results from two case studies from the Southern Levant. In F. Borrell, M. Molist & J.J. Ibanez (Eds.), *The state of stone: terminologies, continuities and contexts in Near Eastern Lithics. Studies in early near eastern production, subsistence and environment 14* (pp. 29–44). Berlin: Ex oriente.

Maher, L. A., Macdonald, D. A., Allentuck, A., Martin, L., Spyrou, A., & Jones, M. D. (2016). Occupying wide open spaces? Late Pleistocene hunter–gatherer activities in the Eastern Levant. *Quaternary International, 396,* 79–94.

Maher, L.A., & Richter, T. (2011). PPN predecessors: current issues in Late Pleistocene chipped stone analyses in the southern Levant. In E. Healey, S. Campbell & O. Maeda (Eds.), *The state of stone: Terminologies, continuities and contexts in near Eastern Lithics*, vol. 13, *Studies in early near Eastern Production, Subsistence, and environment* (pp. 25–31). Berlin: Ex oriente.

Maher, L. A., Richter, T., Macdonald, D., Jones, M., Martin, L., & Stock, J. T. (2012). Twenty thousand-year-old huts at a hunter-gatherer settlement in Eastern Jordan. *PLoS ONE, 7,* e31447.

Maher, L. A., Stock, J. T., Finney, S., Heywood, J. J. N., Miracle, P., & Banning, E. B. (2011). A unique human-fox burial from a pre-natufian cemetery in the Southern Levant (Jordan). *PLoS ONE, 6,* 1–10.

Martin, L., Edwards, Y., & Garrard, A. (2010). Hunting practices at an Eastern Jordanian Epipalaeolithic aggregation site: The case of Kharaneh IV. *Levant, 52,* 107–135.

Martin, L., Edwards, Y. H., Roe, J., & Garrard, A. (2016). Faunal turnover in the Azraq Basin, eastern Jordan 28,000 to 9000 cal yr BP, signalling climate change and human impact. *Quaternary Research, 86,* 200–219.

Mauss, M. (1936). *Sociologie et anthropologie*. Paris: Presses Universitaires France.

McClure, S. B. (2007). Gender, technology, and evolution: cultural inheritance theory and prehistoric potters in Valencia, Spain. *American Antiquity, 72,* 485–508.

Miller, H. (2007). *Archaeological approaches to technology*. New York: Elsevier Academic Press.

Milne, S. B., & Wendrich, W. (2013). Lithic raw material availability and Palaeo-Eskimo novice flintknapping. In W. Wendrich (Ed.), *Archaeology and apprenticeship: Body knowledge, identity, and communities of practice* (pp. 119–144). Tucson: University of Arizona Press.

Minar, C. J., & Crown, P. L. (2001). Learning and craft production: An introduction. *Journal of Anthropological Research, 57,* 369–380.

Moorey, P.R.S. (1999). *Ancient Mesopotamian materials and industries: The archaeological evidence*. Eisenbrauns.

Muheisen, M. (1983). La Préhistoire en Jordanie. Recherches sur l'Epipaléolithique. L'Example du Gisement de Kharaneh IV, l'Université de Bordeaux I.

Muheisen, M. (1988a). The Epipalaeolithic phases of Kharaneh IV. In A. Garrard & H. Gebel (Eds.), *The prehistory of Jordan. The state of research in 1986, BAR international series 396* (pp. 353–367). Oxford: British Archaeological Reports 396.

Muheisen, M. (1988b). Le Gisement de Kharaneh IV, Note Summaire Sur la Phase D. *Paléorient, 14,* 265–269.

Muheisen, M., & Wada, H. (1995). An analysis of the Microliths at Kharaneh IV Phase D, Square A20/37. *Paléorient, 21,* 75–95.

Munro, N., & Grosman, L. (2010). Early evidence (ca. 12,000 B.P.) for feasting at a burial cave in Israel. *Proceedings of the National Academy of Sciences of the United States of America, 107,* 15362–15366.

Nadel, D. (2000). Brush hut floors, hearths and flints: The Ohalo II case study (19 Ka, Jordan Valley, Israel). *Journal of Human Evolution, 38,* A22–A23.

Nadel, D., Danin, A., Power, R. C., Rosen, A. M., Bocquentin, F., Tsatskin, A., et al. (2013). Earliest floral grave lining from 13,700–11,700-y-old Natufian burials at Raqefet Cave, Mt. Carmel, Israel. *Proceedings of the National Academy of Sciences, 110,* 11774–11778.

Nadel, D., Weiss, E., Simchoni, O., Tsatskin, A., Danin, A., & Kislev, M. (2004). Stone age hut in Israel yields world's oldest evidence of bedding. *Proceedings of the National Academy of Sciences of the United States of America, 101,* 6821–6826.

Nadel, D., Weiss, E., & Tschauner, H. (2011). Gender-specific division of indoor space during the upper Palaeolithic?: A brush hut floor as a case study. In S. Gaudzinski-Windheuser, O. Jöris, M. Sensburg, M. Street, & E. Turner (Eds.), *Site-internal spatial organization of hunter-gatherer societies: Case studies from the European Palaeolithic and Mesolithic* (pp. 263–273). Mainz: Verlag des Römisch--Germanischen Zentralmuseums.

Odell, G.H. (2004). *Lithic analysis. Manuals in archaeological method, theory and technique*. New York: Kluwer Academic/Plenum.

Olszewski, D. (2001). My "backed and truncated bladelet", your "point": terminology and interpretation in Levantine Epipalaeolithic assemblages. In I. Caneva, C. Lemorini, D. Zampetti & P. Biagi (Eds.), *Beyond tools: Redefining the PPN Lithic Assemblages of the Levant. Third Workshop on PPN Chipped Lithic Industries, studies in early Near Eastern Production, Subsistence and Environment 9* (pp. 303–318). Ca'Foscari University of Venice: Ex oriente.

Olszewski, D. (2006). Issues in the Levantine Epipaleolithic: The Madamaghan, Nebekian and Qalkhan (Levant Epipaleolithic). *Paléorient, 32,* 19–26.

Olszewski, D. I. (2004). The conundrum of the Levantine late upper Palaeolithic and early Epipalaeolithic: perspectives from the Wadi al-Hasa, Jordan. In A. N. Goring-Morris & A. Belfer-Cohen (Eds.), *More than meets they eye: studies on upper Palaeolithic diversity in the Near East* (pp. 151–170). Oxford: Oxbow Books.

Olszewski, D. I., & al-Nahar, M. (2016). Persistent and ephemeral places in the Early Epipaleolithic in the Wadi al-Hasa region of the western highlands of Jordan. *Quaternary International, 396,* 20–30.

Olzsewski, D. (2011). Lithic "culture" issues: insights from the Wadi al-Hasa Epipalaeolithic. In S.C.O.M. E. Healey (Eds.), *The state of the stone: Terminologies, Continuities and Contexts in Near Eastern Lithics, studies in early near eastern production, subsistence, and environment 13* (pp. 51–65). Berlin: Ex oriente.

Ortner, S. B. (2006). *Anthropology and social theory: Culture, power, and the acting subject*. London: Duke University Press.

Pelegrin, J. (1990). Prehistoric lithic technology: Some aspects of research. *Archaeological Review from Cambridge, 9,* 116–125.

Pelegrin, J., Karlin, C., & Bodu, P. (1988). "Chaînes opératoires": un outil pour le préhistorien. In J. Tixier (Ed.), *Journée d'Études Technologiques en Préhistoire, Notes et monographies techniques No. 25* (pp. 55–62). Paris: CNRS.

Perlés, C., & Phillips, J. (1991). The Natufian Conference-discussion. In O. Bar-Yosef, F. Valla (Eds.), The Natufian Culture in the Levant. International Monographs in Prehistory, Ann Arbor, 637–644.

Pfaffenberger, B. (1988). Fetishised objects and humanised nature: towards an anthropology of technology. *Man, 23,* 236–252.

Pfaffenberger, B. (1992). Social anthropology of technology. *Annual Review of Anthropology, 21,* 491–516.

Pfaffenberger, B. (2001). Symbols do not create meanings—Activities do: or, why symbolic anthropology needs the anthropology of technology. In M. Schiffer (Ed.), *Anthropological perspectives on technology* (pp. 77–86). University of New Mexico Press: Albuquerque.

Pigeot, N. (1990). Technical and social actors: flintknapping specialists and apprentices at Magdalenian Etiolles. *Archaeological Review from Cambridge, 9,* 126–141.

Pirie, A. (2004). Constructing prehistory: Lithic analysis in the Levantine Epipalaeolithic. *Journal of the Royal Anthropological Institute, 10,* 675–703.

Ramsey, M. N., Maher, L. A., Macdonald, D. A., Nadel, D., & Rosen, A. M. (2018). Sheltered by reeds and settled on sedges: Construction and use of a twenty thousand-year-old hut according to phytolith analysis from Kharaneh IV, Jordan. *Journal of Anthropological Archaeology, 50,* 85–97.

Ramsey, M. N., Maher, L. A., Macdonald, D. A., & Rosen, A. (2016). Risk, reliability and resilience: Phytolith evidence for alternative 'Neolithization' pathways at Kharaneh IV in the Azraq Basin. *Jordan. PLoS ONE, 11,* e0164081.

Ramsey, M. N., & Rosen, A. M. (2016). Wedded to wetlands: Exploring late Pleistocene plant-use in the eastern Levant. *Quaternary International, 396,* 5–19.

Richter, R. (2011). Nebekian, Qalkhan and Kebaran: Variability, classification and interaction. New insights from the Azraq Oasis. In E. Healey, S. Campbell & O. Maeda (Eds.), *The State of the Stone Terminologies, Continuities and Contexts in Near Eastern Lithics,* vol. 13, *Studies in Early Near Eastern Production, Subsistence, and Environment (SENEPSE)* (pp. 33–49). Berlin: ex oriente.

Richter, T. (2007). A Comparative Use-Wear Analysis of late Epipalaeolithic (Natufian) chipped stone artefacts from the southern Levant. *Levant, 39,* 97–122.

Richter, T. (2014). Margin or centre? The Epipalaeolithic in the Azraq Oasis and the Qa'Shubayqa. In B. Finlayson & C. Makarewicz (Eds.), *Settlement, survey and stone: essays on near Eastern Prehistory in Honour of Gary Rollefson* (pp. 27–36). Berlin: Ex Oriente.

Richter, T., Garrard, A., Allcock, S., & Maher, L. (2011). Interaction before agriculture: Exchanging material and shared knowledge in the final Pleistocene Levant. *Cambridge Archaeological Journal, 21,* 95–114.

Richter, T., & Maher, L. (2013a). The Natufian of the Azraq Basin: An Appraisal. In O. Bar-Yosef & F. Valla (Eds.), *Natufian Foragers in the Levant: Terminal Pleistocene social changes in Western Asia* (pp. 429–448). Ann Arbor: International Monographs. in Prehistory.

Richter, T., & Maher, L. (2013b). Terminology, process and change: Reflections on the Epipalaeolithic of Southwest Asia. *Levant, 45,* 121–132.

Richter, T., Maher, L. A., Garrard, A. N., Edinborough, K., Jones, M. D., & Stock, J. T. (2013). Epipalaeolithic settlement dynamics in southwest Asia: New radiocarbon evidence from the Azraq Basin. *Journal of Quaternary Science, 28,* 467–479.

Rockman, M. (2013). Apprentice to the environment: Hunter-gatherers and landscape learning. In W. Wendrich (Ed.), *Archaeology and apprenticeship: Body knowledge, identity, and communities of practice* (pp. 99–118). Tucson: University of Arizona.

Sassaman, K. E., & Rudolphi, W. (2001). Communities of practice in the early pottery traditions of the American southeast. *Journal of Anthropological Research, 57,* 407–425.

Schiffer, M. (2011). *Studying technological change: A behavioral approach.* Salt Lake City: University of Utah Press.

Schiffer, M.B. (2001). *Anthropological perspectives on technology.* UNM Press.

Schiffer, M. B. (2003). Style and function: Conceptual issues in evolutionary archaeology. *American Anthropologist, 105,* 428–429.

Schiffer, M. B. (2004). Studying technological change: A behavioral perspective. *World Archaeology, 36,* 579–585.

Schiffer, M. B., Skibo, J. M., Griffitts, J. L., Hollenback, K. L., & Longacre, W. A. (2001). Behavioral archaeology and the study of technology. *American Antiquity, 66,* 729–737.

Schlanger, N. (1994). Mindful technology: Unleashing the chaîne opératoire for an archaeology of mind. In C. Renfrew & E. B. W. Zubrow (Eds.), *The ancient mind: Elements of cognitive archaeology* (pp. 143–151). Cambridge: Cambridge University Press.

Sellet, F. (1993). Chaîne opératoire: The concept and its applications. *Lithic Technology, 18,* 106–112.

Shennan, S. (2020). Style, function and cultural transmission. In H. Groucutt (Ed.), *Culture history and convergent evolution: Can we detect populations in prehistory?* (pp. 291–298). Springer, Cham, Switzerland.

Shipton, C. (2020). The unity of acheulean culture. In H. Groucutt (Ed.), *Culture history and convergent evolution: Can we detect populations in prehistory?* (pp. 13–27). Springer, Cham, Switzerland.

Shott, M.J. (2020). Toward a theory of the point. In H. Groucutt (Ed.), *Culture history and convergent evolution: Can we detect populations in prehistory?* (pp. 245–259). Springer, Cham, Switzerland.

Sillar, B., & Tite, M. S. (2000). The challenge of 'technological choices' for materials science approaches in archaeology. *Archaeometry, 42,* 2–20.

Skibo, J., & Schiffer, M. (2008). *People and things: A behavioral approach to material culture.* New York: Springer.

Snir, A., Nadel, D., Groman-Yaroslavski, I., Melamed, Y., Sternberg, M., Bar-Yosef, O., et al. (2015). The origin of cultivation and proto-weeds, Long Before Neolithic Farming. *PLoS ONE, 10,* e0131422.

Soressi, M., & Geneste, J. M. (2011). The history and efficacy of the Chaîne Opératoire approach to Lithic analysis: Studying techniques to reveal past societies in an evolutionary perspective. *Palaeoanthropology, 2011,* 334–350.

Spyrou, A. (In review). Meat outside the freezer: Drying, smoking and sealing meat in fat in an Epipalaeolithic Megasite in Eastern Jordan. *Journal of Anthropological Archaeology.*

Stark, M. T. (1998). Technical choices and social boundaries in material culture patterning: an introduction. In M. T. Stark (Ed.), *The archaeology of social boundaries* (pp. 1–11). Smithsonian Institution Press: Washington D.C.

Suchman, L. (2001). Building bridges: Practice-based ethnographies of contemporary technology. In. M. Schiffer (ed.) *Anthropological perspectives on technology* (pp. 164–177). Albuquerque: University of New Mexico Press.

Suchman, L., Blomberg, J., & Orr, J. (1999). Reconstructing Technologies as Social Practice. *The American Behavioral Scientist, 43,* 392–408.

Surovell, T. A. (2012). *Toward a behavioral ecology of lithic technology: Cases from Paleoindian archaeology.* Tucson: University of Arizona Press.

Texier, P.-J., & Meignen, L. (2012). Soixante années de technologie lithique: étapes marquantes, apports et écueils. In F. Delpech & J. Jaubert (Eds.), *François Bordes et la préhistoire* (pp. 133–139). Paris: Éditions du C.T.H.S.

Tixier, J. (1963). Typologie De L'Épipaléolithique Du Maghreb. *Arts et Métiers Graphiques, Paris.*

Tixier, J. (1974). *Glossary for the description of stone tools: With special reference to the Epipalaeolithic of the Maghreb. Newsletter of lithic technology: special publication; no. 1.* Pullman: Newsletter of Lithic Technology Washington State University.

Tixier, J. (1979). *Préhistoire et technologie lithique. Journées du 11–13 mai 1979. Centre de Recherche Archéologique du CNRS, publications de l'URA 28, cahiers n°1.* Paris: Centre régional de publication de Sophia Antipolis.

Tixier, J., Inizan, M.-L., & Roche, H. (1980). *Préhistoire de la Pierre Taillée 1: Terminologie et Technologie.* Valbonne: Cercle de Recherches et d'Etudes Préhistoriques.

Tostevin, G.B. (2007). Social intimacy, artefact visibility, and acculturation models of Neanderthal-modern human interaction. In P. Mellars, K. Boyle, O. Bar-Yosef, C. Stringer (eds.),' *Rethinking the human revolution (pp.* 341–358). Cambridge, McDonald Institute for Archaeological Research.

Tostevin, G. B. (2013). *Seeing lithics: A middle-range theory for testing for cultural transmission in the Pleistocene.* Oakville, CT: Oxbow Books.

Valentin, B. (2008). *Jalons pour une paléohistoire des derniers chasseurs, XIVe-VIe millénaire avant J.-C.* Vol. 1: Publications de la Sorbonne.

Wallaert-Pêtre, H. (2001). Learning how to make the right pots: Apprenticeship strategies and material culture, a case study in handmade pottery from Cameroon. *Journal of Anthropological Research, 57,* 471–493.

Watkins, T. (2010). Changing people, changing environments: How hunter-gatherers became communities that changed the world. In B. Finlayson & G. Warren (Eds.), *Landscapes in transition* (pp. 106–114). London: Oxbow.

Watkins, T. (2011). Opening the door, pointing the way. *Paleorient, 37,* 29–38.

Watkins, T. (2013). The Neolithic in transition-How to complete a paradigm shift. *Levant, 45,* 149–158.

Weissbrod, L., Marshall, F. B., Valla, F. R., Khalaily, H., Bar-Oz, G., Auffray, J.-C., et al. (2017). Origins of house mice in ecological niches created by settled hunter-gatherers in the Levant 15,000 y ago. *Proceedings of the National Academy of Sciences, 114,* 4099–4104.

Wendrich, W. (Ed.). (2013a). *Archaeology and apprenticeship: Body knowledge, identity, and communities of practice.* Tucson: University of Arizona Press

Wendrich, W. (2013b). Archaeology and apprenticeship: Body knowledge, identity, and communities of practice. In W. Wendrich (Ed.), *Archaeology and apprenticeship: Body knowledge, identity, and communities of practice* (pp. 1–19). Tucson: University of Arizona Press

Wendrich, W. (2013c). Recognizing Knowledge Transfer in the Archaeological Record. In W. Wendrich (Ed.), *Archaeology and apprenticeship: Body knowledge, identity, and communities of practice* (pp. 255–262). Tucson: University of Arizona Press.

Wenger, E. (1998). *Communities of practice: Learning, meaning, and identity.* Cambridge: Cambridge University Press.

Wiessner, P. (1984). Reconsidering the behavioral basis for style: A case study among the Kalahari San. *Journal of Anthropological Archaeology, 3,* 190–234.

Wilke, P., & Quintero, L.A. (1994). Naviform core-and-blade technology: Assemblage Character as determined by replicative experiments. In H.G. Gebel (Ed.), *Neolithic chipped stone entities of the fertile crescent, studies in early near eastern production, subsistence, and environment 1* (pp. 33–60). Berlin: Ex oriente.

Will, M., Mackay, A. (this volume). A matter of time and space: How frequent is convergence in lithic technology in the African archaeological record over the last 300 kyr? In H. Groucutt (Ed.), *Culture history and convergent evolution: can we detect populations in prehistory?* (pp. 103–126). Springer, Cham, Switzerland.

Yaroshevich, A., Kaufman, D., Nuzhnyy, D., Bar-Yosef, O., & Weinstein-Evron, M. (2010). Design and performance of microlith implemented projectiles during the middle and the late Epipaleolithic of the Levant: Experimental and archaeological evidence. *Journal of Archaeological Science, 37,* 368–388.

Yaroshevich, A., Nadel, D., & Tsatskin, A. (2013). Composite projectiles and hafting technologies at Ohalo II (23 ka, Israel): analyses of impact fractures, morphometric characteristics and adhesive remains on microlithic tools. *Journal of Archaeological Science, 40,* 4009–4023.

Chapter 12
Toward a Theory of the Point

Michael J. Shott

Abstract Points were the tips of prehistoric weapons like darts and arrows. Bifacially chipped stone points all have sharp tips but vary greatly in the size and form of the bases that secured them to the shafts or foreshafts of larger composite tools. Especially in the Americas prehistoric stone points are superabundant, their near-endless forms most beautiful and wonderful. Archaeologists exploit this diversity to order past time, adapting prehistoric tools as chronometric ones. Traditional analysis emphasized type definition among the variation in points, then toolstone acquisition, function and use-wear, distinctions between darts and arrows, and pattern and degree of resharpening. Yet we still treat points and their types as tools that define segments of past time, and merely describe historical changes from one type to another. We also should treat types as subjects of analysis, prompting questions not ordinarily asked. How and where on points do history and timeless function register, and do they compete? How are valid types identified and distinguished? When and why do types originate and end, and how long does either take? What explains the duration and relative popularity of types, and the number present over the duration of a period? How do new types diversify from existing ones? How are historical continuity or discontinuity identified in point sequences? For their prehistoric users points were tools, trivially. For archaeologists point types are tools, trivially. But point types also are subjects, nontrivially, for and about whom we must develop the theory that can explain their origin, development, and ultimate fate.

Keywords Archaeological theory • Function • Paleobiology • Phylogeny • Reduction

M. J. Shott (✉)
Department of Anthropology, University of Akron, Akron, OH 44325, USA
e-mail: shott@uakron.edu

"Archaeology is an undisciplined empirical discipline. A discipline lacking a scheme of systematic and ordered study based upon declared and clearly defined models and rules of procedure. It further lacks a body of central theory capable of synthesizing the general regularities within its data in such a way that the unique residuals distinguishing each particular case might be quickly isolated and easily assessed. Archaeologists do not agree upon central theory, although, regardless of place, period, and culture, they employ similar tacit models and procedures based upon similar and distinctive entities—the attributes, artefacts, types, assemblages, cultures and culture groups. Lacking an explicit theory defining these entities and their relationships and transformations in a viable form, archaeology has remained an intuitive skill—an inexplicit manipulative dexterity learned by rote." (D. L. Clarke, Analytical Archaeology, 1978: xv).

Pardon the extended quotation that precedes this essay. It concerns a problem identified decades ago but substantially ignored ever since, to archaeology's detriment. Clarke's neglected book assayed a comprehensive reformation of archaeological thought and practice. Even in 500+ pages, the effort was ambitious. This essay is not nearly so ambitious, but attempts to follow Clarke's lead in one small respect.

Projectile points are stone tools made at once to create a sharp tip with expanding margins for penetration of prey targets and to connect with the larger armature, shaft or foreshaft, used to deliver them to the target. Not all points are bifaces and not all bifaces are points (see Douze et al. 2020); my subject is bifacial points, "points" henceforth. Points are common enough worldwide; in North America, they occur in numbers almost beyond belief if not counting. In 1859, long before most points had been found, Henry Thoreau could write "some time or other…it had rained arrowheads, for they lie all over the surface of America" (Bode, ed. 1967: 289–290). Nearly a half-century later, Wilson (1899) provided brief glimpses of that abundance, which over a century more of subsequent collecting has only increased.

To some extent, archaeology's treatment of points traces major advances in its intellectual development. Originally

H. S. Groucutt (ed.), *Culture History and Convergent Evolution: Can We Detect Populations in Prehistory?*,
Vertebrate Paleobiology and Paleoanthropology, https://doi.org/10.1007/978-3-030-46126-3_12

points merely were evidence of human presence and undetermined antiquity. Later, as the time-space distribution of particular types emerged, points served as markers of past time and sometimes cultural affinity. Later still, they served as evidence of behavior, usually hunting. More recently, use-wear studies identified a range of specific uses, and degree and pattern of reduction recruited points to emerging theoretical issues like curation rates and their explanation (e.g. Andrefsky 2006; Shott and Ballenger 2007). Separately or together these uses are valid, but do not nearly exhaust points' potential to reveal the cultural past. Yet that fuller potential requires and promotes a reform of archaeological thought along the lines that Clarke sketched, and for which points may be especially suited.

To justify that reform, pardon a necessary digression. Paleobiology circa 1980 forms a crude analogy to archaeology's current dilemma and the prospect that it confronts. Then, paleobiology was a mere adjunct to biology, manufacturing inadequate approximations to the latter's units, imitating its theory and exemplifying its processes. This "passive transference from microevolutionary studies" (Gould 1980: 98) condemned paleobiology to intellectual subservience within the larger field, where it defined the wrong units at the wrong scales that it sought to explain using the wrong theory. The species is a fundamental biological unit, but in synchronic behavior ecology it has only descriptive value; the relevant unit of observation and analysis is the individual, whose anatomy and behavior are governed by microevolutionary adaptation and selection. In diachronic paleobiology, however, species are units of observation and analysis, whose form and behavior but also duration, abundance, origin and fate are governed not by adaptation at the individual level but modes and tempos of change at higher, derived ones. Over long biological time, individuals adapt but only species evolve, and their differential persistence, survival and diversification is explained by inherently paleobiological theory that biology cannot entail. Paleobiology's florescence in the past 40 years amply confirms its "bounded independence" (Gould 1980: 107) as a macroevolutionary field.

Today as then, American archaeology is subsumed beneath anthropology. Its units of observation and especially inference—cultures—and its equally synchronic functional or interpretive theory essentially are anthropological in scale. Contemporary archaeology is characterized by rampant passive transference, and an all-purpose misapprehension of its suitable units, scales and explanations (Perreault 2019). The anthropological model that archaeology adopts treats cultures as integral wholes (yet they are historically contingent types, as Reynolds [2020] notes for broad Paleolithic equivalents), derived from the presence or proportion of artifact and other (e.g. faunal, feature) assemblages. That is, we draw many inferences from many sources of evidence, then bundle together the results as integral "Culture X," which then becomes our unit of analysis. At once, we try to characterize such units by their population size, diet and economy, sociopolitical organization, and systems of meaning. Cultures then change by the appearance or disappearance of artifact types or by dramatic shifts in proportions of artifacts and other materials.

Practically, the way we define archaeological cultures and track their fortunes across time and space denies us a more modular view of our subject—components or units, like point types, which may encompass several cultures at once or persist longer than any single culture—especially when cultures are identified with point or other types assumed not to change over time, merely to come and go in ways and at rates unfathomable. Given the typically coarse time resolution that this practice entails, moreover, we cannot resolve the rate, pattern and direction of culture change as experienced on ethnographic time scales. Over the long periods that we study, this limitation gives our descriptions and explanations of sequences of change a misleading episodic quality noted before (e.g. Frankel 1988; Shott 2003; see Stutz [2020] for a similar view of the Middle-Upper Paleolithic transition).

Like paleobiology compared to biology, archaeology needs methods for the construction and theory for the explanation of the inherently historical (therefore *not* anthropological) units and the phase- and time-pattern regularities (Clarke 1978: 163 and passim) they exhibit. This is a grand task that even Clarke merely foreshadowed and that only recently was taken up again (e.g. Perreault 2019). In Paleolithic contexts, relevant research evaluates rather than assumes the historical behavior of presumed historical units (e.g. Groucutt and Scerri 2014; Monnier and Missal 2014; Reynolds 2020). This essay's far more modest scope concerns points alone. The reorganized thought and practice may require more data, certainly new kinds of data. Chiefly, however, it requires a new way to view points and to derive from them the historical units whose behavior we can describe and must explain. It requires, that is, its own comprehensive theory, some parts of which already exist but key components of which remain undeveloped. Archaeology needs a theory of the point.

A sufficient theory of the point lies far beyond current reach. A first approximation must encompass everything from the dimensions that characterize points and reveal their design, to their use, to the contribution of that use to larger synchronic cultural units and practices, and finally to inherently historical traditions of manufacture and use. Like Clarke in general, therefore, this approach to points progresses by level, from attribute to object to sets of objects in assemblages and finally to types as historical units.

Typology

Point types and their historical properties are among the subjects of an archaeology reformulated along Clarke's lines. A theory of the point requires methods for defining and distinguishing types of points. Most North American point types are defined subjectively by combinations of size, technology but especially outline form, and are distinguished from one another in part by emphasizing modal characteristics thought to differ among them. American archaeology lost its typological innocence by the 1950s Spaulding-Ford debate. At least with respect to points, however, it continued to profess innocence decades later, in the process treating types as revealed kinds rather than constructed units (e.g. Justice 1987). Attempts at typological rigor (e.g. Read 1982) emphasized outline form and neglected the effect of reduction, not original design, upon aspects of that form (Hoffman 1985), as did even more recent typologies constrained by the data requirements of methods like cladistics and innocent of the allometric effects of resharpening. Hoffman (1985) considered what many archaeologists called distinct types as variants of a single original form defined by different degrees and patterns of resharpening (Fig. 12.1). Reduction-sensitive variables are "not appropriate for conducting evolutionary analyses with techniques derived from cladistics" (Goodale et al. 2015: 241; see also Lipo 2006: 106; White 2013: 98; Barrientos 2015: 55; Prentiss et al. 2016: 127). The most systematic approach is a replicable key but it remains descriptive and works best only for where it was designed, the Great Basin (Thomas 1981). We have no general method for defining point types nor distinguishing among them. Instead, we treat points as judges once treated pornography, unable to define them but believing that we know them when we see them.

As described and used, subjectively defined point types are a social fact of archaeological practice, routinely cited in the literature from which in part we must identify the properties of better-defined types. We have no choice but to

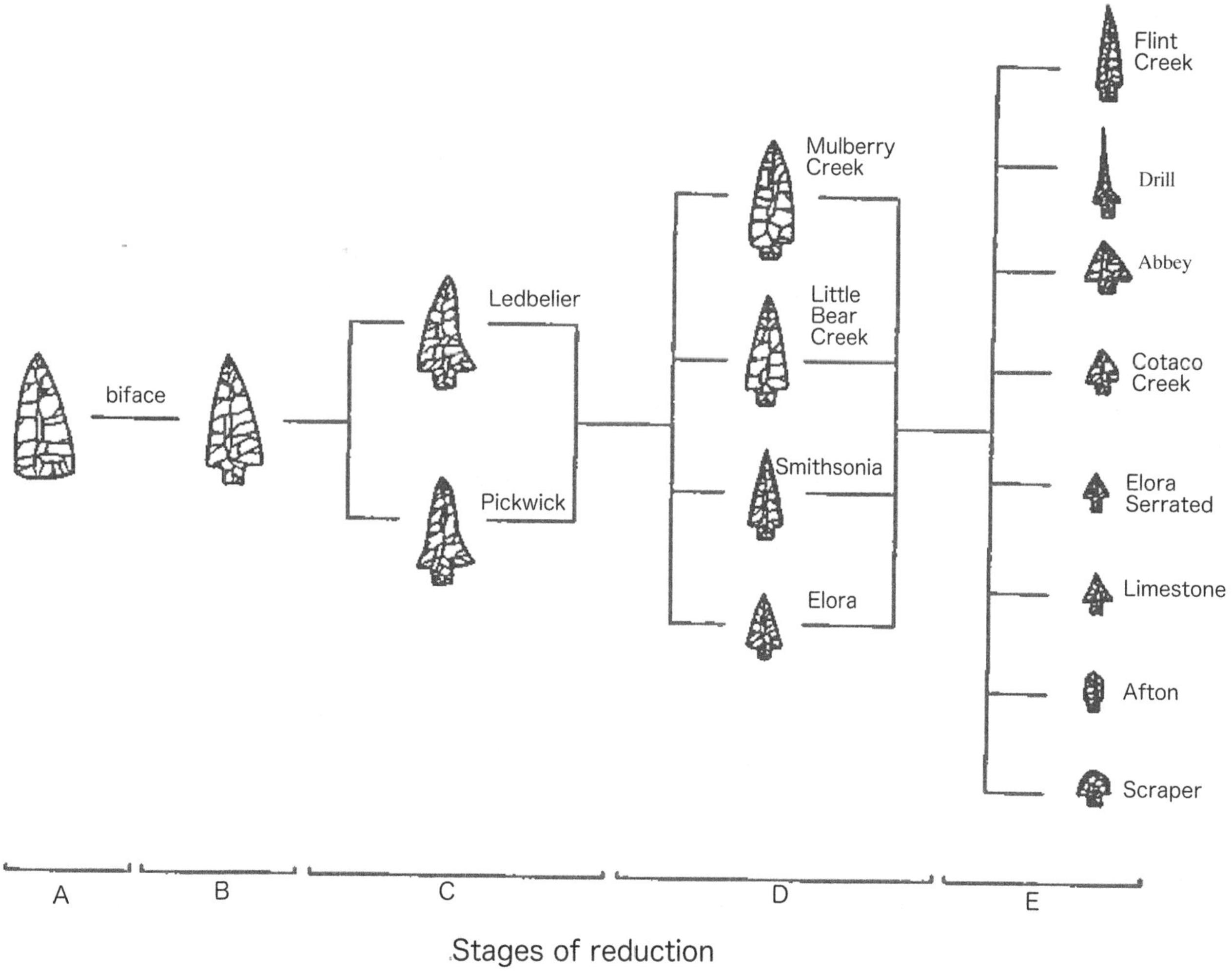

Fig. 12.1 A hypothetical reduction continuum in one point type that links subjectively defined empirical types. *Source* Hoffman (1985: Fig. 18.5)

use them, but only mindful of their serious shortcomings. We need better methods to create types that eventually will replace subjective ones. They must distinguish the most morphologically variable and informative segments of points—their haft elements or stems that articulate with the shaft or foreshaft of the larger weapon of which the point forms a part—from their blades, which are more subject to resharpening effects that compound original design (Hoffman 1985). They must describe in the fullest detail the morphometrics of haft elements. Practically, this requires geometric morphometric (GM) methods applied to two- and three-dimensional (2D and 3D, respectively) models of points (e.g. Thulman 2012), which encode vastly more morphometric information than do orthogonal dimensions; it also enables use of powerful statistical methods not available to conventional dimensional analysis. Better methods also require very large datasets, to fully comprehend types' variation by time, toolstone, curation pattern and rate, and other factors (Barrientos 2015: 56). Whatever types defined will be constructed, not revealed, kinds, units that are constantly arriving but never arrived, "artificial delineations in a continuous evolution of projectile-point morphology over space and time" (Cook and Comstock 2014: 226; see also Bradbury and Carr 2003; Shott 2003), reminiscent of Clarke's (1978: 182) polythetic sets.

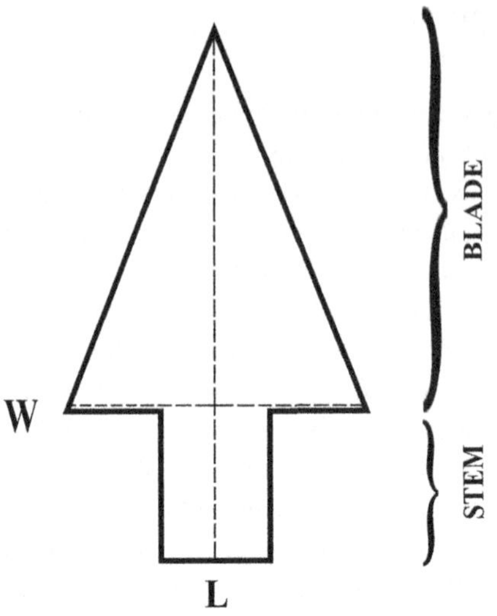

Fig. 12.2 Two-dimensional schematic view of a point, showing separate stem and blade segments and two orthogonal dimensions, length (L) and width (W). Note that values for length, width or other dimensions preserve no information about either their relative positions or overall point shape

Describing Points

Trivially, points are three-dimensional (3D) solids that include at least stem and blade segments (Fig. 12.2). Traditional analysis parses these complex wholes into separate, often orthogonal, dimensions (e.g. length, width, thickness) that measure size and form of the whole or its parts (e.g. blade length, stem width). Dimensions are isolated attributes that lack geometric context within the whole (e.g. a value for width, by itself, says nothing about its value compared to length or thickness, or where along the point's length or thickness profile it was taken), although simple ratios between them or ordination of sets of them can better approximate whole-object form. Still, individual dimensions are as faithful to the fullness of whole-object form as, say, stick-figure drawings are to Leonardo's Vitruvian Man. Other attributes that capture aspects of size (weight, area, perimeter, sectional area [e.g. Hughes 1998: 353]) or form/function [e.g. tip and edge angles]) are less commonly measured, but should be. Besides size and form, points have performance attributes that implicate range, thrust and other ballistic properties but in complex ways not easily derived from individual dimensions (Beck 1998: 24–25; Hughes 1998; Larralde 1990; Ratto 2003: 201–212; Edinborough 2005; Collins 2007; White 2013: 76–77). Points also are durable to varying degree, capable of surviving (or not) more than one firing and impact and large enough to accommodate resharpening (Larralde 1990: 78; Beck 1998; Hughes 1998: 371) experiments suggest much variation in survival depending upon material, targets, weapon systems and other factors (e.g. Odell and Cowan 1986; Cheshier and Kelly 2006; Shott 2016).

Whether used to define types in the first place, size-form variables can track secular trends in sequences of types. For instance, early Holocene midcontinental point sequences can be resolved in part to trends in individual variables that pattern in different directions and rates through time and that implicate complex interactions between weapon technology, hunting methods and environmental structure (White 2013). Viewing subjectively defined normative point types as mere time markers chronicles historical sequences, but detailed attribute studies might identify the underlying causes. Their behavior traced across historical type sequences, attributes also can identify stylistic variation useful for time resolution and functional stasis or change (e.g. Beck 1998; Wilhelmsen and Feathers 2003; Edinborough 2005; Apel and Darmark 2009; White 2013).

Yet recognizing secular trends in types requires controlling for variation by toolstone, and by degree and pattern of use and resharpening. Then, small-scale changes or trends through time within types can be examined, assuming sufficient chronological control. What explains any secular trends identified? Is it toolstone quality or supply? Changing density, body size or behavior of prey species, including people? Social conditions, including high population density and raiding? Changing labor organization of point product or

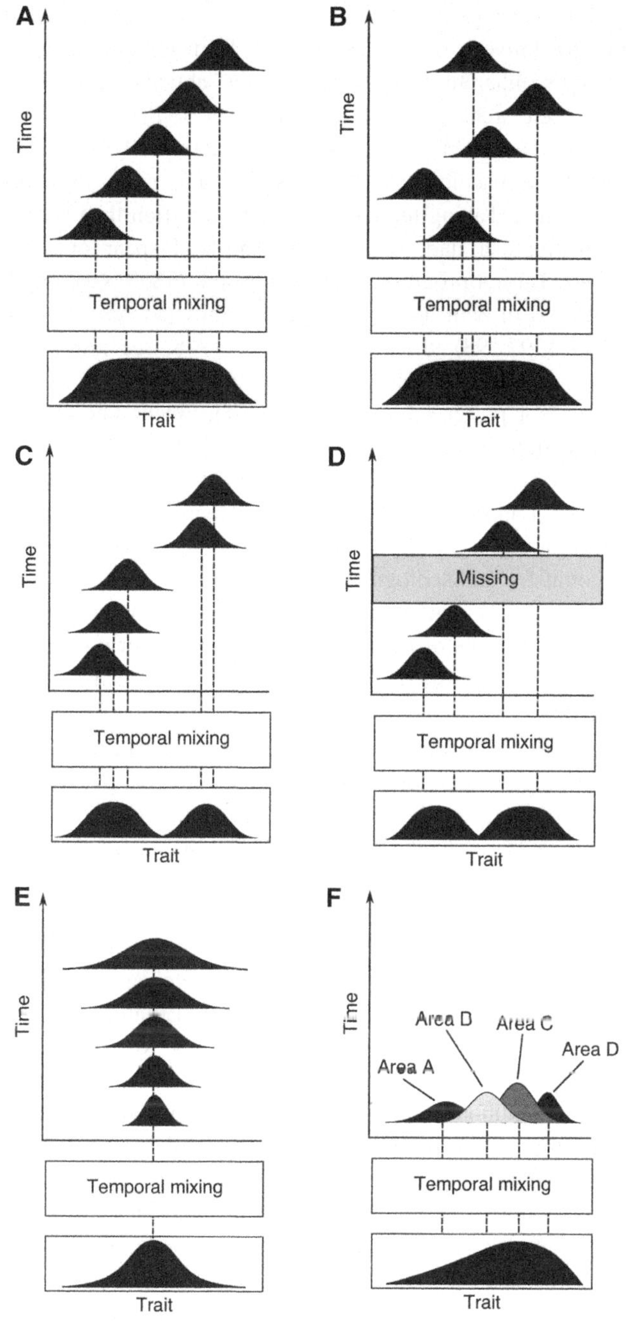

Fig. 12.3 Effects of accumulation ("temporal mixing") upon attribute distributions in time-averaged assemblages. Scenarios **A** and **B** show identical unimodal distributions that result from a steady secular trend (**A**) and fluctuating mean (**B**). Scenarios **C** and **D** show identical bimodal distributions that result from discontinuous trends (**C**) and gaps in accumulation (**D**). *Source* Perreault (2019: Fig. 3.12)

use? Copying error, itself partly a function of population size and number of points produced? Unfortunately, the time scales over which the record accumulates badly compound recognition of secular trends. Trends that vary in rate, magnitude and direction can be swamped or arbitrarily parsed in deposits that themselves vary greatly in scale. The permutations of such pooling effects pose challenges that must be addressed, not ignored (Fig. 12.3).

Moreover, however many individual attributes recorded never will approximate whole-object size and form. What might are GM approaches that characterize 2D and 3D objects by the placement of landmarks (specific functional or morphological coordinate locations like tips, shoulders and base corners) and semi-landmarks. Using landmarks complemented by dense meshes of semi-landmarks, the slope and distance between adjacent locations is reduced to noise, and the configuration of landmarks preserves geometric information and approximates whole-object form. Landmarks can be supplemented with selected attributes (e.g. tip, edge, or notch angles; stem:blade ratio). GM platforms distinguish analyst-defined modules (e.g. stem, blade, tip area) to test for modularity (landmarks in analyst-defined modules covarying significantly more among themselves than with landmarks in other modules), and ordinate morphometric data to characterize ranges of size-shape variation and to measure allometry, none of which is easily accomplished using conventional attribute schemes and manual measurement.

Using GM methods, archaeologists can begin to disentangle complex patterns of variation, for instance in resharpening's allometric effects that alter initial stem-vs.-blade to tip-vs.-rest-of-point modularity (de Azevedo et al. 2014), and measure the complex morphing that describes the transition from one type to another. In this perspective, GM methods and analysis also might explain why individual attributes like base form varied across time both early and late in eastern North American prehistory (e.g. White 2013; Cook and Comstock 2014: 236), perhaps as responses to evolving constraints of shaft width, prey targets or other factors.

Points as Tools

Points are tools, by definition, although what kind of tools they are and whether they are only tools, not also identity-markers, are questions to answer. Use-wear and residues inform on some, certainly last, uses, but treatment here emphasizes function related to morphometrics, mindful of the limits of the approach (Odell 1981). Only extensive experiments, which should be conducted, can identify the ballistic performance requirements of points discussed in the preceding section. Points, especially large ones, may have been designed for use as knives either besides (e.g. Collins 2007: 76–79; Douze et al.

2020) or instead of as one or another type of weapon tip. Discussion here concerns inferring the latter, along with S and L, systemic number and uselife respectively, from Schiffer's (1976: 60) discard equation.

Weapon System

In the limited ethnographic record there are some clear patterns in the use of stone versus other materials for points. Stone is common, but not exclusively so, only for use against larger targets ($\geq$40 kg) (Larralde 1990: 7; Ellis 1997: Tables 1–5). Weapons usually were tipped with organic points for use on smaller targets. If the ethnographic record represents prehistoric practice, then our record of stone points pertains to weapons used only on larger, not all, targets. Yet the further argument (Ellis 1997: 45, 63) that stone point size (or form) is not calibrated to prey size rests on a sample that, despite its considerable size and breadth, is not sufficiently detailed to capture subtle adjustments of point size to prey characteristics—size and others—that may have been salient. The ethnographic sample is dominated by groups which used arrows alone or both arrows and darts or thrusting spears. In these cases, arrows are used commonly against smaller game. That is, weapon system but not point size is calibrated to target size.

Before the adoption of arrows, at least in North America, there was no option to calibrate weapon system to prey size; there were only spears and darts. At that time, the sizes of their points and probably foreshafts and shafts as well were adjusted to some combination of prey size, relative value and hunting method, all within single traditions of point design, manufacture, hafting and use. Similarly, Ellis's comparatively narrow boundary conditions of dart-point use—hafting on light shafts and firing over long distances on open ground (Larralde 1990: 75; Ellis 1997: Fig. 12.2)—may reflect the more circumscribed parameters of dart use when arrows also are available. Detailed studies of arrow design and use (admittedly not of stone) document the careful adjustment of point size to hunting method-, range, and possibly game targets (e.g. Watanabe 1975: 68; Estioko-Griffin 1984: 83); when darts were the sole or chief option, it is not unreasonable to expect that hunters would make similar adjustments of point size and form to relevant considerations like prey size. It is simplistic to suppose a close correlation between point size and prey size, yet declining size of deer, for instance, may help explain declining size of late prehistoric arrow points, along with changes in diet breadth and ecosystem structure (Cook and Comstock 2014: 245).

Either we assume that all stone points were spear or dart tips (or knives), or we reason, both from the ethnographic record's inherent limitations and the abundance of both preserved archaeological (e.g. Thomas 1978) and ethnographic specimens (e.g. Fowler and Matley 1979: 64–66), that the archaeological record also contains many stone arrow points. Taking the latter view, the question then is how to distinguish dart, arrow and hand-held spear or other points. Efforts progressed from Wilson's (1899: 69) length threshold of 3 in, to other simple metrics (sources cited in Shott 1997: 98), then to Thomas's (1978; see also Shott 1997) discriminant analysis of sets of attributes of ethnographic or preserved archaeological specimens known, not assumed, to be dart or arrow tip (see also Ratto 2003: 214–219 for methods that distinguish arrow from hand-held spear points). Latterly there has been a reversion to simple measures based on equally simple assumptions, but they do not account for archaeological data as do multivariate methods (Walde 2014). The next logical step is GM analysis of 3D models of stems of *known* arrow, dart and other points that can be distinguished, for instance, by canonical variates analysis.

Discriminant functions were tested on independent data in original studies (e.g. Shott 1997: 95). Recent discoveries of preserved organic weapon parts that establish either dart or arrow status (e.g. Hare et al. 2012) enable further tests. Together, these data and methods largely confirm Blitz's (1988) scenario of the bow-and-arrow's historical diffusion southward in the first millennium CE. Yet across North America, their application also suggests concurrent use of dart and arrow for significant periods (e.g. Erwin et al. 2005; Rasic and Slobodina 2008; Morrissey 2009; Rorabaugh and Fulkerson 2015; see Dev and Riede 2012 for a European example, although not involving bifacial points), and darts persisting to European invasion in places (Walde 2014: 156). Analysis of radiocarbon data from preserved organic segments found recently in wasting glaciers indicates nearly 200 years of dart-arrow overlap in the Subarctic (Grund and Huzurbazar 2018).

Whether as dart or arrow tip, Cardillo et al. (2016: 49–50) sought but did not find correlations between point form and environmental variables, although they did not control for resharpening effects upon the point-shape axis. Any such correlations that may exist should account for lag effects—possibly centuries in length—between environmental trends and human responses (e.g. Kelly et al. 2013). Fiedel (2014: 88), for instance, timed the apparently abrupt spread of bifurcate-base points to a period about 200 years after an early Holocene environmental shift; the hypothesis is worth testing if used with methods to measure degree of historical continuity between types. As interesting as it would be to correlate change in point size or shape with environmental trends, it would be at least as interesting to document type stasis when environments change. In such cases the question becomes, why do types *not* change as environment does?

Systemic Number S and Uselife L

Beyond noting differences in the number of points in arrow versus dart caches or preserved quivers, one admittedly limited way to test for differences in arrow and dart S is to compare their frequencies per unit time in contexts like wasting glaciers, where preserved organic components make identification reliable and direct dating possible. In sources consulted (Dixon et al. 2005: Table 1; VanderHoek et al. 2007: Table 1; Andrews et al. 2012: Table 1; Hare et al. 2012: Table 6; Lee 2012: Table 1; Lee and Puseman 2017), very few specimens preserved stone points in direct association; anyway, most arrows had organic points. Accordingly, only identifiable arrow shafts or dart shafts/foreshafts were counted; atlatls, bows and other miscellaneous objects were excluded. Obviously, resulting counts are of shafts, not points, so do not directly measure the relative frequency of *points* used per unit time; also obviously, results are limited to high-elevation contexts, assume constant population size or at least hunting rate, and dates are uncalibrated. This is a very coarse estimate, but in consulted sources, 53 darts span a range of 9230–1250 rcybp, or 7,980 radiocarbon years, for a mean figure of 6.6 darts per 1000 years. Excluding two possible specimens that exceed all other arrows by nearly two millennia, 33 arrows span a range of 1710 to 60 rcybp, or 1,650 radiocarbon years, for a mean of 20 arrows per 1000 years. Crudely, arrows occur at three times the rate that darts do, suggesting that their systemic frequency S may have been three times as high. Also, arrows often were made and probably carried in substantially higher numbers than darts were; thus, arrow S probably was higher (Larralde 1990: 177). For the Pawnee, 20–40 arrows was the norm per hunter (Weltfish 1977: 138); burial caches of arrow points sometimes fall in this range (e.g. Wright 2003: 86). If arrows less often than darts were tipped in stone (Ellis 1997), then the notable abundance of stone arrow points *underestimates* the true frequency of arrows and also the difference in S between stone dart and arrow points.

L also is a performance attribute of points. It is futile to try to estimate L in units of time because points were used episodically, not constantly. But if their original size can be estimated (e.g. from cache data) then the reduced size and altered form of points as discarded register degree and pattern of reduction. If a point's utility is measured by the amount of reduction it accommodates as it is used, this in turn relates to its curation (e.g. Shott 1996; Shott and Ballenger 2007), a quantity relevant both to the accumulation rate of point assemblages and to theories of technological organization (Shott 2017) and evolution (e.g. Ugan et al. 2003; Surovell 2009). There is little doubt that some types were subject to extensive repair or resharpening, and therefore curation; their allometric effects are documented in points across a wide contextual and time-space range (e.g. Peterson 1978; Hoffman 1985; Iriarte 1995; Archer and Braun 2010; de Azevedo et al. 2014; Goodale et al. 2015; Lerner 2015; Serwatka 2015). Yet some stone points may have been designed to fracture upon impact in order to increase wound size (Ellis 1997: 51; Engelbrecht 2015). Alternatively, any tendency toward impact fracture may have impaired their functionality (Ellis 1997: 57). Thus, breakability can be a design attribute or a design flaw, depending on circumstances.

Assemblages

In assemblages and their analysis, all points are not discarded alike. Traditionally, we interpret a point as evidence of a unit of activity, usually hunting or perhaps use as a knife. Whatever the particular kinds of use, the *amount* of use represented depends greatly upon the size and condition of the point. For any type designed to accommodate two or more resharpening episodes, *ceteris paribus* the more reduced the point the more use it experienced. If we can estimate—by experiment or comparison of used specimens to cached originals—the number of resharpenings that specimens of a type might undergo and then convert degree of reduction in discarded points to number of resharpenings, then we can estimate the latter (e.g. Shott 2017). Resharpening episodes may encompass two or more different uses, but at least the number of resharpenings might correlate with amount of use and degree of curation. In this view, two points of the same type do not represent equal amounts of use if they differ in amount or degree of reduction from resharpening.

Therefore, discarded points may be counted as discrete units, but must be calibrated to ratio-scale rates of use. Two assemblages of, say, 10 Type-X points each do not necessarily register the same amount or rate of point use, depending upon variation in their curation rates (Shott 1996). Besides their effects upon the size and composition of archaeological assemblages, degree and pattern of reduction and the curation rates implicated thereby have additional value. As assemblages of more types across broader time spans, especially within relatively small areas where toolstone supply and distribution can be held roughly constant, are studied for their reduction patterns and curation rates we can identify patterns in curation that then will require explanation.

Besides intact points that differ in size and reduction, tool fragments pose their own challenges. Broken points are common, as above possibly by design. Methods exist for quantification of original wholes represented by assemblages that combine intact and broken specimens (Shott 2000). Beyond quantification, degree of reduction might be possible to estimate for distal or medial fragments, at least sometimes. If so, these fragments inform on degree of point use as much as do intact specimens. Proximal or haft fragments could break at any stage or degree of use, making it difficult to calibrate the amount of tool use that they represent.

S and L obviously affect the size and composition of point assemblages. Besides calibrating archaeological abundance to past time, their formation effects bear upon inferences to other prehistoric trends. For instance, Bettinger (1999: 69–72) inferred prehistoric Great Basin population trends from changing frequencies of time-sensitive point types. This points-to-people equation assumes, obviously, some constant relationship between the numbers of both. Practically, it assumes that all point types had identical systemic frequencies and use lives, Schiffer's S and L. It assumes, that is, that at different times people used the same number of points per capita that lasted for the same period of time. Otherwise, points-to-people breaks down from the complicating effects of S and L independently of population.

Just in the comparison of arrows and darts, the assumption of constant or constantly proportioned S and L seems questionable. As above, arrow S probably was greater than dart S. There are good reasons to suppose that arrow L was shorter than dart L, again considerably. Arrows typically were thinner relative to their width, which made them perhaps more susceptible to breakage (e.g. Cheshier and Kelly 2006; Engelbrecht 2015). Arrows were fired at considerably higher speed than were darts, upon impact thus placing more stress upon the weapon, not least its point. Arrows could be fired from greater distances, making them easier to lose (Larralde 1990: 62). Thus, more abundant arrows that were more fragile were exposed to greater stresses and higher probability of loss.

Types as Historical Units

Attributes and objects are directly observed, and assemblages are defined by joint patterns of use, discard and deposition. The first two undeniably are "primary historical events" or units (Kitts 1992: 136), of a time-space scale commensurate with observation and experience. Most assemblages probably are time-averaged over at least years and often much longer; strictly they are not such primary events although typically we proceed as though they are, assuming that the size and composition of assemblages that include points characterize synchronic moments of the cultural past. An ethnographer could observe points being made and used, and record their number and context among the many more objects and constructions that typify any culture.

Pompeiis are nice to encounter. But the vast majority of the archaeological record accumulated at time scales orders of magnitude longer than the near-momentary Pompeiian one. We must stop using theory and implicit subjects suitable for very short time scales to explain the time-averaged record. Point types as historical units that persist for decades to centuries are beyond the scale of ethnographic observation. Their salient properties—definition, origins, time-space distribution, changing popularity over that distribution, duration, and fate—must be constructed from the many "primary events" that archaeologists document. Types are secondary historical events or units because they "have no counterpart in the present…[and] are composed of primary events related in a spatial and temporal nexus" (Kitts 1992: 137). As historical units, point types possess properties that are emergent at the lower level of primary events—not deducible from the properties of units at that level—and that require "explanatory principles emergent with respect to" (Kitts 1992: 142) it. No ethnographer can observe a point type in the fullness of its time-space range, or trace its origin, its behavior during its floruit, or its fate.

Yet here lies the gravest shortcoming in both contemporary and past archaeological thought. With rare exceptions (e.g. Perreault 2019) archaeology neglects both point types as units of study and efforts to explain their salient properties. No ethnographer can help us; archaeologists are the only ones who can observe, measure and explain the secondary-level or historical behavior of types over time-space scales that exceed ethnography's. Of course we do not ignore types entirely; we use them as markers that coarsely resolve past time, as clues to function and specific behaviors and, more recently, as contexts in which register technological organizational processes (Shott 2017). Here lies our greatest corollary challenge: developing the method and theory to define and analyze the historical behavior of types. Until we meet it, we are reduced to awkward groping toward a satisfactory account. That groping proceeds from time-space distributions and durations to types' changing abundance across those distributions, and finally to origins and fate together, as linked instances of diversification or extinction.

Time-Space Distributions of Types

Point types often serve as markers of cultures, yet their time-space distributions greatly exceed the equivalent scales of ethnographic cultures. Concerning just time scale, Eighmy and LaBelle (1996) documented point-type persistence on a millennial scale (Old World Paleolithic types, or at least facies defined in part by types, can persist even much longer [Monnier and Missal 2014: 67]), although ceramic types typically persisted for shorter spans. Similarly, the uncalibrated time ranges or durations of 49 eastern North American point types reported by Justice (1987) average 1387 years (although the distribution is right-skewed). Confined to types whose antiquity mid-points exceed 3000 BP—roughly, preceramic or pre-Woodland times—the mean span rises to 1762 years; comparable Plains data yield an average of 1696 years (Eighmy and LaBelle 1996: Table 2).

Overall, types' time ranges and antiquity (measured by range mid-point in years BP) are correlated ($r_s = 0.44$ $p < 0.01$) but a cubic, not linear, model provides the best fit (Fig. 12.4). This result suggests greater precision at the margins of eastern North American prehistory, owing to some combination of archaeological interests attracted to early and late prehistoric cultures and the finer contextual control available in later periods. The slope coefficient of linear regression of $\log_{10}$ type span upon age is 0.75 in Perreault's (2019: 175) deposits that include much earlier and therefore longer Paleolithic contexts. A similar regression of eastern North American point data (although, as above, a linear model fits these data poorly) gives a lower but substantial coefficient—a measure of change in the dependent type span with unit change in independent age—of 0.49. Type longevity is age-dependent, mostly later types lasting for shorter intervals.

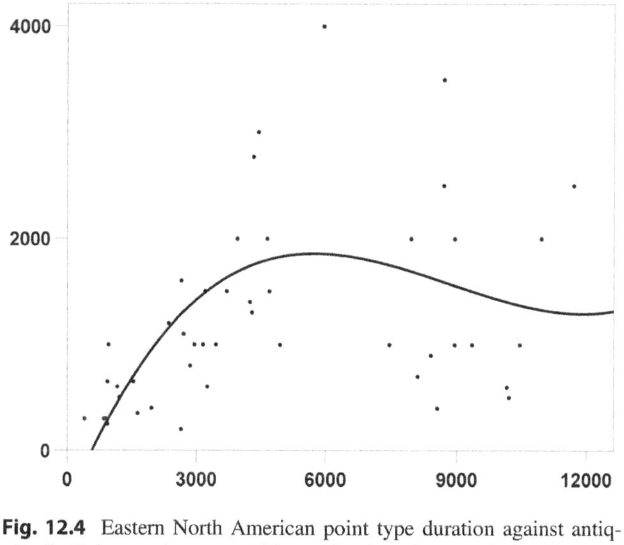

Fig. 12.4 Eastern North American point type duration against antiquity (both in years), measured by mid-point of reported time interval. Curve shows fit to cubic model. *Data source*: Justice (1987)

Properties of Type Floruits

A type's time-space distribution—its floruit—is among its fundamental properties. Fixing time intervals is a matter of adequate sampling, either of stratigraphic sequences, individual closed contexts, or direct dating of points. Considered together, the first two require numerous well dated contexts, for instance in classic alluvial (Coe 1964; Broyles 1971; Dincauze 1976; Stafford and Mocas 2008) or rockshelter (Sherwood et al. 2004) sequences of eastern North America although, as cultures grew progressively more sedentary and depositional regimes stabler through the Holocene, most of these sequences better parse Pleistocene and early Holocene intervals. It also can involve statistical analysis of radiocarbon dates (e.g. Manning et al. 2014; Thulman 2017). Direct dating requires thermoluminescence or other direct methods (Wilhelmsen and Feathers 2003), conceived of but not yet systematically attempted.

A type may exist from t_1 to t_2, but the interval defines only its nominal duration. Its floruit is determined from the interval along with its changing abundance over it. Types that are purely stylistic may exhibit normal distributions, rising gradually from t_1 to reach their maximum abundance at $t_{1.5}$, then declining equally gradually to t_2 (e.g. Manning et al. 2014; Perreault 2019: 237) (Fig. 12.5). Sequences of floruits can overlap only at their tails, forming unbroken sequences over long time periods. But in theory floruits can be skewed, vary in kurtosis, be multi-modal, and overlap in time variably if at all (Fig. 12.5). Over long intervals, some may overlap mostly if not entirely with others, and some intervals of time may be occupied by none.

Unfortunately, summed radiocarbon probabilities (e.g. Thulman 2017: Fig. 12.3) describe samples of radiocarbon dates and points jointly, not the latter alone. They describe the form of floruits over time only by controlling for the manifest biases that reside in radiocarbon samples. In any event, for common types they are derived from vanishingly small fractions of the total point population. Although the contexts are hardly more numerous, types' frequency distributions across dated contexts are somewhat less compounded samples of their changing abundance through time (e.g. Sherwood et al. 2004: Fig. 12.5).

Once the time-space ranges and the forms of floruits are charted, they must be explained. Do time and space ranges correlate with one another, such that more widely distributed types persist longer in time? Do those ranges correlate with their assemblage sizes, such that more common or popular types persist longer or are more widely distributed than others? Do those properties of floruits vary with the length, complexity or failure rates that characterize their production sequences? Do they vary with inferred human population sizes, such that types used by larger populations persist

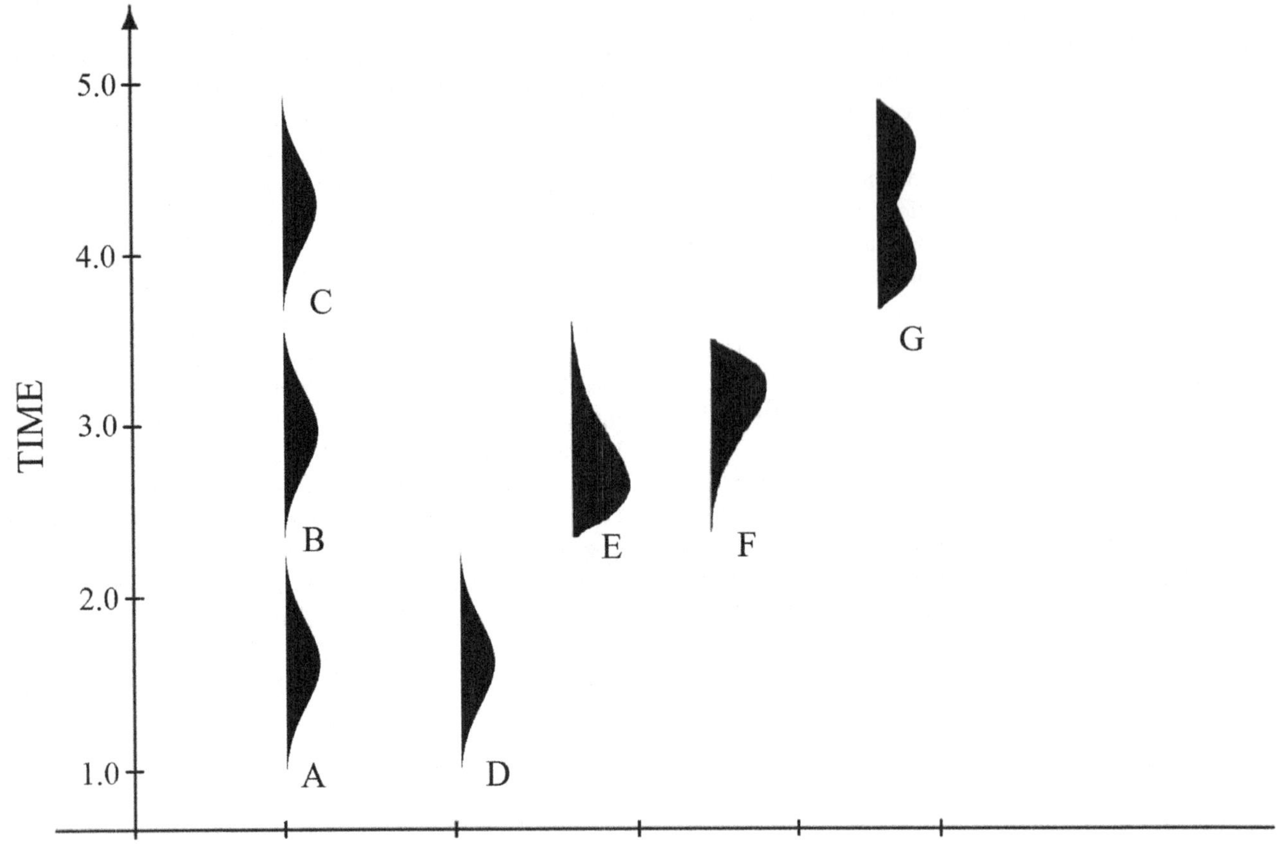

Fig. 12.5 Duration and form of type floruits. Type popularity at any time t_i is proportional to floruit width. **A–C** are similar in all properties save time interval, and all describe ideal unimodal, symmetric floruits of equal duration and popularity. They may overlap one another slightly in time, and together form an unbroken continuum or nearly so. **D–G** also differ in time interval but also in form, duration, popularity and degree of time overlap

longer and wider than do others? Do they vary with scale or type of sociopolitical organization? Do they vary, as Fiedel (2014) argued, with environmental changes?

Stasis and Sensitivity in Types

Stasis describes the tendency of types to change in forms, gradually or otherwise, during their floruits, sensitivity their somewhat opposing tendency to correlate or not with changes in other cultural units or respects. Why do types change at all? Even today, we cannot answer so fundamental a question about so important and abundant a unit of observation. Is type stasis—lack of change over considerable time—a phenomenon to explain or merely the consequence of the absence of sources of change? In analytical terms, types must possess integrity or they become other types, but what range of variation is permissible and exhibited within them? Across time or space, do specimens of a type drift within its morphometric range? Do types persist longer or shorter depending upon their shape, function, or other properties of the individual points? Are types that require lengthier production sequences prone to greater copying error and therefore higher rates of drift or even tendency to diversification? Explaining stasis is "one of the most interesting and potentially revealing aspects of the history of most species" (Gould 1980: 103), and perhaps of point types as well.

Diversification: Origins and Fate of Types

Diversification encompasses both the origin and demise of points. Types originate either *de novo,* as entirely new and original designs, or by change of pattern and degree of ancestral types and their segments. Obviously, the earliest type arises *de novo,* but descendant ones can arise either way. Why do types stop being made? That is, how and why do types end? Types terminate by extinction or by progressive diversification into one or more descendant types. Archaeologists' descriptions of point-type sequences often assume origin *de novo* and termination by extinction, for instance as we speak of LeCroys replacing Kirks. That may occur, but it is an assumption not a demonstrated inference. Valid inferences to origins and fates require methods that both describe type morphometrics and distinguish variation within from variation between types. They also require theory to explain *de novo* origins and termination by either mode.

Types can end in two ways: simple termination or branching diversification. Termination can occur by population replacement (e.g. "The Invasion of the Side-Notched People") or simple abandonment of the type. Either may be treated idiographically, as unique historical events that cannot be explained by general causes. The point-type sequences that underlie typological cross-dating in North America are well known, and involve many apparent terminations. Thus, termination may be fairly common, yet we have no theory to explain it. What, that is, explains one type's end and the next's origin? We cannot even distinguish between replacement and abandonment, never having established or agreed upon the necessary criteria beyond similarity, broadly conceived (Barrientos 2015: 54).

Branching diversification, a typical pattern in biology, has two modes: cladogenesis, in which one ancestral type gives rise to two descendant ones, or anagenesis, in which the ancestor persists as one type branches from it. In the living world, a single taxon at t_1 can yield many descendent taxa by t_{10}. The root taxon may be gone by then, by extinction or speciation. Many descendant taxa may have originated and terminated in the interval. At any time within it, any number of descendent taxa may exist, for varying durations. Yet in general, within phylogenies taxa diversify with time. At t_0 there is only one but at t_2 there may be two, at t_3 seven, and so on to much greater diversity.

In the made world of objects, branching diversification may be an imperfect model for the history of point sequences or any higher-level archaeological units. Or at least the diversification of cultural units like types is constrained compared to the living world. Clovis, say, may be ancestral to any number of types, but rarely to more than a few at any one time. If Clovis lies at t_0, then at any t_x only one or few descendants are apt to exist. Controlling for time span, for instance, Larralde (1990: 67) detected no significant rise in the diversity of early to late Holocene point types on the northern plains, although Lyman et al. (2009) saw evidence for increased type diversity at the dart-arrow transition. No trend, steady or irregular, toward rising diversity is likely to characterize the interval because prehistoric cultures, unlike prehistoric biomes, had limited capacity to accommodate, and limited need for, point types. At t_0 there is only one; at t_2 there may be two or three, at t_3 also two or three, and so on in sequences of relatively fixed typological, if potentially great morphological, diversity.

Cladistic methods commonly are used to generate point-type cladograms, to chart pattern and degree of relationship between ancestral and descendant types (e.g. O'Brien and Lyman 2003). Cladistics is designed to explain patterns of branching diversification, which suits it well to fossil data. But appropriate traits are not merely what are at hand but instead irreducible units that "must…be the result of a process of descent with modification" and that survive tests of unit-transmission integrity (Pocklington 2006: 25). Constrained typological diversity of point types seems better suited to methods that neither assume nor require progressive diversification (e.g. Lipo 2006; Adams and Collyer 2009), which provide merely detailed morphometric descriptions of transformations between point types, and make no assumptions about diversification mode. Our task then is to explain the transformations, their mode, tempo and path, along with the problem of persistence or stasis. All of this requires "in-depth assessment of character hypotheses" (Barrientos 2015: 55), rarely conducted today.

As in other respects, we have little data to catalog the separate occurrences of cladogenesis and anagenesis, and no theory whatsoever to explain the occurrences. Why do some types become two or more descendant ones? Is it determined by human population size or distribution, or perhaps of changing environmental structure and patchiness? Are types of more complex production sequences and perhaps size and shape more likely than others to undergo cladogenesis or anagenesis? If the former, what explains the number of descendant types that form over time? Among diversifying types, are there patterns when viewed across many types from many time-space contexts? Does morphing occur chiefly on haft elements, on blades, or on both at once?

All else equal, presumably more complex production processes and narrowly specified size and shape might limit the potential for diversification. With the historically unique introduction of arrow technology to North America, were dart points "translated" (Hall 1980; see also Clarke's [1978: 228] "transformation types" and White [2003] on the Jack's Reef to triangle sequence in the Great Lakes) by degree into arrow points until it became clear that radical changes—to small triangular forms—were needed? More broadly, and given the functional constraints to which points were subject, does the range of size-shape types produced by diversification over long spans comprehensively sample point phylogenies' theoretical morphospaces, the full range of possible size-shape permutations that they *may* occupy?

Broader Disciplinary Context

Treating point types as units of observation in their own right, seeking to explain the causes, correlates and properties of their time-space distributions, requires great change in our analytical perspective. Yet the shift is not totalizing. It is something to attend to besides, not entirely in place of, what we do now. It does not preclude continued attention to traditional lines of research (e.g. reduction/production processes, attribute analysis, use-wear, curation). Analytical focus upon types as *units* of analysis complements and extends, not replaces, other approaches to points. At the same time, it allows the study of points to contribute to archaeology's maturation as a discipline with units of observation and analysis that are commensurate with the spatial and especially the time scale at which assemblages accumulated.

Approaching this goal requires changes to and improvements in practice. We must amass much larger samples of points to document the morphological and use-related range of variation within types and the fullest time-space range of types (Barrientos 205: 56); practically, this means that we must engage with the larger communities of collectors, who control by far the majority of the known point population. We must compile larger databases of radiocarbon or other chronometric associations with or directly-dated points, and refine their resolution (e.g. Thulman 2017). We must develop and apply systematic methods to identify and distinguish types. We must conduct a wider range of controlled and actualistic experiments to better gauge the functional properties of points as weapon tips and/or as hand-held spears. We must gauge and explain the full range of the pattern and degree of types' resharpening, and their curation rates. Most important of all, to exploit the potential of such improved units of study, we must develop the second-order theory, not of timeless individual or group behavior, but of the behavior of derived types that will explain the pattern that preceding steps described.

Conclusion

As thick as it is with questions, this essay is remarkably thin in answers; in fact, it has scarcely any. Questions are much easier to ask than to answer, but questions of the nature posed here are, I hope, excusable. They arise from rueful acknowledgment that archaeology's units and their scales of accumulation are not commensurate with the theory that it applies, and the resulting conviction about the need to construct units at suitable scales and to explain them using suitable theory. Points are not the only category in which archaeology might seek solutions to our problems, but they certainly are one, and therefore as good as place as any in which to confront the challenges that Clarke identified 50 years ago. Time enough to start the effort.

It is not difficult to substantiate Clarke's claim that archaeology was and remains an "undisciplined empirical discipline." The history of the field in the 1980s and 1990s practically makes the argument for itself. But even Clarke's own time 20 years earlier, when Beatles roamed the earth, shows both the uncritical borrowing of theory and method from other fields and the waxing and waning of fads. At least in the United States, archaeologists then were in the grip of an epistemological fundamentalism. If exaggerated in that case, legitimate concern for grounding inferences always is salutary, but the philosophical agonizing failed to take root. Later, archaeologists conveniently forgot about the need to document their claims in evidence, thus reducing the earlier epistemological rigor to a passing fad. Similarly, the sincere concern for sampling rigor that began to develop in the 1960s was, at length, conveniently abandoned. Still other fads came and went (e.g. trend-surface analysis from geography, factor analysis from psychology, numerical taxonomy from biology).

Thus, in the 1960s archaeologists talked about logical positivism. In the 1980s they talked of praxis. In the past 20 years, they spoke increasingly of agency and identity. If the field does not change, in 20 years they will speak of whatever is then the prevailing intellectual fancy. Not in particulars of course but in the sense that he meant—a passive consumption of other disciplines' method and theory, and a fondness for ungrounded scholarly fashion—archaeology has changed little since Clarke's time. Unless, like paleobiology, we create the truly distinct theory of diachronic pattern and process of units whose time-space scales greatly exceed those typical of anthropology or behavioral ecology, then in 20 years archaeology will remain, as it was before and is now, a parade of passing intellectual fancies without the slightest cumulative progress. It will remain the undisciplined discipline that Clarke deplored. If that happens, we will continue to repeat the errors that paleobiology corrected, defining the wrong units at the wrong scales that we try to explain with the wrong theory.

This is no brief for a crude identification of point types with biological taxa, or an equally crude reduction of archaeology to biology. On the contrary, we deal with material culture that is vastly less constrained in rate, direction and magnitude of change and much more amenable to horizontal transmission. We deal with hierarchies of units —from attribute through type to tradition and cultural phylogeny—and complex patterns of interaction that exceed biology's. The solution is not to transfer our passive transference from anthropology to paleobiology; our challenge is to fashion our own units and theory for our own scales of culture change.

To realize its potential and its rightful place among the historical sciences, it should be clear that archaeology requires unbounded independence from anthropology. The study of points alone, of course, will not make this change but can be an integral part of it. To that extent a theory of the point, put into practice, will make its own modest contribution to correcting the flaws in archaeological thought and practice that Clarke identified so long ago and that, tragically, continue to burden us today.

Acknowledgments Huw Groucutt kindly invited me to write this essay. I alone am responsible for any errors. I thank the University of Chicago Press and Charles Perreault of Arizona State University for permission to use Fig. 12.3. Thanks also to Eric Olson and Calixto Alvarado, who drafted several figures.

References

Adams, D., & Collyer, M. (2009). A general framework for the analysis of phenotypic trajectories in evolutionary studies. *Evolution, 63,* 1143–1154.

Andrefsky, W. (2006). Experimental and archaeological verification of an index of retouch for hafted bifaces. *American Antiquity, 71,* 743–757.

Andrews, T., MacKay, G., & Andrew, L. (2012). Archaeological investigations of alpine ice patches in the Selwyn Mountains, Northwest Territories, Canada. *Arctic, 65*(Suppl. 1), 1–21.

Apel, J., & Darmark, K. (2009). Evolution and material culture. *Current Swedish Archaeology, 17,* 11–28.

Archer, W., & Braun, D. (2010). Variability in bifacial technology at Elandsfontein, Western Cape, South Africa: A geometric morphometric approach. *Journal of Archaeological Science, 37,* 201–209.

Barrientos, G. (2015). On the problem of identifying homologies in lithic artifacts. In M. Cardillo & H. Muscio (Eds.), *Darwin's legacy: The status of evolutionary archaeology in Argentina* (pp. 49–66). Oxford, UK: Archaeopress.

Beck, C. (1998). Projectile point types as valid chronological units. In A. Ramenofsky & A. Steffen (Eds.), *Unit issues in archaeology: Measuring time, space and material* (pp. 21–40). Salt Lake City, UT, USA: University of Utah.

Bettinger, R. (1999). What happened in the medithermal. In C. Beck (Ed.), *Models for the millennium: Great Basin anthropology today* (pp. 62–74). Salt Lake City, UT, USA: University of Utah.

Blitz, J. (1988). Adoption of the bow in prehistoric North America. *North American Archaeologist, 9,* 123–145.

Bode, C. (Ed.). (1967). *The portable Thoreau.* New York, NY, USA: Viking.

Bradbury, A., & Carr, P. (2003). A method for quantifying size and shape dimensions of North American hafted bifaces. In N. Moloney & M. Shott (Eds.), *Lithic analysis at the millennium* (pp. 173–187). London, UK: Institute of Archaeology, University College.

Broyles, B. (1971). *Second preliminary report: The St. Albans site, Kanawha County, West Virginia.* Report of Archaeological Investigation No. 3. West Virginia Geological and Economic Survey, Charleston, WV, USA.

Cardillo, M., Borrazzo, B., & Charlin, J. (2016). Environment, space, and morphological variation of projectile points in Patagonia (Southern South America). *Quaternary International, 422,* 44–56.

Cheshier, J., & Kelly, R. (2006). Projectile points shape and durability: The effect of thickness-length. *American Antiquity, 71,* 353–363.

Clarke, D. (1978). *Analytical archaeology* (2d ed.). New York, NY, USA: Columbia University.

Coe, J. (1964). Formative cultures of the Carolina Piedmont. *Transactions of the American Philosophical Society, 54*(5). Philadelphia, PA, USA.

Collins, M. (2007). Discerning Clovis subsistence from stone artifacts and site distributions on the Southern Plains periphery. In R. Walker & B. Driskell (Eds.), *Foragers of the terminal Pleistocene in North America* (pp. 59–87). Salt Lake City, UT, USA: University of Utah.

Cook, R., & Comstock, A. (2014). Toward more continuous and practical artifact analyses: Defining and learning from key dimensions of Fort Ancient triangular projectile points in the Miami Valley. *Midcontinental Journal of Archaeology, 39,* 222–250.

de Azevedo, S., Charlin, J., & González-José, R. (2014). Identifying design and reduction effects on lithic projectile point shapes. *Journal of Archaeological Science, 41,* 297–307.

Dev, S., & Riede, F. (2012). Quantitative functional analysis of Late Glacial projectile points from Northern Europe. *Lithics, 33,* 40–55.

Dincauze, D. (1976). *The Neville site: 8,000 years at Amoskeag, Manchester, New Hampshire.* Peabody Museum Monographs No. 4. Harvard University, Cambridge, MA, USA.

Dixon, E., Manley, W., & Lee, C. (2005). The emerging archaeology of glaciers and ice patches: Examples from Alaska's Wrangell-St. Elias National Park and Preserve. *American Antiquity, 70,* 129–147.

Douze, K., Igreja, M., Rots, V., Cnuts, D., & Porraz, G. (2020). Technology and function of Middle Stone Age points. Insights from a combined approach at Bushman Rock Shelter, South Africa. In H. Groucutt (Ed.), *Culture history and convergent evolution: Can we detect populations in prehistory?* (pp. 127–141). Cham, Switzerland: Springer.

Edinborough, K. (2005). Evolution of bow-arrow technology. Ph.D. dissertation, Institute of Archaeology, University College London.

Eighmy, J., & LaBelle, J. (1996). Radiocarbon dating of twenty-seven Plains complexes and phases. *Plains Anthropologist, 41,* 53–69.

Ellis, C. (1997). Factors influencing the use of stone projectile tips: An ethnographic perspective. In H. Knecht (Ed.), *Projectile technology* (pp. 37–74). New York, NY, USA: Plenum.

Engelbrecht, W. (2015). Interpreting broken arrow points. *American Antiquity, 80,* 760–766.

Erwin, J., Holly, D., Hull, S., & Rast, T. (2005). Form and function of projectile points and the trajectory of Newfoundland prehistory. *Canadian Journal of Archaeology, 29,* 46–67.

Estioko-Griffin, A. (1984). The ethnography of Southeastern Cagayan Agta hunting. MA thesis, University of the Philippines, Manila, Philippines.

Fiedel, S. (2014). Abrupt changes of climate and of point styles along the Atlantic seaboard of Eastern North America. *Archaeology of Eastern North America, 42,* 77–100.

Fowler, D., & Matley, J. (1979). *Material culture of the Numa: The John Wesley Powell Collection, 1867–1880.* Smithsonian Contributions to Anthropology No. 26. Smithsonian Institution Press, Washington, DC, USA.

Frankel, D. (1988). Characterising change in prehistoric sequences: A view from Australia. *Archaeology in Oceania, 23,* 41–48.

Goodale, N., Andrefsky, W., Osterhoudt, C., Cueni, L., & Kuijt, I. (2015). Cultural transmission and the production of material goods: Evolutionary pattern through measuring morphology. In N. Goodale & W. Andrefsky (Eds.), *Lithic technological systems and evolutionary theory* (pp. 239–252). New York, NY, USA: Cambridge University.

Gould, S. (1980). The promise of paleobiology as a nomothetic, evolutionary discipline. *Paleobiology, 6,* 96–118.

Groucutt, H., & Scerri, E. (2014). Lithics of the late Middle Palaeolithic: Post MIS 5 technological variability and its implications. *Quaternary International, 350,* 1–6.

Grund, B., & Huzurbazar, S. (2018). Radiocarbon dating of technological transitions: The Late Holocene shift from atlatl to bow in northwestern Subarctic Canada. *American Antiquity, 83,* 148–162.

Hall, R. (1980). An interpretation of the Two-Climax Model of Illinois prehistory. In D. Browman (Ed.), *Early Native Americans: Prehistory, demography, economy and technology* (pp. 401–462). The Hague, Holland: Mouton.

Hare, P., Thomas, C., Topper, T., & Gotthardt, R. (2012). The archaeology of Yukon ice patches: New artifacts, observations, and insights. *Arctic, 65*(Suppl. 1), 118–135.

Hoffman, C. (1985). Projectile point maintenance and typology: Assessment with factor analysis and canonical correlation. In C. Carr (Ed.), *For concordance in archaeological analysis: Bridging data structure, quantitative technique, and theory* (pp. 566–612). Kansas City, MO, USA: Westport.

Hughes, S. (1998). Getting to the point: Evolutionary change in prehistoric weaponry. *Journal of Archaeological Method and Theory, 5,* 345–408.

Iriarte, J. (1995). Afinando la puntería: Tamaño, forma y rejuvenecimiento en las puntas de proyectil pedúnculadas del Uruguay. In M. Consens, J. López, & M. del Carmen (Eds.), *Arqueología en Uruguay: 120 años despues* (pp. 142–151). Montevideo, Uruguay: VIII Congreso de Arqueología Nacional Uruguaya.

Justice, N. (1987). *Stone Age spear and arrow points of the midcontinental and eastern United States: A modern survey and reference.* Bloomington, IN, USA: Indiana University Press.

Kelly, R., Surovell, T., Shuman, B., & Smith, G. (2013). A continuous climatic impact on Holocene human population in the Rocky Mountains. *Proceedings of the National Academies of Science, 110,* 443–447.

Kitts, D. (1992). The conditions for a nomothetic paleontology. In M. Nitecki & D. Nitecki (Eds.), *History and evolution* (pp. 131–145). Albany, NY, USA: State University of New York.

Larralde, S. (1990). The design of hunting weapons: Archaeological evidence from Southwestern Wyoming. Ph.D. dissertation, University of New Mexico.

Lee, C. (2012). Withering snow and ice in the mid-latitudes: A new archaeological and paleobiological record for the Rocky Mountain region. *Arctic, 65*(Suppl. 1), 165–177.

Lee, C., & Kathryn Puseman, K. (2017). Ice patch hunting in the greater Yellowstone area, Rocky Mountains, USA: Wood shafts, chipped stone projectile points, and bighorn sheep. *American Antiquity, 82,* 223–243.

Lerner, H. (2015). Dynamic variables and the use-related reduction of Southern Huron projectile points. In M. Shott (Ed.), *Works in stone: Contemporary perspectives on lithic analysis* (pp. 143–161). Salt Lake City, UT, USA: University of Utah.

Lipo, C. (2006). The resolution of cultural phylogenies using graphs. In C. Lipo, M. O'Brien, M. Collard, & S. Shennan (Eds.), *Mapping our ancestors: Phylogenetic approaches in anthropology and prehistory* (pp. 89–107). New Brunswick, NJ, USA: AldineTransaction.

Lyman, R., VanPool, T., & O'Brien, M. (2009). The diversity of North American projectile-point types before and after the bow and arrow. *Journal of Anthropological Archaeology, 28,* 1–13.

Manning, K., Timpson, A., Colledge, S., Crema, E., Edinborough, K., Kerig, T., et al. (2014). The chronology of culture: A comparative assessment of European Neolithic dating approaches. *Antiquity, 88,* 1065–1080.

Monnier, G., & Missal, K. (2014). Another Mousterian Debate? Bordian facies, chaîne opératoire technocomplexes, and patterns of lithic variability in the western European Middle and Upper Pleistocene. *Quaternary International, 350,* 59–83.

Morrissey, G. (2009). Tools and change: The shift from atlatl to bow on the British Columbia Plateau. MA thesis, Simon Fraser University, Burnaby, BC, CAN.

O'Brien, M., & Lyman, R. (2003). *Cladistics and archaeology.* Salt Lake City, UT, USA: University of Utah.

Odell, G. (1981). The morphological express at function junction: Searching for meaning in lithic tool types. *Journal of Anthropological Research, 37,* 319–342.

Odell, G., & Cowan, F. (1986). Experiments with spears and arrows on animal targets. *Journal of Field Archaeology, 13,* 195–212.

Perreault, C. (2019). *The quality of the archaeological record.* Chicago, IL, USA: University of Chicago.

Peterson, R. (1978). Projectile point re-utilization patterns at the Agate Basin site. *Wyoming Contributions to Anthropology, 1,* 139–147.

Pocklington, R. (2006). What is a culturally transmitted unit, and how do we find one? In C. Lipo, M. O'Brien, M. Collard, & S. Shennan (Eds.), *Mapping our ancestors: Phylogenetic approaches in anthropology and prehistory* (pp. 19–31). New Brunswick, NJ, USA: AldineTransaction.

Prentiss, A., Walsh, M., Skelton, R., & Markes, M. (2016). Mosaic evolution in cultural frameworks: Skateboard decks and projectile points. In L. Mendoza Straffen (Ed.), *Cultural phylogenetics: Concepts and applications* (pp. 113–130). Basel, Switzerland: Springer.

Rasic, J., & Slobodina, N. (2008). Weapon systems and assemblage variability during the Northern Archaic Period in Northern Alaska. *Arctic Anthropology, 45,* 71–88.

Ratto, N. (2003). Estrategias de caza y propiedades del registro arqueológico en la Puna de Chascuil (Dpto. de Tinogasta, Catamarca, Argentina). Ph.D. dissertation, Departamento de Ciencias Antropólogicas, University of Buenos Aires.

Read, D. (1982). Toward a theory of archaeological classification. In R. Whallon & J. Brown (Eds.), *Essays in archaeological typology* (pp. 56–92). Evanston, IL, USA: Center for American Archaeology Press.

Reynolds, N. (2020). Threading the weft, testing the warp: Population concepts and the European Upper Palaeolithic chronocultural framework. In H. Groucutt (Ed.), *Culture history and convergent evolution: Can we detect populations in prehistory?* (pp. 187–212). Cham, Switzerland: Springer.

Rorabaugh, A., & Fulkerson, T. (2015). Timing of the introduction of arrow technologies in the Salish Sea, Northwest North America. *Lithic Technology, 40,* 21–39.

Schiffer, M. (1976). *Behavioral archeology.* New York, NY, USA: Academic.

Serwatka, K. (2015). Bifaces in plain sight: Testing elliptical Fourier analysis in identifying reduction effects on Late Middle Palaeolithic bifacial tools. *Litikum, 3,* 13–25.

Sherwood, S., Driskell, B., Randall, A., & Meeks, S. (2004). Chronology and stratigraphy at Dust Cave, Alabama. *American Antiquity, 69,* 533–554.

Shott, M. (1996). An exegesis of the curation concept. *Journal of Anthropological Research, 52,* 259–280.

Shott, M. (1997). Stones and shafts redux: The metric discrimination of prehistoric dart and arrow points. *American Antiquity, 62,* 86–101.

Shott, M. (2000). The quantification problem in stone tool assemblages. *American Antiquity, 65,* 725–738.

Shott, M. (2003). Time as sequence, type as ideal: Whole-object measurement of Woodland biface size and form in Midwestern North America. In M. Soressi (Ed.), *Multiple approaches to the study of bifacial technology* (pp. 251–271). Philadelphia, PA, USA: University of Pennsylvania Museum Press.

Shott, M. (2016). Survivorship distributions in experimental spear points: Implications for tool design and assemblage formation. In R. Iovită & K. Sano (Eds.), *Multidisciplinary approaches to the study of Stone Age weaponry.* Vertebrate Paleobiology and Paleoanthropology Series (pp. 245–258). Dordrecht, Holland: Springer.

Shott, M. (2017). The costs and benefits of technological organization: Hunter-gatherer lithic industries and beyond. In E. Robinson & F. Sellet (Eds.), *Lithic technological organization and paleoenvironmental change: Global and diachronic perspectives* (pp. 321–333). Dordrecht, Holland: Springer.

Shott, M., & Ballenger, J. (2007). Biface reduction and the measurement of Dalton curation: A Southeastern case study. *American Antiquity, 72,* 153–175.

Stafford, C., & Mocas, S. (2008). *The Middle, Late, and Terminal Archaic occupations at the Caesars Archaeological Project, Harrison County, Indiana.* Caesars Archaeological Project Report

Volume 3. Archeology and Quaternary Research Laboratory, Technical Report 38. Indiana State University, Terre Haute, IN, USA.

Stutz, A. J. (2020). The Middle-Upper Paleolithic transition: A long-term biocultural effect of anatomically modern human dispersal. In H. Groucutt (Ed.), *Culture history and convergent evolution: Can we detect populations in prehistory?* (pp. 157–186). Cham, Switzerland: Springer.

Surovell, T. (2009). *Toward a behavioral ecology of lithic technology: Cases from Paleoindian archaeology*. Tucson, AZ, USA: University of Arizona.

Thomas, D. (1978). Arrowheads and darts: How the stones got the shaft. *American Antiquity, 43,* 461–472.

Thomas, D. (1981). How to classify the projectile points from Monitor Valley, Nevada. *Journal of California and Great Basin Anthropology, 3,* 7–43.

Thulman, D. (2012). Discriminating Paleoindian point types from Florida using landmark geometric morphometrics. *Journal of Archaeological Science, 39,* 1599–1607.

Thulman, D. (2017). Dust Cave revisited: A Bayesian reanalysis of the radiocarbon record. *American Antiquity, 82,* 168–182.

Ugan, A., Bright, J., & Rogers, A. (2003). When is technology worth the trouble? *Journal of Archaeological Science, 30,* 1315–1329.

VanderHoek, R., Tedor, R., & McMahan, J. (2007). Cultural materials recovered from ice patches in the Denali Highway region, Central Alaska, 2003–2005. *Alaska Journal of Anthropology, 5,* 185–200.

Walde, D. (2014). Concerning the atlatl and the bow: Further observations regarding arrow and dart points in the archaeological record. *American Antiquity, 79,* 156–161.

Watanabe, H. (1975). *Bow and arrow census in a West Papuan lowland community: A new field for functional ecological study.* Occasional Papers in Anthropology No. 5, University of Otago, Dunedin, New Zealand.

Weltfish, G. (1977). *The lost universe: Pawnee life and culture.* Lincoln, NE, USA: University of Nebraska.

White, A. (2003). Mortuary assemblages and the chronology of the Jack's Reef Cluster. *Michigan Archaeologist, 49,* 43–71.

White, A. (2013). Functional and stylistic variability in Paleoindian and Early Archaic projectile points from Midcontinental North America. *North American Archaeologist, 34,* 71–108.

Wilhelmsen, K., & Feathers, J. (2003). Dating the manufacture of projectile points using thermoluminescence. In N. Moloney & M. Shott (Eds.), *Lithic analysis at the millennium* (pp. 189–202). London, UK: Institute of Archaeology.

Wilson, T. (1899). Arrowpoints, spearheads, and knives of prehistoric times. *Report of the United States National Museum for 1897* (pp. 1–267). Government Printing Office, Washington, DC, USA.

Wright, T. (2003). Madison triangles: There must be a point. In B. Redmond & J. Jones (Eds.), *Facing the final millennium: Studies in the late prehistory of Indiana, A.D. 700 to 1700* (pp. 85–94). Indianapolis, IN, USA: Indiana Division of Historic Preservation and Archaeology.

Chapter 13
Learning Strategies and Population Dynamics During the Pleistocene Colonization of North America

Michael J. O'Brien and R. Alexander Bentley

Abstract Being able to identify individual populations has long been of interest in archaeology, but within the last several decades it has become a specific focus as researchers have linked evolution-based theoretical models of cultural transmission with innovative analytical methods to better understand how groups of agents use culturally acquired information to navigate across fitness landscapes. Other animals learn, but humans have the unique ability to accumulate learned information rapidly and to pass it on to future generations. Nowhere is this interest in applying models of cultural transmission more evident than in the archaeology of the late Pleistocene colonization of North America, where researchers are beginning to identify distinct populations and to trace their movements across complex physical and cultural landscapes.

Keywords Clovis • Emulation • Imitation • Individual learning • Populations • Social learning

Introduction

Archaeologists have long had an interest in being able to identify prehistoric populations (Foley 1987; Lyman et al. 1997; Hermon and Niccolucci 2017; Garvey 2018; Groucutt [Chap. 1] 2020), traditionally using distinctive sets of artifacts—stone tools, pottery, clothing, housing, rock art, fish weirs, and the like—as proxies for the actual groups responsible for making, using, and losing or abandoning the items (McNabb 2020; Reynolds 2020; Shipton 2020; Shott 2020). By the mid-twentieth century, these artifact sets had been used to subdivide much of the North American archaeological record into myriad cultural units such as stages, phases, aspects, foci, traditions, and horizons (e.g. Phillips and Willey 1953; Willey and Phillips 1958; Lehmer 1971). The units contained cores, or sets of artifacts that did not overlap with other sets in either time or space. Extending out from the cultural cores were still other sets that were shared by multiple units. The shared traits were viewed as stemming from common ancestry between populations, from enculturation, and/or from diffusion.

Our goal in this chapter is to offer several alternatives to the standard way of identifying archaeological populations. As examples, we focus on studies that incorporate models of cultural transmission grounded in evolutionary theory and modern analytical methods in order to identify populations and understand their patterns of interaction during the late Pleistocene colonization of North America. The precise timing of the colonization is debatable (see below), but what is not at issue is the point of origin of the colonizing populations. Overwhelming archaeological and archaeogenetic evidence (Waters and Stafford 2007; Goebel et al. 2008; Kemp and Schurr 2010; O'Rourke and Raff 2010; Raff et al. 2010; Morrow 2014; Raghavan et al. 2014, 2015; Rasmussen et al. 2014; Raff and Bolnick 2014, 2015; Hoffecker et al. 2016; Llamas et al. 2016; Blong 2018; Moreno-Mayar et al. 2018; Posth et al. 2018; Davis et al. 2019) indicates that humans moved eastward across the Bering Land Bridge, or Beringia, during the Late Glacial Maximum, perhaps as a result of a shift to warmer/wetter conditions in Beringia between 14,700 and 13,500 years ago, which was associated with the early Bølling/Allerød interstadial (Wooler et al. 2018). Migrant groups then made their way either south along or near the coastline (Fladmark 1979; Erlandson et al. 2007; Gilbert et al. 2008; Braje et al. 2017, 2019) and/or through a corridor that ran between the Cordilleran and

M. J. O'Brien (✉)
Department of Communication, History, and Philosophy, Texas A&M University-San Antonio, San Antonio, TX 78224, USA

Department of Anthropology, University of Missouri, Columbia, MO 65211, USA

R. A. Bentley
Department of Anthropology, University of Tennessee, Knoxville, TN 37996, USA

H. S. Groucutt (ed.), *Culture History and Convergent Evolution: Can We Detect Populations in Prehistory?*, Vertebrate Paleobiology and Paleoanthropology, https://doi.org/10.1007/978-3-030-46126-3_13

Fig. 13.1 Clovis points from various North American sites. Top row (left to right): Townsend Co., Kentucky; unknown county, North Carolina; Williamson Co., Tennessee; Lewis Co., Kentucky (courtesy D. Meltzer); Essex Co., Massachusetts (courtesy J. Boudreau). Bottom row (left to right): Barnstable Co., Massachusetts (courtesy E. L. Bell); Essex Co., Massachusetts (courtesy J. Boudreau); Humphreys Co., Tennessee; Green Co., Kentucky; Columbia Co., Arkansas. All images from Whitt (2010) unless otherwise noted; composite by Matt Boulanger

Laurentide ice sheets that covered the northern half of the continent (Ives et al. 2014; Freeman 2016; Pederson et al. 2016; Potter et al. 2017, 2018). In our view, both scenarios remain equally viable (Potter et al. 2018; O'Brien 2019a).

With respect to timing, colonizing populations could have entered North America before 15,000 years ago (see below), but the earliest widespread human occupation of the continent dates to around 13,400 years ago (Potter et al. 2018), the visible manifestation of which is a tool kit referred to as the "Clovis techno-complex" (Bradley et al. 2010). That techno-complex is marked by a number of distinctive tool types, including bone and ivory rods (O'Brien, Lyman, et al. 2016; Sutton 2018), large prismatic stone blades (Bradley et al. 2010), and bifacially chipped and fluted stone weapon tips, referred to as "Clovis points" (Wormington 1957; Bradley 1993; Morrow 1995; Bradley et al. 2010; Sholts et al. 2012) (Fig. 13.1). The points exhibit parallel to slightly convex sides, concave bases, and flake-removal scars on one or both faces that extend from the base to about a third of the way to the tip. This flake removal, called "fluting," created a thinner base that acted as a "shock absorber," increasing point robustness and the ability to withstand physical stress through stress redistribution and damage relocation (Thomas et al. 2017; Story et al. 2018). Clovis points were hafted to spears that were thrust and/or thrown (Hutchings 2015) and, at least occasionally, functioned as butchering tools (Lyman et al. 1998; Smallwood 2013; Smallwood and Jennings 2016).

As widespread as components of the Clovis tool kit are, they apparently were not the first technological items to appear in North America. Several well-dated sites in Texas, Florida, and Oregon have produced stone-tool assemblages (Waters et al. 2011; Halligan et al. 2016; Williams et al. 2018; Davis et al. 2019) that indicate there was clearly one or more technocomplexes already present by the time Clovis points were first made (see Haynes [2015] for in-depth discussion of other candidate sites). The beginning dates of those technocomplexes are difficult to assess, but it appears that they pre-date Clovis by one or two millennia and perhaps more. Other technologies in the West may have been contemporaries of Clovis (Beck and Jones 2010; Smith et al. 2019), but to us the jury is still out.

Clovis points appear to have originated in the American Southwest (Morrow and Morrow 1999; Hamilton and Buchanan 2007; Meltzer 2009; Beck and Jones 2010; Waters et al. 2011) and spread north and east, including up into the Canadian ice-free corridor (Smith and Goebel 2018). In eastern North America, with a few exceptions the earliest dates from archaeological sites that have produced large numbers of fluted points consistently fall later in time than the earliest fluted points in the West (Haynes et al. 1984; Levine 1990; Curran 1996; Bradley et al. 2008; Robinson et al. 2009; Miller and Gingerich 2013a, b). To simplify a rather complex chronology, we can assign a range of 13,400–12,800 years ago for Clovis in the western half of the continent and 12,800–12,500 years ago in the East, although more restrictive date and spatial ranges have been proposed (e.g. Waters and Stafford 2007, 2014).

The difference in chronological ranges between the East and the West has been explained as the result of Clovis points originating in the West and then spreading eastward as the result of either the movement of populations or down-the-line transmission among established populations (Hamilton and Buchanan 2009; Lothrop et al. 2011, 2016; Smith et al. 2015). It seems highly unlikely, however, that the small sample of radiocarbon dates for the Clovis period has captured the earliest or latest use of Clovis points (Waguespack 2007; O'Brien, Boulanger, Buchanan et al. 2014; Prasciunas and Surovell 2015) in either half of the continent, so we use the ranges above as estimates.

The stone points represent the primary sources of information about the dynamics of Clovis populations, having yielded insights into migration routes, mobility and economics, weapon systems, hunting and domestic activities, and the learning and transmission of technological knowledge (Anderson and Gillam 2000; Cannon and Meltzer 2008; Meltzer 2009; Smallwood 2012; Jennings 2013; Eren et al. 2015). It is the last topic—learning and transmission—that is of particular interest here. *Cultural transmission* encompasses the mechanisms that humans, as well as other primates, use to acquire, modify, and retransmit cultural information in particular instances (Eerkens et al. 2014), whether it be rules concerning the eligibility of potential marriage partners, instructions for how to produce fishing nets, or the proper method of flaking Clovis points. We can refer to the units of transmission as *cultural traits.* After being transmitted, cultural traits serve as units of replication in that they can be modified as part of an individual's cultural repertoire through processes such as recombination, loss, or alteration within an individual's mind. As with genes, cultural traits are subject to recombination, copying error, and the like and thus can be the foundation for the production of new traits (O'Brien et al. 2010). Using cultural traits as general proxies for human behavior might, at first glance, seem straightforward enough, but as we will see, the issue is much more complicated than it appears.

Cultural Units, Transmission, and the Problem of Analogy

Even before Darwin (1859) wrote *On the Origin of Species,* many naturalists made a distinction between what later would be called *analogous* traits and *homologous* traits. Analogous traits—*analogs,* for short—are those that two or more organisms possess that, although they might serve similar purposes, did not evolve because of any common ancestry. Birds and bats both have wings, and those traits share properties in common, yet we classify birds and bats in two widely separate taxonomic groups because birds and bats are only distantly related. This is because these two large groups diverged from a common vertebrate ancestor long before either one of them developed wings. Therefore, wings are of no utility in reconstructing lineages because they evolved *independently* in the two lineages after they diverged. Conversely, homologous traits—*homologs,* for short—are useful for tracking continuity resulting from inheritance because they are holdovers from the time when two lineages were historically a single lineage. As another example, all mammals have a vertebral column, as do animals placed in other categories. The presence of vertebrae is one criterion that we use to place organisms in the subphylum Vertebrata. The vertebral column is a homologous trait shared by mammals, birds, reptiles, and some fishes, and it suggests that at some remote time in the past, organisms in these groups shared a common ancestor.

American archaeologists working in the first half of the twentieth century appreciated not only that there was a distinction between homologs and analogs but that it applied as much to culture as it did to biology. Writing in the 1930s, Kroeber (1931: 152–153) had this to say on the subject:

> There are cases in which it is not a simple matter to decide whether the totality of traits points to a true relationship or to secondary convergence. … Yet few biologists would doubt that sufficiently intensive analysis of structure will ultimately solve such problems of descent. … There seems no reason why on the whole the same cautious optimism should not prevail in the field of culture; why homologies should not be positively distinguishable from analogies when analysis of the whole of the phenomena in question has become truly intensive. That such analysis has often been lacking but judgments have nevertheless been rendered, does not invalidate the positive reliability of the method.

Note that although Kroeber was clear that there are two forms of similarity, one analogous and the other homologous, he was not clear as to how one might distinguish between them. He pointed out that identifying "similarities [that] are specific and structural and not merely superficial … has long been the accepted method in evolutionary and systematic biology" (Kroeber 1931: 151), but he offered no real opinion on how to separate what is "specific and structural" from what is "merely superficial" beyond undertaking a "sufficiently intensive analysis of structure." He was correct: An intensive analysis of structure, especially a detailed *comparative* analysis, is critical to being able to make the distinction, but again, he did not offer any thoughts on how to do that. Thus, Kroeber, and he was by no means alone, landed on the default option: Formal similarities between sets of artifacts must signal *some* kind of relationship, either an ancestor–descendant relationship or one derived through ethnologically documented mechanisms such as diffusion and enculturation (Lyman et al. 1997). Gordon Willey (1953: 363) didn't waffle on the matter, declaring axiomatically that "typological similarity *is* an indicator of cultural relatedness (and this is surely axiomatic to archeology), [and thus] such relatedness carries with it implications of a common or similar history" (emphasis added). This axiom, however, falls prey to a caution raised by paleontologist Simpson (1961), using monozygotic twins as an example: They are twins not because they are similar; rather, they are similar because they are twins and thus share a common history. There is a big difference between the two (O'Brien and Lyman 2000).

The default option—formal similarity signals relationship—continued to dominate archaeology, and the number of articles and monographs emphasizing diffusion and migration as explanatory devices continued to increase throughout the twentieth century. As Rowe (1966: 334) noted, however, most accounts were nothing more than poorly concocted just-so stories: "We are now being subjected in archaeological meetings to ever more strident claims that Mesoamerican culture was derived from China or southeast Asia, early Ecuadorian culture from Japan, Woodland culture from Siberia, Peruvian culture from Mesoamerica, and so forth. In the science-fiction world of the diffusionists, a dozen similarities of detail prove cultural contact, and time, distance, and the difficulties of navigation are assumed to be irrelevant."

One of the studies to which Rowe clearly was referring grew out of the work of Ecuadorian archaeologist Emilio Estrada and two American colleagues, Betty Meggers and Clifford Evans, who saw definite evidence of transoceanic contact between Japan and coastal Ecuador around 5,000 years ago (Evans et al. 1959; Estrada et al. 1962; Meggers et al. 1965). Their claim was based on similarities between some of the pottery they were excavating in Ecuador and pottery they had seen in collections from southern Japan. How did the pottery in Ecuador get there? Estrada and colleagues proposed that Japanese fishermen were blown off course and that Pacific currents carried them to the Ecuadorian coast. It was there that they taught local populations the art of pottery making. It makes an interesting story, but again, similarity does not imply homology.

Style and Function: Not a Simple Dichotomy

Beginning in the 1970s, Robert Dunnell addressed the issue of convergence and divergence with his "fundamental dichotomy" between *style*, which he equated with homology, and *function*, which he equated with analogy (Dunnell 1978, 1980; Shennan 2020). In his scheme, stylistic traits, by definition, are those that are not under selection, whereas functional traits are those that *are* under selection. In archaeology, many examples of this dichotomy come to mind, such as that between a functional canoe paddle versus the stylistic design painted on it or perhaps between more-creative "private" rock paintings in limited-access caves versus tightly regimented and highly visible rock art on a more public landscape (Bradley and Valcarce 1998; Simek et al. 2013).

Although these synchronic distinctions can be used to create hypotheses, Dunnell's (1980) point was to introduce a diachronic distinction between style and function, which could be identified by documenting change through time in the frequencies of artifacts or other proxies for behavior (see Shennan (2020), for an excellent discussion of the use of "style" in archaeology). The frequencies of stylistic traits—those not under natural selection—are expected to change in stochastic fashion, analogous to neutral traits in biology. This creates continuous, unimodal frequency distributions, as things come into fashion, reach their zenith, and then decline, finally disappearing. Conversely, functional traits can display one of several distributions. They might display a sharp rise in popularity followed by a steep decline (O'Brien and Holland 1990) as they are quickly replaced by other functional traits; they might display unimodal frequency distributions similar to those of stylistic traits; or they might display discontinuous, multimodal frequency distributions as a result of convergence or fluctuation in the selective environment.

For some reason, however, some archaeologists began arguing that only stylistic traits, not functional traits, could be used to measure interaction, transmission, and inheritance within and between populations. It was supposed that functional traits were useful only for identifying the presence of selection and measuring its effects. This was incorrect, and the confusion led archaeology down a long, convoluted rabbit hole (O'Brien and Leonard 2001). Put correctly, analogous traits can *always* be assumed to be functional, but the reverse is not always true: Functional traits can be either homologous *or* analogous. In other words, functional traits—those that by definition affect the fitness of the bearer—can show up in two different lineages as a result of either common ancestry or convergence (see Groucutt [Chap. 4] 2020). Let's take a look at an example of misplaced use of functional traits as being unequivocally homologs. We use this particular example because it has a direct connection to our discussion of the early colonization of North America.

In 2012, Dennis Stanford and Bruce Bradley published the book *Across Atlantic Ice: The Origin of America's Clovis Culture* (Stanford and Bradley 2012), the latest version of their proposal that North America was first colonized by groups from southern France and/or the Iberian Peninsula that used watercraft to make their way across the North Atlantic and into North America during the Last Glacial Maximum, some 20,000–24,000 years ago. This 6,000-km journey was facilitated, in their view, by a continuous ice shelf that provided the emigrants with fresh water and a stable food supply. In its initial formulation, the hypothesis was based primarily on similarities between stone tools associated with the Solutrean culture of Western Europe, which dates 23,500–18,000 years ago (Straus 2005), and those associated with the North American Clovis culture, which, as we noted earlier, dates 13,400–12,500 years ago.

Flaws in the "Solutrean hypothesis" were quickly pointed out. The multiple-thousand-year gap between Solutrean and Clovis made an ancestor–descendant relationship highly improbable, meaning that similarities in tool design were instead the result of convergence: unrelated populations of prehistoric flintknappers finding similar solutions to similar adaptive problems (Straus 2000; Will and Mackay 2020). To deal with the large chronological gap, Stanford and Bradley shifted their focus from similarities between the Solutrean and Clovis to supposed similarities among Solutrean, Clovis, and pre-Clovis tool types and production techniques (Stanford and Bradley 2002; Bradley and Stanford 2004). This was an unfortunate modification to their proposal because the pre-Clovis dates used by Stanford and Bradley—all of which are from highly questionable contexts—actually *predate* the Solutrean (O'Brien, Boulanger, Collard, et al. 2014). This would suggest that the traits appeared first in North America and then were carried to Europe. This, of course, is implausible.

That Stanford and Bradley fell prey to the "similarity equals relatedness" principle is not, as we've seen, an isolated incident, and we would be the first to admit that distinguishing between homologous and analogous traits is difficult. As we will show, however, it is not impossible. As an introduction to that issue, we are reminded of a quote from Clarke (1968: 211). What he said was not so different from what other archaeologists had said—Kroeber for example—but it contained an important contrast between two terms, *phyletic* and *phenetic*:

> One of the fundamental problems that the archaeologist repeatedly encounters is the assessment of whether a set of archaeological entities are connected by a direct cultural relationship linking their generators or whether any affinity between the set is based on more general grounds. This problem usually takes the form of an estimation of the degree of affinity or similarity between the entities and then an argument as to whether these may represent a genetic and phyletic lineage or merely a phenetic and non-descent connected affinity.

Both terms, "phyletic" and "phenetic," are grounded in the concept of "similarity," but whereas the former signifies a descent-related affinity—one person or population being related to another one (or more)—the latter has nothing to do with descent. We now have at our disposal a battery of methods and techniques that offer objective grounds for making the distinction. One of them, cladistics, was introduced into biology in the mid-twentieth century (Hennig 1950, 1966) and, in various forms, has become the standard approach in the discipline. It also has seen widespread usage in archaeology, including in research focusing on the Clovis colonization of North America.

Phylogeny and Cladistics

From an archaeological standpoint, if the issue at hand is identifying populations and understanding how they are related—if indeed they are—then the bottom line is, use traits, often referred to as *characters*, that will potentially emit strong *phylogenetic* signals. Phylogenetic—Clarke's "phyletic"—refers to relatedness between or among phenomena, whether they be sets of organisms—including human populations—or sets of stone tools. Conversely, "phenetic" ordering is based solely on similarity. There are several methods of investigating phylogeny, but here we focus on only one, *cladistics*, which defines phylogenetic relationships in terms of relative recency of common ancestry: Two groups—we'll refer to them as *taxa*—are deemed to be more closely related to one another than either is to a third taxon if they share a common ancestor that is not also shared by the third taxon. The evidence for exclusive common ancestry is found in evolutionarily novel, or *derived*, character states. Note that our taxa could be sets of anything that is capable of evolving, including sets of stone tools, manuscripts, and groups of people.

Having said that, we point out that inanimate objects obviously do not breed and reproduce. This narrow view of the Darwinian process doomed early efforts to view the archaeological record in evolutionary terms (Lyman et al. 1997). It overlooked the fact that humans *do* breed and reproduce and that things such as stone tools are part of human phenotypes in the same way that teeth and bones are or that beaks and feathers are for birds. In essence, stone tools are proxies for the human behaviors that create them. All evolution cares about are three conditions being met: (1) variation is present, (2) the variation is inherited, and (3) there is a sorting mechanism that creates differential persistence of variants over generations.

As an example, Fig. 13.2 is a phylogenetic tree that shows relationships among four taxa. It tells us that based on a certain character distribution—more on that below—taxa C and D are more similar to one another than either is to any other taxon. It also says that taxa B, C, and D are more similar to one another than any of the three is to Taxon A. We know that taxa A–D evolved from ancestral taxa, although at this point we know little or nothing about those ancestors except that with respect to certain characteristics, taxa C and D look more like their immediate common ancestor (x) than they do the one (y) that unites them with Taxon B. Likewise, taxa B, C, and D look more like their common ancestor (y) than they do the one (z) that unites them with Taxon A.

In cladistics, convention is to place nodes at the points where branches meet and to refer to the nodes as ancestors that produced the terminal taxa (those at the branch tips). In our tree, taxa C + D, together with their hypothetical common ancestor (node x), form what is termed a *monophyletic group*, or *clade*. Taxa D + C + B and node x, together with their common ancestor (node y), form another, more inclusive clade, and taxa D + C + B + A (and nodes x and y),

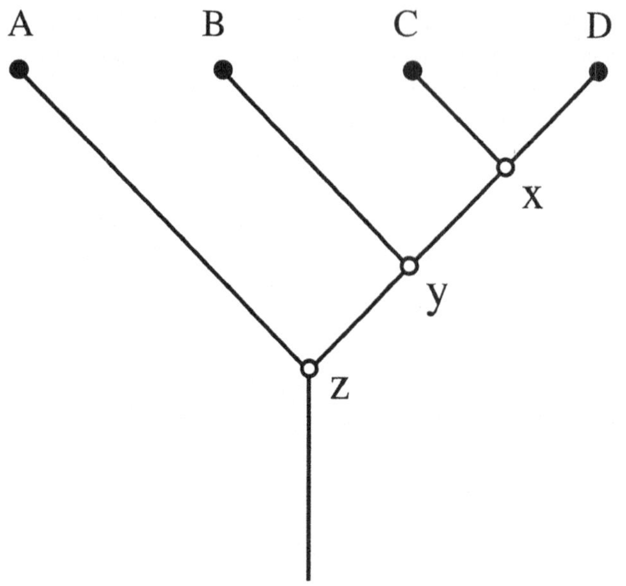

Fig. 13.2 Relationship of four taxa (A–D) and three ancestors (x–z). Based on a certain character-state distribution (not shown), taxa C and D are more similar to one another than either is to any other taxon. Also, taxa B, C, and D are more similar to each other than any of the three is to Taxon A. Related groups and their ancestors form ever-more-inclusive taxa, or clades: C + D + x is one clade; B + C + x + D + y is a second; and A + B + x + C + D + y + z is a third

together with their common ancestor (node z), form yet another, and the most inclusive, clade. A common misconception is that the interior nodes—"ancestors"—are somehow "real." They are not—hence our use of the term "hypothetical" above.

Another series of trees is shown in Fig. 13.3, this time with emphasis on the kinds of characters and character states that one encounters in archaeological phylogenetic studies. The trees show the evolution of a projectile-point lineage that begins with Ancestor A. For simplicity, we are tracking only a single character, fluting, which, again, is the removal of one or more longitudinal flakes from the base of a projectile point in order to thin it. Clovis points, as we noted, are fluted. Here, there are only two character states, fluted and unfluted. Over time, Ancestor A, which is unfluted, gives rise to two lines, one of which, like its ancestor, is unfluted and the other of which is fluted (Fig. 13.3A). Thus the character state "fluted" in Taxon 2 is derived from the ancestral character state, "unfluted." In Fig. 13.3B, Ancestor B (old Taxon 2) gives rise to two new taxa, 3 and 4, each of which carries the derived character state, "fluted." At this point "fluted" becomes a *shared derived* character state, defined as one that is shared only by sister taxa and their immediate common ancestor. Character states in sister taxa that have been inherited from an ancestor more distant than the common ancestor are *shared ancestral* character states. In Fig. 13.3C, in which two descendent taxa have been added, fluting is now a shared ancestral character state relative to taxa 5 and 6 because it is shared by three taxa and two ancestors. But relative to taxa 3, 5, and 6, fluting is a derived character state because it is shared by three taxa and their immediate common ancestor, B. Thus depending on where in a lineage one begins, a trait can be derived or ancestral.

Figure 13.3 does not show a third kind of character, but it is one that occurs on virtually all phylogenetic trees and, if not recognized, creates false positives in terms of similarity resulting from common ancestry. These are analogs, which in cladistics are referred to as *homoplasies*—similarities resulting from processes other than descent from a common ancestor, such as convergence, parallelism, and horizontal transmission between lineages (Sanderson and Hufford 1996; Groucutt [Chap. 1] 2020). Suppose in Fig. 13.3C that the tree is a true depiction of projectile-point evolution. Suppose further that taxa 1 and 6 share a character—say, beveling—that taxa 3 and 5 do not exhibit. We would refer to beveling as a *homoplasious* character—one that arose independently in those two taxa.

Several studies have examined how various Clovis-period and slightly later point types from across North America are related phylogenetically (O'Brien and Lyman 2000, 2003; O'Brien et al. 2001, 2002, 2012, 2013, 2015; Darwent and O'Brien 2006; Buchanan and Collard 2007, 2008; O'Brien, Boulanger, Buchanan, et al. 2014, 2016; Smith and Goebel 2018). Instead of using traditional projectile-point types, several studies used a standardized set of projectile-point classes (taxa) that were defined on the basis of eight characters, including base shape, the shape of the blade, the length/width ratio, and how deeply indented the base was (O'Brien et al. 2001). These characters are shown in the box in the upper left of Fig. 13.4, represented by Roman numerals (I–VIII). Each character has a number of states, and it is the intersection of the states of each character that creates a class (see O'Brien et al. [2001] for the states of each character). The choice of which characters to use was based on expectations as to which parts of a point would change most over time as a result of cultural transmission and thus create a strong phylogenetic signal. Archaeologists, like biologists, lean heavily on experience in selecting characters, and experience has shown that the hafting element—the proximal end of a projectile point (the part that comes into contact with a spear or dart shaft)—is a likely region in which to find characters useful in phylogenetic analyses. Forty of the 41 classes (taxa) used in the latest analysis (O'Brien, Boulanger, Buchanan, et al. 2014, 2016)

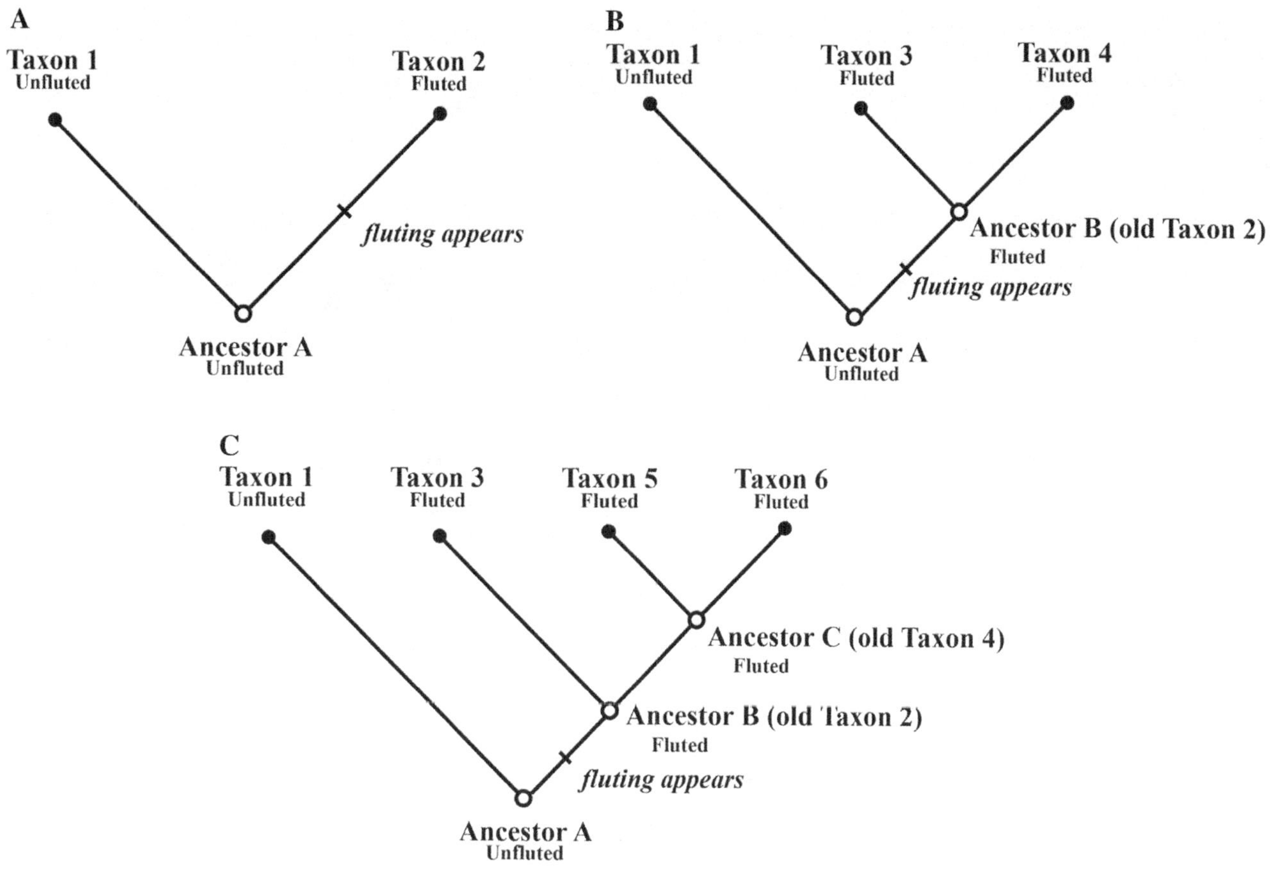

Fig. 13.3 Phylogenetic trees showing the evolution of projectile-point taxa (after O'Brien et al. 2001). In (**A**), fluting appears during the evolution of Taxon 2 out of its ancestral group. Its appearance in Taxon 2 is as a derived character state. In (**B**), Taxon 2 has produced two taxa, 3 and 4, both of which contain fluted specimens. The appearance of fluting in those sister taxa and their common ancestor makes it a shared derived character state. In (**C**), one of the taxa that appeared in the previous generation gives rise to two new taxa, 5 and 6, both of which contain fluted specimens. If we focus attention only on those two new taxa, fluting is now a shared ancestral character state because it is shared by more taxa than just sister taxa 5 and 6 and their immediate common ancestor. But if we include Taxon 3 in our focus, fluting is a shared derived character state because, following the definition, it occurs only in sister taxa 3, 5 and 6 and their immediate common ancestor

are shown at the branch tips, and the class that was used to root the tree—the one predicted to be ancestral to all the other classes (O'Brien et al. 2002)—is at the far left (KDR [12212223]).[1]

The phylogenetic tree shown in Fig. 13.4 is a 50% majority-rule consensus tree, meaning that out of all trees generated during analysis, at least 50% of them had the projectile-point classes in the positions shown. The tree exhibits numerous clades, which, again, are defined as units that consist of two or more related taxa and their common ancestor. Six of the clades in Fig. 13.4 are labeled A–F. Of perhaps more immediate importance are the 48 squares shown on the tree, each of which conforms to one of the three kinds of characters shown at the top of the box in Fig. 13.4. Each square is labeled with a Roman numeral, which corresponds to the characters in the Fig. 13.4 box. The presence of a square indicates that the character has changed states from one generation to the next; the subscript Arabic numeral indicates the evolved character state. For example, the first characters to change were location of maximum blade width (Roman numeral I) and constriction ratio (Roman numeral IV). The former changed from state 1 to state 2, and the latter changed from state 1 to state 3. These changes created an ancestor that then produced Class Kg (22231223) and an offspring class that, with an additional state change, became Class Kj (22232323). White squares on the tree indicate phylogenetically informative changes—shifts that result from descent with modification —-as opposed to changes that result from either adaptive convergence (black squares) or reversals to ancestral character states (half-shaded squares).

Phylogenetic analysis is important because it allows us to track *heritable continuity*—what produced what—as opposed to simply *historical continuity*—what followed what with no reliable knowledge as to whether an ancestor–descendant relationship existed. If we are interested strictly in phylogeny, then our focus is on the white squares in Fig. 13.4 because they are the only ones that resulted from

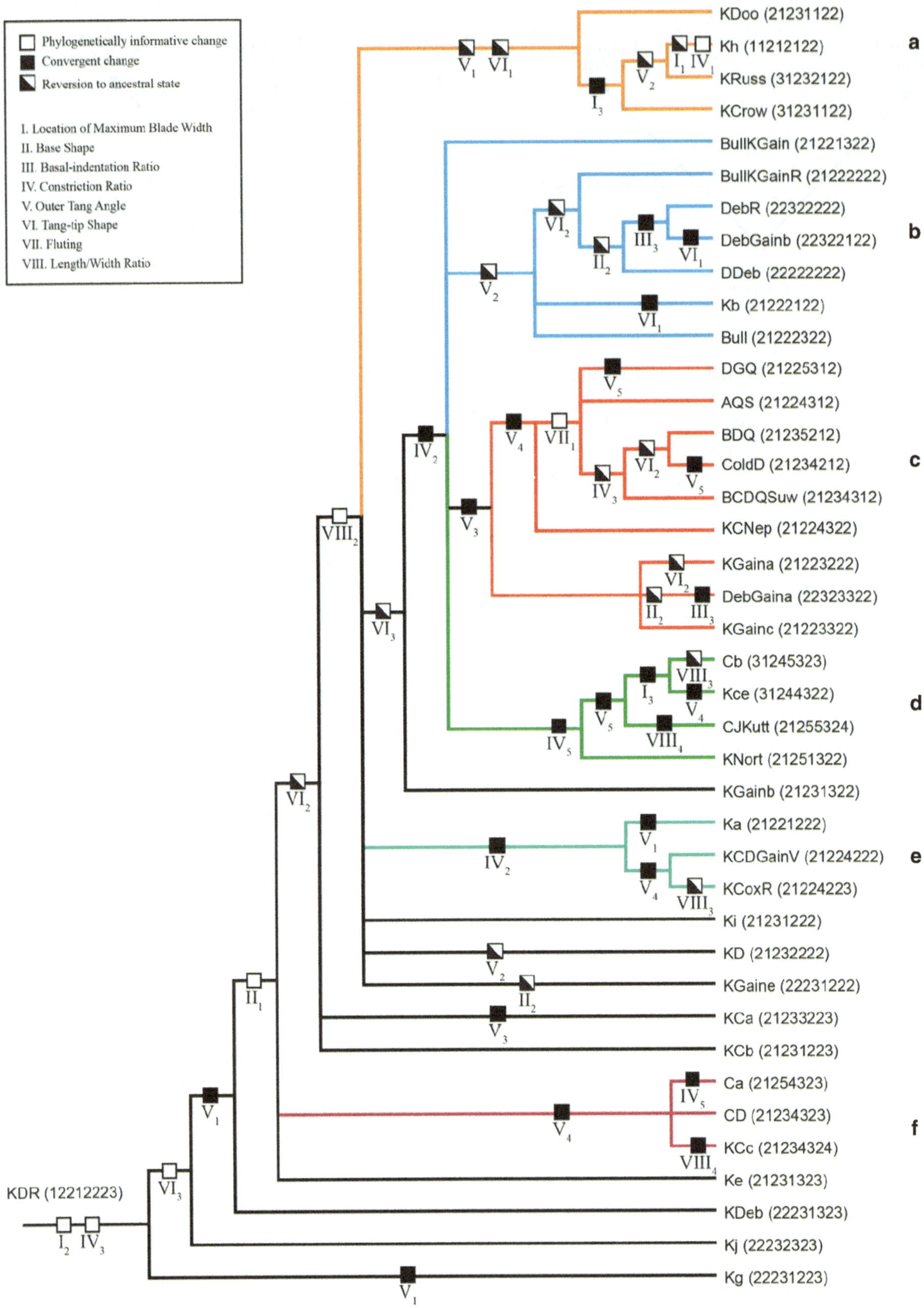

Fig. 13.4 Phylogenetic tree showing 41 classes (after O'Brien, Boulanger, Buchanan, et al. 2014). Roman numerals denote characters, and subscript numbers denote character states. Open boxes indicate phylogenetically informative changes; shaded boxes indicate parallel or convergent changes (homoplasy); and half-shaded boxes indicate characters that reverted to an ancestral state. Six of the clades are labeled A–F. The tree is a 50% majority-rule consensus tree based on 100 replicates

descent with modification. But at a more general scale, all of the morphological changes shown in Fig. 13.4 are important because they give us important insights into how Clovis flintknappers were making decisions about how to manufacture their points. Unless character states were independently invented, the process that led to the traits showing up in the positions they do is cultural transmission. Now, what about the learning processes embedded in the transmission? Do different kinds of learning create different patterns of variation, and at various levels, and can we use the patterns to talk about populations as they move across the landscape? As we will see, the answer to both questions is "yes."

Learning: The Basis of Cultural Transmission

Franz Boas (1904: 522) pointed out at the beginning of the twentieth century that "the theory of transmission has induced investigators to trace the distribution and history of [cultural traits] with care so as to ascertain empirically whether they are spontaneous creations or whether they are borrowed and adapted." Boas (1911: 809) later noted that "we must investigate the innumerable cases of transmission that happen under our very eyes and try to understand how transmission is brought about and what are the conditions that favor the grouping of certain new elements of an older culture." These are excellent points, but again, there was a lack of rigor in producing testable models. It wasn't until the 1970s that Boas's insights led to such models, starting with mathematical work that incorporated cultural information into evolutionary models of differential transmission of genes (e.g. Cavalli-Sforza and Feldman 1973, 1981; Feldman and Cavalli-Sforza 1976), followed by work that brought to the forefront various kinds of learning (e.g. Boyd and Richerson 1985; see Shennan 2020).

For our purposes here, we can subdivide learning into *social learning* and *individual learning* (Cavalli-Sforza and Feldman 1981; Boyd and Richerson 1985; Laland 2004; Mesoudi 2011a; Kendal et al. 2018), keeping in mind that humans are neither purely social learners nor purely individual learners. Rather, certain conditions, perceived or real, dictate which one is used in any particular situation. In fact, there are good reasons to suspect that many species, especially humans, may have experienced selection for reliable social learning, with enhanced individual learning being a by-product (Laland 2017). At the *group* level, social learning is advantageous for most agents, but that benefit relies on the remaining proportion of individual learners and what they know about the environment. Without any individual learners to constantly sample the environment—to produce information useful to the group—social learners cannot track environmental change. Without a source of variation, agents simply copy themselves into stasis—potentially a recipe for disaster in the face of a changing environment. For this to work, however, there has to be an adaptive value for individual learning to occur in the first place. This is achieved by social learning making individual learning less costly (Boyd and Richerson 1995).

Social learning is a powerful adaptive strategy that allows others to risk failure so we don't have to (Henrich 2001; Laland 2004; Aoki and Feldman 2014)—that is, it lets others filter behaviors and pass along those that have the highest payoff (Rendell et al. 2011). This translates into social learning being less costly in terms of energy and/or time (Morgan et al. 2011). Social learning is how individuals acquire their language, morals, technology, how to behave socially, what foods to eat, and most of their ideas from people with whom they come into constant contact. Over generations, the effect is cumulative, as individuals continue to "learn things from others, improve those things, transmit them to the next generation, where they are improved again, and so on" (Boyd and Richerson 2005: 4). As Mesoudi and Thornton (2018: 6) put it, cumulative cultural evolution is "the introduction of behavioural novelty or modification, the transmission of behaviour via social learning, the improvement in genetic and/or cultural fitness or fitness proxies as a result of the learned behaviour and the repeated transmission and improvement of the behaviour over time." This has been referred to as the "ratchet effect" (Tomasello et al. 1993; Tomasello 1999; Tennie et al. 2009). Any number of species exhibit social learning (Hoppitt and Laland 2013; Mesoudi and Thornton 2018), but humans, and a limited number of other species, exhibit an amped-up form of social learning, which we can refer to as "cultural learning" (Dean et al. 2014). Humans excel at cultural learning, which is what makes human minds, not to mention human lives, so different from those of other animals (Heyes 2015).

Learning is the process that ensures what we earlier referred to as heritable continuity—one thing resembling another as a result of transmission (Lyman and O'Brien 1998). Over time, continuity creates what archaeologists refer to as *traditions,* defined as "(primarily) temporal continuity represented by persistent configurations in single technologies or other systems of related forms" (Willey and Phillips 1958: 37). From an evolutionary perspective more explicitly, a cultural tradition "is a socially transmitted form unit (or a series of systematically related form units) which persists in time" (Thompson 1956: 38)—a definition that reflects transmission, persistence by means of replication, and heritable continuity (Lyman et al. 1997; O'Brien et al. 2010).

Copying

Social learning usually involves copying others, which itself is a set of competing strategies. You might, for example, copy someone based on that individual's skill level—perhaps a person who appears to be better at something than you are or someone who appears to be successful—whereas someone else might base his or her decisions on social criteria—copy the majority, copy kin or friends, or copy older individuals (Kendal et al. 2018). The various factors that can affect one's choice of whom or what to copy are often referred to as "social learning strategies" (Laland 2004) or "transmission biases" (Boyd and Richerson 1985)—unique evolutionary forces for the selective retention of cultural variants. The term "biased learning" is commonly used as a synonym for certain social-learning strategies. Given the difference between the effects of copying based on selection for knowledge or a skill level as opposed to copying based on random social interaction, "bias" is used in a statistical sense to indicate some deviation from random, or "unbiased," copying. It is not used in any normative sense, such as "gender bias" or "racial bias."

With respect to model-based transmission—you are picking someone to copy—we might make the underlying assumption that individuals can find a master teacher from whom to learn. Likewise, it might be assumed that individuals can sense how popular a behavior is in the population. These assumptions might be unrealistic for large populations, perhaps where individuals have only local, imperfect knowledge of what models, and hence what behaviors, are optimal (Bentley and O'Brien 2011; Bentley et al. 2014). Thus, we would expect that if individuals are selective and accurate in finding the most skilled model for copying, then the pace of cultural evolution depends strongly on population size, from the Upper Paleolithic Revolution of 40,000 BP (Powell et al. 2009) to the information cascade that confronts us today (Bentley and O'Brien 2017). If, however, learning is relatively unselective, then the pace depends only weakly on population size, if at all, and perhaps more on the level of environmental risk (Collard et al. 2013).

Of course, even with large populations the individual minds involved must communicate in the first place in order to create this "collective-mind" effect. Unconnected individuals are irrelevant to learning and the collective storage/retrieval of information (Bentley and O'Brien 2011). This has been documented time and again, most dramatically in a computer-mediated tournament of learning algorithms held at St Andrews University in 2009 (Rendell et al. 2010). Before the tournament, many expected the winning strategy to be some combination of majority individual learning supplemented by some social learning. In fact, the most successful strategies relied almost exclusively on social learning, even when the environment was changing rapidly. The winning strategy copied frequently and was biased toward copying the most recent successful behavior it observed—an excellent strategy in the face of rapidly changing environments (but see Heyes 2016). Of course, even here there had to be some individuals—a minority—who were creating and updating information for others to copy.

With respect to copying, our view mirrors that of Rendell et al. (2011): Copying confers an adaptive plasticity on populations, which allows them to draw on deep knowledge bases in order to respond to changing environments rapidly. High-fidelity copying leads to an exponential increase in the retention of cultural knowledge. The key term here is "high fidelity" (Boyd and Richerson 1995). What if acquisition costs affect the ability to copy faithfully (Mesoudi 2011b)—a point that applies to all modes of social learning but appears to be especially important for model-based learning? There also is another issue involved with the fidelity of copying, and it involves the difference between *imitation*, copying the form of an action, versus *emulation*, copying the result of an action sequence. This distinction sounds clear enough, but it can be difficult to demonstrate empirically. As an example, let's look at the bearded capuchins that live in the savannah of Brazil. One of the monkeys' economic pursuits involves cracking tough palm nuts using large stones as hammers and stone or log surfaces as anvils. This is no simple task, in that it involves proper stance, proper placement of a nut on an anvil, and a proper striking angle so that the nut doesn't skip away. Adults crack the nuts routinely throughout the year, but juveniles rarely manage to crack a whole nut, even though from a young age and for several years they devote considerable time and effort to watching their elders and practicing pounding actions with bits of nut and small stones.

Can young monkeys learn to crack nuts, or at least improve their technique, from directly copying some aspect of the behavior of others? Some researchers (e.g. Fragaszy et al. 2013) think the answer is no. Beating on a nut because another monkey is pounding on one might increase the copier's skill, but simply pounding a stone on a nut is not sufficient to crack it. Even after a young monkey reliably produces all the relevant actions, and in the correct sequence, it takes another year or more before it succeeds in cracking a whole nut. Does this mean, though, that all nonhuman primates are only good emulators but not imitators? No. Whiten et al. (2009), for example, report results from an experimental study in which a young chimpanzee watching another chimpanzee cracking nuts made repeated and moderately synchronous matching actions, but involving no nut or hammer.

With respect to the manufacture of a Clovis point, there is, as we will see later, a clear distinction between imitation —understanding the actions necessary to produce a point—and emulation—trying to produce a point without understanding the necessary actions (and their correct sequence). Stone-reduction sequences are complex procedures that require a significant amount of investment in terms of time and energy to learn effectively (Geribàs et al. 2010; Stout 2011). Clovis-point production is no exception (Bradley et al. 2010). Fluting can be a challenging technology to master, occurring after a point is already thinned to approximately 7.5 mm (Thomas et al. 2017). That doesn't give the knapper much margin for error.

A Map of Decision Making

Learning, of whatever kind, results in decision making, whether it's how to make a Clovis point, where to find the next meal, or whom to marry. Decisions are affected by two inputs: the kind of learning involved and the costs and benefits related to the knowledge acquired. Figure 13.5 shows a "map" of decision making that is defined by kind of learning along the horizontal axis and by costs and benefits along the vertical axis. Along the left edge, agents are purely individual learners—they use no information from others in making decisions. Along the right edge, agents are purely social learners—their decisions are based solely on copying, instruction, or other similar social processes. In between the extremes is a balance between the two—a flexible measure of the agents represented. The midpoint could represent, for example, a population of half social learners and half individual learners, or each individual giving a 50% weight to his or her own experience and a likewise amount to that of others.

We can compare the kinds of learning against the costs and benefits of acquiring that knowledge. The farther up one goes on the map, the more attuned an agent's decisions will be to the potential costs and payoffs of various decisions. A projectile-point manufacturer, for example, might quickly learn that a certain shape of a base makes a point susceptible to catastrophic failure and thus would likely change the design. Such a decision might be made individually, which places you in the upper left quadrant, or there might be socially identified authoritative experts whom you copy, which places you in the upper right quadrant. As an agent moves down the map, the relation between an action and its impact on performance becomes less clear. At the extreme bottom edge are cases that correspond to total indifference,

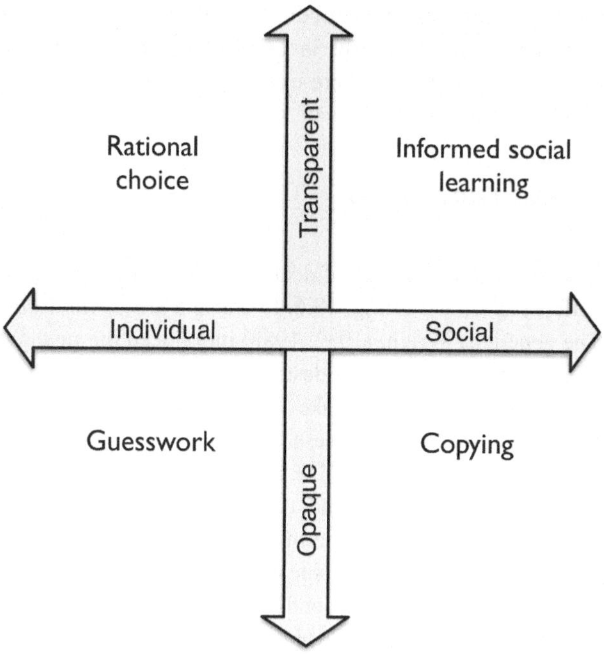

Fig. 13.5 A four-quadrant map for understanding different domains of human decision making, based on whether a decision is made individually or socially (horizontal axis) and the transparency of options and payoffs that inform a decision (vertical axis) (after Bentley et al. 2014)

where choice is based either on randomly guessing among all possible choices (lower left) or copying from a randomly chosen individual (lower right). This area of the cost/benefit spectrum represents cases in which agents perhaps are overwhelmed by decision fatigue—for example, when the number of choices becomes prohibitively large to be processed effectively.

Based on what we see in small nonwestern groups today (e.g. Henrich and Broesch 2011; Muthukrishna and Henrich 2016), we would assume that similarly among Clovis groups, social learning was transparent, as members would have learned adaptive knowledge—tool making, hunting practices, medicinal-plant use, and the like—from respected experts in the group. If learning is nontransparent, then misinformation can invade the social-learning process, such as a misguided panic among a herd of social animals (Couzin et al. 2005). For humans, imagine a case where social influence is strong but transparency is low. This highly social, nontransparent situation might characterize disasters that occur through misguided conformity, such as people remaining in a burning building because they don't yet see anyone else exiting or cult suicides, where everyone drinks the cyanide-laced Flavor Aid, and so on. Although the spread of misinformation is well-documented in modern media-saturated society (Aral et al. 2009; Garcia-Herranz et al. 2014; Vosoughi et al. 2018), we can assume it was much less common in the traditional subsistence societies of prehistory, except perhaps in cases of gossip or deception (e.g. Chagnon 2000), where expertise might not have been transparent to all members of a network.

Fitness Landscapes

We can overlay the map of decision making with peaks and valleys, as shown in Fig. 13.6, to create a *fitness landscape*. The geneticist Wright (1932, 1988) introduced the metaphor of a fitness landscape to describe the possible mutational trajectories that lineages take (evolve) from genotypes that lie in regions of low fitness to regions of higher fitness (Kvitek and Sherlock 2011). We can borrow this metaphorical landscape and turn it into a kind of *design space*, or, in biological terms, a *morphospace* (McGhee 2018). We can also adapt its features so that the highest peak on the landscape corresponds to the optimal design of something—a projectile point, for example—and lower peaks correspond to designs that, although not optimal, are good enough for the intended function at particular points in time. The landscape also contains valleys, which correspond to designs that yield negative fitness. An example of the latter would be a stone spear tip that is so thin that it consistently snaps on the slightest impact—not the best weapon to have when facing a charging animal (O'Brien, Boulanger, Buchanan, et al. 2016).

Note that the bottom half of the fitness landscape contains clouds, which begin to obscure the tops of some of the fitness peaks. Imagine that stone projectile points are variable in design such that some perform better than others for the purpose of, say, hunting mammoth. As the relationship between that variability and the performance for hunting mammoth becomes less clear, it equally becomes less clear as to what changes might be made to increase the performance of a point. Thus, an individual learner is likely to produce variation in design that drifts from one form to the other, but if an agent learns socially, he or she can use the actions of other agents as a guide, although they may be in no better shape to make informed decisions. As the connection between the variation produced and the outcome becomes clearer, agents can make more-informed choices, either singly or collectively (O'Brien, Boulanger, Buchanan, et al. 2016). Again, the key to fitness lies in the effect social learning has on individual learning. Copying can be adaptive *if* it makes individual learning less costly or more accurate. This means that agents use individual learning when it is cheap and reliable and switch to social learning when individual learning is expensive or inaccurate (Boyd and Richerson 1995; Castro and Toro 2004; Kendal et al. 2018).

Fitness landscapes can be simple or complex, depending on the transparency of costs and benefits. A "Mount Fuji" landscape, for example, has a clear solution: The optimum peak is so visible that all you need to do is align your strategy toward the mountain and start climbing. You can get to the top on your own by walking, or you can copy others who are also taking the hike. On more rugged landscapes, however, the highest peak may be over the next ridge, so to prevent getting stuck on a small nearby hilltop, you need to

Fig. 13.6 The four-quadrant map shown in Fig. 13.5 with a fitness landscape superimposed; the view is from the lower left of the figure (from O'Brien, Boulanger, Buchanan, et al. 2016). The presence or absence of clouds corresponds to the transparency of potential costs and payoffs of a decision. Agents, shown as dots, attempt to find the optimum peak, either on their own or with help from other agents. Figure by Matt Boulanger

copy others more and more frequently (O'Brien et al. 2019). Most of those others will also be copying others, who will be copying others, and you hope that somewhere there is someone who actually sees the highest peak. This is why copying only works if at least some people, even if only a minority, are actually looking at the world around them rather than at other people. In other words, we need at least a few producers to supply information to all the scroungers (Mesoudi 2008).

We need to make clear that not all decisions affect fitness, meaning that not all decisions are a matter of life and death. You might, for example, want to buy a mobile phone, but you have no idea of what color to get. Does your fitness rely on which one you pick? Probably not. You could simply look around, point to someone else's phone, and say, "I'll have what she's having," to quote a well-known saying (Bentley et al. 2011). However, just because choices *seemingly* do not have payoff differences with respect to fitness doesn't mean they are *always* unrelated to fitness (O'Brien et al. 2019). Take, for example, carpet designs (Tehrani and Collard 2002), pottery decorations (Neiman 1995; Shennan and Wilkinson 2001), or synonymous words (Bentley 2008; Bentley et al. 2012). It is difficult to think of designs affecting one's fitness. Whether a potter incises triangles or circles into a still-wet ceramic vessel does nothing to affect the ability of the pot to hold and steam food, and the same applies to the designs woven into carpets or whether we say "cop" or "policeman." In the language Dunnell (1978, 1980) used, we would say that designs and synonymous words are stylistic, meaning they have neutral selective value.

Suppose, however, that designs are tightly restricted in terms of social norms, so that you have only a limited number of designs from which to choose. With respect to options 1, 2, or 3, your choice is selectively neutral, but if you pick from outside that range, you could face criticism or even ostracism. All of a sudden, what seemed to be a matter of style becomes a matter of function. Stylistic cultural elements have a payoff based on the particular distribution of choices among other agents, which may favor conformity, anti-conformity, frequency dependence, and so on, none of which depends inherently on the choice itself but rather on its frequency among other agents and their social-learning networks.

Clovis Populations and Patterns of Learning

In a growing and fast-moving population subject to the widespread environmental changes of, for example, late Pleistocene North America, it is understandable why biased-learning strategies, including prestige bias, would have played a key role in fluted-point technologies (Sholts et al. 2012; O'Brien, Boulanger, Buchanan, et al. 2016; O'Brien and Buchanan 2017): When faced with possible weapon failure, especially on an unfamiliar landscape, your safest bet might be to adopt the best model from whom to learn and not change. Under circumstances where ecological conditions change, say, on a generational scale, the mean trait value is often optimal, leading to frequency-dependent bias, or conformism (Henrich and Boyd 1998). In western North America, where Clovis technology apparently began (Beck and Jones 2010; Hamilton and Buchanan 2007; Meltzer 2009; Morrow and Morrow 1999; Waters et al. 2011), point production appears to have been fairly specialized in terms of form, perhaps a result of the focus on fewer prey species in a more stable environment compared to the East (Buchanan et al. 2011). This is consistent with a stronger degree of social learning (biased transmission) in the West relative to the East, as western groups produced fewer point forms overall, and a few particular forms were produced more frequently (Buchanan et al. 2017).

As Clovis groups began moving into eastern North America, they would have encountered environments that were more heterogeneous than those in the West (Thompson et al. 1993), incorporating a greater number of floral and faunal habitat types and greater variability in resource patches (Eren et al. 2015; O'Brien 2019a). A concomitant change in subsistence strategy could have come with a cost to forager time budgets (Buchanan et al. 2017), meaning that populations would have had to invest more time in accumulating knowledge about unfamiliar landscapes in order to understand where productive resource patches were located and in traveling between a greater number of smaller patches. Time available for detailed teaching and learning projectile-point production in the East could have been comparatively diminished, leading to a flourish of individual trial-and-error learning and experimentation, which resulted in higher rates of interregional variation (O'Brien and Buchanan 2017). If this can be demonstrated archaeologically, it says a lot about population dynamics. As we will see below, it *can* be documented.

Models of cultural learning indicate that a mix of social and individual learning is adaptive in environments "that change too rapidly for innate, genetic responses to evolve, yet not so rapid that previous generations' solutions to problems are out-of-date" (Mesoudi 2014: 66). Increasing chronological resolution of the last several thousand years of the Pleistocene has shown that the transition to the Holocene ca. 11,700 cal BP was anything but gradual and uniform, especially in the East (Denton et al. 2010), suggesting this would have been a time when individual learning, at the aggregated group level, might have conferred an advantage, especially if coupled with conformist bias (Hamilton and Buchanan 2009). In other words, information producers took over a larger proportion of the learning process. This appears to account for the significantly greater diversity in Clovis points from the East than in those from the West (Buchanan et al. 2017).

Here, we are using diversity to refer strictly to differences in point shape. Two scenarios have been proposed for the diversity. In one, Clovis groups adapted their hunting gear to the characteristics of prey and local habitat, which resulted in regionally distinctive point shapes (Buchanan et al. 2014; Bement and Carter 2015). In the other scenario, there are no significant regional differences in shape, and any variation is attributable to stochastic mechanisms such as copy error, or drift (Morrow and Morrow 1999; Buchanan and Hamilton 2009). The two scenarios, however, are not mutually exclusive (O'Brien, Boulanger, Buchanan, et al. 2014; Eren et al. 2015). Colonizing populations do not necessarily stay in constant contact with one another, especially as geographic distance between them increases, and thus over time point shapes can begin to drift. Similarly, as they move apart, populations may begin to adapt point shape to regional environmental conditions that differ from those encountered by other groups. In other words, populations begin to explore different local fitness peaks (O'Brien, Boulanger, Buchanan, et al. 2016).

Diversity, however, can also refer to aspects of Clovis points other than shape, including the manner in which they were flaked (O'Brien 2019b). Several recent studies of flaking have shed considerable light on Clovis learning. One study used laser scanning and Fourier analysis to examine flake-scar patterns on a sample of Clovis points from sites across North America (Sholts et al. 2012). This analysis suggested that flaking patterns were similar across the continent, with no evidence of diversification, regional adaptation, or independent innovation. The authors proposed that the lack of diversification was tied to the importance of outcrops of desirable tool stone, where "Clovis knappers from different groups likely encountered each other … [which] would have allowed knappers to observe the tools and techniques used by other artisans, thereby facilitating the sharing of technological information" (Sholts et al. 2012: 3025; see Maher and Macdonald 2020). This sharing created the uniformity in production seen in their sample—a classic case of conformist bias (Sholts et al. 2012), which is a strong form of *stabilizing selection*.

One significant aspect of Sholts et al.'s study was their inclusion of 11 replicate Clovis points made by a modern flintknapper who is well known in the knapping world for his ability to make "superb Clovis points" that are "as thin as anyone could make them" (Whittaker 2004). He copied points from the Drake Clovis cache in Colorado and not only passed them off to highly knowledgeable collectors as authentic but, at least for a while, fooled any number of professional archaeologists highly familiar with Clovis points. How was he able to get away with it? For one thing, he was a master flintknapper and was able to reverse engineer certain aspects of the Drake points (Preston 1999) and then copy them. Until the study by Sholts and colleagues, it was widely believed that the replicas were all but perfectly executed and that his mistakes, which eventually revealed the points' inauthenticity, was his choice of Brazilian quartz as the raw material for some of the replicas (archaeologists assumed the stone was simply from an unknown western North American source) and his use of red clay to buffer the effects of a rock tumbler that knocked off the sharp flake-scar ridges, which would have been sure signs of modern replication.

Analysis by Sholts and colleagues showed, though, that there was another dead giveaway: As skilled a knapper as he was, he could not consistently copy a Clovis knapper's pattern of flake removal. In other words, the modern flintknapper—again, a person widely recognized as one of the best there is—could *sometimes* replicate the flake-removal pattern of a Clovis knapper, but he was inconsistent in his ability to do so. As the flintknapper later told a journalist (Preston 1999), "I just stopped and looked at [a] piece and said, 'That really looks like a Drake-style Clovis if I stop right there.' Until then, I had always kept going, cleaning up the edges, making the point smoother, getting the symmetry dead on, and really dressing the thing up. What I'd been losing was its immediacy, its simplicity." The real reason, of course, for his failure to consistently match the flaking pattern was because he was born 13,000 years too late to have worked side by side with a Clovis craftsman. He was a master emulator but only a so-so imitator (O'Brien and Buchanan 2017).

Eren and colleagues (2015) subsequently used a sample of 115 Clovis points from three chert outcrops in the Upper Midwest as an additional test of the findings by Sholts et al. (2012) that there was no evidence for diversification, regional adaptation, or independent innovation in flaking pattern. Bradley et al. (2010: 177, 106) had proposed that "Clovis flaked stone technology exhibits a bold, confident, almost flamboyant strategy" that "focuses on the removal of large well-formed flakes." Eren and colleagues formulated a straightforward, quantitative measure of "boldness": the number of flake scars on a face divided by the square area of a fluted point. The smaller the value, the bolder a point's flaking pattern. They also used geometric morphometrics to assess variation in shape, but as opposed to the sample used by Sholts and colleagues, which came from scattered regions of North America, the sample used by Eren and colleagues came from a more restricted, environmentally homogeneous region in order to maximize the probability that any patterned variation in point shape should be attributable not to differential adaptation by Clovis groups but rather to decreased social interaction among them. Statistical analysis of flake-scar patterning confirmed that the production technique was the same across the sample—matching the findings of Sholts et al. (2012)—but geometric morphometric analysis also showed distinct differences in point shape associated with the stone outcrop from which particular Clovis points originated.

The dichotomous, intraregional results from the Upper Midwest strongly suggest that Clovis foragers engaged in two tiers of social learning (Eren et al. 2015; O'Brien, Boulanger, Buchanan, et al. 2016; O'Brien and Buchanan 2017; O'Brien 2019b). The ancestral tier, which is an example of *deep homology*, relates to point production and can be tied to conformist transmission of ancestral tool-making processes across the larger North American Clovis population (Sholts et al. 2012), where dispersing Clovis groups were still socially connected across large regions of the continent and directly exchanging technological knowledge, resulting in a low interregional variance in how points were being flaked. The derived tier is tied to point shape, which shows more interregional variance (Eren et al. 2015; Buchanan et al. 2016), which resulted from individual populations spending more time at different stone outcrops. In that tier, the apparent pattern of increased experimentation in shape is what we would expect from guided variation, which is unbiased transmission plus environmental (individual) learning (Boyd and Richerson 1985). In other words, in the absence of strong selection, a population will move toward whichever trait is favored by individual-learning biases (Mesoudi 2011a; O'Brien, Boulanger, Buchanan et al. 2014; Gingerich et al. 2014; O'Brien and Buchanan 2017). Again, this occurs even when the strength of guided variation is weak (Mesoudi 2011a). It should come as no surprise that shape and flake-removal patterns would be driven by different learning and transmission processes (O'Brien and Buchanan 2017). Flaking patterns are a form of "structural integrity," in which key components are more conservative and therefore less likely to change relative to other components—a phenomenon that occurs in other aspects of culture as well (Mesoudi and Whiten 2008).

Over time, the continent-wide method of point manufacture began to shift. In a follow-up study to the one by Sholts et al. (2012), Gingerich et al. (2014) examined flake-removal patterns on specimens of Early Paleoindian eastern fluted-point types that immediately postdate the height of classic Clovis-point manufacture and found more variation and bifacial flake-scar asymmetry than what Sholts et al. (2012) had found among Clovis points. Gingerich et al. (2014: 117) proposed that the differences could represent "a time-transgressive shift, where Clovis interaction and the direct transmission of knowledge responsible for consistent reduction techniques is breaking down, causing biface symmetry to become more variable with greater flake scar variation." In other words, once individual Clovis populations began settling down, and thus encountering other populations on a more limited basis, even the conservative aspects of point manufacture began to dissolve (Sholts et al. 2012; Smallwood 2012; Eren et al. 2015). The resulting regionalization in the East produced a series of morphologically distinct unfluted and fluted forms, reflecting a "relaxation in the pressure to maintain contact with distant kin, a reduction in the spatial scale and openness of social systems, and a steady settling-in and filling of the landscape" (Meltzer 2009: 286).

Conclusion

We would be the last to claim that the theoretical models and analytical methods discussed here can be easily applied to the study of population dynamics generally (Shennan 2020). The dispersal of Clovis groups across North America represents an exceptional case because it occurred within such a short time span and across an area that had at best small resident populations that had not been there very long. Also, Clovis hunters used a stone weapon tip that, despite regional and temporal differences in shape, is a highly visible time marker. The result is that we have temporal resolution rarely seen in archaeology. Compare the resolution available for the spread of the Clovis techno-complex, ca. 13,400–12,500 years ago, to what archaeologists working in the Old World deal with, where resolution can range into the tens of thousands of years, if not more.

The spread of Clovis involved various kinds of learning. Early on, individual populations apparently maintained close social ties as they spread across the landscape, with the result being a pattern of flake removal on Clovis points that was reinforced across generations. Sholts et al. (2012) propose that this reinforcement came about as a result of groups meeting up at chert outcrops, which served as hubs of regional activity (Bradley et al. 2010; Sholts et al. 2012; Smallwood 2012; Waters et al. 2011). For a thinly scattered, mobile population such as Clovis or its immediate descendants, outcrops would have acted as ideal meeting spots because, once found, they would have served as predictable places on an emerging mental landscape map (Gardner 1977; Goodyear 1979; Miller 2016; Miller et al. 2018). Outcrops were places where Clovis groups could not only resupply but also exchange information and the like. This resulted in a low interregional variance in flaking patterns.

Over time, groups began to spend more time at specific chert outcrops (Eren et al. 2015), and although points were flaked similarly across regions, blade shape began to change. This interregional variance could have resulted from drift as well as from adaptation to different environments. At the level of the group, this increased experimentation in shape is what we would expect from individual learning (Boyd and Richerson 1985). Again, in the absence of strong selection, a population will move toward whichever trait is favored by individual-learning biases (Mesoudi 2011a). By the end of the Clovis period in the East, around 12,500 years ago, even the flaking pattern had become diversified (Gingerich et al. 2014), which strongly supports the notion that at the macroscale, social learning had been more or less eclipsed by guided variation.

Future work will be directed toward phylogenetic and morphometric analyses of post-Clovis point assemblages to assess what they might tell us about population dynamics in the resource-rich river valleys of eastern North America. We know, for example, that the Upper Southeast—the modern states of Missouri, Arkansas, Kentucky, Tennessee, North Carolina, and Virginia—contains more post-Clovis projectile-point shapes than any other region in the East (Eren et al. 2016), which is consistent with proposals that the river valleys of the Ohio, Tennessee, and Cumberland were arteries for colonizing populations moving east (Anderson 1990, 1996; Smallwood 2012; Broster et al. 2013). If those diverse type forms are proxies for populations, then they should be useful for tracking various groups that budded off and start moving to the Northeast and Southeast, encountering what perhaps were new fitness landscapes, complete with never-before-scaled fitness peaks and requiring a new mix of individual- and social-learning strategies. This is an exciting prospect for those of us interested in identifying prehistoric populations using items found in the archaeological record.

Note

1. The program used to create the tree was PAUP* (version 4) (Swofford 1998).

Acknowledgments We thank Huw Groucutt for his kind invitation to participate in this volume and Briggs Buchanan, Eric Delson, and two anonymous reviewers for their helpful suggestions for improving the chapter.

References

Anderson, D. G. (1990). The Paleoindian colonization of eastern North America: A view from the southeastern United States. In K. B. Tankersley & B. L. Isaac (Eds.), *Early Paleoindian economies of eastern North America* (pp. 163–216). Greenwich, CT: JAI Press.

Anderson, D. G. (1996). Models of Paleoindian and Early Archaic settlement in the lower Southeast. In D. G. Anderson & K. E. Sassaman (Eds.), *The Paleoindian and Early Archaic Southeast* (pp. 29–57). Tuscaloosa: University of Alabama Press.

Anderson, D. G., & Gillam, J. C. (2000). Paleoindian colonization of the Americas: Implications from an examination of physiography, demography, and artifact distribution. *American Antiquity, 65,* 43–66.

Aoki, K., & Feldman, M. W. (2014). Evolution of learning strategies in temporally and spatially variable environments: A review of theory. *Theoretical Population Biology, 91,* 3–19.

Aral, S., Muchnik, L., & Sundararajan, A. (2009). Distinguishing influence-based contagion from homophily-driven diffusion in dynamic networks. *Proceedings of the National Academy of Sciences, 106,* 21544–21549.

Beck, C., & Jones, G. T. (2010). Clovis and Western Stemmed: Population migration and the meeting of two technologies in the Intermountain West. *American Antiquity, 75,* 81–116.

Bement, L. C., & Carter, B. J. (2015). From mammoth to bison: Changing Clovis prey availability at the end of the Pleistocene. In A. M. Smallwood & T. A. Jennings (Eds.), *Clovis: On the edge of a new understanding* (pp. 263–275). College Station: Texas A&M University Press.

Bentley, R. A. (2008). Random drift versus selection in academic vocabulary. *PLoS ONE, 3*(8), e3057.

Bentley, R. A., Earls, M., & O'Brien, M. J. (2011). *I'll have what she's having: Mapping social behavior*. Cambridge, MA: MIT Press.

Bentley, R. A., Garnett, P., O'Brien, M. J., & Brock, W. A. (2012). Word diffusion and climate science. *PLoS ONE, 7(11),* e47966.

Bentley, R. A., & O'Brien, M. J. (2011). The selectivity of cultural learning and the tempo of cultural evolution. *Journal of Evolutionary Psychology, 9,* 125–141.

Bentley, R. A., & O'Brien, M. J. (2017). *The acceleration of cultural change: From ancestors to algorithms*. Cambridge, MA: MIT Press.

Bentley, R. A., O'Brien, M. J., & Brock, W. A. (2014). Mapping collective behavior in the big-data era. *Behavioral and Brain Sciences, 37,* 63–119.

Blong, J. C. (2018). Late-glacial hunter–gatherers in the Central Alaska Range and the role of upland ecosystems in the peopling of Alaska. *PaleoAmerica, 4,* 103–133.

Boas, F. (1904). The history of anthropology. *Science, 20,* 513–524.

Boas, F. (1911). Review of "Methode der ethnologie" by F. Graebner. *Science, 34,* 804–810.

Boyd, R., & Richerson, P. J. (1985). *Culture and the evolutionary process*. Chicago: University of Chicago Press.

Boyd, R., & Richerson, P. J. (1995). Why does culture increase human adaptability? *Ethology and Sociobiology, 16,* 125–143.

Boyd, R., & Richerson, P. J. (2005). *The origin and evolution of cultures*. Oxford: Oxford University Press.

Bradley, B. A. (1993). Paleo-Indian flaked stone technology in the North American High Plains. In O. Sofer & N. D. Praslov (Eds.), *From Kostenki to Clovis: Upper Paleolithic–Paleo-Indian adaptations* (pp. 251–262). New York: Plenum.

Bradley, B. A., Collins, M. B., & Hemmings, A. (2010). *Clovis technology*. Ann Arbor, MI: International Monographs in Prehistory.

Bradley, J. W., Spiess, A., Boisvert, R. A., & Boudreau, J. (2008). What's the point? Modal forms and attributes of Paleoindian bifaces in the New England–Maritimes Region. *Archaeology of Eastern North America, 36,* 119–172.

Bradley, B., & Stanford, D. (2004). The North Atlantic Ice-Edge Corridor: A possible Paleolithic route to the New World. *World Archaeology, 36,* 459–478.

Bradley, R., & Valcarce, R. F. (1998). Crossing the border: Contrasting styles of rock art in the prehistory of north-west Iberia. *Oxford Journal of Archaeology, 17,* 287–308.

Braje, T. J., Dillehay, T. D., Erlandson, J. M., Klein, R. G., & Rick, T. C. (2017). Finding the first Americans. *Science, 358,* 592–594.

Braje, T. J., Erlandson, J. M., Rick, T. C., Davis, L., Dillehay, T., Fedje, D. W., et al. (2019). Fladmark + 40: What have we learned about a potential Pacific Coast peopling of the Americas? *American Antiquity, 85,* 1–21.

Broster, J. B., Norton, M., Miller, D. S., Tune, J. W., & Baker, J. D. (2013). Tennessee Paleoindian record: The Cumberland and lower Tennessee River watersheds. In J. A. M. Gingerich (Ed.), *In the eastern fluted point tradition* (pp. 299–314). Salt Lake City: University of Utah Press.

Buchanan, B., Chao, A., Chiu, C.-H., Colwell, R. K., O'Brien, M. J., Werner, A., et al. (2017). Environment-induced changes in selective constraints during the peopling of the Americas. *Scientific Reports, 7,* 44431.

Buchanan, B., & Collard, M. (2007). Investigating the peopling of North America through cladistic analyses of Early Paleoindian projectile points. *Journal of Anthropological Archaeology, 26,* 366–393.

Buchanan, B., & Collard, M. (2008). Phenetics, cladistics, and the search for the Alaskan ancestors of the Paleoindians: A reassessment of relationships among the Clovis, Nenana, and Denali archaeological complexes. *Journal of Archaeological Science, 35,* 1683–1694.

Buchanan, B., Collard, M., Hamilton, M. J., & O'Brien, M. J. (2011). Points and prey: An evaluation of the hypothesis that prey size predicts early Paleoindian projectile point form. *Journal of Archaeological Science, 38,* 852–864.

Buchanan, B., & Hamilton, M. J. (2009). A formal test of the origin of variation in North American Early Paleoindian projectile points. *American Antiquity, 74,* 279–298.

Buchanan, B., Hamilton, M. J., Kilby, J. D., & Gingerich, J. A. M. (2016). Lithic networks reveal early regionalization in late Pleistocene North America. *Journal of Archaeological Science, 65,* 114–121.

Buchanan, B., O'Brien, M. J., & Collard, M. (2014). Continent-wide or region-specific? A geometric-morphometrics-based assessment of variation in Clovis point shape. *Archaeological and Anthropological Science, 6,* 145–162.

Cannon, M. D., & Meltzer, D. J. (2008). Explaining variability in Early Paleoindian foraging. *Quaternary International, 19,* 5–17.

Castro, L., & Toro, M. A. (2004). The evolution of culture: From primate social learning to human culture. *Proceedings of the National Academy of Sciences, 101,* 10235–10240.

Cavalli-Sforza, L. L., & Feldman, M. W. (1973). Models for cultural inheritance. I. Group mean and within group variation. *Theoretical Population Biology, 4,* 42–44.

Cavalli-Sforza, L. L., & Feldman, M. W. (1981). *Cultural transmission and evolution: A quantitative approach.* Princeton, NJ: Princeton University Press.

Chagnon, N. A. (2000). Manipulating kinship rules: A form of male Yanomamo reproductive competition. In L. Cronk, N. A. Chagnon, & W. Irons (Eds.), *Adaptation and human behavior: An anthropological perspective* (pp. 115–132). London: Routledge.

Clarke, D. L. (1968). *Analytical archaeology.* London: Methuen.

Collard, M., Buchanan, B., O'Brien, M. J., & Scholnick, J. (2013). Risk, mobility, or population size? Drivers of technological richness among contact-period western North American hunter–gatherers. *Philosophical Transactions of the Royal Society B, 368,* 20120412.

Couzin, I. D., Krause, J., Franks, N. R., & Levin, S. A. (2005). Effective leadership and decision making in animal groups on the move. *Nature, 433,* 513–516.

Curran, M. L. (1996). Palaeoindians in the Northeast: The problem of dating fluted point sites. *Review of Archaeology, 17,* 2–11.

Darwent, J., & O'Brien, M. J. (2006). Using cladistics to construct lineages of projectile points from northeastern Missouri. In C. P. Lipo, M. J. O'Brien, M. Collard, & S. J. Shennan (Eds.), *Mapping our ancestors: Phylogenetic approaches in anthropology and prehistory* (pp. 185–208). New York: Aldine.

Darwin, C. (1859). *On the origin of species by means of natural selection, or the preservation of favoured races in the struggle for life.* London: Murray.

Davis, L. G., Madsen, D. B., Becerra-Valdivia, L., Higham, T., Sisson, D. A., Skinner, S. M., et al. (2019). Late Upper Paleolithic occupation at Cooper's Ferry, Idaho, USA, ~16,000 years ago. *Science, 365,* 891–897.

Dean, L. G., Vale, G. L., Laland, K. N., Flynn, E., & Kendal, R. L. (2014). Human cumulative culture: A comparative perspective. *Biological Reviews, 89,* 284–301.

Denton, G. H., Anderson, R. F., Toggweiler, J. R., Edwards, R. L., Schaeer, J. M., & Putnam, A. E. (2010). The last glacial termination. *Science, 328,* 1652–1656.

Dunnell, R. C. (1978). Style and function: A fundamental dichotomy. *American Antiquity, 43,* 192–202.

Dunnell, R. C. (1980). Evolutionary theory and archaeology. *Advances in Archaeological Method and Theory, 3,* 35–99.

Eerkens, J. W., Bettinger, R. L., & Richerson, P. J. (2014). Cultural transmission theory and hunter–gatherer archaeology. In V. Cummings, P. Jordan, & M. Zvelebil (Eds.), *The Oxford handbook of the archaeology and anthropology of hunter–gatherers* (pp. 1127–1142). Oxford: Oxford University Press.

Eren, M. I., Buchanan, B., & O'Brien, M. J. (2015). Social learning and technological evolution during the Clovis colonization of the New World. *Journal of Human Evolution, 80,* 159–170.

Eren, M. I., Chao, A., Chiu, C., Colwell, R. K., Buchanan, B., Boulanger, M. T., et al. (2016). Statistical analysis of paradigmatic class richness supports greater Paleoindian projectile-point diversity in the Southeast. *American Antiquity, 81,* 174–192.

Erlandson, J. M., Graham, M. H., Bourque, B. J., Corbett, D., Estes, J. A., & Steneck, R. S. (2007). The kelp highway hypothesis: Marine ecology, the coastal migration theory, and the peopling of the Americas. *Journal of Island and Coastal Archaeology, 2,* 161–174.

Estrada, E., Meggers, B. J., & Evans, C. (1962). Possible transpacific contact on the coast of Ecuador. *Science, 135,* 371–372.

Evans, C., Meggers, B. J., & Estrada, E. (1959). *Cultura Valdivia.* Publication, no. 6, Museo Victor Estrada. Guayaquil, Ecuador.

Feldman, M. W., & Cavalli-Sforza, L. L. (1976). Cultural and biological evolutionary processes, selection for a trait under complex transmission. *Theoretical Population Biology, 9,* 238–259.

Fladmark, K. R. (1979). Routes: Alternate migration corridors for early man in North America. *American Antiquity, 44,* 55–69.

Foley, R. A. (1987). Hominid species and stone tools assemblages: How are they related? *Antiquity, 61,* 380–392.

Fragaszy, D. M., Biro, D., Eshchar, Y., Humle, T., Izar, P., Resende, B., et al. (2013). The fourth dimension of tool use: Temporally enduring artefacts aid primates learning to use tools. *Philosophical Transactions of the Royal Society B, 368,* 20120410.

Freeman, A. (2016). Why the Ice-Free Corridor is still relevant to the peopling of the New World. In M. Kornfeld & B. B. Huckell (Eds.), *Stones, bones, and profiles* (pp. 51–74). Boulder: University Press of Colorado.

Garcia-Herranz, M., Moro, M., Cebrian, M., Christakis, N. A., & Fowler, J. H. (2014). Using friends as sensors to detect global-scale contagious outbreaks. *PLoS ONE, 9*(4), e92413.

Gardner, W. M. (1977). Flint Run Paleoindian complex and its implications for eastern North American prehistory. *Annals of the New York Academy of Sciences, 288,* 257–263.

Garvey, R. (2018). Current and potential roles of archaeology in the development of cultural evolutionary theory. *Philosophical Transactions of the Royal Society B, 373,* 20170057.

Geribàs, N., Mosquera, M., & Vergès, J. M. (2010). What novice knappers have to learn to become expert stone toolmakers. *Journal of Archaeological Science, 37,* 2857–2870.

Gilbert, M. T. P., Jenkins, D. L., Götherstrom, A., Naveran, N., Sanchez, J. J., Hofreiter, M., et al. (2008). DNA from pre-Clovis human coprolites in Oregon, North America. *Science, 320,* 786–789.

Gingerich, J. A. M., Sholts, S. B., Wärmländer, S. K., & Stanford, D. J. (2014). Fluted point manufacture in eastern North America: An assessment of form and technology using traditional metrics and 3D digital morphometrics. *World Archaeology, 46,* 101–122.

Goebel, T., Waters, M. R., & O'Rourke, D. H. (2008). The Late Pleistocene dispersal of modern humans in the Americas. *Science, 319,* 1497–1502.

Goodyear, A. C. (1979). *A hypothesis for the use of cryptocrystalline raw materials among Paleoindian groups of North America.* South Carolina Institute of Archeology and Anthropology, Research Manuscript Series, no. 6. Columbia: University of South Carolina.

Groucutt, H. S. (2020). Into the tangled web of culture-history and convergent evolution. In H. Groucutt (Ed.), *Culture history and convergent evolution: Can we detect populations in prehistory?* (pp. 1–12). Cham, Switzerland: Springer.

Groucutt, H. S. (2020). Culture and convergence: The curious case of the Nubian Complex. In H. Groucutt (Ed.), *Culture history and convergent evolution: Can we detect populations in prehistory?* (pp. 55–86). Cham, Switzerland: Springer.

Halligan, J. J., Waters, M. R., Perrotti, A., Owens, I. J., Feinberg, J. M., Bourne, M. D., et al. (2016). Pre-Clovis occupation 14,550 years ago at the Page-Ladson site, Florida, and the peopling of the Americas. *Science Advances, 2,* e1600375.

Hamilton, M. J., & Buchanan, B. (2007). Spatial gradients in Clovis-age radiocarbon dates across North America suggest rapid colonization from the north. *Proceedings of the National Academy of Sciences, 104,* 15625–15630.

Hamilton, M. J., & Buchanan, B. (2009). The accumulation of stochastic copying errors causes drift in culturally transmitted technologies: Quantifying Clovis evolutionary dynamics. *Journal of Anthropological Archaeology, 28,* 55–69.

Haynes, C. V., Jr., Donahue, D. J., Jull, A. J. T., & Zabel, T. H. (1984). Application of accelerator dating to fluted point Paleoindian sites. *Archaeology of Eastern North America, 12,* 184–191.

Haynes, G. (2015). The millennium before Clovis. *Paleoamerica, 1,* 134–162.

Hennig, W. (1950). *Grundzüge einer theorie der phylogenetischen systematik.* Berlin: Deutscher Zentralverlag.

Hennig, W. (1966). *Phylogenetic systematics.* Urbana: University of Illinois Press.

Henrich, J. (2001). Cultural transmission and the diffusion of innovations: Adoption dynamics indicate that biased cultural transmission is the predominate force in behavioral change. *American Anthropologist, 103,* 992–1013.

Henrich, J., & Boyd, R. (1998). The evolution of conformist transmission and the emergence of between-group differences. *Evolution and Human Behavior, 19,* 215–241.

Henrich, J., & Broesch, J. (2011). On the nature of cultural transmission networks: Evidence from Fijian villages for adaptive learning biases. *Philosophical Transactions of the Royal Society B, 366,* 1139–1148.

Hermon, S., & Niccolucci, F. (2017). Formally defining the time–space–archaeological culture relation: Problems and prospects. *Archeologia e Calcolatori, 28,* 93–108.

Heyes, C. (2015). When does social learning become cultural learning? *Developmental Science, 20,* e12350.

Heyes, C. (2016). Blackboxing: Social learning strategies and cultural evolution. *Philosophical Transactions of the Royal Society B, 371,* 20150369.

Hoffecker, J. F., Elias, S. A., O'Rourke, D. H., Scott, G. R., & Bigelow, N. H. (2016). Beringia and the global dispersal of modern humans. *Evolutionary Anthropology, 25,* 64–78.

Hoppitt, W., & Laland, K. N. (2013). *Social learning.* Princeton, NJ: Princeton University Press.

Hutchings, W. K. (2015). Finding the Paleoindian spearthrower: Quantitative evidence for mechanically-assisted propulsion of lithic armatures during the North American Paleoindian period. *Journal of Archaeological Science, 55,* 34–41.

Ives, J. W., Froese, D., Supernant, K., & Yanicki, J. (2014). Vectors, vestiges, and valhallas: Rethinking the corridor. In K. E. Graf, C. V. Ketron, & M. R. Waters (Eds.), *Paleoamerican odyssey* (pp. 149–169). College Station: Center for the Study of the First Americans, Texas A&M University.

Jennings, T. A. (2013). The Hogeye Clovis cache, Texas: Quantifying lithic reduction signatures. *Journal of Archaeological Science, 40,* 649–658.

Kemp, B. M., & Schurr, T. G. (2010). Ancient and modern genetic variation in the Americas. In B. M. Auerbach (Ed.), *Human variation in the Americas* (pp. 12–50). Carbondale: Southern Illinois University Press.

Kendal, R. L., Boogert, N. J., Rendell, L., Laland, K. N., Webster, M., & Jones, P. L. (2018). Social learning strategies: Bridge-building between fields. *Trends in Cognitive Sciences, 22,* 651–665.

Kroeber, A. L. (1931). Historical reconstruction of culture growths and organic evolution. *American Anthropologist, 33,* 149–156.

Kvitek, D. J., & Sherlock, G. (2011). Reciprocal sign epistasis between frequently experimentally evolved adaptive mutations causes a rugged fitness landscape. *PLoS Genetics, 7*(4), e1002056.

Laland, K. N. (2004). Social learning strategies. *Learning and Behavior, 32,* 4–14.

Laland, K. N. (2017). *Darwin's unfinished symphony: How culture made the human mind.* Princeton, NJ: Princeton University Press.

Lehmer, D. J. (1971). *Introduction to Middle Missouri archeology.* National Park Service, Anthropological Papers, no. 1. Washington, DC.

Levine, M. A. (1990). Accommodating age: Radiocarbon results and fluted point sites in northeastern North America. *Archaeology of Eastern North America, 18,* 33–63.

Llamas, B., Fehren-Schmitz, L., Valverde, G., Soubrier, J., Mallick, S., Rohland, N., et al. (2016). Ancient mitochondrial DNA provides high resolution time scale of the peopling of the Americas. *Science Advances, 2,* e1501385.

Lothrop, J. C., Lowery, D. L., Spiess, A. E., & Ellis, C. J. (2016). Early human settlement of northeastern North America. *PaleoAmerica, 2,* 192–251.

Lothrop, J. C., Newby, P. E., Spiess, A. E., & Bradley, J. (2011). Paleoindians and the Younger Dryas in the New England–Maritimes region. *Quaternary International, 242,* 546–569.

Lyman, R. L., O'Brien, M. J., & Dunnell, R. C. (1997). *The rise and fall of culture history.* New York: Plenum.

Lyman, R. L., & O'Brien, M. J. (1998). The goals of evolutionary archaeology: History and explanation. *Current Anthropology, 39,* 615–652.

Lyman, R. L., O'Brien, M. J., & Hayes, V. (1998). A mechanical and functional study of bone rods from the Richey-Roberts Clovis cache, Washington, U.S.A. *Journal of Archaeological Science, 25,* 887–906.

Maher, L. A., & Macdonald, D. A. (2020). Communities of interaction: Tradition and learning in stone tool production through the lens of the Epipaleolithic of Khareh IV, Jordan. In H. Groucutt (Ed.), *Culture history and convergent evolution: Can we detect populations in prehistory?* (pp. 213–243). Cham, Switzerland: Springer.

McGhee, G. (2018). Limits on the possible forms of stone tools: A perspective from convergent biological evolution. In M. J. O'Brien, B. Buchanan, & M. I. Eren (Eds.), *Convergent evolution in stone-tool technology* (pp. 23–46). Cambridge, MA: MIT Press.

McNabb, J. (2020). Problems and pitfalls in understanding the Clactonian. In H. Groucutt (Ed.), *Culture history and convergent evolution: Can we detect populations in prehistory?* (pp. 29–53). Cham, Switzerland: Springer.

Meggers, B. J., Evans, C., & Estrada, E. (1965). *Early Formative period of Coastal Ecuador: The Valdivia and Machalilla phases.* Smithsonian Contributions to Anthropology (Vol. 11). Washington, DC: Smithsonian Institution.

Meltzer, D. J. (2009). *First peoples in a new world: Colonizing Ice Age America.* Berkeley: University of California Press.

Mesoudi, A. (2008). An experimental simulation of the "copy-successful-individuals" cultural learning strategy: Adaptive landscapes, producer–scrounger dynamics, and informational access costs. *Evolution and Human Behavior, 29,* 350–363.

Mesoudi, A. (2011a). *Cultural evolution: How Darwinian theory can explain human culture and synthesize the social sciences.* Chicago: University of Chicago Press.

Mesoudi, A. (2011b). Variable acquisition costs constrain cumulative cultural evolution. *PLoS ONE, 6*(3), e18239.

Mesoudi, A. (2014). Experimental studies of modern human social and individual learning in an archaeological context: People behave adaptively, but within limits. In T. Akazawa, N. Ogilhara, H. C. Tanabe, H. Terashima, T. Akazawa, N. Ogilhara, et al. (Eds.), *Dynamics of learning in Neanderthals and modern humans* (Vol. 2): *Cognitive and physical perspectives* (pp. 65–76). Tokyo: Springer.

Mesoudi, A., & Thornton, A. (2018). What is cumulative cultural evolution? *Proceedings of the Royal Society of London B, 285.* https://doi.org/10.1098/rspb.20180712.

Mesoudi, A., & Whiten, A. (2008). The multiple roles of cultural transmission experiments in understanding human cultural evolution. *Philosophical Transactions of the Royal Society B, 363,* 3489–3501.

Miller, D. S. (2016). Modeling Clovis landscape use and recovery bias in the southeastern United States using the Paleoindian Database of the Americas (PIDBA). *American Antiquity, 81,* 697–716.

Miller, D. S., & Gingerich, J. A. M. (2013a). Regional variation in the terminal Pleistocene and early Holocene radiocarbon record of eastern North America. *Quaternary Research, 79,* 175–188.

Miller, D. S., & Gingerich, J. A. M. (2013b). Paleoindian chronology and the eastern fluted point tradition. In J. A. M. Gingerich (Ed.), *In the Eastern Fluted Point tradition* (pp. 9–37). Salt Lake City: University of Utah Press.

Miller, G. L., Bebber, M. R., Rutkoski, A., Haythorn, R., Boulanger, M. T., Buchanan, B., et al. (2018). Hunter–gatherer gatherings: Stone-tool microwear from the Welling site (33-Co-2), Ohio, U.S.A. supports Clovis use of outcrop-related base camps during the Pleistocene peopling of the Americas. *World Archaeology.* https://doi.org/10.1080/00438243.2018.1461128.

Moreno-Mayar, J. V., Potter, B. A., Vinner, L., Steinrücken, M., Rasmussen, S., Terhorst, J., et al. (2018). Terminal Pleistocene Alaskan genome reveals first founding population of Native Americans. *Nature, 553,* 203–207.

Morgan, T. J. H., Rendell, L. E., Ehn, M., Hoppitt, W., & Laland, K. N. (2011). The evolutionary basis of human social learning. *Proceedings of the Royal Society B.* https://doi.org/10.1098.rspb.2011.1172.

Morrow, J. E. (1995). Fluted point manufacture: A perspective from the Ready Lincoln Hills site, 11JY46, Jersey County, Illinois. *Midcontinental Archaeology, 20,* 167–191.

Morrow, J. E. (2014). Review of "Across Atlantic ice: The origin of America's Clovis culture", by D.J. Stanford & B.A. Bradley. *Lithic Technology, 39,* 76–78.

Morrow, J. E., & Morrow, T. A. (1999). Geographic variation in fluted projectile points: A hemispheric perspective. *American Antiquity, 64,* 215–231.

Muthukrishna, M., & Henrich, J. (2016). Innovation in the collective brain. *Philosophical Transactions of the Royal Society B, 371,* 20150192.

Neiman, F. D. (1995). Stylistic variation in evolutionary perspective: Inferences from decorative diversity and interassemblage distance in Illinois Woodland ceramic assemblages. *American Antiquity, 60,* 7–36.

O'Brien, M. J. (2019a). Setting the stage: The Late Pleistocene colonization of North America. *Quaternary, 2.* https://doi.org/10.3390/quat2010001.

O'Brien, M. J. (2019b). More on Clovis learning: Individual-level processes aggregate to form population-level patterns. *PaleoAmerica, 5,* 157–168.

O'Brien, M. J., Bentley, R. A., & Brock, W. A. (2019). *The importance of small decisions.* Cambridge, MA: MIT Press.

O'Brien, M. J., Boulanger, M. T., Buchanan, B., Bentley, R. A., Lyman, R. L., Lipo, C. P., et al. (2016). Design space and cultural transmission: Case studies from Paleoindian North America. *Journal of Archaeological Method and Theory, 23,* 692–740.

O'Brien, M. J., Boulanger, M. T., Buchanan, B., Collard, M., & Lyman, R. L. (2014). Innovation and cultural transmission in the American Paleolithic: Phylogenetic analysis of eastern Paleoindian projectile-point classes. *Journal of Anthropological Archaeology, 34,* 100–119.

O'Brien, M. J., Boulanger, M. T., Collard, M., Buchanan, B., Tarle, L., Straus, L. G., et al. (2014). On thin ice: Problems with Stanford and Bradley's Solutrean-Clovis hypothesis. *Antiquity, 88,* 606–624.

O'Brien, M. J., & Buchanan, B. (2017). Cultural learning and the Clovis colonization of North America. *Evolutionary Anthropology, 26,* 270–284.

O'Brien, M. J., Buchanan, B., Boulanger, M. T., Mesoudi, A., Collard, M., Eren, M. I., et al. (2015). Transmission of cultural variants in the North American Paleolithic. In A. Mesoudi & K. Aoki (Eds.), *Learning strategies and cultural evolution during the Palaeolithic* (pp. 121–143). Tokyo: Springer.

O'Brien, M. J., Buchanan, B., Collard, M., & Boulanger, M. T. (2012). Cultural cladistics and the early history of North America. In P. Pontarotti (Ed.), *Evolutionary biology: Mechanisms and trends* (pp. 23–42). New York: Springer.

O'Brien, M. J., Collard, M., Buchanan, B., & Boulanger, M. T. (2013). Trees, thickets, or something in between? Recent theoretical and empirical work in cultural phylogeny. *Israel Journal of Ecology & Evolution, 59*(2), 45–61.

O'Brien, M. J., Darwent, J., & Lyman, R. L. (2001). Cladistics is useful for reconstructing archaeological phylogenies: Palaeoindian points from the southeastern United States. *Journal of Archaeological Science, 28,* 1115–1136.

O'Brien, M. J., & Holland, T. D. (1990). Variation, selection, and the archaeological record. *Archaeological Method and Theory, 2,* 31–79.

O'Brien, M. J., & Leonard, R. D. (2001). Style and function: An introduction. In T. D. Hurt & G. F. M. Rakita (Eds.), *Style and function: Conceptual issues in evolutionary archaeology* (pp. 1–25). Westport, CT: Bergin & Garvey.

O'Brien, M. J., & Lyman, R. L. (2000). *Evolutionary archaeology: A systematic approach.* New York: Plenum Press.

O'Brien, M. J., & Lyman, R. L. (2003). *Cladistics and archaeology.* Salt Lake City: University of Utah Press.

O'Brien, M. J., Lyman, R. L., Buchanan, B., & Collard, M. (2016). A review of late Pleistocene North American bone and ivory tools. In M. C. Langley (Ed.), *Osseous projectile weaponry* (pp. 221–235). Dordrecht, Netherlands: Springer.

O'Brien, M. J., Lyman, R. L., Mesoudi, A., & VanPool, T. L. (2010). Cultural traits as units of analysis. *Philosophical Transactions of the Royal Society B, 365,* 3797–3806.

O'Brien, M. J., Lyman, R. L., Saab, Y., Saab, E., Darwent, J., & Glover, D. S. (2002). Two issues in archaeological phylogenetics: Taxon construction and outgroup selection. *Journal of Theoretical Biology, 215,* 133–150.

O'Rourke, D. H., & Raff, J. A. (2010). The human genetic history of the Americas: The final frontier. *Current Biology, 20,* R202–R207.

Pederson, M. W., Ruter, A., Schweger, C., Friebe, H., Staff, R. A., Kjeldsen, K. K., et al. (2016). Postglacial viability and colonization in North America's Ice-Free Corridor. *Nature, 537,* 45–49.

Phillips, P., & Willey, G. R. (1953). Method and theory in American archaeology. *American Anthropologist, 55,* 615–633.

Posth, C., Nakatsuka, N., Lazaridis, I., Skoglund, P., Mallick, S., Lamnidis, T. C., et al. (2018). Reconstructing the deep population history of central and South America. *Cell, 175,* 1185–1197.

Potter, B. A., Baichtal, J. F., Beaudoin, A. B., Fehren-Schmitz, L., Haynes, C. V., Holliday, V. T., et al. (2018). Current evidence allows multiple models for the peopling of the Americas. *Science Advances, 4(8),* eaat5473.

Potter, B. A., Reuther, J. D., Holliday, V. T., Holmes, C. E., Miller, D. S., & Schmuck, N. (2017). Early colonization of Beringia and northern North America: Chronology, routes, and adaptive strategies. *Quaternary International, 444,* 36–55.

Powell, A., Shennan, S., & Thomas, M. G. (2009). Late Pleistocene demography and the appearance of modern human behavior. *Science, 324,* 1298–1301.

Prasciunas, M. M., & Surovell, T. A. (2015). Reevaluating the duration of Clovis: The problem of nonrepresentative radiocarbon. In A. M. Smallwood & T. A. Jennings (Eds.), *Clovis: On the edge of a new understanding* (pp. 21–35). College Station: Texas A&M.

Preston, D. D. (1999). Woody's dream. *New Yorker, 75*(34), 80–87.

Raff, J. A., & Bolnick, D. A. (2014). Palaeogenomics: Genetic roots of the first Americans. *Nature, 506,* 162–163.

Raff, J., & Bolnick, D. A. (2015). Does mitochondrial haplogroup X indicate ancient trans-Atlantic migration to the Americas? A critical reevaluation. *PaleoAmerica, 4,* 297–304.

Raff, J. A., Bolnick, D. A., Tackney, J., & O'Rourke, D. H. (2010). Ancient DNA perspectives on American colonization and population history. *American Journal of Physical Anthropology, 146,* 503–514.

Raghavan, M., Skoglund, P., Graf, K. E., Metspalu, M., Albrechtsen, A., Moltke, I., et al. (2014). Upper Palaeolithic Siberian genome reveals dual ancestry of Native Americans. *Nature, 505,* 87–91.

Raghavan, M., Steinrücken, M., Harris, K., Schiffels, S., Rasmussen, S., DeGiorgio, M., et al. (2015). Genomic evidence for the Pleistocene and recent population history of Native Americans. *Science, 349,* 841.

Rasmussen, M., Anzick, S. L., Waters, M. R., Skoglund, P., DeGiorgio, M., Stafford, T. W., Jr., et al. (2014). The genome of a late Pleistocene human from a Clovis burial site in western Montana. *Nature, 506,* 225–229.

Rendell, L., Boyd, R., Enquist, M., Eriksson, K., Feldman, M. W., Fogarty, L., et al. (2010). Why copy others? Insights from the social learning strategies tournament. *Science, 328,* 208–213.

Rendell, L., Boyd, R., Enquist, M., Feldman, M. W., Fogarty, L., & Laland, K. N. (2011). How copying affects the amount, evenness and persistence of cultural knowledge: Insights from the social learning strategies tournament. *Philosophical Transactions of the Royal Society B, 366,* 1118–1128.

Reynolds, N. (2020). Threading the weft, testing the warp: Population concepts and the European Upper Palaeolithic chronocultural framework. In H. Groucutt (Ed.), *Culture history and convergent evolution: Can we detect populations in prehistory?* (pp. 187–212). Cham, Switzerland: Springer.

Robinson, B. S., Ort, J. C., Eldridge, W. A., Burke, A. L., & Pelletier, B. G. (2009). Paleoindian aggregation and social context at Bull Brook. *American Antiquity, 74,* 423–447.

Rowe, J. H. (1966). Diffusionism and archaeology. *American Antiquity, 31,* 334–337.

Sanderson, M. J., & Hufford, L. (Eds.). (1996). *Homoplasy: The recurrence of similarity in evolution.* New York: Academic Press.

Shennan, S. (2020). Style, function and cultural transmission. In H. Groucutt (Ed.), *Culture history and convergent evolution: Can we detect populations in prehistory?* (pp. 291–298). Cham, Switzerland: Springer.

Shennan, S. J., & Wilkinson, J. R. (2001). Ceramic style change and neutral evolution: A case study from Neolithic Europe. *American Antiquity, 66,* 577–593.

Shipton, C. (2020). The unity of Acheulean culture. In H. Groucutt (Ed.), *Culture history and convergent evolution: Can we detect populations in prehistory?* (pp. 13–27). Cham, Switzerland: Springer.

Sholts, S. B., Stanford, D. J., Flores, L. M., & Wärmländer, S. K. T. S. (2012). Flake scar patterns of Clovis points analyzed with a new digital morphometrics approach: Evidence for direct transmission of technological knowledge across early North America. *Journal of Archaeological Science, 39,* 3018–3026.

Shott, M. J. (2020). Toward a theory of the point. In H. Groucutt (Ed.), *Culture history and convergent evolution: Can we detect populations in prehistory?* (pp. 245–259). Cham, Switzerland: Springer.

Simek, J. F., Cressler, A., Herrmann, N. P., & Sherwood, S. C. (2013). Sacred landscapes of the south-eastern USA: Prehistoric rock and cave art in Tennessee. *Antiquity, 87,* 430–446.

Simpson, G. G. (1961). *Principles of animal taxonomy.* New York: Columbia University Press.

Smallwood, A. M. (2012). Clovis technology and settlement in the American Southeast: Using biface analysis to evaluate dispersal models. *American Antiquity, 77,* 689–713.

Smallwood, A. M. (2013). Building experimental use-wear analogues for Clovis biface functions. *Archaeological and Anthropological Science, 7,* 13–26.

Smallwood, A. M., & Jennings, T. A. (2016). Experiments as analogues: Use-wear analysis of Clovis bifaces from the Gault site, Texas. In M. Kornfeld & B. B. Huckell (Eds.), *Stones, bones, and profiles* (pp. 103–126). Boulder: University of Colorado Press.

Smith, G. M., Duke, D., Jenkins, D. L., Goebel, T., Davis, L. G., O'Grady, P., et al. (2019). The Western Stemmed Tradition: Problems and prospects in Paleoindian archaeology in the Intermountain West. *PaleoAmerica.* https://doi.org/doi.org/10.1080/20555563.2019.1653153.

Smith, H. L., & Goebel, T. (2018). Origins and spread of fluted-point technology in the Canadian Ice-Free Corridor and eastern Beringia. *Proceedings of the National Academy of Sciences, 115,* 4116–4121.

Smith, H. L., Smallwood, A. M., & DeWitt, T. J. (2015). Defining the normative range of Clovis fluted point shape using geographic models of geometric morphometric variation. In A. M. Smallwood & T. A. Jennings (Eds.), *Clovis: On the edge of a new understanding* (pp. 161–180). College Station: Texas A&M University Press.

Stanford, D., & Bradley, B. (2002). Ocean trails and prairie paths? Thoughts about Clovis origins. In N. G. Jablonski (Ed.), *The first Americans: The Pleistocene colonization of the New World* (Vol. 27, pp. 255–271). Memoirs. San Francisco: California Academy of Sciences.

Stanford, D. J., & Bradley, B. A. (2012). *Across Atlantic ice: The origin of America's Clovis culture*. Berkeley: University of California Press.

Story, B. A., Eren, M. I., Thomas, K., Buchanan, B., & Meltzer, D. J. (2018). *Why are Clovis fluted points more resilient than non-fluted lanceolate points?* A quantitative assessment of breakage patterns between experimental models: Archaeometry. https://doi.org/10.1111/arcm.12407.

Stout, D. (2011). Stone toolmaking and the evolution of human culture and cognition. *Philosophical Transactions of the Royal Society B, 366,* 1050–1059.

Straus, L. G. (2000). Solutrean settlement of North America? A review of reality. *American Antiquity, 65,* 219–226.

Straus, L. G. (2005). The Upper Paleolithic of Cantabrian Spain. *Evolutionary Anthropology, 14,* 145–158.

Sutton, M. Q. (2018). Paleoindian-era osseous rods: Distribution, dating, and function. *PaleoAmerica, 4,* 183–201.

Swofford, D. (1998). *PAUP*: Phylogenetic analysis using parsimony *and other methods, version 4*. Sunderland, MA: Sinauer.

Tehrani, J., & Collard, M. (2002). Investigating cultural evolution through biological phylogenetic analyses of Turkmen textiles. *Journal of Anthropological Archaeology, 21,* 443–463.

Tennie, C., Call, J., & Tomasello, M. (2009). Ratcheting up the ratchet: On the evolution of cumulative culture. *Philosophical Transactions of the Royal Society B, 364,* 2405–2415.

Thomas, K. A., Story, B. A., Eren, M. I., Buchanan, B., Andrews, B. N., O'Brien, M. J., et al. (2017). Explaining the origin of fluting in North American Pleistocene weaponry. *Journal of Archaeological Science, 81,* 23–30.

Thompson, R. D., Whitlock, C., Bartlein, P. J., Harrison, S. P., & Spaulding, W. G. (1993). Vegetation, lake levels and climate in the western United States since 18,000 B.P. In H. E. Wright, Jr., J. E. Kutzbach, T. Webb, III, W. F. Ruddiman, F. A. Street-Perrott, & P. J. Bartlein (Eds.), *Global climates since the Last Glacial Maximum* (pp. 468–513). Minneapolis: University of Minnesota Press.

Thompson, R. H. (1956). An archaeological approach to cultural stability. In *Seminars in Archaeology: 1955* (pp. 31–57). Memoirs, no. 11. Menasha, WI: Society for American Archaeology.

Tomasello, M. (1999). *The cultural origins of human cognition*. Cambridge, MA: Harvard University Press.

Tomasello, M., Kruger, A. C., & Ratner, H. H. (1993). Cultural learning. *Behavioral and Brain Sciences, 16,* 495–552.

Vosoughi, S., Roy, D., & Aral, S. (2018). The spread of true and false news online. *Science, 359,* 1146–1151.

Waguespack, N. M. (2007). Why we're still arguing about the Pleistocene occupation of the Americas. *Evolutionary Anthropology, 16,* 63–74.

Waters, M. R., Forman, S. L., Jennings, T. A., Nordt, L. C., Driese, S. G., Feinberg, J. M., et al. (2011). The Buttermilk Creek Complex and the origins of Clovis at the Debra L. Friedkin site, Texas. *Science, 331,* 1599–1603.

Waters, M. R., & Stafford, T. W., Jr. (2007). Redefining the age of Clovis: Implications for the peopling of the Americas. *Science, 315,* 1122–1126.

Waters, M. R., & Stafford, T. W., Jr. (2014). The first Americans: A review of the evidence for the Late Pleistocene peopling of the Americas. In K. E. Graf, C. V. Ketron, & M. R. Waters (Eds.), *Paleoamerican odyssey* (pp. 541–560). College Station: Texas A&M University Press.

Whiten, A., McGuigan, N., Marshall-Pescini, S., & Hopper, L. M. (2009). Emulation, imitation, over-imitation and the scope of culture for child and chimpanzee. *Philosophical Transactions of the Royal Society B, 364,* 2417–2428.

Whitt, E. E. (2010). *Extraordinary fluted points of the Tennessee Valley region*. Florence, AL: Printers and Stationers.

Whittaker, J. C. (2004). *American flintknappers: Stone age art in the age of computers*. Austin: University of Texas Press.

Will, M., & Mackay, A. (2020). A matter of space and time: How frequent is convergence in lithic technology in the African archaeological record over the last 300 kyr? In H. Groucutt (Ed.), *Culture history and convergent evolution: Can we detect populations in prehistory?* (pp. 103–126). Cham, Switzerland: Springer.

Willey, G. R. (1953). Archaeological theories and interpretation: New World. In A. L. Kroeber (Ed.), *Anthropology today* (pp. 361–385). Chicago: University of Chicago Press.

Willey, G. R., & Phillips, P. (1958). *Method and theory in American archaeology*. Chicago: University of Chicago Press.

Williams, T. J., Collins, M. B., Rodrigues, K., Rink, W. J., Velchoff, N., Keen-Zebert, A., et al. (2018). Evidence of an early projectile point technology in North America at the Gault site, Texas, USA. *Science Advances, 4,* eaar5954.

Wooler, M. J., Saulnier-Talbot, E., Potter, B. A., Belmecheri, S., Bigelow, N., Choy, K., et al. (2018). A new terrestrial palaeoenvironmental record from the Bering Land Bridge and context for human dispersal. *Royal Society Open Science*. https://doi.org/10.1098/rsos.180145.

Wormington, H. M. (1957). *Ancient man in North America*. Denver Museum of Natural History Popular Series, no. 4.

Wright, S. (1932). The roles of mutation, inbreeding, crossbreeding and selection in evolution. In D. F. Jones (Ed.), *Proceedings of the Sixth Congress on Genetics* (Vol. 1, pp. 356–366). New York: Brooklyn Botanic Garden.

Wright, S. (1988). Surfaces of selective value revisited. *American Naturalist, 131,* 115–123.

Chapter 14
Culture, Environmental Adaptation or Specific Problem Solving? On Convergence and Innovation Dynamics Related to Techniques Used for Stone Heat Treatment

Patrick Schmidt

Abstract Heat treatment of stone for tool knapping may have been one of the oldest documented transformative techniques of materials. It has been interpreted as shedding light on the technical behavior of past humans, on their time- or resource-management and even on their cognitive capacities. Its earliest invention likely dates to the African Middle Stone Age, but prominent examples are also known from more recent periods on all other continents. In all of these contexts, stone was heated with specific techniques, applying specific parameters, and in many cases these vary between regions. Differences may be interpreted as technical responses to specific problems, as adaptations to environmental factors like climate, or alternatively as more or less random markers of cultural identity. This chapter will consider these possibilities by comparing the techniques and parameters applied during heat treatment in five different archaeological contexts: the earliest known cases from Southern Africa; the European Upper Paleolithic Solutrean culture; the European Mesolithic Beuronien culture; the European Neolithic Chassey culture; and the recent North-American Paleo-Indian period. During these five periods, stone was transformed for purposes that may be interpreted as being similar yet slightly different. The stones themselves were of different nature and strong variability of the used heating parameters can be observed. In the end of this chapter, I will discuss observations on the dynamics of invention, reinvention and technical convergences.

P. Schmidt (✉)
Department of Early Prehistory and Quaternary Ecology, Eberhard Karls University of Tübingen, Schloss Hohentübingen, 72070 Tübingen, Germany
e-mail: patrick.schmidt@uni-tuebingen.de

Department of Geosciences, Applied Mineralogy, Eberhard Karls University of Tübingen, Wilhelmstraße 56, 72074 Tübingen, Germany

Keywords Early transformative technology • Pyrotechnology • Archaeometry • Lithic heat treatment • Invention and re-invention

Introduction

Heat treatment of stone for tool knapping may well be one of the oldest documented transformative techniques of materials. Its invention in the southern African Middle Stone Age (MSA) (Brown et al. 2009) marks a turning point in the cultural evolution of modern humans because stone knappers no longer accepted the properties of available resources, but began to deliberately transform them. Heat treatment is also known from later periods such as the European Upper Paleolithic and Mesolithic (Bordes 1969; Tiffagom 1998; Eriksen 2006), the American Paleo-Indian period (Crabtree and Butler 1964; Wilke et al. 1991) and the European Neolithic (Binder 1984; Léa 2005). It has been interpreted as being a proxy for many archaeological concepts: modern behaviors (Sealy 2009), complex cognition (Wadley 2013), high technical skill (Inizan and Tixier 2001) or non-shared specialized craftsmanship (Léa et al. 2012). The underlying assumption is that heat treatment requires an important investment in terms of cognition, resources and time. This assumption, in turn, is based on interpretations of the actions performed and choices made during the heating process: the heating technique and procedure. Unfortunately, such heating techniques cannot be easily reconstructed from material evidence. This has been possible in the past at sites that preserved intact heating structures (for an example see: Shippee 1963) but evidence of this kind is fairly rare in the archaeological record (Schmidt 2016). In most cases, heating techniques must be understood by reading a set of proxies specific to a particular heating environment or procedure. This has recently been attempted for four chrono-cultural contexts: the southern African MSA (see for example:

H. S. Groucutt (ed.), *Culture History and Convergent Evolution: Can We Detect Populations in Prehistory?*,
Vertebrate Paleobiology and Paleoanthropology, https://doi.org/10.1007/978-3-030-46126-3_14

Schmidt et al. 2015b), the European Upper Paleolithic Solutrean (Schmidt and Morala 2018), the Mesolithic Beuronian of southern Germany (Schmidt et al. 2017b) and the French Neolithic Chassey culture (Schmidt et al. 2013a). These contexts together allow us to appreciate a first picture of the technical solutions to heating stone at different periods and in different natural environments.

The obvious question arising from these first datasets is whether these techniques were similar (either completely, or partially—i.e. containing at least similar elements). The potential implications of similarities observed in a priori unrelated contexts could be of potential importance for understanding the mechanisms of original invention, but also of potentially independent reinvention (see for example: Tennie et al. 2017). Reinvention implies the possibility for partial or full independent convergence that, if observed in two archaeological contexts, raises questions about the role of cultural transmission in explaining similarities (see also, Shennan 2020). In other words, are similar technical behaviors across some or all contexts the result of cultural transmission? Or did similarities arise in the absence of cultural transmission? How may we envision that such independent convergence happen? One potential answer lies in the responses to specific natural environments, another in the predictability of individual reaction to specific technical problems. This chapter will examine these questions by comparing the parameters of some of the archaeologically documented heating techniques. Another possibility to investigate these questions would be to rely on descriptions of ethnographic observations of heat treatment (see for example: Hester 1972; Mandeville 1973), but most reports are short and imprecise in terms of the preformed actions, heating environments, even the nature of the stones that were heated. This discussion will therefore only take into account the archaeological contexts for which explicit material evidence on at least some aspects of the used heating technique is available.

Expectations and competing hypotheses: At first glance, there are two possible explanations of the heat treatment-related patterns observed across contexts: Heating techniques may be [1] culturally transmitted from one context to another, or [2] independently (re-)invented at different places and times. In the first case, transmission, we would expect to observe a relation between techniques used at different times in confined regions. Such a relation might be the absence of change over time, but also gradual not immediately reversible change (local evolution). However, the search for such a relation only allows to make a negative argument: a relation between contexts would neither exclude transmission nor reinvention, but the absence of a relation would plead against transmission. If reinvention had been at play, the question becomes: can we observe independent convergence in heating techniques? The absence of any convergence will most likely lead to the conclusion that it is impossible to explain invention dynamics by external factors. If we can observe (at least partial) convergence, we can try to investigate the mechanisms driving the invention and adaptation of techniques.

The Southern African Middle and Later Stone Age

Archaeologists have only recently discovered the antiquity of heat treatment of silcrete (a relatively coarse-grained pedogenic silica rock) in the archaeological record of South Africa's Cape region. The discovery by Brown et al. (2009) that some artefacts were heat-treated in c. 164 ka old deposits at the site of Pinnacle Point showed for the first time that MSA hunter-gatherers intentionally transformed some of their raw materials with fire. The authors also found that at c. 71 ka, almost all silcrete artefacts had been knapped after being heated, suggesting an important role of heat treatment in local technology. Other datasets from the MSA (Schmidt et al. 2015b; Delagnes et al. 2016) and the Later Stone Age (LSA) (Porraz et al. 2016) also revealed more than 80% of all silcrete to be heat-treated. Hence, heat treatment was used as a standard procedure, applied to all or almost all silcrete before knapping in the MSA and LSA of Africa's Cape region.

The heating technique used for these thermal treatments, and the investment in time and resources they require, have also been the subject of intensive debates, initiating a discussion about technical complexity and cognitive capacity (Brown and Marean 2010; Schmidt et al. 2013b, 2015a; Wadley 2013; Wadley and Prinsloo 2014). For example, if silcrete heat treatment was a time and resource consuming process as suggested by some authors (Brown et al. 2009; Brown and Marean 2010; Wadley 2013; Wadley and Prinsloo 2014), it must have considerably slowed down the lithic reduction sequence and most likely also altered raw material and resource provisioning strategies. This would have a significant impact on the selective context that made such investment worthwhile. However a mineralogical and crystallographic analysis of the transformations taking place in Cape silcrete during heat treatment (Schmidt et al. 2013b) showed that the material itself does not require a specially slow heating procedure, but can be heat-treated in the embers of an open air fire. The temperatures experienced by rocks heated in embers were described as scattering between 350 and 500°C (Schmidt et al. 2015b, 2017b), not imposing the risk of excessive heat fracturing in silcrete. The finding of such great heat tolerance of cape silcrete implied that a specially built heating environment creating slow heating rates (like a sand-bath, see for example: Brown et al. 2009) was not necessary in the MSA. However, this model was challenged by Wadley and Prinsloo (2014) who found in an experimental study that many of their samples heated in open air fires showed signs of heat-induced failure. On this basis, they argued for the necessity of a sand-bath to successfully heat-treat silcrete. A new element in this debate emerged from a study of c. 63–80 ka old artefacts from the Diepkloof Rockshelter. The discovery of a previously undescribed residue indicated that silcrete was indeed heat-treated in open fires during this period (Schmidt et al. 2015b). The residue is an organic wood-tar strictly associated with surfaces that correspond to the outer limits of the silcrete blocks at the time of their heat treatment. It was deposited on the silcrete surface by dry distillation of plant exudations and contains micrometre-sized charcoal inclusions, indicating that it formed in the reducing conditions of a pile of glowing embers (Schmidt et al. 2016a). A second argument in favour of such a fast heating technique came from the finding that up to 10% of the heat-treated lithics from Diepkloof show signs of heat-induced fracturing in a fire after which they were still knapped (Schmidt et al. 2015b). Such heat-induced failure only occurs at high temperatures and fast heating rates and is practically absent when silcrete is heated in a sand-bath (Schmidt et al. 2013b; Schmidt 2014; Wadley and Prinsloo 2014). Since this initial discovery, tempering residue and heat-induced fractures after which knapping continued have been identified in numerous other South African MSA and LSA sites (Delagnes et al. 2016; Porraz et al. 2016; Schmidt and Mackay 2016; Schmidt 2019) and today, it seems to be a secure assertion that at least most of the heat-treated silcrete in MSA and LSA assemblages was heated using a fast and expedient heating technique that relied on the use of open-air fires, perhaps regular domestic fires.

The European Upper Paleolithic Solutrean

The c. 22–18 ka old Solutrean is the oldest European context to have yielded proof of intentional heat treatment of rocks for stone knapping (Bordes 1967, 1969). In this context, relatively fine-grained silica rocks like flint and chert (henceforth only called chert) were heated. In contrast to the African MSA and LSA evidence, Solutrean heat treatment was not universally applied to a large range of artefact types. The artefact class best recognized as being knapped from heat-treated chert comprises the so-called laurel-leaf points or *feuilles de laurier*. Several examples from south-western France (Bordes 1969) and Spain (Tiffagom 1998) document thermal treatment as part of the later stages of the reduction sequence associated with the production of these bifacial points. The production of some of these artefacts also involved a final step of pressure knapping (Aubry et al. 1998). The strict association between pressure flaking and heat treatment in the Solutrean has recently been questioned by a study of the unique Solutrean laurel-leaf points of Volgu (Schmidt et al. 2018). These relatively largest and most skillfully crafted laurel leaf-points known today were not modified by heat, yet some of them benefited of a final step of pressure retouch. Still, the finely crafted laurel-leaf points of the Solutrean document a high technical skill of the knappers of this period and heat treatment was part of this skillset in at least some cases.

Also, the Solutrean was for long considered the oldest culture where heat treatment was practiced (Tiffagom 1998; Inizan and Tixier 2001) before Brown et al. (2009) found the African silcrete evidence. However, together with the Siberian Dyuktai culture (Flenniken 1987), the Solutrean still appears to have yielded the earliest evidences of heat treatment of chert. Such finer-grained silica rocks need to be heated with a procedure that involves relatively low temperature, slow heating rates (Schmidt et al. 2011, 2012) and thus larger investment in time and resources (Schmidt et al. 2016b). This was already noticed by the first experimenters attempting to heat-treat chert (Crabtree and Butler 1964) and the theory of sand-bath heating was used to interpret the heating technique used in the Solutrean (Inizan and Tixier 2001). The technique actually used for heat treatment in the Solutrean was recently investigated by Schmidt and Morala (2018). The authors used a technique based on near infrared spectroscopy (Schmidt et al. 2013a) to investigate the heating temperatures experienced by 44 laurel-leaf points from the Laugerie-Haute site. The underlying assumption behind these analyses was that different heating environments and procedures produce different temperatures. If the pieces had been heated in an open fire, the effective heating temperatures measured in different artefacts could be expected to scatter within a large interval of temperatures, as the embers of open fires were found to produce a wide range of different temperatures (see for example: Bentsen 2013). If, however, Solutrean heat treatment instigators had used a dedicated heating environment like a sand-bath, these temperatures can be expected to fall into a narrower range and be generally lower. The study found that most of the analyzed laurel-leaf points were heat-treated with temperatures between 250 and 300°C, a minor part of the samples between 200 and 250°C and only four samples were heat-treated slightly but insignificantly above 300°C (Schmidt and Morala 2018). The only way such reproducibility of similar heating temperatures can be achieved is by a standardized technique that allows the reproduction of similar conditions during successive heating cycles. A sand-bath or similar underground heating structures allows one to heat-treat stone with a range of temperatures from 200 to 400°C and fairly good standardization (Mandeville and Flenniken 1974; Griffiths et al. 1987; Eriksen 1997; Brown et al. 2009) and therefore appears to be a valid working hypothesis explaining the observed pattern. As it stands, the Solutrean data rule out the possibility of heat treatment in open-air fires and point in the direction of indirect, perhaps underground, heating.

The Mesolithic Beuronian

The Early Mesolithic of south-western Germany, the so-called Beuronian (9600–7100 BC), is yet another period that yielded evidence of stone heat treatment. Its material leftovers are found in the Swabian Jura region, a ~200 km long and ~70 km wide limestone plateau of Jurassic age in south-western Germany. It was a period of important transformations in the way people lived, in their subsistence and in the stone tools they produced. Typical lithic artefacts for this period are small triangular or rectangular microliths that were used as hafted implements on wooden projectiles. The majority of the Beuronian lithic assemblage is made from local chert of Jurassic age, an opaque white and slightly rough-looking chert. Part of this chert was heat-treated prior to tool production (Hahn 1998). Several works explored Beuronian heat treatment, providing the first insights into its relative prevalence in different assemblages (Eriksen 2006), and investigating possibly applied heating techniques experimentally (Eriksen 1997). Some of these studies found that Jurassic chert was particularly heat resistant (Eriksen 1997) and hypothesized a low-investment, cost- and time-effective heating technique, relying on the active part of above-ground fires for this period. A recent study (Schmidt et al. 2017b) on the Beuronian site Helga-Abri investigated the heating environment with the same near-infrared-based technique described above. The authors estimated the heating temperatures of all artefacts that were found to be heat-treated to fall in a relatively large temperature interval, ranging from 350 to 500°C. These temperatures lie significantly above the temperatures determined for other heat-treated archaeological assemblages, namely the Solutrean assemblage described above. The degree of standardization allowed by the Beuronian technique also seems to be considerably lower. The Beuronian temperature ranges of ±75°C are statistically broader than the ±~30°C of the Solutrean (Schmidt and Morala 2018). Standardized heating techniques, such as sand-baths or earth-ovens, are unlikely to produce such great scattering of heating temperatures, precluding the hypothesis of their use in the Mesolithic of south-western Germany. Thus, the study found no indication of a specific heating environment or oven-like structure that would allow to produce, control and maintain a well-calibrated range of heating temperatures in the stones. On the contrary, using open-air fires for heat treatment can be expected to produce a wider range of heating temperatures when the stones are placed at different parts of the embers or ashes and temperatures as high as 550°C have been attained with this technique experimentally (Schmidt et al. 2015b). Schmidt et al. (2017b) conclude in their study that the observed pattern can be reasonably well explained by the hypothesis that Jurassic chert was heat-treated in the above-ground part of camp-fires. This would put the Mesolithic evidence and the African silcrete data on the same page, both documenting the use of fast, expedient and rather opportunistic techniques.

The Neolithic Chassey Culture

The Neolithic Chassey culture of southern France (4100–3500 BC) also documents heat treatment of chert. The treatment was systematically used for producing pressure-flaked bladelets (Léa 2005). It may even have been the reason for the widespread use of a particular type of chert from the French Vaucluse region that can be found at sites in all of southern France, Tuscany (Italy) and Catalonia (Spain). The Chassey reduction sequence included heat treatment of large volumes of this chert shaped into pre-cores (preforms) that attained up to 7 cm in diameter. The discovery of lithic production sites in the Vaucluse region, where these large preforms were heat-treated, shows that the treatment was conducted by specialists who did not seem to have shared their know-how (Léa 2004). Heating large volumes of this chert must be considered a difficult task, as it has to my knowledge not yet been possible to experimentally heat-treat such large preforms of this particular chert in 'actualistic' conditions without thermal fracturing (overheating). Unlike for the African data on silcrete heat treatment, such heat-induced fracturing would render the Neolithic chert preforms useless for further pressure-reduction. One of the reasons for this failure to reproduce Chassey heat treatment is that most of the parameters applied during heating remained unknown until recently. In response to this, two studies aimed at determining the heating temperatures experienced by Chassey artefacts. On experiment used the above described near-infrared analyses (Schmidt et al. 2013a) and the other investigated the pressure in fluid inclusions within heat-treated chert (Milot et al. 2017). Both studies found average heating temperatures between 200 and 250°C for the analyzed flakes and a precision of heating temperatures of

$\pm\sim 25°C$. It thus appears that heat treatment in the Neolithic Chassey culture was an even better calibrated process than in the Solutrean (i.e. producing a slightly narrower interval, and generally lower temperatures). It allowed to produce and re-produce these temperatures in chert during successive heating cycles (Schmidt et al. 2013a; Milot et al. 2017). Similar to the Solutrean, this may be understood as yet another augment for underground heating using sand-baths or similar structures. However, it should be emphasized that, as for the Solutrean, there has not been any other data indicating such a technique in the Neolithic so far. During this period, witnessing a steadily increasing technical know-how and fire-related skills (e.g. the mastering of ceramic firing), it appears prudent to await more detailed data on the techniques used for stone heat treatment before final conclusions can be drawn. At our current state of knowledge, it can only be stated that this technique, similarly to the Solutrean, aimed at producing good temperature control and standardization and that the data support underground heating.

The Paleo-Indian Evidence for Underground Heating

The perhaps most detailed description of an archaeological structure used for heat treatment was made by Shippee (1963). He interpreted an undated feature found in North America as a fire-pit used for heat treatment of chert. He described a $\sim$45 cm-deep pit containing an infill of chert, sediment and ashes. The pit contained at its base a bed of ashes. Chert cores and flakes were placed on top of the ashes. The pit was backfilled with sediment and limestone boulders on top of the chert. This isolated and undated dataset provides a small window onto the North American heat treatment evidence and unambiguously documents the used of underground structures in this context. Although this data is of a very different nature than the above explained examples from Europe and Africa, it can nonetheless be compared with the latter. Similar underground heating techniques have successfully been used to heat-treat fine grained silica rocks like chert in heating experiments (see for example: Mandeville and Flenniken 1974). During these experiments it was noted that the indirect heating in the sand environment allowed good temperature control and slow heating rates. It appears therefore likely that such a technique would allow to produce similar patterns in heated stones as the ones recorded from Solutrean and Chassey artefacts. This indicates that the heat treatment technique used in all three contexts was similar or at least contained similar elements.

Similarities, Dissimilarities, Convergence?

The data detailed above are not all of the same kind, in some cases being precise heating temperatures, in others direct or indirect evidence of heating environments. This is unfortunate and results from the different suitability of silcrete and chert for analysis with analytical techniques (e.g. silcrete is too opaque for the infrared-based method for temperature reconstruction described above). It is nonetheless possible to compare different contexts in terms of the heating environment used (either directly in fires or indirectly in underground or oven-like structures). All heating techniques discussed above are compared in Table 14.1. In summary, two of the above described contexts yielded evidence of stone heat treatment using the above-ground part of fires (the African MSA to LSA and the German Mesolithic) whereas the other three contexts yielded evidence for indirect heat treatment, perhaps in underground structures.

Can cultural transmission explain this pattern? One way of examining this question is by comparing the three techniques throughout the European sequence, from the Solutrean to the Neolithic Chassey culture, where direct or indirect population contact (necessary for cultural transmission) may at least be tentatively assumed. It is not suggested here that there was any type of cultural continuity across the three European contexts. Comparing them will not likely answer the question of whether heat treatment techniques were directly transmitted from one of those contexts to another (for example, heat treatment was not even practiced during the Magdalenian period that separates the Solutrean

Table 14.1 Comparison between the five heat treatment bearing contexts discussed in this chapter. The early date under 'Approx. age/duration' corresponds to the earliest published age for heat treatment within the context and the second date to the end of the context

Context	Approx. age/duration	Heating temp.	Heating environment
MSA/LSA	164–$\sim$12 ka	$\sim$350–500°C	Open-air fires
Solutrean	22–18 ka	$\sim$250–300°C	Indirect heating. Underground?
Beuronian	9600–7100 BC	$\sim$350–500°C	Open-air fires
Chassey	4100–3500 BC	$\sim$200–250°C	Indirect heating. Underground? Ceramic kiln?
Paleo-Indian	Undated holocene	Probably $\sim$200–350°C	Underground heating

and the Mesolithic Beuronian in time; also, there might have been important population turnovers between contexts). However, if such a comparison were to be made and if it would result in the observation of continuity, it might be argued that there were some, not yet understood indirect mechanisms of transmission or perhaps a collective memory of techniques (e.g. via other similar but more regularly practiced fire-based techniques; such as bleeding over of cooking styles). The testing conditions for a relation between contexts would be satisfied if the heating technique practiced in Europe was invariable, or if shifts from one technique to another were gradual, or perhaps if we could observe irreversible changes from one heating technique to another. Neither of these was the case in Europe: the rather well-standardized indirect heating technique of the Solutrean was replaced by an opportunistic camp fire-based technique in the Mesolithic that had no apparent similarities. The following Neolithic yielded evidence for even higher standardization and control that were most likely only possible by indirect heating. Thus, the European data do not provide arguments for the transmission of technical knowledge related to heat treatment. However, again, this sequence is not ideal to test for such transmission. Can the obvious problems of the European sequence be overcome by seeking for transmission in other contexts? The MSA to LSA sequence of Africa's Cape region provides an alternative dataset. There, no archaeological evidence of underground heating has ever been brought forward and, from at least 70 ka (Schmidt and Högberg 2018) to about 20 ka (Porraz et al. 2016), silcrete appears to have been invariably heated in open-air fires. And so, in this case, cultural transmission is a possible scenario. However, this would be in contrast to technological changes in other domains (see also Will and Mackay 2020) and probably also population turnovers during this period. In all non-African cases, heat treatment appears to have been an independent (re-) invention.

The similarities between some of these independently invented heating techniques must then be termed independent convergence. It can be expected that at least some traits of these heating techniques result from inherent processes, necessities or structures within the heat-treating groups (i.e. they were not chosen arbitrarily). For example, building a heating environment that allows temperature control and slow heating rates is cost-intensive (Brown and Marean 2010; Schmidt et al. 2016b) and its re-invention in three distant contexts most likely followed some underlying reasoning. If this was the case, by what factors can the partial convergences be explained and what might have been the reasons for choosing one technique or another?

One possibility is to explain convergence in heating techniques by environmental factors such as climate or the availability of wood fuel. However, such factors are very different in southern Africa during the MSA and central Europe during the Mesolithic, both contexts that documented heat treatment in open-air fires. The same is true for the three contexts that documented underground heating: such techniques are more resource-consuming (Brown and Marean 2010), so that one might expected to find fuel-efficient open-air fires in the arid Last Glacial Maximum (at the time of the Solutrean) and more fuel-consuming underground structures in the temperate and more humid Mesolithic. The contrary was the case. Thus, external factors related to climate cannot explain the observed pattern. Another approach to explaining these convergences comes from understanding the heated rocks themselves. In all three contexts that documented underground heat treatment and good temperature control, it was fine-grained silica rocks like chert that were heated. Such rocks typically have ideal heating temperatures between 200 and 350°C (Schmidt et al. 2012, 2013c, 2017a). Most become even less well suited for stone knapping after heating above these temperatures (see for example: Inizan et al. 1976; Terradas and Gibaja 2001). A similar statement can be made for the speed these rocks can be heated with. If heating rates are too fast, chert may overheat and become un-knappable (Schmidt 2014). Thus, finer-grained silica rocks require slow and low-temperature heating and one way of producing such conditions is by setting up a heating environment (Schmidt et al. 2016b) that relies on indirect, perhaps underground, heating. Other rocks like silcrete do not pose the same problem. Silcrete heated in Africa's Cape region did not require particularly slow or low-temperature heating conditions (Schmidt et al. 2013b). It is therefore not surprising that this context documents the use of above-ground fires for heat treatment. The same is true for the Jurassic chert heat-treated in open-air fires in the German Mesolithic. As detailed above, this chert is unusually heat resistant, not failing when heated rapidly in open fires (Eriksen 1997). The dichotomy between slow indirect- and fast direct-heating may thus be the result of specific problem solving of different groups with access to different types of rock. In other words, it was no coincidence that chert that is susceptible to overheating was carefully heated in unrelated contexts in dissimilar natural environments. This was rather the specific responses to similar technical problems posed by similar materials. It was also no coincidence that more heat-tolerant rocks like silcrete and the Beuronian Jurassic chert were heated in open-air fires because, in the absence of constraints in terms of temperature or heating rate, knappers chose the simplest and most efficient technical solution. Thus, oriented problem solving and the intention not to complicate techniques when it is not necessary appears to provide the best explanation of the partial convergence in techniques used for stone heat treatment in different parts of the world.

Outlook

These observations result in obvious questions about other contexts that document heat treatment of similar types of stone. An ideal case study would be the comparison between silcrete heat treatment in Africa's Cape region and Australia. Both regions are rich in silcrete types that have previously been described as being similar in terms of genesis, mineralogy and structure (see for example: Summerfield 1983). Heat treatment was part of silcrete reduction sequences since at least 25 ka in Australia (Hanckel 1985; Schmidt and Hiscock 2019) and even longer in Africa's Cape region (Brown et al. 2009). Contact or cultural transmission can be confidently ruled out in these two distant contexts.

Where the requirements in terms of heating temperature or heating speed of Australian and southern African silcrete the same? Was the technical response of knappers the same or, in other words, did early Australians heat-treat silcrete in the same way as knappers in Africa? The comparison between both continents would provide ideal conditions to investigate the mechanisms and dynamics of inventions, cultural differentiation and oriented problem solving.

Acknowledgements Financial support for this study was provided by the Deutsche Forschungsgemeinschaft (DFG) (grant number SCHM 3275/2-1). I am thankful to C. Tennie for his advice on the introduction of this manuscript.

References

Aubry, T., Walter, B., Robin, E., Plisson, H., & Ben-Habdelhadi, M. (1998). Le site solutréen de plein-air des Maitreaux (Bossay-sur-Claise, Indre-et-Loire): un faciès original de production lithique. *Paléo*, 163–184. http://www.persee.fr/doc/pal_1145-3370_1998_num_10_1_1135.

Bentsen, S. E. (2013). Controlling the heat: An experimental approach to Middle Stone Age pyrotechnology. *South African Archaeological Bulletin, 68,* 137–145.

Binder, D. (1984). Systèmes de débitage laminaire par pression: exemples chasséens provençaux. In J. Tixier, M. L. Inizan, & H. Roche (Eds.), *Préhistoire de la pierre taillée, 2: économie du débitage laminaire: technologie et expérimentation: IIIe table ronde de technologie lithique. Meudin-Bellevue, octobre 1982* (pp. 71–84). Paris: Cercle de Recherches et d'Etudes Préhistoriques.

Bordes, F. (1967). Considérations sur la typologie et les techniques dans le Paléolithique. *Quartär, 18,* 25–55.

Bordes, F. (1969). Traitement thermique du silex au Solutréen. *Bulletin de la Société préhistorique française*, 66(7), 197. http://www.persee.fr/web/revues/home/prescript/article/bspf_0249-7638_1969_num_66_7_10404.

Brown, K., & Marean, C. (2010). Wood fuel availability for heat treatment drives the rise and fall of silcrete as a raw material in the Middle Stone Age of South Africa, Abstracts of the PaleoAnthropology Society 2010 Meetings. *PaleoAnthropology, 2010,* A0001–A0040.

Brown, K. S., Marean, C. W., Herries, A. I. R., Jacobs, Z., Tribolo, C., Braun, D., et al. (2009). Fire as an engineering tool of early modern humans. *Science, 325,* 859–862. http://www.sciencemag.org/content/325/5942/859.abstract.

Crabtree, D. E., & Butler, B. R. (1964). Notes on experiment in flint knapping: 1 Heat treatment of silica materials. *Tebiwa, 7,* 1–6.

Delagnes, A., Schmidt, P., Douze, K., Wurz, S., Bellot-Gurlet, L., Conard, N. J., et al. (2016). Early evidence for the extensive heat treatment of silcrete in the Howiesons Poort at Klipdrift Shelter (Layer PBD, 65 ka), South Africa. *PLoS ONE, 11,* e0163874. https://doi.org/10.1371/journal.pone.0163874.

Eriksen, B. V. (1997). Implications of thermal pre-treatment of chert in the German Mesolithic. In R. Schild & Z. Sulgostowska (Eds.), *Man and Flint, Proceedings of the VII International Flint Symposium Warszawa-Ostrowiec Swietokrzyski, September 1995* (pp. 325–329). Warsaw: Institute of Archaeology and Ethnology Polish Academy of Sciences.

Eriksen, B. V. (2006) Colourful lithics—The "Chaîne Opératoire" of heat treated chert artefacts in the early Mesolithic of Southwest Germany. In C. J. Kind (Ed.), *After the Ice Age. Settlements, subsistence and social development in the Mesolithic of Central Europe, Materialhefte zur Archäologie in Baden-Württemberg* (pp. 147–153). Stuttgart: Konrad Thesis Verlag.

Flenniken, J. (1987). The Paleolithic dyuktai pressure blade technique of Siberia. *Journal of Arctic Anthropology, 24,* 117–132.

Griffiths, D. R., Bergman, C. A., Clayton, C. J., Ohnuma, K., & Robins, G. V. (1987). Experimental investigation of the heat treatment of flint. In G. D. G. Sieveking & M. H. Newcomer (Eds.), *The human uses of flint and chert. Proceedings of the Fourth International Flint Symposium Held at Brighton Polytechnic 10–15 April 1983* (pp. 43–52). Cambridge: Cambridge University Press.

Hahn, J. (1998). Opportunistic patterns of lithic reduction at the Mesolithic site of Rottenburg-Siebenlinden 1. In N. C. Conard & C.-J. Kind (Eds.), *Aktuelle Forschungen zum Mesolithikum—Current Mesolithic Research, Urgeschichtliche Materialhefte 12* (pp. 251–255). Tübingen: Archaeologica Venatoria.

Hanckel, M. (1985). Hot rocks: Heat treatment at Burrill Lake and Currarong, New South Wales. *Archaeology in Oceania, 20,* 98–103.

Hester, T. R. (1972). Ethnographic evidence for the thermal alteration of siliceous stone. *Tebiwa, 15,* 63–65.

Inizan, M. L., Roche, H., & Tixier, J. (1976). Avantages d'un traitement thermique pour la taille des roches siliceuses. *Quaternaria Roma, 19,* 1–18.

Inizan, M. L., & Tixier, J. (2001). L'émergence des arts du feu: le traitement thermique des roches siliceuses. *Paléorient, 26,* 23–36.

Léa, V. (2004). Centres de production et diffusion des silex bédouliens au Chasséen. *Gallia préhistoire, 46,* 231–250.

Léa, V. (2005). Raw, pre-heated or ready to use: Discovering specialist supply systems for flint industries in mid-Neolithic (Chassey culture) communities in southern France. *Antiquity, 79,* 1–15.

Léa, V., Roque-Rosell, J., Binder, D., Sciau, P., Pelegrin, J., Regert, M., et al. (2012). Craft specialization and exchanges during the southern Chassey culture: An integrated archaeological and material sciences approach. *Colloque international Networks in the Neolithic. Exchange of raw materials, products and ideas in the Western Mediterranean-VII°–III° millennium BC, Février 2011 Barcelone (Espagne), 5,* 119–127.

Mandeville, M. D. (1973). A consideration of the thermal pretreatment of chert. *Plains Anthropologist, 18,* 177–202.

Mandeville, M. D., & Flenniken, J. (1974). A comparison of the flaking qualities of Nehawka chert before and after thermal pretreatment. *Plains Anthropologist, 19,* 146–148.

Milot, J., Siebenaller, L., Béziat, D., Léa, V., Schmidt, P., & Binder, D. (2017). Formation of fluid inclusions during heat treatment of barremo-bedoulian flint: Archaeometric implications. *Archaeometry, 59,* 417–434. https://doi.org/10.1111/arcm.12256.

Porraz, G., Igreja, M., Schmidt, P., & Parkington, J. E. (2016). A shape to the microlithic Robberg from Elands Bay Cave (South Africa). *Southern African Humanities, 29,* 203–247.

Schmidt, P. (2014). What causes failure (overheating) during lithic heat treatment? *Archaeological and Anthropological Sciences, 6,* 107–112. https://doi.org/10.1007/s12520-013-0162-3.

Schmidt, P. (2016). The 'sand-bath' and lithic heat treatment in the South African Middle Stone Age: Myth or reality? *African Archaeological Review, 33,* 99–105.

Schmidt, P. (2019). How reliable is the visual identification of heat treatment on silcrete? A quantitative verification with a new method. *Archaeological and Anthropological Sciences, 11*(2), 713–726. https://doi.org/10.1007/s12520-017-0566-6.

Schmidt, P., Badou, A., & Fröhlich, F. (2011). Detailed FT near-infrared study of the behaviour of water and hydroxyl in sedimentary length-fast chalcedony, SiO_2, upon heat treatment. *Spectrochimica Acta Part A: Molecular and Biomolecular Spectroscopy, 81,* 552–559. http://www.sciencedirect.com/science/article/pii/S1386142511005397.

Schmidt, P., Bellot-Gurlet, L., & Floss, H. (2018). The unique Solutrean laurel-leaf points of Volgu: Heat-treated or not? *Antiquity, Accepted since 18/10/2017.*

Schmidt, P., February, E., Bretzke, K., & Bellot-Gurlet, L. (2016a). Tempering-residue on heat-treated silcrete: An experimental perspective and a potential analytical protocol. *Journal of Archaeological Science: Reports, 15,* 611–619. http://www.sciencedirect.com/science/article/pii/S2352409X16304746.

Schmidt, P., & Hiscock, P. (2019). Evolution of silcrete heat treatment in Australia—A regional pattern on the South-East Coast and its evolution over the last 25 ka. *Journal of Paleolithic Archaeology, 2,* 74–97.

Schmidt, P., & Högberg, A. (2018). Heat treatment in the Still Bay—A case study on Hollow Rock Shelter, South Africa. *Journal of Archaeological Science: Reports, 21,* 712–720. http://www.sciencedirect.com/science/article/pii/S2352409X18302918.

Schmidt, P., Lauer, C., Buck, G., Miller, C. E., & Nickel, K. G. (2017a). Detailed near-infrared study of the 'water'-related transformations in silcrete upon heat treatment. *Physics and Chemistry of Minerals, 44,* 21–31. https://doi.org/10.1007/s00269-016-0833-6.

Schmidt, P., Léa, V., Sciau, P., & Fröhlich, F. (2013a). Detecting and quantifying heat treatment of flint and other silica rocks: A new non-destructive method applied to heat-treated flint from the Neolithic Chassey culture, southern France. *Archaeometry, 55,* 794–805.

Schmidt, P., & Mackay, A. (2016). Why was silcrete heat-treated in the Middle Stone Age? An early transformative technology in the context of raw material use at Mertenhof Rock Shelter, South Africa. *PLoS ONE, 11,* e0149243. https://doi.org/10.1371/journal.pone.0149243.

Schmidt, P., Masse, S., Laurent, G., Slodczyk, A., Le Bourhis, E., Perrenoud, C., et al. (2012). Crystallographic and structural transformations of sedimentary chalcedony in flint upon heat treatment. *Journal of Archaeological Science, 39,* 135–144. http://www.sciencedirect.com/science/article/pii/S030544031100330X.

Schmidt, P., & Morala, A. (2018). First insights into the technique used for heat treatment of chert at the Solutrean site of Laugerie-Haute, France. *Archaeometry, 60,* 885–897.

Schmidt, P., Paris, C., & Bellot-Gurlet, L. (2015a). The investment in time needed for heat treatment of flint and chert. *Archaeological and Anthropological Sciences.* https://doi.org/10.1007/s12520-015-0259-y.

Schmidt, P., Paris, C., & Bellot-Gurlet, L. (2016b). The investment in time needed for heat treatment of flint and chert. *Archaeological and Anthropological Sciences, 8,* 839–848. https://doi.org/10.1007/s12520-015-0259-y.

Schmidt, P., Porraz, G., Bellot-Gurlet, L., February, E., Ligouis, B., Paris, C., et al. (2015b). A previously undescribed organic residue sheds light on heat treatment in the Middle Stone Age. *Journal of Human Evolution, 85,* 22–34.

Schmidt, P., Porraz, G., Slodczyk, A., Bellot-Gurlet, L., Archer, W., & Miller, C. E. (2013b). Heat treatment in the South African Middle Stone Age: Temperature induced transformations of silcrete and their technological implications. *Journal of Archaeological Science, 40,* 3519–3531. http://www.sciencedirect.com/science/article/pii/S0305440312004670.

Schmidt, P., Slodczyk, A., Léa, V., Davidson, A., Puaud, S., & Sciau, P. (2013c). A comparative study of the thermal behaviour of length-fast chalcedony, length-slow chalcedony (quartzine) and moganite. *Physics and Chemistry of Minerals, 40,* 331–340. https://doi.org/10.1007/s00269-013-0574-8.

Schmidt, P., Spinelli Sanchez, O., & Kind, C.-J. (2017b). Stone heat treatment in the Early Mesolithic of southwestern Germany: Interpretation and identification. *PLoS ONE, 12,* e0188576.

Sealy, J. (2009). Modern behaviour in ancient South Africans: Evidence for the heat treatment of stones in the Middle Stone Age. *South African Journal of Science, 105,* 323–324.

Shennan, S. (2020). Style, function and cultural transmission. In H. Groucutt (Ed.), *Culture history and convergent evolution: Can we detect populations in prehistory?* (pp. 291–298). Cham, Switzerland: Springer.

Shippee, J. M. (1963). Was flint annealed before flaking? *Plains Anthropologist, 8*(22), 271–272.

Summerfield, M. A. (1983). Silcrete. In A. S. Goudie & K. Pye (Eds.), *Chemical sediments and geomorphology* (pp. 59–91). London: Academic Press.

Tennie, C., Premo, L. S., Braun, D. R., & McPherron, S. P. (2017). Early stone tools and cultural transmission: Resetting the null hypothesis. *Current Anthropology, 58,* 652–672. https://doi.org/10.1086/693846.

Terradas, X., & Gibaja, J. F. (2001). El tratamiento termico en la produccion litica: el ejemplo del Neoltico Medio Catalan. *Cypsela, 13,* 31–58.

Tiffagom, M. (1998). Témoignages d'un traitement thermique des feuilles de laurier dans le Solutréen supérieur de la grotte du Parpalló (Gandia, Espagne). *Paléo, 10,* 147–161. http://www.persee.fr/web/revues/home/prescript/article/pal_1145-3370_1998_num_10_1_1134.

Wadley, L. (2013). Recognizing complex cognition through innovative technology in Stone Age and Palaeolithic Sites. *Cambridge Archaeological Journal, 23,* 163–183. https://doi.org/10.1017/S0959774313000309.

Wadley, L., & Prinsloo, L. C. (2014). Experimental heat treatment of silcrete implies analogical reasoning in the Middle Stone Age. *Journal of Human Evolution, 70,* 49–60. http://www.sciencedirect.com/science/article/pii/S0047248414000621.

Wilke, P. J., Flenniken, J., & Ozbun, T. L. (1991). Clovis technology at the Anzick Site, Montana. *Journal of California and Great Basin Anthropology, 13,* 242–272.

Will, M., & Mackay, A. (2020). A matter of space and time: How frequent is convergence in lithic technology in the African archaeological record over the last 300 kyr? In H. Groucutt (Ed.), *Culture history and convergent evolution: Can we detect populations in prehistory?* (pp. 103–126). Cham, Switzerland: Springer.

Chapter 15
Style, Function and Cultural Transmission

Stephen Shennan

Abstract Recent evolutionary approaches to the understanding of lithic variability take us back to long-standing issues in lithic studies to do with the claimed contrast between style and function and the Binford-Bordes debate of the 1960s concerning the factors that affect inter-assemblage variation. In fact, the style and function contrast is an unhelpful one, not least when considering the question of convergence. Taking the definition of style as 'a way of doing', all functions are carried out in locally specific ways that have a transmission history, although the extent to which the history of the attributes relevant to the function have been subject to random drift and innovation patterns, as opposed to selection, will vary. Moreover, in a subtractive technology like lithics the extent to which a transmission signal will be visible in an attribute like the angle of a cutting edge is unclear. The contrasting view is that, in the case of lithics, functional requirements will always call into existence the technical innovations to satisfy them, which in any case are not that difficult to find. The paper addresses these and related issues with reference to previous work by Shennan and colleagues on the use of material culture to identify within and between group variation, the extent to which isolation-by-distance in space and time can account for the similarities and differences between assemblages, and the role of phylogenetic methods.

Keywords Lithics • Heritability • Isolation-by-distance • Cultural evolution • Selection • Drift • Phylogenetics • The comparative method

S. Shennan (✉)
Institute of Archaeology, University College London, 31–34 Gordon Square, London, WC1H 0PY, UK

Introduction

The famous Binford-Bordes debate of the 1960s and early 70s (e.g. Binford and Binford 1966; Binford 1973; Bordes 1973) concerning how to explain the pattern of changing Mousterian assemblages in SW France in many ways encapsulated the contrast between the long-standing (European) tradition of culture history and the newly emerging (American) approach of 'new archaeology' (for a recent assessment see Wargo 2009). For the Binfords the patterning was explicable in terms of technical variation between the assemblages, responding to different functional requirements of groups exploiting different resources in different environments at sites that had different roles in mobile settlement systems; in other words, the reasons for the presence of different numbers of different tool types were situational, and by implication convergent. For Bordes they were simply assumed to be a reflection of the social traditions of different human groups, following the long-established interpretive conventions of culture history.

The contrast between the culture history and systemic 'new archaeology' perspectives was also played out, of course, in the study of later periods. Here Binford (1965) was concerned to make a number of important distinctions between different dimensions of variation: the tradition, 'is seen in continuity in those formal attributes which vary with the social context of manufacture exclusive of the variability related to the use of the item. This is termed stylistic variability…; the adaptive area exhibits the common occurrence of artifacts used primarily in coping directly with the physical environment' (pp. 208–9); these are 'technomic' artefacts, or the technomic dimension of artefacts, following Binford (1962). In principle, the commonalities of artefacts characterizing the adaptive area could be the result of independent convergence from different starting points.

Whereas for agricultural societies there may be multiple lines of evidence that can convincingly be argued to relate

H. S. Groucutt (ed.), *Culture History and Convergent Evolution: Can We Detect Populations in Prehistory?*,
Vertebrate Paleobiology and Paleoanthropology, https://doi.org/10.1007/978-3-030-46126-3_15

differentially to these different dimensions, as Binford showed, this is much more problematical for the Paleolithic, where variation in lithic assemblages has had to play multiple roles. In particular, recognizing the multi-dimensionality of the archaeological record meant distinguishing stylistic attributes of artefacts relating to the 'tradition', the ongoing social context of manufacture, from those related to function and adaptation, for which 'tradition' was, by definition, irrelevant. In keeping with Binford, Dunnell (1978) defined stylistic variation as variation not under selection and asserted that stylistic attributes could be defined *a priori*, on the basis of whether or not they involved differential energy expenditure.

However, this is not sustainable. Even decorative attributes on ceramic vessels, the stylistic attribute par excellence, can potentially be under selection for social reasons; for example, pressure to conform to group norms that might have a bearing on people's chances of marriage and reproductive success. In contrast, if we take the definition of style as 'a way of doing', all functions are carried out in locally specific ways that have a transmission history, including adaptive ones, although the extent to which the history of the attributes relevant to the function has been subject to random drift and innovation patterns (see below), as opposed to selection, will vary. Thus, explaining variation in lithic assemblages through time should take all these aspects into account and arrive at conclusions about the factors affecting variation in a set of attributes at the end of a process of analysis, not by *a priori* assumption.

In fact, of course, this was not a line that Binford pursued. His ethnoarchaeological work among the Nunamiut (1978) demonstrated strikingly different situational patterns, for example in the material left behind at different types of site associated with different activities, and the importance of practices such as tool curation in relation to factors such as time stress; in other words, technological organization. More generally, throughout his later career his interest focused on ecological aspects of adaptation, culminating in his 2001 book, *Constructing Frames of Reference*, along lines parallel to those of human behavioral ecology and specifically optimal foraging theory, though these were not approaches he ever accepted. Both exclude culture from consideration, whether tactically or on the basis of an in-principle rejection of the importance of culture in understanding human behavior. From this perspective we can understand the reasons for changing the atlatl for the bow-and-arrow, for example, simply by looking at their effect on the return rates of different prey in terms of the costs and benefits represented in the diet breadth model (e.g. Hames and Vickers 1982) in relation to the environmental conditions, such as the encroachment of forests in northern latitudes at the end of the last Ice Age. The dynamic comes from the environment, not from the cultural system and effectively assumes that as environments change they will call into existence the technical innovations to exploit them successfully. This implicitly presupposes that the innovations concerned lie within what Tennie et al. (2009) call the 'zone of latent solutions', things that are easily inventable by individuals working from first principles, and thus likely to be convergent. This may be true in some cases. It seems that wherever seed exploitation became important it led to the convergent innovation and use of grindstones, but this contrasts with the case of the more complex technology of the bow-and-arrow, for example, whose spread by diffusion can be traced across North America (e.g. Blitz 1988; Angelbeck and Cameron 2014).

If we return to the Bordes side of the argument, it has already been pointed out that the interpretation of the changing Mousterian assemblages as a reflection of changing human communities was no more than an interpretive convention characteristic of the time, based on the assumption that there was some mental template generating the patterns. There is no evidence for it other than the inter-assemblage variation that it seeks to explain and, as Binford pointed out, it seems highly unlikely that there would be a mental template for producing assemblages containing different proportions of different types. In fact, more recent assessments include elements of both interpretations, in keeping with the theoretical principles discussed below. Delagnes and Rendu (2011) argue that the different Mousterian types correspond to different technical principles in *lithic production* (my italics), which have implications for mobility.

Extensive discussions in the 1980s between Sackett (e.g. 1982, 1985), Wiessner (1983, 1985), and others, and later by Carr (1995), addressed the nature of different kinds of artefact variability and the factors affecting technological choices. They provided the theoretical basis for a more sophisticated approach that escaped the conflation of the adoption of different choices with 'ethnic identity' and included the possibility of choices made on the basis of differential efficacy in achieving a goal (summarized in Tostevin 2012, Chap. 3). However, this literature was focused on the choices involved in artefact production, not on the processes that generate assemblages, which are linked to technological organization and its situational use (see e.g. Holdaway and Douglass 2012 for a recent discussion), but are also strongly affected by taphonomy and time-averaging (e.g. Shott 2008), a point to which we will return.

Cultural Evolution and Lithics

With regard to the style and function issues, it has been the development of cultural evolutionary theory, in the sense of a set of ideas and methods for understanding cultural change as a process of descent with modification, since the 1980s (Cavalli-Sforza and Feldman 1981; Boyd and Richerson 1985), that has provided a coherent theoretical framework that can be used to make further advances. This is because it has provided a set of relevant mechanisms for understanding continuity and change through integrating transmission and adaptation. The starting point is the process of cultural transmission, involving a variety of social learning mechanisms and the transmitted environments in which they take place—it may be difficult to distinguish the effects of the one from the other. Innovations, intended or unintended ('copying errors') generate new variation, and various sorting processes, including selection influenced by the environment but also drift, can act on the variation that is transmitted to change the frequencies of different variants. The effects of environmental adaptation on variation in artefacts (the 'technomic' dimension) cannot be considered independently of 'tradition' and the non-selective factors that also affect it. Importantly, the cultural evolution framework has also provided a set of tools for addressing the issues raised by the need to make these distinctions.

While the starting point for psychological or ethnographic studies of cultural transmission processes is the experimental or observational study of the processes themselves, in the case of archaeology it is variation in the artefacts, ecofacts and their spatial-temporal arrangements that is the basis of analysis (Shennan 2011) We need to distinguish the variation related to transmission and the sorting processes affecting what is transmitted from other factors. From the evolutionary point of view, testing hypotheses about convergence in lithic assemblages involves tracing different independent artefactual lineages through time and showing that they arrive at similar solutions from different starting points. Cultural phylogenetics provides a well-established set of methods for making these distinctions, which have been extensively applied to the study of lithics to distinguish convergent characteristics (homoplasies) from features arising from common descent, and specifically shared-derived characteristics (synapomorphies), provided that they are applied to appropriate variables (see e.g. papers in Lipo et al. 2006 or O'Brien et al. 2018). It is the application of these methods that enables us to evaluate the probability in any given case that an innovation is a homoplasy in the 'zone of latent solutions' or builds on a specific set of prior innovations in a specific lineage. . Importantly, it is necessary to recognize that lithic assemblages as such are not the results of transmission processes associated with specific ways of doing, though they are made up of the products of such processes. They are time-averaged outcomes of large number of events affected by many contingent factors as well as evolutionary forces, but also by factors such as artefact use-lives (e.g. Shott 2008). The relevant analogy is paleontological species assemblages. These came originally from ecological communities, made up of many evolving species but varying in response to local variations in temperature, precipitation and edaphic conditions that would have had a selective effect on the components and their relative representation. However, their composition in the paleontological record is likely to be overwhelmingly dominated by taphonomic factors and the scale of time-averaging of different conditions over which they accumulated. However, neither in their original, and even less in their time-transformed, state do they tell us about processes of descent with modification.

Artefact Production

Several recent developments based on adopting a cultural evolution approach to lithic variation contribute to making progress in distinguishing the role of transmission and performance characteristics in *producing* lithic artefacts, for example handaxes (e.g. Key and Lycett 2017), the sphere in which descent with modification becomes relevant. What is emerging from this is that, within broad functional limits where stabilizing selection influenced by the ergonomics of hand-held cutting tools becomes relevant, there is considerable variation that stems from the operation of other cultural transmission processes (Lycett et al. 2016). One of these is drift, chance variation in what is copied within particular transmission chains, depending on who is in contact with whom and therefore on both geographic and temporal distance. But selection also depends on transmission; thus, directional selection will result from the preferential imitation of some specific portion of the available range of variation. For example, if smaller tools are more effective for butchering smaller prey and climate change or an increase in diet breadth resulting from over-exploitation of resources results in increasing exploitation of small prey then the mean size of the tools produced may decrease. This will be spatially and temporally specific, like the fluctuating short-term environmentally-based selection pressures operating on the beaks of Galapagos finches (Grant and Grant 2002) although these time scales may be beyond our levels of resolution. Such pressures may also result in convergence. If the increasing exploitation of small prey is the result of large-scale climate change then the same directional change may occur in a number of local traditions as a result of the operation of the same selection process. This is potentially identifiable by assessing the extent to which tool variation and variation in relevant aspects of faunal assemblages correlate with one another, for example. On the other hand, it is important to emphasize that drift too, resulting simply from copying-error, can also be directional (Bentley et al. 2004; Eerkens and Lipo 2005) and is likely to be greater in a reductive technology such as lithics (Schillinger et al. 2014).

The fact that in finite populations, i.e. in all real world situations, chance processes occurring in the process of cultural transmission can have directional consequences was something never appreciated by the processualists.

This point leads on to another recent development, using what Lycett and von Cramon-Taubadel (2015) call a 'quantitative genetics' approach to distinguishing the role of transmission in generating lithic variation from other factors. The situation is similar to that faced by geneticists trying to understand the factors affecting quantitative dimensions such as variation in height between members of the same species which are the result of complex causality, including the action of multiple genes as well as environmental factors such as diet. We can in principle follow the geneticists in distinguishing between the heritable component of quantitative variation in the cultural phenotype and that produced by other factors as well as random variation. In the case of lithic artefacts, as noted above, in addition to raw material variation there may be variation resulting from re-sharpening. These latter effects are potentially quantifiable and can allow us to obtain the residual heritable variation by subtraction. In any case, as the authors emphasize, so long as there is *any* heritable variation, over the longer or shorter term evolutionary forces will have an effect on the variation concerned as a result of the operation of selection and drift, as discussed. Discontinuities in the heritable component are likely to indicate discontinuities in transmission.

Tostevin (2012) takes a different approach to the same question, proposing a positive approach to characterizing the variation that is culturally transmitted. It is generally agreed that the traditional characterizations of lithic 'industries' cannot be used for this purpose (e.g. Shea 2017), because transmission forces have a limited impact at best on assemblage formation, as noted above. In their place Tostevin proposes a series of variables associated with blank production as well as tool kit selection. These derive from the specific context of the acquisition of the skills of local lithic production in the close observation of flint-knapping episodes, and therefore what is visible in the relevant taskscape. In the light of the close contact implied by lithic learning and the strong evidence for the vertical transmission of craft skills, if not actually from parents then from other close group members of the older generation (Shennan and Steele 1999), continuities and discontinuities through time in the relevant variables reflect continuities and discontinuities in transmission, which are likely to correspond to continuities and discontinuities in gene flow. On this basis, after an analysis of relevant lithic assemblages Tostevin concludes that the appearance of the initial Upper Paleolithic 'Bohunician Behavioral Package' in Central and Eastern Europe and the Levant marked a discontinuity with what went before where it occurred and that it spread through a process of demic diffusion.

Building and Testing Models

Appropriate kinds of empirically and theoretically justified analytical description then potentially enable us to track transmitted variation and the forces that influence it, at the same time minimizing the possibility of mistakenly rejecting the conclusion that the patterns are a result of convergence. Given that this is the case we can define an initial null model to account for spatial and temporal variation in 'ways of doing' that are the outcome of social learning processes. In the spatial domain the model is what geneticists call 'isolation by distance' (cf. Scerri et al. 2018). Cultural transmission depends on interaction, and, for the transmission of skills, often close interaction, as Tostevin (2012) emphasizes. Interaction decreases with distance so, in the absence of other forces, similarity in transmitted variation will also decline in the same way. Similarly in the temporal domain. Other things being equal, change will result from 'drift', the chance loss of variants through time in the course of transmission, and innovation, the generation of novel variation, both dependent on the cultural effective population size, the number of individuals interacting with respect to the specific transmission process in question. When there are departures from such null models the reasons for them can be explored. Spatial and temporal discontinuities may be accounted for by discontinuities in transmission or by shifting selection pressures; continuities by preferential interaction or stabilizing selection. Whether there are indeed departures can be tested by the use of techniques similar to those used for the same purpose in genetics.

Thus Shennan et al. (2015) carried out an analysis to see if spatial and temporal distance were the only factors affecting variation in the sets of attributes describing pottery assemblages and types of ornament at Neolithic sites in Europe. In this case it was postulated that a site's traditional cultural affiliation, based on the characteristics of its domestic pottery, might also have an effect as an indicator of preferential interaction, implying a culturally structured population (cf. Scerri et al. 2018). The results showed that cultural affiliation accounted for significant variation in the similarity between sites in their pottery assemblages even when the temporal and spatial distances between them were controlled. They also showed that variation in the between-group similarity between cultures was strongly associated with time, pointing to the conclusion that there was not a continuum of temporal variation that was arbitrarily divided into different cultures but rather that the through-time patterns were marked by sudden changes. Variation in similarity between sites and cultures in terms of their ornaments did not show the same pattern of variation, with cultural affiliation much less important, pointing to the existence of distinct cultural 'packages' (Boyd et al. 1997) with their own transmission patterns, subject to different biases, as per Binford's argument about the different dimensions of cultural variation.

However, this is not the only possible line of approach. Cultural phylogenetic methods have a major role here in that trees corresponding to specific hypotheses can be constructed and tested, as they have been for later periods using other kinds of data (e.g. Gray and Jordan 2000). In fact, Tostevin could have used such an approach to test his hypotheses although he did not actually do so. However, it is surely no accident that the methods have mostly been successfully applied to rather elaborate types such as projectile points, which have relatively large numbers of distinctive features, some of which have then been shown to be convergent. In contrast, in the case of so-called production flakes, experimental work by Eren et al. (2018) showed that there was an enormous overlap in flake shape even when they resulted from the production of different tools, with different techniques from flint nodules of very different shapes and sizes. However elaborate the description of the objects concerned, they may simply lack information about their transmission history.

Nevertheless, we do not always need such methods to make such inferences. Space and time can themselves be used as independent variables to overcome the problem of lithic assemblages having to play multiple roles in description and explanation. Thus, Moore (2013) uses the differential timing of the appearance of hierarchical reduction sequences in addition to simple chaining sequences in Australia and the Old World to argue that they are convergent trends associated with demographic growth since they are unquestionably independent developments. In a similar vein Clarkson et al. (2018) use the differential timing of the appearance and disappearance of microlithic industries within and between several different world regions, including southern Africa, South Asia and Australia, to argue that they are convergent developments associated with changing mobility. In effect, their invention and use was always within Tennie's 'zone of latent solutions'.

However, a further source of independent evidence to test many Paleolithic hypotheses is now beginning to be provided by aDNA studies. These provide strong evidence, in addition to the rationale advanced by Lycett and colleagues, to believe that some proportion of the variation in space and time observed in lithic assemblages during the Paleolithic would have been the result of variations in interaction that influenced transmission processes. One example is Hajdinjak et al.'s (2018) study of genomic data from late Neanderthal populations in Europe, which showed that their relatedness decreased with geographical distance as a result of decreased interaction over greater distances. Whether this is simply isolation by distance or something more structured is currently impossible to say, but in any event, given the intimate interaction required for the learning of lithic skills, the prediction would be that there is a corresponding decline in similarity in lithic attributes linked to the learning context. Conversely, the genomic evidence from an earlier and a later Neanderthal individual from Mezmaiskaya cave in the Caucasus pointed to population turnover, possibly the result of local extinction and replacement, so the prediction would be that this was also associated with a discontinuity in learned attributes. In any case, the point is that the genomic evidence now provides a new basis for relieving the 'interpretative burden' (Kristiansen et al. 2017) on the archaeological evidence of the lithics themselves, by providing an independent set of data with which the lithic patterns can be compared, just as radiocarbon dating did for later prehistory in the 1970s. In doing so it shows that the kinds of interaction and transmission processes assumed (in a naïve form) by the culture historians can be identified even in the Middle Paleolithic and even though their dating is relatively imprecise.

Similar inferences can also be made for the Upper Paleolithic on the basis of the genomic data. Thus Fu et al. (2016) show that an individual from Goyet Cave in Belgium dating to c. 35 kya and thus corresponding in date to the early Upper Paleolithic Aurignacian complex belonged to a different population group from their Věstonice genomic cluster, which is associated with the Gravettian, and on this basis infer that the spread of the Gravettian was at least partly the result of population movements (see also Sikora et al. 2017). Conversely again, the Věstonice cluster represents a different population from that of the Mal'ta 1 individual from Siberia but examples of the well-known Venus figurines occur with both, suggesting that the relevant cultural process explaining the link is horizontal transmission across populations.

In evolutionary biology the standard way to assess whether traits are the result of common descent or convergent, and therefore by implication adaptive, is the use of the phylogenetic comparative method, in which the occurrence of the traits of interest is mapped onto an independently derived tree characterizing relationships of biological descent; statistical methods are then used to test hypotheses of independence in relation to the tree structure (Harvey and Pagel 1991). In the last 30 years these methods have been extensively used in cultural evolutionary studies of various attributes of present-day societies, for example whether they are matrilineal or patrilineal in descent rules and whether or not these rules are a convergent adaptation. In this case it is a language tree that is taken as the proxy for descent relationships between populations (e.g. Mace and Pagel 1994; Holden and Mace 2003). As Paleolithic ancient DNA data becomes increasingly available it should become possible to go beyond the *ad hoc* inferences made above to map archaeological traits onto the admixture trees being created by geneticists.

However, the ancient DNA evidence also points to other cultural evolutionary factors relevant to understanding cultural variation. Specifically it will enable us to address the much discussed role of population size in influencing cultural change in the Paleolithic (Shennan 2001; Powell et al. 2009), a period for which no other reliable source of information on this is available. In the case of the Neanderthals the genomic evidence of runs of homozygosity from both the Vindija cave individual and, even more so, the Neanderthal individual from Denisova cave in Siberia (Prüfer et al. 2017, 2014 respectively) indicates that the populations were small and isolated. Since drift is a much stronger force in small populations than in larger ones, and can potentially overwhelm selection, one likely inference is that it would also play a significant role in explaining variation within and between Middle Paleolithic lithic assemblages. But it is not just a matter of drift. Hamilton and Walker's (2018) modelling of stochasticity in hunter-gather population dynamics indicates that on average hunter-gatherer populations only continue to exist for a few hundred years, and often less. It is the repeated stochastic patterns of population extinction that produce the long-term outcome of effectively zero population growth in the Pleistocene. This would imply regular loss of cultural features and the need for re-invention, with the likely result again that only relatively obvious features within the zone of latent solutions will be re-invented (cf. Henrich 2004), resulting in a ceiling in the level of cultural diversity (cf. Premo and Kuhn 2010), and also a major role for convergence.

Conversely, as carrying capacity increases the average time to extinction also goes up, although the size of this effect decreases with increasing environmental stochasticity. Thus, evidence from Upper Paleolithic individuals from the well-known site of Sunghir (Sikora et al. 2017) suggests the existence of larger interacting populations with a structure similar to that of known modern hunter-gatherer groups, including low levels of relatedness between the members of co-resident groups. In these circumstances the effects of selection on genetic variation will not be overwhelmed by drift and the same principle should apply to culturally-transmitted variation as well. In other words, there is a greater potential for attributes that improve the efficiency of tools, for example, to increase in frequency. Combined with the fact that populations will on average last longer before they go extinct, there is more scope for the maintenance of cultural traditions, including the development of cumulative traditions that include the recombination of prior innovations (cf. Derex and Boyd 2015; Enquist et al. 2011). This may well be relevant to the increased rate of cultural change during the Upper Paleolithic. It should also lead to lower levels of homoplasy and more robust trees.

Conclusion

The production of lithic artefacts depends on learned behaviors and therefore on cultural transmission, thus the 'ways of doing' concerned have significant heritability, which can in principle be distinguished from the effects of raw material and re-sharpening. Progress has been made in identifying these and describing material in terms of attributes that relate to the transmission process. However, even though they are made up of products of social learning, the composition of lithic assemblages is not determined by transmission in the same way but by situational factors associated with technological organization in local environments (and then, of course, subject to processes specific to the formation of the archaeological record, like time-averaging). Insofar as these situations repeat themselves, there may be strong similarities between assemblages, but they do not tell us anything one way or the other about transmission and its role. Only attributes relating to the production process can tells us this.

With regard to transmitted variation, declining transmission with distance results in decreasing similarity because innovations occurring in one place are less likely to be transmitted to the other and the increasing availability of relevant genomic data provides a new basis for generating testable predictions about the role of population processes like isolation-by distance or expansions and extinctions. Phylogenetic methods have a major role to play in distinguishing isolation-by-distance, the existence of structured populations and the extent of homoplasy. Increasingly too, ancient DNA admixture trees will provide a basis for using the comparative method. For the reasons discussed above smaller effective cultural population sizes are likely to be associated with higher degrees of convergence than larger ones and this is likely to be one of the main factors distinguishing the Middle from the Upper Paleolithic. Here too ancient DNA will play a major role in model testing.

References

Angelbeck, B., & Cameron, I. (2014). The Faustian bargain of technological change: Evaluating the socioeconomic effects of the bow and arrow transition in the Coast Salish past. *Journal of Anthropological Archaeology, 36,* 93–109.

Bentley, R. A., Hahn, M. W., & Shennan, S. J. (2004). Random drift and culture change. *Proceedings of the Royal Society Biological Sciences, 271,* 1443–1450.

Binford, L. R. (1962). Archaeology as anthropology. *American Antiquity, 28,* 217–225.

Binford, L. R. (1965). Archaeological systematics and the study of culture process. *American Antiquity, 31,* 203–210.

Binford, L. R. (1973). Interassemblage variability-the Mousterian and the 'functional' argument. In C. Renfrew (Ed.), *The explanation of culture change* (pp. 227–254). London: Duckworth.

Binford, L. R. (1978). *Nunamiut ethnoarchaeology.* New York: Academic Press.

Binford, L. R. (2001). *Constructing frames of reference.* Berkeley: University of California Press.

Binford, L. R., & Binford, S. R. (1966). A preliminary analysis of functional variability in the Mousterian of the Levallois Facies. *American Anthropologist, 68,* 238–295.

Blitz, J. H. (1988). Adoption of the bow in prehistoric North America. *North American Archaeology, 9,* 123–145.

Bordes, F. (1973). On the chronology and contemporaneity of different Palaeolithic cultures in France. In C. Renfrew (Ed.), *The explanation of culture change* (pp. 217–226). London: Duckworth.

Boyd, R., & Richerson, P. J. (1985). *Culture and the evolutionary process.* Chicago: University of Chicago Press.

Boyd, R., Borgerhoff-Mulder, M., Durham, W. H., & Richerson, P. J. (1997). Are cultural phylogenies possible? In P. Weingart, S. D. Mitchell, P. J. Richerson, & S. Maasen (Eds.), *Human by nature* (pp. 355–386). Mahwah, NJ: Lawrence Erlbaum.

Carr, C. (1995). A unified middle-range theory of artefact design. In C. Carr & J. Neitzel (Eds.), *Style, society and person: Archaeological and ethnological perspectives* (pp. 171–258). New York: Plenum.

Cavalli-Sforza, L. L., & Feldman, M. W. (1981). *Cultural transmission and evolution: A quantitative approach.* Princeton: Princeton University Press.

Clarkson, C., Hiscock, P., Mackay, A., & Shipton, C. (2018). Small, sharp and standardized: Global convergence in backed-microlith technology. In M. O'Brien, B. Buchanan, & M. I. Eren (Eds.), *Convergent evolution in stone tool technology* (pp. 175–200). Cambridge, Mass: MIT Press.

Delagnes, A., & Rendu, W. (2011). Shifts in Neandertal mobility, technology and subsistence strategies in western France. *Journal of Archaeological Science, 38,* 1771–1783.

Derex, M., & Boyd, R. (2015). The foundations of the human cultural niche. *Nature Communications, 6,* 8398.

Dunnell, R. C. (1978). Style and function: A fundamental dichotomy. *American Antiquity, 43,* 192–202.

Eerkens, J. W., & Lipo, C. P. (2005). Cultural transmission, copying errors, and the generation of variation in material culture in the archaeological record. *Journal of Anthropological Archaeology, 24,* 316–334.

Enquist, M., Ghirlanda, S., & Eriksson, K. (2011). Modelling the evolution and diversity of cumulative culture. *Philosophical Transactions of the Royal Society of London. Biological Sciences, 366,* 412–423.

Eren, M., Buchanan, B., & O'Brien, M. (2018). Why convergence should be a potential hypothesis for the emergence and occurrence of stone-tool form and production processes: An illustration using replication. In M. O'Brien, B. Buchanan, & M. I. Eren (Eds.), *Convergent evolution in stone tool technology* (pp. 61–72). Cambridge, Mass: MIT Press.

Fu, Q., Posth, C., Hajdinjak, M., Petr, M., Mallick, S., Fernandes, D., et al. (2016). The genetic history of Ice Age Europe. *Nature, 534,* 200–205.

Grant, P. R., & Grant, B. R. (2002). Unpredictable evolution in a 30-year study of Darwin's finches. *Science, 296,* 707–711.

Gray, R. D., & Jordan, F. M. (2000). Language trees support the express-train sequence of Austronesian expansion. *Nature, 405,* 1052–1055.

Hames, R. B., & Vickers, W. T. (1982). Optimal diet breadth theory as a model to explain variability in Amazonian hunting. *American Ethnologist, 9,* 358–378.

Hajdinjak, M., Fu, Q., Hubner, A., Petr, M., Mafessoni, F., Grote, S., et al. (2018). Reconstructing the genetic history of late Neanderthals. *Nature, 555,* 652–656.

Hamilton, M., & Walker, R. (2018). A stochastic density-dependent model of long-term population dynamics in hunter-gatherer populations. *Evolutionary Ecology Research, 19,* 85–102.

Harvey, P., & Pagel, M. (1991). *The comparative method in evolutionary biology.* Oxford: Oxford University Press.

Henrich, J. (2004). Demography and cultural evolution: How adaptive cultural processes can produce maladaptive losses: The Tasmanian case. *American Antiquity, 69,* 197–214.

Holdaway, S., & Douglass, M. (2012). A twenty-first century archaeology of stone artifacts. *Journal of Archaeological Method and Theory, 19,* 101–131.

Holden, C., & Mace, R. (2003). Spread of cattle led to the loss of matrilineal descent in Africa: A coevolutionary analysis. *Proceedings of the Royal Society of London. Series B: Biological Sciences, 270,* 2425–2433.

Key, A. J., & Lycett, S. J. (2017). Influence of handaxe size and shape on cutting efficiency: A large-scale experiment and morphometric analysis. *Journal of Archaeological Method and Theory, 24,* 514–541.

Kristiansen, K., Allentoft, M., Frei, K. M., Iversen, R., Johannsen, N., Kroonen, G., et al. (2017). Re-theorising mobility and the formation of culture and language among the Corded Ware Culture in Europe. *Antiquity, 91,* 334–347.

Lipo, C., O'Brien, M., Collard, M., & Shennan, S. (Eds.). (2006). *Mapping our ancestors.* New Brunswick: Aldine Transaction.

Lycett, S. J., & von Cramon-Taubadel, N. (2015). Toward a "quantitative genetic" approach to lithic variation. *Journal of Archaeological Method and Theory, 22,* 646–675.

Lycett, S. J., Schillinger, K., Eren, M. I., von Cramon-Taubadel, N., & Mesoudi, A. (2016). Factors affecting Acheulean handaxe variation: Experimental insights, microevolutionary processes, and macroevolutionary outcomes. *Quaternary International, 411,* 386–401.

Mace, R., & Pagel, M. (1994). The comparative method in anthropology. *Current Anthropology, 35,* 549–564.

Moore, M. W. (2013). Simple stone flaking in Australasia: Patterns and implications. *Quaternary International, 285,* 140–149.

O'Brien, M., Buchanan, B., & Eren, M. I. (Eds.). (2018). *Convergent evolution in stone tool technology.* Cambridge, Mass: MIT Press.

Powell, A., Shennan, S. J., & Thomas, M. G. (2009). Late Pleistocene demography and the appearance of modern human behavior. *Science, 324,* 1298–1301.

Premo, L. S., & Kuhn, S. L. (2010). Modeling effects of local extinctions on culture change and diversity in the Paleolithic. *PLoS ONE, 5,* e15582. https://doi.org/10.1371/journal.pone.0015582.

Prüfer, K., Racimo, F., Patterson, N., Jay, F., Sankararaman, S., Sawyer, S., et al. (2014). The complete genome sequence of a Neanderthal from the Altai Mountains. *Nature, 505,* 43–49.

Prüfer, K., de Filippo, C., Grote, S., Mafessoni, F., Korlević, P., Hajdinjak, M., et al. (2017). A high-coverage Neandertal genome from Vindija Cave in Croatia. *Science, 358,* 655–658.

Renfrew, C. (1973). *Before civilization: The radiocarbon revolution and prehistoric Europe.* London: Jonathan Cape.

Sackett, J. R. (1982). Approaches to style in lithic analysis. *Journal of Anthropological Archaeology, 1,* 59–112.

Sackett, J. R. (1985). Style and ethnicity in the Kalahari: A reply to Wiessner. *American Antiquity, 50,* 154–159.

Scerri, E. M. L., Thomas, M. G., Manica, A., Gunz, P., Stock, J. T., Stringer, C., et al. (2018). Did our species evolve in subdivided populations across Africa, and why does it matter? *Trends in Ecology and Evolution, 33,* 582–594.

Schillinger, K., Mesoudi, A., & Lycett, S. J. (2014). Copying error and the cultural evolution of "additive" versus "reductive" material traditions: An experimental assessment. *American Antiquity, 79,* 128–143.
Shea, J. J. (2017). *Stone tools in human evolution: Behavioral differences among technological primates*. Cambridge: Cambridge University Press.
Shennan, S. J. (2001). Demography and cultural innovation: A model and some implications for the emergence of modern human culture. *Cambridge Archaeological Journal, 11,* 5–16.
Shennan, S. J. (2011). Descent with modification and the archaeological record. *Philosophical Transactions of the Royal Society B, 366,* 1070–1079.
Shennan, S. J., Crema, E., & Kerig, T. (2015). Isolation-by-distance, homophily, and "core" vs. "package" cultural evolution models in prehistoric Europe. *Evolution and Human Behavior, 36,* 103–109.
Shennan, S., & Steele, J. (1999). Cultural learning in hominids: A behavioural ecological approach. In H. Box & K. Gibson (Eds.), *Mammalian social learning*. Symposia of the Zoological Society of London (Vol. 70, pp. 367–388). Cambridge: Cambridge University Press.
Shott, M. (2008). Lower Palaeolithic industries, time, and the meaning of assemblage variation. In S. Holdaway & L. Wandsnider (Eds.), *Time in archaeology: Time perspectivism revisited* (pp. 46–60). Salt Lake City: University of Utah Press.
Sikora, M., Seguin-Orlando, A., Sousa, V. C., Albrechtsen, A., Korneliussen, T., Ko, A., et al. (2017). Ancient genomes show social and reproductive behavior of early Upper Palaeolithic foragers. *Science, 358,* 659–662.
Tennie, C., Call, J., & Tomasello, M. (2009). Ratcheting up the ratchet: On the evolution of cumulative culture. *Philosophical Transactions of the Royal Society B, 364,* 2405–2415.
Tostevin, G. B. (2012). *Seeing lithics: A middle-range theory for testing for cultural transmission in the Pleistocene*. Oxford: Oxbow.
Wargo, M. C. (2009). The Bordes-Binford debate: Transatlantic interpretative traditions in Palaeolithic archaeology. Unpublished Ph.D. thesis, University of Texas at Arlington.
Wiessner, P. (1983). Style and social information in Kalahari San projectile points. *American Antiquity, 48,* 253–276.
Wiessner, P. (1985). Style or isochrestic variation? A reply to Sackett. *American Antiquity, 50,* 160–166.

Index

Note: Page numbers followed by *f, t* and *n* indicates figures, tables and footnotes respectively

H. S. Groucutt (ed.), *Culture History and Convergent Evolution: Can We Detect Populations in Prehistory?*, Vertebrate Paleobiology and Paleoanthropology, https://doi.org/10.1007/978-3-030-46126-3

The manufacturer's authorised representative in the EU is Springer Nature Customer Service Centre GmbH, Europaplatz 3, 69115 Heidelberg, Germany. If you have any concerns regarding our products, please contact ProductSafety@springernature.com

Printed and bound by CPI Group (UK) Ltd, Croydon, CR0 4YY
22/09/2026
02221072-0001